ENVIRONMENT
and
ARCHEOLOGY

Karl W. Butzer
University of Chicago

ENVIRONMENT *and* ARCHEOLOGY

An Ecological Approach to Prehistory

second edition

Aldine • Atherton
Chicago and New York

First published 1971 by
Aldine • Atherton, Inc.
529 South Wabash Avenue
Chicago, Illinois 60605

Library of Congress Catalog Card Number 74-115938

ISBN *202-33023-1*

Printed in the United States of America

Second Printing 1973

For

ELISABETH,

HELGA,

and

CARL

Foreword to the Second Edition

In preparing the original version of this book I made an explicit plea for a Pleistocene *geography* concerned with man and the land. This revised edition reflects my concern that dating techniques, geomorphological theory, and regional stratigraphy continue to be overemphasized in the literature and in the classroom, with insufficient stress on the total environment. Much has been said for the ecological approach to archeology, but courses in archeology and prehistory generally impress the student that the natural sciences provide a set of techniques and facts that he should know of. Little effort is made to convey that the ecological approach is a matter of attitude and concept.

Contemporary human groups interact with their environment in many ways and at different levels, depending on their technology and organizational skills. The regional environment provides a resource base that may also be relevant to the development of individual economic traits, primary technology, subsistence patterns, and even social structure. Similarly, on a more local scale, the habitat provides settlement sites and the focus for human activities. Conversely, man leaves his imprint upon the local setting and even on the regional environment. Food-gatherers modify vegetation by fire and accidental dispersal of plants; they exploit and may overexploit food resources, and they impart their mark upon the land. Food-producers leave a far more conspicuous record. Their structures are common and of some permanence. They clear or destroy forest, displace wild game with domesticated animals or mass crops, deplete or destroy the soil mantle, modify or upset the hydrological balance, and initiate the process of pollution that threatens us today.

Prehistoric groups interacted with their environments in similar ways. No single discipline can hope to unravel the story in its entirety, but many can contribute. Paleo-ecology is essentially an approach based on various classes of data. In searching for an understanding of prehistoric groups, their settlements and their way of life, we must not let data-gathering obscure understanding. Instead our work must be more problem-oriented. Each discipline must contribute directly to ecological sythesis, and the archeologist must concentrate on deriving more meaningful results from his own data, rather than attempt to integrate all categories of evidence. More explicitly, I would hope for archeologists who can think as geographers, who can derive more social and economic inference from their materials, and who can interact with other scientists also interested in prehistoric man.

This edition has been completely revised, a difficult task at a time when Pleistocene studies are proceeding apace in both theory and data. Eight chapters have been completely rewritten and considerably expanded, nine more have been intensively reworked and complemented by the addition of new sections and materials. All of the remaining parts were corrected and revised in detail, and some 600 new references worked into the text. Those sections of the book dealing with Pleistocene environments and man-land relationships have been thoroughly redone, and three new chapters dealing with North America and Australia added. As a result, the length of these two parts has been doubled and their scope increased. It is hoped that these applied chapters, which now constitute half of the book, will prove useful to the student as examples of what can and cannot be done with the available information. At the same time, now augmented in detail, they may better serve as a reference for the researcher. This added scope and content warrants the change of subtitle to *An Ecological Approach to Prehistory.*

I am grateful to my colleagues who in reviews or conversation had responded so encouragingly to the first edition. Their comments, and the stimulating multi-authored discussion in *Current Anthropology* (1966, 7:501-513), provided the main guidelines for revision. In the process of rewriting I have been indebted to Glynn Isaac (Berkeley), Leslie Freeman, Daniel Bowman, Marvin Mikesell, and William Hylander (University of Chicago), David Thurber (Lamont Geological Observatory), Stuart Struever (Northwestern University), C. K. Brain (Transvaal Museum), Richard Klein (University of Washington), and Phillip Tobias and Alun Hughes (Witwatersrand) for information, discussion, or criticism. The Division of Social Sciences of the University of Chicago supported part of the work by a research grant, while the Geography and Anthropology departments provided cartographic and typing facilities. John Kirchner deserves particular thanks for the care and interest with which he worked for many weeks on the new maps and diagrams. Last but not least, Alexander J. Morin, my publisher, encouraged me to undertake the task of a complete revision—a unique opportunity that I much appreciate.

Before closing, I should mention that a succession of field seasons in eastern and southern Africa has provided me with fresh stimulus and excitement, opening new horizons that I hope every young student of the Pleistocene may one day share. To those many friends and companions, who made those safaris in the African sun and under African stars so delightful, a special "thank you."

Flossmoor and Chicago, Illinois
October, 1970

Foreword

Pleistocene geology is primarily concerned with stratigraphy and chronology. A more comprehensive study of past environments is needed, a *Pleistocene geography* concerned with the natural environment and focused on the same themes of "man and nature" that are the concern of historical and contemporary geographies. This is a field to which both the natural scientist and the archeologist should contribute—more directly and with greater enthusiasm.

The book grew out of lectures and seminars in Pleistocene or prehistoric geography given at the University of Wisconsin since 1960. Their purpose was to study the contributions of the natural sciences to Pleistocene environmental analysis, and to integrate this information with a cultural-historical perspective. Without a suitable book the widely diverse backgrounds of general students, or majors in anthropology, geography, geology, or botany, required a novel approach. Some collateral reading in elementary physical geography or geology, or in general prehistory, as the case might be, proved helpful. But the interest of the students and the peculiar fascination of interdisciplinary problems were found to transcend any technical barriers.

It seemed to me that the materials assembled could be made useful to a wider audience of students in other courses dealing with either the Pleistocene or with prehistory. In its present form this book is specifically intended as a text in Pleistocene geography for coursework in both Pleistocene geology and geomorphology, and prehistoric archeology. Possibly the book may also encourage closer relations between the natural scientists and archeologists—in the field, the laboratory, and the university.

I owe my interest in paleoclimatology to F. Kenneth Hare (King's College, London, then of McGill University). Carl Troll, Roland Brinkmann, and Paul Woldstedt (all of the University of Bonn) gave me my foundations in geomorphology and geology. My subsequent collaboration with numerous friends, colleagues, and students in the field and on the campus stimulated my enthusiasm for cultural problems. To all these and particularly to Robert Braidwood and Clark Howell of the University of Chicago, Günter Smolla of the University of Frankfurt/Main, Chester Chard, Andrew Clark, and Frederick Simoons of the University of Wisconsin, I am indebted beyond measure. The Wenner-Gren Foundation for Anthropological Research provided me with a unique opportunity to learn at two Burg Wartenstein symposia. Parts of the manuscript were read by several people, including Clark Howell (University of Chicago), Philip Wagner (University of California, Davis), William Denevan, John Emlen, and Jonathan Sauer (University of Wisconsin), all of whom provided valuable criticism. Diagrams and maps were drawn by William Hess.

Madison, Wisconsin
January, 1964

Contents

PART III. INTERPRETATION OF PLEISTOCENE SEDIMENTS

Part VI. Man-Land Relationships in Prehistory

Illustrations

Tables

ENVIRONMENT

and

ARCHEOLOGY

PART ONE

Introduction

ONE

Prehistoric Environment, Geography, and Ecology

INTRODUCTION

Man had lived on the earth for some two million years before the first literate civilizations of the Near East appeared. This long period saw repeated changes of climate with great modifications of the natural environment. At first man was a rather minor element of the biological world. But gradually the development of social organization and technology lent increasing significance to the human species. Then, with the domestication of certain plants and animals, man achieved a unique dominance in the biological world through agriculture. This new symbiotic alliance spread over most of the world, with man fundamentally responsible for widespread and significant modifications of the natural environment.

A study of the environment in prehistory involves more than a complex of interacting physical and biological elements. It adds a new dimension—time. This additional perspective includes natural changes and cultural modifications of the environment.

Methods and techniques in paleo-environmental study cover a wide range of fields including geography, geology, soil science, botany, zoology, meteorology—to name only a few of the natural sciences. The delicate cultural aspects of paleo-environmental work are primarily within the scope of the prehistoric archeologist. This is then an interdisciplinary field of investigation.

Much interdisciplinary work in paleo-environmental matters has of course been done in recent decades, but comprehensive studies have been disappointingly absent. Standard Pleistocene geologies such as P. Woldstedt's *Das Eiszeitalter* (1954–65), J. K. Charlesworth's *The Quaternary Era* (1957), and R. F. Flint's *Glacial and Pleistocene Geology* (1957) have devoted considerable attention to the topic, but have been dominated by stratigraphic considerations and glacial geomorphology and have almost ignored the presence of man. F. E. Zeuner's *Dating the Past* (1958; 1st ed., 1946) specifically attempts to bridge the gap between the natural sciences and prehistory, but is chronologically oriented and, like its more paleo-environmental companion *The Pleistocene Period* (1959; 1st ed., 1945), is based upon pre-1945 literature. Also of relevance are compendia such as A. Laming's *La Découverte du Passé* (1952) or the synopsis *Science in Archaeology* (1963; rev. ed. 1970), edited by D. R. Brothwell and E. S. Higgs. While the subject matter of this book is then not necessarily new, it attempts to chart a new course in emphasizing paleo-environmental study from both theoretical and applied viewpoints, followed by an analysis of man and the environment during prehistoric times.

If only for the sake of convenience, a field of study should have a name. Interdisciplinary research, however, borrows techniques and problems from diverse fields, seldom possessing a methodological literature. This difficulty is particularly acute in the case of paleo-environmental work focused on prehistoric man.

The words "geography" and "ecology" both express the idea of interrelationships. "Ecology," according to M. Bates (1953), is concerned with the external factors that control the survival and abundance of individuals and populations. "Geography," on the other hand, can be defined as the scientific description and interpretation of the earth as the world of man (Hartshorne 1959, pp. 21, 172). Although "ecology" has traditionally been preempted by biologists and anthropologists, both "ecology" and "geography" are applicable to a broad spectrum of studies dealing with man, with the environment and with man-land interactions. For this reason "Prehistoric geography" is chosen as a convenient designation that serves to emphasize both environment and man in contradistinction to "Pleistocene geology."[1]

In view of the many disciplines concerned with the past, there are many different approaches to geography. Basically these approaches are of three kinds:

1. As a matter of convenience the term "Pleistocene" is frequently used as extending to the present, and including the "Recent," "Postglacial" or "Holocene." In this way it would be synonymous with "Quaternary." However, the restricted usage is preferred in the subsequent chapters of this book, since many of the most exciting aspects of prehistory date from the post-Pleistocene.

a) Individual research by the natural sciences, usually carried out independently in the field or laboratory by geologists, geographers, soil scientists, botanists, zoologists, and meteorologists. Although the range of specific goals or interests may vary greatly, most of our basic techniques and paleo-environmental data have been obtained in this way.

b) Interdisciplinary work by natural scientists in collaboration with archeologists, particularly in the field. Pleistocene geology, geomorphology, paleontology, and pollen analysis probably form the most common backgrounds of the individuals concerned. Generally directed toward the study of archeological sites, such interdisciplinary work is particularly valuable in that it contributes ecological as well as environmental information.

c) Paleo-anthropological work by archeologists directed toward a fuller understanding of the cultural ecology (Mikesell, 1967) of prehistoric communities—particularly of the economy, social organization, and interactions with the environment.

EARLY RESEARCH IN THE PLEISTOCENE

Students of the Pleistocene, whether earth or biological scientists, have contributed substantially to an understanding of past environments for over a century.

In the field of glacial geology, the first milestone was set by the work of J. Venetz, J. de Charpentier, and L. Agassiz, who established the idea of a former "Ice Age" in the Swiss Alps (1822–47). Equally significant was the proof of continental glaciation in Europe by O. Torell in 1875, and the recognition of interglacial periods by A. Penck in 1879. J. Geikie provided the first systematic treatment of the "Ice Age" in 1874, while the fundamental study of the geographers A. Penck and E. Brückner *(Die Alpen im Eiszeitalter)* in 1909 marks the beginning of the modern era in Pleistocene geology.

The history of botanical contributions to our understanding of Pleistocene and postglacial climatic history begins with the macrobotanical work of J. Steenstrup in 1841 and A. Grisebach in 1844. A few decades later pollen analysis permitted the first systematic regional interpretations of A. G. Nathorst (Sweden) in 1870, A. Blytt (Norway) in 1876, and E. Engler (Germany) in 1879. Comprehensive studies by C. A. Weber and L. Von Post in 1906 and 1916, respectively, had a similar impact on paleobotany as did Penck and Brückner in their field.

Paleontological work did not lag far behind the advances in the earth

2. Discussions of the historical development of Pleistocene geology are provided by Woldstedt (1954), Flint (1957), and Charlesworth (1957, with references).

3. For general references see Firbas (1949-52, vol. 1) and Faegri and Iversen (1964).

sciences and paleobotany, and studies by L. Ruetimeyer (1862) on the animal remains of the Swiss lake dwellings began an era of ecologically oriented paleontological work.

THE GROWTH OF INTERDISCIPLINARY ARCHEOLOGICAL WORK

Following the establishment of Pleistocene geology, pollen analysis, and paleontology as respectable fields of scientific endeavor about 50 years ago, a later stage of development was more specifically oriented toward prehistoric man. Certainly much of the prehistoric research in France and Switzerland since about 1860 had interdisciplinary overtones, but a new pattern was probably first set by the geological-paleontological investigations of the Grimaldi caves (Boule, Cartailhac, Verneau, and de Villeneuve, 1906–19) and the Grotte de l'Observatoire at Monaco (Boule and Villeneuve, 1927). These represent the first major efforts involving trained natural scientists in direct field association with digging archeologists. Following the same tradition is the association of Elinor W. Gardner (geologist) and Gertrude Caton-Thompson (archeologist) in the Fayum and Kharga oases of Egypt (Caton-Thompson and Gardner, 1929, 1932) and in the Hadramaut of southern Arabia (Caton-Thompson and Gardner, 1939). A similar team, Dorothy M. A. Bate (paleontologist) and Dorothy A. E. Garrod (archeologist), excavated some of the Mt. Carmel caves of Palestine (Garrod and Bate, 1937). Equally significant during the same decade was the work of the botanists and archeologists of the British Fenland Research Committee during 1932–40 (Philips, 1951); the geological-archeological effort of the Cenozoic Research Institute at Choukoutien, China (Teilhard de Chardin, 1941, with references); the geological-archeological teamwork of H. de Terra and T. T. Paterson (1939) in India, of de Terra and H. L. Movius (1943) in Burma; and the paleobotanical work of R. Schütrumpf (1936, 1938) at A. Rust's Meiendorf site near Hamburg. Another important individual study was the pioneer cave sedimentology developed by R. Lais (1932, 1941) in Central Europe.

Following World War II, geomorphological[4] and biological[5] in-

4. Some of the more geomorphologically important studies include sites in sub-Saharan Africa (Cooke, 1946; Bond, 1946, 1957, 1962; de Heinzelin, 1957; Brain, 1958; Haldemann, in Howell *et al.*, 1962; Hay, 1967; Isaac, 1966, 1967; Butzer and Thurber, 1969), in the Near East and North Africa (Hey, in McBurney and Hey, 1955; Wright, 1951, and in Braidwood *et al.*, 1960; Butzer, 1960b; Stearns, in Howe, 1967; Butzer and Hansen, 1968; de Heinzelin, 1968; Farrand, 1969), in Europe (West and McBurney, 1954; Freund, in Zotz, 1955; Movius and Judson, 1956; Butzer, 1965, 1967; Brunnacker, 1963, 1967; Laville, 1964, and in Bordes *et al.*, 1966), and in North America (Bryan, 1950; Judson, 1949, 1953a, 1953b; Moss, 1951; Hopkins and Giddings, 1953; Black, 1959, Black and Laughlin, 1964; Mackay *et al.*, 1961; Malde and Schick, 1964; Haynes, 1968, Haynes and Grey, 1965, Haynes and Agogino, 1966).

5. Including the botanists Walker and Godwin (in J.G.D. Clark, 1954), West (in West

vestigations of archeological sites became frequent, while the number of contributions and new techniques developed by individual natural scientists has grown tremendously.

This survey of interdisciplinary efforts over half a century is necessarily selective. But it serves to illustrate an increasing degree of interdisciplinary interest and a remarkable acceleration of activity during the last twenty years or so. Equally symptomatic of growing interest by a wider group of people was the establishment of institutes of Environmental Archaeology at Groningen and London, a sub-department of Quaternary Research at Cambridge, and Geochronological Laboratory at the University of Arizona; and, more recently, a growing number of jointly appointed professors.

Analysis of the existing substantive literature shows that stratigraphy and climatic interpretation form the most general unifying theme in interdisciplinary work by natural scientists. Petrographical, mineralogical, or out-of-context osteological or vegetable identifications form a surprisingly small segment of the natural science contributions to archeological reports. The lack of distinction commonly made between stratigraphy and environment is understandable since relative dating of sites depends directly or indirectly on the succession of climatic changes. But this slight confusion is serious. Many joint studies neglect a full interpretation of the immediate environment of a site during habitation in favor of discussions of world-wide stratigraphic schemes and absolute chronology. It seems that environmental reconstruction is the higher goal of attainment. Stratigraphy and chronology are indeed important, but they have been unduly emphasized.

PREHISTORIC GEOGRAPHY AND THE NATURAL SCIENCES IN THEORY

Although many natural scientists co-operating with prehistorians in the field, as well as a great number of individual earth scientists and biologists, are implicitly or explicitly interested in man, there is a marked reticence to engage in integration of "natural" and "cultural" data. A glance at the objectives and interests expressed in the limited methodological literature confirms this.

Methodological interest for an integrated "prehistoric" geography has been greatest among geographers. A. H. Clark, in his analysis of the field of historical geography, writes as follows (1954, p. 72):

---

and McBurney, 1954), Helbaek (1959, 1960a), Leroi-Gourhan (1961), Van Campo (1960, 1962, 1964-65, 1966), Hafsten (1961), Martin (1963), and Mehringer (Mehringer and Haynes, 1964); and the paleontologists Bate (in A.J. Arkell, 1949b, 1953, and in McBurney and Hey, 1955), Arambourg *et al.* (1952), Arambourg and Balout (1952), Kleinschmidt (in Tode *et al.*, 1953), Hooijer (1961), Elisabeth Schmid (1958), Reed (1959, 1960), and Lance (1959).

> To insist that historical geography begins where history, as opposed to prehistory, begins would assume some inherent necessity for written records in studying the past geography of an area. Archaeological reconstructions alone have sufficiently demonstrated that no such necessity exists. The reasons for denying the validity of such a division apply with almost equal force to any other. There is, indeed, no logical date or period in time when such studies may properly be said to begin. If physical geography is something more than a summation of geological, climatological, ecological, and similar evidence, then a physical historical geography must exist, which utilizes the kind of evidence that is also studied, often in arbitrarily restricted categories, by the historical geologists, paleontologists, and paleo-climatologists. It is true that for periods before the Pleistocene and for much of that epoch, such studies either do not exist, or have been attempted only by scholars from one of these systematic fields. In practice, "dawn" for the historical geographer rarely antedates the late Pleistocene; he has shown little interest in ages devoid of human culture. In logic, however, his license as a scholar allows him to go back in time as far as he has interest and competence.

Clark further stresses that such an interest on the part of geographers has indeed been demonstrated, for example, by the works of H. J. Fleure (Peake and Fleure, 1927-36) and P. Deffontaines (1930, 1933) which are devoted to a form of "prehistoric" geography. Another striking example of the geographer's contribution to a fuller understanding of prehistory is provided by the studies of C. O. Sauer on the early significance of fire (1947) and on geographical aspects of agricultural origins (1952).

From another theoretical viewpoint, the field of cultural geography (see Wagner and Mikesell, 1962, pp. 1-24) has shown distinct undercurrents that envisage the natural prehistoric or prehuman landscape as the necessary datum line from which cultural "deformations" are to be measured (Gradmann, 1906, 1936; Sauer, 1927). The natural landscape, prior to agricultural colonization, is in effect the background, and understanding it is prerequisite to a full understanding of the cultural landscape. That this goal of reconstructing the natural landscape is indeed attainable is amply illustrated by the prehistoric vegetation studies in Europe by Firbas (1949-52), Iversen (1954, 1960), Godwin (1956), and others. The natural landscape of ancient Egypt has received attention in regional studies by Passarge (1940) and Butzer (1959b). The study of physical conditions of the environment prior to first agricultural settlement is, then, also within the traditional interests of geographical research.

Other more general writings are confined to the possible role of natural scientists in relation to archeological investigations. The geologist's work is outlined by H. E. Wright (1957, p. 50), while W. G.

Reeder (Laughlin and Reeder, 1962, pp. 106–7) describes interdisciplinary functions of the paleozoologist.

ENVIRONMENT AND PALEO-ANTHROPOLOGY

The prehistoric archeologist has long been aware that the natural sciences provide a number of useful techniques. The very fact that specific interdisciplinary efforts have a tradition of over a half century speaks eloquently for the paleo-ecological interest of many excavators. Yet closer inspection shows that few archeologists realize precisely what the natural sciences do have to offer. Speaking as a prehistorian, R. J. Braidwood (1957b) lamented "the almost complete lack of comprehension" on the part of archeologists of *(a)* the necessity of understanding the environment as a functioning entity before cultural interpretation can proceed; and *(b)* the full interpretative potential of nonartifactual materials. So, for example, only a small fraction of excavations at prehistoric sites use the services of a geomorphologist in the field, even for a limited period of time. And although it is a common practice to have bones identified afterwards, the idea of having botanical materials studied is often as remote as having a biologist at the site. Some archeologists who do have a geomorphologist at their excavations, "employ" him as a technician expected to provide ready answers for poorly formulated questions. Corollary to such relationships is a failure on the part of the archeologist to communicate information. Such a one-way flow of information is not particularly productive in an interdisciplinary study, so that the natural scientist can hardly be "problem-oriented."

Despite this occasional lack of perceptiveness and genuine collaboration, outstanding examples of paleo-ecological interest in both practice and theory have been manifest. So, for example, J. G. D. (Grahame) Clark (1957, p. 20) insists that the archeologist wishes to obtain a complete geographical-ecological understanding of a prehistoric community—the paleo-environment, its resource potential and external limitations, and above all, the interactions of man and environment as manifested in the economic sphere. Braidwood (1957a, pp. 15–16) voices an appeal specifically directed at the natural scientist:

> There is a heartening growth of comprehension on the part of a few biological and earth scientists that specialized but fascinating problems exist jointly for them and the archeologists. But these exceptions are few . . . the few biologists and earth scientists of good will who are interested must bootleg the time they invest in projects of joint interest. As a general rule, however, biologists and earth scientists tend to side-step problems which are "culture-linked." Once the hand of man has been laid on the species of their

> concern, an imponderable has been introduced for them, and they seem to find the issue uncongenial.
>
> It seems to me that in reaching for the goal, a new field, perhaps "Pleistocene ecology," might come into focus, allied to archeology (and human paleontology), but setting the archeologist somewhat more free to deal with matters of culture. This field or axis of interrelated disciplines (perhaps "Pleistocene ecology" or "paleo-environment," or "Quaternary geography"—I shall not attempt to name it) would definitely include man as an element in and a factor acting upon the environmental scene.

More recently J. Desmond Clark (1960, p. 308) stated the basic need for a Pleistocene geography particularly well:

> it is essential that the *environment and ecological setting* of cultures . . . be established as accurately as possible, for, without this knowledge, we can hardly begin to interpret the cultural evidence. It is necessary to know the nature of faunas, of vegetation and climate, of kinds and forms of raw materials, available to man and so on. Here, to a very great degree, the archaeologist must rely on workers in other disciplines—geologist, palaeontologist, ecologist, palaeobotanist, soil chemist, and geographer, to mention but a few. It is now fully apparent that unless there is teamwork with other disciplines, we cannot hope to extract more than a fraction of the evidence that in many instances our sites could yield.

Grahame Clark's paleo-ecology, if one may call it so, is clearly not intended for the natural scientist alone, but suggests total interpretation by the archeologist in keeping with the natural science evidence. The approach taken in Clark's *Prehistoric Europe: the Economic Basis* (1952) and *Excavations at Star Carr* (1954) provides a good example of this. Braidwood and Desmond Clark have turned more specifically to the natural scientists. Yet both ask for more than environmental reconstruction (not chronology), namely, for an applied study of the geographical setting to prehistory. If these authors be considered as spokesmen for wider anthropological circles in favor of interdisciplinary collaboration and a geographical-ecological approach, then the broadest scope of "prehistoric" geography could be described as *environmental reconstruction as applied to an understanding of the ecological setting to prehistory.*

The relation of the natural sciences to anthropology has received attention in the anthropological literature. A systematic description of natural science techniques in prehistory by an archeologist was already given by L. F. Zotz in 1951, soon followed by a technical compendium of scientific methods (Laming, 1952). Since then most general texts of prehistory or archeological method have included sections or chapters of this type. A methodological analysis of the position of these studies within anthropological research is due to R. Pittioni (1961), who distinguishes several classes of auxiliary sciences:

1. Field techniques
   a) "Optional" techniques, including phosphate content of sediments, cave studies in general, particle-size analysis, organic and inorganic chemistry.
   b) "Vital" techniques, including Pleistocene and economic geology as well as pollen analysis.
2. Laboratory techniques
   a) Optional: petrography, textile studies, food-chemistry and radiocarbon dating.
   b) Vital: physical anthropology, paleontology, zoology, botany, mining engineering, metallurgy, metallography and spectral analysis.[6]

Pittioni emphasizes that each of these techniques contributes toward an understanding of the natural environment of a site or the economic activities of a prehistoric community. Among the so-called vital techniques, Pleistocene geology and pollen analysis are singled out as studies prerequisite to prehistoric interpretation and considered as "substantive branches of prehistoric research" (Pittioni, 1961, p. 21). But since their immediate object of study is only indirectly associated with man, none of these auxiliary disciplines are considered as "anthropological."

ORGANIZATION AND OBJECTIVES

To rephrase the opening words of this chapter, the purpose of this book is an analysis of the environment and its possible significance for prehistoric man. The name "prehistoric geography" has been suggested for this topic which, understandably, is more a point of view than a scholarly discipline.

The first part attempts to outline a basic scheme of stratigraphy and chronology, intended as background information for the more pertinent topics dealing with prehistoric geography.

The second part deals with the significance of vegetation, soils, and geomorphology as environmental indices. This section also serves as an introduction to the soil and the geomorphic processes so fundamental to paleo-environmental work. A further chapter discusses modern mammalian distributions, both their relationship to the environment and their significance as a food resource for prehistoric man.

Part III attempts a systematic account of Pleistocene sediments and methods for their study. The terminal chapter focuses this information

6. J. Haekel, in comments to this paper, notes that the applicability of this scheme varies according to the culture level, making such distinctions as "optional" or "vital" difficult. Some of the distinctions between "field" and "laboratory" techniques are also questionable. As Haekel indicates, the list is not quite complete.

on the geomorphological investigation of archeological sites. Part IV follows with a briefer outline of biological contributions to environmental reconstruction.

Although none of the methods in Parts III and IV are likely to be applied directly by archeologists, they may provide the student of prehistory with an outline of the more relevant paleo-environmental techniques, as well as an understanding of their possible interpretation. This should facilitate collaboration between archeologists and the natural scientists in the field or laboratory.

The regional descriptions and analyses of Part V give examples of current realizations and limitations in reconstructing the paleo-environments encountered by prehistoric populations. So, for example, the late Pleistocene setting is essential to an understanding of Neanderthal man and of the Upper Paleolithic hunters of Europe and the Mediterranean world. Similarly, colonization of the New World by the Paleo-Indians makes little sense in the context of contemporary environmental resources. Finally, the complexity of "pluvial" conditions in Africa during the later Pleistocene provides a working model for understanding earlier phases of the Pleistocene that witnessed the evolution of man.

The final group of chapters, on man-land interactions in prehistory, forms the core of the book, namely, the explicit study of man and the land. Drawing its materials from four continents, it attempts to consider the ecological aspects of hominization, of cultural innovation and adaptations during the course of the Pleistocene, of the dispersal of man into new environments, of the origins of agriculture in the Near East, and of the impact of food-production and urbanization. The themes have been chosen selectively, to provide documented examples that range through different environments and span the length of prehistoric time. In so tracing the interactions of man and the land, it is possible to see many of our contemporary problems of "environment" in a broader perspective. Our recent awareness of the environment has only reawakened an age-old goal of man, to live within the environment and its resources, rather than to destroy them. Man has always been a member of the ecologic community, and his conscious and unconscious efforts to live in some form of equilibrium with his environment go back over two million years. Despite the ignorance and lack of appreciation that have come through a few decades or centuries of urban civilization, our subconscious predilections developed through the uncounted millenia of prehistory. Surely a fuller understanding of past interrelationships has more than academic relevance today.

TWO

Stratigraphy

THE PLACE OF THE PLEISTOCENE WITHIN THE GEOLOGICAL RECORD

The perspective of time is as important to the culture historian as to the geologist, but concept and dimension differ tremendously for the two disciplines. The Quaternary period, embracing the total time span of the human species, represents the very end of geological history—an interval of two or three million years within a total five billion years of earth history. In fact the Quaternary accounts for only 0.5 per cent of the time that has lapsed since the first appearance of the trilobites during the Cambrian, some 600 million years ago. On the other hand, "man the toolmaker" existed for as much as two million years, only 5,000 of which fall within the outer circumference of historical time. Consequently almost all of the Quaternary represents "prehistory"—for most culture historians a dark age including over 99.5 per cent of man's existence.

Beyond the pale of history, time becomes a relative concept only, dealt with by approximate methods of stratigraphic and "absolute" dating, and with the philosophy of the geologist rather than of the historian. Absolute dating techniques have provided some local chronologies for the last 30,000–50,000 years, and are now suggesting broad approximations for the total time elapsed during the Quaternary. But most of our knowledge of "time" is confined to relative information derived by stratigraphic techniques, i.e., pertaining to regional or world-wide sequences of natural events such as climatic or biologic changes. It is important then, to distinguish between relative stratigraphy and so-called absolute chronology.

The position of the Quaternary within the standard geologic column is shown in Table 1. Opinions vary as to whether the Quaternary should

Table 1. Standard geological column (with the approximate duration of the eras).

CENOZOIC (60–70 million years)	
Quaternary	Holocene (synonyms: Recent, Postglacial) Pleistocene
Tertiary	Pliocene Miocene Oligocene Eocene Paleocene
MESOZOIC (160–170 million years)	
Cretaceous Jurassic Triassic	
PALEOZOIC (360–380 million years)	
Permian Upper Carboniferous (Pennsylvanian) Lower Carboniferous (Mississippian) Devonian Silurian (Gotlandian) Ordovician Cambrian	
PROTEROZOIC (at least 4,500 million years)	

be classified as an era (e.g., the Mesozoic), as a period (e.g., the Cretaceous), or an epoch (e.g., the Pliocene). Some authors would even abandon the name Quaternary in favor of considering all of the post-Pliocene as Pleistocene (Flint, 1957, pp. 282–284). Even though the Holocene (equivalent to Recent or Postglacial) is probably nothing more than a warm interval within the Pleistocene "Ice Age," it is decidedly more convenient to maintain a distinctive name for the last 10,000 years—if only to avoid confusion. But for the sake of convenience, unless otherwise specified, the general term Pleistocene will be here used synonomously with Quaternary.

The Pleistocene is an unusual period in earth history. It is not only broadly contemporaneous with the existence of man, but was witness to one of the rare spasms of extensive and recurrent glaciation affecting the planet. Although very brief by geologic standards, the Pleistocene was marked by numerous, violent changes of climate and environment, so that its over-all effect on the earth's surface was much greater than its duration would suggest. Equally striking have been the effects of the accompanying environmental changes on mammalian (Kurtén, 1968) and perhaps human (Robinson, 1963) evolution.

THE "NORMAL" CLIMATE OF GEOLOGICAL TIME

Throughout the greater part of earth history world climates were warmer and less differentiated than they are today (Nairn, 1961, 1964; Schwarzbach, 1963). Permanent ice was apparently quite absent and higher latitudes were by no means frigid zones. There is good reason to believe that the position of the geomagnetic poles (and presumably of the geographical poles) has changed constantly in the past, while differential movements between the various continents have been clearly demonstrated (Revelle *et al.*, 1969; Nairn and Thorley, 1961, with references). Such changes make interpretation of paleoclimatic features of the Paleozoic and Mesozoic rather difficult. But for the comparatively recent Tertiary period there is abundant evidence that the poles have been approximately at their present position. Apart from the absence of significantly different geomagnetic data, biological distributions suggest latitudinal orientations much like those of today. The *isoflors* of the older Tertiary are already remarkably arranged in broadly concentric circles about the present North Pole (Chaney, 1940; Axelrod, 1960), while the world distribution of corals during the Tertiary and Cretaceous is almost symmetrical about the modern equator (Schwarzbach, 1963).

As a comparatively well-understood example of warmer conditions, the Eocene floral record of the northern hemisphere can be divided into three major latitudinal belts (Chaney, 1940; Axelrod, 1960) (Fig. 1):

a) The planetary belt north of 50° latitude was dominated by the Arcto-tertiary forest, composed predominantly of *Sequoia* species, with pine, fir, spruce, willow, birch and elm. This coniferous-type forest is estimated to have enjoyed a mean annual temperature of 10° C., with January means in the range of 2°-4°, July means, 18°-21° C. A more cool-temperate facies was present in higher latitudes, north of about 70° latitude.

b) In middle latitudes, a subtropical evergreen association of the *Nipa* palm with numerous tropical oaks covered the sites of the modern deciduous or boreal forest belts, while coral growth was possible in

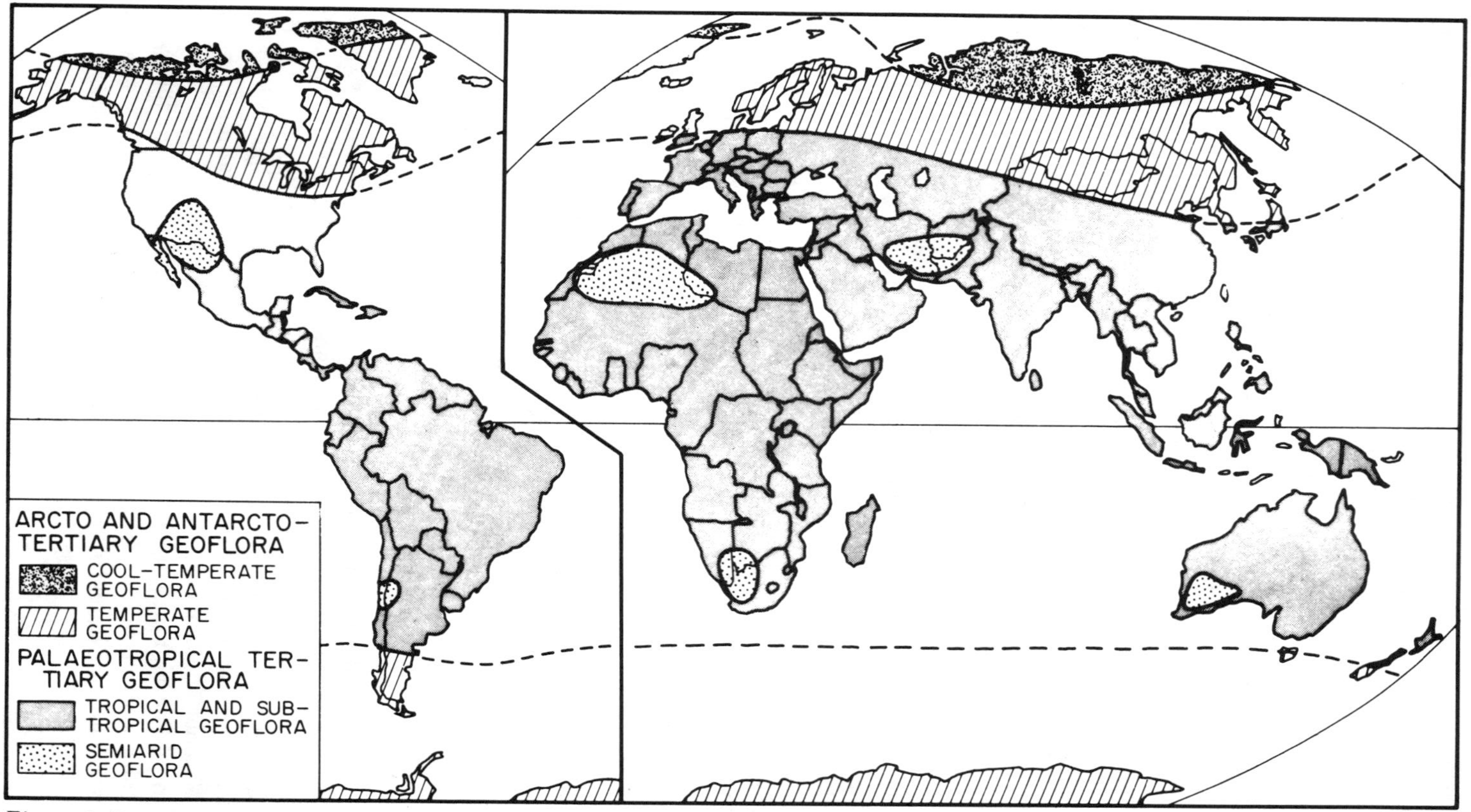

Figure 1. Distributions of geoflora during the Early Tertiary (modified after Axelrod, 1960).

the oceans at the same latitude. Semiarid floras were evolving on the western sides of the continents, in lower middle latitudes.

c) In lower latitudes the presence of typical tropical forests is verified. Tropical and subtropical floras are not subdivided in Figure 1.

From this example it can be deduced that the isotherms of the present temperate zone were displaced poleward by some 15°–20° of latitude. Equatorial temperatures were not necessarily greater, so that the difference is probably more a matter of reduction of the temperature gradient between pole and equator. This small latitudinal gradient, coupled with the lower general relief common to large parts of the geological record, may be related to frequent evidence of widespread aridity in middle latitudes. Such evidence may in part, however, be a result of slow adaptation of plant life to continental conditions, something achieved only in the Mesozoic.

Although climatic distribution was still latitudinal and seasonal in character, the geography and character of the climate were quite distinct from that of today, and even more so from that of the aberrant glacial phases of the Pleistocene. This condition of greater warmth and limited differentiation was called the "normal climate of geological time" by C. E. P. Brooks (1949).

ICE AGES OF THE EARLIER GEOLOGICAL RECORD

The warm conditions prevailing during most of earth history were periodically interrupted by aberrant glacial periods or Ice Ages, recorded by a wide range of geological deposits indicating the presence of great ice masses on several continents.

The first widely recognized Ice Age, the Infra-Cambrian or Eo-Cambrian, is located in the Proterozoic-Paleozoic transition zone. Exact correlation of deposits is impossible, and the duration of the successive stages is unknown. But glacial deposits are widespread in southern Africa, Brazil, and Australia (see L. C. King, 1961, with references) as well as in Greenland, Scandinavia, and the Arctic regions (see Schwarzbach, 1963, with references).

The second major Ice Age, the Permocarboniferous, is a better-defined phenomenon (see Schwarzbach, 1963; L. C. King, 1961, with references). Major continental glaciations affected parts of eastern South America, southern Africa, India, and Australia during the later Carboniferous, circumstances best explained by these continents' common origin in a primeval Antarctic continent then lying adjacent to the now-dispersed southern hemisphere land masses. More localized glacial features are recorded in the same areas in Upper Permian strata. There may have been five major glacial complexes in all during the Permocarboniferous, covering a minimum of ten million years.

HIGHER LATITUDE COOLING DURING THE CENOZOIC

The "normal" climate prevailing since the close of the Permocarboniferous Ice Age underwent temporary oscillations during the Mesozoic and older Tertiary (e.g., see Hamilton, 1968). But a distinctive downward trend of temperatures, heralding the Pleistocene, was first evident during the Miocene.

Illustrations of the magnitude of this cooling are provided by a number of paleobotanical and paleozoological studies (See Fig. 2). The following mean temperatures have been suggested for northwestern Europe (Woldstedt, 1954, pp. 8–9; Schwarzbach, 1963):

Pliocene	14°–10° C.
Miocene	19°–16° C.
Oligocene	20°–18° C.
Eocene	22°–20° C.

They are based upon a number of British, French, and German floras. The present annual mean of these areas is about 9°–10° C. In other words, approximately modern conditions had been reached by the close of the Pliocene, recorded by a progressive change from lowland tropical rain forest to temperate forest (Dorf, 1955). A similar trend is apparent

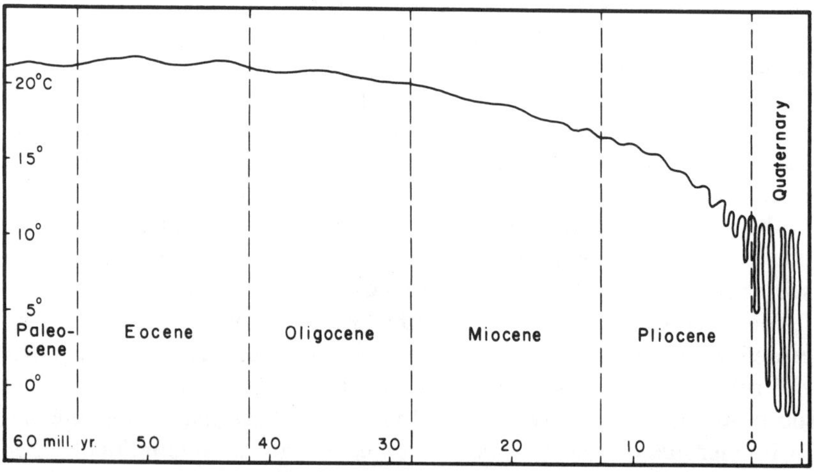

Figure 2. Generalized curve of mean annual temperatures in Central Europe during the Tertiary and Pleistocene (modified after Woldstedt, 1954). The time-scale of the Pleistocene is strongly exaggerated.

in the western United States, between latitudes 40° and 50° (see Nairn, 1964):

Pliocene	8°–5° C.
Miocene	14°–9° C.
Oligocene	18°–14° C.
Eocene	25°–18° C.
Paleocene	14.5° C.
Late Cretaceous	20° C.

On the nearby Pacific coast of America, Durham (1950) estimated February marine temperature for the same latitudes as follows, basing himself on paleozoological evidence:

Recent	10° C.
Pliocene	12° C.
Miocene	18°–11° C.
Oligocene	20°–18.5° C.
Eocene	25°–18.5° C.

Although these studies also reveal oscillations superimposed upon the over-all downward trend, temperature fluctuations during most of the Tertiary—as recorded by deep-sea cores—were of long duration and limited amplitude.

With the beginning of the Pleistocene, a series of brief but intense cold spasms, already heralded by oscillations of increasing amplitude in the later Pliocene, carried the planet into a phase of intense changes of physical environment. Each of the major cold spasms carried world temperatures several degrees below the present mean, and led to migrations and modifications of floras on a continental scale. And at least the last three cold-climate complexes were accompanied by continental glaciations, negative fluctuations of world sea level in a range of 100 to 150 meters, and significant faunal modifications. Compared with the 270 million years elapsed since the Permocarboniferous, the Pleistocene was indeed an abnormal period.

THE PLIO-PLEISTOCENE BOUNDARY

Few of the stratigraphic problems of the Pleistocene are as difficult as a satisfactory delimitation of the boundary between the Pliocene and the Pleistocene. Until 1948 the first of the four conventional Pleistocene glaciations was accepted as the base of the Quaternary. At the International Geological Congress in London during that year, it was widely thought that the continental Villafranchian (with characteristic mammalian fauna) and the marine Calabrian (with the first paleontological evidence of cooler climate) should be added to the Pleistocene (Oakley *et al.,* 1950). These units, best represented in Italy, had previously been referred to the late Pliocene. There was good reason to consider the Villafranchian stage as Pleistocene since it includes biological evidence of cold climate or at least of temporary cold intervals. Whether or not such early cold phases were accompanied by extensive glaciations remains unknown. But from present knowledge, the Villafranchian spans a long period of strongly fluctuating climate which is ultimately responsible for the glacial spasms that mark the climax, rather than the beginning, of the Pleistocene Ice Age.

Three criteria have been suggested for the delimitation of the new Plio-Pleistocene boundary between the Villafranchian and the Upper Pliocene: *(a)* tectonics, *(b)* the appearance of new forms of animal life, and *(c)* first evidence of major climatic deterioration (see the regional review of Flint, 1965). Tectonic disconformities have on occasion proved helpful, particularly in highly disturbed areas such as East Africa. But such "breaks" in the stratigraphic records are not confined to the Plio-Pleistocene boundary, thus limiting their application. The faunal criterion has limitations as well, since the first appearance of the modern genera—elephant *(Archidiskodon),* wild cattle *(Bos),* true horse *(Equus),* and camel *(Camelus)*—was not contemporaneous everywhere. In middle latitudes, wherever long sedimentary sequences are available, paleoclimatic evidence has so far proven most satisfactory. Examples are provided by the immigration of northern mollusca into the Mediterranean, or the abrupt decimation of tropical elements among the flora and fauna of certain Dutch and English beds.

Acceptable Plio-Pleistocene boundaries have since been drawn in a number of Old World countries, although it is probable that a good deal of time-stratigraphic discrepancy remains from place to place.

CONTINENTAL STRATIGRAPHY OF MID-LATITUDE EUROPE

An outline of the stratigraphic sequence of Pleistocene events can be obtained by comparative evaluation of reliable and fairly complete local sequences. Unfortunately few such sequences exist, and much of what is widely held as fundamental is in fact dubious or unclear.

At present, potential stratigraphic sequences are known from several major areas: *(a)* the Alpine zone and northwestern Europe, *(b)* the Mediterranean Sea and Morocco, *(c)* eastern and southern Africa, *(d)* the India-Pakistan borderlands, *(e)* the Irrawaddy Valley of northern Burma, *(f)* the Hwang-ho Basin of China, *(g)* Java, *(h)* the Atlantic coastal region of the United States, and *(i)* the Gulf Coast-Mississippi Valley area. In addition, the deep seas of several oceanic areas have provided valuable information of their own. For the purpose of this study, discussion can be limited to a selection of such stratigraphic columns as now seem most representative of world-wide stratigraphic units. There is little question that the European area is still most fundamental for continental stratigraphy, the Mediterranean region (including the Atlantic coast of Morocco), for littoral stratigraphy.

Temperate Europe—*(a)* the British Isles, *(b)* the Netherlands, northern Germany and Denmark, and *(c)* the Alpine region with southern Germany—provides the fortunate interdigitation of successive glacial deposits with marine or continental beds characterized by distinctive faunas or floras. Although none of these sequences is quite independent or complete, reasonable correlations are often, although not always, possible. Table 2 attempts to sketch the most diagnostic phenomena of each of these three regions, with a suggested nomenclature for wider application. The table is basically arranged with rock-stratigraphic units correlated under headings of "warm" and "cold," and ultimately as a time-stratigraphic sequence. The "cold" intervals later than the Cromerian are either "glacials" or "stadials," the "warm" intervals, "interglacials" or "interstadials."[1] Earlier cold phases were probably not accompanied by continental glaciations, so that the terms "glacial" and "interglacial" are of dubious application in the early Pleistocene.[2] That these earlier cold phases were indeed cold can be readily seen from the European biological record, where each successive cold period eliminated part of the local flora and fauna.[3]

No attempt is made here to discuss the zones of the Würm and Holocene, which are considered in chapters 18, 20, and 29. Absolute chronology is treated in chapter 3.

1. There is no general agreement as to whether, for example, the Riss and Elster complexes represent two or four glacials, depending upon evaluation of the intervening warm phases. If the Würm interstadial fluctuations are used as guides, the Mauer interval would merit interglacial rather than interstadial status. As a result, there may be as many as seven or as few as four major glacials, so that terminologies referring to glacials as "first" or "fourth" are apt to promote confusion and are no longer used by Pleistocene geologists.

2. More general designations such as the German *Kaltzeit* (cold phase) and *Warmzeit* (warm phase) would be more suitable than the current improvisations.

3. According to Kurtén (1960a), only 7 per cent of the mammalian species of the Tegelen faunas are recent, whereas already 31 per cent of the Cromerian and 53 per cent of the Cortonian are. The floral decimations are at least equally impressive: the Cromer Forest Bed flora (Duigan, 1963, with earlier references) contains only 5 per cent extinct or exotic (tropical) species; the Tegelen beds (Vlerk and Florschütz, 1953), 41 per cent; the Upper Pliocene Reuver beds, 79 per cent. The Tegelen-age Schwanheim beds (Baas, 1932) have 28 per cent extinct or exotic species, the nearby Upper Pliocene Frankfurt beds (Mädler, 1939), 83 per cent. Similar observations were made by Szafer (1954) at Mižerna, and Huba and Krośćienko in southern Poland. Much of the outright extinction of species and genera may not have been a result of environmental change, but over half of the floral species in question survive in warmer areas today.

Table 2. Basic stratigraphy of mid-latitude Europe.

Climatic Trend	*British Isles**	*North Germany and Netherlands†*	*South Germany‡*	*Stratigraphic Unit§*	
Warm	Postglacial	Postglacial	Postglacial		Holocene
Cold	Hessle Till	Weichsel Till	Würm Till		Würm
Warm	Ipswich Beds	Eem Beds	Stuttgart travertine		Eem
Cold	Hunstanton Till	Warthe Till	Younger Riss Till		Warthe
Warm	Ilford Beds	Treene Beds	(Soil zone)		Treene
Cold	Gipping Till	Saale (Rehburg, Emmen, Drente) Till	Older Riss Till		Saale
Warm	Hoxne Beds	Holstein Beds	Steinheim *antiquus* gravels		Holstein
Cold	Lowestoft Till	Elster Till	Upper Mosbach Sands; glaciofluvial gravels	Biharian	Elster II
Warm	Corton Beds(?)	(Soil and pollen zones)	Middle Mosbach Sands; Mauer Sands	Biharian	Mauer
Cold	North Sea Drift (?)	Süssenborn Gravels	Glaciofluvial gravels	Biharian	Elster I
Warm	Cromer Forest Bed (Middle)	Bilshausen Beds, Voigtstedt Loams	(Soil zone)	Biharian	Cromerian
Cold	Cromer Forest Bed (Lower)	Kedichem Beds	Glaciofluvial gravels	Biharian	Menapian
Warm	Paston Beds	Kedichem Beds		Biharian	Waalian
Cool	Bavents Beds	Kedichem Beds	Glaciofluvial (?) gravels	Biharian	Eburonian
Warm	Norwich Crag (Lower)	Tegelen Beds	Schwanheim Beds		Villafranchian
Mainly Cool	(Marine beds at Ludham)	Praetiglian Complex			Villafranchian

* The British sequence primarily follows West (1956, 1961, 1968) for the glacial and interglacial deposits, except for the interpretation here proposed for the controversial Hunstanton and Ilford deposits. The warm Cromerian substage is defined by the middle unit (temperate, freshwater muds, and transgressive estuarine silts) of the Cromer Forest Bed Series.

†The continental northwest European glacial sequence is largely based on Woldstedt (1958, 1967), with the exception of the warm Treene (Gerdau) interval, which has in at least some instances been confused with the Eem (see discussion by Frenzel, 1969a, p. 85 ff.; also Ruske, 1965; Picard, 1960; Kunert and Altermann, 1965). Although there is no good evidence for a subdivision of the Elster Till in northern Germany and Poland, the Süssenborn Gravels (see Kahlke, 1961, with references) with their excellent fauna indicate a major cold oscillation, followed by a major period of soil formation. Typical Elster glacial deposits rest on the Süssenborn Beds. The sedimentary and palynological sequence at Voigtstedt also shows a major cold phase as well as a second warm phase, following upon the fossiliferous Voigtstedt Loams and preceding the Elster Till (W. Krutzsch, R. Ruske and K. Erd in Kahlke, 1965). The intra-Elster warm phase is given interglacial status on the basis of the Voigtstedt pollen profile. The Elster I glacier did not advance as far south as the Elster II; the related tills were presumably reworked by the Elster II and subsequent glaciers. The Bilshausen beds (Müller, 1965; Lüttig, 1965) and Voigtstedt Loams (see Kahlke, 1965) are probable equivalents of the estuarine silts and freshwater peat of the middle part of the Cromer Forest Bed Series. The earlier Pleistocene sequence of marine and fluvial beds is confined to the Netherlands and the northern Rhineland and is adapted from Quitzow and Zonneveld (1956), Pannekoek (1956), and de Jong (1967).

‡ The Alpine-South German sequence is much in need of revision, particularly the tills, glaciofluvial beds, and paleosol horizons of the earlier Pleistocene. The scheme adopted here follows Müller-Beck (1957, 1964), Kahlke (1961, 1965), M. Kretzoi (in Kahlke, 1965), Ložek (1964), Adam (1964), Brunnacker (1965), and Kurtén (1968); it is based primarily on the bio-stratigraphic horizons, which can be correlated to other European sites with a reasonable degree of confidence. In particular, the Stuttgart, Steinheim, and Mauer beds have emerged as significant horizons of this type (details in Reiff, 1955; Adam, 1953, 1954; Kahlke, 1961; summaries of Steinheim and Mauer are given by Howell, 1960). The correlation of the so-called Mindel, Günz, and Donau (Danubian) glacials, let alone the identification or subdivisions of their depositional record (see Bartz, 1959; Graul and Brunnacker, 1962; Schädel and Werner, 1963; Fink, 1965; also Richmond, 1970), is so highly controversial that usage of the terms should be discontinued. In this connection the discussion of the French Pleistocene by M.H. Alimen (1967) should be considered with caution; the bio-stratigraphic correlations Alimen proposes for the French and German faunas are grossly incorrect. The Leffe sequence of the southern Alpine foothills (Venzo, 1955; Flint, 1965, with references), although very interesting, requires geological restudy.

§ Obviously none of these sequences is ideal for purposes of nomenclature, and rather than propose a rigid, regional terminology, the names adopted here are selected by the criteria of accurate stratigraphic implication and greatest familiarity. German terms are normally used in the nominative form, which may serve both the purpose of noun and adjective in that language; anglicized adjectival forms such as "Holsteinian" or "Elsterian" are just as ill-sounding as "Illinoisan" or "Sangamonian," apart from being redundant. The time-stratigraphic designation Biharian (see Rónai *et al.*, 1961) is proposed for the broad range of poorly understood time between the Tegelen and Holstein. Lastly, Villafranchian, as a time-stratigraphic term, is here used for the time range covered by the Étouaires, Pardines, St. Vallier, Senèze, Val d'Arno and Tegelen beds (Movius, 1949; Viret, 1954; Lüttig, 1962; Kurtén, 1968). However, in view of the 3.35 million year potassium-argon date from Étouaires (Curtis, 1967) there is some question whether this first stage should not be assigned to the Pliocene (see chapter 3).

Table 3. Coastal stratigraphy in the Mediterranean area and in Western Morocco (Shoreline levels given with respect to modern sea level; continental series in parentheses; intraregional and external correlation of Tyrrhenian stages according to Th^{230}/U^{234} dates.)

Climatic trend	*Italy**	*Mallorca*†	*Morocco*‡	*Tentative correlation*
Warm	Versilian (+2 m.)	+4 m., +2 m.	Mellahian (+2 m.)	Holocene
Cold	(Pontinian)	(Regression)	(Soltanian)	Würm
Warm	Tyrrhenian III (+2 to 8 m.)	+0.5 to 2.5 m	Ouljian II (+5 m.)	Late Eem
Cool		(Minor Regression)	(Presoltanian)	
Warm	Tyrrhenian II (+2 to 11 m.)	+2 to 12 m.	Ouljian I, Kebibatian (+8 to 20 m.)	Early Eem
Cold	(Nomentanan)	(Regression)	(Tensiftian)	Saale
Cool		+4 to 5 m.		Late Holstein
Warm	Tyrrhenian I (+16 to 32 m.)	+16 to 34 m.	Anfatian (+25 to 34 m.)	Early Holstein
Cold	(Flaminian)	(Regression)	(Amirian)	
	Milazzian II		Regressive Maarifian (+40 to 45 m.)	Biharian
Warm	Milazzian I	+50, +62 m.	Maarifian (+55 to 60 m.)	
Cold	(Cassian)		(Saletian)	
Mainly Cold	Sicilian	+110 m.	Messaoudian (+90 to 100 m.)	
Warm	Emilian		(Regregian)	Villa-franchian
Cold	Calabrian		(Moulouyian), Upper Moghrebian	

COASTAL STRATIGRAPHY OF THE MEDITERRANEAN SEA AND MOROCCO

Although terminology derived from the Mediterranean sequence is frequently used in attempts at world-wide correlation in distant parts of the Old and New World (e.g., see Woldstedt, 1960a, 1962a), objective correlation of local sequences is difficult even within the Mediterranean. The basis of correlation is the world-wide fluctuation of ocean level reflecting the growth and wastage of the continental glaciers in North America and Eurasia (see ch. 14). In tectonically undisturbed areas, sequences of high sea levels (transgressions) were, in part, a result of partial deglaciation during warm phases of the Pleistocene. Mainly as a result of long-term changes in the morphology of the earth's crust, these higher levels appear to be temporary stages superimposed upon the over-all lowering of world sea level evident since late Pliocene times. There is reason to believe that regional altimetric correlation is possible within tectonically stable areas. However, in default of faunal criteria or isotopic dating such correlations are seldom convincing.

Three sequences, two in the western Mediterranean Basin, one on the Atlantic coast of Morocco, are presented in Table 3. The data consist of alternating marine deposits and continental beds or series. The older marine deposits were largely laid down at considerable depth, and al-

---

* After Blanc (1957, with references), Castany and Ottmann (1957), Gigout (1962), Ruggieri (1965), Selli (1962; also Emiliani *et al.,* 1961), each with references. Throughout the Mediterranean Basin the Tyrrhenian I, II and III are each subdivided into several substages, in part as a result of local tectonics, in part because of primary differences of sea level or general long-term uplift of the continental land masses; only the range of apparently undeformed shoreline levels is indicated here. The differentiation of the Tyrrhenian II and III has not been systematically attempted in Italy, as it has been in southern France (Bonifay and Mars, 1959; Bonifay, 1962) or Mallorca; Gigout (1962) has, however, demonstrated that such a separation is possible, and a single Th^{230}/U^{234} date of Tyrrhenian III age has been obtained from Italy (Stearns and Thurber, 1967). The Calabrian, Emilian, Sicilian, and Milazzian here refer to the massive sequences of marine deposits published in detail during recent years (Selli, 1962; Ruggieri, 1965), not to shoreline levels frequently labeled with these terms. The Cassian, Flaminian. Nomentanan, and Pontinian are continental series, from the area of Rome, providing convincing evidence of cold episodes (see Blanc *et al.,* 1955; Blanc, 1957). As elsewhere in the Mediterranean region there is a peculiar dearth of continental or littoral deposits from the inter-Tyrrhenian II/III regression. In general, the Italian terminology appears suitable for a time-stratigraphic nomenclature of circum-Mediterranean application.

† After Butzer and Cuerda (1962a, 1962b), Butzer (1962), Cuerda and Muntaner (1960), Cuerda and Sacares (1965, 1966), essentially substantiated by ten published (and five further, unpublished) Th^{230}/U^{234} dates (Stearns and Thurber,

‡ After Biberson (1961a), Choubert (1962), Gigout (1960), and Stearns (in Howe, 1967), in part as coordinated by the Th^{230}/U^{234} dates of Stearns and Thurber (1967). The term Ouljian I is used here for the former pre-Ouljian. The fauna of the upper estuarine beds of the Moghrebian is distinctly Villafranchian despite the controversies surrounding the stratigraphic assignment of the Moghrebian as such.

most universal deformation gives them little or no altimetric significance. The Calabrian beds, for example, commonly occur to levels of up to 300 meters or more. Units accompanied by elevation values, on the other hand, refer to direct shoreline features, generally substantiated by coastal deposits with molluscan faunas. The continental beds, distinguished in the table by parentheses, are usually interdigitated with the marine sequence. In part they pertain to deposition accompanying cooler or moister climates; in part, also, they include coastal accumulations of water and wind-borne materials related to sea levels lower than those of today (regressions). With such a range of features, a strict distinction between rock and time-stratigraphic units is impossible. The Italian terminology has very wide application and may be suggested for the basic stratigraphic sequence.

Although the designations "cold" and "warm" have been included in Table 3, European correlations can only be suggested for part of the column. Direct relationships with the North Atlantic deep-sea sequence discussed in chapter 3 are equally problematical. Despite the theoretical limitations (discussed in chapter 14) and the stratigraphic uncertainties, Pleistocene coastal stratigraphy is a valuable chronological tool of sorts. This is particularly the case in the later Pleistocene, where the associations of continental deposits with transgressions or regressions may be crucial in establishing local stratigraphies.

PLEISTOCENE STAGES

In concluding this outline of the basic stratigraphical (as opposed to chronological) problems of the Quaternary, the question of Pleistocene stages must still be raised. Table 4 presents some of the views on Pleistocene subdivisions. Understandably, Pleistocene subdivision is not an arbitrary matter and has in the past been made on both climatic and faunal arguments, The various limits given in Table 4 each have faunal significance in Europe. According to Thenius (1962) and Howell (1966), the fourfold division (third column, Table 4) into Upper, Middle, Lower, and Basal stages provides the maximum number of distinctive faunal and floral units. A distinction between Basal and Lower Pleistocene is also supported by other criteria. So for example, the australopithecine hominids of the Basal Pleistocene contrast with the new genus *Homo* widespread over the Old World during the Lower Pleistocene. Similarly the Basal Pleistocene, as reaching up to the base of the Eburonian, would include the great complex of cold and warm phases prior to the first

Table 4. Pleistocene stages according to different authors.

Stratigraphic unit	*Second Inqua Congress (Leningrad, 1932)*	*Fink (1960), Thenius (1962), Adam (1964), Howell (1966)*	*Lüttig (1959)*	*Woldstedt (1962b)*	*Kurtén (1968)*
Würm Eem	Upper Pleistocene	Upper Pleistocene	Upper Pleistocene	Upper Pleistocene	Upper Pleistocene
Saale and Warthe Holstein	Middle Pleistocene	Middle Pleistocene	Middle Pleistocene	Middle Pleistocene	Middle Pleistocene
Elster Complex Cromerian Menapian	Lower Pleistocene	Lower Pleistocene	(Erfurt Stage) — Lower Pleistocene	Middle Pleistocene	Middle Pleistocene
Waalian Eburonian Villafranchian	Pliocene	Basal Pleistocene*	(Aachen Stage) — Lower Pleistocene	Lower Pleistocene	Lower Pleistocene

* The term *Basal Pleistocene* was introduced by Howell (1959b) as the logical equivalent for the corresponding German *Ältest Pleistozän*.

major alpine glaciation. Stage designations are quite useful for broad temporal identification, and the Basal-Lower-Middle-Upper subdivisions as defined by Howell (1966) will be used below as time (rather than lithological) units. "Late Pleistocene" will be used specifically for the Würm, while "early" and "later" Pleistocene will be employed in a general way only.

THREE

Dating and Absolute Chronology

INTRODUCTION

Chronology, although not the central theme of this study, is a most important aspect of Pleistocene research. The details of late Pleistocene and Holocene stratigraphy have been clarified by absolute dating as much as by the more conventional methods of stratigraphic research. Similarly the duration of the Pleistocene and the later Tertiary have been made more tangible by isotopic dating. Consequently, both the techniques of "absolute" dating and the present information on Pleistocene chronology complement the available stratigraphic information. And both provide the perspective necessary for an effective study of Pleistocene geography.

The major techniques of isotopic dating as well as the problem of "dating" by the "radiation curve" are discussed in a nontechnical manner. The available data are subsequently applied to the broad lines of late Cenozoic stratigraphy.

RADIOCARBON (C^{14}) DATING

Radiocarbon dating today provides almost the entire structure of geologic and prehistoric chronology for the last 60,000 years. In application since 1949, methods have been so rapidly improved that part of the techniques and dates outlined by Libby (1955), in the one existing comprehensive study, are already obsolete. The physical background of radiocarbon dating has more recently been reviewed by Aitken (1961, pp. 88–99).

The basic principle in radiocarbon age determination is that cosmic radiation produces a small fraction of the radioactive carbon isotope C^{14} in the atmosphere. Biological assimilation of carbon, in the form of carbon dioxide, also includes a certain fraction of the C^{14} isotope. This is then present in the cellular structure of all organic creatures, plant or animal. Similarly, bodies of water absorb atmospheric CO_2, which is exchanged between the atmosphere and hydrosphere, so that inorganic as well as organic sedimentation of carbonates includes a definite ratio of C^{14} to C^{12}.

Immediately after absorption of this carbon isotope, directly by plants and indirectly by animals through the intermediate agency of plant tissue, C^{14} begins to disintegrate. The rate of beta ray emission is such as to reduce radioactivity to a half after about 5,730 years, the so-called "half-life" period. Thereafter the rate of decay slows down and diminishes exponentially. Sensitive geiger counters can trace this radioactivity and so enable a determination of the stage of disintegration of the isotope present. This would theoretically determine the age of the isotope and, indirectly, of the organism of which it forms a part.

As a consequence of this fortuitous isotope, all organic materials such as charcoal, wood, peat, charred antler and bone, hair, hide, leaves, nuts, soil humus, organic lake and swamp beds, freshwater limestones, and spring tufas or travertines can, theoretically at least, be measured for radioactivity and hence age. This is the essensce of radiocarbon dating. The potential of radioactive dating is not unlimited, however.

The practical dating span of C^{14} is limited to about 35,000–40,000 years, since radioactivity is not measurable with accuracy beyond that time. Fortunately, Haring *et al.* (1958) were able to apply artificial isotopic enrichment to samples exceeding 40,000 years in age. Although the increased effect of contamination by modern carbon is, in practice, considerable, the span of dating is theoretically extended to about 70,000 years. This appears to be the outer time range of the C^{14} technique. Only the Groningen Laboratory of the Netherlands is so far capable of dating beyond 40,000 years.

SOURCES OF ERROR IN RADIOCARBON DATING

Errors of three kinds reduce the absolute dating value of the technique: (a) statistical-mechanical errors, (b) errors pertaining to the C^{14} level of the sample itself, and (c) errors related to laboratory storage, preparation, and measurement. These facts may be outlined briefly.

a) A statistical-mechanical error is present as a result of the random, rather than uniform, disintegration of radioactive carbon. This is expressed in the date by a plus-or-minus value in years (e.g., 6,240 ±320 yrs.). The expression conveys that the chances are 67 per cent that the value lies between the stated extremes (e.g., between 5,920 and 6,560 years), and 96 per cent that the value lies within twice that amplitude (e.g., between 5,600 and 6,880 years). This statistical error can be reduced by increasing the time of measurement. It is of no great concern, however, as the practical sources of error inherent to the sample itself are far more serious.

b) Sources of error in the C^{14} content of a sample may be a result of (1) past fluctuations of the C^{14} concentration in the C^{14} exchange reservoir; (2) unequal C^{14} concentration in different materials; and (3) subsequent contamination of samples *in situ*.

1) Fluctuations in cosmic radiation with time may produce slight differences in the C^{14} equilibrium of atmosphere, hydrosphere, and biosphere. Damon *et al.* (1966; also Damon, 1968; and Dyke, 1967) determined the C^{14} content of dendrochronologically dated tree rings for the last 6,300 years, observing a strong increase in the C^{14} content of the atmosphere between 2,000 and 6,000 years ago. As a result, C^{14} dates for this particular period are too low, a fact accounting for serious discrepancies between C^{14} and historical dates for Egyptian materials (see H. S. Smith, 1964). To correct this divergence for the period 2,000 to 6,000 years before the present, Stuiver and Suess (1966) suggested the following correction formula:

$$T \text{ (true age)} = 1.4R \text{ (radiocarbon ``years'')} - 1100.$$

So, for example, a radiocarbon date of 4,000 years should approximate a calendar age of 4,500 years.

It appears that the irregular fluctuations of the C^{14} reservoir during the last 2,000 years are related to fluctuations of solar activity and cosmic-ray intensity, while earlier trends are related to long-term changes of planetary temperature (Damon, 1968). Proir to 6,000 years, crude checks provided by other isotopic dating methods (see Broecker, 1965) indicate that consistent C^{14} dates, on the whole, are of the right order of magnitude. Comparisons with the Scandinavian varve-ring chronology (De Geer, 1940; Zeuner, 1958) do, however, suggest that the level of atmospheric C^{14} was lower during the terminal phases of the last glacial. If Damon's (1968) correlation of atmospheric temperature and C^{14} content proves to be correct, C^{14} ages prior to 9,000 years ago would be too great. However, the potential discrepancies will not be too serious, probably being less than 500 years for the time range 10,000–15,000 years ago, and less than

1,000 for the period of finite dating prior to that. The potential error would, therefore, normally be less than the margins of mechanical error inherent in C^{14} determinations of that age.

The addition of "old" carbon to the atmosphere through burning of fossil fuels (since about 1850) has no direct effect on geologic or prehistoric samples. However, many dating laboratories in the past employed recently grown wood as a measure of C^{14} exchange activity (see Suess, 1955). The first dates published by some laboratories are consequently several hundred years too young. Increased production of C^{14} by H-bomb tests raised the radiocarbon activity of recently grown plants by 25 per cent between 1954 and 1959 (Broecker and Olson, 1960). Although this is not of immediate significance, the penetration of strontium-90 in percolating rain waters may possibly contaminate soils and sediments to considerable depth.

2) Different substances or bodies of water that take up carbon from the atmosphere absorb C^{14} at different rates than C^{12}. So, for example, in the exchange of carbon dioxide between atmosphere and ocean, there is a preferential uptake of C^{14} by the ocean so that C^{14} concentration is 1.2 per cent higher than in the atmosphere (Aitken, 1961, pp. 100–102, 106). There is a similar, but reversed, isotopic fractionation between the atmosphere and plant life, amounting to as much as 4 per cent. As a result, marine shells will be a little too young, organic carbon samples a little too old. Uncorrected fractionation of 1 per cent introduces an error of only 80 years, so that this effect is insignificant for samples of prehistoric age. For historical samples this source of error can be checked by a C^{13}/C^{12} determination: the isotope C^{13} is stable through time, and its concentration permits a good estimate of the fractionation effect.

Quite apart from isotopic fractionation, different samples do not always provide consistent dates. In part, these basic discrepancies involve an imperfect carbon dioxide equilibrium between the original organism and the atmosphere. This effect is important in the case of fresh waters (lakes and rivers), since the equilibrium of CO_2 in atmosphere and hydrosphere is frequently unbalanced by the introduction of radioactively dead carbonates in solution, derived from older rocks rich in calcium carbonate. Consequently, aquatic plants or shells will build "dead" carbon dioxide or carbonates into their organic structures, giving C^{14} dates that may be rather too high (Deevey *et al.,* 1954; Keith and Anderson, 1963). Another fundamental problem is the susceptibility of a substance to later contamination, depending on its porosity and permeability, or its stability with respect to chemical replacement. The suitability of individual substances for C^{14} dating is discussed further below.

3) Geobiochemical contamination of samples *in situ* is the worst and most common offender in C^{14} work. Soil processes are effective to considerable depths in soil and sediment, and humic acids, other organic decay products, and fresh calcium carbonate are carried down by percolating waters. Similarly, modern root networks and microorganisms penetrate deeply into the soil, while earthworms, ants, and burrowing mammals transport younger and older materials bodily—vertically as well as sideways. Many of these problems can be avoided by the judicious selection and removal of samples from undisturbed beds, preferably by an earth scientist. Roots, root hairs, and other macroscopic contaminants can be removed directly, with the help of a hand lens and tweezers. In the case of charcoal, wood, or peat, younger carbonates are washed out with a hydrochloric acid solution while humic acids and microscopic humic contaminants can be partly neutralized by treatment with sodium hydroxide. However, a small and unpredictable residue of younger carbon contaminants almost always remains. This becomes very important in older samples, since only 1 per cent recent humic contaminant in a carbon sample of truc agc 67,000 ycars will yield an apparent age of 37,000 years; for a sample of 23,000 years, 1 per cent of modern carbon increases the apparent age by 5 per cent or 1300 years (Aitken, 1961, p. 97). In the case of shell, bone carbonate, or inorganic carbonates, pretreatment must be limited to acid leaching of the outer layers of the sample. Younger (or older) carbonates precipitated within pore spaces or fissures can hardly be removed, posing a considerable potential source of error for bone and porous carbonate rock. In general, therefore, contamination involves younger substances, so that whenever "dead" carbon can be excluded, aberrant C^{14} dates tend to be too young.

c) In addition to the many sources of inaccuracy already discussed, laboratory processing has introduced a number of further undesirables.

So, for example, the half-life of C^{14} once calculated at 5,570 years, was remeasured to be about 5,760 years by the National Bureau of Standards in 1961. Godwin (1962) suggests use of 5730 $\pm$40 years as a mean of three new determinations. Since laboratories continue to use the 5,570 year half-life in publication, dates may be converted by multiplying them by 1.03.

Early C^{14} work measured the radioactivity of solid carbon, a technique discontinued since about 1955 in favor of more efficient procedures involving less risk of current radioactive contamination. Most widely used are gas proportional counters, for which the carbon is converted into CO_2 and, less frequently, acetylene (C_2H_2) or methane (CH_4). The various techniques of measurement are discussed by Aitken (1961, pp. 110–16).

In addition to the problems of different half-life assumptions, variable efficiency of radiocarbon counters, and variance of results stemming from different techniques, there is the problem of preparatory cleansing of samples. The use of hydrochloric acid or sodium hydroxide or both produces appreciable differences commonly varying by 10 per cent or more. Differences of 20 per cent or more were obtained between some "preliminary" and "final" dates in the 30,000-and-more time range (after special treatment for recent humus contamination) by the Groningen laboratory. In other words, the available dates, running into many thousands, are of variable reliability. So far the unpleasant but necessary reappraisal of dates has not been formally undertaken. Many disciplines will benefit when and if such a survey is made. Publications should always refer to radiocarbon dates by laboratory and number so that readers may have a ready reference to check.

SAMPLE REMOVAL AND SUITABILITY FOR RADIOCARBON DATING

Removal of samples for C^{14} dating is conditioned by the nature of the material. Samples should only be collected when there is no doubt that they are contemporary with the archeological or geological bed to be dated. So, for example, great care should be exercised with materials apparently "weathering out" of a horizon. Materials from superposed horizons should not be mixed, and samples should be carefully labeled to avoid later error or confusion. Organic carbon is susceptible to contamination during removal (by dirty or greasy tools and by younger organic materials falling down section faces) and transport (by poorly sealed or contaminating containers; e.g., plastic, paper, wood, etc.). Clean glass bottles or new aluminum foil should be used to pack samples of organic carbon, avoiding all grease, lubricants, or preservatives. Shell, bone and carbonate rock need only be kept dry and clean, since surface contaminants are removed during superficial acid treatment.

The kinds of material selected for C^{14} dating are normally dictated by what is available. Charcoal and wood are generally considered optimal since they can be readily treated by sodium hydroxide. Peat as well as "carbonaceous soil" and ash, consisting of charcoal powder, carbonized plant tissue, and inert minerals, are more readily contaminated and more difficult to clean, since the sodium hydroxide breaks down all but the larger organic particles. Charred bone and antler are quite suitable (Münnich, 1957), although uncharred bone is problematical. Bone protein or collagen can be dated for bone less than 10,000 years old (see Dyck, 1967), but later contamination by extraneous organic compounds is difficult to eliminate, commonly yielding dates that are too young (Haynes, 1967). Equally difficult is the dating of the organic protein or conchiolin of shells.

The calcium carbonate of bone is also susceptible to C^{14} dating, but is quite prone to removal and replacement by younger or older carbonates carried in soil or groundwater. Shell is normally dated by its calcium carbonate (Rubin *et al.,* 1963). In the case of freshwater shell, older carbonates in solution in rivers, lakes, or ponds are incorporated into the original shell structure, giving dates that may be considerably too old. If ancient limestones are absent from the drainage system, this objection is irrelevant; otherwise, the level of possible contamination can be estimated by dating a sample of modern shell. Subsequent carbonate replacement becomes a serious problem for all shell greater than 20,000 years old, so that if original contamination by "dead" carbonate can be excluded, shell dates in excess of 20,000 years commonly provide only minimum ages. The calcium carbonate of spring and lake deposits (travertine, tufa, etc.) as well as soil horizons (caliche) can also be dated, with reservations similar to those applying to freshwater shell (see Münnich and Vogel, 1959). Porous or noncompact forms are particularly prone to resolution and recrystallization by waters carrying younger lime in solution. Even the groundwater reservoirs of desert regions can be dated, since CO_2 is originally dissolved in rain water (Brinkmann *et al.,* 1960); however, mixing of older and younger waters as well as addition of "dead" carbonates can influence such dates.

The sample size required varies according to the material:

Sample type	*Optimal size* (grams)	*Minimum* (grams)
Charcoal	8–12	1
Peat	10–25	3
Wood	10–30	3
Charred bone	10–30	2
Uncharred bone (carbonate)	50–200	5
Uncharred bone (collagen)	200–500	100
Shell (carbonate)	30–100	5
Shell (conchiolin)	500–2500	200
Inorganic carbonate	30–100	5

Large samples are particularly desirable for older materials.

Other vegetable or animal products such as leaves, nuts, paper, parchment, cloth, skin, hide, or hair can be dated but are seldom or never present in prehistoric associations.

All in all, C^{14} dates are no better than the samples run. Due to the many, often unpredictable, and sometimes major sources of error, isolated dates must be recorded with caution. However, when older and younger levels at a site or from a local stratigraphic scheme provide internally consistent dates, the level of confidence is greatly increased. Despite the fact that C^{14} dates are not absolute, they have revolutionized the field of geochronology and still provide the most useful single tool for stratigraphy in the 5,000 to 60,000-year time range.

Current theoretical and applied developments in the field can be readily followed in the journal *Radiocarbon* (since 1959).

POTASSIUM-ARGON DATING

Next to radiocarbon, the most spectacular results in isotopic dating have been obtained with the potassium-argon method (Evernden and Curtis, 1965; Miller, 1967; Dalrymple and Lanphere, 1969). Fresh volcanic rocks contain small qauntities of the isotope potassium 40; but are free of the heavy gas argon 40. Decay of K^{40} produces Ar^{40} in the mineral; the Ar^{40} is quantitatively retained and can be distinguished from Ar^{40} absorbed from the atmosphere. With a half-life value of 1330 million years, the ratio K^{40}/Ar^{40} can provide ages for a time span that is theoretically unlimited. However, datable materials are restricted to unaltered, potassium-rich minerals of volcanic origin, primarily basalt, obsidian, micas, and feldspars of lava, rock intrusions and, to a lesser degree, ash falls. Apart from mineral alteration, with absorption of atmospheric Ar^{40}, the major source of error is the inclusion of slight quantities of older potassium-rich minerals into a sample. Careful laboratory examination and preparation can normally, but not always, preclude such sources of error.

Although K/Ar dating had already been used to date rocks of great geological age, application of the method to the Pleistocene and Pliocene requires a refinement of techniques that has only been achieved quite recently. There is now a substantial body of dates (for example, Evernden and Curtis, 1965; Evernden *et al.,* 1964; Frechen and Lippolt, 1965) that have added considerably to our understanding of the duration of the Pleistocene, the age of early man, and the correlation of stratigraphic subdivisions from place to place. Unfortunately, a fair percentage of the available dates under 1 million years are erratic, usually too high (see discussion comments in Evernden and Curtis, 1965) so that K/Ar determinations in the late to mid-Pleistocene time range are to be accepted with caution. The technical refinements offered by the new Ar^{40}/Ar^{39} method (Dunken *et al.,* 1958) minimize the inherent inaccuracies of the K/Ar dating and are particularly suitable for younger and small-sized samples.

A useful cross-check for the internal consistency of K/Ar dates is provided by paleomagnetic stratigraphy (Cox *et al.,* 1965; Cox, 1969). Unaltered lavas that remain in place undisturbed preserve a record of the earth's magnetic field at the time of their cooling. Apart from the minor movements of the magnetic poles with time, magnetohydrodynamic processes in the earth's fluid core have repeatedly reversed the positions of the north and south magnetic poles. Consequently, paleomagnetic data shows a bimodal distribution in "reversed" and "normal" fields. K/Ar dating has demonstrated the existence of normal polarity "epochs" from 690,000 years to the present, and again from 3.3 to 2.4 million years ago, of reversed polarity epochs prior to 3.3 million years and again between 2.4 and 0.69 million years ago. Unfortunately, each of these epochs was interrupted by a number of brief "events," characterized by rapid switches of the earth's magnetic field. Although the paleomagnetic record is correspondingly complicated (see Opdyke *et al.,* 1966; Dagley *et al.,* 1967), it nonetheless provides opportunity to check the consistency of K/Ar dates.

A cross-check on the "absolute" calibration of K/Ar dating has been provided by fission-track dating of volcanic glass from Bed I, Olduvai Gorge (Fleischer *et al.,* 1965). This method uses different assumptions and is prone to other sources of error. The number of "tracks" caused by spontaneous fission of U^{238} during the "life" of the sample are counted. Age is determined by obtaining the ratio of the density of such tracks to the number of uranium atoms, which is obtained from the increase in track density produced by fission of U^{235}. The Olduvai cross-check provided a fission track age of 2.0 million years that compares reasonably well with K/Ar dates averaging about 1.8 million years (see Evernden and Curtis, 1965).

URANIUM SERIES DATING

Thorium-Uranium (Ionium-Deficiency) *Method.* Mollusks, marine coral, and freshwater carbonates contain uranium 234 at or shortly after death, but no thorium 230, a daughter element of uranium that is virtually insoluble in natural waters. If the fossil carbonates subsequently remain closed to isotopes of the uranium series, the amount of Th^{230} present will reflect the original concentration of U^{234} and the period of isotopic decay. Since the half-life of U^{234} is 248,000 years, and that of Th^{230} 75,000 years, Th grows to equilibrium with U in about 500,000 years. In the meanwhile, the ratio Th^{230}/U^{234} is a function of age (Broecker, 1965; Veeh, 1966; Stearns and Thurber, 1967). The effective dating range of this ionium-deficiency method is 200,000 to 300,000 years.

The major source of error is the introduction of foreign uranium and its daughter products after the death of the organism. To some extent this type of contamination can be screened out, but isolated age determinations cannot be accepted with any great confidence. The Th^{230}/U^{234} technique has been applied to the study of Pleistocene beaches and lake beds in different parts of the world and there has been sufficient internal consistency as well as consistency with accepted geological correlations to warrant a moderate degree of optimism. In effect, this method has proved crucial for correlating littoral deposits of the last two interglacial periods, both in relation to the radiocarbon-dated portions of the Würm glacial, and to the apparent temperature fluctuations recorded in organic oozes of the deep-sea floor.

U^{234} Method. Uranium is present in carbonate solutions in very small concentrations. It is fixed after sedimentation and, barring possible contamination, is not susceptible to outside addition or loss. U^{234}, the daughter element of U^{238}, is originally present in the carbonate solutions but increases through radioactive decay after sedimentation. If the initial proportions of U^{234} and U^{238} are known for a particular depositional medium, the U^{234}/U^{238} ratio of a fossil carbonate provides an approximate date with a potential dating range of 1 to 1.5 million years (Thurber, 1962). This ratio is fairly constant (1.15) in marine waters but rather variable in fresh waters. Dating of coral has yielded U^{234} ages consistent with independent age formation (Veeh, 1966), although marine mollusks receive an unpredictable contribution of uranium from surrounding sediments, so that they are less reliable. Dating of freshwater carbonates, such as travertines, has been attempted after establishing the U^{234}/U^{238} ratio for modern waters (Cherdyntsev *et al.,* 1966). However, such ratios will vary considerably through time, and the resulting dates are not particularly consistent.

Protactinium-Thorium (Protactinium-Ionium) Method. The decay of uranium (U^{234} and U^{238} in ocean waters is attended by the formation of daughter elements, Pa^{231} and Th^{230}, which accumulate in deep-sea sediments. Being produced by the same element, the ratio of Pa^{231} (half-life, 32,500 years) and Th^{230} (half-life, 75,000 years) should be unaffected by the concentration of uranium in marine waters and should be a function of time only, independent of changes in geological conditions (Rosholt *et al.,* 1961, 1962). These elements are generally but not always resistant to postdepositional diffusion within the sediment (Broecker and Ku, 1969).

A fair number of deep-sea cores, recording glacial and interglacial climatic changes (see ch. 17), have already been dated, providing valuable information on the glacial chronology for a time range of 140,000 years. The dates so far obtained are moderately, although not entirely, consistent with thorium-uranium and U^{234} dates.

Although this method, in its present form, is aimed at deep-sea sediments, some interesting but inconclusive results have also been obtained from freshwater limestones (Rosholt and Antal, 1963).

THE PROBLEM OF MILANKOVITCH'S "RADIATION CURVE"

Through much of the archeological and even the geological literature, the basic concept of Pleistocene chronology is defined by so-called absolute dates obtained from a so-called radiation curve. This curve was originally outlined by M. Milankovitch on the basis of astronomical theory since 1920 and was elaborated in a number of later articles (1941, with earlier references).

The basic concept is one of constant total solar radiation with variations in time of the latitudinal and seasonal distributions. Three periodicities of the earth's orbital geometry are involved: (1) The angle of the ecliptic. Earth's axis to its orbital plane is oblique, varying from 65°24′ to 68°21′ over a period of between 38,000 and 45,000 years. When the obliquity is increased, seasonal contrasts are magnified, and vice versa. (2) The precession of the equinoxes. The date of perihelion, the day when the sun and earth are closest (currently about January 2), shifts around the calendar with a period varying from 16,000 to 26,000 years. About A.D. 1200, perihelion was on December 21 (reducing winter cold on the northern hemisphere), while about 9300 B.C. its date was June 21 (increasing summer warmth in the northern hemisphere). As a result, this precession alternatingly increases or decreases seasonal contrasts, in opposite directions for either hemisphere. (3) The eccentricity of the orbit. Earth's orbit deviates from a true circle to an ellipsoid, with a variable eccentricity in a poorly known periodicity of perhaps 92,000 years. This again influences seasonal contrasts. All together these periodicities of orbital geometry do not affect total radiation received from the sun, but they do affect the degree of continentality, the latitudinal distribution of incoming radiation, and hemispheric contrasts. This is the astronomical background of the radiation curve. Its tenets have been critically reviewed by Carpenter (1955).

Since 1920 these variables have been computed and recomputed into various "radiation curves" that span the last 600,000 years: the most significant of these is for summer at 65° North latitude. The assumption is made that cool summers will favor glaciation with a time lag of several millenia; eventually the growing glaciers will increase the albedo of higher latitudes, providing a powerful self-perpetuation or "feedback" mechanism to increase the heat deficit of the glaciated continents (Wundt, 1933). Since Antarctica has remained glaciated since the end of the Tertiary, and since there is little or no land in the higher mid-latitudes of the southern hemisphere, the opposing hemispheric trends are felt by some to be unimportant, with the alternation of glacial and interglacial phases essentially controlled by changes in the northern hemisphere (see Emiliani and Geiss, 1957). Further climatologic refinements were carried out by Wundt (1944), but have been criticized from the meteorological perspective by Berg (1949). Through the elaborate presentations of Soergel (1937) and his student, Zeuner (1958, 1959), a hypothetical glaciation curve or chronology, built on the foundation of this radiation curve, found its way into the core of chronological thought. Geologists such as Woldstedt (1929, 1954) and Flint (1957; 1st ed., 1947) evaluated the "curve" in the light of geological evidence and found it unsatisfactory. Strong evidence favors the synchronous nature of climatic changes and oscillations in both the northern and southern hemisphere, in conflict with the basic requirements of the "astronomical theory." Equally disconcerting has been the wide range of interpretive variations by different authors who successfully "fitted" the curve oscillations to any number of glacial or interglacial phases.

Despite the almost general reservation of Pleistocene stratigraphers and meteorologists today, there has been no lack of attempts to revive the "curve"—e.g., Van Woerkom (1953) modified the curve itself, while Emiliani and Geiss (1957) and Broecker and van Donk (1969) have attempted to reconcile the "curve" and the geological evidence. Isotopic dating of interglacial coral reefs and deep-sea sediment cores by Broecker *et al.* (1968) does, indeed, show possibilities of background significance of the "curve" for Pleistocene temperature variations. However, recent computations by means of thermodynamic models of the general atmospheric circulation show that the surface temperature changes related to the variations in insolation computed by Milankovitch have been quite small (Sawyer, 1966; Shaw and Donn, 1968), although the slight changes in differential heating of different latitudinal belts may have been sufficient to affect upper air pressure and wind patterns (Kutzbach *et al.,* 1968). Yet without further quantitative analysis of the Milankovitch effect, the recurrent attempts at curve-fitting are premature and without physical foundation. Present information is insufficient to prove or disprove the value of the astronomical radiation curve, and considerably more evidence must be made available before a reliable and satisfactory analysis can be made. Great caution must be exercised in the meantime, and it is not scientifically possible to use the curve for any dating purpose.

OTHER METHODS OF ABSOLUTE DATING

Absolute dating is not confined to isotopic dating. Absolute chronological information has also been obtained by geological methods such as varve counting (see chapter 11), by tree-ring analysis or dendrochronology (see chapter 16), and by pottery-dating techniques such as magnetic dating (Aitken, 1961, pp. 121-55; Cook, 1963) and thermoluminescence (Hall, 1963). Dates have also been estimated from the thickness of weathering films on obsidian artifacts. This obsidian hydration method has a greater dating range than the pottery-dating techniques (Friedman *et al.*, 1963; Katsui and Kondo, 1967), although substantial results must be available before an evaluation can be made.

In a much greater time range, flourine dating of bone (Richter, 1958) and the evolutionary concept of longevity and half-life of mammalian and molluscan species (Kurtén, 1959a, 1960b, 1968) have provided interesting although tentative information.

ABSOLUTE DATING APPLIED TO PLEISTOCENE STRATIGRAPHY

The broad stratigraphic units of the Pleistocene outlined in Tables 2 to 4 become more meaningful in the light of absolute dating. Some of the units or boundaries for which information is available may be discussed briefly.

The Pleistocene-Holocene Boundary. The first absolute date calculated for the Pleistocene-Holocene boundary was the varve-ring count (see chapter 11) of De Geer (1940), namely 7912 B.C. A large number of C^{14} dates, published or compiled in a great variety of sources, suggest a value of about 8300 B.C. or 10,250 years before the present (B.P.)[1] (see Nilson, 1964a, 1964b), assuming a C^{14} half-life of 5,730 years. This discrepancy may be entirely due to fluctuations of the C^{14} reservoir (see Damon, 1968), so that the varve date is probably closer to the truth. Consequently, the conventional date of 10,000 B.P. is best retained for the Pleistocene-Holocene boundary.

1. Most isotopic dates are quoted with B.P. (before the present) rather than B.C. values. For practical purposes B.P. values are readily converted to years B.C., when greater than 10,000 years, by subtracting 2,000 years. For younger periods, however, subtraction of the generally accepted "zero" date A.D. 1950 is bothersome, and the concurrent use of B.P. and B.C. values confusing. A compromise solution is adopted here, quoting B.P. values throughout, except for chapters dealing primarily with Holocene time ranges (chapters 31 to 34).

The Eem Interglacial. The Eem Interglacial, as currently understood from mid-latitude Europe, lies just beyond the theoretical dating range of C^{14} (with the aid of isotopic enrichment), and van der Hammen *et al.* (1967) extrapolate the age of the Eem-Würm boundary at "greater than 70,000 years." The type area of the Eem in North Germany records a marine transgression (see Woldstedt, 1958; pp. 21 ff.), the last high interglacial sea level (see chapter 14 for discussion of glacio-eustatic fluctuations of sea level) recorded in Germany. In the Mediterranean Basin and on the western coast of Morocco, the last high sea level of Pleistocene age is the Tyrrhenian III or Ouljian 2; the eight available Th/U dates (Stearns and Thurber, 1967, also unpublished) on stratigraphically secure samples range between 75,000 and 95,000 B.P., averaging 84,000±5,000 years. On Barbados the youngest interglacial sea level is dated 82,000±2,000 years by the average of four Th/U dates (Broecker *et al.,* 1967). It therefore seems quite reasonable to correlate the termination of the Eemian with the end of the Tyrrhenian III transgression, and to assume an age of 75,000 B.P. for the Eem-Würm boundary.

The last interglacial was a complex affair, however. This is shown by the littoral stratigraphy and Th/U dating of key shoreline sequences, such as those of Barbados (Mesolella *et al.,* 1969; Broecker *et al.,* 1968) and Mallorca (Butzer and Cuerda, 1962a; Stearns and Thurber, 1967, also unpublished). The high beaches of the last interglacial average 115,000 B.P. on the California coast (Veeh and Valentine, 1967), 120,000 B.P. for a variety of islands in the Pacific and Indian Oceans (Broecker and Thurber, 1965; Veeh, 1966), and 122,000 B.P. on Barbados. In the Mediterranean region the complex Tyrrhenian II stage, with its thermophile mollusca, was separated from the Tyrrhenian III only by a minor regression. Several dates cluster about 120,000 B.P., although there are further scatters of uncertain interpretation near 160,000 and 200,000 B.P. Another line of argument is provided by the oxygen isotopic analysis of deep-sea cores (see chapter 17), which possibly provides ocean surface-water paleotemperature records (see Emiliani, 1966, with references) and certainly records glacial-interglacial alternations (Olausson, 1965; Shackleton, 1967; Dansgaard and Tauber, 1969). On two Carribbean cores variously dated by Pa and Th isotopes, the last "warm" zone of the Pleistocene can be approximately dated 125,000 to 75,000 B.P. (Broecker and van Donk, 1969). Similarly, oxygen isotopic analysis of the annual ice increments recorded by the Greenland ice sheet indicate that the last interglacial complex was approximately as long as the last glacial (Dansgaard *et al.,* 1969). Finally, unpublished U^{234}/U^{238} dates recently obtained by Russian workers from the Lower and Upper Travertines at Weimar-Ehringsdorf (H. J. Müller-Beck, personal communication), a type site for Eem faunal and floral spectra, fall into this same time range.

In overview, there seems to be little question that the Eem-type horizons, as well as the last major Pleistocene warm interval of interglacial status, fall within the time span of 125,000 to 75,000 B.P. The duration of the Würm Glacial can be estimated at 65,000 years, that of the Eem Interglacial at 50,000 years (see Table 5).

Middle and Lower Pleistocene Events. Dating of glacial and interglacial phases during the mid-Pleistocene is complicated by problems of nomenclature and correlation, as well as by a much wider scatter of partly contradictory isotopic dates. Available information[2] indicates cold periods of glacial status at ca. 170,000–125,000; at ca. 265,000–225,000; and again at ca. 340,000–300,000 B.P. Protracted warm intervals of interglacial status are indicated at ca. 225,000–170,000; at ca. 300,000–265,000; and again at ca. 380,000–340,000 B.P.

By simple stratigraphic argument, correlating the deep-sea units with littoral or continental dates, these cold phases could be correlated with Elster II, Saale, and Warthe, while the warm phases could be correlated with Mauer, Holstein and Treene. Such a sequence of dates, as shown by Table 5, involves a minimum amount of contradiction, although

2. The least subjective guide to large-scale glacial-interglacial alternations of the middle Pleistocene is provided by Emiliani's (1964, 1966) deep-sea "paleotemperature" curves, as dated and reinterpreted by Broecker and van Donk (1969) and Broecker and Ku (1969). The Tyrrhenian I of the Mediterranean region lies beyond the range of Th/U, with dates variously given as "greater than 200,000" and "greater than 300,000 years" (see Stearns and Thurber, 1967). A U^{234}/U^{238} determination on mollusca from a Tyrrhenian I-equivalent in southern Russia gave "280,000 years or greater" (Cherdyntsev *et al.*, 1966).

Isotopic information on continental sequences is still very fragmentary. Vulcanics related to interglacial incision of the Middle Rhine gave K/Ar dates of 150,000 and 140,000 B.P., while the preceding period of cold-climate alluviation ("middle Middle Terrace") began ca. 220,000 B.P. (Frechen and Lippolt, 1965). The stratigraphic assignment of these events is uncertain since their relationship to the well-established lower Rhine sequence is not quite clear. However, the correlation of the oldest Middle Terrace (base of alluvium) and the younger Main Terrace (upper part of alluvium) in the middle and lower Rhine region are not disputed. These are assigned to the Elster Glacial (see De Jong, 1967; Kempf, 1966; Quitzow and Zonneveld, 1956). The seven K/Ar dates range from 350,000 to 300,000 B.P. (Frechen and Lippolt, 1965), and provide one of the few dated horizons for the mid-Pleistocene. The older Main Terrace, conventionally correlated with the Menapian Glacial (by geologists who do not recognize a major cold phase between the Menapian and "classical" Elster) in the lower Rhine valley, has five K/Ar dates ranging from 420,000 to 390,000 B.P. In the Rome area a single K/Ar date of 270,000 appears to date the warm interval (Tyrrhenian I) preceding the cold Nomentanan (Evernden and Curtis, 1965), with a date of 430,000 for the earlier Flaminian cold period. Finally, Cherdyntsev *et al.* (1966) obtained U^{234}/U^{238} ages of 225,000 and 362,500 years for the Elster-age site of Verteszöllös in Hungary.

Table 5. Stratigraphy and chronology of the Pleistocene (with approximate date for beginning of units).

Holocene			10,000 B.P.
Pleistocene	Upper	Würm	75,000 B.P.
		Eem	125,000 B.P.
	Middle	Warthe	170,000 B.P.
		Treene	225,000 B.P.
		Saale	? 265,000 B.P.
		Holstein	? 300,000 B.P.
	Lower	Elster II	? 340,000 B.P.
		Mauer	? 380,000 B.P.
		Elster I	? 430,000 B.P.
		Cromerian	
		Menapian	
		Waalian	
		Eburonian	
	Basal	Villafranchian	(about 3 million B.P.)

alternative correlations are possible. By implicitly correlating the older Main Terrace of the Rhine with the Flaminian on the one hand and the Elster I on the other, a time span of about 430,000 to 300,000 B.P. is obtained for the Elster Complex. This is reasonably compatible with Kurtén's study of mammalian and molluscan evolutionary rates and extinctions, whereby the Cromerian would have begun about 500,000 B.P. (Kurtén, 1960b, 1968, ch. 19).[3] Similarly, Richter (1958) obtained

3. See Müller (1965) and Shackleton and Turner (1967) regarding the minimum duration of the Cromerian.

interpolated dates of 230,000 B.P. for the Holstein and 640,000 B.P. for the Cromerian by studying rates of fluorine increase in various fossil bone beds. Recalling both the fallibility of isotopic dates, in a time period where few cross-checks are currently available, and the problematical nature of the stratigraphic correlations involved, it nevertheless seems advisable to regard the available dating frameworks for the Saale, Holstein, Elster, Cromerian, and Menapian with great caution.[4]

The Villafranchian. There are a large number of K/Ar dates on fossiliferous strata pertaining to various stages of the Villafranchian (Evernden and Curtis, 1965; Evernden *et al.*, 1964; Curtis, 1967; Howell, 1968). If we accept the first known appearance of primitive archidiskodont elephants in the Lower Omo Basin (East Africa) as a Pleistocene index fossil, these dates would cover a time span from greater than 4.25 million to less than 1.5 million years. Considering the relatively short duration of the Pleistocene glaciations, such a time span of 2.75 million years staggers the imagination. To be sure, Kurtén (1968, 261 ff.) concedes that faunal turnover was only a quarter as fast during the Villafranchian as it was during the later Pleistocene. However, large segments of the Pliocene geological column have already been added to the Villafranchian by use of fossil criteria, and this steady attrition of the Pliocene is bound to continue until a radiometric boundary is drawn and accepted for the base of the Pleistocene. Possibly, in accordance with the spirit of the 18th International Geological Congress in London, 1948, this boundary should be drawn where the first indications of continental glaciation at higher latitudes and of mountain glaciers at high elevations come to light.

In the Sierra Nevada of California, major mountain glaciation is first recorded after 3.1 million and shortly before 2.7 million B.P. (Curry, 1966). Extensive glaciation is recorded on Iceland about 3.1 million B.P. (see Dagley *et al.*, 1967), in the Andes of southern Argentina before 3.2 million B.P. (Mercer, 1969), on New Zealand about 2.5 million B.P. (Stipp *et al.*, 1967), and in the interior of Antarctica prior to 2.7 million B.P. (Armstrong *et al.*, 1968). Large-scale freezing of sea ice around

4. The deep-sea curves of oxygen isotopic data (Emiliani, 1964, 1966) and microfaunal variation (Olausson, 1961a, 1961b; Ericson *et al.*, 1961; Ericson and Wollin, 1968) are not without their controversial aspects. Questions of interpretation of the actual core data (see Shackleton, 1967; Shackleton and Turner, 1967; Dansgaard and Tauber, 1969; Olausson, 1965; Broecker, 1965) are disturbing, as are the problems of dating (compare Broecker and van Donk, 1969, and Broecker and Ku, 1969, with Rona and Emiliani, 1969). These inherent difficulties are distinct from the problems of correlating deep-sea "zones" with continental or littoral stratigraphies.

Antarctica—probably associated with general glaciation of that continent—began between 2 million and 3 million B.P. (Opdyke *et al.*, 1966).

This evidence of a major, general climatic deterioration in higher latitudes suggests that the Plio-Pleistocene boundary could be effectively drawn somewhere between 3.2 million and 3.0 million years ago, i.e., approximately synchronous with the base of the Villafranchian faunal beds in France (see Curtis, 1967). The upper contact of the Villafranchian with the Biharian has not been satisfactorily dated, however, and its temporal position in regard to the first major continental glaciation in North America[5] and in Scandinavia is still obscure.

5. Widespread mountain glaciation is already in evidence for the Miocene, about 13 million B.P., in southeastern Alaska (Bandy *et al.*, 1969).

PART TWO

Vegetation, Soils, and Geomorphology as Environmental Indices

FOUR

The Zonal Concept

INTRODUCTION

Paleo-environmental interpretation is based largely on evidence provided by geomorphologic features, soils, animal, and plant remains. An archeological site in a loess embankment may, for example, show evidence of former soil-frost phenomena and fossil humus horizons, while excavation may produce bones of mammoth and reindeer along with pollen of herbaceous plants. Or, a Pleistocene geologist studying ancient lake beds in a dry subtropical area may find fossilized bones of fish, crocodile, and hippopotamus with various mollusca and traces of different woods. In either case the consideration of all classes of evidence is desirable. Yet the question arises, what environmental significance do the plant or animal remains carry? Under what conditions would the sediments or soils form? If answers can be obtained for these queries, further questions arise: Do the various lines of evidence produce convergent results? If so, does the ecological pattern so obtained have a modern counterpart?

None of the physical and biological data that can be gleaned from a site or exposure speaks for itself. Recognition and identification of phenomena must be followed up by careful interpretation of each line of evidence. Only then can the investigator be reasonably certain that his over-all ecological picture is representative and sufficiently clear. As a consequence, proper understanding of the features and materials commonly found in ancient contexts is vital. And the ultimate key to paleo-ecology is provided by modern distributions of similar features.

Geomorphic evidence may indicate that a site was located by a stream or lake, on a windy plain with conspicuous soil-frost or with waterlogging. This evidence of the *local habitat* or *setting* may be complemented by soil-frost phenomena that are today restricted to certain polar climates. This would suggest that the *regional environment* was a tundra. The wind-borne plant pollen found in the sediment might corroborate the latter evidence, whereas fossilized leaves, stems, or mosses would add more to our understanding of the immediate local scene. In other words, paleo-environmental interpretation requires knowledge of both the setting and the regional environment, evidence that can be obtained only by familiarity with modern processes and distributions.

Fortunately, the elements of the natural environment generally show some order in their location and occurrence on the continents. The most important single distribution affecting the face of the earth is climate. Since all living things require warmth and moisture in varying degree, they ultimately reflect resources primarily controlled by climate. In a broad way the distribution of various species of plants or animals is related to the zonation of climate. In fact, one of the basic principles of vegetation or plant geography is that climatic control is a primary factor determining the general character of world vegetation belts. Animals, unlike plants, can adapt themselves more readily to climatic conditions and often migrate seasonally to avoid unfavorable food or thermal conditions. But in a general way the animal world, when seen as a complex of organisms ranging from bacteria and mollusca to birds and mammals does show recognizable associations related to climatic zonation.

Vegetation and microorganisms are two of the factors influencing the character of soil development in a given area. Climate directly plays an active role in soil development by thermally controlling the rate of chemical reactions. In fact, climate makes such reactions possible in the first place by supplying the necessary water by means of which chemical alteration takes place. Overlooking the nature of the bedrock and the peculiarities of relief and surface configuration, many of the processes of local soil development can, in varying degree, also be related to climate.

Many geomorphic processes are essentially a product of the same factors responsible for soil development. These are the external forces that sculpture and modify features resulting from bedrock lithology or tectonic strucure. Although the broad lines of soil mineralogy or of surface morphology are frequently dominated by bedrock materials and past tectonic activity, the actual modifications underway at any one time are controlled either directly or indirectly by climate. A detailed soil map will reflect an infinity of minor factors, whereas soil types over very large regions show unmistakable analogies. Similarly the surface configuration of the earth shows no latitudinal pattern; yet in its details, regional sculpturing becomes individualistic and discernible.

Broad distributions of biological or physical associations that show latitudinal zonation—ultimately as a response to climate—are called *zonal*. The equatorial rainforest or the podsolic soils of the coniferous forest belt are examples of such zonal distributions. *Intrazonal* features reflect both local factors, such as poor drainage, and regional climate. On the other hand, associations that can occur in any latitudinal belt, such as shoreline forms or tectonic activity, are *azonal*.

The elements of the natural environment are invariably related to both zonal and azonal factors, as Table 6 indicates. Each of these phenomena is interrelated, and the interdependence becomes more versatile as one moves from vegetation to geomorphic processes *(morphogenesis)*, while the importance of climate diminishes in the same order.

The zonal concept is fundamental for the purpose of this study. Characteristic features commonly occur in "fossil" form in prehistoric contexts, in which case they may provide a key to the former regional environment. Unfortunately, the impression of homogeneity readily conveyed by discussion of such large units is misleading, and it is difficult to avoid overgeneralization and oversimplification in a brief review of the pertinent aspects of zonal vegetation, soils, and morphogenesis. Later chapters emphasizing local settings should to some extent restore the balance between generalization and complexity.

Table 6. Relationships of zonal and azonal factors.

Zonal factors	*Phenomenon*	*Azonal factors*
Temperature Radiation Moisture Soils	Vegetation	Mineral nutrients Drainage and relief Other edaphic factors
Temperature Moisture Vegetation	Animal Life	Drainage and relief Population density Predators and parasites
Temperature Moisture Vegetation Microfauna Geomorphic processes	Soils	Bedrock lithology Drainage and relief
Temperature Moisture Vegetation Soils Fauna	Morphogenesis	Tectonic structure Bedrock lithology Drainage and relief

VEGETATION AND CLIMATE

Certain basic requirements of the plant world are a direct function of regional climate: radiation, temperature, and moisture (Cain, 1944; Aario and Janus, 1958; Eyre, 1963).

The chemical life of a plant consists of absorption of water and certain minerals from the soil followed by *photosynthesis,* whereby water, minerals, and atmospheric carbon dioxide are combined to form plant tissue. Radiation is a primary source of energy for photosynthesis, while the rate of chemical reaction as such increases rapidly with temperature. Water is essential as a raw material in molecular structure and as a basic agent in the vital processes of the plant. Consequently the length and intensity of the growing season, as determined by available light, warmth, and water, is a major control of plant life. Another complex determining factor is provided by absolute temperature toleration and optimal temperature ranges.

The free passage and assimilation of soil moisture, the rate of photosynthesis, and the eventual water loss of the plant by transpiration are intimately associated with temperature. Optimal temperatures for these processes and reactions vary for different plants. So do the toleration limits. Few plants tolerate temperatures above 40° C., and no higher plants tolerate values of 50° C. Except for a few arctic plants that may not tolerate maximal temperatures as low as 4° C., most plants do not have significantly different upper toleration limits. The greatest differentiation occurs at the minimum temperatures. Many tropical species cease to grow when temperatures drop to 10° C., while temperate species stop growing at temperatures between 0° C. and 5° C. Many tropical species die at temperature below 5° C. or at the freezing point, so that the complete absence of frost is important for plant distribution. Many plant families seem to tolerate frost of varying intensity or duration by dormancy. Since plants do not have direct physiological protective devices against heat or cold such as thick bark or the like, adaptability is a matter of the organic chemistry of individual species.

Plants do, however, have distinct adaptations to moisture conditions. *Xerophytic* plants are adapted to chronic drought by devices that bring about a reduction of evaporation, greater water storage, and improved root efficiency. Small or needle-like leaves, in part replaced by thorns or scales, reduce evaporation just as do reduced surface area, slick bark, waxy surfaces, or fewer pores. *Hygrophytic* plants are adapted to optimal moisture conditions by broad, thin leaves (providing a maximum assimilation surface with a minimum of material) and thin bark. *Tropophytes* are adapted to alternating dry and wet seasons.

Consequently a particular climatic belt will tend to favor a certain association of plants, grasses, shrubs, and trees known as a *plant formation* or as the *climatic climax vegetation.* These are the highest types of vegetation that can develop under the different aspects of a given climate, and that are in equilibrium with that climate. Obviously climates change, however, and according to Mason (1936) climate is the only significant variable that can stimulate the migration of plant associations. As climatic change is not local, movements of vegetation tend to show a regional parallelism. From North America and Europe we do indeed know of a dramatic succession of flora that moved poleward during the recession of the continental glaciers. Such changes of environment suggest that the plant formations are to some degree temporary features and that they may be complicated by the presence of relict species. Such relicts are valuable evidence for changes in the past.

Factors other than climate do play a part, not so much in determining the large-scale patterns, but in creating innumerable local variations that reflect soil factors such as texture, temperature, moisture, chemistry, available nutrients, and humus type. Well-developed, fertile soils will, for example, support a more luxuriant vegetation with different or more species of plants than will thin soils with limited plant nutrients. Similarly, groundwater available along rivers or lakes may permit local growth of fringing or *galeria* woodlands in otherwise arid or semiarid country. Natural drainage, topography, and slope are also important, so much so, in fact, that a climatic climax community will be limited to well-drained lands with gentle slopes and no disturbance by man.

WEATHERING AND SOIL DEVELOPMENT

Fundamental in understanding the relationship of climate and soil development is the process of *weathering* and its latitudinal variations (see Cotton, 1949; Ollier, 1969). "Weathering" refers to the disintegration and decomposition of exposed rock under the direct influence of the elements. It is prerequisite to, and simultaneously a major process in, soil development. Weathering may be chemical or physical.

One form of chemical weathering is *solution,* whereby soluble salts or calcareous constituents are dissolved and washed away or *leached.* Many salts such as sodas, chlorides, or sulfates can be simply washed out. Calcium carbonate, the basic constituent of limestone, is altered by carbonic acid derived from carbon dioxide dissolved in rainwater. The alteration product, a bicarbonate, is quite soluble in water and is responsible for the chemical decomposition of limestone.

Almost all other rocks can be altered and weakened by *hydrolysis,* whereby percolating waters partially dissociate into differently charged ions that break up minerals into their corresponding bases and acids. In this manner feldspars, silicates, micas, etc. are disintegrated, rendering the rock softer and weaker. Only quartz and some heavy minerals are not generally affected by this agency. Then, also, contact with oxygen molecules leads to the *oxidation* of iron, manganese, and aluminum into alteration products that commonly are weaker.

Each of these agencies requires heat and water, suggesting that chemical weathering will be most intense in the humid tropics and, to a certain extent, in humid temperate lands. Arid zones and cold seasons or cold climates severely reduce chemical weathering, so that there is little of it in arctic and desert regions.

Mechanical agents break up the bedrock into smaller units more liable to chemical attack. Frost is undoubtedly the most important of these forces. Water increases its volume by 9 per cent when it freezes. When this expansion occurs in water held in crevices, fissures, or pores, complex forces and strains work upon the bounding rock surfaces and may eventually shatter the mineral structure. In loose materials with more than 2 per cent fine particles, the freeze-and-thaw of water induces volume changes and relative displacements, together with lateral movements on slopes. On thawing, such materials lose much of their cohesion and become more susceptible to erosion by wind or water. The presence of permanently frozen subsoil *(permafrost)* is linked with frost-heaving, also a potent mechanical agent.

Temperature variations may produce mechanical disintegration without going below the freezing point. Strong daily insolation and rapid nightly cooling in dry regions lead to rapid expansion and contraction of rock surfaces, creating a differential with the interior. Particularly in the case of coarse-grained rocks of variable mineralogy, such as granites, differential expansion seems to produce microfractures, favoring gradual but very slow rock disintegration. All mechanical processes due to temperature variation are included under the designation *thermoclastic weathering*.

Seasonal changes in humidity can produce the contraction of finer materials by dehydration (polygonal cleavage or contraction cracks), Certain clay minerals can expand or contract by several per cent of their volume under the influence of moisture changes, and alternate wetting and drying will tend to split rocks with a high clay content. Many of the alteration products resulting from chemical weathering exert mechanical pressures; for example, the decomposition of feldspars gives an increased volume of clay products. The various geomorphic processes may also aid mechanical weathering by surface impact or friction. Final sources of mechanical weathering are the pressures exerted by the roots of plants and trees, i.e., biological agents. The latter also include incidental items such as burrowing animals.

Summing up, then, the processes of weathering are basically related to the latitudinal distribution of solar energy and the semilatitudinal alignments of precipitation belts.

Soil development involves not only weathering but also biochemical processes associated directly with vegetation and microfauna. These biochemical agencies can be dismissed briefly, as they merely imply an intensification of the chemical processes discussed above. Lower plants such as algae, lichens, fungi, and mosses produce biochemical alteration of the immediate surface materials on which they settle. The removal of nutrients from the incipient soil with only selected minerals, bases, or acids being returned by leaf or needle-fall, has effects similar to hydrolysis. Particularly important here are the carbonic and humic acids of biological origin. Plants also accelerate evaporation from the soil, slowing down the process of solution and leaching. Roots effect an exchange of minerals between the upper and lower soil, while the vegetation itself breaks the impact of the external elements. The kind of humus present, which is determined by the vegetation, may accelerate or slow down decomposition in the soil. A host of bacteria and microbes—whose activity is controlled by the environment—and numerous insects, including beetles and earthworms, are also chemically active.

ZONAL SOILS

As soil development involves bedrock alteration, it cannot show nearly so much zonal arrangement as plant associations, which are independent and spontaneous. Consequently, azonal components dominate in all but mature and well-drained soils. In fact, the climax whereby a soil will ultimately reflect its climate and vegetation rather than its parent material is an ideal case that is never realized. But even immature soils already reflect to some degree the climate and vegetation type under which they developed. Most striking to the eye is the soil color, which often gives significant information on the physical and chemical agencies involved. Gray or blackish colors most frequently reflect humus conditions, while yellowish or reddish shades may reflect chemical alteration. Only in a few cases is the soil color of mature soils directly determined by the parent material. Similarly, the classes and kinds of humus, soil fauna, chemical reaction, and soil texture often permit identification of zonal factors before a mature soil profile has been developed.

Although the properties of poorly drained soils in swampy areas, alluvial soils in river valleys, and stony soils on eroded slopes diverge widely within a region, these azonal or intrazonal soils are still affected by many basically similar weathering processes. Even moderately mature alluvial soils show marked latitudinal differentiation. It is, then, possible to identify large-scale patterns in world soil distribution known as *zonal soils* or *great soil groups* (Glinka, 1927; Kellogg, 1941; Bunting, 1965). These patterns also show some similarity to the world's climatic provinces. But not all soils are mature, and changes in soil development reflecting climatic change may result in soil-forming agencies insufficiently intense to alter older, more mature soils. As a result *relict* soils of various kinds are significant in almost all latitudinal belts.

Zonal and even intrazonal soils can, with due caution, provide information on regional climate, while buried or relict soils *(paleosols)* can provide a clue to former environments. It may happen that the properties of soil profiles recognized in archeological or geological contexts will permit deductions almost as valuable as those derived from paleobotanical evidence.

EXTERNAL GEOMORPHIC PROCESSES AS A FUNCTION OF CLIMATE

Those geomorphic processes reponsible for sculpturing the earth's crust are commonly known as external or *gradational* agencies. They include the work of running water, ice, and wind, and are closely linked with weathering and soil development. Bedrock differences, structure, relief, and topography pose obvious limitations (Büdel, 1963) to zonal characterization of the earth's landforms. But within different climatic areas the gradational agents are balanced differently and show basic, recognizable patterns that ultimately produce distinctive surficial modifications of the crust. As a result several authors such as Büdel (1950a, 1969) and Tricart and Cailleux (1965a) have defined *climatic-geomorphologic zones* in which certain forms of weathering, erosion, sedimentation, and landform sculpture are typical (see Stoddart, 1969). Erosional features or sediments pertaining to specific climatic types are more commonly preserved in a fossil context than are soils or biological evidence. Consequently, the record of former gradational processes is also fundamental for paleo-environmental work.

NATURAL REGIONS

Most, but not all, of the major plant formations, zonal soil groups, and climatic-geomorphologic zones show an approximate overlap with certain first-order climatic provinces. Obviously the phenomena of the earth's surface form a continuum in which regions and boundaries are invariably arbitrary. But certain large-scale *natural regions* can be identified with reservations, even though their number and limits will vary from author to author.

The basic regions employed in this study are adapted from the climatic classification of W. Koeppen, described in many elementary texts in physical geography or climatology (for example, Strahler, 1969). The following oversimplified but useful regions can be recognized. The terminology employs the familiar designation of vegetation types, with climatic overtones:

a) Polar glaciers (Koeppen symbol[1] *EF*)
b) Polar tundras (*ET*)
c) Subarctic or boreal forests (*Dc, Dd*)
d) Continental forests (*Da, Db*)
e) Temperate woodlands (*Cbf*)
f) Dry subtropical woodlands (*Cs*)
g) Moist subtropical woodlands (*Caf, Caw*)
h) Grasslands (*BS*)
i) Deserts (*BW*)
j) Tropical woodland and savanna (*Aw*)
k) Tropical rain forest (*Af, Am*)

1. The Koeppen symbols, with minor modifications by G. T. Trewartha, are defined as follows:

A– Tropical forest climates: coolest month average temperature above 18°C.
B– Dry climates, defined by a linear relationship employing mean annual temperature, annual rainfall, and season of rainfall concentration. Subdivisions include a semiarid or steppe-grassland climate (*BS*) and an arid or desert climate (*BW*).
C– Mesothermal forest climates: coldest month above 0°C, but below 18°C.; warmest month above 10°C.
D– Microthermal forest climates: coldest month below 0°C.; warmest month above 10°C.
E– Polar climates: warmest month below 10°C., including *ET*–Tundra climate with warmest month below 10° C. but above 0° C., and *EF*–Perpetual frost with all months below 0° C.
a– Warmest month above 22°C.
b– Warmest month below 22°C., but at least four months above 10°C.
c– One to three months above 10°C.
d– One to three months above 10°C., but coldest month below −38°C.
h– Hot and dry: all months above 0°C.
k– Cold and dry: at least one month below 0°C.
f– Moisture distributed evenly throughout the year.
s– Dry season in summer.
w– Dry season in winter.

The distribution and location of these climatic regions is shown in Figure 3. Characteristic features of the vegetation, soils, and geomorphologic processes are discussed in subsequent chapters.

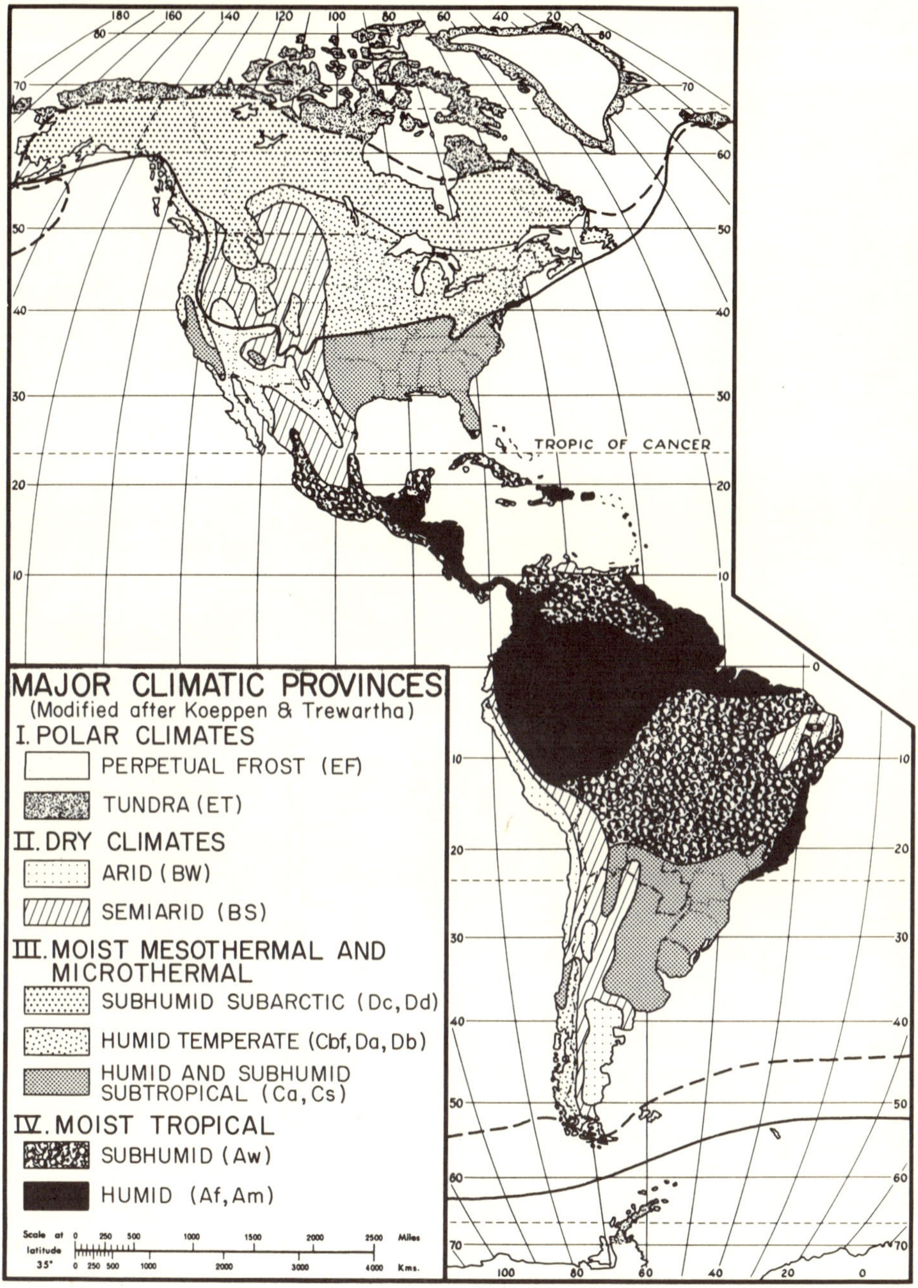

Figure 3. Major climatic provinces (modified after W. Koeppen and R. Geiger [Klima der Welt, 2nd ed., Darmstadt: J. Perthes] and Trewartha et al. *[1967]).*

Figure 3

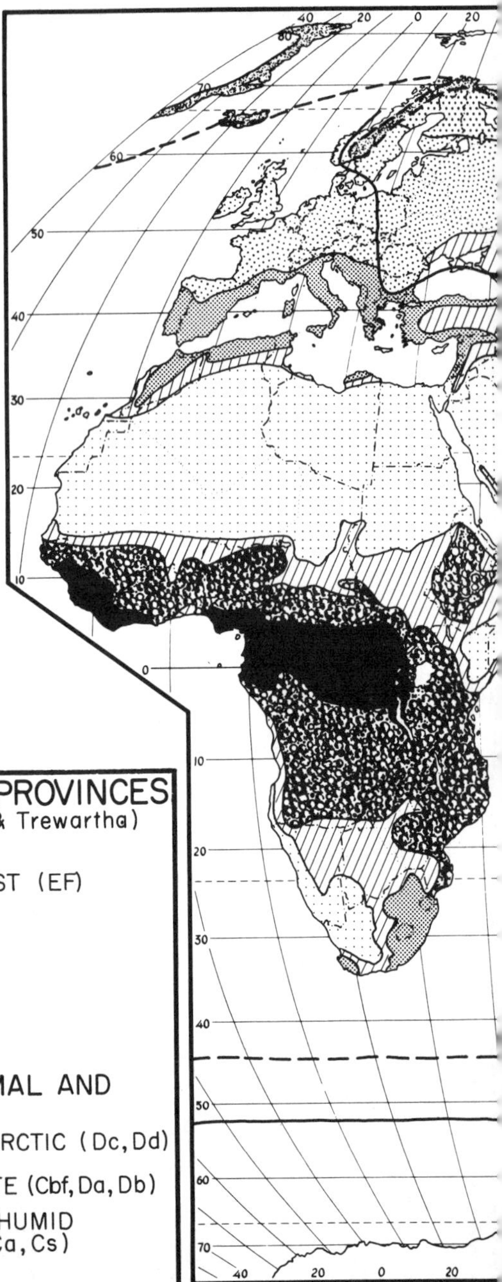

MAJOR CLIMATIC PROVINCES
(Modified after Koeppen & Trewartha)
I. POLAR CLIMATES
PERPETUAL FROST (EF)
TUNDRA (ET)
II. DRY CLIMATES
ARID (BW)
SEMIARID (BS)
III. MOIST MESOTHERMAL AND MICROTHERMAL
SUBHUMID SUBARCTIC (Dc, Dd)
HUMID TEMPERATE (Cbf, Da, Db)
HUMID AND SUBHUMID SUBTROPICAL (Ca, Cs)
IV. MOIST TROPICAL
SUBHUMID (Aw)
HUMID (Af, Am)
Scale at latitude 35°
0 250 500 1000 1500 2000 2500 Miles
0 250 500 1000 2000 3000 4000 Kms.

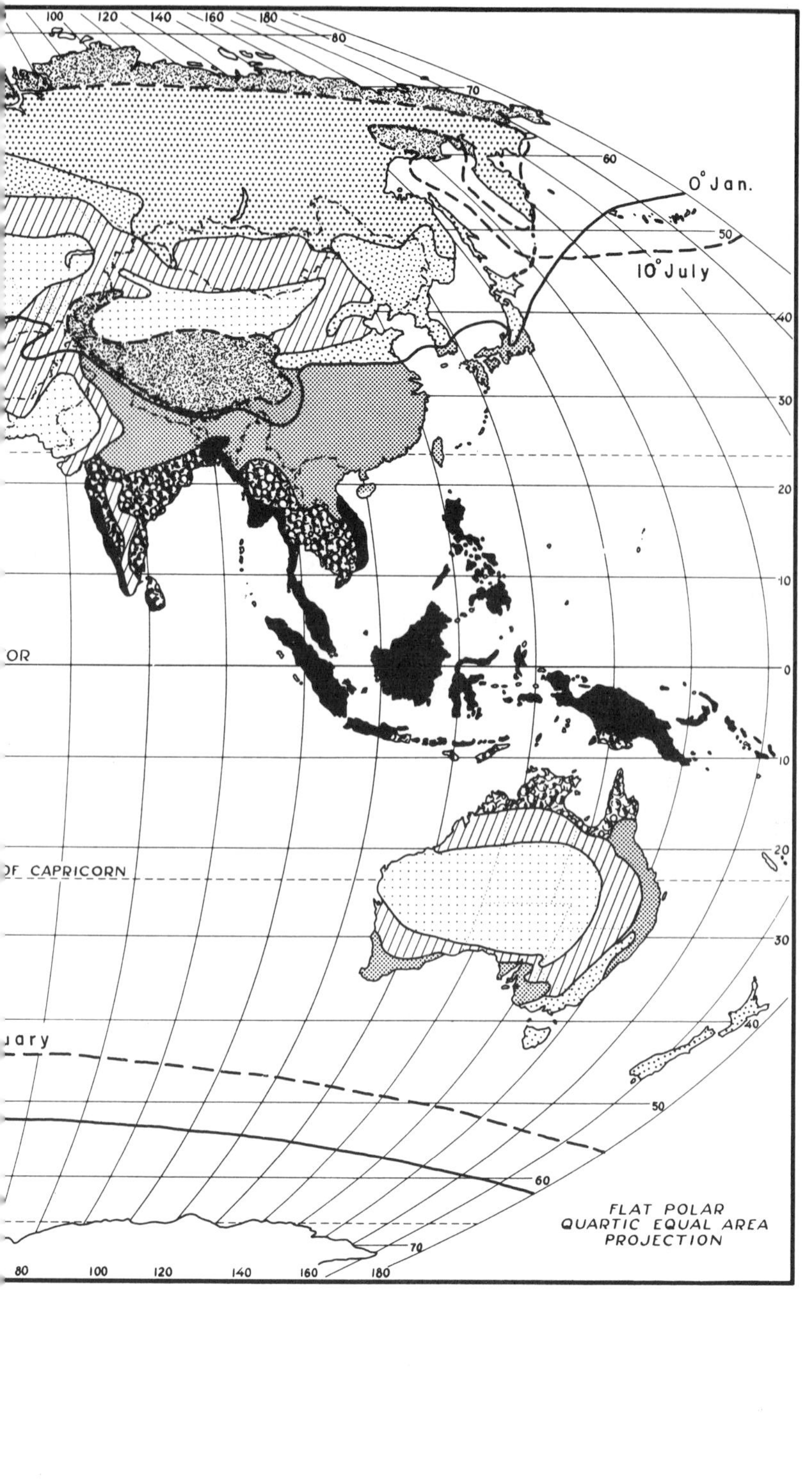

100
120
140
160
180
80
70
60
0° Jan.
50
10° July
40
30
20
10
0
OR
10
20
OF CAPRICORN
30
40
uary
50
60
70
FLAT POLAR
QUARTIC EQUAL AREA
PROJECTION
80
100
120
140
160
180

FIVE

Vegetation and Climate

The major classes of vegetation and their world distribution are shown in Figure 4 and are discussed below in relation to climate. A useful comprehensive outline of vegetation and plant geography is given by Eyre (1963). Ecological requirements and adaptations are more specifically considered by Walter (1960).

POLAR AND ALPINE VEGETATION

Some Characteristics of Climate. The polar regions beyond the limits of tree growth are affected by a harsh, cold climate. In general, winters are cold and long (10-12 months of the year) and the average temperatures of the warmest month commonly are below 10° C. The frost-free season is everywhere less than 50 days and may be nonexistent. Most of the treeless polar tundras have a permanently frozen subsoil *(permafrost)* that is never affected by the summer thaw, which is limited to the uppermost part of the ground.

In the broad belt of open country lying between the northern woodlands and the perennial ice fields, the brief summer vegetative period of one to four weeks is far from benign. Nightly temperatures of −1° to +5° C. alternate with daytime values of 7°- 12° C., while killing frosts are possible at any time. The remaining months of the year do not permit vegetative growth since temperatures remain well below 0° C., and strong, cold, dry winds sweep the surface. Although the snow cover generally lasts more than seven or eight months, the snow cover itself is shallow and frequently incomplete, offering little insulation.

As a consequence, frost may penetrate to depths of tens or even hundreds of meters, and the winter blizzards rapidly desiccate any

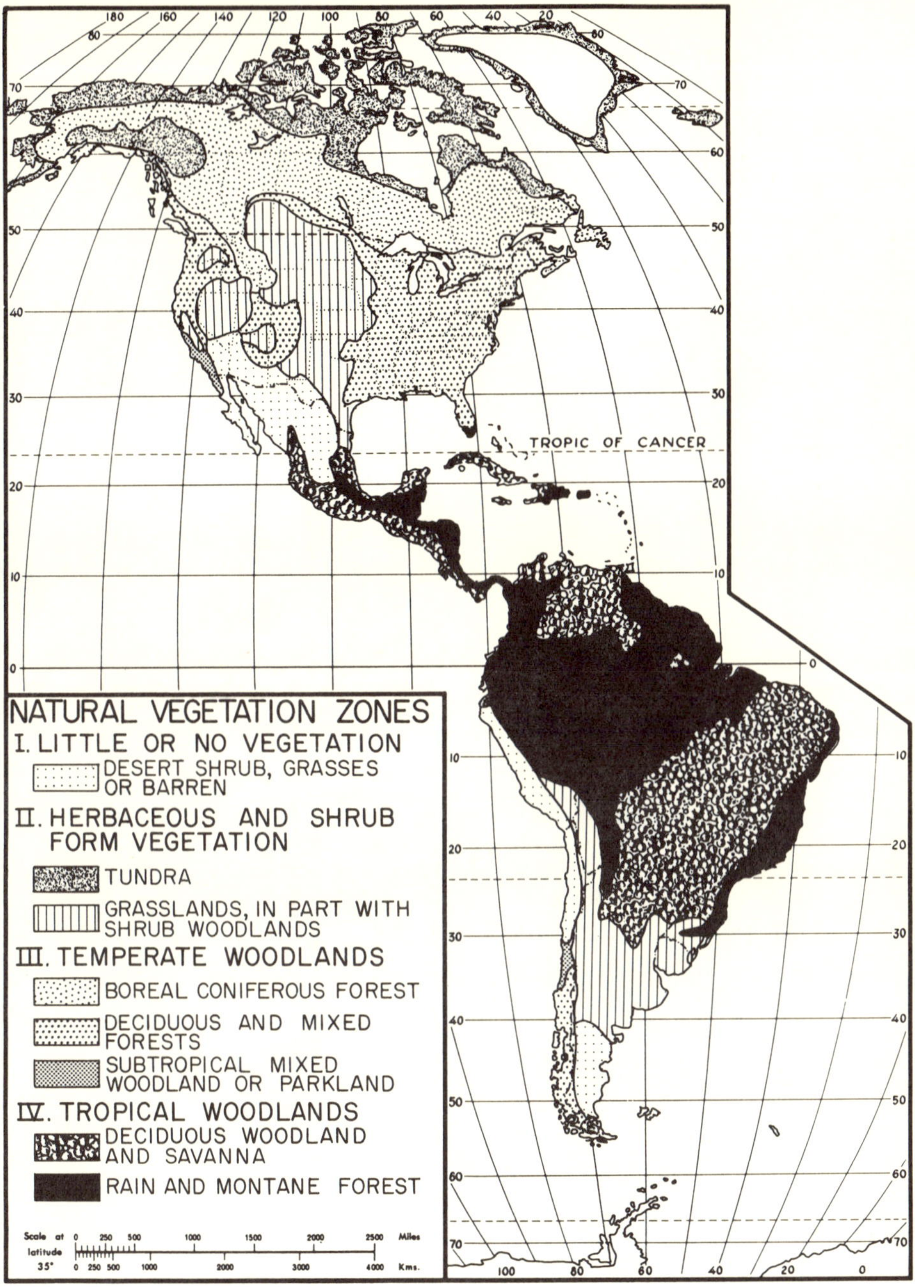

Figure 4. Natural vegetation zones.

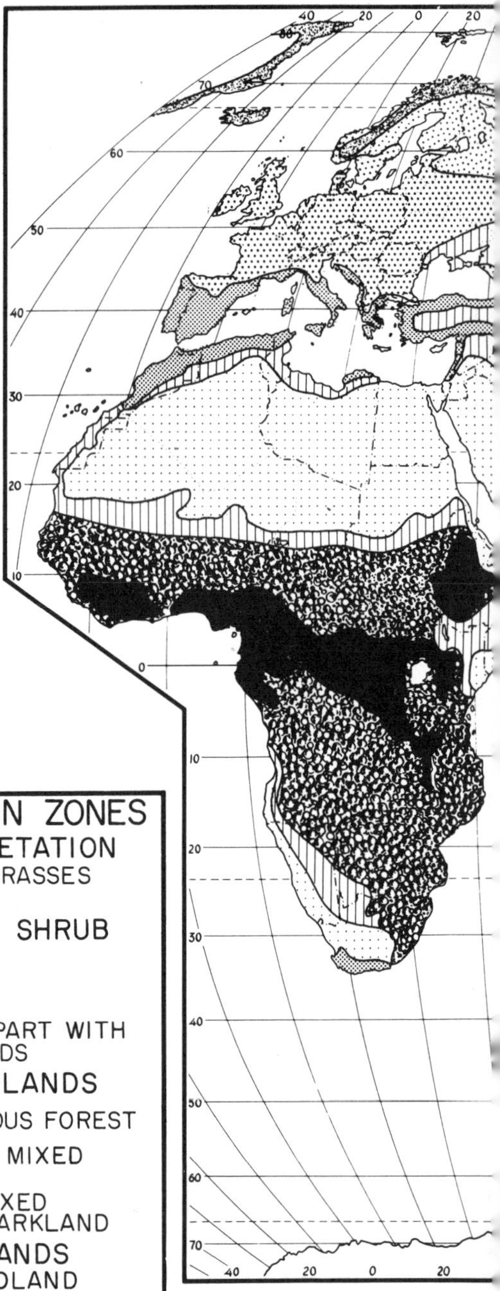

Figure 4

NATURAL VEGETATION ZONES

I. LITTLE OR NO VEGETATION
- DESERT SHRUB, GRASSES OR BARREN

II. HERBACEOUS AND SHRUB FORM VEGETATION
- TUNDRA
- GRASSLANDS, IN PART WITH SHRUB WOODLANDS

III. TEMPERATE WOODLANDS
- BOREAL CONIFEROUS FOREST
- DECIDUOUS AND MIXED FORESTS
- SUBTROPICAL MIXED WOODLAND OR PARKLAND

IV. TROPICAL WOODLANDS
- DECIDUOUS WOODLAND AND SAVANNA
- RAIN AND MONTANE FOREST

Scale at latitude 35°: 0 250 500 1000 1500 2000 2500 Miles; 0 250 500 1000 2000 3000 4000 Kms.

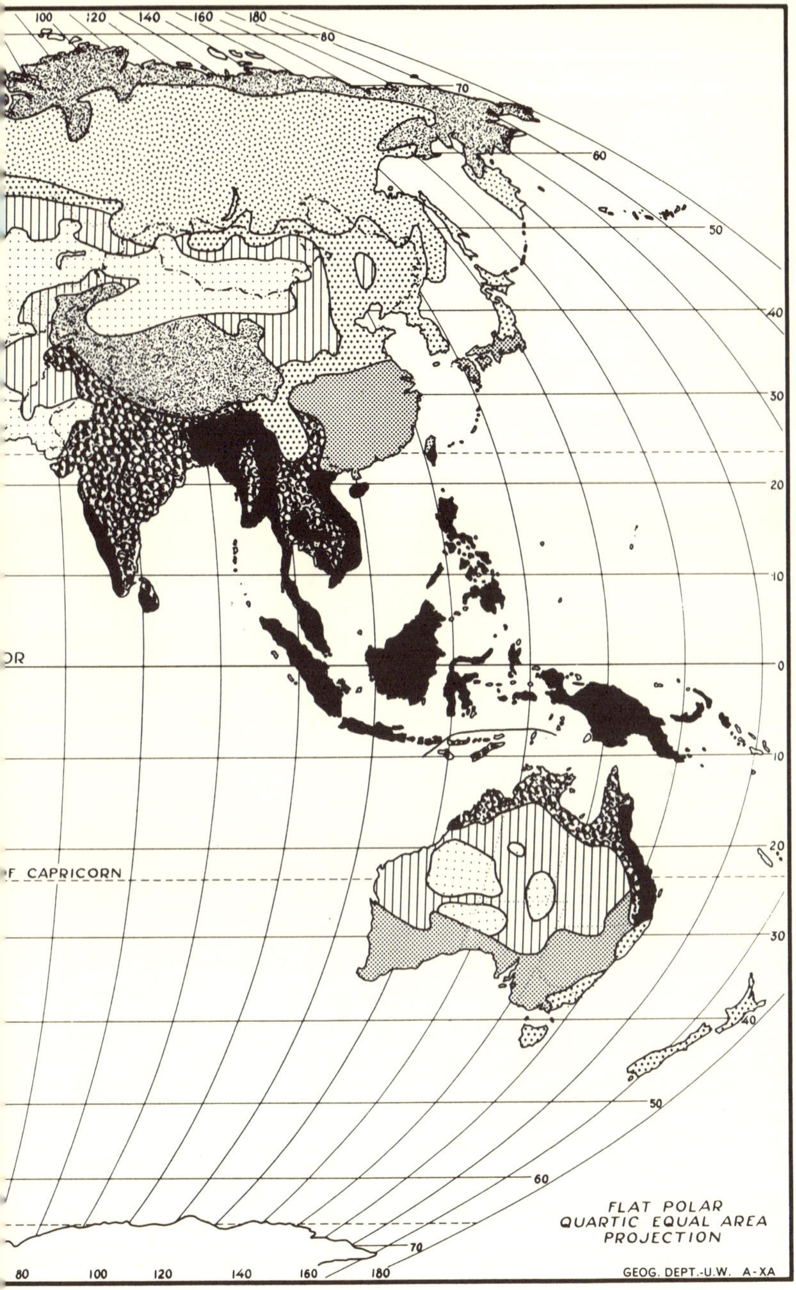

FLAT POLAR
QUARTIC EQUAL AREA
PROJECTION
GEOG. DEPT.-U.W. A-XA
F CAPRICORN
OR

exposed vegetation. Intense physiological drought is the result of transpiration from plant tissues at a time when water replacement through the frozen soil, roots, and stems is impossible. Similarly, cold-air drainage into lower-lying areas may create particularly cold microhabitats. Waterlogged soils have a similar effect, since evaporation of soil moisture results in considerable heat loss from the soil, while the water insulates the permafrost of the subsoil. Another aspect of the microclimate concerns the lowest 30 cm. of the atmosphere, in which normal air mixing is at a minimum. Nightly frosts in summertime may be twice as frequent at the surface than at the standard climatic observation level of 1.5 m. During the day, however, temperatures are 5° C. or more higher at the surface than at 1.5 m. Frost-tolerant plants of dwarf size can best utilize such a microenvironment.

According to Eyre (1963, p. 98 f.) the absence of tree growth in the polar regions reflects several factors, among them winter frost, winter winds, short growing season, and permafrost. At the regional margins farthest from the poles, wind force and the shallowness of unfrozen soil are thought to be most significant, while at very high latitudes near the poles, the growing season is not sufficiently long. Whatever the local cause of tree absence, a tundra vegetation of grasses, sedges, lichens, and mosses replaces the forests of middle latitudes (see Suslov, 1961).

High mountain areas in many parts of the world range into the *alpine zone,* above the altitudinal tree-line. Permafrost is seldom present and temperature patterns vary strongly with latitude. For example, there is no winter season in the tropical highlands; instead, temperature oscillations between night and day may result in regular night frosts. In middle latitude highlands, the thermal patterns more closely resemble those of the polar lands. Although snow cover is more effective and snowfalls are considerable, tree growth is again impeded by very high wind speeds and, probably to a lesser extent, by insufficient opportunity for photosynthesis. The alpine meadows, then, are remarkably similar to the polar tundras in both physiognomy and vegetation composition.

Several subzones of decreasing climatic severity may be recognized within the polar regions. These include:

a) The ice caps of Greenland, Antarctica, and several other areas such as Spitsbergen, Ellesmere, and Baffin islands;
b) The high arctic barrens;
c) The herbaceous tundra (in the strict sense); and
d) The transitional forest-tundra.

The vegetation of the last three subzones will be discussed in a little more detail.

The High Arctic Barrens. In many high arctic areas, vegetation is limited to a discontinuous scatter of mosses, sedges, or lichens, with much bare ground in between. There is no vegetative mat, sod, or turf, and large areas support no plant growth at all. Such desert or semidesert conditions prevail on the Canadian Arctic islands, in the unglaciated parts of northern Greenland and Spitsbergen, on the Soviet Arctic islands, and on the Taimyr Peninsula.

The Herbaceous Tundra. The more typical tundra vegetation, with its comparative luxuriance of plant growth and a continuous, hummocky sod, fringes the northern littorals of North America and Eurasia. Various grasses and sedges (*Carex* spp.) are probably dominant under natural conditions, with a substantial under-story of lichens (such as the reindeer "moss," *Cladonia rangiferina,* as well as the genera *Centraria, Stereocaulon,* etc.) and mosses of the genus *Polytrichum* (Walter, 1954, p. 150 f.; Eyre, 1963, p. 96 f.). A great variety of flowering plants is also present, including several genera—such as the mountain avens (*Dryas* spp.), saxifrages (*Saxifraga* spp.), and gentians (*Gentiana* spp.)—that are more or less confined to polar and alpine environments. Creeping shrubs such as the dwarf willows (*Salix polaris, S. herbacea, S. reticulata*) and dwarf birch (*Betula nana*) are not uncommon, particularly on sheltered slopes and along streams.

Widespread within the herbaceous tundra are poorly drained plains, often with brackish or salty waters. These are areas of bog vegetation, with cotton grass (*Eriophorum vaginatum*), various sedges, bog moss (*Sphagnum* spp.), and at times, a number of salt-tolerating grasses and shrubs.

The Forest-Tundra. As the distance from the poles increases, the tundra belt grades imperceptibly into continental forests, as riverside brush becomes more conspicuous and higher forms begin to replace mosses and lichens on the better-drained soils. A stage of what is often described as low-brush tundra is finally succeeded by terrain with lightly stocked tree groves or river thickets of birch (*Betula*), spruce (*Picea*) and larch (*Larix*). Tundra bogs are still common and extensive in waterlogged areas, while upland surfaces are frequently bare. This fluid transitional zone is known as the forest-tundra. The so-called treeline marks the complex zone where individual trees of the sporadic stands of brush no longer attain more than dwarf size (i.e., maximum heights of 1 m.).

The Polar and Alpine "Tree-Line." The reconstructed position of the tree-line during various Pleistocene or Holocene periods has commonly been used for paleoclimatic interpretations. So for example, Koeppen's boundary between tundra and forest climates (*ET* and *D* or *C*) was designed to approximate the tree-line by the 10° C. July isotherm. In the northern hemisphere, the polar tree-line today generally lies somewhere between the July isotherm of 10° C. and 12° C., depending on the maritime or continental character of the climate.[1]

Unfortunately, the tree-line is neither a "line" nor a true climatic boundary. Instead it is a transitional belt marked by different tree species in different areas. Tree growth is determined largely by exposition, relief, and surface drainage—all of which create specific microhabitats in terms of wind force, air drainage, snow cover, soil temperatures, soil moisture, and depth of annual thaw. There is thus increasing doubt that the polar or alpine tree-lines are a simple function of the macroclimate. In fact there is no question that the tree-line would be located far nearer the poles if summer air temperatures were the only limiting factor.

The position of a Pleistocene tree-line is, then, not simply equivalent to a particular monthly isotherm. In lieu of anything better, the somewhat accidental correspondence to midsummer isotherms of 10°–12° C. cannot be overlooked, but must still be considered with caution.

FORESTS OF MIDDLE LATITUDES

Forest Classes and Climate Types. Trees are sometimes classified on the basis of appearance and physiognomy. They fall into either the broadleaf or needle-leaf (coniferous) class, and may be evergreen or deciduous. Consequently the following types of forest are generally recognized:

a) Broadleaf evergreen,
b) Broadleaf deciduous,
c) Coniferous evergreen,
d) Coniferous deciduous, and
e) Mixed evergreen and deciduous.

Broadleaf evergreen species are confined to the tropics and subtropics, while rather unusual coniferous deciduous woodlands are important in northeastern Siberia where the dahurian larch (*Larix dahurica*) is the dominant tree type. Within the extratropical and subtropical woodlands there does not appear to be any very apparent logical order for the occurrence of the physiognomic forest types. Despite a certain over-all

1. Nordenskjiöld and Mecking (1928) have in fact attempted to give an empirical climatic definition for the tree-line using the formula $W = 9 - 0.1\ K$, where W = mean temperature of the warmest month (in °C.) and K = mean temperature of the coldest month. South of the polar tree-line W is thought to be greater than the right-hand side of the expression. A more recent attempt by Hare (1954) to apply the potential evapotranspiration concept of Thornthwaite is also of interest.

dominance of evergreen conifers in the cool, continental climates, and of mixed or broadleaf deciduous woodlands in the more temperate maritime zones, there are many important exceptions. Consequently, certain characteristic species of limited areal occurrence must also be emphasized in describing broad vegetational groups corresponding to certain climatic environments.

Four major *climatic* provinces may be identified in the middle latitude woodlands. In summary form these can be outlined as follows:

a) The Subhumid Subarctic. The extreme continental climates of Koeppen's *Dc* and *Dd* provinces have extremely cold winters and cool summers (warmest month average 12°–15°C.), excessive annual ranges of temperature (July-January differences of 40°–65°C.), and a brief growing season of one to three months. Although the winters are comparatively dry the snow cover lasts five to nine months and rivers remain icebound for about the same length of time. Permafrost is widespread. Confined to the northern hemisphere, these climates have a forest vegetation of northern or *boreal* conifers.

b) The Humid Temperate Zone. A maritime or oceanic type, corresponding to Koeppen's *Cbf* province, is commonly distinguished from an interior or continental climate (Koeppen's *Da, Db).*

The maritime variety is generally found on the western margins of the continents in middle latitudes. Winters are mild (coldest month 2°–10°C.), summers are warm (warmest month 15°–19°C.), and the growing season lasts five to ten months. Although snow does fall, there is no durable snow cover.

The continental climates with interior and east coast situations are harsher, with cold winters (coldest month −2° to −14°C.) and warm to hot summers (warmest month 16°–22°C.). The growing season lasts four or five months. There is a significant snow cover, amounting to one to four months.

Almost any combination of temperate trees may be found within the humid temperate climates, ranging from deciduous hardwoods to warmth-loving, cold-tolerating conifers. Deciduous or mixed broadleaf-coniferous forests are, however, most common.

c) The Humid Subtropics. The eastern continental margins of lower mid-latitudes fall within the *Ca* climate province of Koeppen. Winters are mild (coldest month 2°–10°C.), summers hot (warmest month 23°–30° C.), with a long growing season of seven to twelve months. There is no snow cover. Moisture is abundant at all seasons, particularly in summer. In southern China and northeastern Australia this environment has favored a subtropical or tropical, broadleaved evergreen forest. In east-central South America a grassland vegetation that poses an ecological enigma prevailed in pre-European times. In the southeastern United States, temperate mixed or coniferous forests are dominant today, presumably as a result of human interference rather than climatic opportunity. Nevertheless the species present are remarkably less thermophile than those of other humid subtropical climates, largely as a consequence of sporadic but exceptionally severe spells of winter cold (to $-10°$ or $-15°$ C.) in the southeastern states.

d) Subhumid Subtropics with Dry Summers. On the western sides of the continents, lower middle latitudes enjoy a transitional climate with distinctly dry summers. These are mediterranean-type climates *(Cs* of Koeppen,) with mild winters (coldest month 5°–12° C.), warm to hot summers (15°–25° C.), and very rare snow or frost. Vegetation growth, although impeded by lack of water much of the time, is possible for nine to twelve months of the year. The vegetation is composed of warmth-loving, subtropical evergreens with some deciduous species. Tree growth is clearly marginal in some of the dry mediterranean borderlands.

In a very generalized manner the *vegetational* aspects of the middle latitude woodlands can be discussed under three topical headings:

a) The boreal coniferous forests,
b) The deciduous and mixed forests, and
c) The subtropical woodlands and parklands.

The Boreal Coniferous Forests. The boreal forests of Canada and northern Eurasia represent a comparatively uniform vegetation type. Densely packed conifers of a limited number of species offer little opportunity for undergrowth or grass sod. Instead the moderately high stands (seldom as much as 20–30 m. tall) of forest are difficult to penetrate.

In northern Eurasia the core of the boreal forests is regionally dominated by species such as the European spruce (*Picea excelsa*) and its Siberian relative (*P. obovata*), by the cembran pine (*Pinus cembra*), and the deciduous European and Siberian larches *(Larix decidua, L. Sibirica).* In regions transitional to the temperate forests of Europe, Scots pine (*Pinus silvestris*), white fir (*Abies alba*), and broadleaved species such as the goat willow (*Salix caprea*) and Eurasian aspen (*Populus tremula*) are characteristic. In the forest-tundra certain deciduous species such as willow, birch, and aspen again assume more importance. Tundra bogs and barren uplands are common in the northern parts of the boreal forest, particularly in central and eastern Siberia.

The Canadian boreal forest is similar to that of northern Eurasia except that deciduous conifers are less common and the dominant species are different. White and black spruce *(Picea glauca* and *P. mariana),* balsam fir *(Abies balsamea),* tamarack *(Larix laricina),* and jack pine *(Pinus banksiana)* are the characteristic forms.

Similar forests form the subalpine zone of many middle latitude highlands.

The peculiar adaptation of conifers to the subhumid subarctic lands is generally thought to be related to two factors (Eyre, 1963, p. 47 f.). Needle-shaped leaves strongly reduce water loss by transpiration during the winter, a time of severe drought stress as a result of persistent dry and cold winds. A second asset is provided by permanent leaves capable of immediate photosynthesis whenever sufficient warmth is available, without having first to grow a new set of leaves in the late spring. Evergreen species can then make the most of a short growing season. However, in the extreme situations of the forest-tundra and of eastern Siberia, very slow-growing deciduous species are even more drought resistant in the winter time. This is thought to account for the larch woodlands of eastern Siberia and for the prominence of certain deciduous forms in the forest-tundra in general.

The Deciduous and Mixed Forests. The forests of more temperate mid-latitude regions are extremely complex in their distribution, but are locally rather uniform. Dense, mixed rain forests thrive on the ultramaritime uplands of southern Chile, Tasmania, and southern New Zealand. The world's tallest forests are found along the Pacific coast of Canada and the United States. These include conifers such as the coastal redwood (*Sequoia sempervirens*) and Douglas fir (*Pseudotsuga taxifolia*). Elsewhere in the eastern United States, Europe, and central eastern Asia, deciduous or mixed deciduous forests are dominant. Evergreens are here largely confined to the higher country and to areas of poorer soil. A few words may therefore be devoted to the broadleaf woodlands of middle latitudes that shed their foliage in winter.

In contrast to the coniferous forest, the elements of the deciduous forest are more widely spaced since most of them require much light. The tree crowns are dense so that little undergrowth is commonly present. These woodlands were originally easy to penetrate and frequently interrupted by glades or areas of fewer trees. Instead of the dense needle littler that floors the boreal forest, there is a discontinuous growth of herbaceous plants.

Among the characteristic European species on rich, well-drained soils are beech *(Fagus silvatica)* and ash *(Fraxinus excelsior).* The pedunculate oak *(Quercus robur)* favors moister lowlands, with alder *(Alnus glutinosa),* aspen *(Populus nigra),* and willow *(Salix fragilis)* characteristic of riverine situations. The sessile oak *(Quercus petraea)* and birch *(Betula pendula)* thrive on shallow, poorer soils. Other members of this association are the hazel *(Corylus avellana),* elm *(Ulmus campestris),* lime *(Tilia cordata),* sycamore *(Acer platanoides, A. pseudoplatanus),* hornbeam *(Carpinus betulus),* and yew *(Taxus baccata).* Many of these species have played important roles in the forest history of Europe as preserved in the pollen record (see Firbas, 1949-50; Walter, 1954). Other species, generally more diverse, are dominant in the deciduous woodlands of North America. Oak, maple *(Acer* spp.), and hickory *(Carya* spp.*)* are leading genera there.

The Subtropical Woodlands and Parklands. Of the two major subtropical woodland types, one is found in the mediterranean, summer-dry climates of the world, the other in southern Japan and China.

The woodlands of the summer-dry subtropics have been largely replaced today by scrub growth or thorn. The natural vegetative cover was very probably a dry, open forest of broadleaved and needle-leaved evergreen species. In the lowland areas of the Mediterranean basin the live oak *(Quercus ilex)* is universal, and the umbrella pine *(Pinus pinea),* Aleppo pine *(P. halepensis),* and maritime pine *(P. pinaster)* are common. The scrubby growth of leathery or thornleaved evergreens—constituting the xerophytic maquis or garrigue of today—probably once formed the undergrowth, particularly in areas of thin or poor soils. All of these species are adapted to the often severe summer drought, but there is no need for them to shed their leaves during the warm, moist winters.

In the upland areas of the Mediterranean, the characteristic submediterranean forms include the black pine *(Pinus nigra)* and a number of deciduous oaks *(Quercus pubescens, Q. cerris),* the chestnut *(Castanea sativa),* the ash-elm *(Fraxinus ornus)* and a dogwood *(Cornus mas).* These species are adapted to cooler winters, and may extend into temperate Europe in favorable warm-dry situations.

Finally, along the dry margins of the subhumid subtropics, the dry Mediterranean woodlands thin out into parklands with groves or scattered stands of increasingly scrublike dry forest. The intervening areas are occupied by garrigue and herbaceous plants. This stage marks the transition to the dry lands.

The moist subtropical woodlands of southern Japan and China can be discussed more briefly. These broadleaved evergreen forests form a unique transition between the temperate deciduous woodlands and the tropical forests. Numerous species of evergreen oak and magnolias

(Magnoliaceae) dominate, with a dense undergrowth of shrubs and small trees, as well as several woody climbers (Eyre, 1963, pp. 84-85.)

VEGETATION OF DRY LANDS

Climatic Aspects. An outstanding natural delimitation of the dry lands is given by the *grassland-forest boundary*. The limits set to tree growth as a result of aridity are in many ways comparable in complexity to those of the polar tree-line, although the boundary is probably more sharply defined. Nowhere is the natural appearance of the grassland-forest boundary preserved today, since human activity (cultivation, herding, fire) and natural interference (grazing, natural fires) have everywhere caused significant modifications. In particular, there is reason to believe that many peripheral grasslands in middle latitudes and much of the tropical savanna would be forested under "natural" conditions. However this may be, the climatic controls are in major part a matter of rainfall (amount, seasonality, reliability, frequency, intensity) and evaporation (temperature, cloudiness, wind speed). Severe sporadic droughts may carry more significance than typical averages. Microclimatic features and frost play a more subordinate role. Edaphic features such as soil moisture, soil water retention properties, and the like are significant. The presence of perennial or seasonal groundwater in depressions or along rivers and lakeshores may account for isolated tree stands or fringing, galeria forests.

The convergence of toleration limits and climatic controls is difficult to express quantitatively. In a very rough way the simple and unsophisticated formulas used to delimit Koeppen's dry *B* climates (see Strahler, 1969) offer an approximation of the grassland-forest boundary. This limit is expressed by $p = 2t$ (for winter-dry climates), or by $p = 2t + 28$ (for summer-dry climates), where p = annual precipitation in centimeters, and t = mean annual temperature (°C.).[2]

Within the dry climates there is, of course, an infinite gradation of

2. Despite their mechanical complexity, empirical formulas by Thornthwaite (1948) provide no significant improvement on Koeppen's results, either in theory or practice. Calculations of the numbers of dry versus humid months per year were made by Lauer (1952) for the tropics and by Jätzold (1960) for North America, and the correspondence of certain critical values with vegetation belts is remarkable, considering the simple basic assumptions. A complicated but physically sound formula by Paterson (1956) sets the grassland-forest boundary at

$$\frac{V}{A} \cdot R \cdot \frac{PG}{12} = 25,$$

where V = mean temperature of warmest month (°C), A = mean annual range of temperature (°C), R = ratio of annual radiation at the pole to that of the location in question, P = mean annual precipitation (in cm.), G = length of the growing season (in months).

moisture. The distinction of arid and semiarid climates is significant from all ecological points of view, although no simple climatic criteria other than Koeppen's will be cited for the purpose of subdivision.[3] The native vegetation belts ranging from prairie or steppe through more arid forms of grassland to semidesert scrub and barren desert speak adequately for themselves.

Climatic elements other than rainfall are also of significance in the dry lands. Temperature contrasts between day and night (12°-20° C.) and between summer and winter (10°-20° C. in lower latitudes, 20°-30° C. in higher latitudes) are exaggerated as a result of low atmospheric humidity. And the middle latitude dry lands of the northern hemisphere (north of about 35° N.) experience very cold winters, with mean cold month temperatures well below the freezing point. This implies that plant growth is impeded during the winter as a result of cold, and during other parts of the year as a result of drought. It is consequently useful to distinguish the "cold" dry climates of middle latitudes (*BWk, BSk* in Koeppen's classification) from the "hot" dry climates of lower latitudes (*BWh, BSh*). Even in the hot deserts sporadic frosts occur as far toward the equator as the tropics of Cancer and Capricorn. Since dry lands extend through fifty-five degrees of latitude on either side of the equator, it is impossible to give representative temperature means here.

Grassland, Prairies, and Steppe. The interdigitation of forest and open country is complex in the middle latitude grasslands of North America ("prairies") and Eurasia ("steppes"). But the grassland vegetation itself is comparatively uniform. The moister parts have a continuous sod of tall grasses exceeding a meter in height. This luxuriant herbaceous vegetation with its dense root network is destroyed in winter, and its decay provides the soil with rich mineral nutrients. In the drier belts shorter bunch grasses, without a continuous sod, dominate, although the root network under the exposed soil is equally dense. Maximum moisutre is available in spring and early summer, and there may also be a secondary rainfall maximum in the autumn. But by late summer, drought conditions are severe, and the vegetation is parched and frequently dormant.

A similar vegetation sequence exists in the subtropical *pampas* of South America, although low latitude grasslands are usually different in physiognomy. They are commonly dominated by semidesert grasses with a scattering of deciduous thorny shrubs or low trees. The xerophy-

3. The boundary between Koeppen's arid *BW* and semiarid *BS* climates is given by $p = t$ (for winter-dry climates), $p = t + 14$ (for summer-dry climates) with p and t defined as before.

tic grasses and deciduous shrubs flourish during the two- to five-month-long summer rainy season, but lie dormant for most of the remaining year. The tropical grasslands consequently provide an environment somewhat different from that of the middle latitude steppes or prairies.

The Desert Margins. The peripheral belt of the semiarid climates and the adjacent desert margins provide a number of intermediate biozones bridging the gap between the bunch grasses or thorn-scrub grasslands and the barren desert wastes. It is here that the dry steppe, the desert steppe and semidesert are found. The herbaceous component of these associations is small in comparison to the evergreen or deciduous thorn shrubs and the spiny, pulpy, drought-resistant succulents. Hairy-leaved shrubs of the genus *Artemisia* are particularly characteristic in North American (sagebrush) and Eurasia (wormwoods). Individual plants are commonly scattered and noncontiguous, and despite the deep and complex rooting systems the soil is not enmeshed in a continuous root turf as in the true grasslands. As the wet season progressively decreases in length from about two months to brief periods following sporadic rains, the spacing of individual plants becomes ever greater. Drainage lines and areas with groundwater near the surface are conspicuously favored. Finally, as the desert interiors are approached, even these edaphically favored patches of episodic life give out.

TROPICAL VEGETATION

Climatic Characteristics. Climatically, the moist tropics are generally subdivided into the hyperhumid equatorial or monsoon climates (Koeppen's *Af, Am*) and the seasonally wet-and-dry tropical climates (*Aw*).

Most of the humid equatorial regions (*Af* climates) experience no marked dry season and are quite wet for ten months or more. Monthly temperatures average about 20°–28° C. all year. In the monsoon climates (*Am*) there is a distinct dry season, but the rainy season is sufficiently long and intense to provide sufficient moisture for growth the whole year through.

In the subhumid tropics, wet summer alternate with dry winters (*Aw* climates). The dry season lasts some three to seven months, with maximum heat occurring just before the onset of the rainy season (daytime temperatures 28°–36° C., nighttime temperatures 15°–20° C.). Temperatures are lower and more uniform during the rainy season (24°–27° C.).

Whereas the evergreen forest of the humid tropics is most luxuriant and dense, that of the wet-and-dry tropics is adapted to seasonal drought and shows a progressive gradation from the equatorial rainforest to the grassland margins. It is dominated by deciduous woodlands and savanna.

Tropical Rain and Montane Forests. The rain forest and its analogous counterpart, the montane tropical forest, is composed of hygrophytic, evergreen, broadleaved forms. Several layers of trees, their canopies forming successive stories within the forest, may reach 50 m. in height. Lianas and other climbing plants are plentiful, while undergrowth and herbaceous vegetation are practically absent. The ground is often bare except for occasional fresh leaves.

When the canopy is incomplete, however, and light penetrates to the floor, a dense, impenetrable undergrowth is commonly present. This is typical for regions with a moderately effective dry season. A number of deciduous trees begin to appear, and a transitional semievergreen or even semideciduous, mixed forest results.

Tropical Woodland and Savanna. The lighter, tropophytic vegetation of the subhumid tropics falls into two phases, a woodland phase and a parkland or savanna phase. Where dominantly woodland, the plant cover grades from mixed tropical forests to deciduous woodland or thorn forest, with an increasing amount of open landscape with both trees and grasses. In the savanna phase, grasslands with interspersed, isolated trees or groups of trees occupy the broad belt between the rain forest and the steppe. Tall, coarse grasses (one to two meters high) with scattered evergreen and deciduous trees occur where the rainy season persists seven to nine months of the year. Shorter, finer grasses set among groves of deciduous thorny trees are found where the rainy season lasts five to six months (Lauer, 1952). It is widely held that savannas of this kind were originally limited to areas with impeded drainage. The present widespread distribution of parkland vegetation is possibly a result of frequent dry-season burning as well as overgrazing—both activities a result of human interference (Richards, 1952; Eyre, 1963). But since the word "savanna" is the commonly recognized name for this belt, "savanna" will be used in a more general sense in the subsequent text.

VERTICAL ZONATION OF CLIMATE AND VEGETATION

Although alpine vegetation and the montane forests of the tropics have been mentioned briefly in the preceding sections, any outline of plant formations tends to overemphasize two-dimensional surface distributions. In addition to continental and latitudinal zonations of climate and vegetation, there also are significant vertical zonations determined by altitude. Mountain country and even hilly landscapes differ materially from the surrounding lowlands, and on the flanks of a sufficiently high mountain the climate may change with increasing elevation from an equatorial lowland to a snowy waste. In other words, temperature and environmental contrasts that span the 40 or 50 degrees of latitude between the tropic and the polar world may be compressed within 4,000 meters vertical distance in a tropical mountain range. Consequently, world patterns of climate and vegetation change through three dimensions, according to both latitude and altitude.

The environmental distinctiveness of highlands is due to several factors:

a) Lower air temperatures. Depending on the mass and irregularity of hill or mountain country, as well as on climate and season, there is a general temperature decrease with elevation, which varies between 0.55 and 0.8° C. per 100 meters. As a result, an upland surface at 1,000 meters elevation may be, on the average, 6 or 7° C. cooler than adjacent lands near sea level.

b) Local contrasts of insolation. The intensity of solar radiation depends on the angle of incidence of the sun's rays. On sun-facing slopes, i.e., those with southerly exposures (in the northern hemisphere), insolation is intensified; correspondingly, shade slopes receive less radiation. Exposure to sun can also modify the number of hours of daylight in areas of rugged topography, with deep valleys and shade slopes experiencing shorter days. In this way sun slopes are warmer and drier, contrasts that are reflected in the soil microclimate as well as in the plant cover. Consequently there is considerable small-scale variation in the vegetation and soils of hill and mountain country.

c) Intensified radiation. Since vapor content, carbon dioxide, and pollutants decrease with elevation in the atmosphere, the absorption and scattering of incoming insolation are reduced. Consequently rapid nightly cooling is as characteristic of mountain environments as is rapid daytime heating. The air is warmed less, yet radiation is stronger than in the lowlands. The soil is correspondingly warmer and vegetative growth favored.

d) Regional contrasts of precipitation. As a general rule, rainfall and snow increase with elevation since rising air cools, favoring condensation and more frequent and more intensive rains. Cloud cover increases correspondingly, favoring cooler summers and a higher relative humidity. However, the orientation of mountain ranges with respect to the prevailing moisture-bringing winds is at least as significant as the altitude. Moist air masses ascend windward slopes, leading to orographic intensification of precipitation. However, air dries as it descends on lee slopes, and much of the original moisture has already been lost on the windward face of the mountain range. As a result, lee slopes are correspondingly dry, frequently creating "rain-shadows." If a range is sufficiently high and continuous, it may create abnormally dry conditions over wide continental lowlands by its interception of moist air masses.

These major factors help explain why hill and mountain country is ecologically complex as well as distinct from nearby lowlands.[4] Despite climatic changes in the past, the basic character of mountain climates did not change. Mountains remained oases of richer vegetation in desert areas, or the foci of cold climate and possible glaciation in temperate and tropical latitudes. Sunny, wind-sheltered valleys may also have served as refuges for thermophile vegetation during glacial periods. In view of the complexity of modern ecological patterns, it is even more difficult to reconstruct past environments from, for example, pollen and macrofloral evidence.

Nonetheless, the broad vertical zonation of climate and vegetation, seen in a larger scale, can also be of assistance in environmental reconstruction. Vertical belts of vegetation migrated up or down, corresponding to latitudinal changes on the continental scale. With an understanding of modern temperature and precipitation gradients with altitude, it may therefore be possible to approximate the magnitude of climatic change responsible for vertical shifts of the vegetation belts (see, for example, Coetzee, 1967).

4. Illustrative examples of vertical zonation and local complexity in mountain country can be cited for Asiatic Russia (Suslov, 1961), the tropical mountains of Latin America and Africa (Troll, 1959), Iran (Zohary, 1963), the Himalayas (Schweinfurth, 1957), and the British Isles (Eyre, 1963, ch. 11).

SIX

Soil Processes and Soil Types

Although it is not yet possible to give an accurate map of world soils, the general lines of regional soil development are understood (Fig. 5). The major processes involved in the genesis of soils are discussed below with relation to the over-all physical environment. Characteristic soil types are outlined on this basis (Figs. 6–8). Paleosols are frequently used as paleoclimatic indicators, making an understanding of analogous modern soils vital.

SOIL PROFILES AND HORIZONS

A *soil* may be defined as a zone of weathering, comprising mineral and intermixed organic matter, directly overlying unweathered parent material. Soil *types* are chiefly identified on the basis of profile, humus, texture, structure, carbonate content, and micromorphology. Before attempting to discuss zonal soil types, some of the essential soil processes or features will be discussed here. Further consideration of soil properties, particularly as observed in the field, is given in chapter 10.

The *profile* of a soil seen in vertical section consists of several subdivisions or horizons, often recognizable on the basis of color. The three major horizon classes are successively known, from top to bottom, as *A*, *B*, and *C*-horizons. The uppermost, *A*-horizon is commonly characterized by an admixture of decomposed or partly decomposed organic matter, the soil *humus*. Many *A*-horizons are also a result of washing out (*eluviation*) of certain fine minerals and humic matter. The *B*-horizon commonly refers to the deeply weathered subsoil, devoid of humus, but with a concentration of fine textured material derived from chemical alteration of the parent material. The *B*-horizon may also be a zone of accumulation or *illuviation* for minerals or humus washed in from an overlying, eluvial *A*-horizon. The *C*-horizon includes the more or less unaltered parent material, which may vary from unconsolidated sediment to resistant rock.

These major soil horizons may or may not occur simultaneously in one soil profile. Frequently horizons will show subdivisions of various kinds, and other, less common horizons may occur in addition to any of the above. Since the profile type of a soil is a major criterion for soil classification, the more important subhorizons are described below.[1]

A-horizons:

A_{00} – undecomposed surface leaf litter (litter horizon);

A_0 – partly decomposed leaf litter (fermentation horizon);

A_1 – humic mineral soil, not noticeably eluviated (humus horizon);

A_{2e} – mineral soil largely eluviated of clay, iron, and aluminum compounds, with little or no humus present; frequently the eluviation is sufficiently intense to bleach the soil (eluvial horizon);

(A) – (called "*A*-bracket") poorly developed organic horizon of shallow, often stony soils, found directly over *C*-horizon; excludes presence of other *A*-horizons (incipient *A*-horizon).

B-horizons:

(B) – ("*B*-bracket") chemically weathered and discolored zone of finer texture than the parent material, but without humus and illuvial materials; precludes existence of an A_2-horizon (color *B*-horizon);

B_h – illuvial horizon with humic matter, clay minerals, and iron compounds derived from an A_2-horizon (illuvial humus horizon);

B_s – illuvial horizon enriched with sesquioxides of iron (Fe_2O_3) and aluminum (Al_2O_3) from the A_2-horizon; although a B_h-horizon may or may not occur when an A_{2e} is present, a B_s is indispensable (illuvial sesquioxide horizon);

1. Needless to say, the symbols used by different authors vary somewhat. The letter designations used here follow the rather general system of Franz (1960, pp. 213-18).

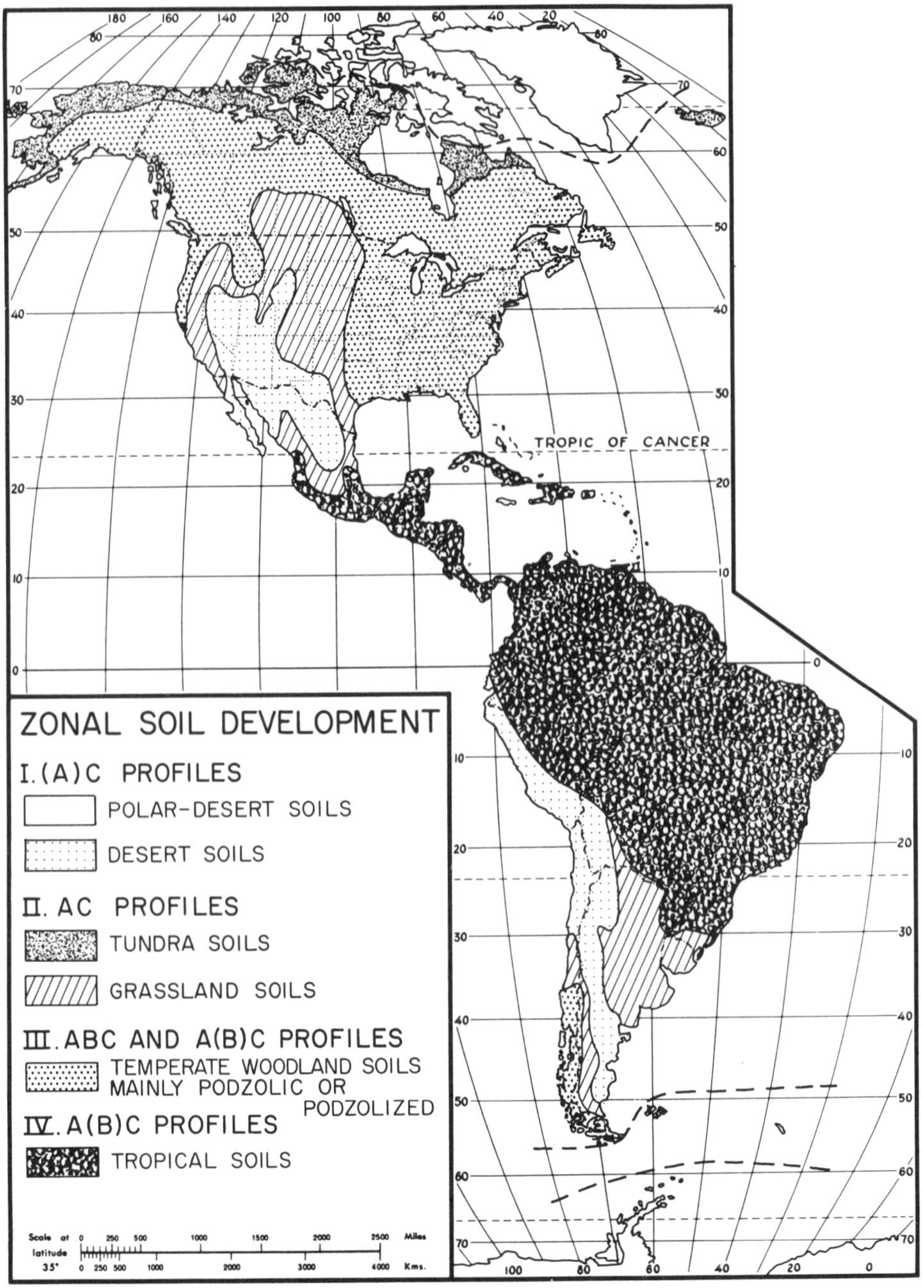

Figure 5. Zonal soil development.

Figure 5

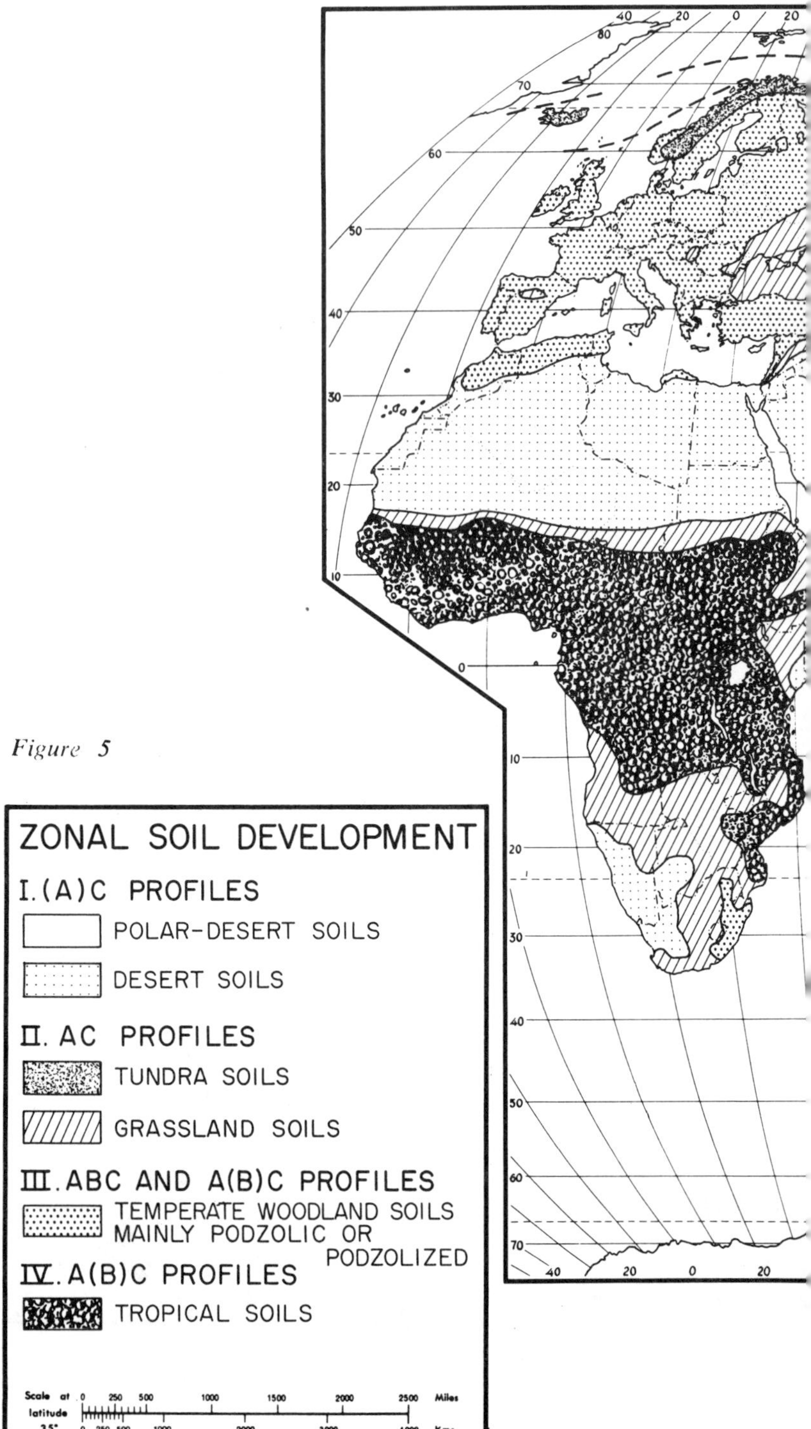
ZONAL SOIL DEVELOPMENT
I. (A)C PROFILES
POLAR-DESERT SOILS
DESERT SOILS
II. AC PROFILES
TUNDRA SOILS
GRASSLAND SOILS
III. ABC AND A(B)C PROFILES
TEMPERATE WOODLAND SOILS MAINLY PODZOLIC OR PODZOLIZED
IV. A(B)C PROFILES
TROPICAL SOILS
Scale at latitude 35°
0 250 500 1000 1500 2000 2500 Miles
0 250 500 1000 2000 3000 4000 Kms.

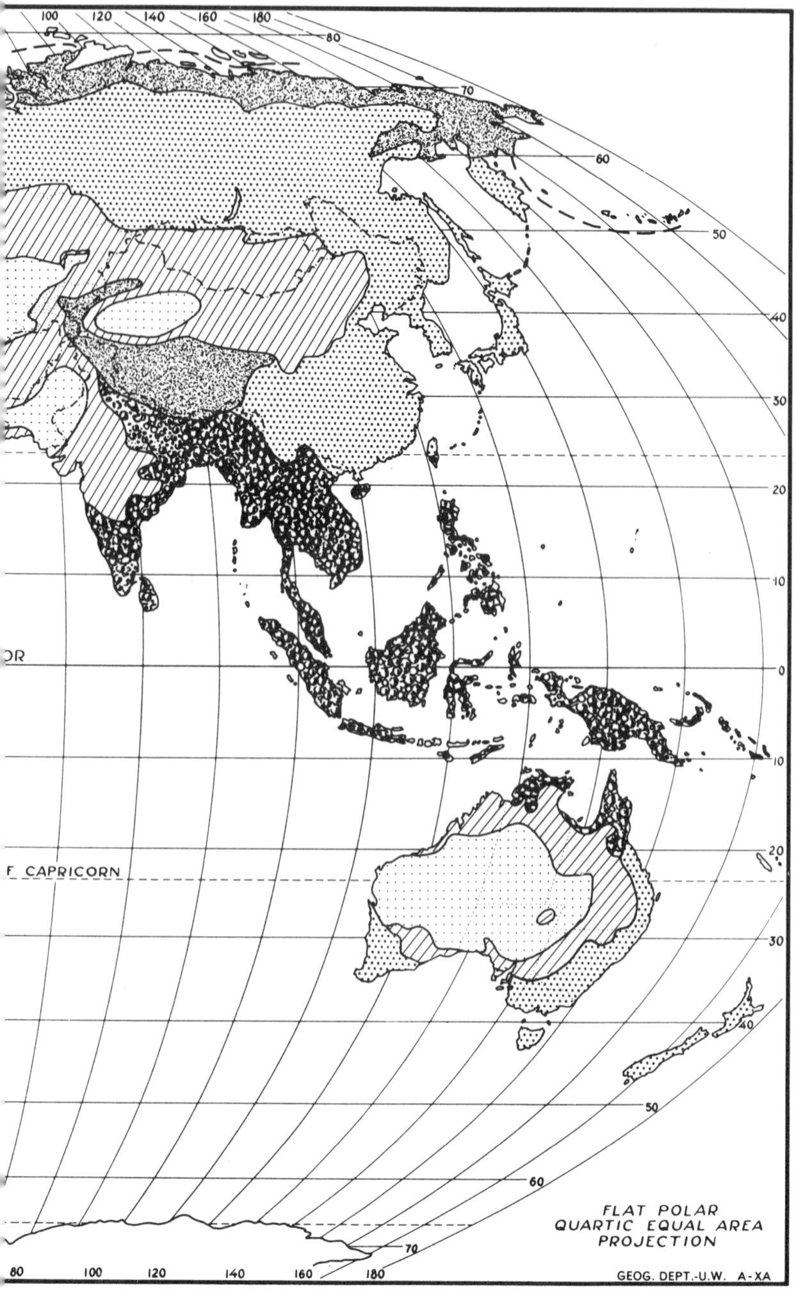

100
120
140
160
180
80
70
60
50
40
30
20
10
OR
0
10
20
F CAPRICORN
30
40
50
60
70
80
100
120
140
160
180
FLAT POLAR
QUARTIC EQUAL AREA
PROJECTION
GEOG. DEPT.-U.W. A-XA

A(B)–transitional horizon with chemical weathering and primary humus enrichment;

BC or *(B)C*–transitional horizon with visible but incomplete physical and chemical weathering.

C-horizons:

C_1– partially disintegrated parent material with some evidence of oxidation;

C_2– intact parent material.

D– underlying bedrock not related to the parent material of the soil.

Other Horizons:

Ca–horizon of diffuse lime precipitate or zone of calcareous nodules, concretions, or crusts found as *A/Ca, (B)/Ca, C/Ca* or as a well developed, distinct *Ca*-horizon just above the *C*-horizon (carbonate horizon);

Sa–halite (NaCl) or gypsum ($CaSO_4 \cdot 2H_2O$) accumulation (salt horizon);

Fe–limonitic iron ($Fe_2O_3 \cdot 2H_2O$) accretion layer (oxide horizon); in the case of tropical soils *Fe* or *Al* is used to designate hardpan concentrations dominated by either iron or aluminum sesquioxides;

P–seasonally waterlogged horizon of mottled reddish and grayish color resulting from seasonal alternations of oxidation and reduction (pseudo-gley horizon);

G–groundwater horizon with chemical reduction or alternating reduction and oxidation (gley horizon);

E–eroded soil sediment; may be used as a suffix, for example, as: *(B)E* indicating material from an eroded *B*-horizon.

Some of the broad classes of soil types are commonly based on profile types. Ignoring the groundwater soils, four more important groups can be identified: *(A)C, AC, A(B)C,* and *ABC*. These are genetic types that are related primarily to vegetation, intensity of chemical weathering, and precipitation-evaporation ratio.

The *(A)C* soils fall into two classes. First, they include the immature, shallow, incipient soils present in all climatic zones. These may be a result of recent surface erosion or, less commonly, recent deposition with insufficient time elapsed for soil development. So, for example, *(A)C lithosols* may be a permanent feature on steep slopes. The zonal soils of the arctic barrens and lower latitude deserts do not develop past the *(A)C* stage as a result of limited chemical weathering and organic matter. Such zonal soils of *(A)C* type reflect a quasi-absence of soil development.

AC soils are commonly found in semiarid or subpolar climates where organic material and biochemical activity are abundant, but where chemical alteration of parent material is impeded by insufficient moisture or low temperatures. Carbonate horizons are frequently found in the semiarid soils of middle and lower latitudes.

A(B)C soils develop when a rich organic environment is combined with intensive chemical weathering. Eluviation and illuviation may be more or less negligible for a number of reasons related to the nutrient cycle, water balance, or parent material of the soil. Carbonate horizons form if there is a dry season. Soils with color *B*-horizons are most common in the subtropical and tropical woodlands as well as in regions of temperate broadleaved forests.

ABC-type soils reflect intensive chemical alteration in acidic environments as a result of vegetation type, parent material, or climate. Although most commonly associated with coniferous forests, *ABC* soils have a limited occurrence in the humid tropics.

The *A(B)C* and *ABC* soils are, then, essentially woodland soils, in contrast to the *(A)C* or *AC* soils of the world's deserts, grasslands, and tundras. Although groundwater modifications and immature soils are common variables to contend with, this basic classification is useful in evaluating paleosols. It can be sufficiently refined to be of considerable paleo-environmental value.

NUTRIENT CYCLE, *p*H, AND HUMUS

Organic plant matter and microfauna are important for the genesis of all soils, but are often crucial in the development of woodland soils.

Although the parent material of a soil provides the raw material for the nutrient cycle, the plant residues themselves are a product of the vegetation. Plant communities demanding large quantities of mineral nutrients from the soil remove potassium, calcium, magnesium, and ammonia which are then employed in building plant tissue. Ultimately these plant products are returned to the soil and party incorporated as humus. Such a soil will probably have an alkaline or slightly acid reaction (*p*H value over 6) as a result of at least partial saturation with bases such as potassium or calcium carbonate (K_2CO_3, $CaCO_3$). More modest plant species incorporate few minerals into their structures and do not afterward provide the soil with nutrient-rich residues. This paucity of bases leads to a rather acidic environment (*p*H under 6).

The nutrients and pH values of a soil are important for the microfauna, which in turn helps determine the humus type and soil structure.

If a soil is moderately well saturated with base nutrients and is not too acidic, a rich fauna of earthworms and beetles is commonly present. Through repeated ingestion and excretion by earthworms, the organic materials are reduced completely to amorphous humic acids in chemical association with iron compounds and clay minerals. The resulting clay-humus complex has few identifiable plant tissues, and the organic and mineral components cannot be separated microscopically. This optimal humus type, derived from earthworm excreta, is known as *mull* or *mild humus* (see Kubiena, 1953, p. 38 ff.) *A*-horizons with mull humus commonly have a shallow A_{00}-horizon over an A_1, without eluvial horizons. Mull humus is confined largely to soils with *A(B)C* profiles and the more favored *AC* soils.

On the other extreme, soils with few nutrients and rather acidic environments have few microorganisms, mainly fungi and mites. Organic materials are broken down physically but not chemically, and plant structures are readily identified with use of a hand lens. The resulting inert residues are not chemically integrated with the mineral soil, so that deep A_{00} and A_0-horizons may form, commonly over horizons with little or no intervening A_1 horizon. This inert, poorly decomposed humus is known as *raw humus* or *mor*. It is typically associated with *(A)C* and many *ABC* soils.

An important intermediate type of humus is known as *moder*. It develops when there is some base saturation with mildly acidic or alkaline conditions, or when an earthworm microfauna is precluded by seasonal drought. Instead the fauna is dominated by beetles, and plant remains are chewed up and combined with organic excreta, but only loosely intermixed with mineral grains. The texture of moder is loose and crumbly, and plant structures can be identified microscopically. The upper soil profile is dominantly a combination of A_{00}, A_0, and A_1 horizons, and moder humus is usually associated with certain *AC* and *A(B)C*-type soils in drier locations or on bedrock with limited base nutrients.

PODSOLIZATION, CALCIFICATION, AND LATERIZATION

The plant communities that favor the development of soils with raw humus and little or no base saturation are, in the main, coniferous forests. The needle litter provides few nutrients, and percolating rain waters combine with the fermenting, resinous needles and become strongly acidic. The resulting environment is unattractive to most microorganisms, and chemical breakdown is slow and incomplete. The acid solutions of penetrating water leach away any soluble bases and further proceed to break down the clay minerals present. These minerals may be eluviated in an unaltered state or decomposed with selective removal of the iron and aluminum sesquioxides. Of the three major components of most clays (two sesquioxides and silica), the silica molecules are released the most slowly. A noticeable horizon of eluviation (A_{2e}), frequently of bleached color, marks this process of *podsolization*. The sesquioxides and other eluviated materials accumulate in the *B*-horizon whose reddish or yellowish hues contrast with the white or light gray color of the A_{2e}-horizon. This, then, is the genesis of *ABC*-type soils.

Not all coniferous forests are associated with extreme podsolization as described above. Previous sandy soils and tree species such as the pine provide optimal conditions for podsolization; denser soils and nutrient-demanding conifers such as spruce may retard podsolization. In such cases the A_{2e}-horizon is brownish and retains a good part of its intact clays as well as of its disassociated sesquioxides. These intermediate soils are called *podsolic* in contrast to the more extreme *podsols* which have bleached A_{2e}-horizons. Soils of the *ABC*-podsolic group are not restricted to coniferous forest, but may also be found under broadleaved forests in wet localities on pervious or nutrient-poor soils.

Soil *calcification* is the converse of podsolization. Base saturation is high or complete, and leaching, even of the soluble salts, is limited or absent. Seasonal drought or over-all semiaridity favors calcification by limiting the amount of percolating rain water. The subsoil is often permanently dry, so that leached carbonates are soon precipitated at moderate depths as the soil water evaporates. Similarly, the dry season may stimulate upward migration of lime-charged soil moisture by capillary action. As this water evaporates, accumulation again follows. The result is carbonate enrichment within the subsoil in the form of carbonate horizons.

Dry soil conditions, however, are only a part of the mechanism of calcification, at least in subhumid or semiarid regions. Without the intervention of the soil humus, occasional protracted rains would easily dissolve the carbonates of many grassland soils. Grasses, largely by virtue of their exceedingly fine, dense, and deep network of roots, provide the greatest quantity of nutrient-rich organic matter to the soil. This organic matter and the favorable alkaline environment attract a teeming microfauna which in turn helps maintain a rich humus type. In this way limited losses through leaching can be compensated for indefinitely.

Most of the semiarid *AC* soils are carbonate soils, and *A(B)C* soils developed on limestone bedrock under subhumid environments commonly show lime accumulation as well. Soils with carbonate enrichment *(pedocals)* are frequently distinguished from *pedalfers,* which have podsolization or intensive carbonate leaching (Marbut, 1936). Although precipitation exceeds evaporation in many areas of pedalfer soils, water balance is not the only factor dominating their genesis. Consequently the pedocal-pedalfer limit does not provide an unqualified index of environmental conditions. Nevertheless, podsolization and calcification are two types of soil process that are of great interest in the study of paleosols.

Laterization is the process leading to hardpan formation in tropical soils. The intensive chemical weathering of the humid tropics leads to the breakdown of clay minerals, with selective release of colloidal silica (SiO_2). This mobile silica is leached downward, leading to a relative enrichment of sesquioxides in the intermediate layers of the soil. These initial processes appear to be common to most tropical woodland soils, and the term *latozation* is sometimes generally applied for intensive weathering with desilicification and sesquioxide concentration. In many latozolic soils the sesquioxides form immobile concretions, and partial desilicification may ensue. Such *lateritic* horizons with concretions or massive concentrations of sesquioxides may be several meters deep, and have been known to reach thicknesses of well over 10 m. As long as these horizons remain in the subsoil, true hardpans do not develop. But when lateritic horizons are exposed at the surface by erosion, particularly in regions of pronounced seasonal drought, irreversible crystallization follows and a durable crust is formed. Concretions may also be concentrated into another form of crust by extended erosion of the unconsolidated soil aggregates. Both of these ferruginous crusts are commonly known as *laterite.*

The significance of laterites and lateritic soils is still not satisfactorily established. Fossil laterites are known from tropical deserts, and some laterites of the subhumid tropics also appear to be fossil. In the traditional view, however, laterization was thought to be characteristic of all mature soils of the seasonally humid tropics. In fact, the sesquioxide accumulation was thought to be similar to the carbonate concentration in semiarid soils, with upward movement by capillary action during the dry season. Field and laboratory studies by W. L. Kubiena (1954a, 1957) suggest a somewhat different genesis, however. Kubiena found that laterization is at least as common in the humid tropics as in the savanna regions, and that, apart from the process of induration, laterization is only possible in permanently moist subsoils. But since laterites exposed and hardened at the surface are known best, an erroneous impression of their distribution and origin is readily possible. In fact, laterization is at least as common within bedrock and parent material as in the soil zone proper. Consequently, Kubiena considers the process as a typical form of tropical petrefaction rather than of soil development. Its prerequisites appear to be the presence of perennial soil moisture and high temperatures (see also Mohr and Van Baren, 1954; G. D. Smith *et al.*, 1960, p. 238 f.).

ARCTIC SOIL TYPES[2]

Soil development in the polar regions is generally impeded by a harsh climate as well as by the potent, ubiquitous erosional forces in the geomorphic sphere (see chapter 7). In view of the wide extent of polar environments during much of the Pleistocene, the soils of the Arctic will be considered in some detail.

The High Arctic Barrens. Within the high arctic wastes, soil development is evidently at a minimum. Biochemical agencies and chemical weathering are almost nonexistent in this inhospitable environment. The snow cover is scanty or absent during most of the dry winters, a result of very low vapor content in cold air. Consequently, frost-weathering acts on the surface rock during much of the year, producing great masses of debris ranging in size from crude rock to medium-grained silts. The presence of permanently frozen subsoil at a few decimeters depth aids in the mechanical breakdown of the bedrock surface. However, the wind, unimpeded by vegetation, sweeps away the silt and fine sand produced by mechanical weathering. The over-all effect is that any incipient soil development is permanently hampered by this persistent removal by wind or water of any fine materials. The climax or zonal soil is a lithosol of *(A)C* type, known as an arctic *ramark* (Fig. 6*a*) (Kubiena, 1953). A discontinuous veneer of raw humus is present over a physically disintegrated C_1-horizon some 25–80 cm. in depth, and largely corresponding to the zone of seasonal thaw and frost-heaving.

The Tundra. The herbaceous tundra provides a more favorable environment for soil development than the arctic barrens. A reasonably complete vegetative mat ensures some rudimentary form of biochemical activity and furthermore helps protect fine mineral grains from deflation by wind. A more durable and effective snow cover also reduces frost-weathering at the surface, although frost-heaving over permafrost is prominent.

2. Although there are several major soil classifications in use, most of them are basically similar and can be readily compared, with exception of a recent revision proposed by the U. S. Soil Survey (G. D. Smith *et al.*, 1960). The classification adopted here is that of Kubiena (1953), which is followed by soil scientists and geologists in many European countries. The text in question has been published in English, German, and Spanish, while the principles have already been introduced to archeologists by Cornwall (1958, chs. 7-9). Apart from these practical aspects, Kubiena's system, with its accompanying detailed soil descriptions, is invaluable for pedological field work and for the study of fossil soils.

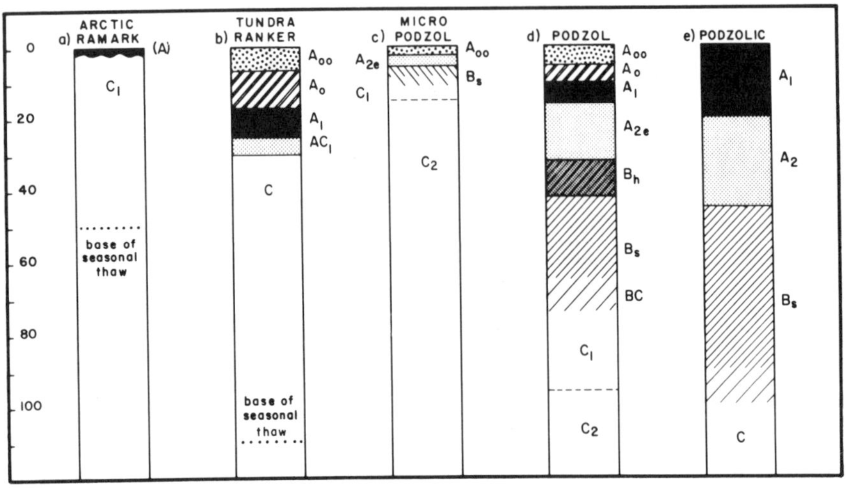

Figure 6. Some higher latitude soil profiles.

Although lithosols occur over wide areas affected by intensive erosion, the greater part of the landscape has both vegetation and soil cover. Drainage is characteristically poor, however, as a result of permafrost and widespread gentle relief. For this reason swamp and groundwater soils are probably most common. Such semiterrestrial soils frequently have a shallow horizon over a gley zone extending down to the base of the seasonal thaw. Permanently oozing with water, the subsoil is not aerated; it is characterized by reduction and sulfide formation which gives the *G*-horizon a pale gray color.

In the case of better-drained terrain, a climax-type soil of *AC*-profile develops in response to a rich but nutrient-poor herbaceous vegetation. The microfauna is limited, and decomposition of organic materials is slow and incomplete, producing raw or moder humus types. Base saturation is very low, reaction acidic. An *AC*-soil with such features belongs to the *ranker* class of Kubiena (1953); the most typical soil of the tundra is a tundra ranker (see also Tedrow and Harries, 1960). Due to the high water absorption of the humus and the almost bacterialess, acidic soil environment, boggy conditions are common, leading to a net accumulation of organic matter in the form of peat. In the better-drained tundra rankers a little leaching of iron compounds may lead to sesquioxide or humus-staining on the surface of rocks in the transitional AC_1 horizon (Fig. 6*b*).

The Forest-Tundra. Conditions in the forest-tundra transitional belt are in good part similar to those of the herbaceous tundra. Gley, peat, and ranker soils are common. In areas of higher vegetation, however, podsolization may be more advanced, with minute but nonetheless typical eluvial and illuvial horizons. The resulting A_{00}-A_{2e}-B_s-C_1 horizon is that of a diminutive or micropodsol as shown in Figure 6*c*(Kubiena, 1953). In locations with unconsolidated bedrock, a deeper *BC*-horizon will occur; in areas with a tundra mat, a deeper *A*-horizon.

Alpine Soils. The soils of high mountain regions show many similarities to those of the arctic. Lithosols are, of course, more widespread as a result of rough terrain, and as a corollary, swamp and groundwater soils less common. Various classes of rankers probably form the most typical soil type on gentler sloping surfaces, at least in middle latitudes.

FOREST SOILS OF HIGHER AND MIDDLE LATITUDES

The soils of the temperate and cold continental woodlands are the product of moderate to considerable chemical weathering during warm to hot summers of one to five months duration. Moisture is available most of the time, and there is no climatic impediment to biochemical activity.

Forest soils of *ABC* or *A(B)C*-type occur as climax type in well-drained areas, although immature *AC*-profiles are frequent on moderate to steeply sloping surfaces and on very resistant bedrock. Such immature soils may be highly base-saturated and calcified in the case of carbonate parent materials *(rendzinas)*, or they may be acidic pedalfers of the ranker group.

The equally widespread groundwater soils deserve brief mention as well. Soils subjected to seasonal waterlogging are known as *pseudo-gleys*. While water-saturated, the subsoil is subjected to limited reduction of iron into its various derivatives, followed by extensive oxidation as the soil dries. The resulting iron and manganese oxides are concentrated in the form of stains, stripes, and concretions that vary in size and shape but are usually of yellowish or orange color, set in a background of gray. The gray tones of pseudo-gley subsoils are more commonly a result of local iron removal than of the presence of reduced iron derivatives. Gley soils are located permanently in the groundwater zone. Gley horizons are therefore characterized by reduction horizons that, depending on the iron compounds present, have dull gray to grayish-green shades. The zone of fluctuating groundwater table undergoes alternating reduction and oxidation, and has many similarities with mottled pseudo-gley horizons, although the products of reduction are more significant.

The Boreal Forests. Frost-weathering is important wherever bedrock is exposed in the boreal forest belt. Similarly, soil-frost phenomena are widespread, particularly where permafrost prevails. The dominant factor of soil development, however, is chemical. During the months when the soil zone has thawed, water percolates freely through the needle litter and partly decomposed raw humus, passing through the mineral soil below. Meltwater is plentiful in the spring and in summer in the wet season, although rainfall is a little irregular and of some intensity. This favors repeated rapid percolation. Thus, in combination with the unfavorable nutrient cycle, podsolization is active. The climax *podsols* have humic horizons at the surface, followed below by a more or less bleached, eluvial A_{2e}-horizon consisting largely of resistant minerals (mainly quartz) in the sand-and-silt size. The illuvial *B*-horizon has a light brown or yellowish-brown zone of redeposited aluminum and iron sesquioxides *(*B_s*)*. There may also be a dark, shallow, intermediate zone with leached clays and humus in addition to iron sesquioxide (B_h) (Fig. 6*d*). The B_h-horizon may form a hardpan, especially beneath cultivated soils.

Podsols are also found under coniferous woodland, particularly on highly pervious subsoils within the marine temperate climates. *Podsolic soils,* with party eluviated A_{2e}-horizons, generally show little bleaching and have no B_h (Fig. 6*e*). These intermediate soils are widespread under coniferous woodland, even within the boreal forests, wherever the nutrient cycle is better. They may also occur on pervious soils in wet localities under deciduous vegetation.

Bog soils and gleys are very extensive in the boreal forest belt, while tundra enclaves are typical in permafrost areas.

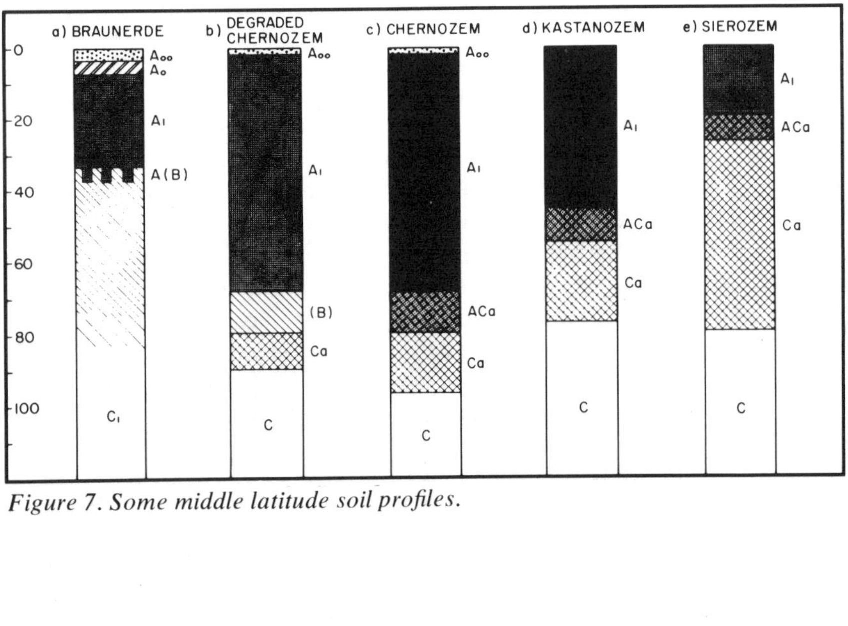

Figure 7. Some middle latitude soil profiles.

The Deciduous Forests. Although podsolic soils are common under deciduous hardwoods, true *A(B)C* profiles are also characteristic.

A protracted warm season with equalized distribution of moisture, provides an opportunity for moderate but prolonged chemical activity. Organic matter rich in base-nutrients is provided by the more demanding, deep-rooted species. All this favors a rich microfauna, and biochemical activity is equally vigorous. The leaf litter is effectively decomposed and distributed through a fairly deep, dark brown A_1-horizon of mull-type humus (Fig. 7*a*). The brown color of the deep *(B)*-horizon is predominantly a result of sesquioxide coatings on the mineral grains (but without apparent illuviation), or of chemical alteration that forms clay minerals and oxides while obliterating the original rock structure. Such brown forest soils or *braunerdes* (Kubiena, 1953) show complex intergrading with similar podsolic profiles.

MID-LATITUDE GRASSLAND SOILS

The soils of the middle latitude grasslands are distinct from other semi-arid and arid soils because of their exceptionally rich humus. Climatically, these regions all experience cold winters (Koeppen's *Bsk* climates), so that vegetation is dormant for several months of the year, during which time little or no biochemical activity takes place. The repeated annual growth provides the soil with considerable quantities of nutrient-rich organic matter every autumn. These nutritious grasses have an exceptionally intricate and deep rooting system. Subsequent decomposition is primarily performed by earthworms, producing a mull humus. But since decay is slow, a considerable accumulation of humus takes place to depths of one meter or more. The soil is frozen in winter, and percolating waters are limited by droughty conditions in midsummer. Leaching is therefore limited, and lime accumulations are commonly found between the *A* and *C*-horizons. Chemical weathering is not sufficient to develop a *(B)*-horizon. A general profile is then A_1-*Ca*-*C* (Fig. 7*c*-*e*). Carbonate horizons are absent in areas with nonbasic silicate rocks, leading to the formation of ranker-type soils.

The *AC*-soil profiles shown in Figure 7 illustrate the most important mid-latitude grassland soils. Under optimal conditions the black *chernozems* (Kubiena, 1953) develop particularly deep, humic A_1-horizons. On the woodland margins there may be leached *AC*-type *prairie soils* of the pedalfer group, or *degraded chernozems* with shallow *(B)*-horizons, with or without some evidence of podsolization (Kubiena, 1953). Degraded chernozems with A_1-*(B)*-*Ca*-*C*-horizons (Fig. 7*b*) are commonly found in Europe wherever forests have invaded former grasslands, partly as a response to climatic change.

On the drier margins the chernozems grade over into less humic soils with shallower A_1-horizons but greater carbonate accumulation. These include the chestnut soils or *kastanozems* (Fig. 7*d*), the brown or *burozems,* and finally the gray semidesert soils or *sierozems* (Fig. 7*e*) (Kubiena, 1953).

OTHER SEMIARID AND ARID SOILS

The lower latitude grasslands do not experience winter cold, but vegetation is dormant and biochemical activity low as a result of seasonal drought. The rapid decay during the hot summer rainy season, combined with a vegetation type producing limited organic matter, lowers the humus content of the soil. Furthermore, tropical grasses are largely poor in mineral nutrients. Consequently many tropical *AC*-profiles are not impressive.

An exception to this are the *tirs* or tirsified soils (which are also known as regur, black cotton soils, tropical black earths, margalitic soils, vertisols) (Mohr and Van Baren, 1954; Kubiena, 1957; Klinge, 1960) which are found mainly in seasonally waterlogged locales on both margins of the lower latitude dry belt, extending well into the seasonally humid tropics. Boggy conditions prevail during the rainy season, mainly over impervious bedrock or clayey subsoil. Some diffuse humus is typical, the reaction alkaline. Although remarkably black when wet, the A_1-horizon commonly 30–150 cm. deep, is light gray when dry. The color is apparently not due to high organic content but rather to some property of the base-saturated clay-humus colloids. The A_1-horizon swells up when wet and contracts strongly when dry, with dehydration cracks penetrating to its base. The clay-humus dispersal through such fissures probably reflects mechanical diffusion rather than the activity of soil microorganisms. Deep *P* or *G*-horizons commonly underlie the *A*, indicating groundwater oxidation and reduction. Also, rather massive *Ca*-horizons may frequently underlie the A_1. Despite their superficial resemblance to chernozems, the tirsified soils are distinctive of seasonally wet climates that experience hot dry seasons with air temperatures exceeding 38° C. at some time. Yet the tirs' peculiar textural properties and relationship to groundwater conditions rule them out as true zonal soils. They appear to be significant as paleosols in some subtropical arid regions (e.g., Morocco and the Sahara).

With little chemical weathering and no vegetation, the arid deserts proper allow very limited soil development, so that $(A)C_1$ lithosols are most characteristic. This is Kubiena's *yerma* or desert soil class. Areas without external drainage have widespread subsoil concentrations, or surface crusts, or precipitates of halite (NaCl), gypsum ($CaSO_4$), or calcium carbonate ($CaCO_3$). This is understandable because any solubles carried by the ephemeral waters will be left behind in the wake of evaporation. Subsurface *Sa* or *Ca*-horizons are also common in well-drained soils, since any available soil moisture will be attracted to the surface by capillary action.

SUBTROPICAL WOODLAND SOILS

Mediterranean Soils. The summer-dry subtropics are a transitional belt in many ways, and the soil pattern is generally rather complex. Only in the Mediterranean Basin are the common soil processes and soil distributions better understood.

Chemical and biochemical weathering are at a low level of intensity as a result of the coincidence between low winter temperatures and the rainy season. Soil development is painfully slow, and on fresh limestone bedrock or other unconsolidated calcareous materials, *AC*-type rendzina soils seem to represent the post-Pleistocene climax soil (Kubiena, 1954b, 1963; Butzer, 1963a, with remarks by Kubiena). On silicate rocks, with low base status, semiarid southern or *meridional braunerdes* (Fig. 8*a*) seem to be the climax type. In contrast, the "typical" red soils of the Mediterranean region no longer develop on fresh surfaces. Various instances of fossil, Pleistocene red soils of this type have been demonstrated (Klinge, 1958; Fränzle, 1959b; Butzer and Cuerda, 1962a; Butzer, 1964a, 1965), and it is now widely accepted that the red soils at the surface are relict, i.e., also paleosols (Durand, 1959). They are in approximate equilibrium with their environment but will not form afresh.

These conspicuous Mediterranean paleosols deserve some attention, since they are closely related to tropical soils, and since similar paleosols of Pleistocene age are found in other middle latitude areas. The most common is the *terra rossa* developed on calcareous parent material. This soil has a shallow A_1-horizon over a deep, red, clayey *(B)*-horizon weathered from the limestone bedrock. The brilliant color is due to the presence of anhydrous iron such as hematite (Fe_2O_3) and goethite ($FeO \cdot OH$), presumably derived from the hydrated limonitic form ($Fe_2O_3 \cdot 2H_2O$) through extreme desiccation during a hot dry season (Kubiena, 1953; Klinge, 1958). On the other hand, the iron compounds are found dispersed within colloidal silica (SiO_2) liberated during breakdown of clay minerals by intensive weathering. There is a little leaching of colloidal silica with relative accumulation of iron. This presumes an alternation of warm-moist and hot-dry conditions, so that the terra rossas are typical products of warm, seasonally humid climates. The anhydrous iron compounds (which do not take up water upon wetting) are stable, so that these red soils are almost irreversible after a change of climatic environment. Although terra rossas are quite decalcified during formation, they have generally been subject to secondary calcification in drier parts of the Mediterranean region today. Also, as a consequence of Pleistocene geomorphic agents and more recent interference by man, most terra rossas are today found as soil sediments, and only rarely can the *(B)*-horizon be traced to the *C* by an intact sequence of *(B)C* and C_1-horizons.

On the cooler margins of the Mediterranean Basin, as well as at higher elevations, the terra rossas are replaced by reddish-yellow *terra fuscas,* the products of equally effective weathering without anhydration of the iron compounds. Analogous soil types on noncalcareous bedrock are *rotlehms* and *braunlehms* (Kubiena, 1953). A general profile type

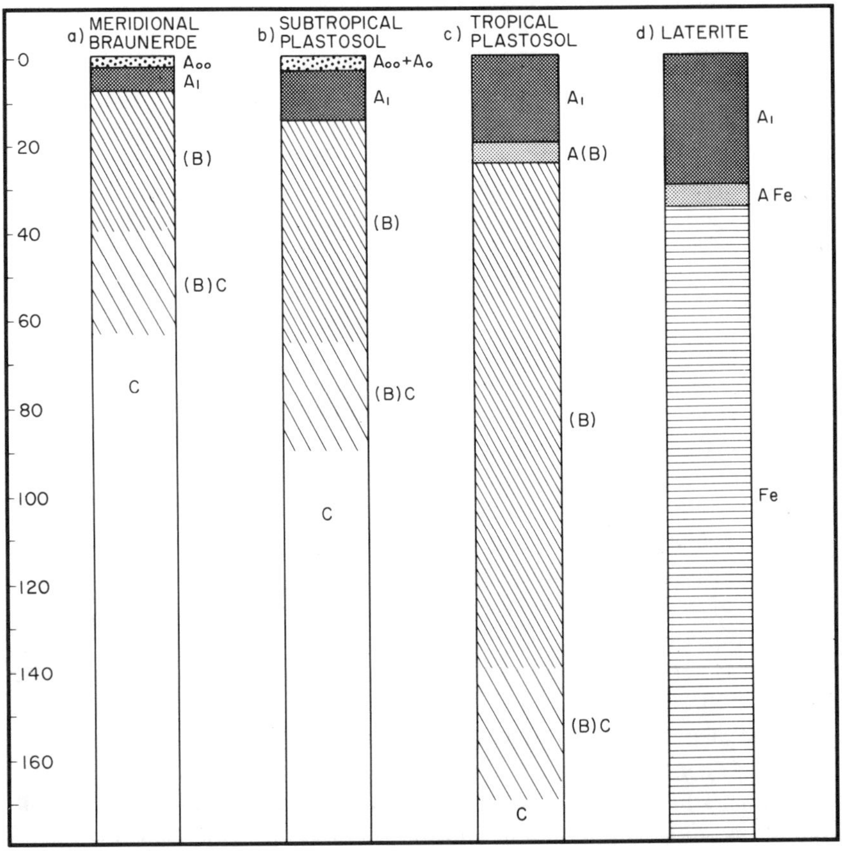

Figure 8. Some lower latitude soil profiles.

applicable to most of these relict, subtropical *plastosols* (*plastic* as opposed to *friable* products of intensive weathering) is given in Figure 8*b*.

The Humid Subtropical Woodlands. The soils of the humid subtropical woodlands are little understood, except for those found in the southeastern United States. There intensively weathered reddish and yellow podsolic soils are dominant, very closely resembling the rotlehms and braunlehms of the Old World.

TROPICAL WOODLAND SOILS

The most characteristic soil development of the seasonally humid tropics is a *rotlehm,* dominated by anhydrous iron. In contrast, *braunlehms* with limonitic iron are found under perhumid rainforests (Kubiena, 1957). Although some latozation and even laterization may be evident, these soils are essentially plastosols (Fig. 8*c*). The A_1-horizon commonly has a fair amount of organic matter, and no A_{2e}-horizons are present. Any lateritic concretions present in the *(B)*-horizon are not crystallized, and the colloidal silica—derived from clay mineral decomposition and re-solution of crystallized SiO_2—is only moderately leached, hence the plasticity. Clays constitute over 25 percent of the *(B)*-horizon. Base saturation is comparatively low, the *p*H acidic. These tropical plastosols (see discussion by D'Hoore, 1960, Birot, 1965, and Tricart and Cailleux, 1965b) differ from the subtropical plastosols only in their intensity of development and depth of profile.

Within the savanna belt, rotlehms vie with *latosolic,* i.e., friable and nonplastic, *roterde* soils (Kubiena, 1953, 1957) which show considerable silica depletion and very low base saturation. The basic profiles are similar, however. Also common in the savanna are tirsified soils in areas of seasonally impeded drainage. Exposed lateritic crusts, frequently under attack by water erosion, help complicate the soils pattern. Soil stripping on all slopes, with redeposition on the flat savanna plains, is probably the most extensive of all these phenomena. The over-all result is that rotlehms or roterdes are more commonly found as derived sediments than as soil profiles *in situ*. Such *(B)E* profiles may attain depths of 30 meters or more.

In the rain forest, groundwater soils are widespread, but their systematics are poorly understood. In many cases humic *A*-horizons overlie deep, mottled, gleylike beds with lateritic concretions or semicontinuous sesquioxide precipitates. Most of these horizons appear to be incipient laterites (Fig. 8*d*) rather than true oxidation and reduction zones. Some true podsols with peaty *A*-horizons are known, but most of the equa-

Table 7. Climate, vegetation, and soil development.

Koeppen climate	*Vegetation type*	*Soil profile*	*Dominant soil process*	*Characteristic soil types (not including ground-water soils or paleosols)*
EF, ET	High arctic barrens	*(A)C*	Mechanical weathering	Arctic ramark
ET	Herbaceous tundra	*AC*	Humification	Tundra ranker
ET, Dd	Forest-tundra	*AC, ABC*	Humification, some podsolization	Ranker, micropodsol
Dc, Dd	Boreal needle-leaved, evergreen and decid-uous forest	*ABC*	Podsolization	Podsols, podsolics
Cbf, Da, Db	Mid-latitude broad-leaved deciduous and mixed coniferous-deciduous forests	*A(B)C, ABC*	Deep weathering and moderate podsoliza-tion	Braunerde, podsolics
BSk	Mid-latitude grasslands	*AC*	Humification, calci-fication	Chernozem, kastanozem, burozem, sierozem

Cs	Dry subtropical, broad-leaved and needle-leaved evergreen woodland	*AC, A(B)C*	Humification, calcification	Rendzina, meridional braunerde
Caf	Moist subtropical, broadleaved evergreen woodland	*ABC*	Deep weathering, with podsolization	Podsolic braunlehm
BWh	Low latitude desert and desert shrub	*(A)C*	Mechanical weathering	Yermas
BSh	Low latitude grasslands	*AC*	Humification, oxidation	(Tirsified soils)
Aw, Caw	Tropical broadleaved deciduous woodland and savanna	*A(B)C*	Deep weathering, latozation, laterization	Rotlehm, roterde, laterite (tirsified soils)
Af, Am	Tropical broadleaved evergreen and semi-evergreen forest	*A(B)C*	Deep weathering, latozation, laterization	Braunlehm, roterde, laterite

torial podsols are found in better-drained situations on base-poor, pervious parent material. Fresh alluvium in the valleys and widespread soil slumping on slopes further reduce the occurrence of the climax braunlehms. Redeposited soil sediments are again widespread and *(B)E*-horizons may attain remarkable thicknesses of 50–75 m. in some areas and several hundred meters in exceptional cases.

VEGETATION AND SOIL DEVELOPMENT

The basic relationships and characteristics of climate vegetation and soil development are summarized in Table 7.

SEVEN

Glacial and Periglacial Geomorphology

GLACIERS PAST AND PRESENT

Depending on the material culture, the environment sets differing limitations on human habitation of the earth. Even today, despite extensive exploration and research on the Greenland and Antarctic ice caps, permanent settlement in glaciated areas remains impossible or at least uneconomical. So too in the past, the glaciated surface of the earth was certainly not inhabited by early man. Consequently, the former extent of permanent snow and ice is important for setting certain limits to the possible distribution of peoples and cultures.

Glaciers past and present are at least equally significant in understanding climate. So, for example, Koeppen's perpetual frost climate (*EF*) has always been at least coextensive with the continental ice caps. Similarly, in mountain areas the altitude of the snowline—that critical limit above which more snow falls than can melt—gives a rough approximation for mean annual temperatures somewhere below the freezing point (0° C.). Pleistocene snowline elevations that differed from those of today may allow broad estimates of local temperature change. In general, the difference in extent of past and modern glaciers provides a valuable quantitative criterion for latitudinal and altitudinal shifts of climatic boundaries. Since glacier location is influenced by moisture sources and the direction of prevailing winds, the presence of Pleistocene glaciers may also provide information about the general circulation of the atmosphere. Finally, glaciers can provide pertinent stratigraphic data permitting identification of glacial or interglacial epochs, either locally or on a much larger scale through meltwater deposits and through the world-wide oscillations of sea level accompanying the growth and wastage of continental glaciers.

The following summary discussion of glacial geomorphology outlines some of the more relevant data treated systematically elsewhere by R. von Klebelsberg (1948-49, vol. 1), Woldstedt (1954), Charlesworth (1957, vol. I), Flint (1957), Tricart and Cailleux (1962), and particularly Embleton and King (1968). A tentative review of paleoclimatic applications of snowline data is also given.

GLACIER REGIMEN AND MOVEMENT

Although observed summer temperatures at the margins of permanent ice fields are often quite close to the freezing point, temperature alone does not determine glacier location. Certain thermal conditions are, of course, prerequisite, but equally important is the relation of snowfall to the annual ice wastage (*ablation*) caused by melting and direct evaporation (*sublimation*). Both temperature and solar radiation are responsible for the rate of ablation, but glaciers may develop in cool areas with heavy snowfall and yet be absent in colder but drier regions. Once formed, however, glaciers create a microclimate of their own. They reflect radiation (increased *albedo*), lower the summer temperatures by contact and radiative cooling, and so lead to significant temperature *inversions* in the lower atmosphere, i.e., a cold skin of air underlying warmer air above. Such inversions often extend far beyond the ice margins. The larger the glacier the greater the environmental modification, and the resulting rigor of the climate is often out of all proportion to the basic thermal character of the region.

Ice fields that form where the snowfall exceeds the annual ablation are the result of compaction and structural alteration from snow to ice. The density of fresh snow is in the order of 0.15–0.16. After settling, removal of part of the pore space, and recrystallization, the stage of granular snow or *firn* (with densities of 0.5–0.8) is attained. Repeated melting and overfreezing, aided by further compaction under pressure of overlying firn and snow leads to complete impermeability to air and densities exceeding 0.82. This is defined as *ice,* which is capable of plastic flow. The resulting ice masses may form either *mountain, valley,* or *piedmont* glaciers in rough highland terrain, or *ice caps* in areas of smoother topography.

The movement of glaciers is a response to gravity and includes two basic components, *internal deformation* and *basal sliding* over bedrock. The relative significance of each varies with slope, ice thickness, and ice temperature. On flat surfaces or reversed slopes, glacier motion is accomplished by a thickening of the ice until gravity again favors downhill flow. In the case of mountain glaciers, movement is directed downhill and makes use of existing topographic channels. The ice cap, on the other hand, expands outward in radial fashion, largely independent of topography. Mountain glaciers advance until ablation exceeds the ice supply from higher elevations. Ice-cap expansion ceases when wastage balances accumulation or supply in the source areas. In the latter case several factors may be involved: increased radiation and higher temperatures in lower latitudes, scanty snowfall in drier continental interiors or high arctic regions, and finally, geographical phenomena such as the termination of ice at the sea coast.

The annual balance of accumulation versus ablation varies, and so ice margins and ice thicknesses fluctuate. These variations of regimen are less noticeable in continental glaciers than they are in mountain ice streams whose tongues may oscillate considerably. Movements are two-

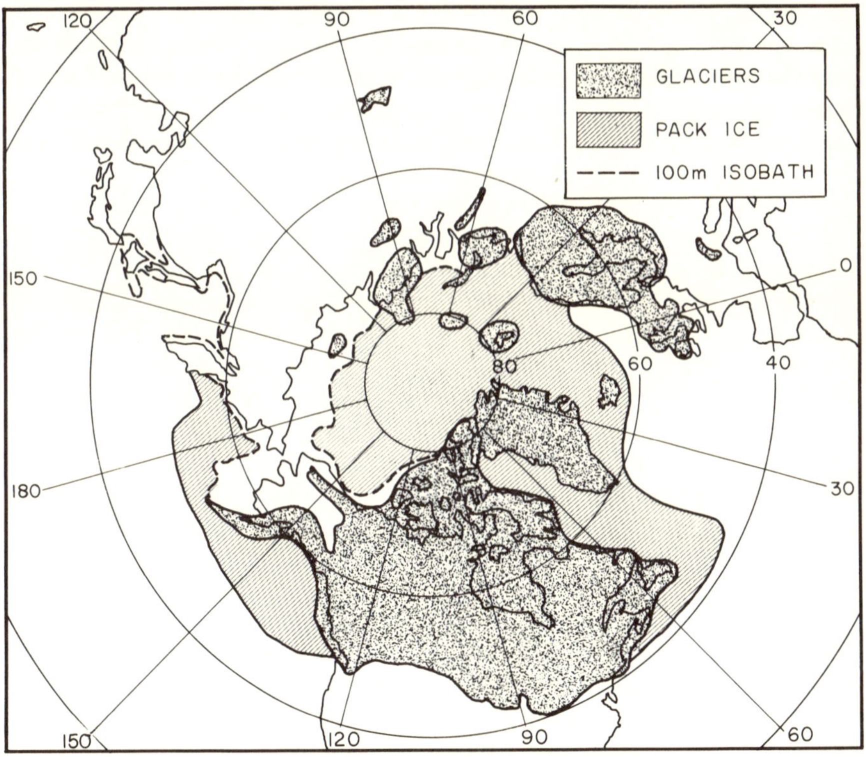

Figure 9. Maximum late Pleistocene glaciation of the Northern Hemisphere (simplified after Flint, 1957; Frenzel, 1959–60; and others).

fold, so that waning may be accompanied by *backwasting,* or retreat of the front, and *downwasting,* or reduction in ice thickness.

DEVELOPMENT OF MOUNTAIN, VALLEY, AND PIEDMONT GLACIERS

Although permanent ice covers only 10 per cent of the world's land surface today, it extended over 32 per cent at the time of maximum Pleistocene glaciation (Fig. 9). Apart from the areal significance of such glacial phenomena, moving ice is also the most powerful agent of erosion and deposition. The rather conspicuous effects of Pleistocene glaciation are apparent both on the broad continental plains of higher latitudes and in diverse highlands at all latitudes.

Snow accumulation in mountain country is greatest wherever snow-drifts can build up in valley-heads and other hollows in the lee of the major snow-bringing winds. Such snowbanks may persist for weeks or months after the snow has melted elsewhere, particularly when they are at least partially in the shade of higher mountain ridges. Such semi-permanent snow fields already exert a peculiar geomorphic influence during the summer season, especially through intensified frost-weathering along the receding snow margins, gravity movements under the snow and firn banks, and removal of debris by meltwaters. This process of *nivation* may produce broad, shallow dimples on gentle slopes or more recessed forms in rugged terrain (Cotton, 1942).

If snow fields persist over several years and evolve into larger ice masses, erosional niches created in the valley-head area will grow through further headward backweathering by freeze-and-thaw as well as by scour beneath the moving ice. Further recession of the steep back face or *headwall* of the valley will increase snow accumulation by favoring drifting and avalanching, and will reduce ablation by increasing the amount of shade. The ice basin itself will be deepened a little by subglacial abrasion and plucking, and broadened by scour and quarrying of lateral rock faces. Such loosened rock is embedded and carried within, on top of, or below the ice. The deeply recessed basin form that results is known as a *cirque;* its *floor* is commonly separated from the downstream valley by a gentle rise or *threshold.* The formative ice is called a mountain or cirque glacier.

Cirque glaciation imparts an individualistic appearance to a highland area. Mature cirques in adjacent basins will cut back the dividing ridges to sawtooth or *serrate ridges,* and reduce radial peaks to jagged, frost-sharpened horns. After deglaciation, whatever its cause, the cirque glaciers leave an amphitheater-like depression, with aprons of loose rock embanked against the headwall; frequently there is also a cirque lake or *tarn* in the center of the basin. The totality of such mountain sculpture by ice is known as *alpine topography* (Fig. 10).

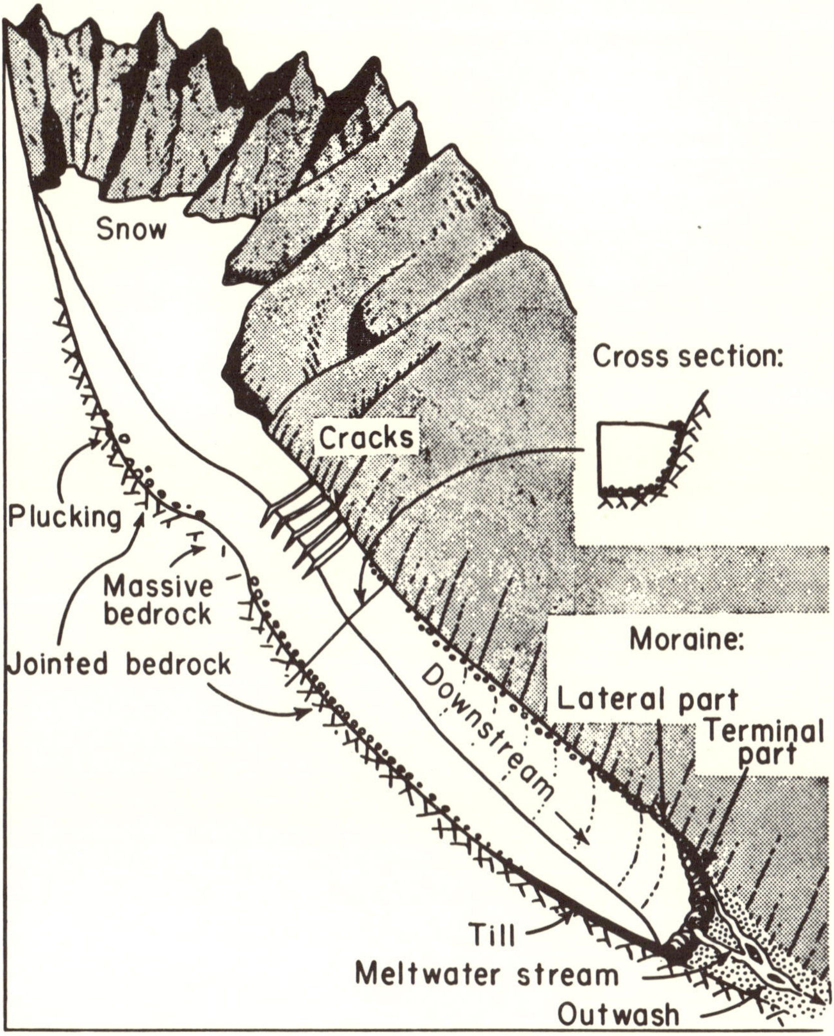

Figure 10. Features related to an Alpine glacier (from Longwell and Flint, 1955, by permission of John Wiley and Sons, Inc., New York).

Cirque growth may lead to the coalescence of several small glaciers, and the enlarged source area may permit the resulting ice tongue to advance down the valley. Such valley glaciers modify existing stream valleys by cutting deep, broad floors flanked by oversteepened cliff faces. These U-shaped, or trough, valleys are commonly several hundred meters deep in the case of mature glaciers, and are conspicuous hallmarks of *valley glaciation.* Toward the terminus of the ice, debris accumulation may form into frontal ridges or *end moraines,* or into subglacial *ground moraines,* or into side or *lateral moraines* which extend back through much of the glacial valley. Coalescing ice tongues may also leave intermediate ridges of rock and dirt known as *medial moraines.* The meltwater deposits of sand and gravel stream-laid ahead of the ice terminus are known as *outwash* (Fig. 10).

Occasionally valley glaciation may be sufficiently extensive so that ice tongues descend through the mountain valleys right out onto the foothill region, where they may spread out to form *piedmont glaciers.* Mountain

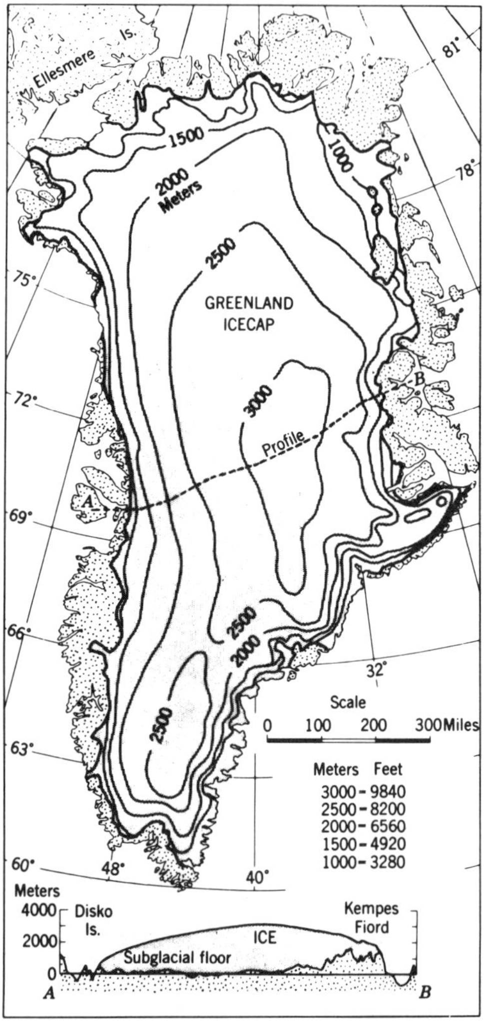

Figure 11. The Greenland ice sheet (after Bauer, 1955, from A. N. Strahler, 1960, by permission of John Wiley & Sons, Inc., New York).

glaciation may also grow to the dimensions of a mountain ice sheet, under which all but a few protruding horns or *nunataks* are submerged.

DEVELOPMENT OF CONTINENTAL GLACIERS

Two large continental glaciations on Greenland (Fig. 11) and Antarctica still provide living examples of the dimensions that the Pleistocene continental glaciers of North America and northern Eurasia once had. Many smaller ice caps on Iceland and several Arctic islands provide firsthand information on processes at work at the ice margins.

An attempt to sketch the development and nourishment of the two extinct continental glaciers has been made by Flint, and his example of the Scandinavian ice sheet (1957, pp. 366–71) can be used here. The Scandinavian ice sheet had its origins as a mountain and valley glaciation in the Scandinavian mountains (crestline elevations 1,200–2,500 m.). With the intensification of glacial climatic conditions the highland glaciation developed into a piedmont glaciation, spreading out eastward over Sweden. To the west it abutted onto deep coastal waters, where the ice terminated as a result of rapid terminal loss of ice through iceberg calving over the open, comparatively warm North Atlantic. Further development of a mountain ice sheet was followed by progressive eastward and then southward expansion into, and ultimately across, the shallow Baltic Sea. The resulting ice sheet (Fig. 12) maintained its symmetry of profile by shifting the radial axis of ice outflow from the

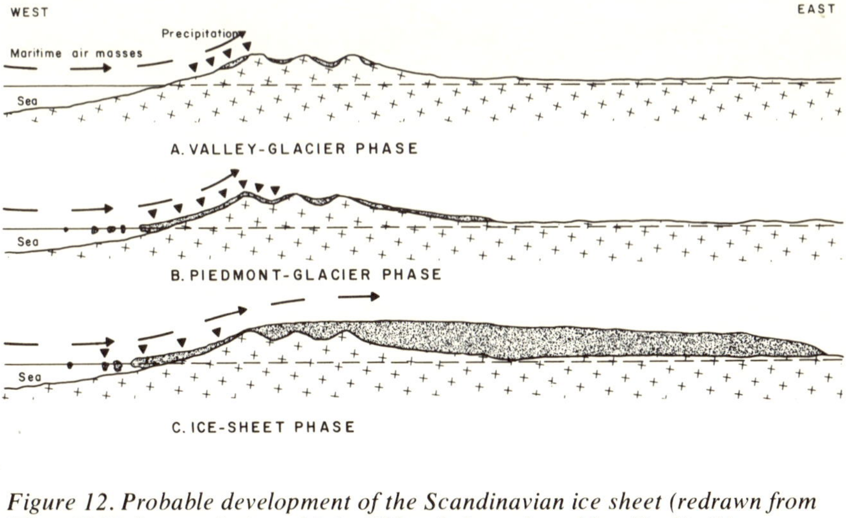

Figure 12. Probable development of the Scandinavian ice sheet (redrawn from Flint, 1957, by permission of John Wiley & Sons, Inc., New York). (Not to scale.)

mountain crests to a point some 400 km. further east. During the glacial maximum, an ice sheet attaining a maximum thickness of 2,500–3,000 m. (Gutenberg, 1941; Niskanen, 1943; Weertmen, 1961) was nourished by moist air masses from the west and southwest, and to a lesser extent from the south and southeast.

The erosion caused by a continental glacier is concentrated in its central area where soil is stripped off, hills are rounded, and bedrock is gouged out in deep grooves, often extending in the former direction of ice movement. In the more peripheral regions, deposition of a variable mantle of ground moraine tends to mask existing irregularities of the topography. And ahead of the ice great sheets of meltwater sweep down broad drainage channels every summer, carrying masses of sand and gravel that eventually remain behind as outwash. After deglaciation, knobby ridges of end moraines, often a kilometer or more wide and with a relief of twenty to thirty meters, extend across the countryside for thousands of kilometers to mark the former ice front. Although they are analogous to the depositional features of mountain or valley glaciers, the moraines and outwash of continental glaciers differ greatly in terms of thickness and extent.

The major features of ice-sheet deposition are summarized in Fig. 13. Characteristic of the glacial bed proper are a lack of horizontal bedding or stratification, and an absence of sorting according to size of the heterogeneous soil products, sand, gravel, and boulders that constitute the *till*. The *glaciofluvial* or meltwater beds ahead of the former ice are stratified through successive horizontal accumulation of materials and are frequently sorted out into dominantly sandy or pebbly beds.

A final feature of interest in the case of continental glaciation is the direction of ice movement. This can be reconstructed from directions of scratches, grooves, or streamline eroded bedrock in areas of dominant

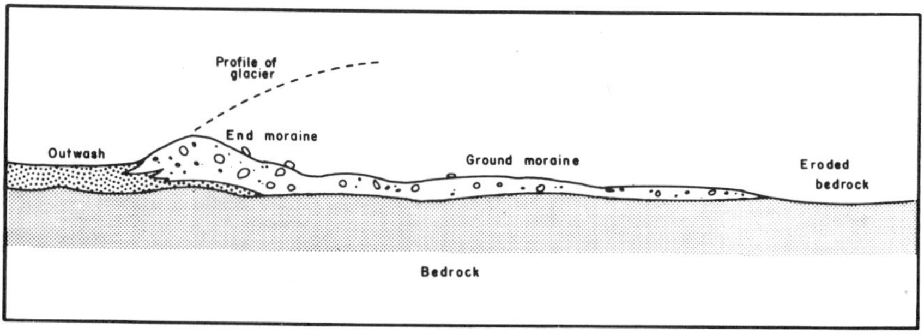

Figure 13. Deposits related to a continental glacier (modified after Flint, 1955). (Not to scale.)

scour. Elsewhere the same purpose is served by the orientation of individual stones or till masses in longitudinal ridges known as *drumlins*. Another method of study is devoted to the dispersal of specific rock indicators derived from localized bedrock areas (Richter, 1956; West and Donner, 1956; Arnemann and Wright, 1959). Such techniques may permit the location of the initial stages of glacier development or of the paths of ice movement to certain marginal areas during different glaciations. Such information is pertinent to understanding the associated climate patterns.

IMPLICATIONS OF THE CLIMATIC SNOWLINE

A controversial although conditionally useful index of climatic change is provided by the altitudinal displacement of the snowline in highland areas having both contemporary and Pleistocene glaciation. In the case of existing cirque and valley glaciers, the modern local snowline can be determined for each glacier in a number of ways, each only approximate. The Kurowski method employs the median elevation of the glacier, the Höfer method the mean elevation between the serrate ridges above and the tongue end of the glacier below. Another method employs the elevation at which the contours of the glacier change from a concave form upstream to a convex form downstream. Each of these techniques, which can also be applied from topographic maps, is inaccurate in varying degree depending on the peculiar bedrock topography.[1] Most dependable are field observations on the lower limit of unmelted firn found on the glacier surface at the end of the summer ablation season. This limit separates areas of net accumulation and net wastage for each individual glacier.

Exposure to radiation and precipitation-bringing winds causes the local snowline to vary considerably, particularly on the northern and southern sides of a mountain range. To offset these local topographic and orographic effects, a mean value can be obtained from the total of local snowline values from a region. This is known as the *climatic snowline*.

Mean annual and summer temperatures at the climatic snowline vary considerably according to latitude (mainly as a result of radiation) and humidity. In a broad way, mean annual temperatures range from 0° to + 1° C. in tropical mountains to about − 5° C. in middle latitude mountains. Mean summer temperatures (June to August) vary even

1. The orographic snowline obtained by averaging the altitude of all perennial snowbanks on a mountain range is of no value for comparison with Pleistocene data, mainly because geomorphic records do not provide comparable information.

more strongly, in the range of 0° to + 4° C. in middle latitudes. Discussions of snowline temperatures are provided by Koeppen (1920), Ahlmann (1924), and Troll (1956).

The local snowline of a former mountain glacier can be determined more accurately than that of an existing one. The median level of the cirque floor may be at or above the local snowline, so that the lowest cirque floor of a group of contemporaneous cirque glaciers will generally give a ±50 m. approximation of the local snowline. Employing the top of the headwall and the end moraine, Höfer's method is also occasionally used to determine Pleistocene snowlines. Median cirque floor levels are preferable, however, and can also be used to estimate regional or climatic snowlines.

If identical criteria are (and can be) employed to determine both the modern and the Pleistocene snowlines of an area, the difference or *snowline depression* is of some importance. The snowline depression of the Würm glaciation generally varies between 600 and 1,400 m., depending on latitude and on maritime or interior location. This would imply a local altitudinal depression of the over-all climatic zonation by about the same value. Temperature is probably the foremost variable, but changes in radiation, particularly as influenced by cloud cover, changes in precipitation amount, ratio of snowfall to total precipitation, seasonal distribution of precipitation, etc., may all be involved. Different mountain topography at different elevations may also play a conspicuous, though more local, role. Consequently, the common practice of singling out temperature as *the* variable involved in snowline depression is unfortunate.

Quantitative estimates of temperature changes responsible for snowline depressions—with the optimistic provision that all other factors remain constant—are made by using the modern temperature decrease with increasing elevation, called the local *lapse rate*. This lapse rate is different within different mountain ranges at different seasons (Koeppen, 1920; Ahlmann, 1924; Baker, 1944; Weischet, 1954; Miller, 1955) and is not identical with lapse rates measured in the free air above nearby lowland stations. Whereas the average lapse rate of the free air is now commonly estimated at 0.55°-0.6° C. per 100 m., empirical surface observations within mountain ranges suggest that average values of 0.6°-0.7° C. per 100 m. are more representative. This type of information is largely lacking in many areas affected by Pleistocene highland glaciation, so that many published estimates of the sort are speculative. Equally disturbing is the fact that, even if we could assume that temperature was the only variable, local Pleistocene lapse rates need not have been identical with those of today. Mortensen (1952, 1957) and

Flohn (1953) have made a good case for widespread, semipermanent temperature inversions in the lower atmosphere between the Scandinavian ice sheet and the Alpine mountain glaciation as a result of the secondary cooling effects of the ice. There is reason to believe that such phenomena were common elsewhere as well. The absolute use of snowline depression values for Pleistocene temperature reconstructions, then, is not justified, at least unless convergent supporting data are available through other methods of investigation.

With more qualified use, however, snowline depression values do contribute towards an understanding of climatic change. For example, unequal regional or local snowline depressions beg explanations. Equally important is the realization that, averaged out by latitude, Pleistocene snowline depression values have been remarkably uniform at all latitudes on both the northern and southern hemisphere (Klute, 1928; Büdel, 1953; Mortensen, 1957). Radiocarbon data are gradually accumulating to prove that the glaciers in question were also contemporaneous, at least on the time scale of $x \cdot 10^3$ years. This world-wide parallelism of glaciation strongly supports the idea of the contemporaneousness of cold or warm climates in both hemispheres. And without a significant change in the water balance of the atmosphere, only a general, world-wide lowering of temperature can explain adequately a more or less similar snowline depression in all latitudinal belts.

CONCEPTS IN PERIGLACIAL GEOMORPHOLOGY

During the Pleistocene glaciations much of Europe and North America experienced polar or subpolar climates with tundra vegetation and permafrost. The European tundras witnessed a succession of remarkable cultures during the Würm Glacial, raising questions about the nature of the environment encountered by late Pleistocene man in these areas. The geomorphic processes of the polar and subpolar world are among the most striking of any, and deserve as much attention as the related vegetation and soils. The totality of information available on the physical environment of the tundras and permafrost lands may permit full interpretation of fossil phenomena, as well as an understanding of the ancient landscape in terms of modern counterparts.

The term *periglacial* is commonly used to describe a group of geomorphic processes dominant in nonglaciated polar and subpolar regions. The use of the concept "periglacial" is a loose one, however. Some authors include any cold-climate phenomena at all; others exclude all but those processes related to permafrost; while still others delimit the area of the periglacial "cycle" to the lands between the glaciers and the polar or alpine tree-lines. An environmental definition of "the periglacial" would

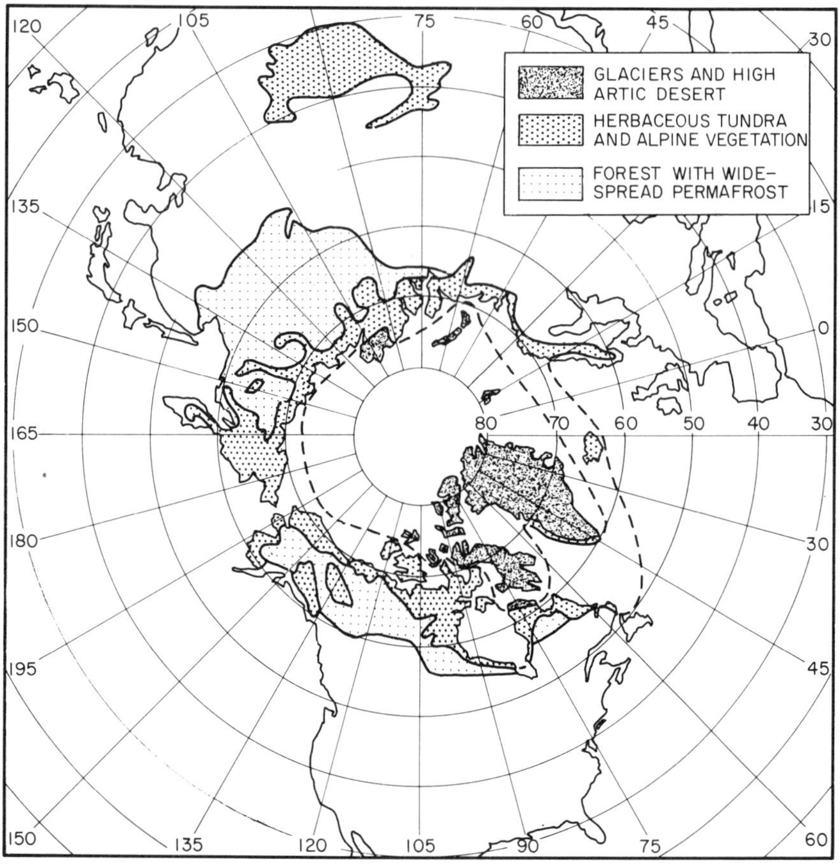

Figure 14. Contemporary extent of glacial and periglacial environments in the Northern Hemisphere.

be most useful, as it would circumvent the necessity of subdividing between "periglacial" and other "cold-climate" phenomena, or of following the unfortunate practice of equating periglacial and cold-climate processes. Use of the word "periglacial" will here be confined to the group of processes peculiar to certain environments:

a) The high arctic barrens;
b) The herbaceous tundra, with or without permafrost;
c) The forest-tundra and coniferous forests with widespread permafrost.

The approximate modern extent of these zones in the northern hemisphere is shown in Figure 14.

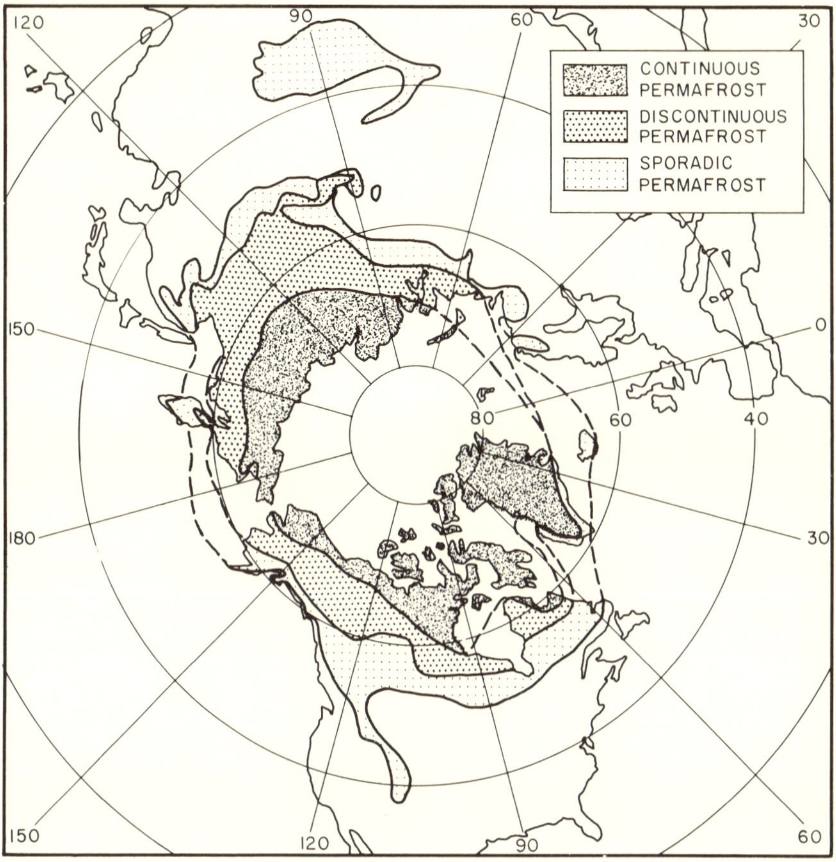

Figure 15. Contemporary extent of permafrost belts in the Northern Hemisphere (after Black, 1954; Brown, 1960; Frenzel, 1959; and others).

The subsequent section will deal with those cold-climate phenomena that occur frequently in one or more of these regions. Less attention is paid to those features that are either rare or not commonly preserved in the fossil state. Unlike glacial geomorphology, periglacial environments have only recently become the subject of accessible, comprehensive studies (see Tricart, 1969; Tricart and Cailleux, 1967; Bird, 1967; Embleton and King, 1968; Davies, 1969).

COLD-CLIMATE PHENOMENA

Permafrost. Permanently frozen subsoil or bedrock accounts for at least 25 per cent of the world's land surface (Black, 1954; R.J.E. Brown, 1960) and is most common in continental areas having long, severely cold winters with brief summers. This belt has been divided into "continuous," "discontinuous," and "sporadic" permafrost zones (Fig. 15). The base of modern permafrost reaches 600 m. in parts of Siberia and 500 m. in North America. The surface horizon which is subjected to annual thaw may range from several centimeters to several meters in depth. This active layer is a zone of considerable geomorphic activity.

Although there is no very close relationship between permafrost distribution and air temperature (Brown, 1960), both continuous and discontinuous permafrost are found only in areas with mean annual temperatures below 0° C. (Black, 1954). The areal extent of permafrost, however, is not necessarily coextensive with all frost-weathering and soil-frost phenomena. Many characteristic cold-climate features are not indicative of any thermal limit and may be found in any area with frost, although the more frequent, larger, and distinctive phenomena are concentrated where frosts are more common.

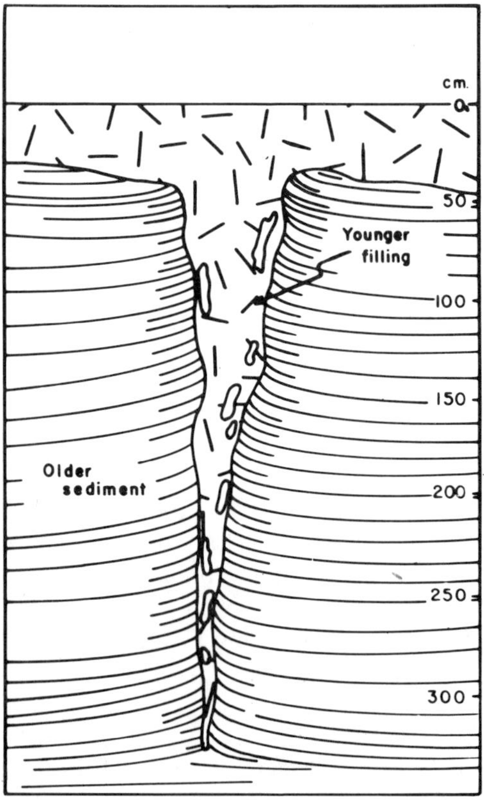

Figure 16. Ice-wedge cast. (To scale.)

Ice Wedges. Perhaps the most typical morphological feature of frost climates are vertical wedges in soil or sediment that in their active state are filled with ice (Fig. 16). Typically some 1 to 5 m. deep, and with a surface width of 30 to 50 cm., these *ice wedges* may range from a few millimeters to as much as 10 m. in width, and up to 10 m. in depth. The cracks are often found in groups intersecting at obtuse angles at distances of 20–30 m. (in extreme cases to 5 km.) to form giant polygonal networks. The centers of the polygons so formed may be either depressed or upraised in relation to the marginal cracks. The difference in height can be as much as 1.5 m. In the more frequent low-centered variety, the strata in contact with the ice wedge commonly bend upward, while they bend downward in the high-centered. Small embankments run parallel to the cracks. The low-centered type may possibly be more typical of swampy terrain, the high-centered of better drained land.

The origin of ice wedges remains controversial. One theory (Leffingwell, 1919; Black, 1954) holds that the crevices are initiated by periodic cracking of sediments as a result of rapid and extreme falls in temperature. During the melt season the fissure is filled with water derived from the supersaturated thawed parts of adjacent beds. When this water freezes on the interior crack surfaces, the increase in volume due to freezing leads to cryostatic pressures that widen the fissure. If the ice wedge does not melt out during the summer, the feature may widen from year to year. Empirical observations tend to support this hypothesis.

Another body of theory (Taber, 1943; Dücker, 1951; Dylik, 1952; Schenk, 1955) accounts for the growth and persistence of these frost cracks—which may be minute originally—by capillary movement of soil moisture from the comparatively warm, adjacent sediments to the dehydrating surface of the crevice walls. Here moisture is said to be precipitated in the form of ice crystals perpendicular to the wall. Such ice needles would grow as long as soil moisture is available, partially reinforced by dew from outside. Cryostatic pressure would continually widen the original crack until the sediments were dehydrated.

The growth of ice wedges is a function of time, sediment type, and subsoil temperature ranges. Wedges hardly, if ever, grow in the sporadic permafrost zone today, while their growth approximates 0.5 to 1.5 mm. annually in the continuous permafrost zone of Alaska (Black, 1954). The width of ice wedges can therefore be used to approximate the time involved in their active growth and thereby provide an estimate for the minimum duration of corresponding climatic conditions. The permanent ice wedge develops exclusively within permafrost, beneath the base of the summer thaw zone (see Büdel, 1961; Lachenbruch, 1962). Wedge depth depends on sediment texture, soil drainage, and climate as well as on the geometry of the intersecting ice-wedge networks.

Today ice wedges are found only in the permafrost zone. In Eurasia larger typical wedges, often forming giant polygons, are found in the poorly drained tundras and to a limited extent in the swampy parts of the forest-tundra. In the latter environment such features occur only on peaty or fine-grained, homogeneous sediments (Frenzel, 1959, p. 75 ff., 1967). Only coarse-grained clastic materials seem to be affected in the permafrost tundra and the high arctic barrens. In the permafrost forest belt, ice wedges are sporadic and found only on similar sediments as in the forest-tundra. They are small and do not form characteristic networks. The ice wedge is, then, a strict indicator of permafrost, and probably indicative of at least former association with the continuous or discontinuous permafrost belts. On this basis it is believed that the former presence of ice wedges indicates mean annual temperatures of below 0° or even - 5° C. at the time of formation (see Poser, 1948; Mortensen, 1952; R. Péwé, 1966; R. J. E. Brown, 1960). Further environmental localization can be made for all sediment types in the arctic barrens and tundra; and also, in the case of highly organic or fine-grained beds, in swampy or tundra enclaves of the boreal forest. Preserved in fossil form, ice wedges are a very specific environmental criterion.

With the gradual disappearance of the ice after climatic amelioration, the crack would be filled with surface wash or slump material, and so be able to preserve its outline. Such fillings of loam, sand, loess, etc., are known as *ice-wedge casts*. Casts are comparatively frequent in former permafrost areas and can be identified in section by distortion of adjacent strata and by the vertical orientation of elongate stones found within the heterogeneous filling. Distortion is absent in the case of fillings produced by decomposed roots, or in the case of contraction fissures due to dehydration of fine sediments.

Frost cracks, or miniature ice-wedge fillings, seldom exceeding 15-20 cm. in length, are relatively common in unconsolidated materials. They form through simple freezing or through dehydration in fine silts, with subsequent filling by needle ice. They are not necessarily associated with permafrost.

Stone Rings. On a very much smaller scale, soil frost may produce ground patterns of a different kind and dimension. These are circular rings or small polygonal structures marked by sorting of materials into fine and coarse classes. On flat ground, stone rings are most common, surrounding centers of finer material. Earth rings, or geometrically arranged earth hummocks, may also contrast with adjacent coarse materials. With gradients of 4 per cent or more, slope movements may elongate such stone rings into nets or garlands, which on steep slopes degenerate to stone stripes running perpendicular to the contours (Troll, 1944; Washburn, 1956). A certain proportion of vertically oriented stones is another hallmark of such patterned ground. The diameter of stone rings may vary from 10–30 cm. in the case of areas with a dominance of daily freeze-and-thaw, to a few meters in areas with protracted seasonal frost or permafrost.

The coarse clastic materials of stone rings are largely due to frost-weathering, and the sorting is a result of soil-frost expansion followed by differential contraction among rock and soil particles of varying particle size and cohesion. Repeated soil frosts of moderate intensity can explain stone rings adequately. The modern distribution of actively developing patterned ground in Alaska (Hopkins *et al.,* 1955) and Eurasia (Frenzel, 1959, p. 80 ff.) shows that, although found most frequently in tundra areas, stone rings are not uncommon throughout the boreal forests. Williams (1961) suggests an equatorward limit in the form of the mean annual isotherm of + 3° C. for all but the largest features, for which he adopts + 1° C.

Occasional fossil stone polygons (e.g., Wortmann, 1956), or the more common fossil stone stripes, are therefore not always indicative of a particular environmental situation. In fact Hastenrath (1960) has been able to show that the occurrence of patterned ground within a considerable range of climatic possibilities is determined above all by the character of the vegetation mat.

Involutions. Another cold-climate phenomenon takes the form of closely folded, highly crumpled laminae of fine-grained sediments, commonly segregated as to color. The resulting pockets or linear contortions are usually found within one or two meters of the surface, and show vertical orientation. Such pocket soils or *involutions* are a common fossil feature (Fig. 17) in regions once affected by cold Pleistocene climates.

Evidence from the modern arctic of exactly analogous phenomena has not been fully satisfactory, since numerous churning or contorting processes occur there (see Hopkins and Sigafoos, 1950; Frenzel, 1959, p. 78-80). Small, often temporary, forms similar to the Pleistocene features have been observed developing in the temperate forest zone today (e.g., Fries *et al.,* 1961).

The most widely accepted theory of development (Steeger, 1944; Schenk, 1955) calls for the presence of permafrost. When, in the autumn, the active layer permanently refreezes, the supersaturated beds

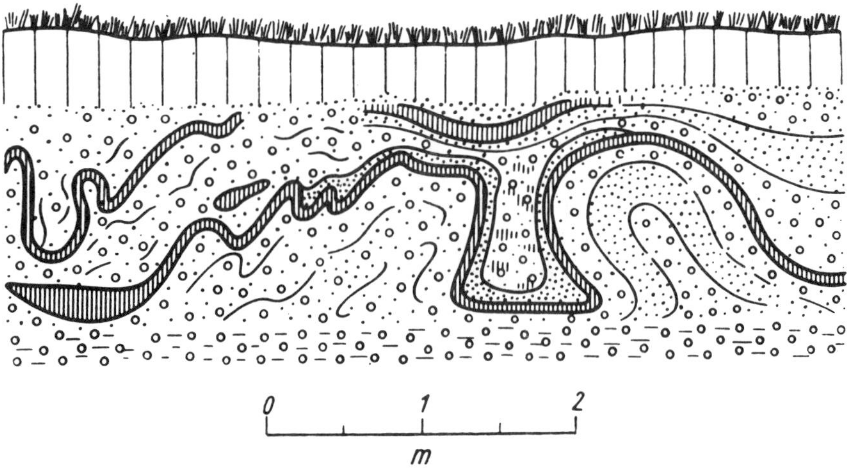

Figure 17. Fossil involutions (after Steeger, 1944, from H. Weber, 1958, by kind permission of B. G. Teubner, Verlagsgesellschaft, Leipzig). The sediments have been intensely folded and partially sorted into fine (dark) and coarse (stippled) laminae through Pleistocene soil frost processes. (To scale.)

are trapped between two impermeable layers and so are subject to complex and intense hydrostatic pressures. Differential expansion of fine and coarse-grained beds may complicate the process. Vertical or horizontal contortions ensue, with frequent segregation of silt bands as a result of frost-sorting. Obviously the same process could take place on impermeable substrata other than permafrost. Tricart and Cailleux (1967) consider convective currents, due to density differences of water during the freezing process, as a likely explanation for the distortions often found in quite homogeneous sediments. The base of the thaw zone (if we accept the presence of permafrost as a prerequisite) will be at 0° C., whereas the water in the saturated ooze near the surface will be several degrees warmer during periods of insolation. Since water achieves its maximum density at + 4° C., turbulence is set up within the thawed layer.

Fossil involutions are generally found in association with ice-wedge casts, so that they are above all a permafrost indicator. But by themselves they must be considered inconclusive, particularly where there is any possibility of horizontal movements of soils and regolith, even on very gently sloping surfaces. Involutions are sometimes referred to as *cryoturbations,* although the term cryoturbation is also widely used to include patterned ground phenomena.

Other cold-climate indicators involve slope deposits partly related to gravity action.

Solifluction. Solifluction in its strict sense refers to slow, almost imperceptible soil flow under the influence of gravity, i.e., a general downslope displacement of masses of soil and rubble saturated with water. It is common in areas with permafrost, which provides an impermeable substratum under an active layer of saturated soil (Büdel, 1944, 1961). Analogous conditions are provided in springtime where some subsoil is still frozen without the intervention of permafrost. Tricart (1969) emphasizes that an impermeable substratum is not necessary. Instead, he considers that the fundamental mechanism proceeds in two stages. First, freezing breaks the colloidal structure of silts and clays, so that they lose their cohesion when thaw sets in. The thawing material has a high water capacity, and the abundance of seepage water provides a lubricated ooze capable of mass movement even on gentler slopes. In fact, C. Schott (1931) showed that solifluction still occurs in warm-temperate climates today, depending on lithological conditions.

Movement by solifluction is slow and irregular, varying from a few centimeters to a few decimeters a year. Solifluction materials are unstratified, locally unsorted, and quite angular; they commonly occur in smooth, extensive sheets (*congeliturbate mantles*) not confined to topographic channels but rather covering broad slopes to a depth of two to

Figure 18. Coarse Pleistocene solifluction deposits. The angular limestone slabs are set in a loamy matrix, oriented downslope, in part grading into intact bedrock. Sorted and contorted lenses also occur. (To scale.)

three meters, possibly grading down into stream valleys below. When solifluction materials overlie the parent strata, various stages between intact, deformed, and removed or reoriented bedrock can be observed (Fig. 18). Solifluction sheets themselves display a crude sorting into lenses or pockets of finer and coarser materials as a result of differential velocity according to particle mass. However, coarse blocks may be found at random in finer beds. In practice, many solifluction deposits are badly contorted and consequently similar to involutions. In fact, solifluction and involutions have indistinguishable transitional forms.

Practically all slope deposits on inclinations of less than 25 to 30 per cent in middle latitudes are considered to be "fossil" and of Pleistocene age. They often contain blocks transported many miles across areas now forested. Most of the finer materials are, however, derived from local slope bedrock. Glacial deposits are often reworked by solifluction, and hence foreign stones may be common at times. Also, the borders between till and solifluction are sometimes obscure.

Solifluction on slopes without vegetation is much more rapid, and the deposits have less relief than features resulting from solifluction under a vegetative mat. The second process is slower and may develop distinct surface forms on more symmetrical slopes. Such features, known as *terrassettes,* are the result of tear-slumping or overturning of sod. Terrassettes may follow the contours or have a lobate, tongue-like configuration. In conclusion, treeless situations are generally necessary for solifluction, but not permafrost. Nevertheless, the greater part of the low-angle slope deposits in western Europe is probably due to solifluction over permafrost.

Solifluction phenomena cannot be used to delimit the permafrost zone in marginal areas. There are many gradations between true fossil solifluction phenomena and superficially similar but water-laid colluvial deposits in the subtropics (see Butzer, 1964b), and slumping of lubricated soil on slopes also may produce analogous features.

Block streams. Numerous coarse deposits of chaotic masses can be found localized in valleys and extending as mantles on slopes of 5–45 per cent (Büdel, 1944; Tricart, 1963). Many such *block streams* are simply eluviated fans of coarse congeliturbate material due to solifluction. Other block streams are coarse by origin, having formed below extensive rock outcrops. Block stream movement has often been observed well below the tree-line, and modern distributions in Eurasia show a concentration in the forest permafrost zone and in cold, mountainous country (Frenzel, 1959, pp. 82-86).

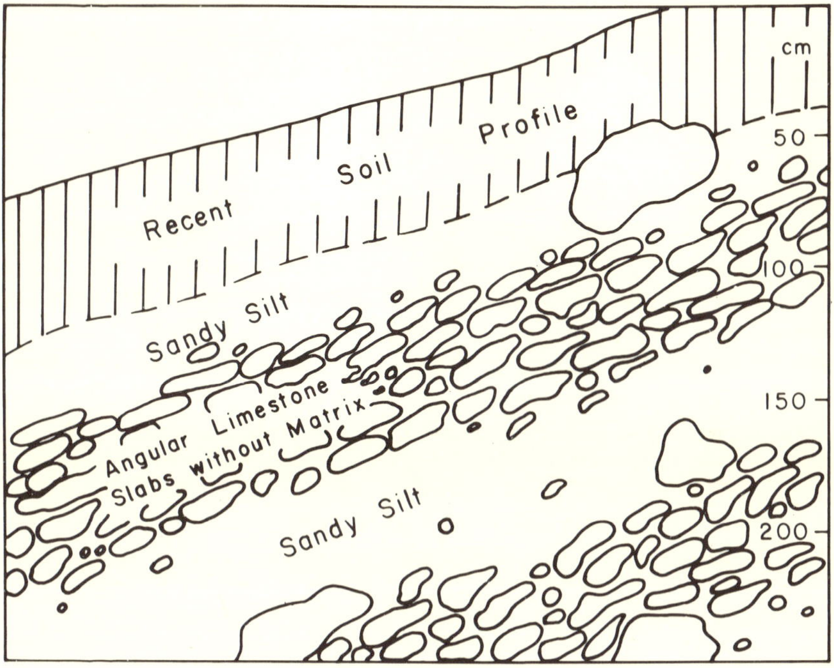

Figure 19. Pleistocene éboulis ordonnés. (To scale.)

Sorted Talus (Éboulis ordonnés, Grèzes litées). Another common fossil feature, due to frost-shattering on slopes and subsequent downhill sliding of materials, is beds of assorted talus, often found on slopes of less than 17 per cent but occurring at angles of as much as 60 per cent. Their inclination decreases noticeably downslope. The beds show distinct alternations of fine and coarse materials, invariably quite unrolled and angular, with the coarse beds well stratified and containing little or no interstitial materials (Fig. 19) (Tricart, 1963). The individual beds may have thicknesses ranging from 5 to 50 cm. The coarse particles are thought to accumulate during the first thaws of spring, by sliding over still-frozen soil. Thereafter, solifluction is responsible for transportation and deposition of fines over the coarser bed (Tricart, 1956a). Permafrost is obviously not a prerequisite, and the modern distribution of these phenomena in Eurasia indicates that they occur at least as frequently outside the permafrost or tundra belts as within these areas.

Other Cold-Climate Phenomena. Many additional cold-climate processes could be enumerated here, but most of these carry less paleo-environmental importance. So, for example, the large mushrooms of ice or mud known as *pingoes* have been recognized in fossil form (e.g., Pissart, 1963; Wiegand, 1965). The remaining circular ridges of dirt may be as much as several hundred meters in diameter, and are mainly found over permafrost. Only small pingoes of 10–15 m. diameter and up to 2 m. high are known from outside the permafrost zone (Frenzel, 1959, pp. 88-90, Fig. 13).

Another widely cited cold-climate phenomenon is the so-called *asymmetric valley,* which in cross-section has a gentle slope on one side, a steep slope on the other. The various hypotheses for development of asymmetric valleys have been discussed by Maarleveld (1951) and Ollier and Thomasson (1957), and there can be no doubt that only some of these forms are unequivocally related to specific cold-climate processes.

Cryoplanational or *goletz terraces* are more or less horizontal surfaces cut into the bedrock of mountainous terrain by frost-weathering and solifluction (Tricart and Cailleux, 1967; Frenzel, 1959, pp. 82 ff.). Such features show no relationship to bedrock structure or stream valleys. They appear to form today only within the permafrost or tundra zones, but have been recognized in fossil form in western Europe.

Features due to stream and wind action will be discussed later.

A SYNTHESIS OF PERIGLACIAL GEOMORPHOLOGY

The High Arctic Barrens. The geomorphic processes in the extreme periglacial environment of the rocky wastes of the high arctic have been described most effectively by Büdel (1950a, 1960, 1961, 1963). These are among the basic prerequisites to landform evolution:

a) Little or no vegetation, so that geomorphic processes are in no way impeded by a plant cover.
b) Permafrost, with an active layer of only 20–80 cm. depth. The permafrost is an agency of deep mechanical weathering, particularly at the base of the active layer where soil frost and massive ice in rock joints and cracks produce abundant, fresh rock rubble every summer.
c) Lack of permanent groundwater percolation, due to the presence of impermeable permafrost near the surface, and minimal evaporation, due to low temperatures (seldom over 10° C.). Consequently the summer snow meltwaters, the meltwater from the active layer, and any summer precipitation must all flow off the surface or through the supersaturated layer.
d) Intensive frost-weathering of all exposed rock faces.
e) Extreme mobility of the active layer during the warmer season,

through the churning and heaving of various forms of soil frost, downhill solifluction of supersaturated ooze, and a remarkable subsurface drainage of water over the base of the active layer.

In this way the high arctic wastes provide enormous masses of mechanically disintegrated rubble as well as potent agents for its removal. In addition, the meltwaters of neighboring glaciers may add further masses of outwash.

Surface denudation of slopes proceeds through subsurface drainage and solifluction. Stone stripes or garlands are common on slopes of 3–25 per cent, and on even steeper slopes runoff drainage and solifluction combine to allow rill-cutting or slope dissection.

The major stream valleys are overloaded with detritus and are invariably underlain by ever-fresh, mechanically disintegrated bedrock. During the early summer runoff spates, stream incision is rather rapid, and broad, steepsided but smooth-sloped valleys (Fig. 20) are excavated by the combined action of stream cutting and slope movements. Broad, coarse gravel beds cover the stream channel, commonly extending well back into the headwater areas. Resistant rock strata are vigorously attacked both on the valley slopes and in the stream bed, and so do not give rise to noticeable breaks in gradient.

While the waters are held in the frozen state and a snow cover is absent, deflation by wind is unimpeded and removes finer grain-size particles from the frost-rubble as well as any outwash areas present. Coarser sands are deposited in parabolic or blowout dunes, and fine dustlike materials of the loess category are usually deposited as veneers

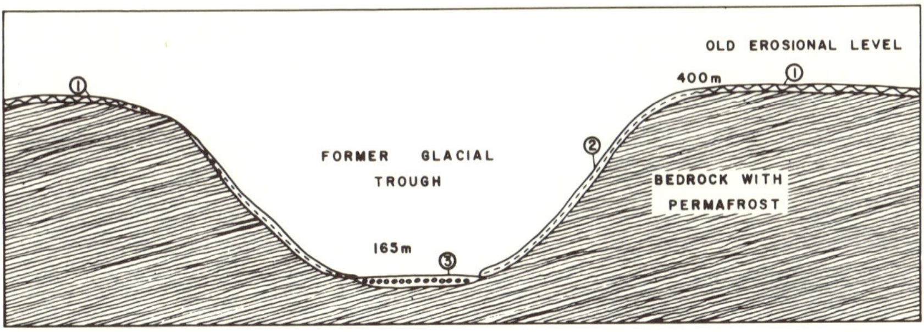

Figure 20. Valley profile in the periglacial zone of Spitsbergen (generalized after Büdel, 1963): (1) cryoturbated active layer on upland surface, (2) congeliturbate mantle with subsurface drainage, (3) gravel bed of stream. The original Tertiary erosional surface, cut as a Pleistocene glacial trough, has been much modified by valley-deepening and slope retreat.

or mantles, possibly in distant, more vegetated zones of the tundra or forest-tundra (see chs. 8 and 12).

The Tundras. Many of the more maritime tundras do not have permafrost, and surface frost-weathering is a little reduced by the effective mat of vegetation and a more durable snow cover in winter. Soil-frost activity and the production of debris in general are consequently less pronounced than in the barren high arctic wastes. The rate of gravity movements is severely checked by the plant cover, and this restricted (German *gebundene,* literally "bound") solifluction leads to the formation of terrassettes or lobes perpendicular to the contours on slopes with gradients of 3 to 35 per cent. The terrassettes and vegetation in turn slow down surface runoff, reducing the potential of streams for both erosion and deposition. As a result, upstream drainage channels often deteriorate into block streams or dells choked with detritus.

Wind erosion is confined to dry river beds and block streams, although deposition of wind-borne loess from high arctic zones of deflation may be very prominent. The greatest part of the European and North American loess was accumulated under a rich herbaceous vegetation as innumerable vertical root zones indicate, in fact, the vertical cleavage of loess has been attributed to such root structure.

The alpine zone above the tree-line and below the snowline conforms in many ways to the various physical attributes of the true tundra, and in part to that of the high arctic wastes. Frost-weathering, various soil-frost mechanisms, and solifluction are present, although they are less intense than in the polar world and are adapted to terrain largely with steep or moderate slopes.

The Permafrost Forests. The forest-tundra and, even more so, the coniferous forest belts with widespread permafrost are in all ways a transitional part of the periglacial zone. Frost-weathering, cryoturbation, solifluction, surface denudation, stream cutting and deposition, and even some aeolian processes are admittedly still in effect. But all forms of geomorphic activity are rather subdued in both spatial and temporal dimensions, particularly as a result of the protective vegetation and winter snow cover.

Synthesis of Periglacial Geomorphology. Following Troll (1948) and Büdel (1950a, 1963) the peculiar group of formative agents at work in the periglacial zone are summarized by the following:

a) Frost-weathering, including disintegration of surface rock by frost action in joints and fissures, the attack of bedrock by ice-wedging, etc.
b) Congeliturbation, the total of soil-frost activity through bedrock dis-

integration, soil churning and heaving, particularly in the active area over permafrost.

c) Solifluction, the slow flow of the water-saturated mantle of soil and loose rock by gravity and frost action, aided by subsurface water drainage.

d) Stream incision and gravel deposition, as a result of accelerated seasonal runoff and overloading with materials derived from frost-weathering, congeliturbation, and solifluction.

e) Deflation and deposition by wind.

f) Slope retreat and surface denudation by the total of these processes.

EIGHT

Humid, Arid, and Tropical Geomorphology

STREAM EROSION AND DEPOSITION

Glacial erosion is probably the most powerful individual agent of gradation, but the work of running water is certainly the most universal. Before attempting to synthesize the geomorphic processes of the nonpolar world, a simplified outline of "normal" stream erosion and deposition may be of use to the general reader.

The residual mantle produced through mechanical and chemical weathering of the earth's crust may subsequently be attacked in a number of ways. The direct mechanical impact of raindrops on unprotected surfaces is the smallest unit of erosion by water. It is remarkably potent in cultivated fields and a prime agent in man-induced soil erosion. Elsewhere, under natural conditions, *raindrop erosion* and mud-spattering are important wherever rainfall is heavy and irregular and the vegetative mat incomplete. The next step is corrasion through shallow, poorly defined rivulets following channels of temporary drainage. These may transport both suspended mud and a certain amount of sand. Under more extreme conditions, thin sheets of water may spread out over much of the surface, moving within the vegetation. Such *sheetfloods* may, in the course of time, contribute to an appreciable lowering of the general surface (three-dimensional or areal *denudation*). All of these mechanical effects of rainwash are intensified and accelerated on steeper slopes where they are aided by gravity in a number of ways, thus favoring *slope retreat*.

Rainwash is also important as a chemical agent of denudation, acting through the rapid solution of salts and carbonates and the surface corrosion of loose or intact rock. An appreciable amount of material may thus be carried in solution. Equally important are solution and corrosion by percolating waters which pass on through the groundwater before they eventually reach the stream network.

Streams are fed by intermittent surface runoff after rains, and by protracted underground percolation through seepage or springs. They may be permanent, or otherwise seasonal or even sporadic, depending upon the climate and, possibly, local factors. Stream transport accounts for a *load* of:

a) Chemical solvents carried in solution;
b) Clays, silts, and fine sands carried in suspension;
c) Coarse sands and small pebbles alternately pushed and flipped along the stream bed through irregular saltation; and
d) Pebbles and cobbles rolled and pushed along by simple traction.

Since the specific gravity of most rocks in the earth's crust varies around 2.6–2.7, size is the major factor determining mode of transportation at a given stream velocity. Although rapidly moving clear water will occasionally tear away loose particles from the stream bed or banks, mechanical stream erosion is performed mainly *by the load* through simple abrasion. The greater the load, the greater the capacity for stream cutting. The greater the turbulence and total velocity of the waters, the greater the transport capacity and the greater the erosive force. Stream erosion leads to downcutting of the stream bed by vertical incision, and to undercutting of its banks by lateral erosion. The upper ends of a stream or its tributaries may also work their way backward by headward erosion. The major local effect of stream erosion is *dissection* (essentially two-dimensional or linear) as opposed to surface denudation and slope retreat by rainwash and allied phenomena.

Materials that have been transported are ultimately laid down. Such stream deposition, or alluviation, ensues when the turbulent velocity decreases, thereby reducing the transport competence. The coarse load is deposited first, followed by the fines. There commonly are well-defined zones within a stream system where effective velocities are greater or less. So, for example, water movement is faster near the bottom of the stream near its center, than out near its banks. Further, stream gradients are greatest in the headwater area, so that turbulent and eddying motions of the water are most violent there. As the gradient decreases downstream, the turbulent energy and total velocity of the stream are reduced, even though the usual increase in volume, as a result of tributary convergence, leads to an increase in net downstream velocity or laminar flow. But effective velocity, in terms of transport competence, is usually reduced in the downstream parts. Generally speaking, the regions of predominant erosion or deposition within a river fluctuate in the course of the year. Almost all streams are affected by strong seasonal variations of discharge, in response either to rainfall maxima, to a spring snowmelt, or to a pronounced seasonality of evaporation. Erosion is generally most effective during the onset and height of the runoff maxima, while deposition is most widespread during the period of subsiding water level. If surface runoff and stream discharge maxima are particularly violent, a stream system will be affected by correspondingly greater erosion and deposition.

The lower courses of most well-developed streams are situated in low-lying level country, and seasonal variations of discharge may lead to regular or sporadic inundations of the surrounding flats. Such *floodplains* are the areas of most characteristic stream deposition or *alluviation*. The aggradation proceeds by bedload sedimentation along the bed of the slowly shifting channel as well as by release of the suspended load through bank overflow.

The various features of the floodplain are significant for an understanding of prehistoric settlement in river valleys. In the case of sluggish rivers, the actual river bed is accompanied by banks or *levees* which tend to lie a few meters above the general elevation of the periodically inundated alluvial flats. The lowest ground is frequently found on the outlying alluvial flats where surface or groundwater may be present throughout the year *(backswamps)*. The gradual decrease in elevation from the river banks to the outer limits of the floodplain is a result of the heavier, coarser materials being deposited in the area of swiftly moving water, at the bottom or margins of the meandering low-water stream channel. The slower flood waters carry only finer materials to the margins of the floodplain, so that these smaller-size particles raise the level of the land at a slower rate; hence the higher-lying banks or levees of many larger rivers. Levees, alluvial flats, and backswamps have always had a direct functional significance for river valley settlement by both food-collectors and agricultural populations.

For a more detailed account of fluvial activity the reader may consult Leopold *et al.* (1964) and Morisawa (1968).

HUMID ZONE GEOMORPHOLOGY

General Features. The external geomorphic forces responsible for landscape sculpture in humid, temperate environments are far better understood than those of the periglacial zone (see Büdel, 1950a, Tricart and Cailleux, 1965). Mechanical and chemical weathering are both moderately significant since winters are cool or cold, and considerable moisture is available. The only important gradational agent is running water, aided by localized gravity movements. This is the climatic-geomorphic province that W. M. Davis' "normal cycle of erosion" seeks to describe. Vegetation cover, except in the mediterranean regions, is characterized by closed woodlands that privide considerable protection for the soil. Consequently soil profiles and weathering zones are well developed and subject to serious erosion only when man removes the tree cover. Geomorphic processes are of moderate intensity, and the dynamism of periglacial sculpture is absent. Instead, surface denudation is limited under the natural plant cover, and soil profiles are deep.

It is therefore not surprising that "fossil" features of Pleistocene vintage have suffered little erosion and only moderate weathering. In fact, ancient glacial deposits and erosional features are rather well preserved. Where not obscured by subsequent deposition, even early Pleistocene glacial forms are commonly recognizable. In contrast to the periglacial zone, the limits of the former continental glaciers can be reconstructed with accuracy in the humid-temperate zone. Equally important is the excellent preservation of glacial-age cold-climate phenomena.

Within the humid-temperate zone there are moderate differences between the temperate-continental forests, with seasonal frosts and winter snow cover, and the marine-temperate woodlands with only occasional frosts and no durable snow cover. Differences also exist between the humid and subhumid subtropics.

The Continental-Temperate Forest Zone. The cool-temperate *Da* and *Db* climates are subject to long, often severe winters. Frosts are of some duration and intensity, penetrating to as much as a meter or more into the soil or bedrock. Late winter or spring thaws lead to a marked seasonality of stream discharge, and larger rivers develop extensive floodplains. Mechanical weathering of any exposed rock, moderate stream dissection of slopes exceeding 25–35 per cent, and gradual, grain-by-grain denudation of slope surfaces all provide a moderate load of fine and coarse materials for the rivers. As a result, stream erosion and deposition are of some significance although in no way comparable to that of the periglacial zone. Gravity movements only assume primary importance in areas of steep slopes exceeding 60 per cent.

"Fossil" features are not only preserved but often dominate the landscape. Paricularly impressive is the disrupting effect of glacial deposition and erosion on drainage. Widespread swampy conditions and innumerable lakes within the boreal forest may reflect the local inefficiency of modern fluvial activity in providing an effective integrated drainage network.

The Temperate-Marine Woodland Zone. In the warm-temperate *Cbf* climates winters are mild and frost action limited. As the winter snowfalls are sporadic and ephemeral, there is no period of seasonal thaw. Rainfall is distributed rather evenly throughout the year, and is of less intensity and greater frequency than in the continental interiors. Stream discharge is consequently marked by only moderate seasonal fluctuations and is generally ineffective in moving coarser materials. The complete mat of vegetation remains effective even in winter, thus restraining surface rainwash and stream dissection. As a result, soil mantles are deep and fluvial sculpture is almost at a standstill wherever deforestation and agricultural misuse have not accelerated soil erosion.

The Moist Subtropical Woodland Zone. In moist subtropical regions such as the *Caf* climates of China and the southeastern United States, winter frosts are also of only limited importance. But the seasonality of rainfall may be pronounced, and individual downpours are of considerable intensity, favoring rainwash and rill-wash. Chemical weathering is intensified by high summer temperatures and by the coincidence of maximum warmth and moisture. Soil stripping is important, and stream erosion and deposition are active, so that in general both denudation and dissection are comparatively great, with corresponding deposition on lower slopes and in the lower stream valleys. Understandably, less is known of "fossil" Pleistocene landforms in these zones.

Geomorphic agents comparable to those of the modern *Caf* climates appear to have been prominent in the summer-dry Mediterranean lands during several phases of the Pleistocene.

The Summer-Dry, Subtropical Woodland Zone. Frost is also a negligible factor in the summer-dry *Cs* or mediterranean-type climates today. But the degree of mechanical disintegration is partly compensated for by an incomplete mat of herbaceous vegetation, permitting weathering by means of insolational heating as well as by the direct attack of rainwash on soil or bedrock. Seasonal desiccation of soils prepares loose, clastic material for running water to remove. Rainfall is also irregular and intense.

Vigorous soil stripping, headward erosion, and stream dissection are then theoretically possible. In practice, however, only the moister uplands with considerable moderate or steep slopes are typically affected by such phenomena. Numerous exceptions occur, of course, where land misuse is rampant, particularly in areas of soft, erodible sediments. But, depending upon the area, many or most of the features loosely ascribed to "historical" erosion (e.g., Büdel, 1950a) are in fact of Pleistocene age (Butzer, 1961c, 1964a). In brief, under natural conditions, geomorphic equilibrium would be typical of the mediterranean lands, and even under cultivation, geomorphic equilibrium is characteristic of level terrain (see Vita-Finzi, 1969). The Pleistocene processes of this ecozone will be described in a later chapter.

ARID ZONE GEOMORPHOLOGY

General Aspects. Among the common traits of the arid zone, in the broad sense of the word, is a dominance of mechanical over chemical weathering. The vegetative cover decreases with intensified aridity, favoring mechanical disintegration while restricting chemical reactions. The major agents of mechanical attack are as follows (see Tricart and Cailleux, 1960–61; Peel, 1966):

a) Frost-weathering, favored by considerable daily and annual variation of temperature, but somewhat limited by reduced rock moisture. The fragments detached are usually small, promoting grain-by-grain disintegration. Frost is common and severe in middle latitudes, but only sporadic in lower middle latitudes. It is largely absent between the two tropics (Fig. 21).

b) Thermal expansion and contraction, as a result of insolational heating and rapid cooling at night or of sudden rainfalls, which put repeated stress on the mineral structure of exposed rock. In finer-grained, homogeneous rocks, microfractures are produced and small splices are eventually detached. Granular disintegration is promoted in heterogeneous rock. Although slow and almost certainly concomitant with superficial chemical attack, this form of weathering is important in dry zones without frost.

c) Salt hydration (Evans, 1970)—the expansion of anhydride salts after odd rain or dew falls as well as in response to heating.[1] Such salts are due partly to weathering, partly to atmospheric transport. They are concentrated at or near the surface by capillary movement to the surface due to evaporation. In response to temperature oscillations and alternate wetting and drying, these salts expand and contract considerably and thereby loosen the individual grains or slabs of material at the face of the rock.

1. This is both a chemical and mechanical agent, and is commonly described in the literature as a form of chemical weathering. There seems to be just as good a case for considering it a mechanical agent because of its analogies with various forms of thermoclastic weathering.

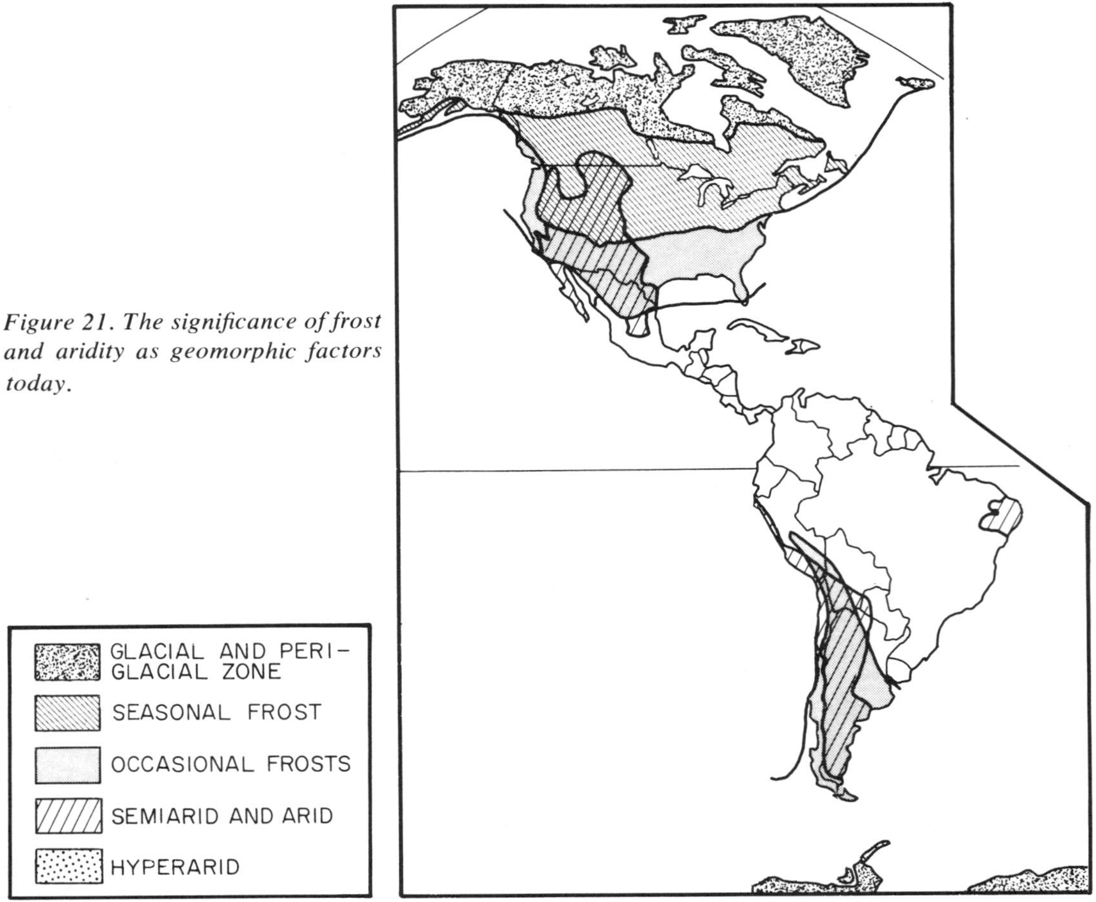

Figure 21. The significance of frost and aridity as geomorphic factors today.

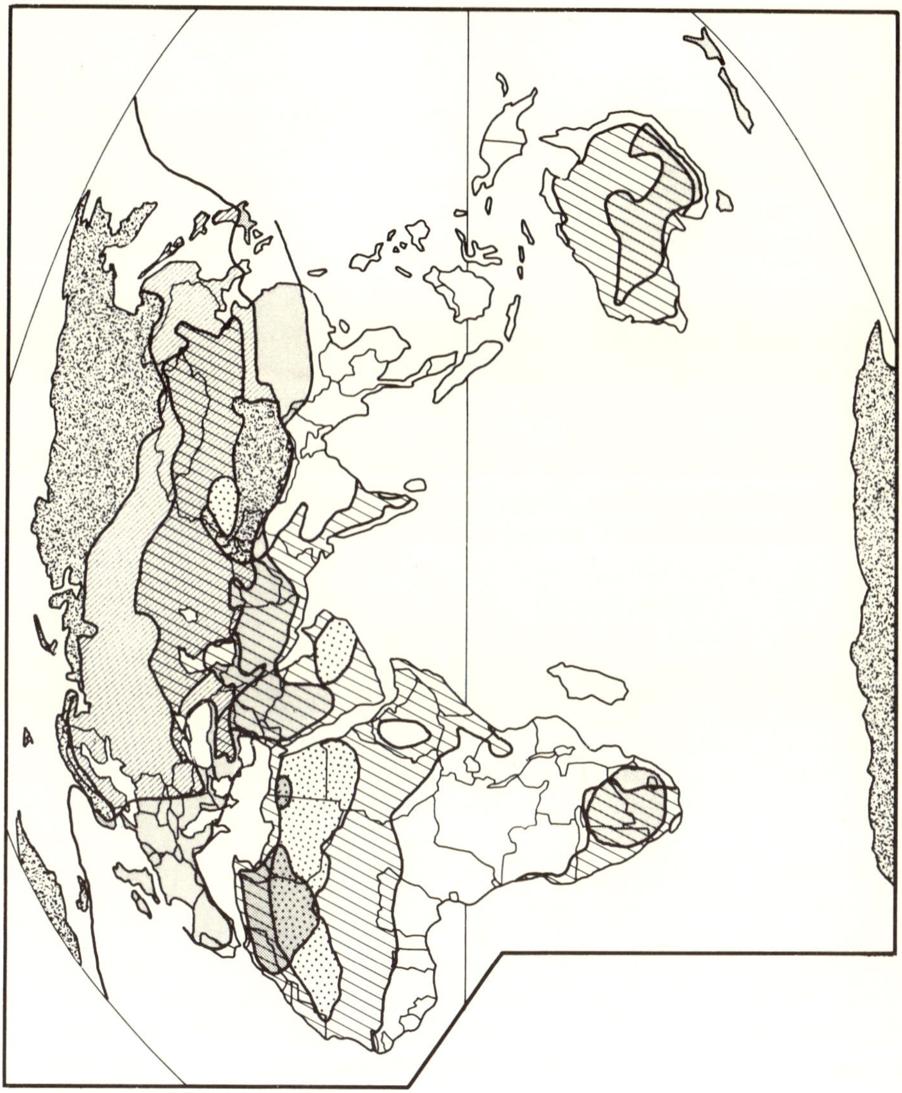

The over-all mechanisms and effects of all three of these mechanical agents are analogous, and they are difficult to isolate. Both forms of thermoclastic weathering probably assume dominance in the middle latitude arid zone, while alternating contraction and expansion through thermal or chemical processes will dominate in lower latitudes.

The products of mechanical weathering accumulate at the base of rock outcrops in the form of loose slope talus, largely as a result of gravity. Consequently slopes will tend to weather back uniformly in uniform bedrock, maintaining constant angles. Part of the debris will be removed by running water, but although evidence of periodic or sporadic water action is manifest everywhere, there frequently is unsufficient fluvial activity to remove all of the coarse debris. In many areas the sparse vegetation permits denudation as well as vertical erosion, particularly in unconsolidated sediments. Fine materials may be attacked by wind erosion, and aeolian deposition may take the form of sand dunes or loess mantles. But throughout the arid zone, running water is and usually has been the major single agent of landform sculpture.

Within the arid zone the presence or absence of frost and the amount and intensity of geomorphic rainfall are responsible for differences in the kind and intensity of geomorphic activity. Consequently the mid-latitude grasslands, the marginal environments of the semidesert, and finally, the deserts proper will be considered individually. General references to arid zone geomorphology are provided by Büdel (1950a), Tricart and Cailleux (1960–61), Peel (1966), and Birot (1968).

Mid-latitude Grasslands. The sod and dense rooting network common to undisturbed mid-latitude grasslands offers effective protection against mechanical weathering. Only on steep slopes, where soils are thin and the herbaceous vegetation scarce, does frost-weathering produce accumulations of detritus. Chemical weathering is not important, although still present.

During the winter the soil is frozen to some depth and there may be a snow cover. For a brief period during the spring, conditions are favorable for rapid runoff; as any snow cover melts, and the soil is still incompletely thawed out, drainage will be concentrated at the surface with little or no percolation. The most effective erosion takes place at this time (Schmidt, 1948), which usually coincides with the major rainy season. Serious denudation is prevented by natural vegetation, but headward stream dissection is conspicuous in areas of poorly consolidated sediments. Steep-sided canyons develop along the fringes of drainage lines. Without human interference, however, such erosional forms remain localized. Corresponding to the seasonality of stream discharge, most river beds are broad and sandy, although the runoff is confined to a small, shallow channel for most of the year.

Wind deflation may attack dry alluvial sands in the river bottoms, and silt-size loess of local or rather distant origin may accumulate on the upland surfaces.

The Desert Borderland Zone. With the sparse vegetation and lack of sod usual in the desert borderlands, mechanical weathering and erosional processes are barely impeded. Much detritus of all particle sizes is provided for removal by running water. Rainfall is seasonal or sporadic and rather intensive so that soil percolation can only account for a fraction of the water, the remainder sweeping the surface as rill-wash or as sheetfloods. But precipitation is localized and commonly limited in duration so that local spates dissipate rapidly and seldom attain a well-defined, channeled flow in a major drainage course. Diffuse runoff is also encouraged by scattered semidesert shrubs. This sporadic, unintegrated, and brief form of fluvial activity promotes denudation rather than stream incision. In combination with the steep slopes maintained by thermoclastic weathering and salt hydration, landform sculpture favors the development of *pediments*. These semiarid forms are characterized by extensive rock-cut plains offset against the irregular fringes of original, higher surfaces by steep rock faces, often making sharp angles at their base (Fig. 22) (see Tator, 1952, for a general discussion). Only in very rough terrain is the drainage strongly channeled, developing steep-sided although still rather flat-floored canyons. When such gorges open onto level plains or major stream systems, alluviation ensues as the periodic waters spread out with reduced gradients, rapidly losing velocity. Radial alluvial *fans* are formed in this manner.

Despite the modern significance of water action, the greater part of the landforms of the arid zone are probably "fossil" and owe their origin

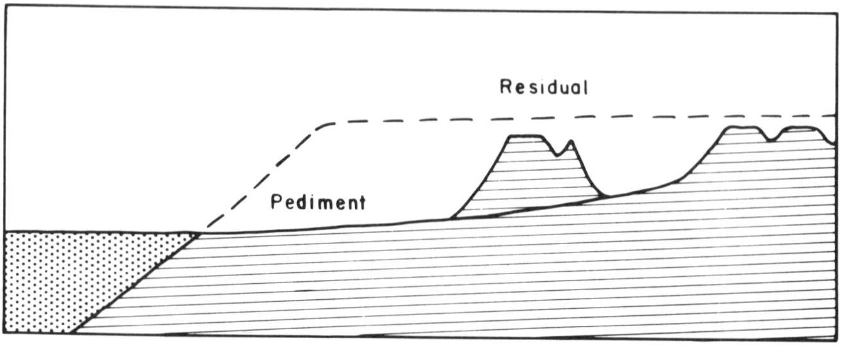

Figure 22. A pediment profile. The original land surface is indicated by the broken line. Sedimentation has been dominant in the depression at the left.

to the moister climates of the Pleistocene. So, for example, many pediments of the lower latitude arid zone are quite defunct today, and pediment development now seems to be confined to areas having at least sporadic frosts. Frost action is by far the most rapid agent of mechanical weathering; the low latitude desert borderlands show little talus development while cliff faces weather back at painfully slow rates. The resulting lack of coarse material for stream transport conspicuously reduces the erosive force of running water. Consequently semidesert landforms are similar in both low and middle latitudes, even though the significance of Pleistocene processes differs, and the present intensity of sculpture is much greater in middle latitudes.

Wind erosion plays little part in slope retreat, but deflation of fine materials from disintegrating level surfaces or alluvial plains can be important. Shallow depressions or blowouts may be created, and the over-all effects for denudation may be appreciable. The dust removed may be swept through the air for hundreds of kilometers before it is laid down in the form of loess mantles elsewhere. In contrast, sand removal and transport is limited, and the resulting dunes and sand sheets are seldom found far from their source. A good part of the aeolian features of the arid zone are more or less immobile and appear to be "fossil" today. Wind deposits will be discussed in more detail in chapter 12.

The Deserts. In the hyperarid and largely barren desert interiors (Fig. 23) geomorphic processes are infinitesimally slow. The rare rainfalls are effective for neither genuine surface sculpture nor true chemical weathering. Even the salt hydration of the desert borderlands becomes infrequent, and practically all landforms appear to belong to the museum of the Pleistocene (see Butzer and Hansen, 1968).

Slow physical disintegration, occasional brief floods at long intervals, and the slow but ever present deflation of fine materials by wind are characteristic. The selective removal of fines may produce surface concentrations or *pavements* of pebbles or crude rock fragments. Sand dunes occupy a role that is rather important areally, but not functionally.

TROPICAL GEOMORPHOLOGY

Tropical Weathering. Several general aspects of tropical weathering are basic to understanding geomorphic activity in the humid and subhumid tropics (for general discussions see Büdel, 1950a, 1958; Tricart and Cailleux, 1965b; Birot, 1968; Ollier, 1969; also Semmel, 1963; and Doornkamp, 1968):

a) There is a complete absence of frost, the most effective single agent of mechanical breakdown of rock in middle and higher latitudes.

b) Daily and seasonal temperature oscillations are moderate or negligible.
c) Complete mats of grassy vegetation or dense forest protect soil and rock surfaces from erosion, and help rule out mechanical weathering as a potent agency.
d) Considerable or excessive moisture is available part or all of the year, while temperatures are persistently high. Chemical weathering is correspondingly intense and effective.

As a result, soil development is remarkably prominent, and smooth or moderately inclined slopes are entirely enveloped in soil or derived soil sediments of appreciable depth (Fölster, 1964; Ollier, 1969; Ruxton and Berry, 1960). Crude rock fragments are seldom exposed at or near the surface. Clays and silts are the dominant particle sizes, and even coarse sands are rare since quartz is rapidly broken down in size to a fine sand component. Bedrock is exposed only on slopes steeper than 20–35 per cent in the subhumid savannas, and on slopes exceeding 120–150 per cent in rainforest areas. But in view of the absence of significant mechanical weathering, slope detritus is still rather scarce. In the seasonally dry savannas, exfoliation, a gradual concentric peeling off of rock slabs, is partly a result of salt hydration. Any talus that may accumulate is rapidly decomposed as soon as it comes into contact with the soil zone.

The over-all result of this overwhelming preponderance of chemical weathering is that sand and gravel, the best tools of stream erosion, are rare in the watercourses of the tropics (Bakker, 1957). This fact is of great significance for landform sculpture.

Savanna Planation. Chemical weathering, which leads to over-all surface decomposition, and insignificance of stream dissection lend a peculiar aspect to the landscapes of many tropical regions. Denudation is effective while downcutting is not, with the result that river valleys remain shallow but grow to the dimensions of great plains. These are the so-called savanna plains, commonly bounded by higher plains of similar origin from which they are separated by moderately steep, convex bedrock slopes (Fig. 23). Isolated remnants of higher ground stud the plains in the form of *inselbergs* (German for "island mountains"). The related process of savanna planation has been developed by Büdel (1958) and Semmel (1963), and reviewed by Cotton (1961).

The prerequisites to erosion are provided by a nonconsolidated mantle of deeply weathered residual products[2] which continue to form

2. It is uncertain whether or not the examples of deep soil profiles of some tropical areas cited by Büdel (1958, 1963) represent true *(B)*-horizons over gley or pseudo-gley horizons

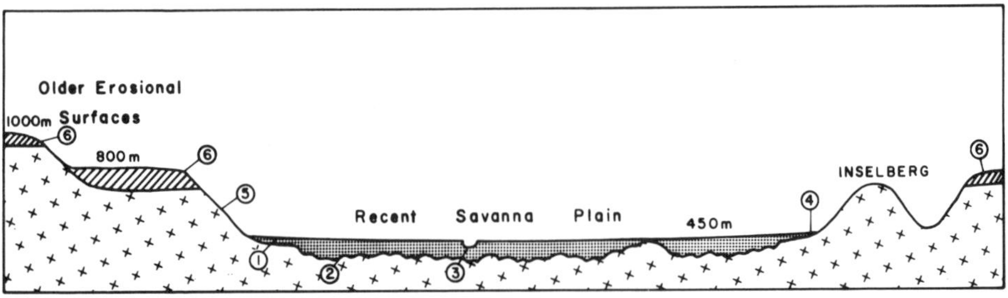

Figure 23. Savanna plain in Guinea (generalized and modified after Büdel, 1958): (1) 10–30 m. clay soil or sediment, (2) base of weathered zone, (3) river channel, (4) pediment-like foot slope, (5) hillslope, (6) massive lateritic crusts on older, raised savanna plain surfaces. General vertical exaggeration 2.5x, depth of deposits strongly exaggerated.

under optimal conditions of chemical alteration. Rainfall is at its world maximum of intensity in the subhumid tropics, and the violent downpours often fall on parched, hard soils. Denudation is therefore effective as clays and silts are removed. But coarse materials are lacking, so that stream beds are broad, shallow, and poorly defined as they wander across the plains, unable to cut vertically. Resistant banks of rock are not sawed through but remain prominent as cataracts or waterfalls in the stream's profile for an indefinite period of time. Streams do not then develop steeper gradients, but continue to lie on the level of the plain. Excess runoff does not drain away through a network of drainage channels, but spreads out over the surface in the form of sheetfloods. The effects of such washing are extensive, gently sloping plains developing at the level of the major stream, partly by lateral planation, partly by significant washing-together of *colluvial* sediments (as opposed to channeled, alluvial stream deposits).[3] Wind deflation of fine alluvial or collu-

in situ—attaining the depths of 10–30 m., with even deeper horizons showing clay mineral formation as a result of breakdown of all but quartz grains. There is reason to believe that parts of this material may have been accumulated through water action. Much remains to be clarified about the actual depth of typical tropical weathering profiles, although the outcome would not seriously affect Büdel's geomorphic arguments. Detailed local work has seldom failed to substantiate the fact that the residual mantle under such savanna plains is in fact deep, regardless of whether the material is redeposited.

3. Unlike the pediments of the arid zone, savanna plantation proceeds on the surface of a residual mantle, rather than on a rock-cut platform. Such platforms are limited to the margins of savanna plains. Pediments, formed by mechanical processes that take place in climates not conducive to the destruction of rock fragments by chemical weathering, are therefore only superficially similar to savanna plains, a point demonstrated by Ruxton and Berry (1960) and Cotton (1961). Obviously the phenomenon of soil denudation (and colluviation) is not confined to the so-called savanna belt and to the humid subtropics, but common to all regions with accelerated soil erosion—particularly as a result of human disturbance of the natural equilibrium. But geologically significant deposits do not develop elsewhere today.

vial deposits may be of local interest during the dry season.

Remnants of analogous landforms of Tertiary age ("peneplains") are widespread in middle latitudes, together with fossil soils pertaining to tropical environments. Even during certain Plesitocene interglacials, analogous subtropical plastosols repeatedly formed in the Mediterranean region, and soil denudation and colluviation were a geomorphically significant factor at times. Savanna-like processes are therefore of considerable interest in the Pleistocene record of temperate or subtropical latitudes.

The Rainforest Zone. The intensity of chemical weathering and the insignificance of mechanical agents are exaggerated in humid equatorial regions, both in the tropical rain forest and in montane forests. Dense vegetation of tree type clings to steep slopes of as much as 150 per cent gradient. This, together with a less intensive rainfall, slows down surface runoff. The seasonality of rainfall is also limited, and stream discharge is more uniform. It is consequently difficult to imagine that an analogous form of sheetflooding, soil denudation, and colluviation could be as prominent here as in the savanna belt. Instead, moderate fluvial activity, equally impeded by a lack of coarse bed materials, probably accounts for rather slow but different landform sculpture.

Only in rough terrain do geomorphic activities gain considerable significance. Under the influence of gravity movement, earth flows of water-saturated soil masses may periodically create slope scars under steep, forested hillsides. Tributary streams establish gully networks through headward erosion, making use of such slump scars. Coarser materials may enter the streams in this manner. As a result, fluvial dissection is greatly exaggerated in the hill and mountain country of the humid tropics (Ruxton, 1967).

The major climatic-geomorphologic regions and features are summarized in Table 8.

Table 8. Climatic–geomorphologic regions and features.

	Zone	*Koeppen Climate*	*Dominant Form of Weathering*	*Dominant Agents of Erosion and Deposition*	*Type of General Erosion*	*Characteristic Sediments*
	Glacial	*EF*	Frost-weathering	Ice; running water		Moraines; coarse glaciofluvial beds
Periglacial	High Arctic Barrens	*EF, ET*	Intensive frost-weathering and permafrost	Running water solifluction, wind	Intensive denudation and dissection	Coarse alluvial beds, gravity/soil-frost deposits, soil-frost features, dunes, loess
	Tundra	*ET*			Denudation and dissection	
	Forest with Permafrost	*Dc, Dd*		Running water	Dissection and denudation	Coarse alluvial beds, soil-frost features
Humid	Temperate-Continental	*Da, Db, Dc*	Frost and some chemical weathering	Running water	Dissection	Fine to coarse alluvial beds
	Temperate-Marine	*Cbf*	Moderate chemical weathering	Running water	Limited dissection	Fine alluvial beds

Humid	Moist Subtropical	*Caf*	Chemical weathering	Running water	Dissection and denudation	Alluvial and colluvial beds
	Summer-dry Subtropical	*Cs*	Limited mechanical and chemical weathering		Moderate dissection	Alluvial beds
Arid	Mid-latitude grasslands	*BSk*	Frost-weathering	Running water and wind	Dissection	Coarse alluvial beds, loess
	Desert margins	*BS, BW*	Thermoclastic weathering and salt hydration		Denudation and dissection	Alluvial beds, dunes
	Desert	*BWh*			Denudation	Alluvial beds, dunes
Tropical	Savanna	*Aw, Caw, Am*	Intensive chemical weathering	Running water	Denudation	Fine colluvial beds
	Rainforest	*Af, Am*			Limited denudation and dissection	Fine alluvial beds

NINE

Mammalian and Human Distributions

THE CONCEPT OF ANIMAL REGIONS

The distribution of mammalian faunas today is of particular interest for Pleistocene geography in two ways. First, most of the bone commonly preserved in ancient deposits belongs to larger mammals whose identical genera or species may still survive today. Then also, the kinds and relative abundance of game in various environments were of fundamental importance for early hunting and gathering populations.

Animals, like vegetation and soils, show peculiar distributions related to broad zonations of the natural environment. Mammals, for example, are partly adapted physiognomically to the natural vegetation of their habitats: climbers, springers, runners, breakers, burrowers, etc.

They are also adapted to the particular food and water resources provided by their environment. So for example, herbivores prefer certain combinations of grasses, shrubs, leaves, bark, roots, etc., while carnivores prey chiefly on certain plant eaters.

In considering animal distributions, zoogeographers commonly employ two different terminologies for different purposes. Zoogeographical, or animal, *regions* are differentiated from one another by the assemblages of mammals and birds they contain (see Darlington, 1957; George, 1962). They represent geographical localizations that are determined by physical barriers to dispersal in the course of geological history. Understandably, these continental regions bear only a loose relationship to the zonations of vegetation or climate. *Bioregions*—reflecting the ecological conditions of the moment—have also been identified. These are superimposed on the animal regions and coincide rather closely with other zonal distributions (see Tischler, 1955; Hesse *et al.*, 1951; also Newbigin, 1936; Troll, 1950).

In the traditional zoogeographical approach, the following major animal regions are identified;

a) Holarctic, including Nearctic (North America) and Palearctic (Eurasia, excluding India and Southeast Asia, but including North Africa) subregions;
b) Ethiopian, including tropical Africa and southern Arabia;
c) Oriental, including India, southern China, and southeastern Asia;
d) Australian, including Australia, New Guinea, and most of the Pacific islands;
e) Neotropical, including Central and South America.

Some of the more typical Palearctic animals include bear, fox, wolf, hyena, deer, ibex, sheep, wild cattle, bison, horse, camel, hare, beaver, mole, and vole. Typical of the Ethiopian region are wart hog, hippopotamus, giraffe, okapi, elephant (of the genus *Loxodonta*), rhinoceros (*Ceratotherium* and *Diceros*), zebra, gnu, baboon, gorilla, and chimpanzee. The Oriental region includes the gibbon, elephant (genus *Elephas*), rhinoceros (genera *Rhinoceros* and *Didermoceros*), dwarf deer, tree shrews, and tarsiers among its typical species.

For our purposes the mammalian distributions of the bioregions are more useful. Most mammals, understandably, range widely and are seldom strictly confined to a single ecological habitat. Similarly, the habits of different genera or species of the same family may differ considerably. Consequently, the association of "typical" species with particular environments is a loose one.

MAMMALIAN DISTRIBUTIONS IN THE MAJOR NATURAL REGIONS

The Tundra. The modern polar regions support a specialized fauna largely adapted to conditions of extreme cold and reduced food resources that last as much as nine to ten months a year (Tischler, 1955, p. 238 ff.). With a few exceptions, the bird fauna is migratory. Even the larger endemic mammals, with the exception of the musk ox (*Ovibos moschatus*), may all frequent the adjacent forest-tundra and boreal woodland to some degree or other during the long, snowy winter.

The ecological balance is based upon the herbivorous reindeer or caribou (*Rangifer tarandus* ssp.), rodents such as the lemming (*Lemmus* spp.) and several voles (*Microtus* spp.), the arctic hare (*Lepus arcticus*), snowy partridge (*Lagopus* spp.), geese, and various song birds. These in turn support endemic carnivores such as the arctic fox (*Vulpes lagopus*) or wider ranging species such as the wolf (*Canis lupus*) and the ermine (*Mustela erminea*), as well as various birds of prey. In the coastal areas, seals and polar bears (*Thalarctos* spp.) obtain much or all of their nourishment from fish. All of the above animals are essentially circumpolar in the northern hemisphere, or at least were so during the Pleistocene.

During the Pleistocene the European tundra species were reinforced by great herds of now-extinct woolly mammoths (*Elephas primigenius*), less specialized open-country denizens such as the extinct *Bison priscus*, and a number of subspecies of the wild horse (*Equus caballus*). These were supplemented by smaller numbers of cold-loving species such as the extinct giant elk (*Megaceros* spp.) and woolly rhinoceros (*Tichorhinus antiquitatis*), and the saiga antelope (*Saiga tatarica*), still present in the Central Asian grasslands today. All of these animals roamed widely between the tundra and the adjacent, lighter forest belts, and sometimes into the cool grasslands.

There can be no doubt that these Pleistocene low latitude tundras had a tremendous carrying capacity, somewhat greater than the recent, higher latitude tundras. Human interference is probably involved in the modern paucity of gregarious herbivores in the tundra. But some environmental differences must have existed, possibly related to different drainage and radiation conditions between high and low latitude tundras. This point has never been raised by animal ecologists, but certainly deserves attention. So, for example, the extended periods of hibernal darkness in high latitudes may carry significance for some species, while intensive summer radiation in lower latitudes would favor plant growth out of proportion to the actual air temperatures.

The ecological environment of the forest-tundra is optimal insofar as cold environments are concerned. The migrating gregarious herds of the open country seek their winter refuge here where many species, faced with a deep snow cover, can augment their diets by consuming tree shoots and bark. Arboreal groves provide windbreaks during storms, while the immediate environment provides all that the tundra otherwise offers. For the human occupant, an additional hunting element of forest species is provided.

In the analogous alpine meadow environments, the marmot (*Marmota* spp.) occupies an ecological niche like that of the lemming, while the chamois (*Rupicapra rupicapra*) and certain mountain sheep take over the role of the reindeer.

The Boreal Forest. The higher latitude coniferous forests (see Tischler, 1955, p. 251 ff.) are rather dense and provide little herbaceous vegetation for gregarious herbivores. The larger herbivores are comparatively few and include the elk or moose (*Alces* spp.), deer or wapiti (*Cervus* spp.), the extinct Old World aurochs (*Bos primigenius*), and woodland bison *(Bison bonasus)*.[1] Seasonally, considerable numbers of tundra reindeer range through the boreal forest, while the endemic forest reindeer are present all year. Plant-, bark-, and root-eating rodents, including the beaver (*Castor* spp.) and various mice, rats, and rabbits, are ground-dwellers; others are climbers of one sort or other. The latter group includes squirrels. The rodents provide the chief food source for carnivores or omnivores such as foxes (*Vulpes* spp.), marten (*Martes* spp.), the wolverine (*Gulo gulo*), the lynx (*Lynx* spp.), certain Holarctic bears (*Ursus* spp.), as well as buzzards and owls. Several of the carnivores also prey upon the insect-, and plant-eating bird life, while the omnipresent wolf and the wolverine prey upon larger herbivores as well.

During the winter most bears and many rodents hibernate, while the deer and many birds migrate south. Reindeer and elk, however, obtain sufficient nourishment from bark, dry leaves, and lichens.

All in all the variety of big game is small, and their number is limited.

The Deciduous and Mixed Forests. The more lightly stocked woodlands of the warm-temperate and even the subtropical zones offer considerably better grazing facilities for large herbivores during all or most of the year (Tischler, 1955, pp. 223 ff. and 284 ff.). These include the red deer or wapiti, the aurochs, and the bison of the boreal forests, as well as the boar (*Sus scrofa*) and several additional deer: the white and black-tailed deer of the genus *Odocoileus* in North America, the roe deer (*Capreolus capreolus*) in Europe, and fallow deer (*Dama* spp.) in the Near Eastern area. The principal rodents and carnivores are similar to those of the boreal forest, although their number is greater, corresponding to a higher carrying capacity.

Mid-latitude Grasslands. The great mid-latitude grasslands and steppes until recently supported a limited variety of swift-footed, gregarious herbivores, some of which were present in exceptionally great numbers. Varied, rich, and nutritious grasses are mainly responsible for the excellent carrying capacity. Foremost of the open-country herbivores in

1. The bison, reindeer, and horse are three major genera originally present as both woodland and open-country forms, identical on the specific level. Thus even today the North American caribou has woodland and tundra forms, while the woodland form of the almost identical Eurasian reindeer seems to be almost extinct. Similarly, the wild horse of Eurasia belonged to two open-country subspecies—the western tarpan *(Equus caballus gmelini)* and the eastern Przewalski horse *(Equus caballus feral)*. A rather heavy, woodland species was also apparently present *(Equus caballus silvestris)*. For discussions of both horse and reindeer see Zeuner (1963, pp. 299 ff. and 112 ff.). A similar situation seems to exist for the North American and European bisons.

North America was the American bison (*Bison bison*), once totaling over 60 million animals (with grassland and woodland subspecies). Possibly equally significant in Eurasia, prior to human interference, was the wild horse of the tarpan and Przewalski subspecies. Occupying similar ecological niches in the New and Old World respectively are the pronghorn (*Antilocapra americana*) and the saiga antelope (*Saiga tatarica*). Other large herbivores are not significant in either area, although several gazelles and the central and southwest Asian onager *(Equus hemionus)* are of some interest in the Eurasian Pleistocene. All of these fast-moving open-country forms vacated the snowy plains in winter, seeking out green pastures in southerly areas.

The rodents are mainly burrowers finding protection underground from carnivores and from the cold, winter season. Certain species of voles, prairie dogs, ground squirrels, pocket gophers, hamsters, kangaroo rats and mice, as well as species of the ubiquitous rabbits and hares belong to this group.

Preying upon the mammalian herbivores are the wolf, coyote (*Canis latrans*), and several smaller carnivores in the New World; and the wolf, jackal (*Canis aureus*), the East European steppe fox (*Vulpes corsac*), and several weasels (*Mustela* spp.) in the Old.

The Lower Latitude Arid Zone. The grasslands and desert shrub areas of lower latitudes are by no means as luxuriant or attractive as those of higher latitudes, and the plant species are remarkably less nutritious. The carrying capacity in terms of larger herbivores is limited, and leading Old World species include a number of swift gazelles and antelopes, the wild ass (*Equus asinus*) of northern Aftica, and the wild ancestor of the dromedary (*Camelus dromedarius*)—all able to survive with little water. Here too there are numerous specialized rodents. Large carnivores in the same areas include the jackal, hyenas (*Hyaena* spp.), lion (*Panthera leo*), and leopard (*Panthera pardus*).

The Tropical Parklands and Woodlands. Many of the savannas and light tropical woodlands, particularly those of eastern and south-central Africa, have a high carrying capacity exceeding even that of the middle latitude grasslands. Others, such as those of West Africa or South America, have less nutritious or less palatable grasses and are rather inferior. The characteristic African "savanna" fauna with its once endless herds of antelopes, gazelles, zebra, giraffe, and elephant needs little emphasis. Some species, such as the giraffe (*Giraffa camelopardalis*) and the non-mammalian ostrich (*Struthio camelus*), are thought to be open-country forms as are the zebras, antelopes, and gazelles. Others, such as the African and Asian elephants and rhinos, are adapted both to forest and open parkland. The rodents play an insignificant role by comparison.

Among the carnivores the great cats, including lion, leopard, tiger, and cheetah, are best known. All range through grassland and woodland.

The dense rainforests, like the boreal forests, provide little grazing, and there are few large herbivores and few carnivores (Tischler, 1955, p. 212 ff.). Instead, a great variety of insectivores, partly mammalian, take advantage of the particularly rich insect life. The few typical species of larger herbivores include the tapirs, forest buffalo, the duikers (*Cephalophus* spp.), the okapi (*Okapia johnstonii*), the chevrotain (*Tragulus* spp.), and several anthropoid apes (gorilla, chimpanzee, orangutan). Elephants were at one time relatively common, for example, the Congo dwarf form (*Loxodonta africana cyclotis*). Game resources are, however, generally sporadic and limited.

An approximate index of differential carrying capacities between grassy tropical woodlands and rainforest is given by biomass statistics of hoofed mammals (ungulates) in several reserves and parks in Africa (Bourlière, 1963). In East African game reserves—without conspicuous overgrazing—the biomass of savanna woodland environments varies from 5,000 to 20,000 kilograms per square kilometer. In some overgrazed parks it exceeds 35,000 kg./sq.km. On the other hand, a dense rainforest environment in Ghana, the Tano Nimri forest reserve, has less than 6 kg./sq.km. These statistics speak for themselves.

A few tropical genera were of considerable importance in higher latitudes during parts of the Pleistocene. Most of them were represented by extinct species, some less specialized, others adapted to rather different environments.

CULTURE, TECHNOLOGY, AND THE HUMAN RESOURCE BASE

Turning from mammalian distributions to the dispersal of the human species, cultural and technological achievements strongly determine the settlement potential of particular environments. In other words, although the nonhuman ancestors of man shared some sort of tropical distribution such as that of the anthropoid apes, man has progressively occupied and culturally adapted to numerous diverse environments. From the viewpoint of prehistoric geography and ecology it is important to assess the availability and assets of different environments *at different cultural levels*. We commonly tend to interpret the implications of a particular environment for early man from a contemporary perspective. Yet for humans possessing neither fire nor clothing, middle latitudes would provide a rather marginal environment. The grassy tropical woodlands or savannas would provide a far better resource base for a general hunting economy than would the temperate forests. And the tundra was not a peripheral environment for the mighty hunting clans of the late Pleistocene.

During most of the Pleistocene the cultural-technological status of man was that of an *unspecialized food-collector*. A great variety of edible roots and plants, as well as many animal foods, were used. At first man may have confined his carnivorous attentions to smaller mammals and to birds, approaching bigger game on a more incidental, possibly scavenging basis. Later, as the archeological evidence attests, big game was hunted on a large scale, possibly eclipsing the importance of vegetable foods in many areas. By this time, in the later Lower Pleistocene, fire was in widespread use, and some sort of protective clothing was presumably worn where necessary.

The terminal stages of the Pleistocene witnessed a number of *specialized, hunter-gatherer economies* in both the Old and the New Worlds. Considerable advances were made in hunting technology, so that the great clans wreaked considerable havoc among the gregarious herbivores (see ch. 28). A certain proficiency in the making of clothing is also evident. Some hunter-gatherer groups specialized further still by exploiting fish and other seafood resources.

By the beginning of the Holocene a new cultural and economic level of subsistence had been established in parts of the Near East, where plant and animal domestication are first recorded. Primitive agriculture led to a renewed emphasis on plant foods, although hunter-gatherer activities still continued to play some economic role in these new communities. Agriculture led to the exploitation of new niches in the environment, particularly of the soil. Soil availability, texture, fertility, and moisture rapidly became new criteria for settlement potential. And the adaptability of available seed crops and household or herd animals to new environmental conditions further influenced the dispersal and establishment of food-producing economies. Primitive mineral extraction was begun, lending some interest to regions with accessible mineral resources. With time, big game and edible wild plant foods were no longer vital to farming populations, casting a completely different perspective on the settlement potential of different environments.

Unfortunately there is no systematic body of information concerning the human ecology of contemporary "primitive" groups.[2] Also, most of

2. Good monographs exist for a number of cultures, although their emphasis is commonly upon social rather than ecological attributes. The one existing attempt at a more comprehensive study of habitat and economy (Forde, 1934) is rather incomplete and in some ways obsolete. Although emphasizing other attributes, the *Man the Hunter* conference (Lee and deVore, 1968) provides much useful data in this direction.

the present-day food-collecting populations have been driven into undesirable refuges, particularly in the humid tropics. Hence the optimal patterns of resource exploitation by Pleistocene food-collectors can only be inferred today (see Bartholomew and Birdsell, 1953). Similarly, cultural contact with higher economies has influenced all modern "primitive" groups to some degree. The ethnological parallels for an ecological understanding of early man are, then, incompletely preserved and often inadequately understood.

In the domain of the biological and physical sciences, more information is needed concerning (*a*) the animal-carrying capacity of various grasses and vegetation types, (*b*) the range and dietary value of edible wild plant resources in different environments, and (*c*) the working and productivity of different soils under primitive agriculture.[3] In other words, the biological foundations for a fuller ecological study are also lacking.

The subsequent outlines of environmental resources for food-collecting and primitive agricultural economies are not more than provisional and rudimentary. Possibly they will serve to emphasize a need for further systematic work.

ENVIRONMENTAL RESOURCES FOR EARLY FOOD-COLLECTORS

The wild vegetable foods eaten by food-collecting economies are obviously most abundant in warmer latitudes, where a cold season does not impede or slow down plant growth. Long, intensive dry seasons are also undesirable. The products used vary from roots and tubers to seeds, nuts, acorns, and fruits; the aborigines of central Australia, for example, are known to have exploited more than seventy-three plant species.[4] Sources of firewood also deserve consideration (Heizer, 1963).

In addition to these vegetable products, insects, lizards, snakes, birds, small rodents, birds' eggs, and fish may also have been eaten. In fact, primitive fishing, particularly in shallow inland waters, may have been of considerable importance locally. Whether or not the fish traps used by various "primitive" groups today were known in the Pleistocene is not established. But fish gorges and harpoons were well known during the terminal stages of the European Pleistocene, while fish bones are archeologically verified from earlier sites. Freshwater fishing resources are not known to vary conspicuously wherever inland waters are available.

3. A good example of cattle-carrying capacities according to grass types in Africa is given by Rattray (1960). A valuable pioneer study on soils and primitive agriculture has been written by Nye and Greenland (1960).

4. For a very general discussion of plant utilization see Schery (1952).

Salt-water fishing requires far greater skills, and was probably not systematically followed until the early Holocene. The equally protein-rich shellfish of the seashore, including oysters, cockles, mussels, lobsters, crabs, crayfish, and shrimp, probably already provided a valuable food resource for people with limited technological skills. Bivalves, gastropods, and crabs are also available in many inland waters.

The greatest individual source of high-calorie, protein-rich food for all the larger hunter-gatherer populations was almost certainly animal flesh. It is here that the great regional differences in carrying capacity asssume paramount importance. The hoofed mammals became the best meat provider for early man since carnivores were rarely attacked, and smaller mammals, such as rodents, provided less meat in proportion to the efforts and skill required to hunt them. It is on the big herbivores that primary attention must be focused. A review of the potential of various environments is given by Table 9, which lists ungulate biomasses recorded in a number of game and forest reserves. Biomass and carrying capacity are by no means identical, but in view of the limited information available about the latter, biomass statistics provide some sort of index of potential utilization by all resident species. Since the number of ungulate species available increases the potential resources of an environment due to fuller exploitation of ecological niches, species data are included where available. It is also of interest from the hunter's point of view.

Table 9. Ungulate biomasses of certain environments (from Bourlière, 1963. Data from overgrazed reserves omitted).

Vegetation type	*Locality*	*Species Number*	*Biomass (kg./sq. km.)*
Rainforest	Ghana	3	5.6
Thorn Forest	Southern Rhodesia	15	4900
Savanna Parkland	Congo; Uganda	5–11	5950–19540
High Grass Savanna	Kenya; Transvaal	17–19	1760–16560
Low Grass Savanna	Kenya; Tanganyika	over 15	5250
Semidesert Grassland	Chad	4	83
Desert Shrub	Rio de Oro; Mauretania	2	0.3–189
Temperate Grasslands	Eurasian steppe; Great Plains		350–3000
Deciduous Forest	Scotland		ca. 1000
Mixed Forest	Carpathians		ca. 500
Tundra	Northern Canada		ca. 800

It is obvious from Table 9 that the tropical savannas stand out as optimal areas. Since many savannas are thought to be artificial, resulting from human interference with former grassy, open woodlands, the potential utilization under "natural" conditions may have been less. Yet even the thorn forest has a remarkably high biomass. Consequently, the tropical parklands and the mid-latitude grasslands remain optimal areas. The fauna of the Canadian tundra has been decimated by intensive hunting and competition by domesticated reindeer (Fitting *et al.*, 1966); moreover, Pleistocene lower latitude tundras may have been far more suitable. The deciduous woodlands also have a high biomass. The boreal forest (no data), the tropical rainforest, and truly arid situations provide the lowest carrying capacities of all.

In summary, vegetable foods were probably optimal in the humid and subhumid tropical and subtropical climates. The relative suitability of various environments for early hunter-gatherer populations can be suggested by using the following three categories based primarily on animal food resources:

a) Optimal. The grassy, tropical deciduous woodlands and savannas; the mid-latitude grasslands; the lower latitude Pleistocene tundras.

b) Intermediate. The temperate and subtropical deciduous and mixed woodlands; the high latitude tundras.

c) Marginal. The tropical rainforest; the boreal forest; the semideserts and deserts.

ENVIRONMENTAL RESOURCES FOR PRIMITIVE AGRICULTURISTS

Primitive agriculture in prehistoric times was largely based on cereal or vegetable cultivation, and remains so even today. The following comments will then be confined entirely to relevant problems of crop cultivation.

Early working of the soil was done with digging and planting sticks or hoes made either of wood, bone, or stone. Animal traction was not used for cultivation in prehistoric times prior to the invention of the plow. Stone tools for forest clearance were present in larger numbers in some prehistoric cultures, but most forest (and grassland) clearance appears to have been done by burning. Consequently, the immediate technological problems were not presented by clearing off vegetation but by breaking up hard soils and combating densely rooted sods. Weeding was presumably another problem to contend with in many areas.

Hard soils are most common where heavy, clayey soils are subject to seasonal drought. The first is a matter of soil texture, the second a matter of climate. Dense soils may occur almost anywhere. Seasonal drought, however, is largely confined to the arid zone and to the lighter tropical and subtropical woodlands. Forest soils are therefore not necessarily more friable than grassland soils. All areas with dry seasons of some importance will experience soil drying and hardening. As already pointed out, lateritic soils become indurated only when stripped by soil erosion and exposed at the surface. Lateritic soils still in process of development do not necessarily make primitive agriculture more difficult than nonlateritic soils.

Densely rooted sods are largely confined to the middle latitude grasslands and, to a lesser extent, to some tropical grasslands. Dense rooting is no problem in most forest locales.

If a soil can indeed be managed mechanically, the next problems are posed by available moisture, sufficient warmth, and soil fertility.

In a general way, primitive agriculture is possible in the humid tropics and subtropics most or all of the year; in the subhumid and semiarid tropics during the summer rainy season; in the subhumid subtropics during the moist winters; in the middle latitude grasslands during the spring and early summer; in the temperate mixed woodlands during the summer. The boreal forest and tundra belts have too short a growing season for any of the genetically unspecialized crops of primitive agriculturists. The arid lands were too dry without irrigation and, therefore, soon attracted communities that abandoned planting but emphasized herding, giving rise to pastoral economies in very late prehistoric times.

Soil fertility is difficult to assess as a resource problem for primitive agriculturists, even though animal dung as a fertilizing agent was, and often remains, unknown. In the absence of animal dung, variable periods of fallow, lasting up to thirty years or longer, help the soil to regain its natural humus content, structure, and base saturation. Some of the principal factors affecting soil fertility are as follows:

a) *Humus content and type*. Rich mull humus is very much more productive than a moder or raw humus, retains more moisture, and provides good aeration. Similarly the clay-humus complex in mull A_1-horizons is optimal. Chernozems, kastanozems, tirsified soils, degraded chernozems, braunerdes, and podsolics generally have the most favorable humus environment. Podsols, rendzinas, rankers, and tropical and desert soils commonly have poor quality humus and lack a good clay-humus complex.
b) *pH and base saturation*. Soil environments with an intermediate *p*H (5–8) are desirable for most crops, and base saturation reflects directly on the availability of the principal nutrients (calcium, nitrogen, phosphorus, potassium.) This is partly a matter of vegetation

type and climate, and partly a question of bedrock. The middle latitude grassland soils, braunerdes, and terra rossas rate best in this sense, the podsols and leached tropical soils least favorably.

c) *Degree of podsolization or latozation.* Podsols and latosolic soils are not only poor or lacking in the principal nutrients, but often also poor in some micronutrients (boron, cobalt, copper, iodine, iron, magnesium, manganese, sulpher, zinc).

d) *Texture and structure.* Moderately fine to fine-grained soils (silt and clay-size particles) are preferable to those with sandy texture. They are water-retentive, less readily leached, and provide more nutrients made readily accessible to the roots. Some fine-textured soils are rather dense by nature, i.e., have a rather compact, nonporous structure. Among these are the tirsified soils, most tropical plastosols, and many groundwater soils. Other fine-textured soils, such as the temperate grassland *AC*-types or the braunerdes, maintain a well-aerated so-called crumb or spongy structure through their rich mull humus. But such soils can deteriorate rather rapidly through overuse. In fact, structural deterioration may occasionally be more important than nutrient depletion for reduction in soil fertility by overuse.

e) *Local factors.* A number of local intrazonal factors are also of importance for soil fertility. They include drainage, bedrock texture, and chemistry. Free-draining soils are generally desirable, as is bedrock with a high base content. Certain bedrock types may also provide too permeable or impermeable a base, or may weather extremely slowly due to high resistivity. These are all part of the great mosaic of innumerable local variations.

In overview, weighing all factors of mechanical soil cultivation and of soil fertility and productivity, a few generalizations can be attempted. Probably the most undesirable environments of all are the deserts, tundras, and boreal woodlands. Either there is no water, or the growing season is insufficiently long, and the soils are poor or rudimentary. The optimal environment seems to be provided by the moist and temperate deciduous woodlands with their fertile braunerdes. The runners-up each have some negative aspects, despite an over-all positive balance. They include the temperate grasslands, the podsolic woodland soil zone, and the terra rossa lands. Regions of other grassland soils, of immature *AC*-type soils, and above all, the greater part of the tropics occupy an intermediate position. With exception of the humus content, which seems to be lower in the grassland soils, there seems to be no generally valid reason why tropical woodland soils should be better than tropical grassland soils when all factors are considered.[5] Nor are subtropical woodland soils inherently more suitable than temperate grassland soils when all factors are weighed.

In terms of resource potential for primitive agriculturists, the various habitats can tentatively be classified as follows:

a) Optimal. The warm-temperate and subtropical woodlands, and the temperate grasslands.
b) Intermediate. The humid, subhumid, and semiarid tropics.
c) Marginal or unsuitable. The deserts, semideserts, boreal woodlands, and tundras.

Contemporary population densities, despite the impact of industrialization, run remarkably parallel to these estimates of resource potential for primitive agriculture. This serves to emphasize the rather different criteria applied to the environment by hunter-gatherer populations.

5. According to Denevan (1966) the savanna and forest regions of the northeast Bolivian lowlands and of the *Llanos* of the Orinoco are about equally easy to cultivate under shifting agriculture—even though the savanna soils are less productive and were probably only used when population pressure was great.

PART THREE

Interpretation of Pleistocene Sediments

TEN

Field and Laboratory Study of Sediments

INTRODUCTION

When investigating a sediment exposure or interpreting an archeological site with geological stratification, the earth scientist commonly studies certain features in the field while removing select samples of material for subsequent laboratory study. Part of the field study involves the surficial expression or morphology of the land. But detailed study of the sediments (and possible soil zones within it) is equally important. Such sedimentological work is crucial for an understanding of the depositional environment, which in its turn may be the key to understanding the local and the macrosetting of an archeological site or to interpreting stratified biological remains.

The importance of carrying out morphological and sedimentological work simultaneously can hardly be stressed enough; one is of only limited meaning without the other. The subsequent chapters are therefore devoted to a brief, systematic outline of the major Pleistocene deposits preceded by a general outline of field and laboratory techniques and interpretations. A final chapter in this group applies this information to a survey of different types of geologically stratified archeological sites.

A useful and detailed description of mechanical and chemical laboratory techniques is given by Cornwall (1958, chs. 10–17). The various aspects of sand and gravel sedimentology are discussed in detail by Cailleux and Tricart (1963). General texts on sedimentation are strongly oriented toward marine environments and consolidated rocks, although

Pettijohn (1957) also brings in general information applicable to the Pleistocene.

SEDIMENT DEPOSITION

Stratification. The *stratification* or disposition of beds or materials is a fundamental characteristic of any deposit laid down by water, wind, ice, or mass movements. A sediment may lack defined beds or horizons, although closer inspection might show that the sand-sized particles are disposed as inconspicuous laminae or that individual pebbles are bedded, i.e., laid down on their flatter faces. Such a sediment can conveniently be called "moderately stratified," reserving the use of "unstratified" to quite unbedded deposits. When both the individual materials are aligned in parallel planes and conspicuous beds are present, "well stratified" is an appropriate designation.

Stratified beds may be classified as horizontal, inclined, cross-bedded or undulating (Fig. 24). Horizontal beds are commonly associated with standing waters, stream gravels, or water-laid sands deposited during laminar flow of uniform velocity. Sheets of aeolian sand or loess may also be horizontally stratified.

Inclined beds may be found in certain slope deposits such as in *éboulis ordonnés*. They are more common however in sand dunes, delta beds, and stream banks. Sand dunes commonly show gently inclined *backset* or *topset* bedding planes dipping to the windward with steeply inclined *foreset* beds in the lee (Fig. 25). The backset and topset beds are derived from particles rolling or bouncing up the slope, the foreset bed from particles blowing or falling down the steep ice face. Delta beds are laid down by streams into standing waters at the edge of a sea or a lake. Foreset beds are most conspicuous, although lenses of topset beds, dipping gently seaward, are frequent (Fig. 26). Seen parallel to the direction of stream movement, stream bank deposition also produces backset and topset beds inclined toward the center of the channel and commonly wedging out into alternating fine and coarse strata at the margins (Fig. 27).

Cross-bedding refers to complex patterns of discontinuous foreset beds with different inclinations, truncated by erosion and interstratified between topset beds. Cross-bedding results from alternations of deposition and erosion, with changes of velocity or direction of movement. It is confined to marine, stream and aeolian deposits. In turbulent streams, rapid changes of local stream velocity produce localized erosional hollows which may be soon refilled by advancing foreset beds. Topset beds may be built up against raised surfaces present in the stream bed. Such

Figure 24. Common types of stratification: (a) cross-bedding, (b) undulated bedding, (c) inclined bedding, (d) horizontal bedding.

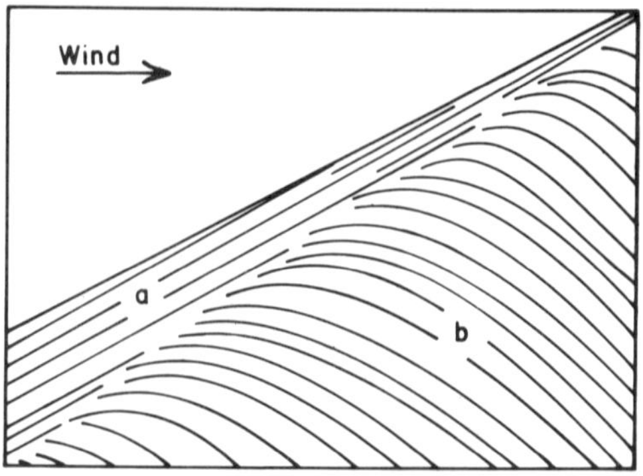

Figure 25. Aeolian bedding: (a) backset, (b) foreset.

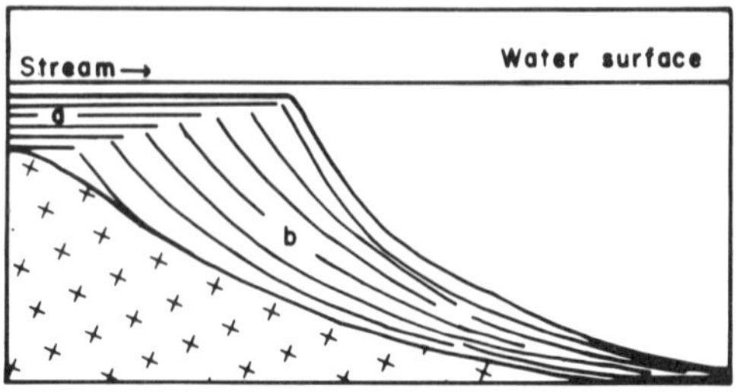

Figure 26. Deltaic deposits showing: (a) topset and (b) foreset beds.

alternating horizontal and inclined strata are a common product of swift, irregular flow, and may be called *current bedding*. Cross-bedded sand dunes are mainly confined to situations with several major wind components, and are rare where one wind direction is dominant.

Finally, *undulating* beds are most commonly exposed in sections

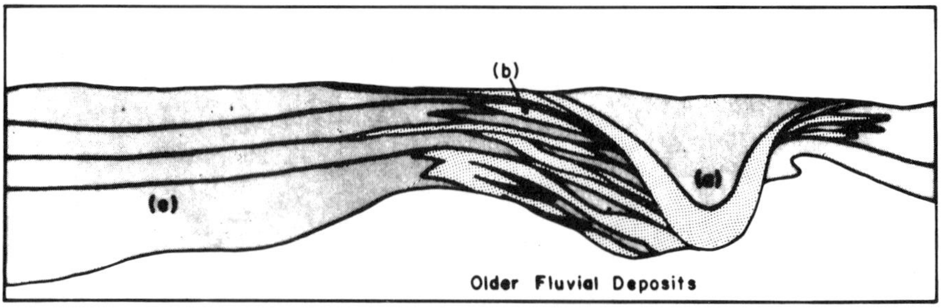

Figure 27. Ancient floodplain deposits: (a) channel bed filling, (b) levee backset and topset beds, and (c) fine, mainly horizontal beds of the alluvial flats. Older fluvial beds below. Sandy beds stippled, silt and clay dark.

perpendicular to beds with ripple marks. Only those found in stream deposits are of interest here. According to Cailleux and Tricart (1963, pp. 324 ff. and 334), undulating ripple bedding is characteristic during phases of rapid increase in stream velocity. Beds are inclined downstream as a product of advancing sand bars. With further increase of velocity, horizontal beds are laid down by laminar flow of water. If the velocity is great enough, pebbles may also be laid down horizontally, gently inclined upstream. As the stream decelerates, following a flood peak, undulating ripple beds may form once more.

The thickness of distinctive stream beds further provides information on stream velocities (see Cailleux and Tricart, 1963, p. 334): thin, extensive beds—up to 2-5 cm. thick—are commonly laid down during periods of quiet or moderate flow; thick, localized beds or lenses—exceeding 5-10 cm. in depth—during times of violent, torrential flow.

Sorting. The various size components of clay, silt, sand, gravel, and cobbles are seldom present in equal proportions. Strong winds will sweep up both sand and dust, depositing the coarser grades first and closer to the source of deflation. The finer particles will be carried farther. Streams will deposit coarser materials at a particular locality during periods of strong flow, while the fines are swept further downstream. During periods of gentle flow, fines may be laid down exclusively. Only glacial till is laid down without any regard to size.

Sorting as to dominant or average particle sizes may shed further light on the medium of deposition. Till commonly shows rocks of all sizes chaotically intermixed in a finer matrix. Many slope deposits are equally unsorted. Aeolian deposits are remarkably homogeneous on a local scale, due to careful wind sorting. The same applies for most deposits in standing water. Stream beds are more intermediate. Although seldom unsorted, stream-laid deposits may lack distinctive horizons of clayey, sandy or pebbly beds if the rate of deposition is moderate and subject to little variation. Strong episodic or seasonal fluctuations of waterflow commonly produce noticeable sorting into conspicuous beds of different particle size and color. On this basis it is possible to distinguish between "unsorted," "moderately sorted," and "well-sorted" beds (Fig. 28).

A particular bed is *graded* bedding, where a marked and progressive decrease in particle size—usually from pebbles to sand or silt—is apparent between the base and top of a bed. Such graded beds have abrupt contacts, and result from flood-level erosion followed by deposition in waters of gradually diminishing velocity (Fig. 29).

Orientation. From a different dimension, orientation of stream channels or individual pebbles provide useful information. In the case of aeolian deposits, the orientation is given by the direction of dip of particular beds. This compass orientation is an average value for the responsible wind direction.

Pebble orientation may give equally significant information for stream, slope, coastal and glacial deposits (see Cailleux and Tricart, 1963, p. 289 ff.). During transport, individual pebbles may be rolled, in which case their major axis will be oriented perpendicularly to the direction of movement. In the case where pebbles are pushed by sliding, they tend to point nose downstream or downhill. All pebbles are affected by both kinds of movement, but the orientation of a pebble at any one time dominantly reflects either the one or the other form of motion. The *compass orientation* of the major axis of 100 pebbles in a horizon can be plotted in the form of a rose diagram (Fig. 30). The results may help determine contemporary flow direction at a point, and may permit differentiation of slope and stream deposits in marginal cases. In archeological sites this technique is equally useful for recognition of random scatters of rocks, possibly artificial, or in plotting stone patterns that may be cultural or natural.

Inclination or dip of pebbles has also been studied by Cailleux and Tricart (1963, p. 304 ff.) and others, but the results are less diagnostic (see Pettijohn, 1957, p. 250 f.).

Degree of Consolidation. Consolidation of most Pleistocene sediments is limited, except where calcareous cement is available to fill the pore spaces and interstices. Ferruginous and siliceous cements may also be present, either derived from without or through alteration of the

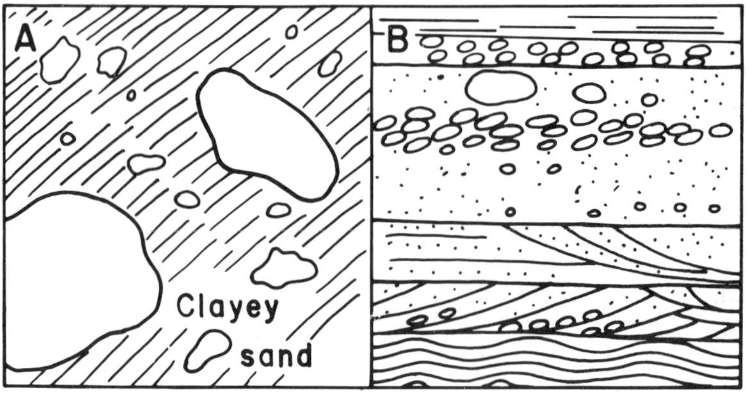

Figure 28. Till (A) and fluvial (B) deposits. The till is unstratified and unsorted while the stream beds are stratified and sorted, and partly current-bedded.

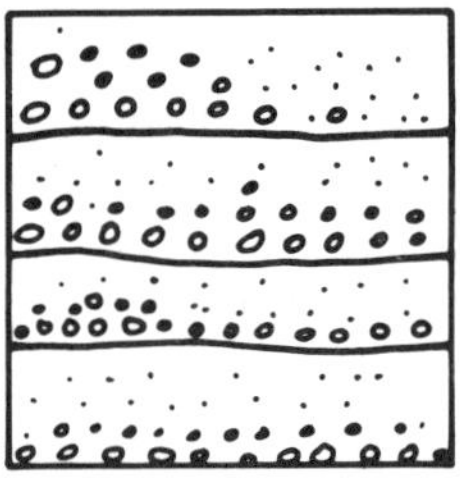

Figure 29. Graded bedding.

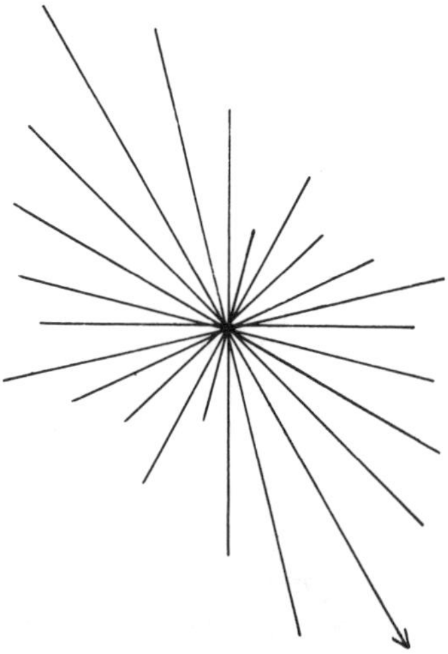

Figure 30. Pebble orientation diagram of pebbles displaced by solifluction. Slope inclination corresponds to vector with arrow.

sediment itself. The degree of consolidation attained can be described as follows:

a) Loose. Sample noncohesive in both wet and dry states.
b) Unconsolidated. Sample cohesive but soft or pliable in wet state.
c) Consolidated. Sample not soft or pliable in wet state, but edges fairly brittle.
d) Cemented. Sample can only be broken up by use of hammer, even in wet state.

SIMPLE QUANTITATIVE ANALYSES

Sand Size Analysis (wet-sieving). Loose, unconsolidated, or semi-consolidated sediments can be readily analyzed for over-all texture or particle size either in the field or laboratory.

A set of standard brass sieves with different mesh sizes, a balance sensitive to 10 mg., a pan, a source of water, and a dispersal agent (a detergent or mild hydroxide will usually do) are required. A 100 gm. sample is weighed in the dry state and often set in solution overnight to ensure separation of the fine silts or clays from the sand grains. It is subsequently washed through the set of sieves. Each sieve component is weighed after drying (in the sun or by use of hot plate or drying oven). The difference between the cumulative totals and the original sample weight belongs to the component finer than the closest mesh sieve. The coarsest sieve normally used is one with openings of 6.0 or 6.4 mm., since coarser sieves lack accuracy. The coarser components are generally considered separately. The finest sieve that can be realistically used for simple wet-sieving is a 0.063 or 0.060 mm. mesh. Many semimentologists prefer dry-sieving, without use of a dispersant, but employing a mechanical sieve-shaking device. Although simpler, accuracy is reduced if a considerable fine component is present.

The various size components used by sedimentologists vary, although two major classifications are rather well known. The modified Wentworth grade scale (Wentworth, 1922; see Pettijohn, 1957, p. 19) is most widely used in North America. It has the following logarithmic subdivisions:

boulders	over 256 mm.
cobbles	64-256 mm.
pebbles	2-256 mm.
sand	0.064-2 mm.
silt	0.004-0.064 mm.
clay	under 0.004 mm.

The nonlogarithmic, modified Atterberg scale widely used in Europe has slightly different nomenclature and size units. It has become the standard of the International Soil Science Society and the British Standards Institution. Fortunately the basic terms of both classifications can be used more or less interchangeably. The major classes are as follows:

cobbles	over 60 mm.
coarse pebbles	20-60 mm.
medium pebbles	6-20 mm.
fine pebbles	2-6 mm.
coarse sand	0.2-2.0 mm.
medium sand	0.06-0.2 mm.
fine sand	0.02-0.06 mm.
silt	0.002-0.02 mm.
clay	under 0.002 mm.

Angular materials coarser than sand may be classified differently, with grit (2-20 mm.) distinguished from detritus (over 20 mm.)

Depending on the importance of the gravel, sand, silt, or clay components, the textural classification of the sediment can be made according to Table 10 (after Wentworth, 1922). There is very little standardization of textural classes however.

In the case of consolidated or cemented deposits, thin sections are commonly examined on a calibrated grid by binocular microscope. This procedure is tedious and complicated.

With a little practice it is possible to approximate sediment or soil texture in the field. A little of the material is rubbed between the fingers in the dry state, examined with a hand lens and then tested for pliability and plasticity by rolling a small quantity in moistened condition (Table 11, modified after Franz, 1960, p. 103.)

Gravel Size Analyses. Mechanical analyses of gravel size can be made using round-meshed sieves or, for components greater than 6.0 or 6.4 mm., by simple measurement of major axes on graph paper. This measurement of pebble length can be used statistically or grouped according to classes. It can also be used to sort out gravel components for subsequent weighing.

In practice, gravel coarser than 6 mm. is best considered independently since much larger and rather heavy samples are needed for a representative count. The component below 6 mm. can be removed by simple dry-sieving. The coarser gravel can be simultaneously analyzed for shape while measuring.

Table 10. Textural classes of sediments.

Component	*Percentage*		*Class Term*
Gravel	>80		Gravel
Gravel	>sand	> 10, others> 10	Sandy gravel
Sand	>gravel	> 10, others>10	Gravelly sand
Sand	>80		Sand
Sand	> silt	> 10, others >10	Silty sand
Silt	> sand	> 10, others> 10	Sandy silt
Silt	>80		Silt
Silt	>clay	> 10, others >10	Clayey silt
Clay	> silt	> 10, others>10	Silty clay
Clay	>80		Clay

Table 11. Field approximation of texture (not valid for latosolic soils).

Texture	*In Dry State*	*Under Hand Lens*	*In Wet State*
Clay	Fine, homogeneous and very hard; greasy appearance	No sand grains visible	Very sticky and pliable, can be rolled into wire form
Silt	Not quite homogeneous; hard	Sand grains visible	Plastic but not very pliable (no wire is formable)
Sandy silt	Heterogeneous and somewhat brittle (clay mixed with sand)	Sand grains visible	Slightly plastic
Silty sand	Sand grains predominant		Too friable to be rolled out on hand
Sand	Almost exclusively sand grains		Does not stain hand

Morphometric Gravel Analysis. The pebble load of a stream or the detritus of a slope deposit is mechanically worn down and modified in shape during transport. Material moved down a slope will be largely rough and angular in shape, whereas material carried along a stream bed for several kilometers will be smoothed and rounded. Gravel rolled across the bed will tend to be squat; gravel pushed along the bed by sliding motions will tend to be flat.

Various indices have been devised for quantitative expression of gravel shape. Most widely known are those of A. Cailleux (see Tricart and Schaeffer, 1950; and Cailleux and Tricart, 1963, p. 259 ff.).[1] Two formulas are most commonly used:

$$\text{Index of rounding} = \frac{2r \,.\, 1000,}{L}$$

$$\text{Index of flattening} = \frac{(l + L),}{2E}$$

where r is the smallest radius of curvature on the circumference of the pebble expressed in cm., measured from the smallest, smoothed convex segment of the circumference:

l is the minor axis (width) of the pebble;
L is the major axis (length) of the pebble; and
E is the height of the pebble (Fig. 31).

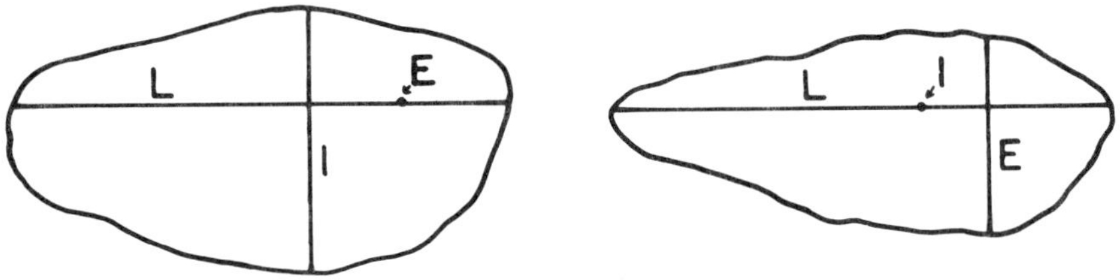

Figure 31. Length, width, and breadth in pebble measurement.

The index of rounding varies from 0 (completely angular) to 1,000 (a true sphere); the index of flattening ranges from 1 (a square cross-section) through 10 (extremely flat).

These classifications can be performed in the field and should be applied to samples of 100 pebbles, although smaller samples of 50

1. Other, rather different methods of analysis include those proposed by Wadell (1932) and Krumbein (1941).

pebbles can be effectively used for simple comparative analyses. Marianne Blenk (1960), providing the most intensive evaluation and overview of such techniques to date, has shown that statistical improvement by use of 200 rather than 100 pebble samples is negligible. Samples must be selected from materials of similar or identical bedrock, which must be specified in the description. Pebbles should be between 2 and 10 cm. in length.

Several modifications to Cailleux's system have been suggested. Blenk (1960) has reviewed their applicability and found that none of them provide a significant improvement. An analogous but different system, however, was suggested by Lüttig (1956, also 1962). It provides (*a*) greater mechanical simplicity, reducing the time required in measurement by almost a half, and (*b*) considerably reduced error and individualistic differences of measurement by different workers. The writer has used both techniques and considers the technical accuracy of Lüttig's index of rounding to be greater than Cailleux's. Its theoretical base is sounder and its possibilities of mechanical differentiation greater. Indices of Lüttig applied to flattening are as accurate although much simpler than Cailleux's. The Lüttig indices are as follows:

Index of rounding (ρ), expressed as per cent of smoothed, convex circumference of a pebble, obtained by careful visual estimation;

Index of flattening (π), as E/L, expressed in per cent, which may be read directly from a graph-paper chart. According to suggestions by Blenk (1960), the ratio E/l, also expressed in per cent, can be used as an auxiliary tool to express the degree of flattening.

Making use of the basic Lüttig indices, the following procedure of morphometric gravel analysis is suggested here:

a) Mean sample value of ρ, employing the following classes:

(ρ) 0-10% – angular
11-20% – subangular
21-40% – subrounded
41-60% – rounded
over 60% – well-rounded

b) Homogeneity of the ρ *values of the sample.* The coefficient of variation of the sample $CV = 100\sigma/\text{mean}$, where σ is the standard deviation, can be introduced as follows:

(CV) 0-25% – very homogeneous
25-50% – homogeneous
50-75% – heterogeneous
over 75% – very heterogeneous

c) Detrital component. The percent of pebbles that have undergone very little transport (ρ values$\leq$8 per cent) provides useful information on the significance of slope or local rubble in the sediment.

d) Transport motion. The sample averages of *E/L* or *E/l,* or both, provide information on mechanical transport by sliding as opposed to rolling motions. The following limits may be suggested:

E/L	*E/l*	
under 50%	under 65%	sliding motion dominant
50-60%	65-75%	both sliding and rolling
over 60%	over 75%	rolling motion dominant

e) Average pebble length. The average value of *L* provides basic reference data for the above measurements, as well as quantitative data of gravel size distribution.

f) Mechanically fractured pebbles. Originally subrounded to well-rounded pebbles with fresh fractures may be counted. More often than not they imply frost-weathering, although such pebbles may in part be the result of transport or weathering *in situ.*

The lithography of pebbles considered for morphometric analysis requires a little more attention. Quartz, quartzite, chert, and flint are next to useless, since they are commonly derived from older deposits and hardly amenable to effective shape modifications during reasonably brief periods of transport. Absolute lithological uniformity of material is not required, and it is more important to differentiate according to the degree of induration of sedimentary rocks rather than to their type, e.g., limestones or sandstones. Most igneous and moderately metamorphosed rocks may be safely grouped together, as they behave rather similarly, with exception of the fissile metamorphics.

As a critical evaluation it should be remembered that morphometric gravel analyses do not provide absolute results. Rounding, for example, is a function of transport distance, and only indirectly of stream competence or climate. Yet angular gravels are largely confined to slope deposits and to dry stream beds of the arid zone. Rounded gravels are found in most large rivers and in the smaller rivers of humid lands in general. Surface corrosion of exposed limestone fragments may also produce rounded edges together with vermiculate ridges and pock-marks. Interpretation of a gravel analysis from a Pleistocene bed can best be made after comparative analyses of modern bed materials at the same locality. It is a comparative technique, designed to contrast past and present stream transport conditions in the same stream.

The homogeneity of a gravel sample is useful in understanding the regularity of stream flow and possible lateral intermixture of colluvial or slope components in the bed load. Further information on local slope derivatives is provided by the detrital component.

The major mechanical component of pebble transport is less indicative of climate than of stream bed features. Sliding motions will be insignificant in the case of colluvial gravels, but will dominate on the beds of larger streams, particularly with strong and uniform stream velocities. Insignificant streams of dry regions can motivate considerable rolling of gravel during rare floods. A sandy stream bed will impede rolling, whereas a rocky stream bed favors it. Finally, rocks with pronounced bedding planes or fissile cleavage will flatten rather more rapidly than massive rock types.

Pebbles freshly fractured after initial stream transport may be predominantly, although never entirely, attributed to frost action. Comparative statistics on such fractured pebbles in beds of differing age, but within the same stream valley, may allow crude approximation of comparative frost significance.

In overview, morphometric gravel analysis is by far the most significant single quantitative technique in stream sedimentology. It may permit:

a) accurate quantitative description of sediments;
b) comparative analysis of transport capacity and, indirectly, of precipitation effectiveness;
c) insight into the mechanics of bed transport;
d) differentiation of fluvial, colluvial, or slope components within heterogeneous beds; or, also, identification of the dominant transport agent in the case of dubious beds;
e) rough estimation of the comparative significance of frost.

Gravel Petrography. Hydrographic changes within a drainage basin frequently produce changes in the petrographic composition of gravels. Study of gravel petrography may therefore help determine the former dimensions of the catchment area or the hydrological significance of different tributaries within the drainage system. One or two hundred pebbles may be classified and counted out. A procedure has been described by Zeuner (1932).

FIELD DESCRIPTION OF PALEOSOLS

General. Many Pleistocene sections expose horizons of humification or weathering well below the modern soil profile. Such buried or *fossil* soils deserve particular attention. They may have developed during periods of slow or interrupted deposition, or they may represent a long period marked by a hiatus in the sedimentary record. Many fossil soils have been "truncated," i.e., have lost part of their upper profile through erosion.

Relict soils are the product of a somewhat different environment, but they are exposed at the surface. They may or may not show evidence of more recent pedogenesis. Most of the braunlehms and rotlehms found in temperate Europe are examples of such relict soils.

Other vestiges of ancient soil development are provided by soil *sediments*—older soil materials that have been eroded and redeposited by stream, gravity, or wind action. They may be buried or exposed at the surface, analogous to relict soils. A good number of alluvial and colluvial sediments are in fact partly derived from older, nonfunctional soils. Their interpretation is more difficult, although possible (see Kubiena, 1954b).

The concept of ancient soils or *paleosols* includes fossil and relict soils as well as soil sediments. Each provides paleo-environmental information, so that every zone of discoloration or abnormally fine texture in a sedimentary profile should be examined for evidence of weathering *in situ* or from derived soil products. Much of the necessary analysis and a provisional interpretation can already be made in the field by the qualified earth scientist. Further laboratory studies are commonly required for final interpretation. Persons not familiar with soils can remove samples at selected vertical intervals from the top of a particular stratum downward to the "normal" material at its base. A color photo with a scale object can provide other relevant data for a specialist consulted later on.

Pertinent information in soil documentation has been outlined by Franz (1960). It includes—as far as possible—a description of both the location (including terrain, soil-moisture, climate, vegetation, and bedrock) and the soil profile. Profile description includes (a) soil horizons and general profile, (b) soil color, (c) texture, (d) carbonate content, (e) humus type and amount, (f) structure, and (g) concretions, stains, flecks, etc. Most of these characteristics have been discussed in chapter 6 but color, carbonate content, structure, and concretions deserve further attention here.

Soil Color. Accurate color description is not only important for soil horizons but for archeological layers and sediments as well. The *Munsell Soil Color Charts* have rapidly attained international status and are to be recommended. Colors are compared between a quantitatively organized scheme of color chips and natural, fresh soil or sediment surfaces. Moist or dry samples may be used, although these should be specified, as the results are by no means identical.

Colors diverging from those of the parent material are commonly due to soil development. Grayish to blackish colors are often produced by organic materials such as humus and charcoal. Black flecks in the *B*-horizon frequently indicate the presence of manganese oxide, a result of moderate seasonal waterlogging. Grayish or greenish horizons over parent material may indicate reduction within the permanent water table (*G*-horizon). Whitish colors indicate the presence of carbonates, gypsum (calcium sulfate), salt, or bone ash (calcium phosphate). They may also record a bleached A_{2e}-horizon. Brownish to yellowish colors commonly suggest limonitic iron compounds due either to weathering in (*B*) or *P*-horizons, or to illuviation in sesquioxide horizons. On the other hand, brownish surface horizons may be the result of both humification and weathering. Lastly, reddish colors may indicate the presence of anhydrous iron or hematite in the soil profile.

In review, the depth and color of blackish or brownish surface horizons commonly reflect on the intensity of humus accumulation, while the depth and intensity of yellowish or reddish discoloration of the subsoil commonly indicate the intensity of weathering or illuviation. Color is frequently used to distinguish *A*- and *B*-horizons in the field: the *B*-horizon must be one Munsell unit redder and one unit brighter than the *A*-horizon.

Carbonate Content. Although carbonate content cannot be determined quantitatively in the field, a reasonable estimate is possible through spraying a vertical column with a 20–25 per cent solution of hydrochloric acid (HCl). In the case of a dry profile it may be preferable to wet the section first, so that released soil air does not simulate effervescence. High calcium carbonate contents are recorded by increased reaction. The following qualitative description of reaction has been suggested by Franz (1960, p. 233):

a) no reaction;
b) audible reaction only;
c) brief visible reaction;
d) conspicuous, persistent reaction;
e) strong effervescence.

Since the reaction of different aggregates in a single sample may be different (e.g., finer or coarser grained materials, concretions, oxidation stains), it is sometimes useful to check whether reactions are uniform. Differences of carbonate reaction in a sedimentary or archeological profile may help locate buried weathering horizons or occupation levels.

A representative sample fragment or two submerged in a vial with dilute HCl can provide further information. Many samples break down into their various particle-sized components in HCl. If this is so, the texture can be estimated with a little practice. Other samples may or may not break down, regardless of their reaction. Among the types of materials that do not break down are salts, noncalcareous silts or clays, and siliceous or ferruginous cements.

Structure. The mineral grains and humus of soils and finer grained sediments may be combined in various ways with respect to pore space and interstices. The resulting *structure* is largely a function of humus type and the kind and amount of clay minerals and is often preserved in paleosols. Some of the major structural types (Nikiforoff, 1941; G. D. Smith *et al.*, 1960, pp. 256–57) are as follows (Fig. 32):

a) Platelike: geometrical arrangement into horizontal laminae or sheets. This structure is common in poorly drained soils and generally attributed to periodic presence of groundwater.

b) Prismatic or columnar: breaks up into columnar units with well-defined vertical faces and angular or rounded caps. This type is commonly a result of dehydration of fine materials. Salts, gypsum, or carbonate precipitates may occur on the face of such cracks.

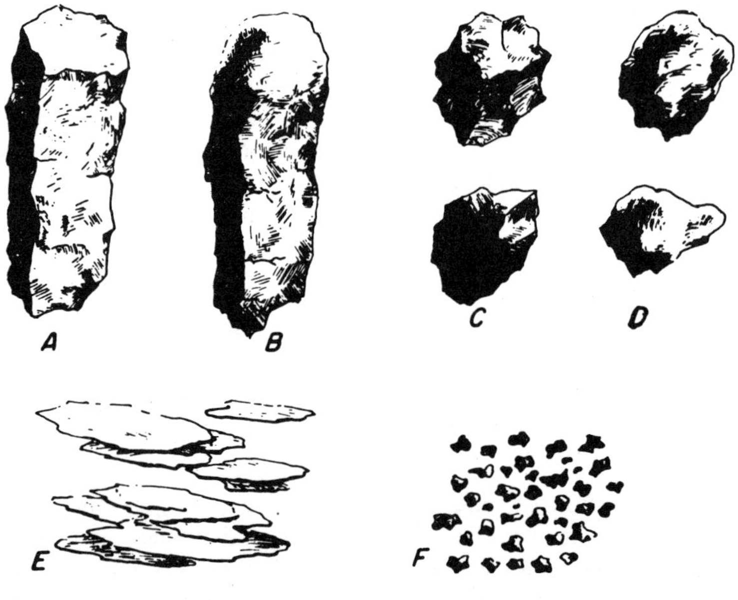

Figure 32. Some major types of soil structure (from G. D. Smith et al., 1960, by permission of the author): (A, B) prismatic and columnar, (C, D) angular and subangular blocky, (E) platy, (F) granular.

Prismatic structure is frequent in arid and semiarid soils, particularly when saline or alkaline.

c) *Block:* breaks up into variable blocklike, polyhedron-like or spheroidal aggregates, with all three dimensions of the same order of magnitude. Several important subtypes are distinguished:
   1) Blocky or *polyhedral:* densely fitted, regular, geometric fragments that may be refitted, having angular or rounded vertices. This structure is common in dense, nonhumic *B*-horizons, where it results from contraction of fine textured material. In other cases polyhedral structure is due to eluvial pore spaces in podsolized soils.
   2) *Granular:* loosely fitted, irregular, spherical fragments with relatively little pore space and little cohesion. Found mostly in friable soils, particularly in sandy materials and moder humus horizons.
   3) *Crumb:* loosely fitted, irregular, spherical fragments, conspicuously porous, and consisting mainly of earthworm excreta. It is commonly associated with mull humus and with most woodland agricultural soils.

Dry sediments or soils generally exhibit their natural structure when plowed up, exposed in vertical sections, and the like. Removal of samples by hand or pick generally leads to disintegration into the desired aggregates. Aggregate size of the individual structures varies considerably. In the case of platelike, granular, and crumb structures, the individual aggregates range from less than a millimeter to a centimeter in diameter; in the case of polyhedral structure, from less than 5 mm. to over 50 mm., in the case of prismatic structure, from under 10 mm. to over 100 mm. The dimension of structures in paleosols or fine-grained sediments may provide additional information for the soil scientist.

Concretions and Staining. Concretions are consolidated aggregates of rather irregular shape, with rough or jagged surfaces. They may be distinguished from smooth-surfaced, usually ellipsoidal *nodules*. Whereas nodules are usually found in bands in relation to the groundwater table, macroscopic concretions are more dispersed and may form as a result of upward or downward migration and localized concentration of salts, carbonates, and oxides of iron or manganese. Both may provide useful information, and should be recorded as to frequency, size, and constituent material.

Stains or flecks of color, mainly of iron oxides, may record seasonal or perennial waterlogging (*Fe*-, *P*-, or *G*-horizons).

Interpretation. Field study of the features discussed above, as well as of the humus and general profile (see ch. 6), can permit tentative recognition of soil horizons and the profile type of a paleosol. Identification of the soil type is, however, hazardous without further laboratory study. For a discussion of paleosol identification and interpretation, see Simonson (1954).

MECHANICAL ANALYSES IN THE LABORATORY

General. Although detailed discussion of a specific laboratory technique is well beyond the scope of this book, a brief outline of the requirements and purpose of certain basic analyses will be pertinent. For a general discussion of mechanical analyses the reader is referred to Kilmer and Alexander (1949), while various methods are described by Cailleux and Tricart (1963), Cornwall (1958, chaps. 10-13) and Thun *et al.* (1955, Part II).

Pipette and Hydrometer Techniques for Particle-Size Study. The accurate determination of all particle-size components (including the fine sands, silt, and clay) is imperative for paleosol examination and may be desirable for general sedimentology as well. The two most widely used techniques are the pipette and hydrometer methods.

In both cases the materials coarser than 2 mm. are removed, and the sample dispersed in a solution by use of distilled water and a dispersant such as sodium pyrophosphate ($Na_4P_2O_7 \cdot 10\ H_2O$). Contrary to the specifications of some published procedures, the organic and calcareous components should *not* automatically be removed.

The pipette method requires small sample sizes (10 gm. of material under 2 mm.), a decided advantage in some cases. The components under 0.06 mm. are determined through sampling a suspension at specified depths and times by means of an inserted pipette. The coarser grades are studied by wet-sieving.

The hydrometer method is somewhat simpler and also accurate, but requires a 50 gm. sample. The grade components are determined through measurement of changing density of the suspension as the sediment settles out, the coarsest materials first. A succinct description for the general reader is given by Cornwall (1958, pp. 128-130).

Heavy Mineral Analysis. The mineral components of a sandy sediment include a number of rarer and therefore often useful minerals, particularly those with a specific gravity exceeding 2.89. These are commonly rather resistant and may remain statistically representative even after the weathering of the sediment itself.

Heavy minerals are isolated through immersion in bromoform ($CHBr_3$), which has a specific gravity of 2.89. The "heavies" are then removed through filtering and studied by a mineralogist (see Cornwall, 1958, pp. 133-36; and Cailleux and Tricart, 1963, pp. 38-49).

The one-hundred-odd heavy minerals are in each case present in certain combinations in certain rocks only. As a result, bedrock source regions may be localized for alluvial or lacustrine sediments. Since the heavy mineral composition in different stratigraphic units is commonly distinctive, their study is sometimes vital for stratigraphic correlation over wide areas.

Quartz Grain Micromorphology. Microscopic analysis of the sand-size quartz grains after removal of other materials in a sample may disclose information on the history of the individual sand particles. Cailleux (1942) and Cailleux and Tricart (1963, pp. 54-103) have shown that quartz grains transported by moving waters in streams or at the beach are commonly well-rounded in shape and glossy or polished in appearance. In contrast to such *water worn* grains, materials that have been transported by wind action are crudely rounded and characteristically dull or frosted in appearance. These are *wind worn. Unworn,* fresh quartz grains are fully angular and rough in shape (Fig. 33). Sand grains 0.3-1.5 mm. in diameter are most frequently used for the purpose. One or two hundred grains are classified, and the relative composition of a sample may provide information on the genesis or the derivation of the sediment.

Quartz grain micromorphology has been overemphasized in the French sedimentological literature, and interpretations can only be made with reservation. Quartz sand is rather resistant and has commonly been modified and redeposited several times. The mystifying statistics provided by some unequivocal aeolian sands bear this out. In granite areas, where coarse quartz grit is abundant, many kilometers of stream transport are necessary to round off and polish quartz grains. Commonly this is only accomplished after several generations of transport, i.e., after

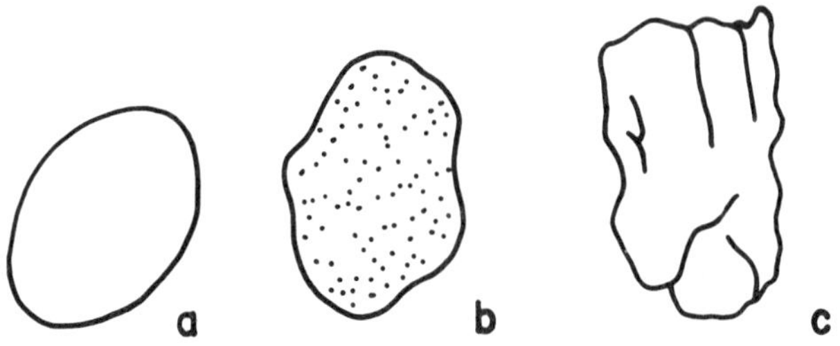

Figure 33. Micromorphology of quartz and grains: (a) water worn, (b) wind worn, and (c) unworn (modified after Cailleux and Tricart, 1963).

repeated derivation from older deposits. The resulting spectra are therefore not too meaningful.

Even more serious are the conclusions of recent experimental work by Kuenen and Perdok (1962), that frosting of quartz grains is due only in minor degree to mechanical action. Instead, chemical action by corrosive solution, or by alternate solution and deposition, is shown to be dominant, and true desert frosting is thought to be the result of alternating precipitation and evaporation of dew. In other words, only in some environments will frosting form, so that its relationship to wind abrasion is dubious (Butzer and Hansen, 1968, Appendix C).

Microscopic study of fabrics. Examination of soil or sediment aggregates under a binocular microscope may be done either with direct light or through transmitted light in a *thin section.* Thin sections are prepared through impregnation of incohesive materials, after which a thin, transparent section is cut out and mounted in glass (see Cornwall, 1958, pp. 141-51, for preparation procedures).

Microscopic examination of samples may provide the expert with a wealth of information on mineralogy and fabrics. The *fabric* refers to arrangement of minerals or microconcretions and to the absence or presence of colloids and precipitates in the intergranular spaces and conducting channels. Such micromorphological work has been developed by Kubiena (1938, with updating in subsequent books and articles) and is often basic in soil identification. Cornwall (1958, chap. 17) provides a brief but useful discussion.

CHEMICAL ANALYSES IN THE LABORATORY

The sample size required for most chemical analyses in the laboratory is small, and 50 gm. will usually be adequate for general purposes. Several chemical analyses are described by Cornwall (1958, chs. 14-16), while Jackson (1958) and Thun *et al.* (1955, Part III) provide detailed accounts of all the standard procedures. Only a few such techniques are of direct interest here.

*p*H values should be determined for paleosols and for the various strata of complex stratigraphic profiles. Samples are suspended in distilled water or a potassium chloride (KCl) solution and then measured by means of a electrometer. Colorimetric tests can also be made with indicator papers or fluids, but they lack accuracy and are at best useful for tentative approximations in the field.

Absolute calcium carbonate content can be readily determined in the laboratory by a number of similar devices which measure the volume of carbon dioxide released after application of sulfuric or hydrochloric acid.

Total carbonate content can also be determined by weighing sample loss after boiling with HCl and removing the solubles. The results may be valuable for recognizing weathering or secondary carbon accumulation in a soil or section. The amount of organic matter present in the soil is often of considerable interest, but analysis is difficult and the margins of error may be sizable.

Clay mineral determinations, now commonly made by X-ray diffraction (M. L. Jackson, 1964), may provide information as to the source of a fine sediment or about the intensity of chemical weathering. The principle involved is that none of the major clay minerals such as illite, montmorillonite, and kaolin have identical origins. Finally, the amounts of iron and aluminum sesquioxides and of colloidal and total silica may also be determined.

In concluding, some or all of the qualitative and quantitative, mechanical, and chemical techniques discussed and evaluated here may be employed to interpret and understand the depositional environment of a sediment, the genesis of a soil, or the setting of an archeological site. The sedimentological work can only be performed by a qualified earth scientist, but it is essential that the problems, possibilities, and techniques of study be familiar to both the biologists and archeologists with an interest in Pleistocene geography.

ELEVEN

Stream and Lake Sediments

STREAM TERRACES

The intensity and extent of alluviation in a stream valley may vary considerably in different environments. In the arctic barrens and tundra, streams are overloaded and deposit sediments along the length of their courses. Progressive water loss through evaporation and seepage along arid zone watercourses may also lead to alluviation. In the boreal forests large floodplains are characteristic, while in the temperate and tropical woodlands the rate and extent of downstream alluviation are comparatively limited. The savanna lands are somewhat exceptional through significant colluviation.

Seen in the perspective of time, the rate of alluviation—or the relation of downcutting to alluviation—also varies appreciably. On the annual basis, major deposition generally follows the flood-season discharge maximum. On a longer-term basis, many streams of the world also show distinct evidence of past periods of considerably greater stream alluviation. Largely responsible for this were Pleistocene climatic changes inducing major latitudinal shifts of climatic-geomorphic regions. During glacial intervals, tundra climates prevailed in many mid-latitude regions, and river cutting and alluviation were understandably accelerated. The effects remain conspicuous in the landscape today. In arid lands, shifts from drier to wetter climates led to overloading of streams with subsequent alluviation.

When a stream has accelerated its activity due to changes in sediment load, water volume, or river gradient, readjustment sets in the moment the aberrant impetus is removed. The stream re-establishes a form of equilibrium related to its gradient and transport ability, and will frequently cut down its bed to a lower and smaller floodplain. The older floodplain becomes obsolete and is separated from the new, functional floodplain by vertical escarpments forming *terraces*. Such alluvial terraces consist of benches, built of river deposits, remaining at the level of defunct higher floodplains (Fig. 34).

Assuming no changes of base level, the following idealized and simplified sequence may take place. At the start, the floodplain has a certain elevation and rate of deposition. Increased floodstage discharge with greater transport ability and load will lead to *(a)* more extensive flooding and consequent enlargement of the floodplain, with undercutting of nearby hill slopes and *(b)* a higher floodplain level due to accelerated deposition. The new floodplain, across which the river migrates horizontally, is broader and higher, and characterized by deposition of more and larger sized materials. When the volume and rate of deposition decrease to their original level, the stream will attempt to maintain its velocity—despite a decreasing volume—by shortening its course and thereby increasing the gradient. A straighter course is adopted, usually associated with a predominance of downcutting. The new floodplain will be smaller, and will be cut out as a limited section of the greater floodplain. In this way alluvial deposits are built up at various elevations and with a distinctive morphology. For general discussion of river terraces due to various origins see Cotton (1945) and Leopold *et al.* (1964).

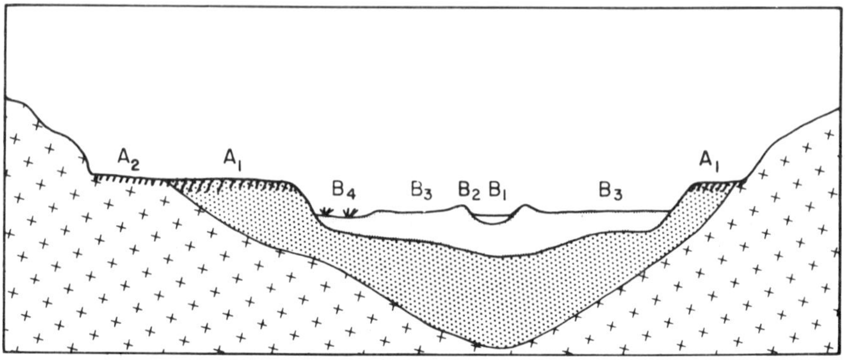

Figure 34. River floodplain with terraces: (A_1) alluvial terrace, (A_2) rock bench belonging to A_1, (B_1) low-water channel of modern floodplain, (B_2) levee, (B_3) alluvial flats, (B_4) backswamp. (Not to scale.)

ALLUVIATION DUE TO GLACIAL ADVANCE

A spectacular Pleistocene phenomenon of higher middle latitudes was the creation and advance of continental and mountain glaciers. One impact of these glaciers was stream alluviation ahead of the ice front.

During the summer ablation period tremendous water masses leave the glacier by watercourses below, within, or on top of the ice. As these issue in front of the ice they lose velocity because of the decrease in gradient and the diffusion of waters over a broad, level area. The considerable load is then rapidly deposited ahead of the ice front and eventually concentrated along drainage lines at greater distance. In the case of continental glaciers, broad fans of outwash contrast with the more localized valley alluviation downstream from mountain glaciers. The materials are coarse—mainly sand, gravel, and cobbles—and conspicuously bedded into alternating, sorted strata of finer or coarser materials.

Glaciofluvial deposits may extend for hundreds of kilometers ahead of the ice front, forming broad floodplains conspicuous along rivers such as the Danube, Rhine, and Rhone in Europe, and the Ohio and upper Mississippi in North America. The protrusion of a glacier into the headwaters of a stream is of more than landform interest. The outwash deposits can be employed as stratigraphic markers over wide areas. The bed materials often reflect rock types not present in the modern drainage basis, a criterion useful in distinguishing periglacial alluviation from glaciofluvial deposits. Foreign rocks or minerals may, of course be reworked from older glacial drift. Also the material size, stratification, sorting, or degree of water-rounding of pebbles in glaciofluvial and periglacial terraces are difficult to distinguish. In fact such deposits were often contemporary, intergrading at stream confluences. The only way, then, to identify glaciofluvial deposits with certainty is to associate terraces with moraines. This is possible through careful study of the geomorphology and sedimentology.

Once identified, glaciofluvial terraces may provide valuable chronological aids. Meltwaters actively deposit ahead of the ice front during both the advance and standstills of a glacier while deposition declines, sometimes to be replaced by downcutting, during glacial retreat. Consequently such terraces date glacial periods with considerable precision. Local geomorphic events, possibly relevant to a particular archeological site, may be stratigraphically associated with such a glaciofluvial terrace. This remains the principal application of glaciofluvial terraces to prehistory, since practically all of the sporadic implements or animal fossils found within them are derived from older deposits bulldozed by the glacier. No occupational levels have yet been recognized in glaciofluvial beds.

TERRACES DUE TO PERIGLACIAL ALLUVIATION

Throughout middle latitudes and at higher elevations, stream basins were repeatedly subjected to the processes of the periglacial zone in the course of the Pleistocene. The immediate results, apparently, were slope denudation by solifluction and subsurface washing, thus injecting great quantities of frost-weathered debris and soil into the overloaded streams (Büdel, 1944; Tricart, 1969). Reduced evaporation, stronger seasonal concentration of runoff in late spring or early summer, and no water percolation into the impermeable, frozen sub-soil each increased the vigor of seasonal stream discharge. Alluviation of gravels and sands was equally accelerated, while large boulders were often transported great distances by ice floes. Most humid mid-latitude stream terraces can be attributed to periglacial alluviation. The environmental implications are clear.

Stratigraphically, periglacial stream gravels are as important as glaciofluvial deposits. Theoretically, alluviation should continue for the duration of cold climate, i.e., the time span of the glacial advance and maximum. There is, however, good reason to believe that periglacial alluviation was limited to the periods of glacial advance and, in a general way, characteristic of each early glacial interval. So for example, the periglacial terraces of streams draining northward across Germany were overrun shortly afterward by the southward-moving glacier (Soergel, 1921; Grahmann, 1955, with references). Schaefer (1950) and Fink (1962) have convincingly demonstrated this stratigraphic relationship. It is assumed that the change of climatic environment, rather than the persistence of tundra conditions, initially provided great masses of frost-weathered debris and solifluction materials. Their production was apparently reduced during the glacial standstill (Büdel, 1950b). At any rate, less material was conveyed to the streams, and these reduced their rate of deposition or ceased alluviation altogether.

Periglacial terraces are of limited importance in lower latitude highlands. They are difficult to isolate from alluviation due to increased moisture and fluvial action.

PLUVIAL ALLUVIATION IN ARID REGIONS

Pleistocene alluviation was rather important in the lower latitude arid lands. Increased cold during glacial intervals certainly did not pass unnoticed in the dry subtropics. But fluctuations of the hydrological balance—related to precipitation amount, seasonality, intensity, and the rate of evaporation—had an even greater effect on stream equilibrium. The following discussion is modeled according to conditions in the Mediterranean lands and northern Africa. Interpretation of alluvial deposits in these areas has been attempted by the author (Butzer, 1963b, Butzer and Hansen, 1968, ch. 2).

The dry subtropical woodlands of the Mediterranean region have no complete mat of grassy vegetation today, and much bare soil is exposed everywhere. An increase in aridity in such areas (with 500-1,000 mm. annual rainfall) would not significantly increase the area of bare soil. A trend to drier climate would only reduce the stream runoff and the potential for erosion and deposition. In fact, modern Mediterranean streams originating in lowland catchment areas carry nothing but a few irregular waterflows per year. These seldom suffice to fill the dry stream valleys or *torrents* from end to end. In other words, lowland drainage basins with little gradient are almost defunct today. The active watercourses, responsible for accelerated erosion as a result of human interference, invariably have strong gradients and obtain their waters from highland drainage basins. These cannot, therefore, be considered fully characteristic.

An increase of rainfall in the lowland Mediterranean region would lead to flood erosion capable of transporting water-saturated soils as well as ready, mechanically disintegrated detritus. These would be carried into drainage channels and lowland basins to be deposited there. The stratigraphic evidence indicates that this was the case during the early glacial intervals (Butzer, 1963a). Pleistocene river terraces of small, "typical" Mediterranean torrents invariably contain much coarse material (indicating greater stream competence) and better rolled gravel (indicating longer-distance transport than today).

In semiarid grasslands the vegetative mat is commonly composed of contiguous sod or bunch grasses providing fairly efficient soil protection. A reduction of plant cover on hillsides during drier climate would permit accelerated runoff, resulting in soil stripping and headward erosion or gullying. Deposition would be localized in the lower valleys of major rivers (Bryan, 1941; Antevs, 1952). During moister climates an increase in the vegetative mat would reduce the intensity of runoff, and fine eroded materials would soon be deposited by slowly moving waters in the upstream parts of the drainage basin, leading to general alluviation.

In the deserts, stream activity is limited today and the vegetation can do little to prevent soil erosion. A rainfall decrease would therefore have no serious effect on the vegetative mat but would simply reduce the stream potential. An increase in rainfall would produce similar results as a change to moister climate in the Mediterranean woodlands (Cotton, 1945; Tricart and Cailleux, 1960–61, vol. 2, p. 142 ff.; Butzer and Hansen, 1968).

Interpretation of alluvial deposits in dry lands must be studied in relation to contemporary environments in each particular case. Greater rounding of gravel samples indicates greater transport distance and consequently, not only greater or longer waterflow or both, but also more runoff and a greater availability of moisture. Greater pebble size indicates greater erosive or transport capacity. Better stratification and moderate over-all sorting of beds may indicate perennial or seasonal, rather than episodic, flow. The polemic as to whether alluviation or downcutting indicates greater aridity or greater humidity, or whether alluviation upstream and incision downstream suggest semiaridity as opposed to aridity, or vice versa, is unnecessary. The deposits themselves reflect the conditions of deposition and tell their own story in relation to the contemporary balance of vegetation, runoff and erosion. On the basis of sediment analysis it seems that the greatest stimulus to arid zone stream alluviation is an increase in moisture in hyperarid regions essentially devoid of vegetation, or a decrease of rainfall amount or seasonality in semiarid regions with a closed mat of vegetation. In either case, alluviation is a temporary result of disruption of the delicate balance of erosion and deposition through a change from one form of geomorphic equilibrium to another. Soils, residual mantles, and detritus provided by one climatic balance may, with a trend to more effective or more violent runoff, be available for large-scale denudation and resulting alluviation of the entire stream channel.

It is symptomatic of stream deposits of northern Africa and of the Mediterranean area, for example, that terraces usually accompany the whole or most of the length of rivers, thus allowing no differentiation of "erosion upstream, deposition downstream" or vice versa. Instead, the differentiation of areal from linear erosion and deposition is more significant. The fact that many coastal streams aggraded their beds even during falling sea levels suggests that absolute loads are more significant than longitudinal distribution of complementary agencies.

Detailed sediment analyses of alluvial sands and gravels may then provide good paleo-environmental evidence. Various generations of alluvial deposits can provide local relative stratigraphies, especially when found in association with prehistoric assemblages, faunas, or floras. Particularly desirable, however, direct linkage of alluvial beds with the sequence of world sea-level fluctuations. If sufficient care is taken not to confuse the two effects, stratigraphic dating may then be extended to wider areas.

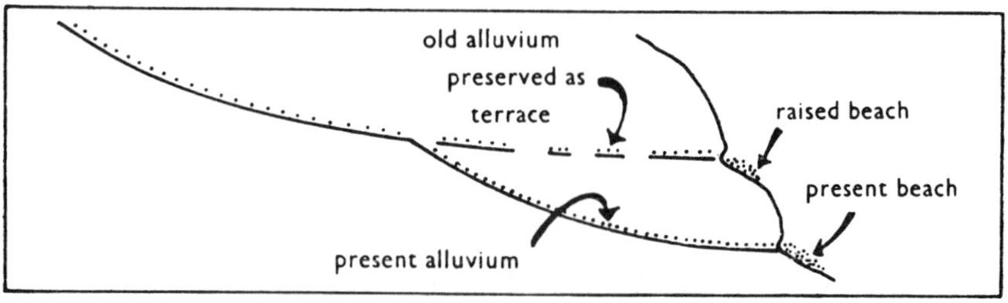

Figure 35. The effect of sea-level fluctuations on stream profiles (from G. H. Dury, 1959, by permission of the author).

COASTAL STREAMS AND SEA-LEVEL FLUCTUATIONS

Integrated stream systems drain onto a base level, either an inland lake or sea, or the ocean. If the base level is raised, the gradient of the lower stream is reduced together with the stream velocity and transport capacity. The theoretical result is increased deposition near the stream mouth. A gradual lowering of base level would increase gradients and velocities and hence favor downcutting. As discussed in chapter 14, world sea level fluctuated appreciably during the Pleistocene. During the glacials, sea level was 100–150 m. below that of today; during some interglacial periods, apparent sea level was somewhat higher. The complications for river alluviation around the world's coasts are considerable (Fig. 35) (see Baulig, 1935; Dury, 1959, p. 79 ff; Sparks, 1960, ch. 9).

Larger streams with weak gradients rapidly alluviated their lower courses and left terraces of fine alluvium graded to high interglacial sea levels (see Zeuner, 1959). Then during glacial-age cold climates, mid-latitude streams alluviated in response to tundra climates or the protrusion of glaciers into the stream headwaters. Woldstedt (1952) showed that cold-climate streams resorted to downcutting downstream in response to falling sea levels. This calls for reversals of erosion or deposition, upstream or downstream, between glacials and interglacials (Fig. 36). In the case of large streams, such factors are extremely difficult to separate. Plant and animal remains, fossil soils, or geomorphic phenomena are therefore necessary to distinguish "warm" and "cold" alluvia in some mid-latitude coastal regions.

In lower latitudes, however, pluvial rainfall during periods of falling sea level induced alluviation to well below modern sea level, even while the stream adjusted to a lower sea level. At any rate, frequent examples can be cited of coastal streams in the Mediterranean and Red Sea regions which actively alluviated during periods of low sea level (see Butzer, 1964a; Butzer and Hansen, 1968, ch. 8).

Generally speaking, complications in lower stream courses resulting

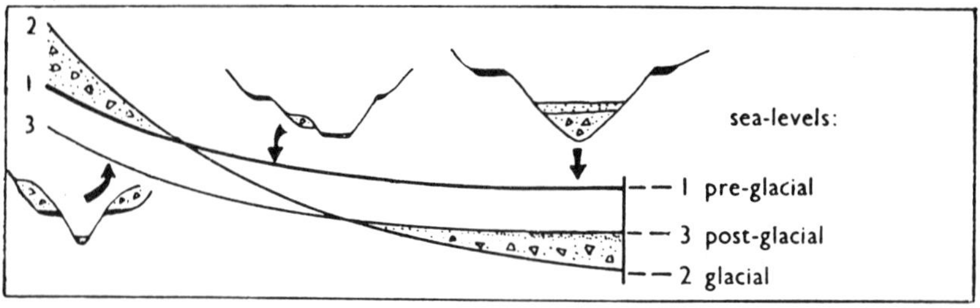

Figure 36. Profiles controlled by changes of sea level and of climate (from G. H. Dury, 1959, by permission of the author).

from Pleistocene sea-level fluctuations are common, although interpretative difficulties are surmountable in all but the larger streams. The precise interpretation of alluvial deposits of the coastal segments of larger rivers is a most difficult geomorphic problem (e.g., for the complex problem of Nile and wadi terraces in Egypt, see Butzer, 1959a; for the Mississippi terraces, see Woldstedt, 1960a).

Despite these difficulties, valuable stratigraphic information may be provided wherever alluvial deposits can be associated with high or low sea levels. Hence the Pleistocene terraces of many smaller rivers draining to the sea may be of great stratigraphic importance for correlation with the continental interiors.

CRUSTAL INSTABILITY AND STREAM ACTION

Earth movements, involving uplift or subsidence of mountains or coastal plains, tilting or other modes of crustal deformation with a stream basin, provide another possible stimulus to alluviation or downcutting. Although most geomorphic treatises of older vintage attribute river terraces to such tectonic disturbances of base level and stream equilibrium, only a few alluvial terraces have been convincingly demonstrated to be the direct result of crustal deformation. Such alluviation must be recognized in the individual local case. Although it is unjustified to minimize the significance of tectonic activity on stream profiles and equilibrium, it is reasonable to assume that few stream terrace sequences can be primarily attributed to local tectonic activity.

LAKE AND SWAMP BEDS

Lake and swamps beds have been laid down in standing waters and are more generally known as *lacustrine* deposits. They include:

a) evaporites, usually gypsum or salts;
b) calcareous beds, including chalk;

c) marls;
d) silts and clays;
e) sands;
f) organic deposits.

Some of the more typical physical aspects of these sedimentary facies may be outlined below.

Evaporites consist mainly of gypsum (calcium sulfate) and other salts such as sodium, magnesium, and potassium chlorides or sulfates. Such beds frequently indicate desiccation or lake shrinkage—periodic shrinkage during the dry season or long-term reduction of a larger lake to a lagoon or salt pan. So for example, the Pleistocene ancestor of the modern Dead Sea deposited some 50,000 banded alternations of silts (rainy season influx) and carbonates, sulfates, or chlorides (dry season evaporites) (see Butzer, 1958a, p. 78, with references). Evaporites, with the exception of open coastal lagoons, are indicative of some degree of aridity or at least of a high ratio of evaporation to precipitation.

Lacustrine chalks usually indicate perennial lakes which are not subject to very great seasonal fluctuations of oxygen content. Lacustrine chalks are common in many climatic zones. In temperate Europe they may be deposited organically by pond weeds; in dry areas such as the Sahara, inorganic precipitation is more important. Plant and animal remains are common in such beds.

Marls or calcareous silts are deposited both in lakes and swamps. The lime content may be derived through plant or inorganic agencies; the clays and silts represent soil products carried in by streams and rainwash. Common in humid and even semiarid lands, freshwater marl sedimention is commonly confined to comparatively small water bodies.

Silts and clays are generally carried into standing waters in suspension by local streams. They may occur wherever finer weathering products are available, from glacier-fed lakes in the arctic to spring-fed lakes in the Saharan oases. Lacustrine silts and clays are, however, most common in moister climates. In some cases, such as the "lacustrine loess" of the Persian Lut Desert (Huckriede, 1962), similar beds may be at least partly of aeolian origin.

Sands of lacustrine deposition are found most widely in areas with limited vegetation. Glacial meltwater streams feed sandy products to ice-margin lakes. In lower latitudes the widespread lacustrine sands of the Sahara were largely derived from sandy wadi deposits in the course of the Pleistocene. The prehistoric Chad and Fayum lakes of northern Africa are striking examples of lacustrine sands derived from direct stream influx as well as lake wave action on local sandstone bedrock.

Organic deposits, of many different kinds and complex origins, are most common in cooler latitudes although they are not quite unknown in the tropics and subtropics. The various facies will be considered in detail further below.

INTERPRETATION OF MINERAL OR SEMIORGANIC LACUSTRINE BEDS

Simple mineral sediments, such as evaporites, marls, silts, clays, and sands, as well as semiorganic deposits, such as chalks, may be of considerable archeological or paleo-environmental interest. The direct origin of such standing waters in localities which are dry today is chiefly the result of (*a*) moister climate (in the arid zone), (*b*) poor drainage (in humid lands), or (*c*) a rise of sea level (in coastal areas). Sedimentation may vary considerably from place to place and is necessarily determined by local conditions. In higher middle latitudes, lacustrine beds are mainly found in poorly drained areas of ground moraines dating from after the retreat of the continental glaciers.

Apart from particle-size and certain chemical analyses (carbonate content, *p*H, organic matter), biological studies usually provide significant paleo-ecological information. Combined with study of any macroremains of plant leaves, fruits, stems, or wood, the pollen spectrum offers an excellent picture of the regional setting. Pollen diagrams can be further studied from different strata to yield chronological information. Depending on the sediment, various other studies directed towards ecological interpretation of mammalian fauna, snails or mollusca, algae or diatoms, etc., may yield results of interest.

Each of these laboratory approaches provides vital complementary evidence to the more standard geomorphologic field investigation of the lacustrine sediments in a wider setting.

VARVE ANALYSIS

General. A particular study of interest to both archeology and geochronology is that of *varves*. These are annual, graded. bands of sediment laid down in glacier-fed lakes contiguous with the margins of continental glaciers. Detailed work by G. de Geer (1912, and later authors) on such annual sediment layers shows that a new load of sediment enters the lake in the wake of each spring's thaw. The coarser materials (mainly silts) settle down first while the fines (clays) gradually settle during the course of the summer. In larger lakes, wave motion may impede fine sedimentation until autumn when the lake surface freezes over. In numerous cases, fine sedimentation continues under the ice throughout the winter. When course silts or fine sands are deposited again during the succeeding spring, a sharp contact zone is formed, so enabling clear identification of the annual increment.

Further seasonal distinctions are provided through biological evidence. The coarse springtime accretion is generally dark and rich in organic matter, while the fine summer sediment is light-colored due to calcium carbonate precipitation. The late summer and autumn sediments are dark again. Pollen examinations of the upper dark layers have shown pollen sequences according to the time of blooming, while micro-organisms such as diatoms are concentrated in the light, summer segment.

The thickness of the annual deposit or varve varies from year to year depending on the course of the annual weather and its influence on the ablation of the nearby glacier. A warm year produces large varves, a cold year narrow ones. A requisite to the regular laminar sedimentation is the temperature contrast of warmer, inflowing waters and cold lake waters, whereby the sediment is distributed evenly over the lake bed. Such conditions are best met in ice-margin lakes. Attempts have been made to study annual, varve-lake evaporite sediments in lower latitudes, e.g., the Saki salt lake of the Crimea (Shostakovitch, 1936), the Dead Sea (see references in Butzer, 1958a, p. 78), and some lake beds of Kenya in East Africa (de Geer, 1934).

Teleconnection of Varves. De Geer first recognized that varve sequences were very similar between nearby lakes – within a kilometer of each other – on account of the similarity of local climate. On this basis sequences were correlated and extended in time from area to area. By following the various stands of the retreating ice front, De Geer established an almost complete sequence covering 15,000 years from the late Upper Pleistocene well into historical times. This provided a true chronology whereby glacial features related to the retreat and dissipation of the European glacier could be more or less precisely dated. For example, the close of the Pleistocene was fixed by the event of the draining of the Baltic ice lake, which, according to the varves, occurred in 7912 B. C. Radiocarbon cross-dating suggests that this date may be at most a few centuries off. During four decades De Geer's varve-chronology of Scandinavia remained an invaluable tool the significance of which for prehistory and geochronology is all too easily overlooked today.

Difficulties in the Varve-Chronology. Within Fennoscandia the varve-chronology, as established by De Geer (1912, 1940) and Sauramo (1929), has in part remained a respectable body of evidence. It has been shown, however, that storms create multiple varves annually in *shallow* lakes through addition of extra influx and the stirring of sediments (Hansen, 1940). As most of the lakes south of the Fennoscandian moraines, dating about 9000 B. C., are shallow, the earlier chronology is now considered doubtful. Most of the lakes north of this line, extending across south-central Sweden and southern Finland, are quite deep and, combined with Sauramo's more conservative approach to the Finnish chronology, there is little ground for serious criticism here.

The establishment of varve-chronologies outside Scandinavia, as attempted by Antevs (1925) in North America, has not been very successful. A major reason for this failure has been extrapolation of sequence segments over hundreds of miles. World-wide correlations of a frivolous type were attempted later whereby reversed seasons in the northern and southern hemispheres, of nonglacial characteristics of varves, have been simply ignored. These attempts have discredited the varve method and, generally speaking, other techniques have now replaced the varve-chronologies everywhere except in Fennoscandia.

SPRING DEPOSITS

Although rather localized in their occurrence, many springs in limestone areas have promoted a class of particular calcareous sediments occurring in direct association with the spring or in adjacent stream beds or lacustrine basins. Two major kinds of deposit are formed by evaporation of, or precipitation in, lime-charged waters:

a) *Tufas,* formed through precipation of cryptocrystalline calcite on growing plants, leaving an inhomogeneous, spongy, porous and often brittle rock. The stems, grass blades, and leaves are preserved as open casts, as a dense calcite replacement, or as a partial cast-filling of lime sand. The filling between the plant structures is commonly soft and very porous, consisting of cemented calcite sands.

b) *Travertines,* dense, banded cryptocrystalline calcite occurring as dripstone in caves, or as horizontal beds within and outside of cave environments. Cave travertines are further discussed in chapter 13. The external travertines can be further subdivided into two types:

1) "True" travertines, precipitated as horizontal, undulating, or bulbous bands, commonly alternating from dense calcite, crystallized with columnar structure, to porous calcite with little or no macroscopic crystalline structure.
2) Sedimentary crusts (the *croûtes zonaires* of the French authors), consisting of fine, wavy laminations of cryptocrystalline calcite. These are frequently confused with *Ca*-horizons of the soil zone.[1]

1. Variously called calcareous crusts, tufaceous or travertine crusts, and caliche. For detailed discussion of these controversial features in Algeria, see Durand (1959, pp. 75–136); for Morocco, Gigout (1960, pp. 91–129); for the Balearic Islands, Butzer (1963b); for the Libyan Desert, Butzer and Hansen (1968, ch. 7). For a general analysis, see Tricart and Cailleux (1960–61, vol. 2, p. 147 ff.).

Fossil crusts of this kind are widespread in the arid, semiarid, and subhumid subtropics. Their occurrence may indicate either moister or drier paleoclimates, depending on the situation. In subhumid climates with acid rocks, sedimentary crusts presumably record drier conditions. In arid or semiarid limestone areas, they probably indicate greater spring activity and more abundant moisture. Such travertines should preferably be associated with defunct springs or lacustrine beds when such an interpretation is made.

INTERPRETATION OF ORGANIC SEDIMENTS

Standing waters, partly closed in by plant growth and not subject to mechanical water turbulence, tend to develop an oxygen deficiency. This may be the case for deep waters with limited vertical stirring or where the water supply is either poor in mineral plant nutrients or acidic as a result of base deficiency in the catchment area. In such moderately acidic waters, organic precipitation of humic solutions known as gel mud or *dy* ("sedimentary peat") takes place. In extreme cases of oxygen and nutrient deficiency, anaerobic conditions and reduction produce ferric sulfide, hydrogen sulfide, and methane through lack of, or limited oxidation of, plant materials. This bad-smelling, clayey humus is known as *sapropel.*

Where partly overgrown water bodies are sufficiently well aerated and supplied with nutrients, sediments are rich in organisms such as diatoms, and the plant and animal remains are partially decomposed. The resulting gray to gray-black, occasionally brown, sediment recalls a highly humic marl, and is known as a *gyttja.*

The above forms of organic deposits are essentially lake types. Emergent plant growth along the margins of such water bodies is limited to the shores and shallower waters. Plants with submerged roots—reeds, horsetails, and water lilies—inhabit water to about a meter in depth. On the immediate shore, plants such as sedges and rushes are found on wet ground, but with their roots out of the water. Since the supply of dead organic matter is great in the reed bank zone, and since wave action and oxygen distribution are effectively impeded by root and stem networks, oxygen is insufficient to enable complete decomposition. Consequently, a net accumulation of organic materials progressively narrows the open water surface, and is followed by colonization of the peaty shores by swamp plants. A botanist can approximate the depth of water associated with peaty swamp deposits, and may be able to reconstruct the general ecological setting in some detail.

When such a lake has finally been reduced to a swamp, the ensuing deposition may be entirely organic, without mineral matter. This is known as *peat.* A distinction is usually applied between a *low moor,* at or under the water level, and *high moor,* above the water level. The low moor harbors reeds, sedges, horsetails, and water-tolerant arboreal species such as alder and willow. When peat accumulation rises above the groundwater table, the moisture supply is largely derived from rainwaters, which are notably lacking in nutrients. The resulting highly acidic environment leads to rapid growth of sphagnum or bog moss, heather, and cottongrass, while the deciduous trees are replaced by pine and birch. Ultimately the pines, as the last tree species, are also replaced by peat mosses.

The colonization and ultimate disappearance of innumerable late glacial lakes and swamps on the Würm-age ground moraines of northern Germany have been systematically studied by Overbeck (1950). According to Overbeck the lacustrine sequence typically began with deposition of sands, silts, and clays in topographic depressions by the meltwater streams of the retreating continental glacier. As herbaceous vegetation recolonized the areas just abandoned by the ice, simple mineral sedimentation was gradually replaced by deposition of thick beds of *dy* or *gyttja.* Rapid increase in warmth favored a more luxuriant vegetation whereby the lakes were reduced to bogs, low moors, and eventually high moors.

The present world distribution of ancient organic beds or contemporary bogs is largely confined to *(a)* areas overrun by ice during the last glaciation, and consequently suffering from disrupted drainage, *(b)* low-lying coastal areas with a correspondingly high water table, and *(c)* poorly drained mountain localities with high rainfall. This limits the distribution of older organic deposits to higher middle latitudes, with the exception of occasional coastal or highland bogs in subtropical latitudes. More recent bogs are also widespread in the tundras.

The interpretation of organic deposits for the purposes of archeology and Pleistocene geography is only to a limited extent a matter of geomorphic investigation. Detailed analyses are mainly concerned with organic materials and are therefore performed by the biologist. On account of their acidic, anaerobic environment, which enables them to preserve most organic materials—ranging from flesh and bone to plant materials—bogs are of exceptional paleo-ecological value. Paleobotanical studies consequently provide detailed environmental and chronological information. Geomorphic investigation is confined to interpretation of the lacustrine setting, as well as to sedimentation features such as disconformities, burnings or peat cuttings in the depositional record.

TWELVE

Wind-borne and Slope Sediments

WIND ACTION

Erosion by wind is limited to dry, loose, and fine-grained sediments not protected by a plant cover. Under natural conditions, wind erosion will be more or less limited to the arid zone and high arctic barrens, except for locally favorable areas: broad sandy beaches and exposed stream or lake beds during low water. Particles in the silt or fine and medium sand size (under 0.2 mm.) are carried in suspension by stronger winds. Coarse sands are moved by saltation. Such grains are picked up by local micro-turbulence and then deflected downwind. As they strike the surface their impact may cause them to rebound or to start saltation by other grains (Fig. 37). Wind-driven coarse sands are responsible for any and all wind abrasion of the surface. Such abrasion is limited to the lowermost meter of the atmosphere, although rarely significant above 50 cm. A secondary effect of saltation is general disturbance of the surface sands which are motivated to a steady forward motion through rolling, sliding, or low saltation. This is known as surface creep.

Transport of the suspended load, consisting of silts and finer sand grades, is effected over long distances. During strong dust storms, great masses of aeolian materials may be carried over hundreds of kilometers, only to be deposited very slowly in response to decreasing wind velocities, or more rapidly by being washed down by rain. Extensive aeolian sedimentation of silt and fine sand may then occur well outside of those environments suitable for wind erosion. The coarser sands of the bed load can move only along the ground, migrating as sand ripples, ridges, or dunes. These materials will ordinarily be confined to the general source region, with the exception of smaller dunes that migrate from the coast or along other local sources of sand.

Corresponding to the mode of transport of the different particle sizes, wind-borne sediments may consist of striking coarse-grained sand mounds or dunes, or of smooth, extensive sheets or mantles of fine-grained materials. The morphologically conspicuous, coarse-grained types are confined largely to the world deserts and the arctic barrens, whereas the sand or dust (loess) sheets may be deposited almost anywhere, although they only retain their structure and other characteristics when laid down in open country.

The localization of aeolian features in different environments is summarized in Table 12. General discussions of wind processes are given by Bagnold (1954), Thornbury (1969, ch. 12), Flint (1957, ch. 10), Sparks (1960, ch. 11) and Tricart and Cailleux (1960-61).

FEATURES OF WIND EROSION AND ABRASION

Pavement. One of the more common forms of aeolian erosion is the gradual deflation of fine materials, leaving increasing concentrations of heavier grit, detritus, or gravel at the surface. Such a *desert pavement* or *lag deposit* forms extensive desert surfaces today—the rocky *hamadas* of disintegrating, angular bedrock and the pebbly *serirs* of deflated alluvial fans or weathered old conglomerates. On a much more local scale, buried stone concentrations in various Pleistocene deposits have frequently been identified as pavements, particularly if other evidence of aeolian action is also present.

Blow-outs. Wind scour and deflation may locally achieve more than simple denudation. Small depressions may be excavated repeatedly in dune fields or other types of sand accumulation. Such *blow-outs* may also develop in areas with poorly consolidated bedrock, such as the Great Plains (Judson, 1950; Wendorf, 1961) or southern Africa (Flint,

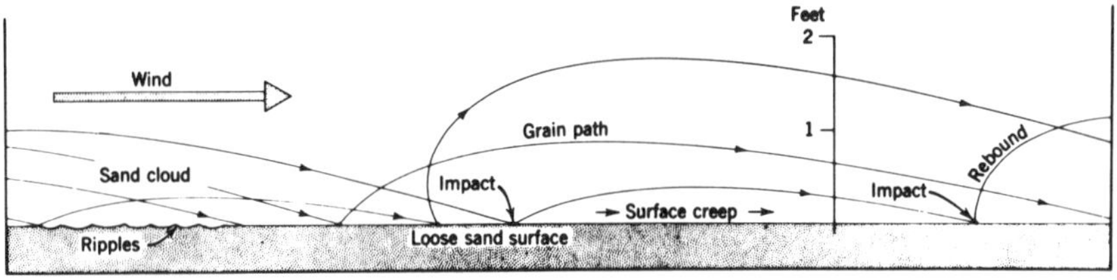

Figure 37. Aeolian transport (after R. A. Bagnold, from A. N. Strahler, 1960, by permission of John Wiley & Sons, Inc., New York).

Table 12. Wind activity in different environments.

Vegetation	*Wind erosion*	*Wind deposition*
Desert and arctic barrens	Active	Dunes and sand sheets
Grassland and tundra	Localized	Loess and sand sheets, some local dunes
Forest	None	Negligible
Littoral zone	Localized	Local dunes and sand sheets

1959), or even in consolidated bedrock, as in the Libyan Desert (Pfannenstiel, 1954). Whatever the origin of the desert hollow at start, or its dimensions today, a combination of chemical weathering and deflation is generally postulated for its development. Salt hydration or various forms of chemical weathering produce fine residual products in hollows where surface drainage collects and percolates or where the water table is high. Solution may also have been an important auxiliary factor in limestone bedrock. Periodic or long-term drying permits repeated deflation of accumulated silt and sand, leading to steady deepening of the hollows. The dimensions of such blow-outs in the Great Plains area are rather modest—as much as a few square kilometers in size, and up to 50 m. deep. The oasis depressions of the Libyan Desert may be a thousand square kilometers or more in area, and several hundred meters deep. Despite their size, the excavation of the Libyan depressions cannot be explained except through wind removal of loosened materials.

Yardangs. Whereas blow-outs are probably more or less exclusively due to deflation, a less common but equally conspicuous abrasional form occurs in nonconsolidated fine sediments. Elongated, **U**-shaped grooves or furrows, oriented with the prevailing wind, are separated by jagged hillocks or ridges of sand-scoured clay, silt, or fine sand. Known as *yardangs,* these forms are a product of both abrasion and deflation and may have a local relief of several meters. Uncommon in mid-latitude Pleistocene contexts, they play a prominent role in the archeology of the Sahara, where various prehistoric cultures are related to deflated and scoured fluvial, lacustrine, or spring-deposited silts (Caton-Thompson and Gardner, 1932; Butzer and Hansen, 1968, ch. 7).

Ventifacts. Wind does not erode only by deflating, pitting, grooving, scouring, but also by polishing and faceting. Pebbles may be polished and faceted on one or more sides by wind-driven sand. The *ventifacts* so formed require only a few decades, given strong winds, plentiful sand, and no vegetation. Ventifacts are common in lag deposits and provide corroboration for their identification. Cailleux (1942) has shown that ventifacts are common in many European Pleistocene beds, and Cailleux and Tricart (1963, pp. 216-41) give a number of techniques for systematic analysis of ventifacts. Multifaceted ventifacts are far more common in the periglacial environment than in lower latitude deserts. Cailleux attributes this to soil-frost heaving, whereby ventifacts may be repeatedly overturned and faceted on new faces.

BED LOAD DEPOSITS: SAND DUNES

Dune types. Dunal forms include migratory "free" dunes, whose existence is independent of topography, and "tied" dunes, related to some permanent wind obstruction. The free dunes include several types:

a) *Longitudinal* dunes or *seifs* occur in groups of long, parallel ridges, with many peaks and sags. They may be 100 km. long and over 100 m. high, lying parallel to the direction of strong winds. Their formation may be aided by local turbulence, leading to accumulation now on one side, now on the other.

b) *Crescentic* dunes or *barchans* are, as the name implies, crescentic in plan, the horns and steep concave slopes facing downwind (Fig. 38). Barchan dunes may attain 30 m. in height and 400 m. in width and length. They develop with unidirectional effective winds.

c) *Transverse* dunes form irregular, wavelike ridges at right angles to the effective wind direction, sometimes merging or occurring simultaneously with barchan fields. Some authors do not recognize transverse dunes as an independent type; others consider them as identical with the barchan. There are, however, several types of complex dune fields (as opposed to single dunes) of controversial origin that are quite distinct from merging barchan fields (see Tricart and Cailleux, 1961, vol. 2, p. 77 ff.).

d) *Parabolic* or **U**-shaped dunes are superficially similar to a barchan, but are more elongated and slightly asymmetrical, with the gentle, concave slope facing windward, the steeper, convex face down-wind (Fig. 39).[1] Merging parabolic dunes may form very narrow ridges running parallel with the effective winds, producing *hairpin* dunes.

The "tied" dunes include a number of types of which *lee* dunes are probably most conspicious. Often nondescript in appearance, they

1. Discussions of parabolic dunes are conspicuously absent in some basic textbooks.

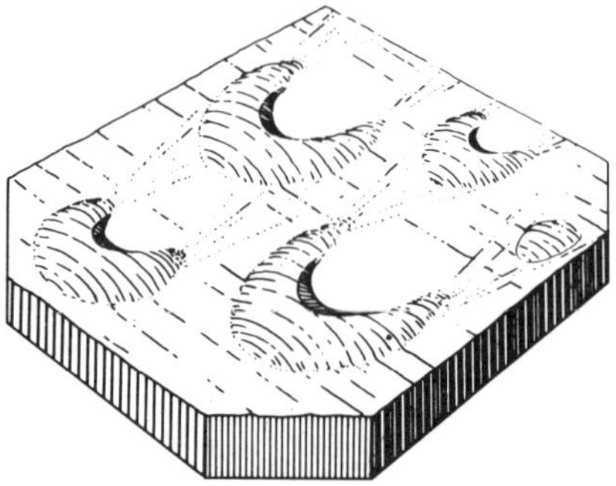

Figure 38. Barchan dunes (from A. N. Strahler, 1960, by permission of John Wiley & Sons, Inc., New York).

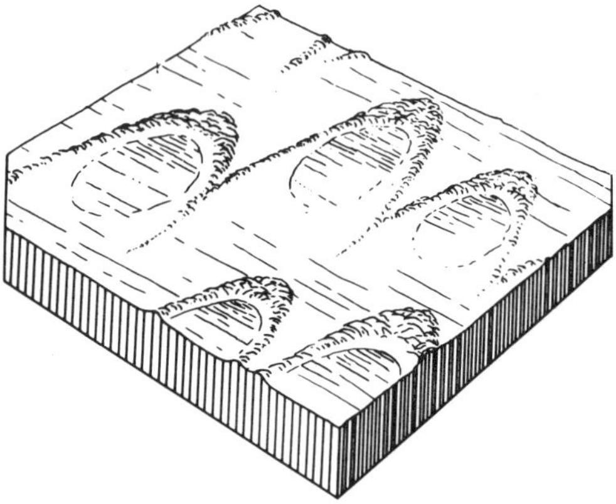

Figure 39. Parabolic dunes (from A. N. Strahler, 1960, by permission of John Wiley & Sons, Inc., New York).

develop in the lee of obstacles, frequently growing into longitudinal forms.

All of the above types belong to the class of interior or *continental* dunes. The somewhat different coastal or littoral types are discussed in chapter 14.

Sedimentology. Particle-size analyses are particularly useful in distinguishing various classes of aeolian deposits. Although there is usually a concentration of coarser sand around the dune crests, the grade size of continental dune sands is remarkably uniform. Medium sands (0.06–0.2 mm.) provide 20–90 per cent of the material, and there is an almost total absence of silts or clays under 0.02 mm. (see Tricart and Cailleux, 1960–61, vol. 2, p. 89 ff.). The constituent material is almost exclusively quartz, although a calcareous component may be present locally and rather exceptional gypsum dunes are known. Only in transitory small dunes is there, on occasion, a high silt or clay component.

In the case of fossil dunes, particularly where the morphology has been obscured or destroyed, dune and wind directions may be obtained from bedding. The compass bearing of bedding-dips determined from a number of localities gives the effective wind direction responsible for sedimentation. The *prevailing* winds may be gentle and from another direction than the *effective* winds which are sufficiently strong to shape the sand. The effective wind is then a storm wind that may or may not coincide with the average wind. Whereas effective wind directions can be obtained from the orientation or bedding directions of a dune, prevailing winds can be deduced from the position of a dune with respect to the source area of sand. Uniform layer-by-layer bedding suggests uniform effective wind directions, while cross bedding implies winds of variable direction. Uniformity of grain size throughout the sediment suggests that the effective winds had a relatively constant velocity, while sporadic coarse sand laminae may record periodic gales. Information on former effective winds can be compared with modern climatic records of strong winds and, if present, with recent dunes or wind-deformed vegetation.

Environmental interpretation. Although the exact origins or development of most dune types are still controversial or uncertain, the formative environment of the major types is clear.

Mobile longitudinal, barchan, and transverse dunes of moderate or great size are now limited to areas without vegetation. Localized fields of small, active dunes may move through vegetated terrain along the sandy beds of some streams in semiarid zones. But if no ready source of sand is available, these types indicate aridity and a lack of vegetation. The same applies to lee dunes.

Parabolic dunes more commonly indicate a subhumid climate with some, although sufficiently modest, herbaceous vegetation. They were once widespread in the European tundras and on the Great Plains of North America (Poser, 1950; H. T. U. Smith, 1965). They commonly form when the vegetation cover of extensive sand deposits is degraded by man or by increasing aridity (Tricart and Cailleux, 1960–61, vol. 2, p. 76 f.).

Barchan and parabolic dunes are generally thought to indicate rather uniform effective wind directions, whereas opinions diverge in the case of longitudinal dunes. Bagnold (1954) has given good reason to believe that the typical seifs with irregular, knifelike crests form where the effective winds are of different direction than the prevalent winds. Some of the simpler and smaller longitudinal dunes may, however, be a product of unidirectional winds.

In addition to providing information on vegetation, sand sources, and wind directions, dunes may also record effective wind velocities. So, for example, longitudinal dunes may be due to stronger winds than are any of the other types. Of greater potential are the parameters governing wind speed and sand size, a relationship differing according to whether the effective winds are unidirectional or not. Studies by Poser (1950),

Dubief (1952), and Hastenrath (1967) are of interest here, although much more experimental work is required.

SUSPENDED LOAD DEPOSITS: LOESS AND SAND SHEETS

Loess. Loess is a pale yellowish, unstratified silty sand, rich in vertical capillary structures. The material typically consists of quartz (60–70 per cent), carbonates (10–30 per cent), and clay minerals (10–20 per cent). The particle-size distribution shows a distinctive maximum in the silt and fine sand grades, with 70–95 per cent below 0.06 mm. in size and 97–99.5 per cent below 0.2 mm., i.e., nothing coarser than medium sand. For comprehensive studies see Guenther (1961) and Lugn (1962).

Two classes of loess are recognized: *(a) periglacial* loess, deflated from outwash (Leighton and Willman, 1950), from freshly exposed till, and from the rocky surfaces of the tundra and arctic barrens;[2] and *(b) desert* or continental loess, derived from desert areas. Grahmann (1932) believes it possible to distinguish the two on the basis of particle-size distribution. Periglacial loess should show a smaller grain-size range due to double selection—first, sorting by meltwater streams or selective frost-weathering, followed by selective wind deflation. Desert or continental loess is only wind sorted and contains more clay and medium sand particles with a less striking silt or fine sand maximum.

Pleistocene loesses, largely of the periglacial type, are rather widespread in Europe, China, and the central United States, as well as in the Pampas of Argentina. With a subdued topography they mantle hills and valleys to depths varying from 50 cm. to over 50 m. Much or most of this periglacial loess was laid down in open grassland or tundra, as is suggested by little or no evidence of soil development (see Schönhals, 1951, 1953), by the contemporary snail faunas (see Ložek, 1964; Schultz and Frye, 1968), and by the nature of the pollen (see Frenzel, 1964, 1965). This does not mean that loess was never laid down in forest areas, and contemporary cases of sedimentation under woodland are indeed known (see Péwé, 1951). But it has not yet been proven that any of the extensive loess beds preserved from the Pleistocene were deposited under forest vegetation. Soil development with its leaching, humification, and biological mixing would continue to destroy the characteristic aspects so that a theoretical forest loess would normally be assimilated soon after deposition—unless the depth and rate of accumu-

2. There is a remarkable concentration of fines between 0.01 and 0.1 mm. among the frost-weathered surface materials of the polar regions (Dücker, 1937). This loess-like size is readily deflated in summer, while a constant supply of materials is assured by continued frost-weathering. Apparently weathering of this type does not produce material of clay or coarse sand size.

lation were excessive. The typical columnar structure and vertical cleavage of true loess are compatible with herbaceous vegetation but not with a woodland soil environment. The absence of calcified root-casts and impressions or secondary root-fillings is of interest in the same connection.

Much of the European and North American loess was subsequently redeposited on a local scale through rainwash, a feature sometimes noticeable through fine wavy laminations, lenses of fine pebbles, and horizons of stratified mollusca. These *Schwemmlöss* (German for "colluvial loess") characteristics do not, however, influence the over-all interpretation of periglacial loess, since such washing may also have taken place during deposition in rainy weather.

Periglacial loess is also significant for stratigraphical purposes. Generally indicative of glacial phases, the loess of western and west-central Europe is attributed mainly to the glacial maxima (see Büdel, 1950b), that of the more continental parts of central and eastern Europe to the glacial advances and maxima (see Ložek, 1964; Frenzel, 1965). Similarly loesses, with their buried weathering profiles, molluscan faunas, etc., have provided the greatest impetus for stratigraphic controversies concerning the details of European glacial stratigraphy (see Fink, 1965; Frenzel, 1964; Ložek, 1964).

Lastly, the periglacial loesses of several Old World areas have assumed importance in another way. Their functional coincidence with highly fertile, well-drained soils, and open or lighter-stocked woodland vegetation destined the loess plains for the earliest agricultural colonization in middle latitudes.

The status of "continental" loess is more problematical, in that little conclusive evidence can be marshaled for loess derived from desert areas. The massive loess of North China can probably be attributed to deflation from the floodplains of the Hwang-Ho and the Yangtze, rather than from the Gobi Desert (Gellert, 1962). The loess of the central Great Plains, sometimes ascribed to deflation in the desert Southwest, is probably derived from local fine-grained sediments (Lugn, 1962). In Tripolitania and southern Tunisia, widespread fine-grained sands with loessic structure were probably deflated from the Mediterranean littoral during glacio-eustatic regressions. A similar origin can be postulated for frequent but very localized patches of poorly sorted, water-reworked silts of loessic structure found in lowland areas of eastern Spain (Butzer, 1964a) and southern France (Aumen *et al.,* 1965). All in all there is a large, heterogeneous class of loess and related deposits that formed outside of the periglacial zone, but the environmental distinction of "periglacial" and "continental" loess is an arbitrary one. Very probably the genetic overtones common to loess definitions should be abandoned and replaced by strictly textural and structural criteria. Furthermore, loess-like deposits that fit the structural but not the textural definition should be excluded from the concept of loess.

Sand Sheets. Sheets of fine to medium-grained sands with incoherent or subdued morphology are rather less common than loess. Included here are a variety of features comprising all classes of drifted sand, excluding dunal forms, coastal features, and specific loess deposits. Particle sizes are commonly in the silt to medium sand grade. Since this is a heterogeneous intermediate class in all ways, few environmental generalizations can be made other than that a ready source of sand must have been available at not too great a distance. Of particular interest are the widespread sand sheets of the southern margins of the Sahara and the "redistributed" Kalahari sands of Angola and the Congo, as well as the "cover sands" of the Low Countries, northern Germany, and Poland. The first of these is primarily aeolian, the second are both aeolian and water-laid (see ch. 20); the "cover sands," consisting of poorly sorted medium sands with a silt admixture, are attributed to deposition with the co-agency of blown snow during winter blizzards (Maréchal and Maarleveld, 1955; Tricart and Cailleux, 1967, p. 398 f.).

AEOLIAN DEPOSITS OF VOLCANIC ORIGIN

The destruction and burial of Pompeii and Herculaneum by the eruption of Mt. Vesuvius in A.D. 79 was the result of volcanic ash and other, coarser materials ejected violently from the crater. The exceptional preservation of the archeological features in these towns illustrates the potential significance of wind-borne volcanic deposits. Although vulcanism is not restricted to any particular environment, volcanic activity does become an environmental characteristic or even a determinant in regions such as East Africa, Japan, and Iceland. In addition, wind-borne volcanics frequently provide direct environmental information in past contexts.

Vulcanism involves magmas or molten rock that are either injected into the upper part of the earth's crust from below or forced out, or extruded, onto the surface as lava or ash. The latter category includes great lava domes, craters of mixed lava and ash composition, small ash-and-cinder cones, and wide, level tracts of lava and ash. Magmas may be of two fundamental types: basic or acidic. *Basic* volcanics are rich in ferromagnesian minerals and soda-lime feldspars; they commonly weather into fertile soils. As lavas, they have a low melting point and are comparatively fluid, flowing readily from volcanic craters or from fissures in the earth's crust. The most widespread basic lavas are *basalts,* which form massive spreads over much of Ethiopia, the Deccan Plateau of India, and parts of the New World. *Acidic* volcanics consist primarily of potash feldspars and quartz, so that the weathered residue is comparatively infertile. Rocks of this kind have a high melting point and are stiff and viscous as lavas; because of the high temperatures, the contained gases frequently promote violent explosions during the course of eruption. Much of the molten rock is thereby ejected in the form of bombs, pellets, and ash that fall to the ground or are carried away by wind. These ejecta, which normally cool and solidify in midair, form *pyroclastic* rocks. Not all pyroclasts are acidic, just as violent eruptions are not confined to acidic magmas, but explosive extrusions and ash falls are nonetheless more typical of this category. Acidic lavas, belonging primarily to the group of *felsites,* are of limited areal extent. Their surface is very rough and irregular, compared with the smooth, ropy surface more characteristic of basalts.

Pyroclasts are subdivided on the basis of particle size. *Tuffs* consist of compacted or cemented ash and dust in the silt and sand grade. Volcanic *breccias* are dominated by lapilli (gravel size) or bombs (cobble to boulder size) in a matrix of ash.

Tuffs are of particular interest here since they are disseminated over great distances by wind and, possibly, by secondary agencies. At the same time they may bury and preserve or embed bone, shell, organic impressions, and artifacts. In terms of material, tuff consists of fresh magma, blasted into froth by the expanding gases, intermixed with fragments of older rock torn from the sides of the vent. Quartz and feldspars are the dominant minerals, although ferromagnesians are prominent in basic tuffs. Such volcanic ash may settle out in thin films or sheets of wind-borne dust.[3] Frequently, too, ash of silt and sand grade is immersed in stream or lake waters and then water-laid. In fact, even on land surfaces, tuffs are frequently reworked into colluvial spreads or alluvial beds, admixed to some degree with older weathering products. In this way they contribute a diversified and informative sedimentary record, frequently present in successive stratigraphic units and, on a more local scale, comparable in many ways to the loess mantles of former periglacial environments.

Tuffs and other volcanic ash may be sufficiently distinctive on textural or mineralogical grounds to serve as local or regional stratigraphic mark-

3. A special type of tuff is produced when a rapidly-moving cloud of red-hot ash settles out: temperatures may be sufficiently high to allow partial remelting of the ash at the ground, forming a *welded* tuff with glassy texture, readily confused with a lava.

ers, e.g., the Pearlette ash of the Great Plains (see Wilcox, 1965), while fresh, primary feldspars from tuffs are susceptible to K/Ar dating.

SLOPE DEPOSITS

Slope deposits range from landslide masses to hillwash. involving highly variable volumes of material and equally different rates of movement or accumulation. Although gravity movements, possibly aided by soil frost, are most characteristic in the origin of slope deposits, rainwash and colluvial action play a large if not dominant role in many types.

Certain kinds of slope deposits are most characteristic in mountain country where slopes are steep and relief is great. After preparatory weathering, isolated stones or large masses of rocks are detached from cliff faces by frost, rain, wind, or earthquake shocks. The resulting *rockfalls* crash down the hillside, with rock fragments accumulating at or near the foot of the cliff. At the same time a constant dribble of loose stones along the footslopes leads to the development of *talus* cones or aprons. Such talus subsequently moves internally (on slopes of 45 per cent or more) as the individual rock particles "settle" a little downslope in response to frost, daily temperatures, and renewed rock impact from above. Of a different nature are sliding motions, whereby single or complex blocks of rocks slip downslope by simple, planar *gliding* or rotational *slumping*. Such sliding, usually across lubricated soil or clay, is characteristic of intermediate slopes in cold or humid environments. The individual block may be moved in a single slide or intermittently over long periods of time. *Landslides,* involving large masses of rock and soil, are preconditioned by unstable rock structures and steep, undercut slopes. These rapid, catastrophic movements may be triggered off by spring thaws (in cold environments), by heavy rains, or by earthquakes. Landslides and their related breccias appear to be fairly common in arid regions during periods of increased rainfall (see Wright, 1951; Butzer and Hansen, 1968, p. 362 ff.).

The gravity movements typical of intermediate slopes with limited topographic relief commonly involve soil mantles and weathering products. The solifluction, block streams, and sorted talus characteristic of cold environments have already been discussed (ch. 7). Far more widespread is the almost universal grain-by-grain downhill displacement of soil particles (on slopes of as little as 8 per cent) known as *soil creep*. This almost imperceptible movement may result from soil frost, temperature changes, and alternate wetting and drying of soil, as well as from raindrop impact on bare surfaces. Creep is most effective in areas with frequent, intensive frosts, grading over in solifluction. In areas of limited vegetation or very intensive rainfall it grades into colluvial deposits. Soil mantles can also be disrupted by slumping, in cold environments prone to solifluction or in humid environments where overgrazing has destroyed the vegetative mat on hillsides. More rapid movements, comparable to landslides in their mechanics but not in scale, include *earthflows*. Finally there are the *mudflows* of semiarid regions and bare volcanic slopes, where large accumulations of silt-sized materials may be set in motion as liquid mud by intensive rains.

In practice, the Pleistocene record reveals an equally great variety of colluvial deposits, due to rainwash and creep, found at the base of gently sloping hills, beneath the edge of alluvial terraces, loess embankments, and the like. Understandably, such colluvial beds may bury archeological sites, or they may include fossils or artifacts derived from upslope. Such deposits may or may not be susceptible to geomorphologic interpretation, depending on the nature of the present slope equilibrium under modern vegetation. Of particular interest are the so-called *stone lines* of many tropical environments, which are horizons of crude rubble found between fine soil layers. Such surface debris suggests the denudation of upper slopes, with accelerated washing of lower slopes, under an incomplete mat of vegetation (see Tricart and Cailleux, 1965b, p. 289 f.).

For a more detailed account of slope sedimentation the reader is referred to Sharpe (1960).

THIRTEEN

Cave Sediments

INTRODUCTION

Caves were first "discovered" for science by archeologists, and despite the enthusiasm of amateur cave explorers, caves and archeology remain almost synonymous in the public mind. The earth sciences have also shown considerable interest in caves and subterranean caverns. In fact, the various processes of groundwater solution and cave formation are a part of *karst* geomorphology (Grund, 1903; Davis, 1930. For general discussions see Birot, 1954; Thornbury, 1969, ch. 13). Practically all true caves have developed as a result of solution in limestone, and the term "karst" refers to landscapes noticeably modified through the dissolving agency of underground waters. Karst is an intrazonal feature in much the same way as is vulcanism. Many of its processes are individualistic. And the geomorphologists who first studied cave phenomena, often in relation to archeological sites, had to interpret a set of rather peculiar sediments.

Man and animals have sought shelter in caves since the beginnings of prehistory, and some of the most interesting cultural sequences have been derived from cave sites. Almost as a by-product, biological evolution and changing geomorphic environments have also been studied successfully in cave strata. Today certain sequences of cave sediments, faunal assemblages, and pollen are as vital for Pleistocene stratigraphy as the cultural horizons are for Stone Age archeology. Cave environments and processes consequently deserve special attention (for outlines, see Trimmel, 1968, and Poulsen and White, 1969).

CAVE ENVIRONMENTS AND SEDIMENTATION

Two major kinds of caves are distinguished: *exterior* caves and niches, and *interior* passages and caverns (Schmid, 1958, 1963; Trombe, 1952). The exterior type may vary from simple overhangs and shelters to shallow caves. Most of these have been dissolved or eroded near the watermark by streams or wave action at the coast. Sometimes they are produced by hollowing out of softer rock strata. Although generally found in limestone, exterior caves may also occur in other bedrock. The interior type is limited to limestone country and is mainly the result of karst activity. Underground corrosion along rock joints and bedding planes leads to enlargement by removal of dissolved material in subterranean waters. Irregular, vaulted caverns and narrow, cleft-like passages follow underground streams or natural rock fissures. Some caves of this kind extend for several kilometers, often at several levels. Locally the surface drainage may be swallowed up in sinkholes, only to emerge at some distance from a hillside spring.

Cave environments are highly variable. Direct sunlight is reduced or eliminated entirely. Relative humidities are high, particularly in deep, shaded caves. Mean temperatures are lower than outside, but an almost unlimited range of possibility exists: deep caves maintain relatively constant temperatures; shallow caves may be heated by the sun in day, and cool off rapidly at night. Except at the very entrance, temperatures are usually too low for soil development, and chemical weathering is practically limited to carbonate solution. Schmid (1958) consequently distinguishes several cave environments (Fig. 40):

a) The *entrance* is located under the cliff face and is exposed to external weathering agents such as frost, rainwash, wind, and possibly sunlight. Plant life is present, soil development is possible, and talus commonly falls in from the cliff face. Sediments at the entrance consequently accumulate rapidly and are rather complex in nature. Many of the rock overhangs or *abris* of prehistory are of this type.

b) The *front part* of a cave is still affected by the external weathering agents but to a limited degree only. Rainwater may wash the floor locally, but wind activity is very subdued, ventilation limited, and temperature contrasts more restricted. Plants are generally absent, snails are few and uncommon, and soil development ceases. Talus is absent. Instead, rocks fractured off the ceiling and walls of the cave accumulate slowly on the floor. Soil products may be washed in from outside through large joints in the cliff, and soluble limestone precipitates may accumulate on the ceiling, walls, or floor. Exterior caves and the external portions of interior caverns or passages may be

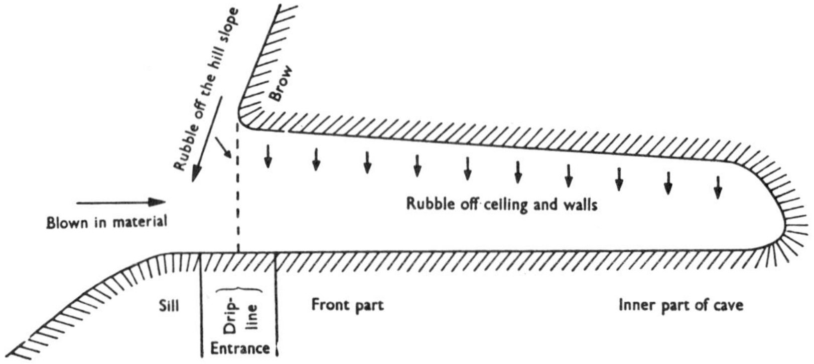

Figure 40. Sedimentary environments of a cave (from E. Schmid, 1963, by permission of Thames and Hudson, London).

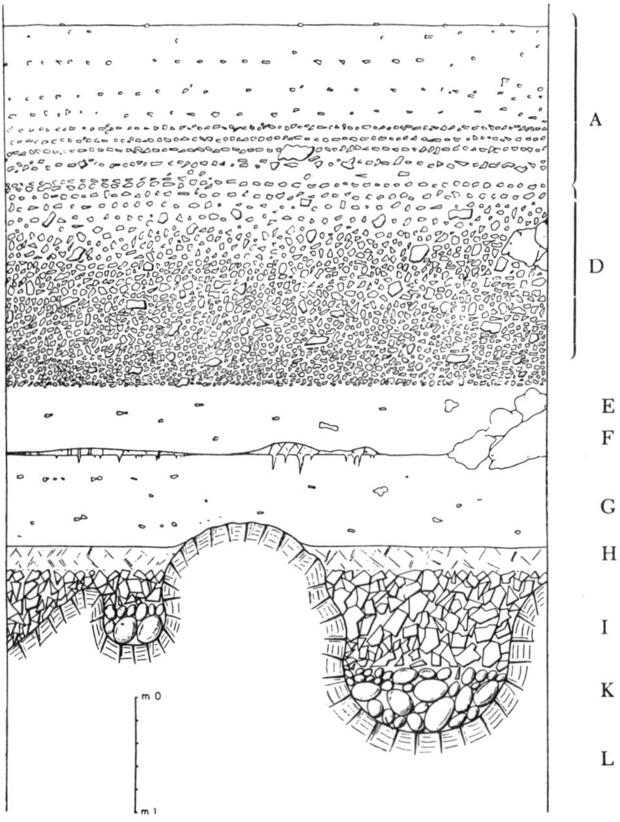

Figure 41. Sedimentary profile of the Grotta Romanelli, southern Italy (from G. A. Blanc, 1921): (A–D) éboulis secs beds, with "cold" fauna, (E) brown earth, partly aeolian, with cold fauna, (F) discontinuous stalagmitic layer, (G) reddish earth, partly aeolian, (H) dripstone and flowstone, (I) limestone detritus with "warm" fauna, (K) beach conglomerate of Tyrrhenian II stage, (L) limestone bedrock.

listed as part of this environment. Most prehistoric cave living-floors belong here.

c) The *inner part* of the cave is permanently dark. Temperatures are practically constant all year, commonly around 5°–15° C. in midlatitudes. Humidities are high. Sediment production is restricted. Habitation by men or animals is sporadic and less common.

The geological layers found in caves are therefore partly of external origin, partly internal. The extraneous materials may be washed in by rainwash, drawn in by gravity, blown in by wind, moved in through solifluction, or washed through rock joints by percolating soil and groundwater. In addition, man and beast may carry in a variety of inorganic objects or materials, deliberately or inadvertently. Local weathering within the cave is limited to thermoclastic agencies and solution. Oxidation is almost entirely absent so that weathered products such as "cave earths" are either extraneous or simply the clayey residuals left after solution and removal of the carbonates constituting the bulk of the bedrock. There are then no true cave *soils,* only *sediments.*

The principal forms of mineral cave sediments therefore include (see also Fig. 41):

a) Extraneous mineral sediments or materials introduced from the outside by water, wind, gravity action, man or animals;
b) Limestone detritus derived from mechanical or chemical attack on the walls and ceiling of the cave;
c) Limestone precipitates derived from solution of limestone products;
d) Residual products, mainly clay minerals, derived from the dissolved limestone.

In addition to these general geological strata, Schmid (1958, 1963) distinguishes further biological horizons:

1) Fossil layers: animal bones, carcasses, feces, etc. included naturally within a sediment;
2) Archeological layers: the occurrence of individual proofs of human presence with or without fossil remains, but little or no alteration of the sediment;
3) Cultural layers: sediments strongly influenced by human activities such as fire and toolmaking, and with many imported objects such as stones, bones, shells, plant matter, etc.

Major attention will here be devoted to mineral sediments. Unfortunately the exact interpretation of such deposits still leaves much to be desired. Systematic analyses were introduced by R. Lais (1932, 1941) and followed by those of Freund (in Zotz, 1955), Bonifay (1956), Schmid (1958), Vértes (1959), and Laville (1964). Yet there still is little comparative material on cave microclimates or contemporary cave sedimentation processes, or on differences within different parts of a single cave or between caves in different regional environments. Much of the outline given here must be considered as tentative. It is only an attempt to sketch and discuss the information available.

LIMESTONE DETRITUS

Limestone fragments are separated from the ceiling or walls either by corrosion and solution (chemical weathering) or by thermoclastic fracturing (mechanical weathering).

a) Chemical Agencies. Subsurface waters contain atmospheric CO_2 or humic acids in solution, and therefore act corrosively on limestone bedrock. Passing through joints, bedding planes, and fissures, a certain amount of calcium carbonate is dissolved and carried away in solution, while minute quantities of residual clay or sand are washed out. The cracks are gradually enlarged until, one by one, corroded fragments of rock break off and fall to the floor. Rocks derived in this fashion are superficially corroded, have somewhat rounded edges, and are generally embedded in residual clays, coarse calcite sand, or crystalline carbonate precipitates (travertines). Both Lais (1941) and Bonifay (1956) consider coarse calcite sand (0.2–2.0 mm.) as primarily due to chemical decomposition.

Production of limestone detritus by chemical agencies is limited to environments that are at least moderately moist and either temperate or warm.

b) Mechanical Agencies. Thermoclastic weathering may be effective as a result of expansion and contraction following rapid temperature changes in abris or cave entrances, or in the interior of inhabited caves as a result of fires. Frost-weathering is far more significant in the front part and entrance of a cave. With even a little percolating moisture, ice forms in joints and fissures, creating pressures because of the increased volume. Repeated freeze and thaw will lead to physical disintegration of superficial rock fragments which drop to the floor. Known as *éboulis secs* (French for, literally, "dry talus") such detritus is sharp-edged, angular, uncorroded, and often flattish in shape. When washed, the faces and rims look quite fresh, unlike those of other talus. The characteristic size range is that of detritus 4–10 cm. in length (Laville, 1964; Laville and de Sonneville, 1967), although coarse detritus and large tabular blocks may be common in the resulting formations.

Although deposition may take place either while the waters are frozen or after thawing, the breccia-like sediments so formed are usually – although not necessarily – unconsolidated and porous, lacking a matrix of fines. *Éboulis secs* are almost mutually exclusive with travertines.

It is generally accepted that typical *éboulis secs,* as defined above, are due mainly to frost-weathering. A more specific climatic interpretation is difficult however. Repeated, strong, penetrating frosts will probably be most effective in producing larger quantities of *éboulis secs,* whereas long protracted frost periods, or frequent, ineffective freeze-and-thaw alternations will be less productive. Subtropical and tropical climates with little or no severe frost are definitely excluded, whereas persistently cold high arctic climates would provide little opportunity for thaw. Another variable is introduced by the degree of water saturation of the rock. Even though frost-weathering is effective on cliff faces in western Europe today, *éboulis secs* are comparatively rare in Holocene cave deposits of central Europe (Lais, 1941) and France (Movius, 1960, reply comments), presumably as a result of the reduced temperature amplitude of cave environments. Although *éboulis secs,* by argument, do not demand extremely cold conditions, they will presumably be most favored by a cold-temperate climate, not too dry, but with appreciable daily and seasonal temperature ranges.

In general, the concentration of *éboulis secs* in Würm-age cave horizons speaks its own story for the case of temperate Europe (Lais, 1941; Movius, 1960, reply comments). But interpretation must proceed with due caution as too little quantitive work has been done on frost-shattering intact limestones.

LIMESTONE PRECIPITATES: TRAVERTINE

Waters percolating through the joints of limestone bedrock will inevitably acquire some dissolved calcium carbonate through corrosion and solution. This effect is most evident when *(a)* soil and vegetation are present at the surface over the cave, so increasing the acidity and CO_2 content of the water, and *(b)* abundant water percolates through into the cave. This water may drip from the ceiling or flow over the cave floor. The distance, speed, and amount of water percolation will determine the degree of saturation of the waters leaving the ceiling, and the rapidity or likelihood of precipitation of the dissolved lime salts will depend on temperature, humidity, ventilation, and carbon dioxide content of the cave air (Cornwall, 1958, pp. 35-36). It is therefore not surprising that the resulting precipitates show a considerable range of variation. For example, both slow evaporation of saturated waters and rapid evaporation of unsaturated waters may produce dense precipitates, while very rapid evaporation of saturated waters may only produce an unconsolidated calcareous grit.

The general class of deposits is grouped as *travertine,* and occurs as:

a) Dripstone, including icicle-like forms hanging from the ceiling *(stalactites),* upward-growing *stalagmites,* and *columns* in cases where stalactites and stalagmites have merged.

b) Flowstone, including a variety of horizontal beds ranging from hard, translucent, banded, or vertically crystallized strata to soft, opaque chalky, gritty, or clayey deposits. Flowstone may assume bizarre shapes or spread in fine, laminated crusts or veneers. Depending on the amount of embedded clay, oxides, sand, and organic deposits, colors may range from clear to white and yellowish-brown or even reddish-yellow.

Horizons of flowstone or fractured dripstone are comparatively frequent in the Pleistocene beds of many mid-latitude caves, and travertines are also known as surface deposits in some subtropical areas. Whenever carbonate precipitation is important in a cave, clastic deposits on the floor may be cemented into resistant *breccias.*

Precise paleoclimatic interpretation of fossil travertine strata is again difficult. Protracted frost will, of course, impede solution processes; severe cold will limit vegetation at the surface and reduce the CO_2 content of the waters; extreme aridity will not provide the requisite water. Consequently travertines pre-eminently suggest a temperate or warm climate and a humid or subhumid moisture regime. In arid and semiarid regions, caves without modern dripstone formation will, by fair judgment, produce such features only with an increase of moisture. In contemporary humid environments, however, the same implication of "wetter climate" becomes highly debatable. Consequently the only reasonable deductions possible from fossil travertine horizons found in caves of temperate mid-latitudes are that the climate was fairly moist and not subarctic in character.

RESIDUAL CLAYS AND EXTRANEOUS SEDIMENTS

The commonest deposits of most caves are soft, moist, plastic cave earths, mainly of silt or clay size, but frequently containing coarser inclusions. The calcareous component of such earthy deposits is sometimes rather small. Yet, even in the deep interiors of dark caverns and passages, only a part of this material is a local residual from dissolved limestone. Foreign materials from the surface trickle in through cracks and crevices, even where washing, blowing, or sliding of sediments from the entrance is excluded. Kerekes (1951) calculated the residual materials to be expected by solution of limestones corresponding to the volume of certain central European caves. In each case the noncalcareous sediment present was greater than the calculated residual materials. Obviously much foreign matter had been intermixed.

Residual clays may be more or less uncontaminated when directly embedded in travertine during its formation. Aluminum silicates are most common, although many minerals are present, including a certain amount of limonitic iron.

Wind-borne loess or sands are probably the most important extraneous sediments. They are either deposited directly at the entrance or in the forepart of the cave, or washed in through the entrance or through the overlying rock. When caves lie at river level in valley gorges, stream deposition may take place. Elsewhere subterranean waterflow may carry foreign sediments far within a cave. And solifluction can induce sliding of extraneous material into a cave from the entrance (see Black, 1959). The interpretive value of such complex extraneous mineral sediments necessarily lies in their association with processes outside.

Materials introduced naturally through the cave entrance generally reflect on the external environment at the time a particular stratum was formed. Thus, loess deposition or a stream terrace may be correlated and used for interpretation. In the case of materials washed through the walls or ceiling, older sediments or paleosols may be derived from the surface above. As a result, such materials are suspect.

ORGANIC DEPOSITS

Biological deposits may assume considerable importance in caves once occupied by cave bears, cave hyenas, and cave lions, or in caves frequented by owls or bat colonies. Birds and bats leave masses of dung, and there may also be considerable quantities of feces of larger mammals. The bones and other tissues of small mammals, particularly rodents, may be added to the osteological remains of their carnivorous hunters. And, repeatedly, the carcasses of dead mammals partially decompose and are added to the accumulating sediment. In addition to this animal matter, plant parts that adhere to the animals or are excreted through the intestinal tract are added to the "fossil layer."

Decomposition of these biological remains produces various compounds, the most important of which are the reddish-brown phosphate beds corresponding to many fossil layers in middle latitudes. Schmid (1958) emphasizes that these have, on occasion, been confused with climatically significant mineral sediments. Humic acids may also be derived from decomposition of excreta and carcasses.

Human occupation, even in cultural layers, is not evidenced by a similar intensity of organic accumulation. Instead, a great deal of mineral ash, partly decomposed or charred plant matter, and foreign rock are introduced.

Plant materials are carried in by man and animals and, together with wind-blown pollen grains, are remarkably well preserved, even in the subtropics (e.g., the Cueva del Toll, Catalonia; see Donner and Kurtén, 1958). Bone preservation is also good in the calcareous, alkaline sedimentary environment. Snail faunas, where present, are also well preserved in most environments (e.g., the Haua Fteah Cave, Cyrenaica; see Hey, in McBurney, 1967).

SPECIAL SEDIMENTARY ANALYSES

A few specialized sedimentological techniques deserve brief mention here, in addition to the more standard techniques (particle size, carbonate content, clay minerals, heavy minerals, etc.). Lais (1941), Bonifay (1962), Brain (1958), Laville (1964), and Brunnacker (1967b) have devised mechanical analyses to help estimate the relative importance of mechanical and chemical weathering. These include *(a)* simple sieve analysis of silts and clays (under 0.06 mm.), sands (0.06–2.0 mm.), fine grit (2–10 mm.), and coarser detritus (over 10 mm.); *(b)* classification of detritus by weight into size classes, e.g., 10–40 mm., 40–100 mm., over 100 mm.; and *(c)* particle-size analysis of the noncalcareous residual under 2 mm. Such basic analyses normally serve to illustrate the proportions and composition of coarse, clastic components—reflecting primarily on mechanical weathering—and the significance of the quartz and clay residue—reflecting in part on extraneous sediments, in part on limestone decomposition within the cave. Such studies can be followed up with level-by-level determination of *p*H values, and of calcium carbonate content in the fine fraction under 2 mm., as well as by computation of a variety of indices expressing sediment porosity or compaction, corrosion of detritus, "gravel" rounding and flattening, and proportions of travertine detritus and primary concretions. In this way refinements of our understanding of both the original sedimentation and of postdepositional changes may be obtained.

Further insights into the nature and intensity of chemical weathering, inside or outside of the cave, may be obtained by X-ray diffraction of clay minerals (see Laville, 1964) or by analysis of trace elements (see Sokoloff and Carter, 1952; Martin *et al.,* 1961; Sabels, 1960). A last specialized cave sediment technique is determination of the phosphate content of various strata (see Schmid, 1958). However, Schmid cautions that phosphate content varies according to the size and number of the cave inhabitants and the rate of mineral sedimentation, so that phosphate contents do not provide a simple correlation with the frequency or importance of animal or human occupation.

CONCLUSIONS AND PROBLEMS

Cave sediment analysis by the earth scientist, in collaboration with the biologist, can provide good paleo-environmental evidence. Although the existing body of comparative information on modern cave sedimentation (as opposed to Pleistocene cave sediments) leaves much to be desired, the basic processes are recognized and tentatively understood. With the many analytical methods now available, recent studies of cave sediments in France, Germany, Yugoslavia, and South Africa have served to show that an objective evaluation of such deposits is indeed possible. Add to this the ecological interpretation of fauna and flora, and it is evident that a careful cave study can, potentially, provide an exceptionally detailed and valuable key to understanding Pleistocene environments and events.

Due caution must, however, be maintained in generalizations about cave sediment interpretation:

a) Each cave must be assessed and interpreted individually, since cave ventilation, temperature amplitudes, and water regime vary considerably even within local areas. Even within a single cave there are significant differences of microenvironment at the entrance, in the interior, and at different elevations within the deep passages and fissures.

b) Sedimentary analyses in the laboratory, no matter how sophisticated, cannot substitute for a thorough investigation of the over-all geomorphology. Changes in cave hydrology commonly accompany geomorphic evolution of the outside landscape, and local soil mantles are often the key to understanding sediments inside a cave. An evaluation of what sediment components are extraneous and by what means they were introduced is vital to correct interpretation. In fact, mechanistic sediment studies can provide a false sense of confidence, and this author is concerned about the paucity of "external" studies reported or evident in the spate of recent cave publications.

c) Postdepositional changes in cave strata are more common than is usually appreciated, and no mechanistic technique can hope to assess them accurately. Limestone grit and detritus corrode rapidly in wet soil environments, sometimes to the point that they are reduced to a soft, chalky substance that rapidly breaks down in the course of mechanical analyses. When sufficiently corroded, derived grit due to thermoclastic weathering can no longer be distinguished from dripstone pellets once precipitated *in situ*. Cementation of strata frequently takes place long after sedimentation, and several strata at different levels can be impregnated *at the same time* by vertical and lateral movement of carbonate-saturated waters moving selectively through permeable, inclined beds. In fact, postdepositional cementation, well beneath the cave floor, can simulate flowstone development when undulating laminae of caliche accumulate in sandy or gravelly strata. Moisture or hydrological changes are commonly inferred from the proportions of limestone precipitates. In reality, however, they are equally well recorded by postdepositional cementation at depth and by oxidation-reduction phenomena that may effect the entire sedimentary column at one time. Limonitic mottling and staining due to a high local water table, although present in many mid-latitude cave sediments, are surprisingly seldom reported. A final postdepositional factor of note is human or animal occupation. Organic acids of different origins will inevitably leach carbonates and may eluviate clay minerals and organic compounds into otherwise sterile, often sandy strata below.

d) Protracted cave sedimentation will gradually modify the local microhabitat by reducing the size of the cave entrance, by changing the configuration of the ceilings, and by obstructing and altering cave drainage lines. As a result the ventilation, temperature amplitudes, humidity, and hydrology may also change through time.

e) Lastly, the possible impacts of man should not be underestimated. Artificial structures, such as tents or shelters constructed inside the cave, or windbreaks erected at the entrance, modify the cave microclimate. Evidence of structures is available from Paleolithic strata in several French and Spanish caves, and modern cave-dwellers are known to seal off cave mouths. Such changes would reduce temperature amplitudes while a greater influx of extraneous "dirt" carried in by man would give the impression of greater chemical weathering. On the other hand, fires in the cave would, locally, accelerate thermoclastic weathering. Finally, alternating settlement and abandonment of a cave site can simulate cyclical changes of the weathering balance that in reality reflect nothing more than periodic temperature modification by man.

FOURTEEN

Coastal Phenomena and Sea-Level Fluctuations

INTRODUCTION

Coastal phenomena give little information about local climate, except in the case of mangrove swamps or coral reefs. Mangroves are largely confined to the humid and subhumid tropics, while coral reefs are restricted to warm seas with water temperatures never dropping below 18° C. Coasts that develop entirely as a result of inorganic processes show some, although minor, features of zonal character. But generally speaking, the significance of Pleistocene coastal phenomena lies in somewhat different spheres.

For one, the Pleistocene experienced a series of remarkably rapid fluctuations of world sea level—from a maximum of about 180–250 m. higher than that of today to a minimum of about 150 m. lower. The stratigraphy of these oscillations has been outlined in Table 3, chapter 2. The application of sea-level stages, some of them world-wide, has been largely stratigraphic. Excellent possibilities for correlating continental sequences with glacial or interglacial phases are provided, and cold or warm molluscan faunas in beach deposits may contribute direct paleo-ecological evidence and possibilities for isotopic dating.

The other contribution is more topographical in nature. The location of the coastal zone varied considerably in areas of gentle gradients, and land bridges were periodically opened, permitting, for example, the settlement of the Americas and Australia by late Pleistocene man. Numerous other archeological applications are also of interest here.

This chapter consequently diverges from the pattern of chapters 11–13 in presenting data on Pleistocene sea-level variation and on fossil coastal forms—partly stratigraphic, partly topographic and environmental in application. Unless otherwise stated, reference is to marine coasts (rather than to lakes or landlocked inland seas).

MECHANISMS OF WORLD SEA-LEVEL FLUCTUATIONS

Two major factors are involved in the complex pattern of Pleistocene sea level. The over-all trend has been downward, ever since the later Pliocene when sea levels were apparently found in the range from +180 to +250 m. Quite probably these values are misleading since slow, epeirogenic uplift on a continental scale and local tectonic displacements have all obscured the record. But the Pliocene generally conformed to a transgressive phase in geological history, leaving marine deposits well inland from the present coasts in most parts of the world. Whatever its true dimensions, a general lowering of sea level is admitted for the Pleistocene, and it is attributed to possible changes in the water-holding capacity of the great oceanic basins (Baulig, 1935; Zeuner, 1952, 1959; Valentin, 1952). Such fluctuations of sea level may have been responsible for a good part of the shoreline changes evident in the geological past.

A number of violent, rapid oscillations were superimposed upon the downward trend. These short-term features had a range of at least 150 m., and are attributed to actual changes in the amount of ocean waters due to the alternating glaciation and deglaciation of the northern hemisphere continents. They are designated as *glacio-eustatic* fluctuations.

The range of possible glacio-eustatic sea levels depends on the calculation of *(a)* how much water is still retained in the solid state in Greenland, Antarctica, and other smaller glaciated areas, and *(b)* how much more water was held in the glaciers that existed during the various Pleistocene glaciations. For general treatment and a discussion of the problem of isostatic adjustment, see Flint (1957, p. 258 ff.).

Ablation of the existing ice caps and mountain glaciers would provide a considerable volume of water to the oceans. Various estimates prior to 1955 (see Valentin, 1952, with references) were first superseded after the results of the French polar expeditions to Greenland (1948–49) determined the thickness of the Greenland ice cap. Based on these results and guess estimates for the Antarctic, Bauer (1955) set the possible rise at +54 m. Preliminary reports from the International Geophysical Year expeditions to Antarctica (1957–58) indicated that the thickness of the Antarctic ice cap had previously been underestimated. The rise in sea level has now been estimated at 66 m. (Thiel, 1962).

Allowing complete deglaciation during the earlier Pleistocene warm intervals, glacial-eustasy alone could never explain the high sea levels of the Pliocene or earlier Pleistocene. On the contrary, there is as yet no proof that the Greenland and Antarctic ice sheets were appreciably smaller at any time during the Pleistocene. It is consequently unproven that even the modest higher sea levels of the later Pleistocene were primarily a result of greater ablation of existing ice caps.

Turning to the negative side of the glacio-eustatic ledger, estimates of theoretical sea-level lowering depend on an accurate calculation of the volume of the Pleistocene ice sheets.

Although the areal extent of the defunct Würm-Wisconsin glaciers is known accurately, the extent of former glaciation in Greenland, Antarctica, and several smaller highland areas is difficult to assess. Whereas sources of error are presumably small in the case of the Würm, they may be rather more appreciable in the case of older glaciations outside of North America and Europe, about which we know next to nothing.

The average thickness of the former continental glaciers is more controversial. Estimates vary strongly, although the current calculations of 1,400 m. for the Scandinavian glacier and 2,000 m. for the North American glacier carry some degree of conviction. Donn *et al.* (1962) arrive at slightly higher values by assuming these glaciers had thickness-area ratios comparable to those of existing ice sheets. None of the estimates for thicknesses of pre-Würm glaciers are based on direct geophysical or geomorphical evidence, and they can be rated only as reasonable assumptions.

As a result, estimates of glacial sea levels vary considerably. For the Würm-Wisconsin maximum Valentin (1952) gives a value of −95 to −100 m.; Woldstedt (1954, p. 293) gives −90 to −100 m.; Donn *et al.* (1962) give −115 to −134 m. For the maximal pre-Würm glaciation (Riss complex?), Valentin (1952) and Woldstedt (1954) suggest −115 to −120 m.; Donn *et al.* (1962), −137 to −159 m. There are a great number of submerged shoreline features that can be freely correlated with any one of these values, but they do not prove the reliability of the one or the other. It should be realized that precise estimates are impossible at present, and a general estimate of −100 to −150 m. is quite sufficient for all practical purposes. Greater precision would be misleading, both theoretically and in specific application, since any local area may have since been affected by small or large tectonic movements.

In overview, the positive or negative trends of Pleistocene sea-level fluctuations can be understood by means of both ocean volume and glacio-eustatic changes, but neither of these mechanisms carries a meaningful absolute value. Instead, the stratigraphy of high and low Pleistocene sea levels is a relative one, and even where altitudinal correlations appear to be possible they do not necessarily have a theoretical explanation.

EVIDENCE OF HIGH PLEISTOCENE SHORELINES

General. The high Pleistocene shorelines forming the basis for glacio-eustatic stratigraphy can best be understood by means of contemporary coastal geomorphology. Numerous classifications of shorelines or coastal types have been made. That of Guilcher (1958, p. 60 f.) is possibly the most simple and useful. Guilcher recognizes four groups of littoral forms:

a) Cliffs with rocky platforms;
b) Sandy beaches with coastal dunes;
c) Tidal marshes, stream estuaries, and deltas;
d) Coral reefs.

Each of these phenomena can and has been recognized in the fossil state. With the exception of coral reefs and tidal marshes, most of these features may be found on both lake and seashores. But the lakeshore phenomena are, in general, only poorly developed, and the subsequent discussion is focused on sea coasts. For further descriptions see Cotton (1949, p. 396 ff.), Valentin (1952), Guilcher (1958), Thornbury (1969, ch. 19), Sparks (1960, ch. 8), and Butzer (1962).

Cliff Coasts. One of the most common coastal types is the cliff coast, marked by a cliff, a basal notch or knock-point at the water mark, and a sand- and pebble-strewn abrasional platform at its foot. The *cliff* may have a gradient anywhere from 30 per cent to an almost vertical face, and its height may vary from 3 m. to 50 m. or more. Wave erosion at the watermark continually undercuts the cliff and thereby widens the platform. Overhangs may be created at the *notch,* and *sea caves* excavated by mechanical abrasion in rocks with pronounced structure or differential resistance, by mechanical and chemical attack of homogeneous limestones. Repeated collapse of undercut cliff sections will periodically destroy the notch or individual sea caves, but their distinctive association with the high watermark (in the case of the open ocean) or mean sea level (along seas or lakes with little or no tidal amplitude) remains clear. In some cases the cliff may plunge directly to well below sea level without an abrasional platform and only a small notch at the high watermark (Fig. 42*a*). In areas with rather heavy surf, the watermark may be located away from the foot of the cliff, somewhere on the platform. Both of these divergent situations may also be a result of sea-level fluctuations or relative movements of the land.

The *wave-cut platform* itself commonly has a gentle gradient, and may attain a width of several hundred meters or more. The materials eroded at the cliff base are sooner or later swept out by the undertow and deposited as a *wave-built terrace* at the seaward edge of the abrasional platform (Fig. 42*b*).

A shoreline dominated by cliffs and abrasional platforms is commonly interrupted by prominent headlands and by coastal indentations with sandy beaches. The headlands may have plunging cliffs; they may also be fronted by small islets or *stacks*—resistant erosional remnants of the retreating cliff face.

Abandoned sea cliffs and marine terraces are among the best known vestiges of higher shorelines. Notches and sea caves may still be recognizable and may allow precise determination of the former high watermark (see Zeuner, 1961). Approximate mean sea level can then be obtained by deducting half the local tidal amplitude from this value. Where notches are unavailable, lines of holes produced by rock-boring organisms may be substituted. Marine beds are seldom preserved *on* abrasional platforms, except where rough seas are rare. They may however, be preserved on the wave-built platform. In the case of cliff coasts, caution should therefore be applied when using marine beds to determine former sea levels.

Nip and Shingle-Beach Coasts. Not all contemporary or fossil coastal forms can be grouped as either cliff or sandy beach types. Many coastal sectors, for example, exhibit an incipient cliff or *nip,* with a face on the order of 0.5–2 m. height. This nip is incised into a moderately sloping marine platform with a veneer of beach sands and gravels (Fig. 42*c*). Many of the fossiliferous marine sediments of the Mediterranean Sea pertain to former nip coasts. If the nip can be identified, sea-level determination can be made with a fair degree of accuracy.

If the original seaward slope of the land is gentle, the "cliff" may be limited to a step some decimeters high. Marine beach sands mask the platform on which coarse blocks and other wave-eroded detritus or gravel are thrown up and accumulated as a *shingle* beach or ridge in the rear of the nip by high seas (Fig. 42*d*). Shingle ridges may also develop as exceptional *storm beaches* above nip or low cliff coasts.

Sandy Beaches. Sandy beaches develop in gently sloping, shallow-water bays interrupting cliff coasts; elsewhere they may be found fringing low-lying coastal plains. In some instances such low, sandy, shallow-water coasts develop broad sandy beaches fringed by semi-aeolian sand ridges or coastal dunes to the landward (Fig. 42*e*).

When gradients are extremely gentle, a coastal type develops, with

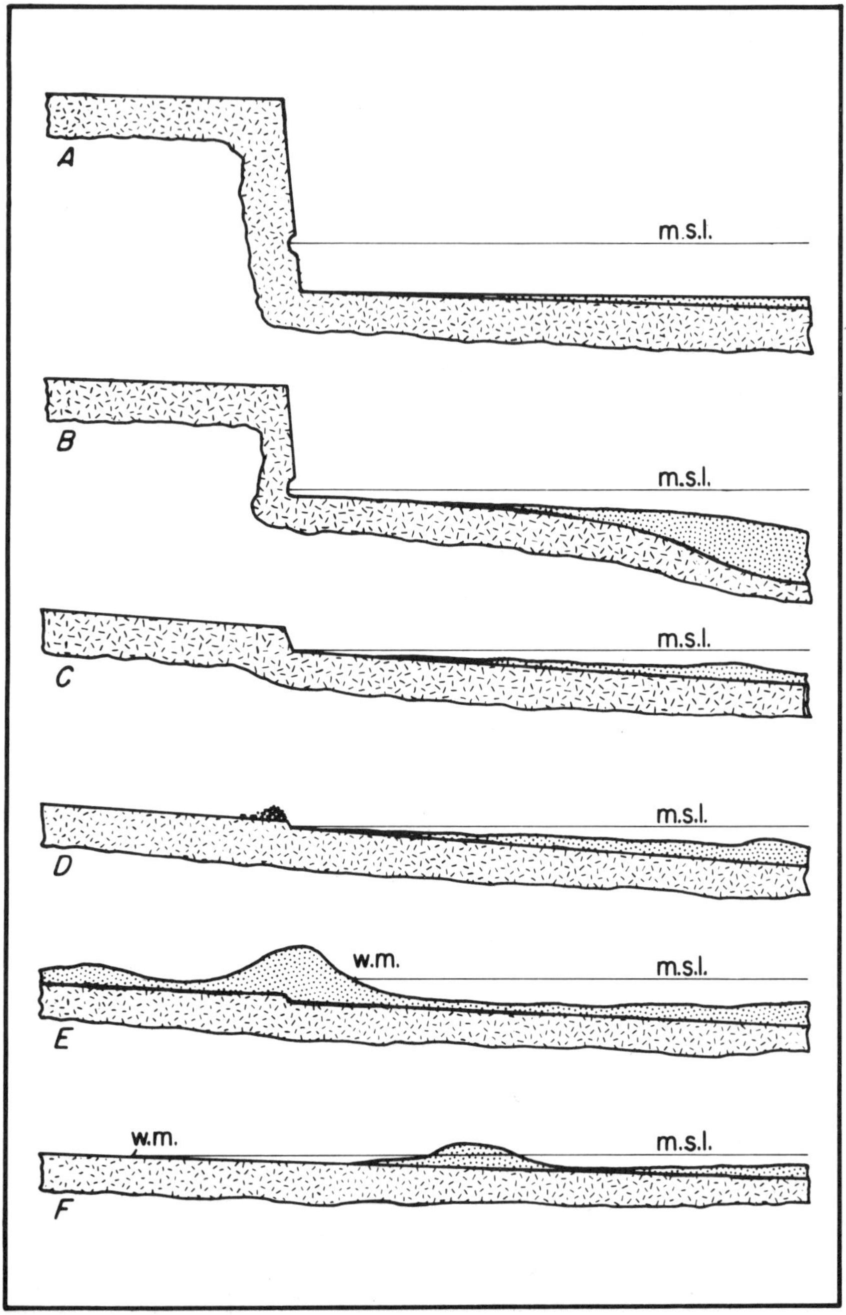

Figure 42. Coastal profile types: (A) plunging cliff coast, (B) cliff coast with notch, wave-cut, and depositional platforms, (C) nip coast, (D) shingle-beach coast, (E) sand-ridge coast, (F) bar-and-lagoon coast (w.m. = watermark).

wave action barely affecting the immediate shoreline. Instead, waves break, erode, and deposit offshore, forming submarine or offshore sand bars (Fig. 42f). These are separated from dry land by a shallow-water zone, a lagoon, or a tidal marsh, possibly characterized by mixed marine and continental deposition. If the bar becomes large enough, the coast may develop a semi-aeolian beach ridge or a dune cordon. Such sandy beach features are often difficult to distinguish in the fossil state.

In the case of the bar-and-lagoon coast, former sea level lies somewhere between the level of the lagoon floor and the mean elevation of the offshore bar. It can be precisely determined only by an examination of the lagoonal sediments: the altitude of the transition between aquatic-lagoonal and terrestrial facies in the center of the former lagoon closely approximates the former mean sea level (Zeuner, 1961). Exact sea-level determinations are difficult for other sandy shoreline features.

Marine-Littoral Sediments. Sands and gravels are among the most common sediments of the marine-littoral environment. Coarse sands, sometimes with a scattering of pebbles, are commonly found on the submerged platforms of sandy beaches or nip coasts. Moderately stratified, they commonly occur as topset beds inclined seaward at 5 to 10 per cent or more. Molluscan shells may be present, and if so, they seldom show the broken edges and badly scoured surfaces usually found with derived shell in coastal dunes or semi-aeolian beach deposits. The pebbles of the sandy beach are well-rounded, rather flat, and homogeneous. Various aspects of beach-sand sedimentology are outlined by Shepard and Young (1961).

Pebbly beach deposits, with little or no sandy matrix, are probably characteristic for the abrasional platforms of cliff coasts. Rounding indices are very high and homogeneous except where detritus is in plentiful supply at the shoreline. Flattish pebbles are usual except in potholes carved into the platform. Pothole churning produces more spheroidal shapes. Pebble size varies from beach to beach according to the bedrock and the roughness of the sea during stormy weather. Beds are stratified, sometimes with comparatively steep inclinations.

The possible environmental information provided by fossil faunas contained in such sediments will be discussed in chapter 17.

Coastal Dunes. Coastal dunes are common along many modern sandy beaches. A classification proposed by H. T. U. Smith (1954) includes the four major continental type dunes (parabolic, barchan, transverse, and longitudinal) and the nondescript beach foredunes that lie adjacent and parallel to the shore, forming mounds up to 10 m. high. Shepard and Young (1961) have discussed various sedimentological criteria for distinguishing between beach and littoral dune sands.

Estuarine and Deltaic Features. Deposition at the mouths of rivers may be rather significant. It may take the form of foreset-bedded deltas built out into a lake or into the sea, particularly in areas with little or no tidal variation. In areas of marked tidal amplitude, broad tidal marshes may develop along the river banks near the mouth with mixed fluvial-marine or estuarine deposition.

In Pleistocene contexts, the surface of uneroded estuarine beds gives a fair estimate of the high watermark (Zeuner, 1961). In the case of delta formations, sea-level approximation may be possible when the uppermost sediments are formed by very gently inclined topset beds.

Coral Reefs. Coral reefs are limited to tropical salt waters, with a salinity of 2.7–4.0 per cent, with temperatures never dropping below 18° C. or exceeding 36° C. Light requirements restrict major coral growth to the upper 25 m. of the water, and growth is usually impossible at depths greater than 60 m. Nourishment for the reef-growing organisms is obtained from the open sea, so that the reef grows seaward, extending up to about mean sea level. Coral is absent near the mouths of muddy streams or larger rivers in general, partly because of the presence of fresh water, partly because of the reduced light.

The shorelines built by coral reefs may take the form of *(a) fringing* reefs, attached directly to the shore as platforms; *(b) barrier* reefs, separated from the shore by a flat-floored lagoon of variable width; or *(c) atolls,* "islands" consisting of more or less circular reefs enclosing a lagoon, but without a land surface inside.

Some of the more important general features of modern coral coasts include these (see also Fig. 43):

1) The steep external slope, formed of both coral and talus, often dropping off for many decameters with gradients of as much as 100 per cent;
2) The reef crest adjacent to the external slope, partly constructed of coral-building algae, partly of detrital coralline sand, and usually attaining about mean sea level, although sand ridges may be higher;
3) The reef surface itself, composed of both dead and living coral, often several hundred meters wide, extending no higher than mean low-water; and
4) The lagoon, if present, with a sandy floor, interrupted by sporadic coral growths reaching up to the mean low-water mark.

Since fossil coral reefs are rather widespread and conspicuous in lower latitudes, they provide valuable information both on shoreline levels and on the local marine environment.

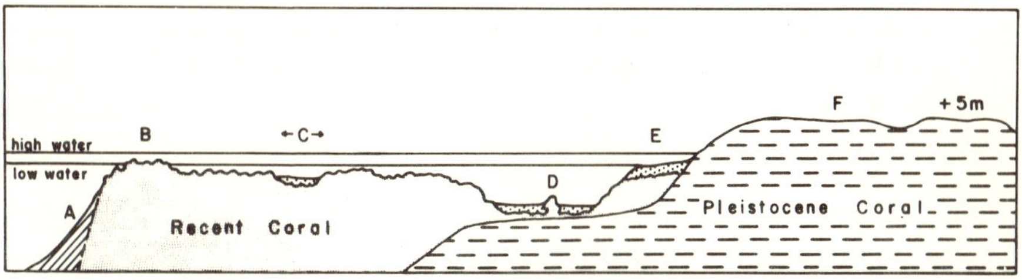

Figure 43. A fringing coral reef: (A) talus slope, (B) reef crest, (C) reef surface (D) lagoon, (E) sandy beach, (F) high Pleistocene interglacial reef. (Not to scale.)

EVIDENCE OF SUBMERGED PLEISTOCENE SHORELINES

Submerged Shorelines. The coastal types discussed above can be used to recognize most high Pleistocene shorelines, generally of interglacial date. They may also record higher sea levels of the Holocene.

Field evidence for "submerged" shorelines, pertaining to glacial-age regressions of world sea level, is a little different. Well-developed cliff coasts and other extensive beach platforms can be recognized as extended platforms with marginal steps or ridges in the submarine topography. Submerged offshore bar-and-lagoon or coral reef topography may also be identifiable from good hydrographic maps. Submerged fluvial phenomena such as stream deposits, estuarine beds, and deltas may also be traced onto shallow marine shelves or may be tapped in bore profiles below modern coastal plains.

Generally speaking, however, little detailed evidence on submerged shorelines is available to date. Even where such shorelines have been indisputably recognized, correlations with particular regressions are difficult or impossible. Stratigraphically more important is the indirect evidence of regressions provided by wind-borne sediments of the coastal zone.

Aeolianites. Coastal dunes functionally attached to sandy beaches play only a subordinate role in the Pleistocene. Far more important are the subtype of consolidated *regressional* dunes, generally called *aeolianite*. These are widespread in lower and lower-middle latitude littoral zones, and were blown up from freshly exposed marine sediments during the regressions of world sea level that accompanied the advance of continental glaciers. They consequently provide valuable stratigraphic information when present in relation to fossil shorelines or continental deposits.

Two major arguments support the fact that well-developed aeolianites pertain specifically to marine regressions rather than to sandy shorelines in equilibrium. For one part, they most often extend to well below modern sea level; on the other hand, they frequently occur along coasts that have no exposed beach sands today. Detailed discussion of some Mediterranean aeolianites is given by Butzer (1962, 1963b).

Aeolianites developed in typical aeolian facies may be found either as *(a)* steeply inclined, uniformly bedded dunes of transverse type embanked against coastal cliffs with typical seaward dip values of 40-60 per cent, and landward dip values of 60-80 per cent; *(b)* free longitudinal dunes of subdued morphology on coastal plains, where they may form littoral cordons—the relief of one dunal generation may be on the order of 5-25 m. while slopes are gentle and seldom exceed 25 per cent; or *(c)* undulating sand sheets with subdued topography of longitudinal affinities, concentrated in the face and lee of minor surface irregularities. These sheets are found beyond the rims of coastal cliffs and well inland on coastal plains or level uplands.

The constituents of aeolianite are mainly lime sands, overwhelmingly derived from the rubble of calcareous marine organisms. In areas of bedrock other than limestone the importance of quartz and other minerals may jump from less than 1 per cent to as much as 90 per cent.

Grain size distributions of the typical coastal facies of aeolianite include a 70 per cent coarse sand (0.2-2.0 mm.) component, while in aeolianites of interior facies this proportion may be reduced to an average of 40 per cent. Clay and fine silt (up to 0.006 mm.) components almost always account for 5-15 per cent. Grain size characteristics of continental dunes are quite distinct: coarse sand components are small, while medium sands (0.06-0.2 mm.) average 20-90 per cent (over double that of aeolianites), and there is an almost complete absence of any component under 0.003 mm. Littoral and continental dunes are, then, both morphologically and sedimentologically distinct.

Such aeolianites are largely indicative of a marine regression actually in progress, for once regression ceases or a renewed rise in sea level occurs, no new sands are exposed to deflation, and consequently sedimentation stops (Wright, 1962*a*). Interruptions of aeolianite deposition may indicate world-wide halts or oscillations of the continental glaciers. Aeolianite deposition will more or less cease when the maximum of a glacial regression has been attained. True aeolianites are therefore stratigraphically equivalent to the phases of glacial advances in much the same way as are periglacial terraces.

Paleo-ecological information may also be available. Poor or absent stratification is usually associated with frequent calcified roots or root-casts of shrubs or conifers. Such aeolianites were probably deposited under vegetation, there intermingling with the needle litter derived from coniferous woodlands. Unbedded littoral dunes form in this fashion on the Balearic Islands today. On the other hand, well-bedded coastal plain or interior aeolianites without root-casts are suggestive of sparse vegetation and prevailing aridity. Further paleoclimatic information may be derived from bedding directions, as in the case of continental dunes.

Other Evidence of Glacial Regressions. During the glacial regressions, continental processes were responsible for erosion or deposition in areas subsequently submerged. Valleys cut by streams or glaciers may now be drowned, forming inlets known as rias or fiords. Stream or glacier deposits may also be submerged, possibly forming islands of peculiar shape. Such drowned deposits may be of considerable stratigraphic importance.

GLACIO-EUSTATIC STRATIGRAPHY

Even the best glacio-eustatic stratigraphy, that of the Mediterranean Basin and western Morocco (Table 3), is not without controversy or problems. Promising sequences also exist along the Gulf-Atlantic coastal plain of the United States, in southern Australia, and on the North Sea coasts of Europe. Useful summaries and tentative correlations of these particular areas have been given by Woldstedt (1958, 1960a, 1960b, 1962a), Richards (1962) and Thom (1969), while Fairbridge (1961), Farrand (1964), and particularly Milliman and Emery (1968) discuss radiocarbon data for sea-level fluctuations during the last 35,000 years or so (Fig. 44). Seen objectively however, the possibilities of world-wide correlation have not been fully explored.

Numerous difficulties exist, even though Pleistocene sea-level fluctuations were a world-wide phenomenon. Local tectonic activity has

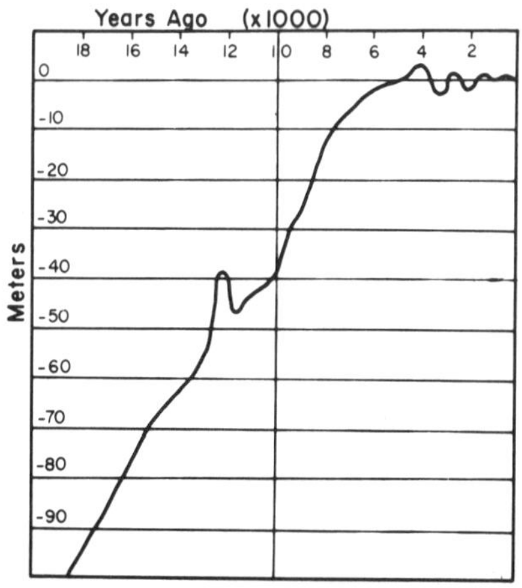

Figure 44. Curve of recent sea-level changes (after Farrand, 1964, and Lind, 1969).

hopelessly complicated the basic picture of eustatic sea levels in some coastal sectors. In those regions overrun by the continental glaciations, compression of the earth's crust under the weight of ice has been matched by isostatic uplift since deglaciation. And elsewhere, slow continental-scale epeirogenic uplift or downwarping has spread the altimetric range of eustatic sea levels over a wider amplitude than was originally the case. In particular, the Sicilian and Calabrian stages are impossible to employ on the basis of elevation. World-wide correlation of the Eemian high sea levels may, however, prove feasible in tectonically stable areas.

The stratigraphic value of glacial eustatic sea-level fluctuations plainly does not yet lie in world-wide altitudinal correlation of high Pleistocene shorelines. The only fact of general applicability is that the Elster, Riss, and Würm glaciations corresponded to major regressions, while the intervening interglacials experienced sea levels at least as high as the present. Excellent local sequences can theoretically be constructed upon these premises. But prior to the Eem there is little possibility so far of specific correlation with higher-latitude continental stratigraphy. And prior to the Cromerian there is little certainty as to whether the glacials corresponded to regressions or the interglacials to transgressions.

Nonetheless, Pleistocene shorelines are significant from a number of perspectives. Firstly, the world-wide rhythm of glacial regressions and interglacial transgressions—at least during the second half of the Pleistocene—provides an invaluable stratigraphic guide to fixing local relative continental stratigraphies. Then again, local sequences of higher sea levels and intervening regressions can be established as regional stratigraphic horizons. Some coastal deposits such as coral, aeolianite, or molluscan faunas may provide direct paleoclimatic information. And lastly, the topographic significance of Pleistocene sea levels for the distribution of human settlement and the paths of human dispersal need hardly be stressed.

FIFTEEN

Geomorphological Study of Archeological Sites

GEOLOGIC CLASSES OF ARCHEOLOGICAL SITES

Introduction. Archeological sites, depending on the cultural time range involved, may represent former houses, villages, or towns; they may pertain to temporary or seasonal camps, or to killing or butchering sites. Other sites may have little or no ecological meaning, but consist only of scattered artifactual material, possibly redeposited within a river terrace. Sites dating from historical or late prehistoric times are commonly found at the surface, possibly buried under cultural debris or a little blown dust, and altered by a weak modern soil profile. Many sites of all time ranges are found exposed at the surface. A great number of prehistoric sites, however, are found in direct geologic context, within or underneath sediments deposited by some geomorphic agency.

For the archeologist a "stratified" site is one with distinct archeological horizons, with or without a gelogic context. The term "surface" site might be used to describe a variety of things, such as an archeologically unstratified surface find or even an ancient open-air encampment—as opposed to cave site—now buried by a meter or two of loess or marl. From the archeologist's point of view, then, sites may be classified according to their cultural-ecological meaning, or according to digging criteria. The earth scientist, interested in providing a stratigraphic date or a geographic-ecological meaning for a site, would naturally use different criteria of classification.

"Stratified" and "Surface" Sites. The most interesting kind of archeological site for the earth scientist is one found in a direct geologic context, i.e., geologically stratified or geologically *in situ*. This should not imply that the cultural materials have not been derived, but only that their present location is geologically circumscribed. For the sake of convenience, "stratified" will here be used in this geologic sense only. "Surface" site will be restricted to materials found at the surface, without geologic context.

According to the basic geomorphological situation or the type of deposits involved, archeological sites can be geologically classified as follows:

a) Alluvial sites: artifacts, fossils, occupational floors, and the like found within former stream deposits.
b) Lacustrine sites: archeological materials found in former lake beds, in ancient bogs, swamps, or spring deposits.
c) Aeolian sites: archeological materials found in or under wind-borne sand or loess, or found in relation to features resulting from deflation or wind scour.
d) Slope sites: archeological materials found within or under deposits due to mass movements motivated by gravity action, including congeliturbate beds, slope wash, reworked loess, or talus.
e) Cave sites: archeological materials found in caves with some form of geologic or archeological stratigraphy.
f) Coastal sites: archeological materials found in direct relation to erosional or depositional phenomena associated with former coastlines.
g) Surface sites: the great mass of scattered archeological materials found at the surface, with little possibility of direct association with any geomorphic event.[1]

Identification of the depositional medium contemporary with or subsequent to an archeological site is vital to the earth scientist. The specific relation of a cultural horizon to a geomorphic event can provide direct paleo-environmental information. This local environmental setting may in turn be stratigraphically linked to regional or world-wide changes of climate. Both environmental reconstruction and dating of sites or

1. In many cases sites can definitely be associated with two different geologic categories, e.g., cave sites found along former shorelines. A further category could also be set up for volcanic sites buried by lava or ash falls. However, almost all volcanic sites would be related to ash falls which, as discussed in ch. 12, are primarily wind-borne. Consequently, the majority of volcanic sites can be best classified as aeolian, the others as alluvial or lacustrine.

occupation levels may then be possible through geomorphological investigation.

Pleistocene and Holocene sites. The frequent statement that only pre-Holocene sites have geologic context is not completely true. Numerous late prehistoric sites have been found in swamp or cave deposits or have been buried by aeolian sediments. On the other hand, there are countless pre-Holocene surface sites, some of them still recognizable as occupation sites. Although there are more sites with meaningful geologic contexts in the Pleistocene (in the restricted sense) than there are in the Holocene, many late prehistoric and even historical sites are buried by sediments of greater interest than most archeologists would suppose.

A distinction of greater validity between Pleistocene and Holocene sites concerns an understanding of the original topography. A completely different perspective is required in studying a Pleistocene site. The great changes of climate and environment introduced rather different local settings in many areas: shifts of stream channels, valleys swept by glacial meltwaters, swamps or lakes that no longer exist, loess plains where forests prevail today, shorelines well inland. Many geologically stratified sites were once a part of valley floors that remain as dissected terraces high above the present stream. The environment and topography of post-Pleistocene sites has changed in degree only. One has little difficulty in envisaging the site in its setting, a factor of more than psychological advantage.

ALLUVIAL SITES

Alluvial sites rank second only to cave sites in the early history of archeological excavations. Excavations or borings in river valleys have frequently struck alluvial sands or gravels of various ages containing animal remains or human artifacts. Natural exposures in terrace faces have also revealed archeological materials. Many such sites have little more to offer than sporadic, water-rolled stone implements and possibly a little bone of dubious association. Other sites, however, may represent occupation floors with rich associations of undisturbed tools and fossils. Interpretation of such sites can, with enough effort, be carried to a satisfactory stage of environmental and stratigraphic understanding.

The periglacial stream terraces of the Old World were probably first studied by Paleolithic archeologists, and the well-known Somme River succession of northern France was established as a sequence of intergraded solifluction beds, loesses, and periglacial stream deposits (Breuil and Koslowski, 1931-32). Once assumed to be the framework of Paleolithic cultural stratigraphy, many of the sites in question are of limited importance since they were mainly collections of derived artifacts rather than occupation floors. Of far greater significance today are well-studied sites such as that of Salzgitter-Lebenstedt, near Braunschweig, Germany, which is situated along the banks of a late Pleistocene tundra stream at the base of a slope affected by solifluction (Tode *et al.,* 1953). The Middle Pleistocene site of Markkleeberg, near Leipzig, was similarly located in periglacial gravels of a northward flowing stream subsequently overrun by the Saale moraines (Grahmann, 1955).

The alluvial terraces of the arid zone have played a significant role archeologically in both the Old and New World, even in rather late prehistoric times. The twin sites of Torralba and Ambrona, in the Spanish province of Soria, were situated on the marshy floodplain margins of a stream valley during a cold, moist phase of the Lower Pleistocene (Butzer, 1965; Howell, 1966). In a drier environment, the terminal Paleolithic cultures ("Sebilian") of Kom Ombo, Egypt, were concentrated along the river banks of several defunct Nile branches (Butzer and Hansen, 1968).

Swanscombe is an example of a significant site associated with downstream valley alluviation during a high, Middle Pleistocene sea level (see Howell, 1960). The site was occupied on a Thames floodplain almost 30 m. above that of the present. Although not wholly undisturbed by stream redeposition, the contemporaneousness of the human and animal fossils was established by fluorine tests.

Geomorphological investigation of alluvial sites should concentrate on two aspects of detailed work, apart from the more general question of external correlations and regional setting:

a) Have the materials been derived? Cornwall (1958, pp. 23-24) raises the problem that archeological objects may be carried into a stream bed by floodwaters or abandoned on the floodplain by man. In either case implements could be part of deposits reworked several times through shifting stream channels. Consequently, Cornwall suggests that a "rolled" artifact need not be derived from deposits older than those in which an unabraded artifact is found. The difference in condition could be due entirely to the distance the artifacts traveled before once more coming to rest. Since animal bones are far less resistant, Cornwall believes that they will rarely survive redeposition, so that the relation of fossils to artifacts in alluvial beds raises a serious problem. If the artifacts are quite unworn, bone and tools may be strictly contemporary. If the tools are water worn, however, it is reasonably certain that a considerable time distance separates them (Cornwall, 1958, p. 24).

Some alluvial sites clearly represent occupation floors that are *in situ* in every way, with no question that the entire cultural assemblage is contemporary. Even here the excavator or the geomorphologist can contribute information by making orientation studies of tools, bones, and the like, to see whether some reorientation has taken place through stream action. Careful attention should also be given to possible water wear on artifacts or fossils. The natural mineral sediments in the cultural layer may also provide information about whether running water affected the site during occupation or only afterward.

When a site does not constitute a clearly defined cultural floor, derivation problems may be more serious. Stream rolling of fossils produces noticeably worn joints. Different degrees of rolling can be recognized in stone implements. Sharp, fresh edges suggest little or no water wear. Sharp edges that are smooth or blunt to the touch should be checked with a hand lens. The edges may show minute rounding while scratches may be evident on the faces. Such an artifact is probably *water worn.* If the edges are conspicuously blunted, sometimes beyond recognition, the implement is clearly *rolled.* In the case of dubious assemblages, a representative sample should be analyzed by an improvised classification of wear characteristics. If the assemblage is not largely homogeneous, caution should be exerted. Gravel analyses applied to all apparently unworked stone may show that a part of the rock present was deliberately fractured or is foreign to the bed load of the stream. Implement orientation and, finally, sedimentological study of the mineral beds should permit an opinion as to the significance of the site.

b) What part of the stream valley was occupied? As important as derivation and redeposition is the topographic location of a site within the former floodplain. Sedimentological studies should establish whether the occupation floor was located on the backset beds of the river bank, in the intercalated backset and topset beds of the levee, or out on the alluvial flats. Particle sizes and sorting should reveal whether stream flow was strongly seasonal or uniform, whether the waters were fast- or slow-moving, what their direction was, and finally, what conditions of waterflow ultimately overcame and buried the site?

All in all, in the ideal case, stream terrace sequences—with archeological assemblages in geologic context or in place on the surface, with surface paleosols or derived soil sediments at the base of terrace alluvium—can be of exceptional interest for both the prehistorian and the earth scientist.

LACUSTRINE SITES

Prehistoric settlement was common around the banks of lakes. So for example, the early Holocene site of Star Carr, Yorkshire, was situated next to a now extinct lake, and subsequently buried by bog deposits

(J. G. D. Clark, 1954). In the Fayum depression of northern Egypt, high Nile floods were responsible for the creation and maintenance of several late Pleistocene and Holocene lakes (Caton-Thompson and Gardner, 1929, 1934). Various Paleolithic and Neolithic populations occupied the fringe vegetation of these lakes, leaving cultural and animal remains along the former shorelines or within the sands of the beaches. At the Lower Pleistocene site of Ternifine, in western Algeria, a rich fauna with skeletal remains of the hominine *Atlanthropus (Homo) mauritanicus* is exposed in clays and spring deposits of a former lacustrine basin (Arambourg, 1955). And last but not least are the *Zinjanthropus* and "pre-Zinj" sites found in mixed lacustrine and volcanic ash beds of Bed I, at Olduvai Gorge, Tanganyika, dating from the Basal Pleistocene (Hay, 1963).

In higher-latitude Europe, lacustrine beds were mainly found in poorly drained gound moraine areas abandoned by the continental glacier. So for example, the Lower Pleistocene occupation level at Hoxne, near Ipswich, is located in clayey silts of Holstein interglacial age. The beds record a former lake within a depression in Elster till (West and McBurney, 1954). The interesting Middle Paleolithic spear of Lehringen, near Hanover, was found with an intact elephant skeleton in lacustrine marls of Eem interglacial age, overlying the Saale ground moraine (Adam, 1951).

Swamp and bog deposits, some of them postdating sites, have long enjoyed considerable archeological interest in northern Europe. They have produced potsherds, plowshares, house or village foundations, and even fully intact corpses.

The general environmental setting of lacustrine beds may be a task for both the earth scientist and biologist. Why, for example, was a lake or swamp there in the first place? The more detailed geomorphological site investigation revolves around the site's association with the lake or swamp in question: how was it located in relation to this lake; did the lake exist at the time; was it shallow or deep, seasonal or perennial; was the site flooded soon after occupation? Sedimentological investigation may contribute in many ways to a full understanding of both the local and the regional setting.

AEOLIAN SITES

Specific archeological associations with aeolian features are mainly of three kinds:

a) Occupation floors or scattered artifacts found under or on top of sand dunes;

b) Archeological materials found under, within, between, or on the surface of loess;

c) Archeological materials exposed by wind deflation or scour.

One of the best examples of an archeological site related to a complex sequence of stream and wind erosion and deposition is the Holocene San Jon site of eastern New Mexico (Judson, 1953a). Most of the terminal Pleistocene Sebilian cultures of the Kom Ombo plain, Egypt, were deflated and are now found partly on yardangs scoured out of old Nile deposits (Butzer and Hansen, 1968).

In the late Pleistocene, innumerable loess sites in central and eastern Europe are examples of occupation during or after loess sedimentation. Possibly among the most famous are the Moravian Upper Paleolithic stations of Předmost (see Žebera *et al.,* 1955) and Unter-Wisternitz (Dolni Vestoniče) (Klima *et al.,* 1962, Klima, 1954).

Geomorphological investigation of aeolian sites is concerned primarily with whether aeolian activity was contemporary with occupation, and whether it preceded or followed occupation. Evidences of soil development in the stratigraphic profile are important, and other indications of sedimentary breaks may be obtained from vertical curves of particle sizes, carbonate, or humus content. With due caution, pollen studies may also be possible in the humic horizons of an aeolian profile. In general, the exact stratigraphic correlation of sediments and archeological levels can almost always be determined, and careful examination may possibly reveal both the contemporary environmental setting of the site and the changing environmental patterns of the period.

SLOPE SITES

A large number of late Pleistocene sites in Europe and Holocene sites in the modern arctic have been seriously affected by slope solifluction or cryoturbation of artifacts and fossils, leading to disruptions of archeological stratigraphy. Other gravity movements in nonarctic regions may also be significant for sites. The various "slope sites" (in the broad sense) require more detailed attention.

a) Sites subsequently buried by solifluction can be studied by means of pebble orientation and morphometric gravel analyses. So for example, at Salzgitter-Lebenstedt (Tode *et al.,* 1953) orientation studies in a complex stratigraphic sequence at the base of a slope determined the varying amounts of lateral downslope movement caused by solifluction as well as the fluvial sedimentation related to stream bedding. Gravel analyses can be applied to determine comparative indices of flattening between the weathered coarse source material still *in situ* and similar materials that have suffered short solifluid transport. Transported detritus is slightly flattened because of dominantly sliding motions. Repeated frost-splitting during transport also increases the angularity of the sample. By combining both techniques at Salzgitter-Lebenstedt, it was possible not only to define the physical conditions contemporary with the cultural horizon, but also to determine the position of the site in the climatic stratigraphy of the Würm.

Another application of solifluction phenomena to archeology is that of the "stone lines" running perpendicular to moderate slopes in many former tundra regions. Most of these are nothing but fossil stone stripes, a patterned ground phenomena.

b) Sites disturbed by solifluction or involutions during or after deposition may show complex vertical and horizontal movements or artifactual materials—bedded in sediments of differing density and grain size, with different rates of frost-penetration and frost-heaving. Organic and mineral matter in particular tend to behave differently. As a result, horizons of fossils and artifacts may be mixed or even inverted. A surface site may also be distributed through many decimeters of soil, giving the impression of a long-term site with geologic context, or even of complex archeological stratigraphy. The difficulties involved in interpreting the late prehistoric Eskimo site at Engigstciak, Yukon Territory, Canada (Mackay *et al.,* 1961) illustrate this well. Any indication of contortion or other nonhorizontal disturbances should be carefully studied for possible effects of cryoturbation.

c) Slope and gravity phenomena of middle and low latitudes may effect archeological sites in different ways. They carry little or no environmental significance but are nonetheless important. Some of the more common categories can be enumerated:

1) Talus accumulation and rock falls have buried or destroyed many sites located near steep cliffs or under former rock overhangs. There are few stratigraphic or environmental problems here.

2) Creep and slope wash often lead to accumulations on or at the foot of even moderate slopes (over 8 per cent). In this way scattered artifacts may be found in geologic contexts with no reason other than subsequent mass movement. Many *Schwemmlöss* beds contain former surface tools embedded during redeposition by slope wash.

3) Pseudostratigraphic situations can be produced by earth slips or earth flows, the notorious south German site of Lengfeld, near Regensburg, being a case in point. Three so-called cultural levels

of final Paleolithic aspect proved to be nothing but a random collection of Paleolithic, Mesolithic, and Neolithic surface materials stratified by three successive earth slips in late prehistoric times (see Zotz, 1956). Geomorphic investigation of such a site would have saved the original excavator both money and reputation.

4) Related to problems of surface materials buried by slope redeposition is the widespread occurrence of Middle Paleolithic implements several decimeters within the soils of tropical Africa. This is attributed to activity of microorganisms, particularly ants and termites, which inadvertently displace such artifacts from the surface to the base of the biologically active subsoil (J. D. Clark, 1960). Similar phenomena have been observed in England, although Cornwall (1958, p. 53) emphasizes that relative stratigraphic order is always preserved. Burrowing animals may also disturb archeological context (Marx and Reed, 1957).
5) Another problem of some relationship to gravity movements and redeposition is the possibility that implements may "sink" into well-lubricated, unconsolidated beds. This form of movement is not confined to swamp or lake beds but apparently can occur in river floodplains as well. So for example, Predynastic potsherds occur at the base of the somewhat older Nile Valley mud. Sandford (1934, pp. 107–8) believes they migrated downward through gravity action.
6) Finally, volcanic lavas, although not typical gravity phenomena, may be mentioned at this point. They frequently have stratigraphic and dating applications, but need no further discussion.

CAVE SITES

Cave sites and associated sediments have been considered in detail in chapter 13. The only point requiring further comment is the type of use that early man made of caves. In all but the rarest cases, occupation was limited to the foreparts or entrance area of a cave. Deep interior caverns were widely used for ritualistic or artistic purposes in some areas, but such damp, lightless vaults would hardly have proved attractive for dwelling purposes. If certain modern ethnological analogies have bearing on Paleolithic cave-dwellers, it may be mentioned that the Australian aborigines of the northern Lake Eyre area, the Shoshones of the Great Basin, and the Kalahari Bushmen are all known to have occupied caves or overhangs on occasion. In each case leaves, branches, grass, moss, bark, etc. were used to line the walls. Together with the use of fire this emphasizes that cave microhabitats were actually modified by man during his occupation, possibly sufficiently to affect the geomorphic environment. This question needs further checking in the field.

Cave sites have assumed importance at many times and in many areas, ranging from the australopithecine caves of South Africa to the crevice breccias of Peking, from the Upper Paleolithic caves of southern France and adjacent Spain to the terminal Pleistocene cave cultures of the southwestern United States. It would be superfluous to give specific examples here.

COASTAL SITES

Fluctuating Pleistocene sea levels are obviously of more than passing interest for archeology. For one thing, they were a physical reality for prehistoric man to cope with, although not in any dramatic way such as "watching the waters rise" (e.g., Pfannenstiel, 1944). Shorelines have always set limits to the habitable earth, especially during the Pleistocene when large coastal areas were alternately submerged and exposed. Glacial regressions provided land links between islands and continents, a factor of fundamental importance in Paleolithic migrations. The natural distribution of insular floras and faunas was also partly the result of glacial regressions that allowed temporary passage over exposed continental shelves. Finally, shoreline stages may permit stratigraphic dating of coastal sites.

Man has often been attracted to the coast for economic reasons, so that there are specific coastal sites related to modern or Pleistocene shorelines. In other cases man has also occupied caves abandoned by the sea during glacial regressions.

Coastal sites may consist of *(a)* occupation sites located in former sea caves, *(b)* occupation sites located on former beaches and often incorporated into marine or aeolian sediments, and *(c)* transient occupations of a littoral zone, recorded by scattered artifactual materials found in geological context or at the surface of coastal erosional features. In other cases, marine transgressions destroyed former continental sites and reworked their artifactual materials, embedding these in beach deposits.

Foremost among coastal sites are former sea caves, usually occupied during the sea's regression from a particular shoreline, and occasionally submerged anew during a subsequent transgression. To mention only one example, the renowned cultural sequence of Sidi Abderrahman near Casablanca (Biberson, 1961a, 1961b) is preserved largely with infillings of what were a number of ancient sea caves. Methods of study are in part similar to those of cave sites in general, although in part they also involve gravel or paleontological studies of marine beds and mechanical analyses of aeolianite strata.

A number of important surface occupation sites may be found adjacent to former shorelines. These belong to *strandloper*-type cultures devoted to seafood consumption. It is possible, although difficult to prove, that comparatively rich archeological assemblages found but little rolled within marine sediments of the beach platform may actually have belonged to an immediate shoreline settlement. Stratigraphic dating possibilities are again excellent at such localities, although environmental aspects are more difficult to reconstruct. Organic materials, apart from shell and bone, are most likely to be absent.

Scattered artifacts found in coastal deposits should be regarded with caution. They may be derived from destruction of interior sites by marine transgression, so that the marine sediments in question only provide a chronologically younger limit or *terminus ante quem.* In the case of scattered surface implements, as opposed to concentrated occupational sites on the surface, the shoreline itself more often than not serves only as a *terminus post quem.*

SURFACE SITES

Not least in either significance or interest are the great mass of surface finds—scattered implements or true occupation sites—not found within or under geologic deposits. Geologists engaged in archeologic-geomorphic work generally shrug their shoulders at surface sites, with the implicit observation that there is nothing for them to do here. Admittedly, in terms of chronology, a time span of anywhere from the present to the Würm or the Paleozoic may be all that is indicated.

However, even such *termini post quem* can often be of importance. A site located on last glacial till is probably of Holocene age. One example that such information is not useless was provided by seven Acheulian-type bifaces found at the surface near Tocra, Cyrenaica, in 1943. Investigation at the site indicated that these implements of pre-Würm appearance were found on top of Würm-age pluvial gravels (McBurney and Hey, 1955, pp. 172-74), suggesting possible human interference.

Chronology is, however, far from being the only point of interest. The geographic setting of surface sites can, in part, be studied successfully by geomorphic methods. Siting with relation to regional landforms can be particularly significant for archeological surveys. When the typical geomorphic situation of certain occupation sites is understood, regional landforms can be evaluated as to their possible role in the pattern of prehistoric settlement. A case in point is the Neolithic-Predynastic archeological "gap" between Upper and Lower Egypt, which is in fact accidental rather than cultural: any sites within this area have either been buried by Nile alluvium or drifting sands, or destroyed by expansion of the cultivated land (W. Kaiser, 1961; Butzer, 1960b, 1961b).

The geographic setting and environmental evaluation of a site should also be attempted by a study of outside evidence from contemporary neighboring sites, or by the biological investigation of organic remains obtained from the site itself.

Obviously there are no fixed rules for geomorphological work applied to surface sites. It is here that the personal intuition, improvisation, and interest of the earth scientist begins to be decisive, and the problems to be solved or formulated must grow out of discussion with the archeologist. Depending upon physical and human factors, the problems involved will vary from country to country or culture to culture. To recognize the problems in the first place, the geomorphologist must have some familiarity with archeology and must actively exchange ideas and notions with the anthropologist. In other words, the "straight" geologist with little direct interest in the cultural aspects of a site cannot apply himself fully to problems that can only be formulated in interdisciplinary discussion.

GEOMORPHOLOGY, SURFICIAL GEOLOGY, AND ARCHEOLOGICAL SITES

Through study of the geomorphology and sediments of a site the earth scientist may be able to:

a) Reconstruct the local habitat or setting of a site, including
   1) terrain type and geographic location with respect to terrain features;
   2) water resources, if available, and whether permanent or not;
   3) ground water conditions and likelihood of flooding.

b) Reconstruct the regional environment—preferably with contributions from the biological sciences.

c) Establish a local stratigraphy that may be integrated into the chronology of a wider area.

The specific relationships of sites to particular geomorphic situations or Pleistocene deposits may also be of value in archeological surveying, either in the field or through use of aerial photographs or geologic maps.

The field work and environmental interpretation involved can be undertaken only by a qualified earth scientist. But the archeologist must also understand something of what the geomorphologist or Pleistocene geologist can do and what he is about. This is essential if there is to be any profitable interdisciplinary discussion. The archeologist need not become a field geomorphologist, but it is vital that he be able to understand the full implications of the earth scientist's findings. In many reports, archeological and earth science evidence remain quite unintegrated and of correspondingly limited value. Possibly this chapter will be of some use, in connection with the preceding section, in outlining principles and possibilities of such geomorphological investigation, thus helping to sponsor the necessary synthesis of cultural and physical data.

PART FOUR

Contributions of the Biological Sciences

SIXTEEN

Palynology and Paleobotany

GENERAL PRINCIPLES OF POLLEN ANALYSIS

Pollen analysis, or palynology, is by far the major botanical technique in paleo-ecological work, although examination of plant macroremains, where available, is equally vital. The results achieved by palynology are truly spectacular in their detail, interpretive value, and record of rapid time-change when compared, for example, with geomorphic investigation. But palynology may be limited both areally—for want of suitable sediments—and temporally—through poor representation in certain strata. It is consequently only one of several major fields of study and an excellent source of complementary evidence in sedimentological work. Comprehensive studies of theoretical palynology and its application have been made by Erdtmann (1954), Firbas (1949–52, vol. I), Faegri and Iversen (1964), Overbeck (1950), and Felix (1961).

The basic principle of pollen analysis is that most wind-pollinated trees, shrubs, and grasses emanate pollen in great quantities. The particle size of pollen is on the order of 0.01–0.1 mm., and the absolute weight less than 10^{-9} grams. Consequently, pollen grains are readily removed by wind and widely dispersed in the lower atmosphere where the grains are carried in suspension. Distances of 100–250 km. are crossed readily by traveling pollen, and grains may be found up to several kilometers in the lower atmosphere. Pollen density is greatest at elevations of 200 to 500 meters above the ground, and the density remains appreciable to elevations of 2 km. Pollen accumulations in any one locality will, therefore, provide a regional rather than a local cross-section of the pollen-emanating plants present.

The annual pollen "rain" in a vegetated area amounts to several thousand grains per square centimeter. A part of this pollen may be preserved indefinitely if oxidation is limited or absent, particularly in dense, poorly aerated sediments or in acidic environments such as provided by bogs or many lake beds. Year after year, stratified laminae of sediments, including a small cross-section of the year's pollen that is preserved, may be laid down under various conditions at a number of localities. Each of these sediments, then, preserves its own chronological and environmental record.

Of great importance to the botanist is the fact that the pollens of different plants are quite individualistic and can in many cases be identified easily as to genus (Fig. 45). On the specific level, some trees and most nonarboreal species are rather more difficult. Only in rare cases can subspecies be identified accurately, and then only with difficulty.

FIELD REMOVAL OF SAMPLES FOR POLLEN EXAMINATION

Field sample selection and removal may be carried out by earth scientists and archeologists, as well as by palynologists.

Generally speaking, organic lacustrine sediments and semiorganic beds such as marls are excellent for pollen preservation. Clays and dense silts, where not conspicuously weathered, are generally favorable. Sandy sediments are usually poor, unless dense impermeable strata "seal" the beds in question. Beds discolored by oxidation are also poor.

Samples should be removed from stratigraphically meaningful horizons or sections at vertical intervals of 5, 10, 20, or 30 centimeters as the case warrants. Some 20 cubic centimeters of material will suffice in the case of all but sands or deeply weathered sediments. The sample should be removed with great care so as to avoid contamination from the atmosphere or the removing tool, preferably after cleaning and discarding the surface layer. Short test tubes with a cap are most convenient for storage. In the case of general pollen studies applied to nonexposed strata, such as lake or bog sediments, core borings are made using various hand-operated devices.

In the case of open sections, and as far as possible in core profiles as well, careful study should be devoted to the possible presence of sedimentary disconformities, erosional surfaces, traces of fire or peat cutting, and the like.

LABORATORY PREPARATION OF SAMPLES

Although the laboratory preparation and ultimate analysis of samples is strictly a task for the qualified specialist, it is useful for the persons concerned to be aware of the different results possible, depending on the

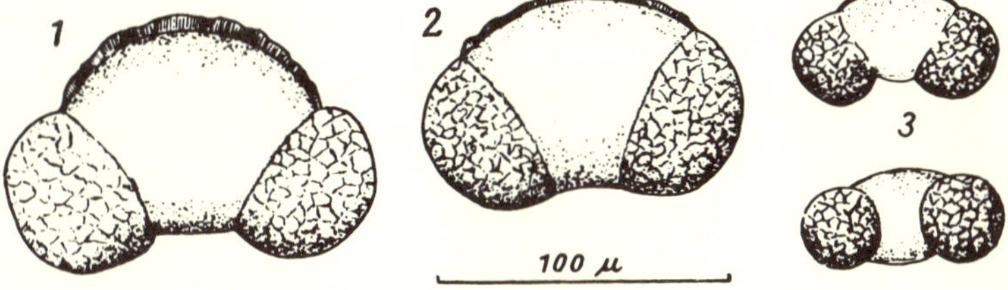

Figure 45. Some pollen types (from F. Overbeck, 1950, by permission of the Niedersächsisches Institut für Landeskunde und Landesentwicklung, Göttingen): above (1) Abies, (2) Picea, (3) Pinus; below (1–2) Alnus, (3–4) Betula, (5) Corylus, (6) Carpinus, (7) Quercus, (8–9) Ulmus, (10-11) Tilia, (12) Fraxinus, (13) Salix, (14) Fagus, (15) Juglans, (16) Castanea.

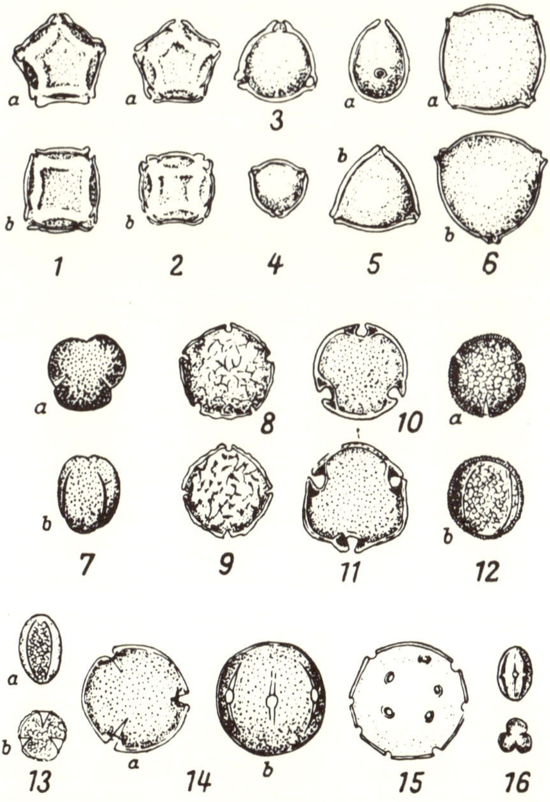

method of preparation used. There is, at present, no standard method of preparation. Many authors specify their methods, others do not. Some authors have obtained pollen from certain samples by using one technique; other authors employing another method may fail to find any pollen. Some techniques are widely considered acceptable, while others are frequently considered of dubious validity. Other less frequently mentioned but equally serious grievances are directed against overintensive preparation of samples with massive destruction or mutilation of pollen. Since pollen has now been widely and successfully studied from rather "unorthodox," nonacidic, sedimentary environments in the arid zone and humid tropics, new preparation methods have necessarily been introduced to preserve from wanton destruction such pollen as is present. The writer, not in any way qualified to evaluate these methods, cannot comment specifically on the topic. But it is highly recommended that geomorphologists and archeologists carefully ascertain the methods whereby their samples are studied, and if need be, seek outside advice on their reliability.

Until recently the only evaluation of the problem, unfortunately not well suited for the nonspecialist, was a detailed compilation by C. A. Brown (1960). The revised text book of Faegri and Iversen (1964) consequently fills a long-felt need.

Only one, widely employed technique is briefly described here, in order to illustrate the stages of "cleaning." Three undesirable substances may be present and may be removed in the following manner:

a) Calcium carbonate is removed with cold, diluted (25 per cent) hydrochloric acid.
b) Silica is removed by letting the sample stand for 48 hours in 40 per cent concentrated hydrofluoric acid, after which the sample is washed and then heated with 10 per cent hydrochloric acid.
c) Unwanted organic matter is destroyed by first boiling in 10–15 per cent hydrogen peroxide and then, after washing, boiling the sample a second time in 10 per cent potassium hydroxide.

All three techniques may have to be applied to clays or marls, whereas only *(c)* may be required in the case of peat, lignite, or coal. When the various undesirables have been so removed, the final residue of pollen is mounted in glycerine jelly on a permanent slide or suspended in liquid glycerine for immediate investigation under the microscope with 300× to 1000× magnification.

THE POLLEN SPECTRUM AND POLLEN DIAGRAM

Either 100 or 200 grains are identified, the total pollen assemblage of

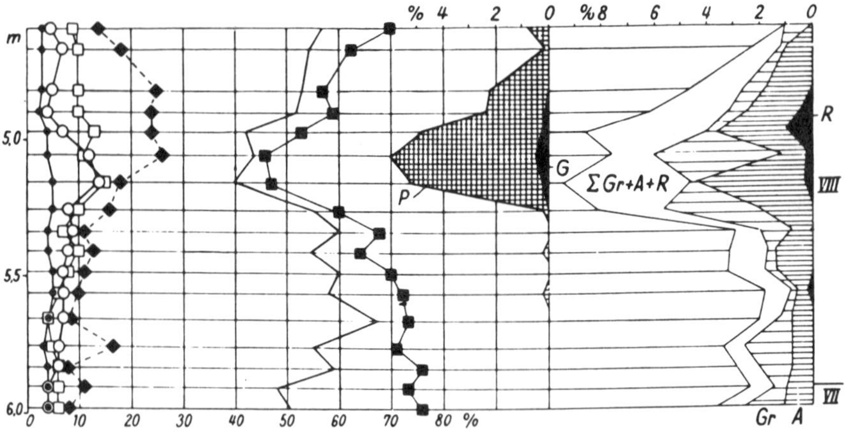

Figure 46. Partial pollen diagram from Körup Sö, Denmark (after Iversen, 1949, from Walter, 1954, by permission of J. Iversen and of Eugen Ulmer Verlag, Stuttgart). The tree pollen symbols are as follows: dark squares, oak, plus, elm and lime; drawn line, oak only; dark rhomboids, hazel; dark dots, pine; open squares, alder; open circles, birch. The NAP (on the right) include Plantago (P), cereals (G), grasses (Gr), Artemisia (A), and Rumex (R).

which is known as a *spectrum*. When samples from successive stratigraphic layers are studied, data are presented in the form of a stratigraphic *diagram* representing changing spectra through time (Fig. 46). Such pollen diagrams generally show the absolute level, the sedimentary facies of the stratigraphic column, and the percentage of each genus with respect to the total count of arboreal pollen (AP) grains. Modern analyses invariably include a single or specified curve of the nonarboreal pollen (NAP) present, expressed as a percentage of the total AP. The pollen diagram may consist of a single composite diagram, including all of the AP, with separate recording of NAP in one or more vertical columns. Other authors prefer separate vertical columns for each pollen species (see Fig. 75).

DIFFICULTIES IN THE INTERPRETATION OF POLLEN PROFILES

The validity of the pollen spectra and diagrams is never absolute, even with proper identification. Errors are sporadically or systematically introduced by a variety of factors:

a) Differential representation of pollen due to
  1) differing surface receptivity of lakes, swamps, bogs, etc.;
  2) differential preservation of pollen under different environmental conditions.

b) Primary over- or underrepresentation of species:

1) some species are systematically overrepresented on account of excessive production: *Corylus* (hazel), *Pinus* (pine), *Alnus* (alder), and *Betula* (birch);
2) some species tend to be present in more representative proportions: *Abies* (fir), *Carpinus* (hornbeam), *Picea* (spruce);
3) some species are underrepresented on account of small pollen production or insect pollination: *Fagus* (beech), *Quercus* (oak), *Ulmus* (elm), *Tilia* (lime) and *Salix* (willow);
4) some species are little or not at all represented on account of easily decomposed pollen: *Populus* (poplar or aspen), *Acer* (maple), *Fraxinus* (ash), *Castanea* (chestnut), *Larix* (larch), and *Juniperus* (juniper).

c) Long-distance transport of pollen by wind. Pollen represented by less than 1–2 per cent of the AP spectrum may be a result of long-distance transport. It is thought by some that pollen from distances exceeding 100 km. represents less than 1 per cent of the spectrum, provided there is no local source (Firbas, 1949–52, vol. I, pp. 23–24).

d) Redeposition of pollen from older sediments.

e) Long-distance transport by streams.

f) Truncated, interrupted or incomplete profiles due to

1) fire or peat cuttings leading to destruction of sections;
2) natural interruptions of sedimentation;
3) lateral differences of sedimentation within the water body.

The implications of the first source of possible error above are presumably small, although no comparative material exists for an objective assessment. "Errors" introduced by (*b*) and (*c*) are rather substantial, but do not provide an objection to the use of pollen data. The palynologist is fully aware that a pollen spectrum does not provide an absolute or relative picture of forest composition. True *forest composition* can only be estimated by, first, full evaluation of available macrobotanical evidence, such as preserved leaves, leaf impressions, wood, seeds, fruits, and second, by comparative study of the pollen spectra associated with various forest or grassland associations today. Relative *forest density* can be estimated by, first, the relative percentages of AP and NAP pollen in a diagram, and second, the relative pollen density per unit of sediment compared with pollen density in similar sediments within the pollen diagram. In other words, palynology entails considerably more than mechanical identification and enumeration of pollen. It requires that the analyst be a knowledgeable plant geographer or plant ecologist.

Finally, the possible errors introduced by factors (*d*) and (*e*) can be evaluated systematically by careful site examination. Redeposited pollen must be derived from a source whose direct accessibility should be considered. In case of question, sedimentary strata possibly affected by erosion may be checked for pollen content, if any. Disturbed profiles may be readily detected in section, although this would not be easy in the case of simple core-bore sampling.

APPLICATIONS OF POLLEN ANALYSIS

Pollen analysis may be applied to a broad range of paleo-environmental problems.

a) Reconstruction of local vegetation. Careful interpretation of contemporary pollen spectra from neighboring sites may provide a good picture of local vegetation and ecology, a technique that may be extended to particular regional settings (Firbas, 1949–52, vol. II).

Certain floral elements are characteristic of certain environments, although most genera are distributed rather more broadly. Species identification is usually required, and just this is difficult or impossible through palynology alone. The direct paleoclimatic interpretation of many spectra is consequently limited. If species identification is possible and supported by some macrobotanical evidence, the presence of several tundra elements[1] may, for example, establish the vegetation as tundra or forest-tundra. The additional presence of steppe genera such as *Artemisia, Hippophae* (sea buckthorn), and *Helianthemum* (rock rose) would contribute a further environmental note. Vegetation reconstruction is commonly based on such combinations of ecologically significant species within the over-all pollen spectrum obtained, for example, from a geological exposure or from an archeological horizon.

b) Regional pollen maps. Plotting of data of approximately contemporary pollen spectra over wider areas cannot, of course, assume the validity of a vegetational map. But absolute "pollen maps" can be constructed as, for example, those carried out by Firbas (1949–52, vol. I) for central Europe and by Godwin (1956) for Britain. Such geographic patterns of pollen spectra are highly informative for environmental reconstructions (Fig. 47).

c) Climatic change. Although with considerable qualification, it may be said that specific changes in pollen spectra with time may indicate climatic or ecological changes at a locality. However, plant successions, during colonization of an area by pioneer plant communities, probably

1. Some of the most common Eurasian species of indicator value (see Firbas, 1949–52, vol. I; Iverson, 1954; Godwin, 1956) have been discussed in Chapter 5.

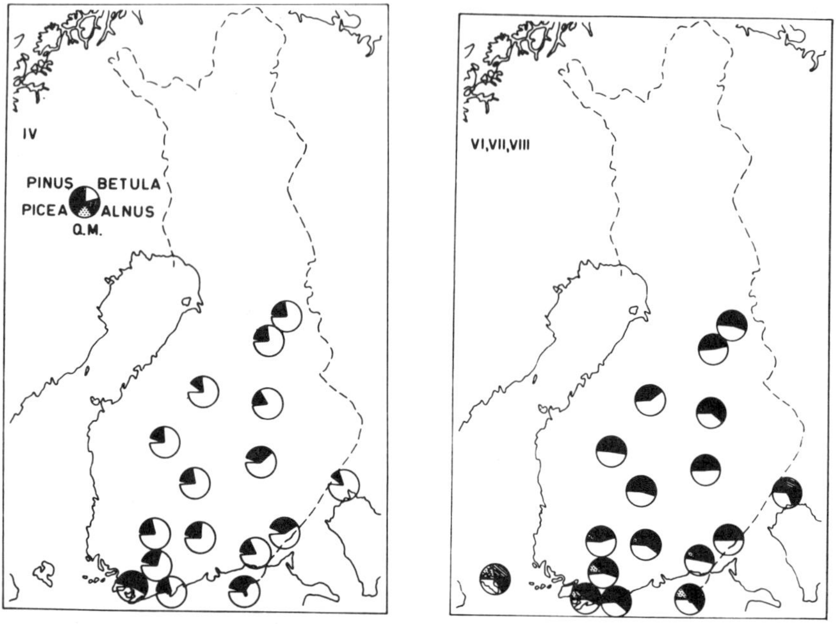

Figure 47. Pollen spectrum maps of Finland (left) during the Preboreal, and (right) during the Atlantic/Subboreal (from J. J. Donner, 1963, by permission of the author). The symbols show the percentage of tree pollen as explained on the left diagram, with "Q.M." indicating mixed oak woodland.

account for many of the assumed climatic changes recorded in pollen diagrams (Iversen, 1960) (Fig. 48).

d) Stratigraphic dating. Characteristic pollen diagrams have been described for certain interglacial periods or for the Holocene period in temperate Europe (Woldstedt, 1954, p. 220 ff). Such standard profiles are frequently used as dating tools, either within the span of a certain diagram, or as fossil assemblages referring to a particular interglacial interval. Artifactual materials in bogs can occasionally be dated according to their position within pollen profiles.

e) Prehistoric settlement. Forest clearance, burnings, and agricultural colonization are dramatically recorded in pollen profiles by the sudden abundance of NAP, the appearance of weed or cereal pollen, and the like. In fact the earliest agricultural settlements in temperate Europe frequently have been first recognized by pollen diagrams, as in the case of Denmark (Iversen, 1960) (Fig. 46).

MACROBOTANICAL REMAINS

Macrobotanical remains are usually indispensable as an auxiliary form of evidence in a palynological interpretation.

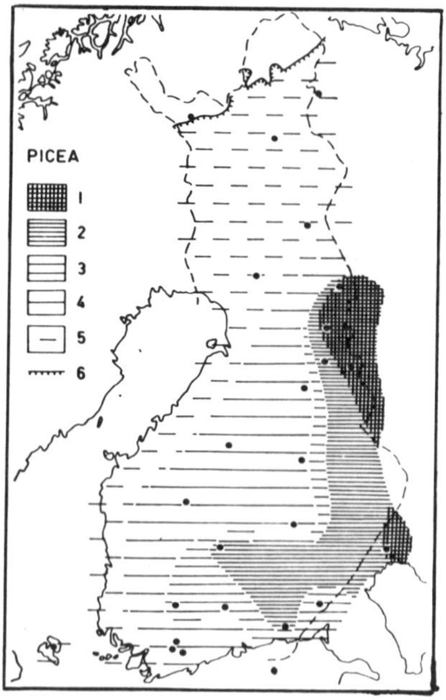

Figure 48. The immigration of Picea into Finland, during (1) the Boreal, (2) the early Subboreal, (3) the middle Subboreal, (4) the late Subboreal, (5) the Subatlantic. The barbed line (6) shows the present northern limit of Picea abies. Dots indicate pollen sites. (From J. J. Donner, 1963, by permission of the author.)

Detailed analysis of macrobotanical remains, particularly of seeds and fruits, offers statistical possibilities that rival the quantitative value of the pollen spectrum (see Watts and Winter, 1966; Wasylikowa, 1967; Watts and Bright, 1968). In fact, a well-preserved macroflora, such as that of the late Pleistocene volcanic ash of the Laacher See (Schweitzer, 1958), near Andernach, Germany, may well exceed a pollen spectrum in interpretive potential.

TREE-RING STUDIES

Tree-ring analysis is a botanical technique with strong analogies to varve study. The underlying principle is that nontropical trees add an annual growth increment to their stems. Particularly in "stress" zones, along the polar and grassland tree limits, annual radial growth fluctuates widely, depending on the fluctuations of the growing season climate. In warm semiarid regions, available moisture largely controls the rate of radial growth of trees: the tree ring of a moist year is wide, while that of a dry year is narrow or, on occasion, missing entirely. In subpolar regions, rainfall is less significant since the late spring snows keep the water content of the soil sufficiently high. Instead summer, and particularly

July, temperatures show the most significant correlation with radial growth.

Tree-ring studies have been almost exclusively carried out in the semiarid American southwest (Schulman, 1956), the western arctic of North America (Giddings, 1954), and the Finnish subarctic (Sirén, 1961). Despite all optimism, areal correlation of tree-ring growth is not very convincing over distances of more than 200 km. Nor are the correlation coefficients between radial growth and single variables, such as growing season precipitation or midsummer temperatures, really satisfactory. A statistically significant approach, drawing different variables into account, is in the process of development by Fritts (1963). Fritts' work has yielded excellent correlation results in detailed, local studies at the grassland tree limit of Illinois. Such methods, applied to the classical areas of tree-ring analysis, promise to revolutionize the field.

The major application of tree-ring studies has been directed toward chronology. The method involved is a simple teleconnection of rings from different logs, partly in archeological context. These chronological aspects, known as *dendrochronology,* have proved most useful in North America when confined to sufficiently small regions and to a time span covered by many long series with appreciable overlap. For outlines, with references, see Fritts (1965) and Bannister (1963).

The second application of tree-ring studies, to paleoclimatology, has been frequently misused in the past by simplistic interpretations of limited data in terms of single variables. Several authors have further abused tree-ring information by attempting to recognize climatic periodicities, such as the sunspot cycle. By rigorous application of variance spectrum analyses to a large number of long tree-ring series, Bryson and Dutton (1961) were able to show that (*a*) there is little evidence of important periodicities, although several weak periodicities of widespread distribution (at 2+ and 3+ years) may be indicated; (*b*) there is some evidence of local "ecological" periodicity; and (*c*) there is no reflection of sunspot periodicities in the tree-ring series. It was also found that a single varve-ring series examined was more like the spectrum of a series of random numbers than like the tree-ring spectra. Responsible interpretations of tree-ring thickness series according to climatic fluctuations can be expected only from refined mathematical analyses of all the ecological and growth variables involved. Fritts (1966) has discussed these factors and reviewed the problems and potential of paleoclimatic interpretation for semiarid regions in western North America. Among certain coniferous species, tree-ring widths represent the integrated effect of climate on food-making and food accumulation in the tree crown throughout the 14 to 15 months previous to and including the period of growth. Trees in warm, low-elevation sites may utilize winter moisture most efficiently; trees in cool, high-elevation, or more northern sites utilize early summer and early autumn moisture more efficiently. Other variations are introduced by local edaphic and microclimatic differences, variability among and within individual trees, and by the compounding effects of occasional fires, of insect or other infestations, and of recurring years of high seed production. Yet even with these multiple complications, a significant amount of climatically controlled variance is found to be common among tree-ring series from a wide range of sites, species, and environments. Fritts *et al.* (1965) provide a particularly fine example of such a paleoclimatic analysis from tree-ring studies at the ancient Pueblo Indian settlements of Mesa Verde, Colorado.

SEVENTEEN

Paleontology and Paleotemperatures

INTRODUCTION

Fossil bone and shell may be obtained from a number of natural and cultural sedimentary environments:

a) stream, lake, swamp and spring beds;
b) beach and estuarine beds;
c) loess and volcanic ash;
d) "fossil," "archeological," and "cultural" cave strata;
e) artificial situations such as kitchen middens, burial pits, etc.

The study of such materials by paleontologists or paleozoologists may yield data of considerable environmental and stratigraphic importance. And for the archeologist, the findings may be indispensable for proper understanding of a prehistoric site—whether the animals present were wild or domesticated, and whether the site belonged to agriculturists or food-collectors.

Yet, generally speaking, paleontological data lack the precision and reliability of palynological work. Statistical samples are usually small and seldom representative of the former animal community because of highly selective sedimentation, hunting, or preservation. Stratified sequences are seldom present in any one locality. Living species today commonly have wide geographic ranges and are seldom rigidly confined to any one environment. In addition, many fossil species and even whole genera are extinct today, so that their ecology is at best imperfectly understood (see Ager, 1963). Despite these qualifications, paleontology is vital for prehistoric geography and ecology, as well as for stratigraphic purposes.

The following discussion will be devoted mainly to mammalian faunas, including a discussion of some characteristic Pleistocene faunas of Europe. Birds and mollusca are considered very briefly, while domesticated animals are relegated to chapter 32. Finally, the paleo-environmental background to deep-sea paleotemperature work is outlined in relation to the stratigraphic data of chapter 2.

FOSSIL PRESERVATION

Dehydrated animal bone consists of about two-thirds mineral matter and one-third organic matter. The mineral component is mainly calcium phosphate with some calcium carbonate and other salts. The organic components—including fat, citric acid, organic carbon, nitrogen, and amino acids—are largely combined in proteins and fats such as the marrow filling the shafts of long bones (Cook, 1951; Cook and Heizer, 1952). Cornwall (1956, p. 205) indicates that soaking in dilute soil acids will dissolve the calcium compounds, leaving only the soft, pliable, organic tissue of the bone. On the other hand, rapid oxidation of the organic matter, which can be produced by heating, converts bone to a brittle structure of light, porous mineral ash. In humid climates, therefore, bones left on the surface will break down rapidly into the constituent components of carbon dioxide, water, ammonia, and mineral salts under the impact of chemical weathering.

Depending on the conditions of sedimentation or the chemical environment, rapid burial of bone or shell may preserve either the mineral or organic matter. Lacustrine, cave, and aeolian sedimentation are least likely to exert mechanical friction on bone and thus destroy it during deposition. Peat, spring deposits, cave strata, and loess consequently supply some of the best fossiliferous deposits. Fluvial sedimentation (except perhaps in the floodplain) and beach deposition take a far greater toll of destroyed bone. Yet only a small fraction of that which is buried intact is preserved for the paleontologist.

Alkaline or calcareous environments are favorable for the preservation of the mineral components of bone. The organic materials are largely decomposed and carried away in solution. In this *fossilized* condition the bone is characteristically light in weight, porous and brittle. Soil waters may percolate freely through fossilized bone, carrying oxides and carbonates in solution. When the soil dries out, a film of mineral precipitates is left in the pore network of the bone. Eventually these spaces are refilled, and mineral replacement of bone material by calcium carbonate, sequioxides, or silicates may take place. The *mineralized* end product is considerably heavier and harder than the original bone. In an arid soil or sediment, however, bone may remain in the fossilized state.

In an acidic, waterlogged environment, the mineral component may be largely dissolved, while various organic materials, including fleshy parts, horn, hide, hair, or wool, may be preserved. These are the conditions responsible for remarkable conservation in some bogs. Bone material is very soft under such conditions and must be removed with great care.

The intermediate case of noncalcareous, neutral, or acidic soils with permanent or seasonal aeration and oxidation is a rather common sedimentary environment. Under such circumstances, all of the bone will eventually be destroyed.

A few exceptional cases of preservation of animal carcasses may be mentioned. For example, the artificially mummified animal fauna of historical Egypt (Lortet and Gaillard, 1903) is a case in which an exceptionally arid climate has helped preserve hair, hide, and bone and, aided by mummification, fleshy tissue as well. Elsewhere in the Sahara, surface fossils are rare and preservation of buried bone is not exceptional. In other words, the artificial case of careful burial and mummification is not a typical sedimentary environment. The case of some Siberian and Alaskan mammoths, removed from permafrost quite intact in flesh and bone, is different. Such continuous refrigeration is an important zonal feature for carcass preservation. Other examples of exceptional conservation in petroleum-saturated beds in central Europe and California are also of interest.

METHODS OF STUDY

Field collection and preservation of fossil bone has been described by Heizer (1958). Subsequent study of fossil bone from geologic strata or archeological sites involves a number of steps (Thenius, 1961b):

a) Taxonomic identification, for which purpose skull, dentition, antlers, horn cores, and long bones are particularly useful;
b) Quantitative analysis, i.e., determination of the minimum number of individuals for a species present, for which the quantity of the most frequent diagnostic skeletal part is used;
c) Age, sex, and size composition;
d) Ecological interpretation, based on comparison of the morphology, behavior, and ecological relations for a living species, or comparative anatomical traits and assemblage composition for an extinct species.

Animal remains, including bone and a wide range of organic refuse pertaining to dietary habits, are invariably richest in occupation sites of man or "sedimentary" predators. The latter include the cave-dwelling bears, hyenas, lions, and owls of the European Pleistocene. The prey of owls in particular includes a valuable cross-section of the local rodent population. More relevant information is provided by human occupation sites, particularly kitchen middens. Kitchen middens are probably as old as the Pleistocene, although they only assume an important role towards its close. They are most numerous along the seashore or on river banks where a molluscan and fish component could be added to the diet. These are the richest sites for mammalian or molluscan faunas.

Archeological salvage of occupation sites is increasingly concerned with the physical analysis of habitation residues. As Heizer (1960) amply illustrated, such refuse may provide invaluable information on dietary economy, settlement pattern, human activities, and environment. The paleontologist should be at the site during the excavation and removal of animal bones, since only then can the context and conditions of deposition be interpreted correctly (Thenius, 1961b).

RELATIVE DATING OF BONE BY CHEMICAL ANALYSIS

An often raised question concerns the possibility of relative or absolute dating of prehistoric bone by techniques other than isotopic analyses. The breakdown or removal of proteins or amino acids, the removal of hygroscopic or hydrated water, the degree of mineralization, and the like have been discussed. So far, with exception of the fluorine method, it has not been possible to provide a reliable and objective method (see Cook, 1960; also Baud, 1960). The major difficulty lies in the variable conditions of the soil or sediment environment.

The so-called fluorine dating method, although a relative technique intimately associated with the peculiarities of soil water, has already provided rather useful information (Oakley, 1963). According to Richter (1958), bones must have been permanently immersed in groundwater ever since fossilization. Fluoric apatite gradually enters the bones from the groundwater and accumulates at a relatively constant rate. The absolute fluorine content depends on local groundwater conditions, and varies from place to place. The value of the method lies in the fact that the bones of an assemblage may be tested for stratigraphic equivalence since contemporary bones at one locality should normally have a comparable fluorine content. This test was largely responsible for the exposure of the Piltdown Man fraud and has proved extremely useful in assessing a number of other critical fossil assemblages (Oakley, 1963, with references).

With further study of the chemistry and physics of soil processes and fossilization, valuable advances in such chronological techniques may be expected.

GLACIAL AND INTERGLACIAL FAUNAS OF THE EUROPEAN UPPER PLEISTOCENE

The environmental significance of the European Upper Pleistocene faunas is better understood than that of any other Pleistocene fauna. Only three of the genera are extinct, and two of these, the woolly mammoth and the rhino, have been found more or less intact at certain localities, so that their diet and cold adaptations are well known. A half dozen further species became extinct at the close of the Pleistocene, but allied species of the same genera are still present. All in all, these Upper Pleistocene faunas can be fully evaluated in terms of their modern (or historical) environmental distributions. They are therefore an interesting case in point. For excellent syntheses see Hescheler and Kühn (1949) and Thenius (1962). Monographs on the mammoth are provided by Pfizenmayer (1939), on the reindeer by Soergel (1941) and by Degerboel and Krog (1959), on the musk-ox by Soergel (1942).

The characteristic mammalian species of the interglacial (Eem) fauna are the extinct, straight-tusked woodland elephant (*Elephas* [*Palaeoloxodon*] *antiquus*), the extinct woodland rhino (*Dicerorhinus mercki*),[1] the African hippopotamus (*H. amphibius major*), the boar (*Sus scrofa*), the fallow deer (*Dama dama*), and the roe deer (*Capreolus capreolus*). In mid-latitude Europe these animals are rarely, if ever, found in glacial age deposits. They do, however, occur in the Mediterranean lands during part or all of the Würm. In addition to these species, a few dozen mammals of the temperate and boreal woodlands are also found in mid-latitude Europe during glacial periods. These include elk (*Alces alces*), red deer (*Cervus elaphus*), aurochs (*Bos primigenius*), woodland horses ancestral to *Equus caballus silvestris,* lynx, wild cat (*Felis silvestris*), fox (*Vulpes vulpes*), wolf, wolverine, sable (*Martes zibellina*), and brown bear (*Ursus arctos* ssp.).

The glacial (Würm) fauna is far more complex (Fig. 49). In part it includes temperate and boreal woodland forms, particularly in the southern parts of France and in southeastern and southern Europe. In part it consists of the "typical" tundra fauna: reindeer (*Rangifer tarandus*), musk ox (*Ovibos moschatus*), snowshoe and arctic hares (*Lepus timidus, L. arcticus*), the mountain lemming (*Lemmus* [*Myodes*] *lemmus*), and the

1. The correct name for Merck's rhino is *Dicerorhinus kirchbergensis,* but the more widely known form is retained here.

Figure 49. Characteristic Upper Pleistocene mammals of mid-latitude Europe (from an original kindly made available by E. Thenius; see Thenius, 1962).

arctic fox (*Vulpes* [*Alopex*] *lagopus*). Alpine forms such as the steppe ibex (*Capra ibex prisca*), the chamois (*Rupicapra rupicapra*), the alpine marmot (*Marmota marmota*), and the alpine vole (*Microtus nivalis*) were found well outside of their high mountain haunts.

In addition to these "expectables," a cool, mid-latitude steppe fauna was also present, ranging through Hungary into southern France (Astre, 1937; Jánossy, 1961). Included here are the saiga antelope (*Saiga tatarica*), the wild steppe horses of tarpan and Przewalski types (see Zeuner, 1963, p. 299 ff. for a useful discussion of the late Pleistocene subspecies problem), the steppe fox (*Vulpes corsac*), the steppe polecat (*Putorius putorius eversmanni*), the steppe marmot (*Marmota bobak*), the hamster (*Citellus citellus*), and a gerbil (*Allactaga saliens*).

Some of the best-known "cold" elements are surprisingly enough not quite as specialized as commonly believed. These include the woolly mammoth (*Elephas* [*Mammonteus*] *primigenius*), the woolly rhino (*Tichorhinus* [*Coelodonta*] *antiquitatis*), the steppe bison (*Bison priscus*), and the giant elk (*Megaceros giganteus giganteus*). Thenius (1961b) classifies these species as steppe-and-tundra forms. The woolly mammoth was a huge creature standing up to 3.5 m. tall with curved tusks as long as 4 m. It was obviously adapted against the cold with a 10 cm. layer of fat under its skin, a woolly undercoat (10–12 cm. long), and a hairy overcoat (30–70 cm. long). Stomach contents examined from well-preserved carcasses (Farrand, 1961; Polutoff, 1955) suggest that grasses formed the basic diet. In its heyday, during the height of the Würm-Wisconsin, the woolly mammoth ranged from northern Spain across Eurasia into Alaska, extending eastward to the New England area and southward into Florida. The woolly rhino was not as universal, but it was also found throughout the colder parts of Eurasia. This rhino was as much as 3.5 m. in length, standing 1.6 m. at the shoulder, with a forehorn up to 1 m. long. A woolly undercoat, up to 6 cm. thick, was complemented by a long hairy overcoat, as in the case of the woolly mammoth. Less clearly adapted to cold were the steppe bison and the giant elk. *Bison priscus* was probably an open-country grazer, with considerable cold tolerance, living on both the tundra and the loess steppes. The giant elk, supporting antlers 3.5 m. broad, seems to have preferred open habitats for practical reasons, but has also been found in association with interglacial woodland faunas (see Mitchell and Parkes, 1949).

The characteristic cave faunas of the European Pleistocene included the cave bear (*Ursus spelaeus*), the spotted cave hyena (*Crocuta crocuta spelaea*), and the cave lion (*Felis* [*Panthera*] *leo spelaeus*). Each of these species was cold tolerant but rather intermediate in its requirements. They are not "cold" indicators by any means. Whereas the cave bear was distinct from the brown bear (*Ursus arctos*) on the specific level—probably due to isolation in western Europe during the Elster glaciation (Kurtén, 1959b)—the cave hyena and cave lion were not very much different from the living spotted hyena (Kurtén, 1957) or the now extinct lion of the Balkans.

STRATIGRAPHIC ASPECTS OF PLEISTOCENE MAMMALIAN FAUNAS

The comparatively rapid evolution of mammalian faunas during the Pleistocene (Kurtén, 1968) is of great stratigraphic value. Particularly useful are the genetically rather differentiated elephants and rhinos of Europe. It seems that these two genera, as well as several other tundra or boreal elements, specialized only during the course of the Pleistocene (Thenius, 1961a, with references).

A few words may be devoted to the evolution and paleo-ecology of the extinct European elephants and rhinos (see Soergel, 1940, 1943; Adam, 1953, 1954, 1961; Kurtén, 1968; Thenius, 1961a, 1962). The earliest Pleistocene elephant in Europe was the forest type *Elephas (Archidiskodon) meridionalis*. Specialization took place during the earlier part of the Elster glacial complex when the steppe or grassland form *Elephas (Archidiskodon) trogontherii* diverged from the *meridionalis* stock, and the more specialized temperate-woodland type *Elephas* (Palaeoloxodon) antiquus replaced *E. meridionalis*. During the Riss glacial complex, *E. trogontherii* specialized further into a cold steppe or tundra form, the woolly mammoth. With exception of the straight-tusked loxodon group, all of the Pleistocene elephants of Europe belonged to the curved-tusked archidiskodon or mammoth groups. Mastodonts, primarily distinguished from the true elephants on the basis of their dentition, became extinct in Europe in the Basal Pleistocene.

In the rhinocerotid family each of the Pleistocene genera in Europe was two-horned. The earliest species, *Dicerorhinus (Opsiceros) megarhinus* and *D. etruscus,* were probably intermediate, warm, or warm-temperate woodland forms. A bifurcation similar to that of the elephants marked the evolution of *D. etruscus* during the Elster, with the appearance of a more specialized woodland species, *D. mercki,* and of the steppe form *D. hemitoechus* (Zeuner, 1935). At about the same time the rather specialized Asiatic woolly mammoth makes its first appearance in Europe.

Both the woodland elephants and the rhinos evacuated mid-latitude Europe during the early Würm and became extinct in southern Europe before the end of the same glaciation—not much earlier than the final disappearance of the more northerly woolly mammoth and rhino at the close of the Würm. Table 13 summarizes the stratigraphy of the European proboscidians and rhinocerotids.

In a more general way, the European Villafranchian was characterized by a Pliocene fauna (see Thenius, 1962, with references). Despite the appearance of certain Pleistocene guide fossils, world-wide faunal affinities existed. Widespread moist, subtropical, or temperate climates did not favor the development of geographical variants, and most of these extinct mammals were ecologically less specialized (Thenius, 1961a). Great faunal and floral changes accompanied the close of the Villafranchian, and the biological array of the Cromerian is largely one of modern genera. Most of the Pliocene elements had disappeared when more specialized woodland and steppe forms first appeared.

At the beginning of the Upper Pleistocene the faunal picture of Europe is divided into the characteristic interglacial and glacial assemblages discussed above.

The last major faunal zone, the Holocene, is characterized by a fully modern fauna. Most of the great Pleistocene mammals, such as the mammoth, woolly rhino, steppe bison, giant elk, and cave bear, died out at the close of the Pleistocene, while the musk ox withdrew from temperate Europe, eventually becoming extinct in the Old World. Other tundra elements still persist in higher latitudes today, while the wild horses, the aurochs, and a woodland bison survived into recent centuries, when they were destroyed by man.

The problem of extinctions at the close of the Pleistocene is discussed in chapters 28 and 29 in relation to the environmental changes and hunting activity of the time.

The stratigraphy of other continental areas is less completely understood. The African faunas have been carefully synthesized by H. B. S. Cooke (1963), while Hibbard *et al.* (1965) may be consulted on the North American fauna.

NONMAMMALIAN FAUNAS

Despite the paleo-environmental potential of nonmammalian faunas, little systematic knowledge is available for such classes as birds, fish, mollusca, and the like. The value of nonmigratory birds for understanding past climatic and vegetational changes has been demonstrated for sub-Saharan Africa by Moreau (1963). Certain larger bird genera are sufficiently numerous in archeological contexts of the late Pleistocene and Holocene to offer good possibilities for local environmental interpretation. Particularly well-known tundra or northern denizens in the Würm record of mid-latitude Europe include snowy partridges (*Lagopus* spp.), the snowy owl (*Nyctea scandiaca*), and the great auk (*Alca im-*

Table 13. Stratigraphy of the extinct European elephants and rhinos (after Kurtén, 1968, and others).

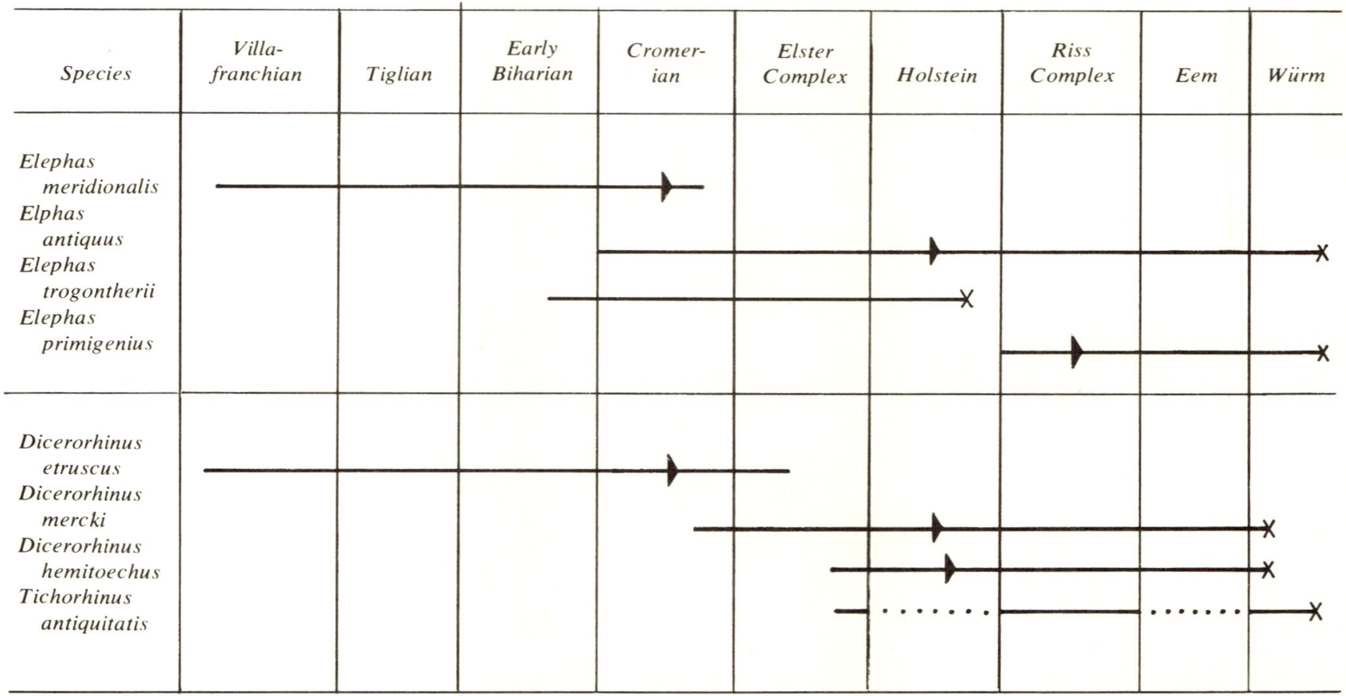

pennis). A bibliography of bird remains in archeology is given by Dawson (1963).

Fish are also relatively scarce in the archeological record, although they may attain some importance for paleo-ecology. For references see Ryder (1963), J. G. D. Clark (1952), and Miller (1965).

Marine mollusca have, in some areas, been studied to the same degree as mammalian faunas. But most marine mollusca are almost indifferent to their climatic environment within ranges of 30 degrees of latitude or more. Ecologically significant species are largely confined to the level of local assemblages and have less general validity than mammalian indicators. So, for example, the relative frequency of cold-loving species such as *Yoldia arctica, Saxicava arctica,* and *Tellina calcarea,* and of thermophile species such as *Lutricularia ovata, Gastrana fragilis, Mytilus lineatus, Lucina divaricata,* and *Haminea navicula,* may mean the difference between glacial and interglacial periods in the North Sea area (Woldstedt, 1954, p. 242). Similarly, in both western and eastern North America (Valentine, 1961; Richards, 1962), there have been ecological shifts in the marine-littoral environments, judging by the regional appearance of faunal elements composed of species that are now markedly "northern" or markedly "southern" in their distribution. However, local effects such as sandy or rocky shores, inshore water depths, and patterns of currents may commonly mask regional climates.

The case of the Mediterranean molluscan faunas is unique. Here the Straits of Gibraltar have repeatedly admitted cold-loving or warmth-loving species during colder or warmer climatic fluctuations. When the climate changed once more, these exotic species again became extinct in the Mediterranean Sea. Colder northern mollusca must have been moving down the west coast for some time before it was possible for an invasion of the Mediterranean to take place, so that northern immigrants such as *Cyprina islandica, Mya truncata,* and *Buccinum undatum* would reflect a long-term change to colder conditions. Similarly, thermophile west African species first have to move through the cold, upwelling waters characteristic of the Moroccan Atlantic coast, so that their arrival in the Straits area would be possible only after an appreciable warming of the deep oceanic waters. Once inside the Mediterranean Sea, many Senegalese elements would find conditions congenial even today. The most thermophile fauna is that of the Tyrrhenian II ("Eutyrrhenian" of Bonifay and Mars, 1959), contemporary with the early Eem Interglacial (see chapters 2 and 3). These Senegalese or "Tyrrhenian" species include *Conus testudinarius, Tritonidea viverrata, Strombus bubonius, Natica lactea, Mytilus (Brachidontes) senegalensis, Arca plicata,* and *Cardita senegalensis* (Cuerda, 1957).

Rigorous faunal stratigraphic studies on the Balearic Islands indicate that *Strombus, Natica,* and *Cardita* are absent from the "impoverished" Tyrrhenian III (late Eem, "Neotyrrhenian" of Bonifay and Mars, 1959) assemblages (Butzer and Cuerda, 1962a, 1962b; see also Stearns and Thurber, 1967). This generalization seems to apply for most other areas where *Strombus* is found and the local stratigraphy is properly understood (Provence, Italy, Syria, Egypt, Tunisia), although isolated and almost certainly derived Strombidae have been found in younger deposits. Only in southeastern Spain is there unequivocal evidence of *Strombus* in beaches dated as Tyrrhenian III (see Stearns and Thurber, 1965). Presumably conditions were unfavorable for renewed dispersal through the Mediterranean Sea following a cold interval during the mid-Eem interglacial. The dispersion of the Senegalese species into the northern and eastern reaches of the Mediterranean Sea was unequal and not uniform during the Tyrrhenian stages, and the moderately thermophile and salt-loving Mediterranean mollusca *Cardium tuberculatum* and *Tapes calverti* play the role of interglacial index fossils in the Black Sea (Pfannenstiel, 1944). The Tyrrhenian I of the Balearic Islands clearly lacks any Senegalese elements (Butzer and Cuerda, 1962a; Cuerda and Muntaner, 1960; Cuerda and Sacares, 1965, 1966; also Stearns and Thurber, 1967), a generalization that applies to all other Mediterranean littorals. However, certain endemic warm-temperate species such as *Purpura (Thais) haemastoma, Patella ferruginea,* and *Pectunculus (Glycymeris) violascescens* were more frequent and often proliferated in Mediterranean littorals of Tyrrhenian I and earlier age. Similarly, no thermophile elements of Senegalese origin entered the Mediterranean during the Holocene, and alleged Versilian beaches with *Strombus* in southeastern Spain have proved to be either of Tyrrhenian II or III age (see Stearns and Thurber, 1965).

Oxygen isotopic analyses of Tyrrhenian I, II, and III mollusca from Provence, Italy, Morocco, and Portugal confirm the paleontological and geological inferences that these are interglacial faunas (Emiliani and Mayeda, 1964); by the same technique, a regressive fauna with *Cyprina islandica,* from depths of 80 to 100 m. off the Provence coast, was shown to be of glacial age.

Terrestrial mollusca, particularly when studied as a general spectrum, provide more specific evidence than marine-littoral assemblages since they reflect their local habitat more accurately. Interglacial and glacial indicators are comparatively well understood for Great Britain (Sparks, 1963), central Europe (Ložek, 1964, 1967), and France (Germain, 1923), while useful molluscan studies in North America are outlined by Taylor (1965). Local studies in the Sahara are due to Gardner (1932,

1935), Sparks and Grove (1961, 1964), Llabador (1962), Martin (1968) and E. G. Leigh (in Butzer and Hansen, 1968, appendices G and H). The immobility of some terrestrial snails may make them particularly suitable for statistical analyses in stratigraphic profiles, as has been impressively illustrated from the Pleistocene of Czechoslovakia (Ložek, 1964). In this way land mollusca may carry both stratigraphic and ecological implications.

PALEOTEMPERATURES AND DEEP-SEA SEDIMENT CORES

Deep-Sea Cores. Sedimentation on the more level parts of the deep-sea floor is limited to very slow accumulation, consisting partly of fine organic oozes. Some of the organisms found in these sediments are calcareous skeletons of one-celled Foraminifera. While alive, a few species of foraminifers inhabit the photosynthetic zone of the ocean's surface waters. Consequently they are adapted to ocean surface temperatures which are, in their turn, in equilibrium with regional climates. After death, these planktonic foraminifers accumulate on the deep-sea floor. The faunal composition of this accumulation will reflect environmental conditions, particularly temperature, so that deep-sea sediments may record stratigraphic successions of microfauna capable of paleoecological study.

Specially equipped ships can operate a so-called "piston-corer," which is able to remove from the sea floor thin, columnar sedimentary sections twenty meters or so in length. These deep-sea cores are subsequently studied and evaluated in the laboratory. The number analyzed is now sufficiently great that sedimentary interruptions or disturbances can be recognized (Ericson *et al.,* 1961). These can be eliminated through careful micropaleontological layer-by-layer cross-correlation between cores from different localities within the same sedimentary basin. The collections of several deep-sea expeditions have now been evaluated and are sufficiently numerous to inspire considerable confidence (Ericson *et al.,* 1961; Olausson, 1961a, 1961b, 1965; Emiliani, 1964, 1966).

Temperature Evaluation. Several different approaches have been followed to determine paleotemperatures as recorded by deep-sea cores:

a) Identification and statistical analysis of "warm" and "cold" foraminifers;

b) Calcium carbonate content and foraminiferal productivity, which will vary according to temperature or, in some oceanic areas, according to oceanic circulation and nutrient level (Arrhenius, 1952; Olausson, 1961a; Wiseman, 1966).

c) The ratio of the oxygen isotopes O^{18}/O^{16} employed in the skeletal development of foraminifers is thought by some to vary according to water temperature, and could, therefore, be converted into isotopic temperatures by use of appropriate equations (Emiliani, 1961);

d) Faunal zones characterized by the presence or absence of certain indicative foraminifers in faunal assemblages (Ericson *et al.*, 1961).

The determination of paleotemperatures from planktonic foraminifers must be considered with caution. So, for example, the "absolute" isotopic temperatures obtained by Emiliani's technique were seriously questioned by Broecker (1965) and Olausson (1965), while Shackleton (1967) and Dansgaard and Tauber (1969) have presented convincing arguments that oxygen isotope analyses do not provide an index of ocean water temperatures. Instead the changes in isotopic composition of the foraminifers examined are almost certainly a result of changes in salinity and water density, leading to readjustment in the depth habitats of various species. Latitudinal shifts of the marine planktonic faunal provinces are similarly to be understood as a response to changes in salinity rather than to changes in water temperature: during glacial periods, water density increased, with increasing salinity, as a great volume of water was removed from the ocean reservoirs to feed the growing continental glaciers; this water was returned rather rapidly during deglaciation, decreasing salinity and promoting readjustments of the foraminifers. In fact, ocean bottom waters now lie only 2° to 3° C. above the freezing point, so that it is difficult to postulate an appreciable reduction in the energy reservoir of the oceans during the glacial periods. On the other hand, if we accept that the faunal and isotopic changes observed are primarily a result of the extraction of large quantities of water from the oceans with storage in the form of glacier ice, their relationship to glacial events is clearcut. It is simply necessary that every faunal or isotopic curve be reinterpreted, rephrasing "cold" by "extensive continental glaciation," and "warm" by "glaciers reduced to their present level" (Shackleton, 1967).

Consequently the deep-sea cores assume an invaluable role in recording the time-sequence and gross stratigraphy of Pleistocene glacials and interglacials. In other words, the faunal and isotopic curves provide a stratigraphic rather than a direct paleo-ecological tool. However, as emphasized in chapter 3, considerable disagreement and uncertainty still shrouds the absolute dating and nomenclature of glacial events in the early and mid-Pleistocene time range.

PART FIVE

Some Pleistocene Environments of the Old and New World

EIGHTEEN

Mid-Latitude Europe during the Late Pleistocene

INTRODUCTION

Whereas the preceding sections attempted to outline the methods and theory of environmental reconstruction, the chapters in this section will be devoted to a sketch of some Pleistocene environments. Three major aspects of Pleistocene geography appear to warrant reconstruction. These are, first, the environmental changes attendant to glaciations in higher latitudes; second, the phenomenon of pluvial phases in lower latitudes; and third, environmental conditions during interglacial periods. Reconstructions must necessarily be selective on the basis of available information and representative situations. The late Pleistocene, in particular the Würm, provides a good example of both glacial and pluvial conditions. Spatially, the only "documented" areas are Europe, the Mediterranean borderlands, North America, and certain parts of Africa. Our knowledge of the Pleistocene of eastern and southern Asia, Australia, and South America is fragmentary and inadequate for regional synthesis. Even in Europe the detail and reliability of Pleistocene research is only satisfactory in a qualified way. But the available data at least permit tentative reconstruction of glacial environments.

EXISTING RECONSTRUCTIONS OF LAST GLACIAL CLIMATIC ZONES

Serious attempts to reconstruct the geographical distribution of ecozones in Europe during the last glacial have been made by Poser (1948, 1950), Büdel (1951a), and Frenzel (Frenzel and Troll, 1952; Frenzel,

1968a, 1968b). Their work forms the basis for the subsequent discussion on late Pleistocene climatic zonation of temperate Europe. Cold-climate phenomena are also discussed by K. H. Kaiser (1960), while Wright (1961) has provided a general survey of the criteria for environmental reconstruction.

Poser's regional reconstruction (1948) is by far the most original and stimulating of the three approaches, although not all of his criteria are considered valid today. Poser attempted to delimit the southern boundary of permafrost (a very crude approximation for the −2° C. mean annual isotherm) and the position of the arctic tree limit (an approximation for the July isotherm +10° C.). Of the three criteria he employed to determine the extent of permafrost, only fossil ice wedges are generally

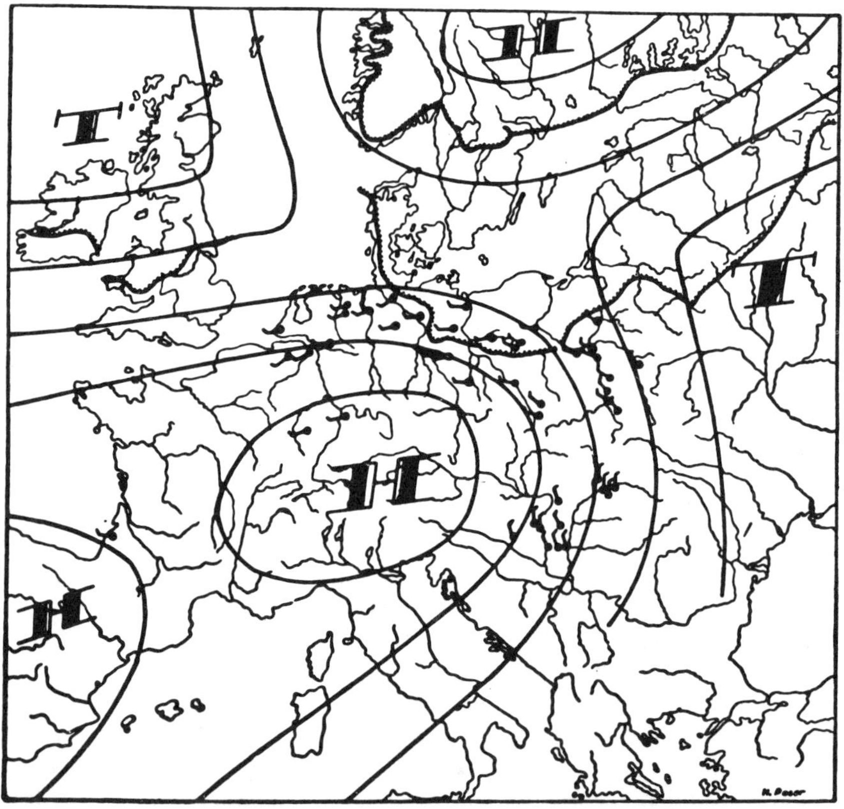

Figure 50. Late glacial summer circulation in Europe (from H. Poser, 1950, by permission of Erdkunde): H = high pressure, T = low pressure center. Barbed lines indicate ice margins during the full and late glacial period. Wind arrows are based on dune orientations.

considered valid. Involutions and asymmetrical valleys must be questioned, although this uncertainty would not significantly alter Poser's picture on the basis of the then available information. Poser further attempts to estimate the degree of summer warmth by drawing isolines of equal depth of involutions, meant to indicate the depth of the active thaw layer. Although this assumption is open to question, the consistency of his findings suggests there may be considerable validity to the concept. It is more difficult, however, to follow Poser's arguments that the vertical depth and surface width of ice wedges is proportional to the severity of winter cold (rather than to sediment texture and drainage features). Overlooking such details, however, the primary importance of Poser's reconstruction is that the essentials are sound, even though subsequent work on periglacial phenomena in France and on palynology in Poland and Russia impel modifications in detail. Poser further attempted isobaric reconstructions of pressure distribution during the Würm maximum using the unfounded argument of the glacial anticyclone (see Matthes and Belmont, 1950). That particular theoretical sketch (Poser, 1948, Fig. 3) is then of no scientific value. A later study by Poser (1950) is devoted to a reconstruction of mean summer wind directions during the later Würm glaciation based on the alignment of fossil parabolic and longitudinal dunes (Fig. 50). This very significant reconstruction will be discussed further in chapter 23.

Büdel's reconstruction (1951a) has curiously found the greatest access into the general literature, presumably on the basis of its cartographic suitability for reconstruction. Yet its criteria are open to question, despite a considerable degree of intuitive accuracy. The basis of much of Büdel's analysis is a shrewd evaluation and extrapolation of data derived from modern isothermal distributions. There is no evidence that paleobotanical data was employed in either the delimitation or classification of forested areas. Büdel's attempt is not a factual reconstruction but a hypothetical sketch, certainly interesting, but to be regarded with due caution.

Frenzel (1968a, 1968b) has contributed a wealth of material on the U.S.S.R. derived from an exhaustive study of the Russian literature. The information employed is largely paleobotanical, partly paleontological, partly geomorphic. The amount of data available for Russia and western Siberia was appreciable, although it is less satisfactory for other parts of the U.S.S.R. where the reconstruction sometimes becomes hypothetical. Frenzel, however, rates the reliability of his work objectively, and does not delude the reader with unfounded covering statements. Unfortunately, Frenzel uncritically employed a speculative sketch of the China area by Wissman (1938), and his approximately

1:37,500,000 map (Frenzel, 1968a, Plate 10), despite its over-all value and broad validity, is of variable reliability from region to region. His reconstruction for the Saale glacial is equally useful but unfortunately groups the Saale and Warthe records.

BASIC CHRONOLOGY OF THE LAST GLACIAL IN MID-LATITUDE EUROPE

Before attempting to review existing information on full glacial climatic conditions in Europe during the Würm, it is essential to emphasize the oscillating character of the ice front and the fluctuating nature of climate during the 65,000-year time span involved. A fairly good radiocarbon chronology[1] is available for the period, primarily based on Groningen dates (Van der Hammen *et al.*, 1967; Zagwijn, 1961; Fink, 1965; Ložek, 1964; Movius, 1960). The basic pattern can be tentatively outlined as follows:

a) ca. *75,000* B.P. End of Eem Interglacial (see ch. 3) and onset of Early Würm. Inception of Scandinavian glacier. Cold climate.

b) ca. *65,000* B.P. and ca. *60,000* B.P. Temperate phases of the Early Würm, designated as the Amersfoort and Brörup, respectively. Almost total deglaciation in Scandinavia (?) (see Frenzel, 1968a, Plate 16).

c) ca. *58,000–40,000* B.P. First cold maximum of the Würm with full glacial conditions *(Lower Pleniglacial).* Ice front south of the Baltic Sea (?).

d) ca. *40,000–29,000* B.P. Complex interval of cool-temperate climate, the Würm *Interpleniglacial.* Once incorrectly called the Gottweig, now generally designated as Paudorf Interstadial. Two brief temperature maxima ca. 37,000 and ca. 30,000 B.P. (Hengelo and Denekamp phases), interrupted by a longer, cold interval. Partial deglaciation.

e) ca. *29,000–13,000* B.P. Second cold maximum of Würm *(Upper Pleniglacial).* Maximum glaciation during Brandenburg interval, ca. 20,000 B.P., followed by recessional halts at the Frankfurt (ca. 18,000 B.P.) and Pommeranian (ca. 15,000 B.P.) moraines.

f) ca. *13,000–12,100* B.P. Rapid glacier retreat initiated at beginning of *Late Glacial;* relatively warm oscillation ca. 12,500 B.P. (Bölling interval). Palynologically the Upper Pleniglacial is known as the Oldest Dryas, while the cold oscillation between the Bölling and the Alleröd is called the Older Dryas.

g) ca. *12,100–11,300* B.P. Important oscillation of temperate climate,

1. Isotopic dating is usually expressed in terms of years B.P. (before the present) rather than B.C. See footnote 1, chapter 3.

the Alleröd. Ice retreat into central Sweden and south-central Finland.

h) ca. *11,300–10,200* B.P. Last phase of the Late Glacial, the Younger Dryas interval. Very cold but only minor glacial readvance.

Conditions to be outlined here refer in a broad way to the full glacial or Pleniglacial phases. Wright (1961) has raised the objection that the very coldest intervals were also the driest, and consequently unfavorable to ice wedge development. There is, however, no serious discrepancy between maximum areal extent of permafrost indicators and typical climatic conditions during the full glacial. The latter probably represented a complex of moister glacial advances and drier glacial standstills or retreats—both the advances and standstills during a time of intense cold.

THE SCANDINAVIAN, BRITISH, AND ALPINE GLACIERS

At the maximum of the Würm (Brandenburg-Valdai moraines), the Scandinavian glacier attained an area of about 4,250,000 sq. km., and an average thickness of 1,900 m. (Grahmann, 1937; Niskanen, 1943). The ice sheet was centered at about 21° E., 59° N. over the Gulf of Bothnia, where the maximum thickness may have amounted to 3,000 m. The ice merged with the British glacier over the North Sea (Valentin, 1957), then extended down the length of Denmark following the Elbe valley to south of Berlin, reaching farthest south in the Oder valley (about 52° N.) (Figs. 51 and 52). Beyond the Oder the ice margin recurved slowly northeastwards toward the Volga River headwaters and thence due north to the White Sea (S. A. Jakovlev, 1957, cited after Woldstedt, 1958).

The British glacier was not only considerably smaller (about 370,000 sq. km.) but also much shallower than the Scandinavian one. Directions of flow outwards from several major highland centers were controlled by the underlying topography. The glacier formed through the merging of several ice caps on the Scottish (Grampians and southern uplands), Welsh, and Irish highlands. All of these mountain areas are of moderate elevation (1,300 m. in Scotland, 1,000 m., in Wales and Ireland), but receive considerable precipitation (to over 2,000 mm.). Louis (1934) estimates the snowline in western Ireland at 400 m., and the general snowline depression at 1,200 m. At its greatest extent during the Würm, the ice extended eastward from Limerick across Ireland to southern Wales (about 51°30′ N.), recurving up to about Manchester and passing on to the North Sea at Hull (Charlesworth, 1957). A local glaciation existed in the Kerry Mountains (1,041 m.) of southwestern Ireland. During the late glacial, the British glacier again disintegrated into its discrete parts.

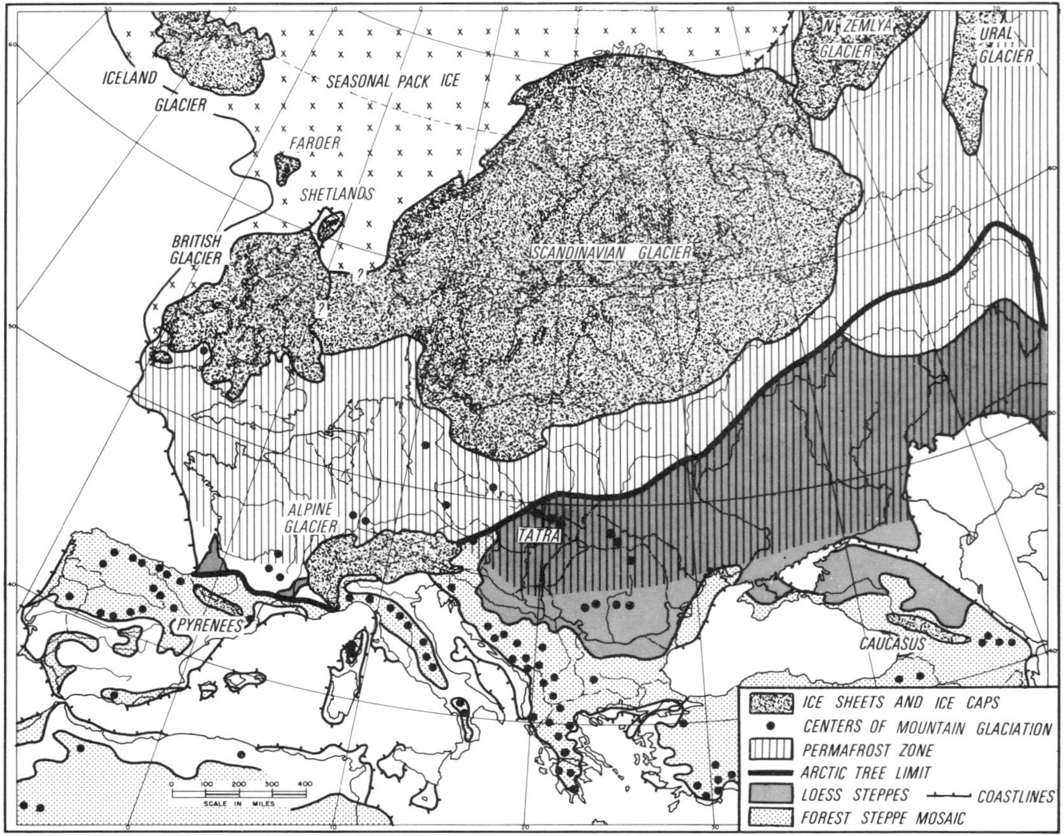

Figure 51. Environmental zonation of Europe during the Würm Pleniglacial (ca. 20,000 B.P.*).*

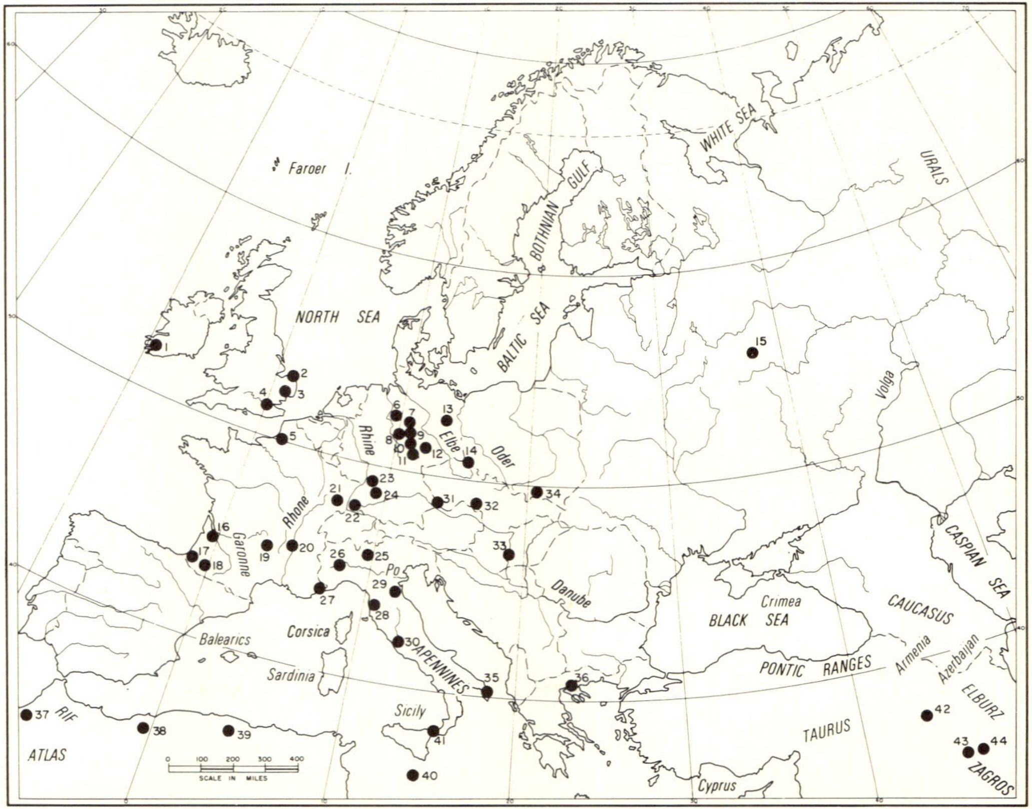

Figure 52. Location Map of Europe: (1) Kerry Mountains, (2) Cromer, (3) Hoxne, (4) Swanscombe, (5) Somme Valley, (6) Lehringen, (7) Lebenstedt, (8) Bilshausen, (9) Harz, (10)Voigtstedt, (11) Weimar, (12) Markkleeberg, (13) Berlin, (14) Sudeten Mountains, (15) Moscow, (16) Bordeaux, (17) Biarritz, (18) Lourdes, (19) Massif Central, (20) St. Vallier, (21) Vosges, (22) Black Forest, (23) Mauer and Mosbach, (24) Stuttgart and Steinheim, (25) Leffe, (26) Villafranca, (27) Nice and Monaco, (28) Pisa, (29) Ravenna and Forli, (30) Rome, (31) Bohemian Forest, (32) Unter-Wisternitz and Předmost, (33) Verteszöllös, (34) Kroščienko and Mižerna, (35) Otranto, (36) Salonika, (37) Sidi Abderrahman, (38) Ternifine, (39) Jurjura Mountains, (40) Malta, (41) Messina, (42) Shanidar. (43) Palegawra, (44) Marivan.

The Alpine glacier occupied an area of about 25,000 sq. km. and consisted of overgrown, coalescent valley glaciers, up to 2,000 m. thick, spreading out in a chain of small piedmont glaciers beyond the edge of the mountains (see Klebelsberg, 1948–49, vol. II). The ice extended down to elevations of about 500 m. on the north, 200 m. on the west, and 100 m. on the south of the range. The Würm snowline elevation varied between 1,500 and 2,000 m., rising eastward and into the interior of the mountains. Snowline depressions amounted to about 1,000–1,200 m. Considering that the 1,000 km. range has crest elevations of 2,000 to 4,000 m. and a modern precipitation of 1,000 to 3,000 mm., the extent of glaciation at this latitude is not surprising. The advance of glaciers right onto the edge of the Po Valley suggests that a major source of moisture was from the south, derived from the warm waters of the Mediterranean Sea.

Further extensive glaciations existed on Iceland and the Faroer Islands, on Novaya Zemlya and adjacent parts of northern Russia, on the northern Urals, the Caucasus, and the Pyrenees.

Climatically these areas of permanent ice must have had midsummer temperatures little above the freezing point, and presumably belonged to the class of frost *(EF)* climates of Koeppen. Flohn (1953) has attempted to evaluate the paleoclimatic implications of the 1,200 m. snowline depression. Provided the lapse rates, radiation, and precipitation remained constant, the evidence possibly suggests a temperature lowering on the order of 6° C. As a test case, Flohn illustrates the example of the Faeroe Islands, which today receive solid precipitation (snow) during the months December through March. A lowering of the mean temperature by 5° C. would reduce the mean temperatures of November and April to below 0° C., and thereby increase the number of days with snow from 44 to 218 days per year, out of a total of 281 days with precipitation. More snow would then fall than could melt in summer, and glaciation would ensue. Nonetheless, snowline deductions should be considered with caution (see ch. 7), since precipitation did not remain the same and cloud cover must also have varied. Temperatures estimated from other geomorphic criteria are substantially lower (see below).

NONGLACIATED REGIONS OF WESTERN EUROPE

Turning to the nonglaciated areas of France and the southern British Isles, much of this region experienced both permafrost and tundra conditions during the main Würm. The 100 m. lowering of world sea level produced an important change in topography, whereby the southern North Sea, the English Channel, and much of the Atlantic continental margin were exposed. This may have affected the continentality of local climate.

Geomorphic evidence from southeastern and south-central England includes ice wedges, frost cracks, involutions, and widespread solifluction features (see West, 1968, with references). This zone was certainly well within the discontinuous permafrost area, although information is not yet available from the more maritime regions of southern Ireland. As the mean annual temperatures range about +10° C. today, a temperature lowering of at least 12° C. is implied for southeastern England.

The pollen evidence of the Oldest Dryas, prior to the Bölling oscillation, indicates the dominance of tundra or open park-tundra vegetation (Godwin, 1956; Suggate *et al.,* 1959). Nonarboreal pollen (NAP) is almost exclusive, including considerable *Artemisia* and *Thalictrum,* both indicative of at least seasonally dry soil conditions. At least as far back as the Older Dryas, however, copses of dwarf and arctic willow were to be found in Cornwall and of willow and dwarf birch in Ireland (County Monaghan) (Godwin and Willis, 1959). This implies that even during the colder parts of the late glacial, forest-tundra occurred in southwestern Great Britain, suggesting mean July temperatures around the 10° C. mark (+10 to +14° C. today). Whether or not this was also so during the full glacial is not known.

In France, geomorphic evidence relating to Würm-age cold-climate phenomena is relatively well dispersed (Tricart, 1956a, 1956b). Unfortunately ice wedges are poorly recorded. Scattered localities with ice wedges occur through much of the northern half of the country, and a single such site is recorded at Bordeaux. As there is no ready way to judge the validity of the latter evidence, the question must remain open whether or not discontinuous permafrost did extend south of latitude 46°.

Interesting comparative information is available from the glaciated highlands. For example, a number of valley glaciers existed in the Mt. Dore (1,886 m.), Cantal (1,858 m.) and Cevennes (crest elevations at 1,700 m.) mountains of the Massif Central (Boisse de Black, 1951). The Würm snowline was found at about 1,200 m.

The Pyrenean glaciation was considerably more important. Crest elevations run about 2,000–3,000 m. (Pico d'Aneto is 3,404 m.), and the range receives 1,000–2,000 mm. precipitation. The present snowline rises from 2,500 m. in the west to 2,800 m. in the east, whereas the Würm snowline was located between 1,500 and 2,300 m. Except for small piedmont glaciers in the upper Pau and Neste valleys, glaciation was limited to valley glaciers (Nussbaum, 1928; Alimen, 1957, with references). These extended down to elevations of 500 m. on the northern flanks, and to 1,000 m. on the southern flanks of the range.

Full glacial paleobotanical evidence is scanty from France. Lemée

(1954) has outlined Older Dryas pollen records from elevations at 1,100–1,350 m. in the Massif Central. Nonarboreal pollen (NAP) include *Artemisia* and *Helianthemum,* two "steppe" species, although no statistics are given. The arboreal pollen (AP) recorded at the Cantal site is dominated by pine and dwarf birch *(Betula nana),* without oak or hazel. At Mézenc, in the mountains of Vivarais, pine is dominant, with some birch and with traces of *Quercus* and *Corylus*. Lemée believes that an oak and hazel, mixed deciduous woodland may have been found in the lowlands at this time. Elsewhere, on the northern flanks of the Pyrenees, considerable work has been done by Florschütz and Menéndez Amor. Published to date is only one profile from Poueyferré, near Lourdes, at 420 m. (De Vries *et al.,* 1960). Several Groningen radiocarbon dates leave no doubt as to the stratigraphic position. Twenty-four samples antedating 13,000 B.P. (and the Bölling oscillation) have AP under 10 per cent. Pinus is dominant, with traces of *Betula* in 16 samples, *Quercus* in 6, and *Corylus* in 5. *Artemisia* forms 10–25 per cent of the NAP. From this one must conclude a (mountain) tundra vegetation with scattered stands of wood at a little distance, and of mixed deciduous woods at greater distance in lower elevations. There is little likelihood that *Quercus* and *Corylus* pollen were derived from beyond the Pyrenees, so that these species must have occurred somewhere in the lowlands of the Gascogne. Yet, results by Oldfield (1964) in the coastal area of Biarritz show less than 3 per cent AP for a full glacial deposit.

The exact position of the polar tree-line in France cannot be determined from this scanty information, but it must have been well south of the modern 20° C. July isotherm (at about latitude 46° N. in the lowlands). But further pollen studies may still reveal that forest-tundra existed in the Garonne and Rhone lowlands during the Oldest Dryas.

NONGLACIATED REGIONS OF CENTRAL EUROPE

The Scandinavian and Alpine glaciers bordered central Europe (the Low Countries, Germany, Czechoslovakia, and Poland) on two flanks. The severe climate of this area pinched between two cold centers is not surprising.

Ice wedges are common throughout (Poser, 1948; Maarleveld, 1964; Paepe, 1965; K. H. Kaiser, 1960; Sekyra, 1960; Dylik, 1956). Discontinuous or continuous permafrost seems to have been universal, so that a mean annual temperature depression of *at least* 11° C. must be assumed for the region as a whole. Needless to say, numerous other cold-climate indicators are equally frequent.

Local mountain glaciers were present on the Vosges (1,423 m.), in the Black Forest (1,493 m.), and on the Tatra (2,663 m.), and in a more limited way on some smaller highlands such as the Harz and the Sudeten ranges (data summarized by Woldstedt, 1958). The Würm snowline elevation was at 800-900 m. in the Vosges, in the Black Forest at 850-950 m., and in the Tatra at 1,450-1,650 m. (2,500 m. today). The snowline depression of these intermediate highlands was also on the order of 1,000-1,200 m. (see maps 1 and 2 in Frenzel, 1959-60, vol. I). This is well out of key with the permafrost evidence, and may reflect both the presence of a surface temperature inversion (Flohn, 1953) and a reduction in precipitation due to greatly increased continentality (see also Weischet, 1954). In this connection Poser's (1948) deductions on the limited depth of the active thaw layer in the east-central Europe (70-150 cm.) are relevant.

In some of the areas close to the ice front, high arctic barrens with little or no vegetation prevailed during the maximum of the Würm (e.g., the Netherlands; see Zagwijn, 1969). Throughout the central European loess belt, and probably in other areas as well, herbaceous tundra was, however, more characteristic. Firbas (1949-52), Andersen (1961), and Iversen (1954) consider that a herbaceous tundra with wild grasses, sedges, herbaceous plants—and steppe indicators such as *Artemisia*—is generally indicated by Oldest Dryas pollen profiles of Germany and Denmark. *Salix herbacea* is locally present, and Firbas supposes the existence of a scrub tundra with *Salix herbacea, S. polaris,* and *Betula nana* in the warmer Upper Rhine lowland and the Bohemian plain at about the same time. According to Frenzel (1968a), the arctic tree limit crossed the Danube at Vienna and extended across Moravia into southern Poland, where Klimaszewskii *et al.* (1950) found pollen of *Larix* sp., *Pinus silvestris,* and *P. montana* at 228 m. near Krosčienko, while pollen of *Salix reticulata, S. retusa, Betula nana,* and numerous tundra elements were found at a neighboring site in 437 m. This places the contemporary tree limit very approximately at the modern July isotherm of 19° C., suggesting a main Würm midsummer temperature depression in the order of 7°-9° C. Andersen (1961) assesses this value at 10° C. for Denmark during the full glacial. Curiously enough however, *Armeria maritima,* which does not now occur in areas with mean cold-month temperatures below − 8° C. is present in contemporary Danish materials (Iverson, 1954). This would imply a maximum January temperature depression of 8° C. for Denmark. As this is not totally compatible with the geomorphic evidence, the limitations of indirect evidence suggest a need for caution.

SOUTHEASTERN EUROPE

The lands of southeastern Europe (Hungary, Romania, Yugoslavia, and Bulgaria) have a rather varied relief dominated by the Hungarian and Wallachian plains on the one hand, and by the Carpathian, Dinaric, and Balkan mountain systems on the other. The only substantial information for much of the area concerns numerous localities of limited mountain glaciation. The Carpathians (crest elevations 1,200–2,500 m.) had a Würm snowline rising from 1,550 to 2,100 m. north to south; while that of the Dinaric-Balkan ranges (crest elevations 1,000–2,500 m.) rose from 1,300 to 2,300 m. west to east.

Cold-climate phenomena are recorded only from Hungary, where ice wedges occur in the northern half of the country. The exact southern limit of discontinuous permafrost is unknown but is still thought to run across southern Hungary and northern Wallachia (see Poser, 1948; Pecsi, 1964; Brunnacker, 1967b; Frenzel, 1959). Loess deposition was prominent in both the Hungarian and Romanian lowlands.

Pollen information is a bit better. According to Serčelj (1966) a subarctic boreal forest occupied much of Slovenia in the 800–1,200 m. elevation range. It was dominated by pine, with a little birch, willow, and alder present on occasion. In Hungary an open brush woodland of *Larix decidua, Pinus cembra,* and *P. montana* occupied the foothills between 400 and 900 m. elevation, while a forest steppe, with galeria woods of pine and larch, was located on the Hungarian plain (Zólyomi, 1953). In the lower Carpathians of Romania, Pop (1957) analyzed a large number of profiles showing that pine was dominant during the full glacial, almost exclusive during drier phases, and present with spruce, oak, and hazel during moister (and slightly warmer?) phases. Charcoal of deciduous hardwoods found at various Hungarian sites as well as in northern Yugoslavia presumably dates from warmer interstadial episodes.

Information from this sector of Europe is still incomplete and does not permit any definitive reconstruction. Frenzel (1968a) provides a reasonable although partly hypothetical sketch of vegetation belts.

NONGLACIATED EUROPEAN RUSSIA

The full glacial geography of European Russia has been well investigated and is carefully summarized by Frenzel (1968a, 1968b). Separate from the Scandinavian glacier in the northwest, a smaller ice cap centered over Novaya Zemlya (crest elevations 200–1,000 m., Würm snowline under 200 m.) invaded a small area of the Arctic coast, while a mountain glaciation enveloped the northern Urals (crest elevations 1,000–1,600 m., Würm snowline at 500–800 m.). Further east, glaciation in Siberia was limited to certain highland areas as a result of continentality and insufficient moisture. Frenzel (1959, Fig. 4, p. 102 f.) indicates that ice wedges and involutions have been found as far south as the northern shores of the Black and Caspian seas. The southern limit of discontinuous permafrost was then probably located at about the 46th parallel where mean annual temperatures today range between +8° and 10° C.

Three major physical belts are distinguished by Frenzel in the unglaciated parts of the country: *(a)* A broad expanse of herbaceous tundra was found north of the main Würm tree-line, which crossed east-northeastward from southern Poland to the southern Urals. *Artemisia* as well as a number of salt-tolerant plants were prominent in this vegetational association. Saline soils are not unknown over permafrost in the arctic today, although the origin of the salt is obscure. The tree-line itself was located close to the modern +19° C. isotherm for July, again suggesting a summer temperature depression of 7°–9° C. *(b)* South of the tree-line a broad belt of forest-tundra or forest steppe gave way to small woodland areas around the Carpathian foothills, on the central Russian uplands, in the Crimea and the southern Urals, and along the shores of the Caspian Sea which at that time was expanded to about twice its present size, *(c)* As a third area, loess tundra extended over southern Russia, characterized by an association of NAP similar to that of the herbaceous tundra adjacent to the ice sheet, but with widespread loess deposition. Galeria forests, partly with coniferous species, accompanied the moister river lowlands.

The dominance of steppe and tundra through most of the country, despite the presence of a parkland vegetation with scattered arboreal tree growth, can only be explained by a dry, extremely continental climate.

FULL GLACIAL TEMPERATURE CONDITIONS

In the previous sections the approximate distribution of ice wedges and, by inference, of discontinuous or continuous permafrost was discussed for Europe. On the existing evidence it would be unjustified to draw a precise "line" designed to represent the limit of either the permafrost or the ice wedges. In some areas, the ice sheets set one of the few more convincing limits, for it is assumed that the insulation provided by the ice would enable heat from the earth's interior to remove permafrost conditions previously established. Consequently, the ice sheets set definite limits to the north, while the Alpine glacier apparently set a comparable limit in the south, since ice wedges have not yet been recorded from the Po Valley. Because of the variation of local climatic and soil conditions in mountain areas, there is little chance that a reliable permafrost limit could be drawn through mountainous terrain. In three major lowland areas – western France, the Balkan region, and the southern Ukraine – information is far too incomplete. The extent of full glacial lowland permafrost as shown in Fig. 51 is therefore rather tentative.

The location of the arctic tree-line is better known, although no exact data are available from France. Information concerning altitudinal zonation of climate is generally unavailable from the highland areas. The zone between the ice front and the tree-line can, with some reliability, be placed in the tundra *(ET)* climates of Koeppen. It serves no obvious purpose to distinguish climatic subtypes within this *ET* zone, such as Poser (1948) suggests. The available evidence is insufficient, and the quantitative significance of such subdivisions would be obscure. Neither does the distinction of "frost rubble" and loess tundra in Büdel (1951a) or Frenzel (1959–60) have any legitimate inference for either vegetation or climate. Much, if not the greater part, of this "frost rubble tundra" was in reality clad by herbaceous vegetation.

Reviewing the values of suggested temperature depressions, one can hardly avoid the conclusion that mean annual temperatures throughout the main Würm permafrost zone were on the order of 10° to 12°, or more, lower than today. There is no reason to assume that these temperatures were confined to Germany between the Alpine and Scandinavian glaciers, as Poser (1948) and others have implied. In the light of evidence elsewhere, it is difficult to reconcile such a value with the general planetary temperature reduction of about 4° C. suggested by Flohn (1953). At the Würm tree-line, the inference for July temperature depressions is at least in the order of 7–9° C., possibly as much as 8–10° C. (Frenzel, 1967, p. 148), suggesting that winters rather than summers were anomalously cold.

Mortensen (1952) was probably the first to suggest the presence of a secondary cooling effect in the areas adjacent to the continental glaciers. In this way a cold air dome (or surface temperature inversion), considerably cooler than the overlying atmosphere (at 2,500 m. according to Mortensen [1952], at 1,000 m. according to Flohn [1953]), would persist during much of the year. The area affected by such an inversion must have extended 1,000 to 1,500 km. beyond the major continental glaciers, however, something not realized by either Mortensen or Flohn at the time. Modern, semipermanent analogs from the Greenland ice cap (Flohn, 1952) and the Antarctic ice sheet (Sabbagh, 1962) are convincing. Seasonally, the present-day winter surface inversions over North America and Siberia (Willett and Sanders, 1959, Fig. 13; Flohn, 1952, Fig. 1), during well-developed high pressure or anticyclonic conditions, are easily comparable in both dimensions and mechanics to the surface cooling effects suggested for Würm-age Europe. The temperature inversion responsible for the continental dimensions of the former European inversion must, however, have been at an elevation closer to the 2,500 m. mark cited by Mortensen, rather than the 1,000 m. elevation suggested by Flohn. It may be assumed that a semipermanent winter anticyclone, similar in intensity and depth to the modern Siberian anticyclone, was centered over Europe during the full glacial. During the greater part of the winter half-year, local surface temperatures may have been on the order of 20° C. "below normal." Averaged out for the year as a whole, this would more or less account for the available evidence.

FULL GLACIAL MOISTURE CONDITIONS

Of apparent significance for evaluation of moisture conditions in glacial Europe are the widespread loess mantles (Fig. 51) south of the polar tree-line (see Frenzel, 1959, pp. 109-15). Accepting the presence of permafrost in most of these regions, it is difficult to weigh the effects of physiological drought due to extreme winter cold and the effects of possible climatic aridity. Widespread "steppe indicator" plants and steppe faunal elements may only indicate that the physiological processes of vegetative life were frequently interrupted in summer by freezing or near-freezing soil temperatures. Similarly, the halophytes present, capable of exerting exceptionally great osmotic pressures in the root zone, were possibly better adapted to physiological (as opposed to climatic) drought. There may certainly have been saline soils in the permafrost belt, but the salt-tolerating halophytes, the steppe indicators, and the absence of forest growth through much of the southeast European lowlands are all a part of the general problem of physiological versus climatic aridity in very cold environments. Paleobotanists have now begun to attack this problem (Frenzel, 1964, 1965, 1968a).

The available physical evidence in favor of increased aridity in middle latitude Europe is expressed in the following arguments: *(a)* that early glacial solifluction and rising groundwater tables (under moister, oceanic conditions?) were frequently replaced by full glacial loess deposition (under drier, continental conditions) (Büdel, 1950b; Brunnacker, 1962a; Reich, 1953; Fink, 1961; Richter, 1968); *(b)* the occurrence of loess steppes in areas forested today; *(c)* the occurrence of steppe floras and faunas in both the tundra and steppe regions; and *(d)* the fact that the snowline depression of the central European highlands (1,200 m.) is not "sufficient" in view of a temperature depression of at least 11° C. The question of physiological drought renders arguments *(a)* to *(c)* inconclusive. Point *(d)* is open to dispute. Klein (1953) argues that the "deficient" snowline depression was a result of reduced precipitation.

Klein suggested a precipitation reduction of 150 mm. for every 100 m. of difference between the observed Würm snowline, and the snowline "to be expected" by a temperature depression of, for example, 12° C. Although a part of the difference may well be due to reduced precipitation, Klein's argument as such is fallacious. A surface temperature inversion, such as postulated for Europe at the time, would have been much less evident at higher elevations, i.e., at the critical snowline altitude (Flohn, 1953).

The geomorphic processes recorded do not all support an argument for climatic aridity. The Aral and Caspian seas stood 12 m. and 75-77 m., respectively, higher than today (see Frenzel, 1967, p. 185 ff; Butzer, 1958a, pp. 95-97). The latter overflowed into the Black Sea across the Manych depression during the interpleniglacial, although it appears that the Caspian Sea level dropped to 26-28 m. above the present during the full glacial. The major control of Caspian Sea fluctuations today is summer temperature over the Volga drainage basin (Butzer, 1958d). Evaporation over the watershed as well as over the sea itself was considerably reduced during the Würm, while some of the rivers draining northward to the Baltic and White seas were diverted southward to the Volga system. This can adequately explain the improved hydrological budget of these Asiatic lakes, but at the same time it does not support the hypothesis of greater aridity. A similar argument is offered by the general alluviation of streams in what is now temperate Europe. This cold-climate aggradation, discussed in chapter 11, does not prove greater humidity but does not necessarily support greater aridity either. Thomé (1958) has convincingly illustrated the impressive summer discharge of the Rhine River during the full glacial.

The most convincing argument in favor of increased aridity is climatic. Severely reduced air temperatures over Europe necessarily brought a considerable decrease in atmospheric moisture content; reduced evaporation over the oceans (by somewhat over 20 per cent, [Flohn, 1953]) would decrease average world precipitation in general; and finally, a cold air dome over Europe during the winter half-year would impede the passage of rain-bringing cyclonic disturbances. These theoretical arguments are convincing even though not as yet supported by sufficient empirical evidence.

THE GEOMORPHIC LANDSCAPE OF THE WÜRM TUNDRAS OF EUROPE

To complete the picture of the physical landscape of full glacial Europe—as a potential environment for man and animal—a brief sketch of geomorphic processes may be of interest.

Firstly, the general topography of the continent was altered by the glacio-eustatic sea-level lowering of 100 m., as well as by the extension of the Scandinavian and the British glaciers. Not only was the land surface of these areas obliterated, but the existing drainage patterns were disrupted and new streams of glacial meltwaters complicated the hydrographic picture.

Ahead of the ice, meltwater streams deposited outwash over wide areas inundated periodically during early summer. At intervals, extensive fans of sandy deposits were alluviated along the ice margins. In central Europe these meltwaters eventually drained into the great river valleys, known as glacial spillways, which carried the diverted waters of the Vistula (Weichsel) and Oder (and during older glaciations, of the Elbe, Weser, and Rhine as well) westward along the front of the ice. The outwash beds were liable to wind deflation in late summer, as were also the fresh till beds gradually exposed by the ice during the glacial retreat.

Further south, rivers and streams alluviated disproportionately wide floodplains. River channels were braided and characterized by great floods from spring to early summer, following the annual snow thaw. Smaller valleys ("dry valleys"), with little or no surficial stream discharge today, harbored watercourses because of the impermeability of the permafrost subsoil.

Between the rivers, aeolian deposition of loess was common in many of the lowlands and in topographic hollows within more rugged terrain. Frost-weathering was prominent on exposed bedrock, and solifluction processes were active on moderate and steeper slopes. Talus accumulation was rapid at the foot of rock ledges or cliffs. Flat-lying surfaces were probably poorly drained as a result of permafrost in the subsoil.

Generally speaking, the dynamism of geomorphic processes was both similar to and different from that found in the modern Arctic. The primary contrast lies in the widespread deposition of loess, a matter of considerable ecological relevance.

THE NATURE OF THE OPEN GLACIAL-AGE VEGETATION

The preceding regional overviews have indicated that, during the full glacial, an open vegetation prevailed over most of the unglaciated terrain of mid-latitude Europe. Unfortunately, recoverable pollen is scarce in the loess beds and other mineral sediments that generally record periods of extreme cold. It has long been assumed that a typical tundra vegetation, with herbs, grasses, lichens, sedges and shrubs, prevailed west of longitude 20° E. This opinion had not been palynologically verified because of the extremely sparse pollen and macrobotanical record for the full glacial. In this connection, the great biomass of gregarious herbivores seemed incongruent with that of contemporary high-Arctic tundras in North America and Eurasia; and Thenius (1962) had already emphasized that most of the characteristic Würm-age mammals of Europe were open-country grazers, many of them better adapted to the cold dry environment of present-day Central Asia than to the Eurasian tundra. The abundance of steppe rodents in the Würm has supported this view.

Recent work by Frenzel (1964, 1965, 1968a, 1968b) provides rather new perspectives on the true nature of the open vegetation during full glacial times. A new technique of processing pollen samples by flotation in a mixture of potassium and cadmium iodide was developed and the results amply tested (16,000 pollen counts) in a wide range of sediments and sections for possible sources of error.

Applying this technique to Pleistocene loess deposits and paleosols of eastern Austria, Frenzel (1965) presents fascinating patterns of spatial and temporal variation, both in the proportion of arboreal pollen and in the composition of the nonarboreal components. During the periods of maximum cold, coinciding with loess deposition, tree pollen in the moister foothill country ranged from 20 to 70 per cent, as compared with less than 5 per cent in the drier lowland plains. Consequently, scattered trees or parklands of pine, spruce, alder, and birch are postulated for the foothill zone. The open vegetation of the foothills was dominated by *Artemisia* (recalling the related sagebrush plains of the intermontane basins of Utah and Nevada) and by herbaceous plants. The plains were dominated by a grassy steppe, rich in herbs, with a component of semidesert or salt-tolerant Chenopodiaceae. Although tundra and alpine elements are present in all spectra, this was not a true tundra vegetation. For example, there is a great abundance of plants favoring dry but eutrophic environments. This corroborates the geomorphic and pedologic evidence indicating that the Pleistocene loesses bear little analogy to the highly acidic and generally waterlogged soils and surficial sediments of modern Arctic tundras. Certain floristic analogies of full glacial Austria with the steppes and semideserts of Central Asia are apparent, but Frenzel stresses that this late Pleistocene vegetation cannot be described adequately in terms of modern plant associations.

Considerable temporal change of vegetation is apparent within the Würm Glacial itself (Fig. 53). The earlier periods of maximum cold appear to have experienced a *relatively* moister climate, and *Artemisia* is a significant component of spectra in the foothill country. However, the final pleniglacial was drier, a conclusion borne out by the molluscan faunas (Ložek, 1964); and the chenopods attain their greatest prominence. This basic loess-steppe pattern was interrupted by several more temperate interstadials. These were characterized by soil formation and

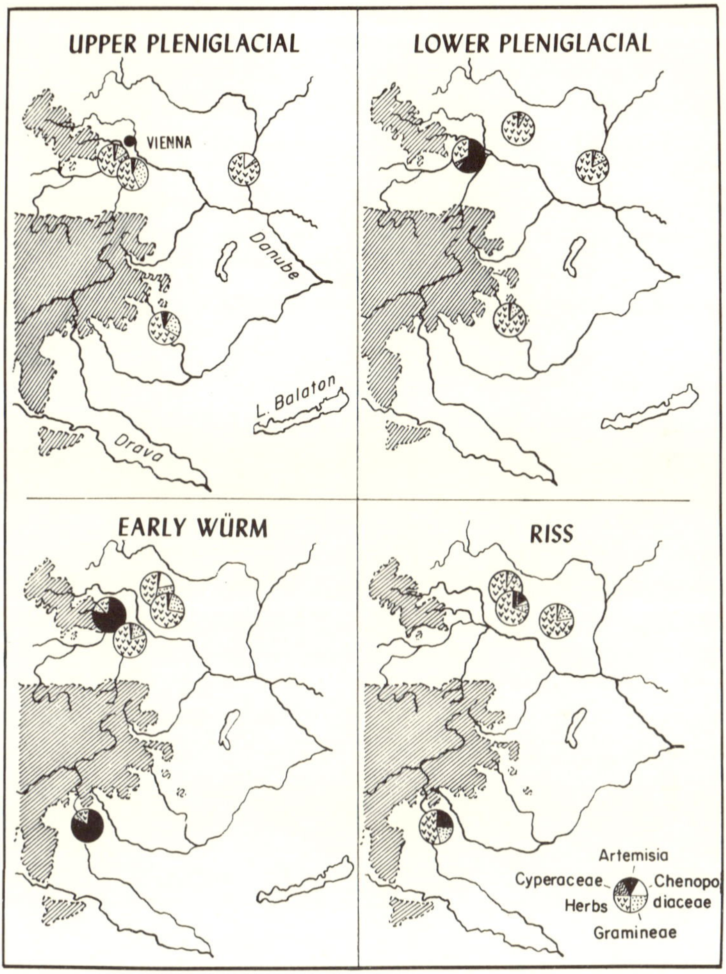

Figure 53. Composition of nonarboreal pollen spectra for several Würm loesses and the Riss Loess in eastern Austria. After Frenzel (1965).

a vegetation of subalpine-coniferous and riverine-deciduous woodlands along the major valleys, with steppe formations retaining their dominance on the drier loess plains. The prospect of further regional studies of Pleistocene mid-latitude loess-steppes in Europe and North America promises to elucidate the true nature of the resource base and ecology of the Pleistocene mid-latitude "tundras."

PHYSICAL CONDITIONS DURING THE INTERSTADIAL PERIODS

In a broad sense, the improvement of physical conditions during the

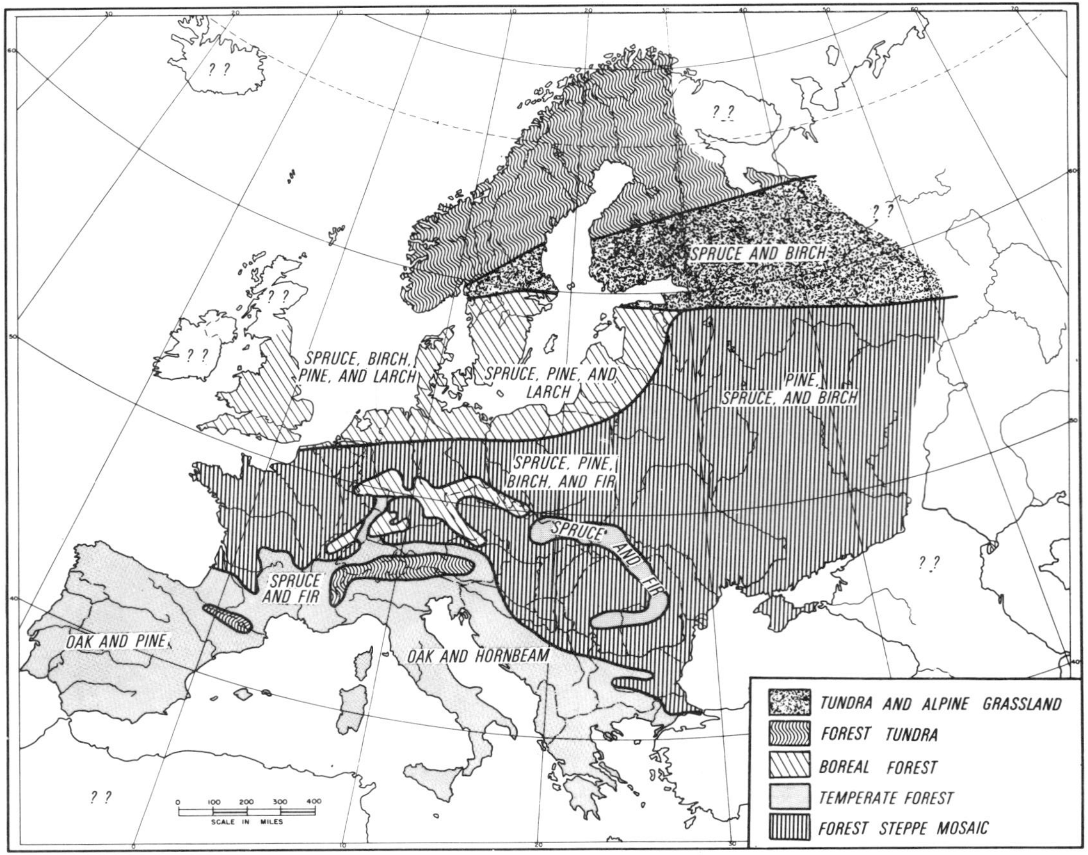

Figure 54. Environmental zonation of Europe during the Brörup Interstadial (ca. 60,000 B.P.*). Simplified after Frenzel (1968a), with corrections in southwestern Europe.*

many warmer spells, characterized by oscillations of the ice margins and the vegetation zones of Europe, was at a level intermediate between contemporary and full glacial conditions. Forest-tundra, scrub, or boreal woodlands reoccupied most of the former tundra belt in a rather short time.

The nature and stratigraphy of the European interglacials has received much attention since the 1950's, and substantial progress towards a fuller ecological understanding has been made since 1960, when Danish and Dutch workers (Andersen *et al.*, 1960) established the existence of the Brörup Insterstadial. Much of the confusion concerning the dating of interstadial records has been eliminated since Gronigen radiocarbon dates have separated phenomena pertaining to the Eem and to the Amersfoort, Brörup, and Paudorf interstadials.

More than 70 sites with pollen records have been identified with the Brörup Interstadial thus far (Frenzel, 1968a), allowing a reconstruction of environmental zones (Fig. 54). From the nature of interstadial pollen spectra at Biarritz (Oldfield, 1964, 1968) and in Catalonia (see Butzer and Freeman, 1968), woodlands can be inferred for southwestern Europe. Oak and hornbeam forests can be suggested for northern Italy, Yugoslavia, and Albania (see Frenzel, 1968a). Most of the lowlands of northern France, central Europe, and European Russia carried a complex mosaic of steppe, with groves or open woodlands of pine, spruce, and birch, as well as small percentages of more thermophile elements. Boreal woodlands, with spruce, birch, pine and larch bordered the North Sea, with similar forests, lacking much birch, around the Baltic Sea. A forest-tundra of spruce and birch is postulated for southern Finland and northern Russia, with tundra or alpine vegetation in Norway, northern Sweden, and the remainder of Finland. Forests of spruce and fir, with some pine, clad the Carpathians and the Tatra, while cooler forests of spruce, pine, birch, and fir covered the intermediate highlands of central Europe.

A large number of Brörup paleosols are known, and their occurrences have been summarized by Frenzel (1968a). Most of the profiles north of the Alps and the Black Sea pertain to grassland soils with *AC*-profiles, many showing signs of groundwater-logging. Braunerdes and podsolic soils are present locally, however, and assume importance in northern Italy (Fränzle, 1965).

The Brörup Interstadial may have lasted some 6,000 years, but this period was marked by considerable fluctuations of vegetation and climate (Andersen, 1961; Zagwijn, 1961). On the basis of cold-tolerant and thermophile species in Denmark, July temperatures may have been 2 to 3° C. lower than today, although summer temperatures in the Netherlands during the thermal optimum of the Brörup were similar if not identical to those of today. Winter temperatures in northern Germany and central Russia appear to have been 4° C. lower (Frenzel, 1968a). Since there is no reason to assume that the climate was unduly dry, the existence of a forest-steppe mosaic over most of mid-latitude Europe may have been due to incomplete adjustment of optimal vegetation types (Frenzel, 1968a). The northward migration of species, involving differential rates of movement from diverse refuge areas, was inhibited by several reversals of temperature and complicated by immature soils, widespread poor drainage, and local areas of coarse, mineral sediments. Similar problems are met in evaluating the early Holocene vegetation of Europe (see chapter 31).

The record of the Amersfoort Interstadial is too incomplete to allow any generalization; temperatures appear to have been lower than during the Brörup and forest development rather less complete. The Paudorf Interstadial is better understood, but again the data does not allow any general reconstruction. Vegetation zonation appears to have been intermediate between that of the Brörup and those conditions already described for the pleniglacial.

A classification of interstadial soils has been proposed for central Europe by Brunnacker (1965), ranging from gley soils (*AGC*-profiles) in poorly drained sites to tundra rankers or micropodsols at drier locales. Humification and some degree of subsurface waterlogging, recorded by limonitic mottling and platy structure, seem to have been typical of each warmer oscillation over most of the plainlands of western and central Europe. Further east, chernozems were characteristic of the cold steppes. Detailed descriptions of interstadial soils are given by Brunnacker (1959, 1960) and Fink (1961).

SUMMARY OF PLENIGLACIAL CLIMATES IN MID-LATITUDE EUROPE

In retrospect, the present state of knowledge about the Würm pleniglacial climate in mid-latitude Europe permits several tentative conclusions.

a) A frost climate, comparable to the Koeppen *EF* climates, was characteristic for three broad areas occupied by the Scandinavian, British, and Alpine glaciers. July mean temperatures certainly did not exceed + 4° C., and were probably generally less than 0° C.

b) A tundra climate, comparable to the Koeppen *ET* climate, occupied most of western and central Europe, as well as the unglaciated northern parts of Russia. This broad belt had continuous or discontinuous permafrost. The absence of tree growth and the presence of permafrost suggest a July mean under + 10° to 12° C., and an annual mean under − 2° C., most probably well under − 5° C. This implies midwinter temperatures averaging below − 20° C.

c) A forest-tundra or loess-tundra belt with permafrost, possibly comparable to Koeppen's winter-dry *Dcw* and *BSk* climates was found in Hungary, Romania, and the southern half of European Russia. At a guess, by modern analogy, July means were possibly around the 12° to 13° C. mark, with 30–40 frost-free days a year. Midwinter temperatures were about as severe as in the tundra belt *(b).*

(d) Cold-temperate woodlands, without permafrost and possibly comparable to those of Koeppen's subhumid *Dc* and *Db* climates, were probably encountered in the Mediterranean basin and along the southern shores of the Caspian Sea. Temperature and ecological transitions were probably rather rapid, and there is abundant evidence in favor of open vegetation or forest-steppe mosaics even here (see chapter 19).

These conditions appear reasonable as a reconstruction of the ecozones of the Würm pleniglacial in mid-latitude Europe, and probably also give a fair impression of conditions during the slightly colder (and drier?) Riss and Elster glaciations. Prior to the late Pleistocene, however, documentation of cold-climate phenomena is still—and may well remain—hopelessly inadequate.

NINETEEN

The Mediterranean Region during the Late Pleistocene

INTRODUCTION

Lower middle latitudes experienced considerable changes of climate during the course of the Pleistocene, although the geomorphic imprint of these paleoclimates on the landscape is less apparent than in temperate Europe. Among the reasons for limited landform sculpture is the absence of cold-climate processes except in the uplands. But changes in moisture regime and precipitation intensity are recorded.

The interpretation and documentation of Pleistocene features and paleoclimates in the Mediterranean lands are unfortunately incomplete. The pollen record known to date is fragmentary: a few profiles are recorded from Spain and Italy, with little or nothing elsewhere. Faunal evidence is available and may occasionally assume an important role. Geomorphic evidence is to some extent confused by features indicating greater cold in the highlands (presumably of full glacial age) and others indicating pluvial conditions in the lowlands (during the early glacial in particular). Although a rough qualitative picture of the environment may be obtained, accurate quantitative estimates of temperature or precipitation anomalies are not possible at the moment. Nor is a regional outline possible on account of the fragmentary nature of the evidence. Consequently, the discussion will be arranged topically. As comprehensive studies are lacking for the area, individual categories of evidence are often outlined in some detail.

EVIDENCE OF MOUNTAIN GLACIATION

Mountain glaciation, particularly during the maximum of the Würm, was common, although areally insignificant, in most of the Mediterranean world. Practically everywhere confined to small cirque or valley glaciers, glaciation's importance lies solely in its paleoclimatic implications.

Beginning in the west, the Iberian peninsula experienced local glaciations in the Cantabrian Mountains adjacent to the Bay of Biscay (crest elevations 1,500–2,650 m., Würm snowline at 1,350–1,750 m.), in the Central Sierras (crest elevations 1,900–2,600 m.), Würm snowline rising west to east from 1,650 to 2,000 m.), and in the Sierra Nevada (to 3,478 m., Würm snowline at 2,350 m.). The snowline depression of the Cantabrian Mountains was 800–1,200 m. (Lotze, 1962; Messerli, 1967); it was 1,300 m. on the northern face of the Sierra Nevada (Messerli, 1965), compared with 600–700 m. on the southern flanks and 1,400–1,500 m. on the northern face of the Pyrenees (see Solé Sabarís *et al.*, 1957, pp. 15–17). In a more general way the regional climatic snowline depression can be estimated at about 1,200.

In Morocco the Rif ranges (2,000–2,450 m., Würm snowline at 2,500 m.) and the Middle and High Atlas (3,300–4,150 m., Würm snowline at 2,900–3,500 m.) had localized glaciers (summarized by Messerli, 1967; Raynal, 1956). The regional snowline depression is estimated at 1,200 m. in the maritime Rif, at 1,050 m. in the continental interior. The low Jurjura Mountains (2,308 m.) southeast of Algiers were also glaciated, with a snowline estimated at 1,900 m. (Büdel, 1952).

The now unglaciated Apennine ranges of Italy (crest elevations 1,600–2,900 m.) have dispersed glacial forms in the highest regions, suggesting a Würm snowline at 1,200–1,500 m. in the Italian Riviera region, at 1,700–2,000 m. in central Italy, at 1,900 m. in the southern part of the ranges (Messerli, 1967; Klebelsberg, 1948–49, vol. II). As the highest mountains of the Gran Sasso reach 2,921 m., the snowline depression presumably exceeded 1,200 m. In Corsica (to 2,707 m.), the Würm snowline is estimated at about 1,800 m. (see Klaer, 1956). The other large Mediterranean islands (Mallorca, 1,445 m.; Sardinia, 1,843 m.; Sicily, 3,263 m.; Crete, 2,456 m.; and Cyprus, 1,953 m.) show no convincing evidence of glaciation.

Further east, the highlands of Albania and Greece (crest elevations 2,100–2,900 m.) were locally glaciated, indicating a Würm snowline rising from about 1,300 m. in the northern coastal range to 2,300 m. in the interior (Messerli, 1967; Klebelsberg, 1948–49, vol. II). Anatolia was likewise affected with numerous local glaciations in the major ranges (see Messerli, 1967; Birman, 1968). The Pontic or Black Sea ranges, rising from crests of 2,400–2,500 m. in the west to 3,700–3,900

m. in the east, had Würm snowline elevations of 2,300–2,600 m. In the western Taurus of southwestern Anatolia (2,300–3,000 m.) the snowline varied from 2,300 to 2,600 m.; for the central Taurus (crestlines 2,700–3,900 m.) this value is 2,600 to 2,900 m. for the southeastern Taurus (crestlines 3,400–4,100 m.), 2,700 to 3,000 m.

In the Armenian Plateau conditions were similar, while the important Caucasus range (crest elevations 3,600–5,600 m.) was extensively glaciated (see Klebelsberg, 1948–49, vol. II). The modern snowline here lies at 2,800–3,500 m., rising from west to east; the Würm snowline is estimated at 2,000–2,800 m. The climatic snowline depression for eastern Anatolia and the Caucasus can be estimated at 800 to 1,200 m.

Beyond the immediate confines of the Mediterranean area, evidence of glaciation is recorded from the northern Zagros, in Kurdistan, where Wright (1926b) obtains a phenomenal snowline depression of 1,200–1,800 m. for the 3,000–3,600 m. ranges. H. Bobek, however, finds it impossible to confirm these conclusions on the basis of air photo surveys (personal communication). Elsewhere in Iran the Würm snowline in Persian Azerbaijan (Seidan, 3,615 m.; Savalan, 4,812 m.; Sahend, 3,690 m.) was located at 3,200–3,500 m.; in the Elburz range (crest elevations 4,000–5,600 m.) at 3,300–4,150 m.; in the southern Zagros (Zardeh Kuh, 4,286 m.) at 3,350–3,400 m. (Butzer, 1958a, pp. 46–47, with references). The snowline depression amounts to at least 700–800 m. in northern Iran, and 600–650 m. further south. Glaciation of the High Lebanon (3,088 m.) and Mt. Hermon (2,814 m.) indicate a snowline of 2,700 m., over 1,000 m. lower than that of today (Messerli, 1966, 1967).

Late Pleistocene snowline depressions have been discussed in detail by Messerli (1967): 1,300–1,600 m. in the southern Alps; 1,200 m. in Iberia, northern Morocco, peninsular Italy, the southern Balkans, and the peripheral mountains of Anatolia; 800–1,000 m. in central Morocco and interior Anatolia or Iran. If a lapse rate of 6.5° C. per 100 m. is used, then, all other conditions being equal, the pleniglacial was 8° to 9.5° C. colder in the northern Mediterranean area but only 5.5° to 6.5° C. on the margins of the Sahara and within the great high plateaus of the Near East. If the pleniglacial was drier than today, these estimates should be increased. Comparatively higher temperatures in the southern and eastern Mediterranean borderlands may reflect greater aridity, both today and in the past, in the Sahara and the interior plateaus of the Near East; they may also indicate comparatively warmer temperatures in the subtropics. Both factors were probably involved. More elaborate climatological deductions, such as those of Kopp (1963), are theoretically unwarranted.

LATE PLEISTOCENE SOLIFLUCTION IN THE MEDITERRANEAN HIGHLANDS

Reliable evidence of Pleistocene permafrost is lacking in the Mediterranean region. However, cold-climate phenomena are relatively frequent in the highlands and were of considerable areal extent during the late Pleistocene. Generally the evidence consists of solifluction mantles, occasionally with cryoturbations or with grading over into block streams. True involutions are very scarce. Such evidence of periglacial activity is a result of short-term soil frost on a daily or seasonal basis. Its significance is mostly indirect: the lower limit of cold-climate solifluction gives a crude approximation of the altitudinal forest limit and so helps to delimit the zone of alpine meadows, as well as to provide some information on vertical zonation of climate during the Würm.

Features due to late Pleistocene solifluction are best known from the Iberian peninsula, the French Midi, and Morocco. Similar features have been little studied elsewhere. A major difficulty is stratigraphic dating. Raynal (1956) has indicated that in the Rif of Morocco, cold-climate solifluction sheets grade over imperceptibly into pluvial slope breccias of colluvial origin, which in turn extend down to and below modern sea level. It has been shown that such colluvial deposits are of early glacial age in eastern Spain (Butzer 1964a, 1964b) so that probably many of the solifluction deposits in question do not date from the period of greatest cold during the full glacial. They may instead reflect comparatively cool, moister conditions during the earlier Würm. In many areas there is evident confusion concerning the identification of cold-climate solifluction and simple slope breccias due to colluviation ("pluvial solifluction").

Starting with Spain, the lower limit of modern solifluction in the central southern Pyrenees appears to lie at about 2,200 m. (Solé Sabarís *et al.,* 1957, pp. 70-71), whereas Würm (?) solifluction mantles occur down to 800 m. in the same area (see Butzer and Fränzle, 1959). On the northern coast of Spain, colluvial deposits with evidence of some soil frost are locally found down to sea level (Llopís-Lládo, 1957, p. 32; Butzer, unpublished; Fränzle, 1959a, p. 64). In northwestern Spain, in the province of Pontevedra, evidence of cold-climate solifluction is lacking below 400 m. (Butzer, 1967). In Mallorca, late Pleistocene solifluction is wholly absent below 950 m. (Butzer, 1964b) and does not occur down to 500 m. (as previously reported by Mensching, 1955b). An "ice-wedge" reported from an elevation of 200 m. near Zaragoza (Johnsson, 1960) cannot be accepted as valid evidence, but must be interpreted as colluvial filling of a crack due either to microtectonics or to dehydration. In the central cordillera, where solifluction features occur today above about 2,000 m. (Fränzle, 1959a, p. 34 ff.), late Pleistocene block streams occur to below 1,100 m. elevation (Fränzle, 1959, p. 62), while colluvial slope breccias with limited evidence of solifluction (Würm?) occur to a lower limit of 1,000 m. (Butzer, 1964b). Finally, in the Sierra Nevada, Pleistocene solifluction features occur to a lower limit of 1,000 m., about 1,100 m. lower than today (Paschinger, 1961). In summary, the upper limit of *closed* forest for the Iberian Peninsula can be estimated at about 1,000 m. in southern, eastern, and central Spain, at 800 m. in the southern Pyrenees, and probably near sea-level north of the Cantabrian Mountains. As shown below, this is substantiated by critical pollen studies in Catalonia and northwestern Spain as well as by faunal evidence. The depression of the lower limit of solifluction amounted to perhaps 1,400 m. in the far north, to 1,000-1,100 m. elsewhere in the country. This exceeds the value of the regional snowline depression. The exceptionally high value in the Pyrenean and Cantabrian regions may indicate the presence of strong horizontal temperature gradients in these areas, which separated the subarctic and temperate parts of Europe during the glacial periods.

In Morocco, the lower limit of Würm-age solifluction in the western Rif lies at 800 to 1,000 m. (Mensching, 1955a), in the eastern Rif at 800 m. (today 1,500-1,550 m.), in the central Rif at 900 to 1,000 m. (today 2,000 m.) (Raynal, 1956). In the Middle Atlas, Würm solifluction deposits of semifluvial character are typical to elevations as low as 1,600 m. and occur to 1,400 m. (1,900-2,000 m. today) (Raynal, 1956, 1960b). In the High Atlas such features occur above 2,700 to 3,100 m. today (Raynal, 1956), but extended down to 1,800-1,900 m. during the late Pleistocene (Raynal, 1960b). The lower limits of Pleistocene cold-climate solifluction in Morocco seem to lie at about 800 to 1,000 m. in the northwest, rising gradually to 1,800-1,900 m. in the southeastern ranges. The Würm-age depression varied regionally between 700 and 1,100 m., averaging about 850 m. This value is less than the one for Spain, a fact which may reflect either greater aridity or smaller Würm temperature depressions at lower latitudes.

In southern France, Würm solifluction with evidence of cryoturbation is found to elevations of 200-300 m. in the western French Alps (Raynal, cited by Mensching, 1955a), and possibly to about sea level in Languedoc and in the French Riviera (Tricart, 1956a). In Corsica, similar features occur to elevations no lower than 500 m. (Büdel, 1953; Klaer, 1956; see also Tricart, 1956a). Information for Italy is almost limited to some cursory remarks by Nangeroni (1952) and by Tricart and Cailleux (1956) about Würmian (?) cold-climate deposition of vague implications at 550 m. in the Umbrian Apennines. Büdel (1951b) reports fossil solifluction deposits from about 1,000–1,200 m. in southern Italy (today above 2,000 m.).

In the eastern Mediterranean Basin the lower limit of Würm solifluction is estimated at 400 m. on the Adriatic coast of Yugoslavia, at 300 m. in Southern Yugoslavia, and at 800 m. on Crete (Brunnacker, 1967b; Poser, 1957). Evidence in the Near East is scanty and summarized by Butzer (1970) and Klaer (1963). Fossil slope breccias, possibly related to solifluction, occur down to 1,500 m. in the central Taurus of Anatolia, and to 2,300 m. in Iran. No general conclusions can be suggested for either the central or eastern Mediterranean areas.

PALEOBOTANICAL EVIDENCE FOR THE WÜRM PERIOD

Palynological and macrofossil study of late Pleistocene vegetation in the Mediterranean region is still in its infancy and practically confined to Spain and Italy. Its significance is almost limited to corroboration of the upper forest limit estimated from geomorphic evidence, and to broad identification of the forest types in the northwestern Mediterranean borderlands.

A pollen profile from the Laguna de las Sanguijuelas at 1,000 m., near the Portuguese border in northwestern Spain (Zamora Province), has been studied by Menéndez-Amor and Florschütz (1961) and radiocarbon dated. The lowest part of the profile, prior to a C^{14} date (Gro. 705) of 13,700 B.P., is characterized by 65–85 per cent nonarboreal pollen (NAP), and 15–35 per cent pine, together with traces of birch (8 out of 8 samples), willow and oak (each in 6 samples), hazel and alder (each in 2 samples). Grasses, *Artemisia* (to 20 per cent), and other steppe indicators such as *Ephedra, Helianthemum, Thalictrum,* and *Hippophae* account for the NAP. Macrofossils of birch have been identified in the same profile. The interpretation of an open parkland with scattered groups of pine suggests a montane forest-steppe or steppe with some few thermophile species growing at lower elevations. A temperate woodland was already found in the area during the later Alleröd, while oak and pine dominate the vegetation today. Almost identical results have been obtained from Padul (740 m.) on the southern Sierra Nevada slopes (Menéndez-Amor and Florschütz, 1962).

Equally valuable is a pollen profile studied from the Cueva del Toll at 750 m. near Moyá, 50 km. north of Barcelona (see Butzer and Freeman, 1968). In what is faunistically the late Pleistocene, pollen spectra recording forests dominated by *Pinus silvestris* (NAP 15–34 per cent) alternate with spectra suggesting open herbaceous vegetation (NAP 49–51 per cent) with a few deciduous trees. Donner and Kurtén (1958) interpret the former associations as "temperate-humid," the latter as "warm-dry." In one case of the high NAP assemblages (51 per cent, layer *h*), however, the geologic deposits (Serra Ráfols *et al.*, 1957) consist of a breccia of uncorroded, noncemented, very flat and angular *éboulis,* with a mammalian fauna containing three alpine species as well as *Tichorhinus antiquitatis.* Admittedly, half the fauna of this stratum is woodland in character, while among the 49 per cent of arboreal pollen, oak accounts for 4 per cent, elm 1 per cent, hazel 1 per cent, with birch 2 per cent and pine 41 per cent. Despite these apparent contradictions, layer *h* may represent the Würm maximum. The remainder of the profile leaves no doubt that cool-temperate, subboreal forests dominated the area during most of the late Pleistocene. The present vegetation consists of evergreen oak *(Quercus lusitanica)* and beech, and is submediterranean in character.

In northern Italy, pollen has been obtained from some peaty clays present in deep bores near Ravenna. Two sections studied at −22 and −26 m. at Forli (Firbas and Zangheri, 1954; Dubois and Zangheri, 1957) showed an average of 85 per cent pine, 4 per cent spruce, 4 per cent birch, 4 per cent willow, 1 per cent larch, 1 per cent aspen. A single pollen of oak is the only thermophile representative, leaving no doubt that this was a time of boreal forest. Judging by the modern distribution of oak in Europe (Walter, 1954, Fig. 135) this vegetation assemblage belongs in the *Db* or *Dc* climates of Köppen, possibly suggesting a mean annual temperature under +6° C. (today +16° C.) and a July mean on the order of +16 to 18° C. (today +25° C.) for the lower Po Valley. This is analogous to the Würm temperature depression suggested for middle latitude Europe. A last glacial age is suggested for these spectra by the presence of an interglacial type assemblage, with 49 per cent pollen of deciduous species, found at − 75 m. in the same area. A similar glacial spectrum was studied underlying a postglacial pollen sequence near Este in the Po-Adige lowland by F. Lona (cited by Bonatti, 1966). However, here the NAP was counted, indicating that the full glacial vegetation was open and dominated by *Artemisia.* A late Würm profile from 880 m. elevation near Lake Garda (Beug and Firbas, 1961) has 18 per cent *Pinus,* 18 per cent *Juniperus,* and 62 per cent NAP, mainly *Artemisia,* further supporting the notion of a steppe vegetation in northern Italy during much of the Würm time span.

On the coast of central Italy, two further successions document a decrease of humidity and an increase in cold during the course of the Würm. In the Agro Pontino, southwest of Rome (see A.C. Blanc, 1959, with references; Tongiorgi, 1938; Blanc *et al.*, 1957), a series of peat strata overlie a typical Tyrrhenian II beach with the mollusca *Strombus bubonius* and *Conus testudinarius* at + 5.1 m. At the base of the terrestrial deposits, species such as wild grape *(Vitus vinifera), Quercus robur, Cornus mas,* and *Carpinus betulus* indicate submediterranean to warm-temperate conditions. At higher levels, further warm-temperate species, such as hazel, yew, and beech, as well as white fir *(Abies alba),* a cool-temperate indicator, appear in a stratum dated at 59,000 B.P. by C^{14} (comment by A. C. Blanc on Movius, 1960). *Abies alba* dominates the top of these peats. The fauna of the peats includes typical sylvan forms such as *Elephas antiquus, Dicerorhinus mercki, Cervus elaphus, Capreolus capreolus,* and at another contemporary site, hippopotamus. The sequence is followed up by regressional dunes containing bones of woolly mammoth and showing evidence of frost-weathering, but with no plant remains. In the Bassa Versilia, northwest of Pisa (see A. C. Blanc, 1936; Tongiorgi, 1938), *Abies alba* forests were eventually replaced by cool-temperate forests with *Pinus montana mugo* and *Pinus silvestris;* and a comparable *Abies* stage appears to have preceded the *Pinus* maximum at Ravenna on the Adriatic coast (see Dubois and Zangheri, 1957).

The aridity and cold of the main parts of the Würm are most apparent from a core taken at 507 m. from a crater lake 50 km. northwest of Rome (Frank, 1969). Here the pleniglacials were characterized by a cold *Artemisia*-gramineae steppe, with some stands of pine or juniper. The Brörup-Amersfoort interval, however, was moist and cool with a closed pine-spruce-oak forest, while the mid-Würm interstadials were only slightly less severe than the remaining pleniglacial time range. Comparable patterns were found from two nearby lakes at only 200 m. elevation (Bonatti, 1966), where an *Artemisia* steppe of ca. 25,000-15,000 B.P. was replaced by a grassland and ultimately, between 12,000 and 11,000 B.P., by hazel, oak, and fir woodlands.

Open vegetation, possibly representing a forest-steppe mosaic (see Fig. 51), was also characteristic of the southern Balkans during most of the Würm. In Macedonia, at about sea level near Salonika, a deep core indicates 80-95 per cent NAP for the pleniglacial (van der Hammen *et al.,* 1965), with the NAP spectra including between 35-75 per cent *Artemisia,* 15-50 per cent grasses, and 10-15 per cent chenopods. These spectra are similar to those found today near the forest limits of the west Siberian steppes (see Frenzel, 1968a, p. 231 and plate 12). The driest conditions in Macedonia were reached about 14,500 B.P. A more detailed sequence from northwestern Greece comes from a marsh near 400 m. elevation, 15 km. northwest of Ioannina (Bottema, 1967). A dry and cold steppe, with 60-80 per cent NAP (mainly grasses and *Artemisia*) and 10-20 per cent *Quercus,* is indicated for a brief period ca.

65,000 B.P. and for an extended time from ca. 35,000 to 8000 B.P. However, an open oak woodland or forest-steppe with 40 per cent *Quercus* and abundant *Artemisia* can be inferred ca. 60,000–50,000 B.P., while a closed forest of fir, oak, hornbeam, and beech ca. 50,000–40,000 B.P. suggests cool and very humid conditions. The dating is extrapolated from a radiocarbon date of 40,000±1000 (GrN-4793), so that exact correlations with the Brörup-Amersfoort and the Lower Pleniglacial are difficult. Nonetheless, it is clear that parts of the early Würm sequence in Greece were cool and wet, just as in west-central Italy.

Evidence from the eastern Mediterranean basin and its borderlands is rather sparse. In the Derna area of coastal Cyrenaica, in tufa deposits immediately postdating a 6 m. Upper Pleistocene beach, McBurney and Hey (1955, pp. 109–110) found the Canary laurel *(Laurus canariensis)* and the Aleppo pine, suggesting a mediterranean-type woodland. Later deposits lack pollen but suggest dry conditions. In the Lebanon Mountains, at 2,000 m., tufa impressions of apparent late Pleistocene age include warm-temperate species such as oak, beech, elm, and hazel, rather than submediterranean forms as are present there today (see Butzer, 1958a, p. 81). In northeastern Iraq preliminary pollen studies are available from Shanidar Cave at 650 m. (Solecki and Leroi-Gourhan, 1961). A spectrum at −4 m. (C^{14} date 31,000 B.P.) had 90 per cent NAP. More convincing are several cores studied in detail by Wright and his associates (Zeist, 1967; Wasylikowa, 1967; Megard, 1967; Wright *et al.*, 1967), in a broader context, in the Zagros country. An open vegetation of steppe and mountain tundra (see discussion in Butzer, 1972) is indicated here from before 22,000 to 12,000 B.P., with 80–95 per cent NAP, mainly Chenopodiaceae and *Artemisia*. The local tree-line was 1,000 m. lower than today. A gradual recolonization by oak woodland is indicated between 12,000 and 6000 B.P.

The above evidence indicates that the Mediterranean basin was cool and wet during parts of the early Würm, probably during the Eem-Würm transition and again during the time interval of the Brörup-Amersfoort. However, parts of the early Würm were already cold and dry, and such conditions prevailed from mid-Würm times until the onset of the Holocene. It is still uncertain whether the Würm-age steppes were primarily a result of climatic aridity or of physiological aridity in relation to great winter cold, complemented by summer drought. In either eventuality, mediterranean or submediterranean woodlands were very rare during the pleniglacial—they may have been confined to favorable topographic location in the southern half on the Mediterranean Basin, probably at intermediate elevations.

FAUNAL EVIDENCE FROM THE LATE PLEISTOCENE

In lieu of further paleobotanical information, a selection of faunal assemblages from different parts of the Mediterranean region will provide a better ecological picture.

In Spain, three particular faunal assemblages are of interest. They include, first, an annotated bibliographic study of archeologically associated fossils from the north coast province of Asturias (Oviedo) by Fraga Torrejon (1958), specifically Aurignacian to Magdalenian in date (ca. 35,000-12,500 B.P.). It provides a fair picture of the fauna, as paleontologically verified, for the environment of the 0-500 m. elevation range of the northern Spanish coast. This picture coincides remarkably well with that of the contemporary cave art representations of the Cantabrian region (J. M. Leverenz, unpublished research paper, 1960). From northeastern Spain, the rich fauna of the Cueva del Toll (750 m.) (Serra Ráfols *et al.*, 1957) is equally interesting. In Table 14 dominantly woodland animals are indicated with (w), open-country forms (tundra, grassland, mountain) with (o), and more or less indifferent types with (i).

Of the 31 species identified for the Asturian and Cantabrian area, 10 are essentially woodland forms and 10 nonwoodland types, while 11 are more indifferent in their ecological requirements. Of the 21 species identified in Catalonia, 9 are woodland types, 6 open-country forms, 6 indifferent. In each case specific tundra or alpine types are present. But the high percentage of sylvan forms implies that scattered woodlands must have persisted in the 0-500 m. elevation range of Catalonia throughout even the coldest phases of the Würm. The Asturian area was presumably a forest-tundra, while in Catalonia one might expect a subalpine steppe above the lower limit of solifluction (800-1,000 m. in northeastern Spain). Similar conditions must have persisted elsewhere in the Spanish uplands, as suggested by the mixed fauna depicted in Castilian and Andalusian cave drawings in the 500-1,000 m. elevation range, or by contemporary fossils found at cave sites in Gibraltar (Garrod *et al.*, 1928).

The main Wurm faunas of Italy are as important as those of Spain. The information provided by the Monte Circeo caves (A. C. Blanc, 1942), the Grotta Romanelli near Otranto (G. A. Blanc, 1921), the Grotto San Teodoro near Messina (Vaufrey, 1928), the Agro Pontino (A. C. Blanc, 1935), or the Ghar Dalam cave of Malta (Vaufrey, 1929) corroborates the sketch of geomorphic and paleobotanical features above, and is faunistically quite analogous to the Iberian peninsula. Similar mixed faunal associations occur in the Riviera caves (Boule *et al.*, 1906-19; Boule and De Villeneuve, 1927; de Lumley, 1963).

In northwestern Africa, an Ethiopian rather than a Palearctic fauna

Table 14. Main and late Würm faunas in Spain
[(w) = *woodland,* (o) = *open country,* (i) = *indifferent forms*].

Species	*Asturias (Fraga Torrejon, 1958)*	*Cantabrian Cave Drawings*	*Cueva del Toll (Serra Rafols* et al., *1957)*
Ursus spelaeus (cave bear) (i)	*x*	*x*	*x*
Ursus arctos (brown bear) (w)	*x*		
Canis lupus (wolf) (i)	*x*		
Vulpes vulpes (red fox) (w)	*x*		*x*
Crocuta crocuta spelaea (cave hyena) (i)	*x*	*x*	*x*
Felis silvestris (wildcat) (w)	*x*		*x*
Felis pardina spelaea (cave lynx) (i)	*x*	*x*	
Felis leo spelaea (cave lion) (i)	*x*		*x*
Meles meles (badger) (i)	*x*	*x*	*x*
Mustela erminea (ermine) (i)	*x*		
Putorius putorius (polecat) (i)	*x*		
Lutra lutra (otter) (i)	*x*		
Erinacaeus europaeus (hedgehog) (w)			*x*
Marmota marmota (marmot) (o)	*x*		
Oryctolagus cuniculus (rabbit) (i)			*x*
Lepus timidus (snowshoe "rabbit") (o)	*x*		
Castor fiber (beaver) (w)			*x*
Apodemus silvatica (field mouse) (i)			*x*
Arvicola amphibius (water vole) (i)	*x*		
Microtus nivalis (alpine vole) (o)			*x*
Talpa europaea (bat) (w)			*x*
Elephas primigenius (o)	*x*	*x*	
Dicerorhinus mercki (w)	*x*		
Tichorhinus antiquitatis (o)	*x*		
Equus caballus spp. (wild horse) (o)	*x*	*x*	*x*
Equus asinus hydruntinus (ass) (o)	*x*	*x*	
Sus scrofa (boar) (w)	*x*	*x*	*x*
Cervus elaphus (red deer) (w)	*x*	*x*	*x*
Capreolus capreolus (roe deer) (w)	*x*	*x*	*x*
Dama dama (fallow deer) (w)	*x*	*x*	
Alces alces (elk) (w)	*x*	*x*	
Rangifer tarandus (reindeer) (o)	*x*		
Rupicapra rupicapra (chamois) (o)	*x*		*x*
Capra ibex pyrenaica (ibex) (o)	*x*	*x*	*x*
Bison priscus (o)	*x*	*x*	*x*
Bos primigenius (w)	*x*	*x*	*x*

remained dominant through most of the Pleistocene. However several new Palearctic forms appeared during the Würm regression (Arambourg, 1952a, 1962), presumably via Asia and the southern Mediterranean littoral. The Palearctic species include *Bos primigenius,* several deer *Megaceroides algericus, Cervus elaphus),* the brown bear *(Ursus arctos),* a bear *(Ursus faidherbi)* intermediate between the brown and the cave bear, Barbary sheep *(Ammotragus lervia),* boar *(Sus scrofa),* the extinct woodland rhinoceros *Dicerorhinus mercki,* and a dwarf elephant *(Elephas* [*Palaeoloxodon*] *iolensis).* These temperate woodland species, presumably indicative of a cooler climate and a southward faunal migration, occur alongside of Ethiopian (or Saharan) species such as the hippopotamus, the white rhino *(Ceratotherium simum),* an elephant closely related to the modern African species, wart hog, zebra, camel, and a number of antelopes and gazelles. Both the hippo and Merck's rhino had disappeared by the time of the main Würm.

The contemporary fauna of the Upper Paleolithic strata of two caves in Cyrenaica (Bate, 1955; Higgs, 1967), seem to suggest that a number of Palearctic species passed through the area during the Pleistocene. Together with the bat and such rodents as the shrew, gerbil, field mouse, and dormouse, the fox, an extinct Asiatic buffalo (*Homoioceros*), aurochs (?), barbary sheep, and an unidentified rhinoceros occur in association with Ethiopian forms. The latter include hyena, lion, antelope, gazelle, and zebra. The fauna of the Egyptian Nile Valley included few Palearctic species, e.g., at Kom Ombo (see Butzer and Hansen, 1968, p. 113).

No comparable faunal studies are available from southeastern Europe or Anatolia, but the Levant area provides much interesting information. Exceptionally rich faunas have been studied at the Mt. Carmel caves of Palestine (the Aurigacian beds *D-F* of Mugharet el-Wad, about 70 m. elevation, by Garrod and Bate, 1937) and at Ksar Akil on the Lebanese coast (about 25 m., by Hooijer, 1961). The fauna includes rodents and other small mammals such as porcupine, squirrel, hare, and hedgehog; carnivores such as marten, badger, wildcat, spotted hyena, wolf, brown bear and its subspecies the Syrian bear *(Ursus arctos mediterraneus),* leopard *(Felis* [*Panthera*] *pardus),* red fox, and the mustiline *(Vormela peregusna);* herbivores such as boar, red deer, roe deer, fallow deer *(Dama mesopotamica),* wild cattle, gazelles, wild goat *(Capra aegagrus),* the onager *(Equus hemionus),* and wild horse. Of a total of no fewer than twenty-seven species present, nine are more or less indifferent forms, twelve are preferably associated with woodland environments, six with open landscapes. Numerically, the woodland form *Dama mesopotamica* dominates the fauna of both sites during what presumably represented the main Würm. Merck's rhinoceros disappeared from the area at the beginning of the Würm. The fauna of the Umm-Qatafa cave in Judaea, studied by Vaufrey (in Neuville, 1951), is quite analogous to that of Mt. Carmel. No strictly contemporary faunas are known from other parts of the Near East, although late Würm levels at Palegawra (about 1,000 m.) and Shanidar (about 650 m.) also have a comparable fauna (see Braidwood, Howe *et al.*, 1960, pp. 58–59, 168–70). The complete absence of truly cold forms in North Africa and the Near Eastern lowlands or foothills is suggestive, while the appearance of numerous temperate zone species points to a southward extension of submediterranean woodlands.

All in all, the faunal evidence of the Mediterranean basin well substantiates the fragmentary pollen record of extensive open vegetation or parklands in the now subtropical Mediterranean lowlands. Furthermore, the migration of woodland species of Asiatic derivation into northern Africa during the early Würm probably suggests the presence of temperate woodlands or, more probably, forest-steppe mosaics fringing the northern coasts of Libya.

LATE PLEISTOCENE FLUVIAL PHENOMENA IN THE MEDITERRANEAN LOWLANDS

Throughout the mediterranean climate zone of the Mediterranean Basin, the early Würm was heralded by intensive sheet erosion and deposition of a heterogeneous class of colluvial silts and alluvial gravels. The characteristically "mediterranean," fossil colluvial silts of this morphodynamic phase are largely derived from reddish interglacial soils (terra rossas in limestone bedrock areas), hence the French technical term of *limons rouges*. One group of these sediments is located in caves, where Arambourg (1952b) first outlined their stratigraphic peculiarity in association with a Mousterian-type industry and an Upper Pleistocene fauna of warm-temperate affinities. Another group is largely fluvial; for example, the great spreads of red silts associated with the late Pleistocene of Morocco, first interpreted by Choubert (1948a, 1948b). Probably by far the most characteristic are slope breccias of colluvial origin, such as described by Wiche (1961) for southeastern Spain. Local aeolian components have also been recognized in Catalonia (Virgili and Zamarreño, 1957; Butzer, 1964a), in southeastern Spain (Brunnacker and Ložek, 1969), and in Tripolitania (Hey, 1962).

Colluvial silts occur through most of the mediterranean woodland zone. For example, they are of considerable importance in the Balearic Islands and to a lesser extent in Catalonia. Analogous, though sporadic, phenomena can be observed in the interior of Spain where they intergrade with solifluction deposits. The following generalizations are based on field and laboratory studies by the writer in these areas (Butzer, 1963b, 1964a, 1964b; also Durand, 1959).

The morphology of these deposits is mainly one of relatively thin areal sheets (to 1 m.), attaining greater thicknesses only at the foot of slopes or in original topographic hollows (to 5 m. or more). Surface slopes seldom exceed 15 per cent, and characteristic examples can be cited where the bedrock topography has slopes of 1 to 3 per cent. In drainage channels these beds grade laterally into alluvial fill with rounded gravel.

Sedimentology is highly variable. Stratification of individual beds rather than individual coarse components is usual; sorting is less common. At the base of steeper slopes, beds may be detrital in character. The amount of coarse component over two millimeters diameter is usually confined to coarse, angular to subrounded gravel embedded in finer sediments. The fines show a grain-size spectrum distributed more or less uniformly among the clay, silt, and fine, medium, and coarse sand fractions. Often a moderate maximum may be found in the coarse silt and finer sand fraction. The writer knows of no spectra over 70 per cent by weight in the silt-fine sand range, so that typical loess deposits are unknown. However, semiaeolian beds are not uncommon in association with regressional dunes—specifically, aeolian materials bedded by rainwash (see chapter 12 on loess). Most frequently the grain-size distribution is analogous to that of the source materials, i.e., soils and weathered aeolian beds.

Frequently the colluvial silts are interupted by, interbedded with, or capped by calcareous crusts of the *croûtes zonaires* type (see chapter 11), which occasionally grade into true travertines.

The stratigraphy of the silts is well defined in the Mediterranean littoral regions (Butzer and Cuerda, 1962a, 1962b; Bonifay, 1962). The silts are often interbedded with marine beds of the Tyrrhenian III and were gradually superseded by regressional dunes during the major negative movements of sea level accompanying the advance of the Würm glaciers. Terrestrial silt beds can be found to well below modern sea level, and laterally conformable alluvial fill shows abnormally steep gradients at the coast. Several generations of silts are recorded in Spain and in Provence, where two or three positive interruptions of the Würm regression were followed by colluviation and subsequently by renewed aeolian deposition. This leaves no doubt that several major oscillations of the continental glaciers in higher latitudes had direct repercussions in the Mediterranean area, i.e., the readvances were associated with secondary maxima of the peculiar climatic regime responsible for silt colluviation.

Geomorphological interpretation of the *limons rouges* must account for extensive denudation accompanied by colluviation and valley alluviation. The primary fact of greater stream competence and accelerated denudation is quantitatively confirmed by morphometric gravel analysis. Comparative statistical analyses of recent and various Pleistocene gravels in identical stream valleys showed that during much of the early Würm, fluvial processes were more effective than during the Holocene. Colluvial spreads on nearly flat surfaces indicated a fair degree of rolling of the contained pebbles, often more so than in modern torrent beds. No assumption of perennial stream flow is warranted for torrents which now enjoy only episodic or seasonal waterflow. But the periodic flash floods must have been more frequent or capable of carrying a load over greater distances.

The paleoclimatic interpretation is then one of *(a)* greater rainfall intensity to permit effective erosion;[1] *(b)* rainfalls of sufficient duration to permit thorough soaking of the soil and extensive transport of soil-wash, as well as greater stream competence; and *(c)* a pronounced dry season, to account for the incomplete nature of the vegetation mat. Whether or not total annual precipitation was different from that of the present cannot be decided from the geomorphological evidence alone. It is abundantly clear that a simple decrease of temperature will not reduce the proportion of evaporational losses for heavy rainfalls of the type that produce localized sheetwash; instead, reduced evaporation will only increase long-term stream discharge by increasing the base flow derived from springs and subsoil seepage. In fact, lower temperatures would favor a more complete vegetation mat, tending to reduce the rapidity of runoff and, as a corollary, stream competence. In other words, the accelerated geomorphologic processes indicated by the colluvial silts of the Mediterranean basin require a change of rainfall regime, with more frequent rainstorms of high intensity and considerable duration.

These colluvial silts are by far the most characteristic Pleistocene sediments of the Mediterranean lowlands. They are well developed in western Morocco, northern Algeria and Tunisia, northern Tripolitania, Israel and Lebanon, western Greece, Provence, and eastern and northwestern Spain. Laterally such silts grade from slope wash to extensive piedmont alluvia, rich in red soil sediments, or true alluvial beds of variable caliber. In hilly or mountainous terrain they are interdigited with mixed colluvial and slope screes involving a larger proportion of mass movements. In northern latitudes and at higher elevations, accelerated frost-weathering provided much of the coarser detritus, while creep,

1. Quantitative work by Barat (1957) confirms that erosional activity and sheetflood level are proportional to rainfall intensity and raindrop size.

slumping, and earthflows as well as soil frost appear to have become major agents of transport (Butzer, 1964b, 1967). Thus the Mediterranean silts find transitional forms to the coarse footslope detritus and hillwash developed in parts of western Europe and in southern Britain, where they are known as *head*. Toward the arid zone there also are affinities with piedmont mantles or coalescent fans, while a genetic relationship with colluvial spreads in the tropical savannas, or slope wash under rainforest, is unmistakeable. Nonetheless, the colluvial silts are a distinctive regional phenomenon.

The intensive rainfalls suggested by the colluvial silts do not span the duration of the Würm Glacial. Several generations of massive regressional aeolianites can commonly be recognized in coastal areas, bedded alternately with colluvial silts. The aeolian deposits are not always restricted to the shore zone but may extend far inland, as during one or more mid-Pleistocene regressions when dune sands swept right across the island of Mallorca. In other coastal areas the regressional dunes are closely related to the origin of the semi-aeolian silts, with loess structure, as already discussed in chapter 12. Considered in conjunction with the pollen evidence from Italy and northern Greece outlined above, the aeolian components well inland suggest increasing aridity and the existence of open country, presumably forest-steppe mosaics, in other lowland areas. Consequently the pleniglacials must have been quite dry, a fact substantiated by the limited development of synchronous alluvial deposits in Spain and Provence and the increasingly angular nature of the detritus, as well as the finer nature of the stream deposits in many areas.[2]

During the earliest parts of the Würm Glacial, paleosols periodically developed prior to renewed colluviation, commonly on a parent material of calcareous silt or aeolianite. Such soils suggest braunerde and terra fusca profiles in southeastern Spain and Greece (Brunnacker and Ložek, 1969). After aeolianite deposition came to a halt, during the late Würm Glacial, soil development was and has remained of limited intensity until the present day. Terra fuscas and terra rossas did not develop on fresh

2. Vita-Finzi (1969) postulates a single generation of "Older Fill" encompassing the 60,000 time span of the Würm Glacial in the Mediterranean lowlands. Detailed regional studies demonstrate that this is a gross oversimplification, possible only as a result of the fragmentary preservation and poor dating possibilities of nondescript alluvial fills. One of Vita-Finzi's examples, the Cyrenaican, can be cited to this point. Many of the "Younger Gravels" of Hey (1962, 1963; McBurney and Hey, 1956) are developed as coastal alluvial fans under regressional aeolianites. However, at various localities, such gravels may be found over beach conglomerates, under or over aeolianites, or present as multiple strata within aeolianites. Inevitably, a detailed stratigraphy would recognize different generations both of the aeolianites and of the fluvial units (e.g., Bonifay, 1962; Butzer and Cuerda, 1962a; Hilly, 1962).

late Pleistocene sediments; instead *AC*, *A-Ca-C*, or weak *A(B)C* profiles are typical, depending on the nature of the parent material.

RETROSPECT OF THE MEDITERRANEAN ECOZONE DURING THE LATE PLEISTOCENE

In comparison with the many regional or synthetic studies of the late Pleistocene geography of mid-latitude Europe or certain parts of Africa, no systematic attempt has been made to discuss the major features of the Mediterranean lands during the late Pleistocene. Consequently this discussion has necessarily been detailed, the presentation topical rather than regional in approach. Without a proper understanding of this area, the interplay of diverse geomorphologic processes at different latitudes and elevations can hardly be deciphered. Similarly, the interrelationships of moraines, solifluction mantles, slope screes, colluvial silts, piedmont and stream alluvia, semi-aeolian silts, regressional dunes, and beach deposits must be explained if the concepts of "periglacial" realms in the north and of "pluvial" belts in the south are to be handled rigorously, separating fact from fiction or polemic.

It is not yet possible to present a sequence of radiocarbon-dated or, for that matter, stratigraphically dated late Pleistocene events of broad, circum-Mediterranean validity. However, the broad contrasts between the early glacial (and the interstadials?), the pleniglacials, and the late glacial can be recognized in many areas. Similarly, certain generalizations can be suggested for the different categories of information:

a) Localized mountain glaciers were developed in numerous highlands, indicating somewhat lower temperatures in the mountains. The Würm snowline was 1200 m. lower than that of today around most of the Mediterranean Basin, although the snowline depression on the margins of the Sahara and amid the vast mountain country and plateaus of eastern Anatolia and Iran was as little as 800–1,000 m.

b) Cold-climate solifluction, in what was probably alpine meadow or subalpine parkland, affected most of the highlands of southern Europe above elevations of 1,000 m. or so, and the highlands of North Africa and the Near East at considerably higher elevations (900–1,900 m. in Morocco, with scattered data in the Near East suggesting 1,500–2,300 m. [?]). The Würm depression of this "lower limit of solifluction" and, possibly, of the altitudinal limit of closed forest amounts to 1,000–1,200 m. in Spain; 800–1,000 m. in Italy; 1,000 m. in Yugoslavia and Crete; and 850 m. in Morocco. Many of the "periglacial" phenomena described in the literature are of questionable interpretation, and many "solifluction" deposits are nothing but colluvial silts or screes. The "lower periglacial limit" described by some authors, such as Tricart (1966), is a fiction—there can only be lower limits to specific cold-climate phenomena—and the definition of any "periglacial" environments in the Pleistocene Mediterranean region is open to serious criticism. The acceleration of geomorphologic agencies and the frequent ruptures of geomorphologic equilibrium were more often than not related to processes now peculiar to arid or tropical environments. There is no acceptable evidence of permafrost from any part of the Mediterranean basin.

c) Biological information leaves no doubt that subtropical woodlands were largely replaced by temperate species in southern Europe, and that submediterranean or even temperate genera were of major importance along the coasts of northern Africa and in the Levant. Similarly, alpine meadows, montane steppes, cold-temperate woodlands, or forest-steppe mosaics were widespread in the hill country of southern Europe and were frequented by a good number of cold-tolerant mammals. Temperate woodland faunas were important in the Levant and spread westward along the littoral of northern Africa. An increasing body of palynological evidence, as well as geomorphic criteria, indicate that montane steppes and forest-steppe mosaics were characteristic of what is today the submediterranean zone, extending down to sea level in Italy, the Balkans, and the Aegean area (see Fig. 51). In part this was a direct response to cold, as in the highlands; elsewhere, it was primarily the effect of a relatively dry climate. Despite the reduced evaporation, in response to temperatures (6–9° C. [?]) lower than today, many areas now experiencing a dry season of four to six months had a dry season of at least five to eight months at the climax of the Würm pleniglacials.

d) The evidence of fluvial geomorphology, ranging from alluvial terraces, fans, and piedmont plains over colluvial silts and soil sediments to slope screes, indicates several periods of accelerated runoff, effective sheetwash and, presumably, more frequent, high-intensity seasonal rains of considerable duration. In coastal areas such deposits are interdigited with or rest upon the Tyrrhenian III beaches, and they distinctly preceded several phases of aeolianite accumulation. Widespread aeolian beds in coastal areas during the Würm pleniglacial, together with limited evidence of fluvial activity, point towards a semiarid climate—with widespread forest-steppe—in the Mediterranean lowlands. Semiarid soil development during the late glacial and also during most of the Holocene suggest that dry conditions have persisted, to some degree or other, into more recent times.

TWENTY

Some African Environments of the Late Pleistocene

THE PLUVIAL PROBLEM: INTRODUCTION

Pleistocene environments in lower latitudes have been the subject of speculation and study since 1884 when E. Hull first postulated that the Pleistocene Ice Age—considered as a single unit—corresponded to a period of moist climate, a "Pluvial Age," in the Palestinian desert. In 1896, for the Jordan Valley, and in 1901, for the Nile Valley, M. Blanckenhorn outlined a sequence of lake and alluvial deposits to support a number of Pleistocene pluvial episodes. In this way the concept of "pluvial" climates in lower middle latitudes was born, much in the same way that the concept of glacials and interglacials was conceived in Europe during the 1870s. Pluvial lake beds were subsequently identified in the American Southwest and in Central Asia, while J. W. Gregory postulated a great pluvial lake in the Kenya Rift Valley following field studies in 1893 and 1919. Between 1922 and 1934, E. J. Wayland had defined a succession of "pluvials" and "interpluvials" in East Africa. Largely under the impetus of L. S. B. Leakey, the first Pan-African Congress of Prehistory at Nairobi, 1947, defined four Pleistocene pluvials (Kageran, Kamasian, Kanjeran and Gamblian), separated by three interpluvials. The implicit parallelism to the obsolete fourfold glacial sequence of Günz, Mindel, Riss and Würm is obvious. In addition, two Holocene moist intervals, the Makalian and Nakuran, were recognized. These designations were suggested as time-stratigraphic terms for use throughout Africa.

The enthusiasm with which geologists, geomorphologists, and prehistorians surveyed the Pleistocene record of Africa, the Near East, and southern Asia during the decades before 1950 had tended to create a false impression of understanding. Sweeping generalizations adopted by workers in one area were accepted at face value elsewhere, so that regional studies were freely interpreted in terms of a scheme of external Pleistocene events, i.e., either the East African "pluvial" or the European glacial sequence. In fact, however, Pleistocene studies outside of Europe and North America were quite unsatisfactory. There were few institutions, such as universities, museums, or government surveys, organized to study the vast surfaces of Africa, Asia, Australia, and South America. Those scattered local institutions that did devote their energies to geological or archeological work were hampered by limited budgets, ridiculously small staffs, and an excessive diversity of goals. Sporadic fieldwork by foreign-based scientists was equally inadequate to resolve the basic problems at hand. It is not surprising, then, that our knowledge of the African Pleistocene as of 1947 was limited essentially to scattered reconnaissance studies in Morocco, Algeria, and Tunisia, to an exploratory survey of the Nile Valley in Egypt and the northern Sudan, to scattered work in East Africa, and to some incipient studies in the Vaal River basin. In western Asia there were some good studies in what are now Israel and Jordan, as well as scattered observations on Pleistocene glaciers of the high country. In southern and eastern Asia there had been some regional reconnaissance in northwest India and Burma, while local studies were available from Japan, northern China, and Java. In South America nothing was known except for the location of the major Pleistocene glaciers of the Andes.

Despite twenty years of rapidly accelerating work in the subtropics and tropics, regional information is still seldom adequate and frequently contradictory. The quality of work remains variable, and it is often impossible to judge the validity of data or inference without field rechecking. Interpretation of sediments, paleosols, and erosional features can hardly be generalized from one environment to another and must be undertaken on the basis of detailed field recording, laboratory analysis, and comparative observations on modern geomorphologic processes. Unlike glacial features, on which there is some unanimity, the distinctive facets of arid and tropical geomorphology pose many problems so that few phenomena have unequivocal explanations. Fluvial deposits do not provide convenient stratigraphic horizons, so that relative dating has again and again been prone to gross error. Only the advent of isotopic dating has created some semblance of order and veracity for certain regional sequences. And finally, the pseudo-astronomers have not hesitated to add to the confusion by applying that panacea of armchair geologists, the Milankovitch radiation curve, to "resolve" matters of dating and paleoclimatic interpretation in lower latitudes (see Zeuner, 1959, ch. 8; Bernard, 1962; Fairbridge, 1965).

Clearly, the most exciting Pleistocene discoveries of the second half of the twentieth century are being made in lower latitudes. But precisely because so little is known from immense areas, it is premature to attempt regional syntheses except in a few select areas. The most suitable "case" studies can be made for parts of Africa,[1] where the local density of intensive recent studies is good and where several broad zonal environments can be "tested" at the same meridian of longitude as Europe and the Mediterranean Basin. Nonetheless, even here the quantity of information is not comparable with that of Europe. Palynological work, reviewed in successive volumes of *Paleoecology of Africa* (since 1966) by E. M. van Zinderen Bakker, has only been initiated but is proceeding apace. Mammalian faunas, although well-publicized, are unavailable for many areas or time spans, and generally lack the specific environmental implications of high latitude faunas (see Cooke, 1963). Isotopic dating is largely restricted to parts of the Saharan area and to eastern and southern Africa.

The subsequent discussion on the nature and chronology of late Pleistocene events will be focused on several of the distinct geographical environments so well illustrated by the *Atlas of African Prehistory* (J. D. Clark, 1967): the northern and southern margins of the Sahara, the equatorial "savanna" setting of East Africa, the southern savanna margins of the Congo Basin, and the Vaal River basin on the southeastern fringes of the Kalahari. In this way some insight may be gained into the vicissitudes of the low latitude desert belts and the tropical savannas, while the questions of subtropical versus tropical "pluvial" climates may be explored. The pluvial concept can be reexamined only after the

1. Zonneveld (1968) has recently reviewed the abundant recent literature for northwestern South America and the adjacent Caribbean, concluding that the value of the information is unequal and that the relations between local climatic changes and the sequence of glacials and interglacials are poorly understood. Except for the glacial sequences in southern Argentina (see Woldstedt, 1965, with references), little systematic information is available elsewhere in South America. In India, interesting studies are currently underway in the Deccan region, but there is next to no radiometric information and publication is incomplete. Little new field work has been done in Indonesia since the 1930s (see van Heekeren, 1957; Andel *et al.*, 1967), while much of the spate of recent work in China and Japan (see overview in Woldstedt, 1965) remains to be published in detail. Despite a wealth of excellent geomorphologic investigations in Australia and New Guinea (e.g., Jennings and Mabbutt, 1967), there is little palynological data (Walker, 1966) except for New Zealand. Here again a synthesis of Pleistocene environments is premature, as several recent surveys (see Galloway, 1965; Woldstedt, 1965) have demonstrated (see chapter 30).

evidence has been reviewed in sufficient detail. Originally the word "pluvial," used by Hull and Blanckenhorn, was vaguely applied to the evidence of relatively moist climates indicated by high lake levels and accelerated fluvial activity. Subsequently, some authors have restricted the meaning to long-term increases of mean annual precipitation of sufficient duration and intensity to leave a geological or organic record (see Flint, 1959). Others have extended the term to the whole range of phenomena indicating moister conditions, whether a result of lower temperatures, of intensified rainfall, or of greater overall precipitation. In the latter sense, "pluvial" would be a counterpart to the widest interpretation of the concept "periglacial." At the conclusion of this chapter, the different kinds of pluvial phenomena will be considered and a more precise definition suggested.

LATE PLEISTOCENE SEQUENCES OF THE NORTHERN SAHARA

The nature of late Pleistocene events along the northern peripheries of the Sahara can be elucidated on the basis of studies in western Morocco, central Algeria and Tunisia, northern Tripolitania, and Cyrenaica. Geomorphologic sequences can be established and, in many instances, dated with respect to the Tyrrhenian II or III beaches of interglacial age. Environmental changes here are presumably related to the westerly circulation of the Mediterranean basin and, in fact, the physical record is remarkably similar to that already discussed in chapter 19. Pollen or mammalian faunas are recorded locally, and help illustrate the nature of the environments inferred from geomorphologic criteria.

In western Morocco the late Pleistocene is recorded by the Tyrrhenian III (Ouljian II) beaches and the succeeding Soltanian sedimentary units (Gigout, 1960, ch. 3; Biberson, 1961, ch. 12; Choubert, 1948a, 1948b). In the coastal sector of Sidi Abderrhaman, the Soltanian began with red colluvial silts that rest disconformably on Ouljian beach deposits or erosional forms; they were followed by aeolian sedimentation (regressional dunes). A period of red soil-formation ensued, with a second generation of colluvial silts subsequently sealed by calcareous crusts. This sequence is quite similar, although not as detailed, as that of the western Mediterranean basin. Further inland, the Soltanian is extensively developed in a variety of facies: alluvial fills, colluvial silts, slope breccias, and red soils—both *in situ* and derived—while corresponding erosional surfaces (pediments) are found among the mountain foothills. Two stratigraphic units are recognized. The Older Soltanian deposits, some 15 m. thick, presumably relate to the red silts under the Soltanian aeolianite. They are terminated by *croûtes zonaires*. The Younger Soltanian, seldom more than 1 m. thick and also capped by calcareous

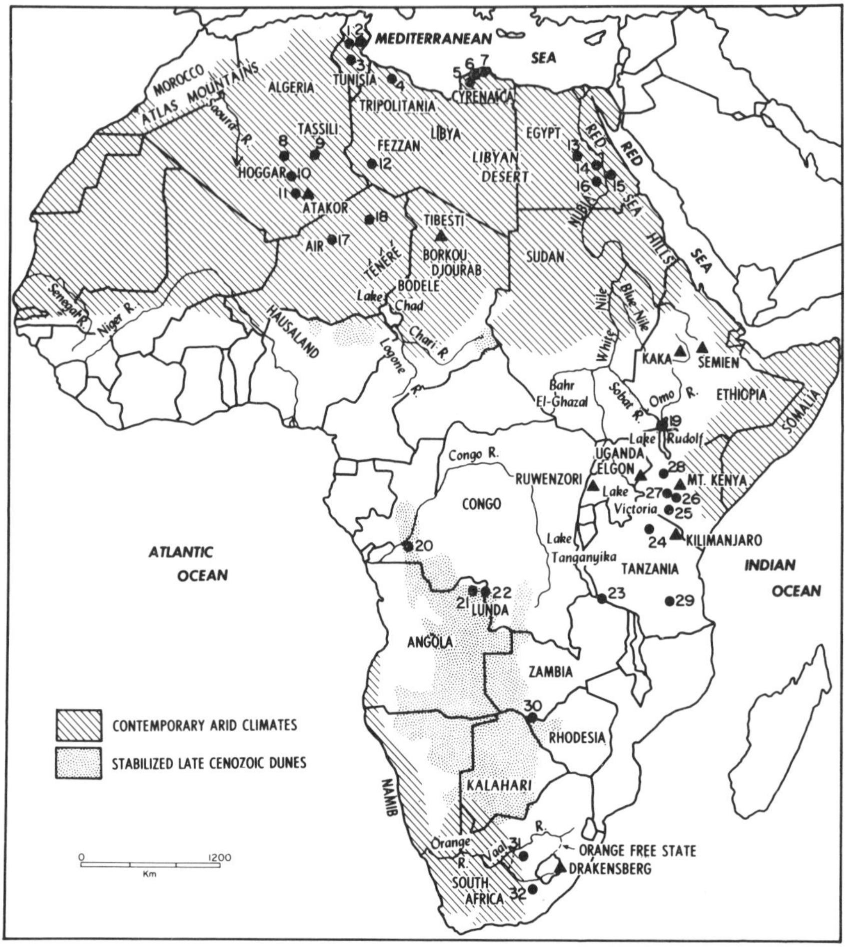

Figure 55. Contemporary arid zones of Africa (as definied by Thornwaite) and fossilized dunes or aeolian sands of Pliocene to Holocene Age (shown only outside of the modern arid zone). Sites or locations discussed in text: (1) Guettar, (2) Akarit, (3) Ain Brimba, (4) Wadi Gan, (5) Hagfet et-Dabba, (6) Haua Fteah, (7) Derna, (8) Meniet, (9) Tihodaine, (10) In Eker and Wadi Outoul, (11) Tarhenanet, (12) Uan Muhuggiag, (13) Kharga, (14) Hierakonpolis, (15) Kom Ombo, (16) Kurkur, (17) Adrar Bous, (18) Djado, (19) Lower Omo Basin, (20) Leopoldville (Kinshasa), (21) Dundo, (22) Mufo, (23) Kalambo Falls, (24) Olduvai, (25) Olorgesaillie, (26) Naivasha, (27) Nakuru-Elementeita, (28) Cherangani, (29) Isimila, (30) Victoria Falls, (31) Florisbad, (32) Aliwal North.

crusts, appears to postdate the major aeolianite. The Soltanian deposits are contemporary with a Palearctic fauna, including *Rhinoceros mercki,* two subspecies of brown bear, and *Sus, Cervus,* and *Megaceroides.* The deposits in general are believed to record periods of accelerated morphogenesis, greater fluvial activity, and a moist climate. Stratigraphic relations with cold-climate phenomena, solifluction mantles and colluvial screes in the high mountains are postulated (Raynal, 1956); however, the highland phenomena need not have been strictly contemporaneous, since the typical Soltanian units are unquestionably of early Würm age. There is no good information regarding the environment of the lowlands during either the pleniglacial or the late glacial. The post-Pleistocene of Morocco is characterized by formation of dark floodplain soils, followed or accompanied by two stages of silt alluviation in the major river valleys (the Rharbian Formation). The earlier alluvial beds are Holocene, the later ones post-Roman (Gigout, 1960, chs. 4 and 8; also Vita-Finzi, 1969, ch. 5).

In central Algeria, the Saoura-Guir drains the Saharan Atlas of eastern Morocco, debouching on the Saharan lowlands in a maze of ancient dune fields. Here Chavaillon (1964, ch. IV.1) identifies a Saourian cycle of erosion, followed by deposition, thought to represent the late Pleistocene. Maximum fluvial activity is related to an initial period of fill dissection (10–12 m. in the Atlas foothills, some 25 m. in the Saharan lowlands). Subsequently, some 25 or 30 m. of sandy beds were deposited in five sedimentary units. The facies include water-reworked aeolian sands; marls, travertines, organic tufas and weakly humic alluvial soils related to marshy sites or ponded waters in intradunal depressions or old channel hollows; fluvial sands and gravels; and some primary aeolian beds. The Saourian sedimentary complex suggests rapid climatic changes, with environmental conditions repeatedly oscillating between hyperarid and subarid. Pollen (see Table 15) analyzed from organic beds or swamp facies (Saourian unit II) included a bare 6 per cent tree pollen (primarily *Salix*), with abundant sedge and *Typha,* reflecting wet conditions (Beucher and Conrad, 1963). The bulk of the pollens pertain to desert plants. Pollen from organic tufas (Saourian unit IV) have not been published but appear to be strongly dominated by NAP (grasses, chenopods, *Artemisia*) with traces of ten genera of mediterranean or submediterranean trees (Beucher, 1963).

Peat from Saourian unit II has given C^{14} dates of "greater than 39,900 years" (I-1787) and "greater than 38,000 years" (I-1761). *Limnaea* shells from the top of unit II gave 33,900 ± 1900 B.P. (T-429); from the base of unit IV, 32,700 ± 1700 B.P. (T-428). Carbonate beds from the top of unit IV, 16,300 ± 350 B.P. and 14,500 ± 190 B.P.

(I-1651, I-1991) (see Conrad, 1969, for dates and interpretation; Chavaillon, 1964, for stratigraphy). This suggests that the Saourian sedimentary cycle represents the greater part of the Würm time range. The period of fill dissection offers little convincing evidence for the long periods of moist climate postulated by Chavaillon (see Butzer and Hansen, 1968, p. 78 ff.), and it may well be that some of the terminal, massive gravels of the preceding Ougartian sedimentary sequence (units VI and VII) represent the earliest Wurm-age deposits. These deposits indicate a much greater stream competence, grade laterally into colluvial screes, and contain water-worn, late Acheulian artifacts.[2]

The evidence for three moist lacustrine periods during the mid-Holocene – 11,500–8500 B.P., ca. 6500–4,200 B.P., and possibly ca. 1,000 B.P. – at least in part associated with the Guirian cycle of Chavaillon (1964), is discussed further in chapter 33.

In central Tunisia, north of the great saline *chotts,* Coque (1962, chs. I.3, III.2–3) identifies a last major pluvial with planation of a great pediment ("Glacis 3") by rillwash, sheetwash, and occasional sheetfloods under moister, yet semiarid conditions. Subsequent alluviation of this pediment by subangular, poorly sorted detritus, under comparatively dry conditions, was contemporaneous with Mousterian and Upper Paleolithic blade industries. This pediment alluvium is sealed by a laminated gypsum crust, attributed to intermittent aeolian deposition of saline dust (obtained by deflation under arid conditions), subsequently cemented by infiltrating rain waters. Pediment-cutting (and alluviation?) was approximately contemporary with a great brackish lake that filled the chott basin to a depth of 65 m. and was fed primarily by artesian waters – without any connection to the Mediterranean Sea (surface at 45 m. above sea level). Salt-tolerant mollusca, such as *Cardium glauca, Hydrobia,* and *Melanoides tuberculata,* are dominant, although lateral detrital beds with freshwater shells (*Corbicula fluminalis*) indicate a considerable influx of surface discharge. This lacustrine phase obviously records optimal moisture conditions.

Pollen and mammalian fossils are available to elucidate the environment contemporary with the pediment-alluviation. At Ain el-Guettar (75 m. elevation), in the foothill country north of the chotts, organic spring beds near the base of the fill include Palearctic (*Rhinoceros mercki, Bos primigenius, Ovis tragelaphus, Camelus dromedarius)* and Ethiopian faunal elements (white rhino, gazelle, antelope, and striped hyena), with

2. Despite his meticulous publication of sections, maps, and sediment data, the stratigraphic correlations and depositional environments suggested by Chavaillon (1964) are sometimes open to question, and are now superseded by the study of Conrad (1969).

Table 15. Pollen from late Pleistocene deposits of the western and central Sahara (excluding samples with less than 25 grains).

Locality	*No. pollen identified*	*Percentage NAP*	*Source*
ALGERIA			
Wadi Saoura (unit II)	366	94	Beucher and Conrad (1963)
In Eker (upper bed)	28	60	Pons and Quézel (1957)
Tarhenanet	164	88	Van Campo *et al.* (1964)
Wadi Outoul (upper bed)	1,965	83	Van Campo *et al.* (1967)
Wadi Outoul (lower bed)	324	85	Van Campo *et al.* (1967)
TUNISIA			
El-Guettar	1,285	90	Van Campo and Coque (1960)
Ain Brimba	3,240	96	Van Campo and Coque (1960)
El-Akarit	2,413	98	Van Campo and Coque (1960)
CHAD			
Yogoum (Djourab)	129	84	Quézel and Martinez (1958)
Kaortchi (Borkou)	118	71	Quézel and Martinez (1958)

a Mousterian industry. The pollen is 90 per cent nonarboreal, with abundant *Cedrus* and *Cupressus* grains (88 and 35 respectively), suggesting that open cedar-cypress woodlands had replaced the juniper scrub now found in the local hill country. Near the Mediterranean coast, at Wadi (Oued) el-Akarit (20 m. elevation), clayey and organic lenses with Mousterian and blade tools have a strictly Ethiopian fauna, including white rhino, zebra, and gazelle. The pollen (see Table 15), is 98 per cent NAP, with a fair number of grains (27) belonging to Aleppo pine. At Ain Brimba (155 m. elevation), south of the chotts, thin organic beds, interbedded with silty sands of the pediment alluvium, include 96 per cent NAP, with substantial Aleppo pine (35 grains) and some northerly elements such as *Erica scoparia* and *Thymelia,* neither of which is prone to long-distance transport. A pine woodland is suggested for the adjacent chain of hills (200–400 m. elevation), in the place of a halfa-grass steppe, while 15 per cent sedge reflects local spring activity. As Van Campo and Coque (1960) emphasize, these three pollen spectra do not suggest a radical climate change, since open vegetation was dominant adjacent to each site. However, the presence of different tree pollen than might be expected today, as well as a number of cooler elements, including elm, linden, deciduous oak *(Quercus suber),* and olive, suggest slightly moister and slightly colder conditions. Differences among the spectra may reflect local ecology and the fact that they are not quite contemporaneous.

The age of the pediment (Glacis 3), of the *Cardium* lake, and of the pediment-alluvium is partly fixed by the 5 meter *Strombus* beaches of the adjacent littoral sector: the pediment was cut *after* the beach deposits were laid down. However, with abundant *Strombus,* it is more probable that these mark the Tyrrhenian II stage,[3] rather than the Tyrrhenian III, so that this fascinating evidence from central Tunisia is not dated precisely within the late Pleistocene. There is limited evidence for accelerated fluvial activity, with pediment-cutting in older fill, from the mid-Holocene.

A late Pleistocene depositional sequence has been outlined by Hey (1962) for northern Tripolitania, where loess-like aeolian silts mantle the limestone uplands. These Plateau Silts may represent interior facies of aeolianites younger than interglacial shorelines at + 6.5 m. and + 3 m. (Tyrrhenian II and III?). In Wadi Gan the aeolian beds overlie about 25 m. of unsorted cobble conglomerate (Mudflow Series), thought also to be of late Pleistocene age. Younger than the Plateau Silts are well-developed, fine alluvial deposits (Higher Silt Terrace), as much as 50 m. thick, mainly silty sands with some gravel and Middle Paleolithic or Aterian artifacts. More frequent but far less violent rainfalls are suggested. It is tempting to relate the Higher Silt Terrace to Chavaillon's (1964) Saourian sedimentary cycle, while, on the other hand, the Mudflow Series and the Higher Silt Terrace occupy a stratigraphic position similar to the Older and Younger Soltanian beds of western Morocco. More precise conclusions can hardly be drawn from the Tripolitanian evidence. A period of post-Roman alluviation is well substantiated (Vita-Finzi, 1969, ch. 2), but its implications are poorly understood.

An interesting geomorphologic sequence has been established in Cyrenaica by Hey (1963, 1968; McBurney and Hey, 1955). All of the deposits postdate a period of intensive wadi erosion as well as the accretion of Tyrrhenian II or III beach deposits, at + 6 m., lacking thermophile elements. Initially, greater spring activity and slow-moving, ponded streams are recorded by massive organic tufas and stratified

3. The Mousterian at Tangier is dated between the Tyrrhenian II and III (Howe, 1967; also review by Butzer, *American Anthropologist,* 70: 809–810).

calcareous marls, with some gravel lenses. The deposits contain leaf, stem, or cone impressions of brambles, reeds, Aleppo pine, and Canary laurel, as well as bones of a zebra, an extinct buffalo *(Homoioceros* sp.*)*, Barbary sheep, and a small land tortoise (Bate, 1955). This "Tufa-Marl Terrace" suggests wetter conditions, little or no colder than the present. It contains a good Mousterian site *in situ,* closely resembling the "typical Levalloiso-Mousterian" of the Haua Fteah cave, level BII (see McBurney, 1967), which dates ca. 50,000 B.P. Thus in Cyrenaica the early Würm was moist.

Following an erosional break, a massive body of well-rounded gravels were laid down in the form of wadi fills (20 m. or more thick) or alluvial fans. At the coast these gravels are complexly interdigited with regressional aeolianites; laterally they intergrade with massive slope screes attributed to accelerated frost-weathering. Lenses of red silt and clay, with isolated pebbles, are interpreted as mudflows, possibly derived from terra rossa soil mantles. Terminal beds may be calcreted. The "Consolidated Wadi Gravels and Screes" are attributed to a colder climate, with relatively moist winters. Another generation of frost detritus, the "Unconsolidated Screes," was deposited on hill slopes following a period of erosion. The volume of scree is small, suggesting that possibly the second cold period was too dry for intense frost-shattering. The earlier screes contain Mousterian artifacts (comparable with level BIV, with massive development of *éboulis secs,* at Haua Fteah, ca. 40,000 B.P.), while the younger screes include Upper Paleolithic (Dabban industry, level C at Haua Fteah, ca. 35,000–15,000 B.P., coinciding with a major horizon of *éboulis secs* in its uppermost level; see McBurney, 1967). Locally, in the cave environment of Haua Fteah, *éboulis secs* remain prominent in late glacial and early Holocene deposits, as late as 8,000 B.P. The Cyrenaican sequence, quite comparable with that of western Morocco, is complemented to a limited degree by the fauna of Haua Fteah[4] and Hagfet et-Dabba (see chapter 19). All in

4. Higgs (1967) has grossly exaggerated the ecological implications of the Haua Fteah fauna. Although 5,000 bone fragments were identified at the family or generic level, a bare 74 were at the specific level. Only 16 genera and 7 species are recognized, and no list of numbers of individuals was attempted per level. The identified bone *fragments* of the early archeological levels are far too few to have ecological significance: Pre-Aurignacian, 29; Mousterian levels, 97; Dabban levels (2 m. of deposit representing 20–25,000 years!), only 297. Strongly dominant in the Pleistocene levels are Barbary sheep (*Ammotragus* sp.), a wide-ranging form, while small numbers of gazelle bone in all strata indicate that open country was always within hunting range of the cave, as it is today. Higgs contends that the large bovines reflect dry, open conditions and that, together with the gazelles, the bone counts can be presented in histograms showing climatic changes through time. Unless these large bovines are identified on the generic level (at least!), such speculation is highly misleading, however. Equally misleading is the mechanical collation of the graphs in Fig.

all it provides clear-cut evidence of an early Würm moist phase, with two intervals of major cold corresponding to the pleniglacials. Continued cold, with little sedimentation and essentially dry conditions, extended through the late glacial. An incomplete vegetation mat, with a relatively dry climate, must be postulated for the periods of scree formation, as elsewhere in the Mediterranean basin (see chapter 19). Post-Roman alluviation is also evident in Cyrenaica.

In conclusion, the northern margins of the Sahara experienced a late Pleistocene climatic sequence analogous to that of the Mediterranean lands. Relatively moist conditions, with accelerated geomorphologic processes, are recorded at the onset of the Würm, with cooler and relatively dry conditions characteristic for the pleniglacials and the late glacial. Temporary phases of soil formation and of accelerating fluvial activity are at least hinted at during the first half of the Würm, presumably earlier than the last pleniglacial. There are no indications whatsoever of a true pluvial broadly corresponding to the Würm Glacial. The comparatively brief phases of accelerated fluvial processes left a striking geomorphologic record; however, like their Mediterranean counterparts, these phenomena do not provide unequivocal evidence for *greater* annual rainfall.

THE LATE PLEISTOCENE HISTORY OF THE NILE VALLEY

The Nile Valley provides a stratigraphic link between the summer-rainfall provinces of Ethiopia and the Upper Nile, and the winter-rainfall belt of Egypt. Today the tributary wadis of Egypt contribute practically no sediments to the Nile floodplain; the rare rainstorms and local wadi floods, related to westerly frontal-cyclonic disturbances (November through March), perform little more than local redeposition of wadi bed materials. Sedimentation in the Nile Valley is essentially confined to flood silts and channel sands primarily derived from the summer monsoon rains (June through September) over the catchment of the Atbara, the Blue Nile, and the Sobat rivers.

This balance of geomorphic activity was not always so, and as recently as 5,000 years ago, the wadis of southern Egypt were considerably more active, their deposits interfingering with nilotic flood silts

---

II.6. So, for example, there is no gazelle bone in 10 of the 20 strata of Phases B and C. Yet Fig. II.6 shows a constant 10 per cent level for the Middle Paleolithic strata, of 5 per cent for the Dabban. Such lumping obscures the paucity of the evidence. Equally fallacious in Fig. II.6 is the detailed averaging of isotopic "paleotemperatures" from marine shells (as originally determined by C. Emiliani)—let alone the validity of oxygen isotopic analyses as temperature indicators (see Shackleton, 1967; Dansgaard and Tauber, 1969).

along the margins of the Nile Valley. Such interrelationships of nilotic and wadi sediments can be readily determined in southern Egypt, allowing systematic study of the comparative vicissitudes of summer Nile floods and winter wadi spates: Such investigation has considerable paleoclimatological implication, since it can establish whether moister paleoclimates in Egypt (related to winter rains of the planetary westerlies) were or were not synchronous with periods of higher and more vigorous Nile floods (related to summer monsoon rains over the Ethiopian highlands).

The early and middle Pleistocene were characterized by net erosion and downcutting in Egypt and Nubia, with temporary aggradation of gravels of local origin: mainly quartz, together with igneous and metamorphic components or ferricrete sandstone at the mouths of larger wadis draining the Red Sea hills. There is no clear evidence of materials exotic to Egypt and the northern Sudan (Butzer and Hansen, 1968; de Heinzelin, 1968). However, important hydrological changes occurred in the Nile basin during early Upper Pleistocene times; and the late Pleistocene and Holocene of the Nile Valley in southern Upper Egypt and Lower Nubia are characterized by a complex sequence of nilotic sediments—primarily derived from Ethiopia and the southern Sudan in the wake of the summer flood regime—and of lateral wadi deposits—derived from sporadic local rains. This suite of intercalated nilotic and wadi fill is younger than intensively rubefied and locally ferricreted gravels. Subsequent to a period of Nile dissection, the following sequence of events can be outlined in some detail (after Butzer and Hansen, 1968):

a) Wadi Floor Conglomerate. At the base of the late Pleistocene sequence, up to 5 m. of ferricreted, well-rolled, cobble conglomerates overlie the bedrock floors of several Nubian wadis. These gravels were graded to a floodplain level at least as low as today; they suggest major wadi activity and a true pluvial climate. Deposition of the Wadi Floor Conglomerate was followed by a period of erosion and consolidation.

b) Korosko Formation. An extensive spread of coarse quartz sands and marls was deposited by a rapidly aggrading, braided Nile to + 34 m. near the Sudanese border and to + 20 m. on the Kom Ombo Plain. Local wadis injected great quantities of sand and gravel into temporary lacustrine environments along the valley margins. Coarse, well-rolled wadi conglomerates immediately underlie the Korosko Formation, and similar gravels are interdigited in the middle of the sequence. These gravels record the terminal phase of late Pleistocene "pluvial" conditions in Egypt. Swift and turbulent summer floods of

Ethiopian origin introduced silt, clay, and solubles, while local materials—made available by wadi discharge during the winter months—were reworked and redeposited in the Nile Valley. The heavy mineral and clay mineral spectra suggest a greater influx of Bahr el-Ghazal waters and sediments, relative to those of the modern Nile. Middle Paleolithic flakes and scrapers appear to be contemporaneous with these deposits. A very approximate age of 50,000 to 25,000 years can be extrapolated from terminal radiocarbon dates. The subsequent period of Nile and wadi incision lowered the local base level to below the modern floodplain, prior to renewed aggradation.

c) *Masmas Formation.* Extensive horizontal flood silts, rich in Ethiopian heavy minerals, filled the Nile Valley to + 33 m. in Nubia and to + 22 m. on the Kom Ombo Plain. The sedimentary environment indicated pertains to a floodplain (alluvial flats or backswamps), and only rarely to channel or levee beds. A broader floodplain was regularly inundated and levee-breaching was common, indicating somewhat greater discharge. The clay mineral composition again suggest more Bahr el-Ghazal waters. However, in Egypt, wadi activity was minimal throughout this period. The Masmas beds attain a thickness of over 40 m. at Kom Ombo, extending from Upper Nubia into Lower Egypt. They suggest an intensified summer flood regime. A time range from 24,000 to 18,000 B.P. can be inferred from radiocarbon dates. The Masmas Formation is equivalent to the Dibeira-Jer (Khor Musa) Formation identified in Sudanese Nubia by de Heinzelin (1968). During a subsequent period of Nile downcutting (to below present floodplain level) the local wadis dissected these older floodplain silts. At about this time or shortly thereafter, a floodplain soil developed under arid climatic conditions.

d) *Gebel Silsila Formation.* The third and final episode of nilotic alluviation includes a sequence of fine gravels, silts, and sands related to channel and levee environments of a rather more vigorous and competent Nile. Horizontal flood silts are rare; instead, former shoals of bed gravel or sands interfinger with laterally embanked topset and backset strata. Relative proportions of clay minerals and heavy minerals are, for the first time, almost identical to those of today. The Gebel Silsila Formation includes three major periods of nilotic aggradation, separated by periods in downcutting:

Darau Member: 17,000–12,500 B.P. (+ 22.5 m. in Egyptian Nubia, + 13 m. on the Kom Ombo Plain). The deposits exceed 18 m. in thickness and include geologically stratified Sebilian and other Late Paleolithic sites, formerly located along the banks of the Nile. Contemporary with accelerated fluvial activity in the local wadis, leading to accumulation of as much as 9 m. of fill. The stratigraphic interdigitation indicates that more vigorous Nile floods (summer) alternated with fairly frequent local rains of moderate intensity (winter). Equivalent to the Ballana and Sahaba Formations of de Heinzelin (1968).

Arminna Member: 11,200–10,000 B.P. (+15 m. in Egyptian Nubia, absent at Kom Ombo). Accelerated fluvial activty, with accumulation of at least 12 m. of fill. Wadi aggradation continued until ca. 8000 B.P., and was followed by development of a minor red paleosol with a 30 to 50-cm. deep *(B)*-horizon. Oxidation and decalcification were accompanied by formation of kaolinitic clays, indicating biochemical weathering, some form of vegetative mat, and more frequent, gentle rains. The age of this almost ubiquitous paleosol in southern Egypt may be about 7,000 B.P.

Kibdi Member: ca. 6000–5000 B.P. (+6–7 m. in Egyptian Nubia). Accelerated fluvial activity in the wadis (over 6 m. of fill), accompanied by extensive sheetwashing of hillslopes, as a result of sporadic, but protracted, torrential rains.

The nilotic deposits of southern Egypt were derived from the sub-Saharan drainage basin of the Nile, and indicate that flood amplitudes and velocities were significantly greater than today for most of the period between about 50,000 and 5,000 years ago. This can only be explained by greater precipitation and discharge over Ethiopia and the southern Sudan, a matter of interpretation corroborated by de Heinzelin's (1968) study of similar deposits in Sudanese Nubia. From this we can infer three complex pluvial periods for the summer monsoonal belt of the Nile Basin. If we compare these fluctuations of Nile discharge with the activity of local streams in southern Egypt, several facts may be noted:

(i) Maximum "pluvial" conditions in southern Egypt, as recorded by the Wadi Floor Conglomerate (ca. 60,000 B.P. [?], probably correspond to the early Würm in Europe. There is no evidence for higher Nile floods at the time. Egyptian climate was still "pluvial," although subarid in nature, during the first half of the Korosko sedimentary hemicycle, in this case correlated with much more vigorous Nile floods. But later during the Korosko stage, local wadi activity was limited. Consequently there was only a broad, incomplete correspondence between Egyptian and sub-Saharan pluvial oscillations during most of the time span of the first pleniglacial and the interpleniglacial (ca. 60,000–25,000 B.P.).

(ii) It is notable that the maximum cold of the second pleniglacial (ca. 24,000–18,000 B.P.) was dry in Egypt, although comparatively wet in the sub-Saharan drainage basin of the Nile.

(iii) Between 17,000 and 5000 B.P. greater Nile discharge of Ethiopian origin coincided very closely with accelerated wadi activity in Egypt. It could be shown in the field that, despite the synchronism of events, local winter spates and summer Nile floods were, in fact, seasonally out of phase. In other words, the evidence of late Pleistocene and Holocene wadi activity in northern Egypt can be largely, if not entirely, attributed to winter rains and not to an intensification and northward extension of the sporadic summer shower activity evident in Sudanese Nubia today.

(iv) The accelerated activity of the Egyptian wadis at about 10,200–8000 B.P. and again about 6000–5000 B.P. suggests very minor "subpluvial" episodes, which were contemporary with higher Nile floods. The intervening red paleosol (ca. 7000 B.P.) has no obvious nilotic counterparts. These three episodes account for most of the time represented by the Younger Dryas, the Boreal, and the Altantic phases in Europe (see chapter 31). Counterparts to these late alluvia in the northern Sudan are discussed along similar lines by de Heinzelin (1968).

During the 3rd millenium B.C. Nile flood levels dropped and wadi activity was reduced to a minimum, while sand dunes were periodically activated around the margins of the Nile Valley (Butzer, 1959a, 1966; de Heinzelin, 1964). The climate of Nubia and Egypt has remained hyperarid ever since.

In conclusion, there is clear-cut evidence for a modest increase of annual precipitation in southern Egypt and the northern Sudan during the late glacial and the early to middle Holocene. The primary evidence is provided by fairly homogeneous wadi sands, sometimes with abundant root impressions of xerophytic shrubs, or with profusions of terrestrial shells (*Zootecus insularis, Pupoides sennaariensis, P. coenopictus*), now extinct in Egypt. All this would suggest a vegetative mat along the wadi floors and periods of fairly frequent, moderately intense winter rains. The coarse wash deposits and colluvium contemporary with the Kibdi unit, by exception, indicate violent sheetfloods, suggesting a period of frequent torrential rains. Finally, the reddish paleosol of mid-Holocene age can best be explained by biochemical weathering, some form of vegetative mat, and more frequent gentle rains. Nonetheless, all of these pluvial phenomena, excepting the Wadi Floor Conglomerate, are indicative of arid or subarid conditions. Consequently the amplitude of these moister intervals should not be overestimated, despite their obvious ecological significance. Furthermore, subpluvial conditions were frequently interrupted by extended interludes of hyperarid climate—ca. 12,000 B.P., ca. 8000 B.P., and again ca. 6000 B.P.

Analogous nilotic deposits are known from Upper Nubia (de Heinzelin, unpublished) and the central Sudan (Berry and Whiteman, 1968; see discussion of sedimentary and stratigraphic interpretation in Butzer and Hansen, 1968, p. 323 ff.).

LATE PLEISTOCENE ENVIRONMENTS OF THE SAHARAN HIGHLANDS

Theoretically, the high country of the central axis of the Sahara could receive either winter rains (of Mediterranean origin) or summer rains (of tropical of "monsoonal" type) during a pluvial period, or there might be an overlap between both rainfall types. The late Pleistocene record of the Hoggar Mountains (2,918 m.) is perhaps most informative in this regard, and Rognon (1967) identifies four successive climatic "types," following upon the penultimate cold-wet period (Riss age?):

1) *Savanna climate.* Deposition of lake beds with diatomaceous clays in small basins within the elevation range 1,100–1,950 m. The diatoms, which indicate shallow waters, include no extinct or cold-loving species, while the mammalian fauna is typically Ethiopian (*Elephas recki,* hippopotamus, white rhino, antelopes and gazelles). The limited pollen (unpublished), mainly NAP, indicates the presence of some mediterranean trees, including Tertiary relicts. An early Upper Acheulian industry has been identified from such lake beds at Tihodaine. Pediment-cutting in the high country completes the picture of warm, semiarid conditions in the Hoggar, during either the late Middle or the early Upper Pleistocene. To Rognon this is a "tropical pluvial," although the primary source of the rainfall is obscure.

2) *Sahelian climate.* Accumulation of aeolian beds in the lowlands and piedmont alluvia of sands or gravels in the uplands suggest warm, arid conditions, such as are now found in the southern Sahara.

3) *Humid-pluvial climate.* Development of brown or dark gray organic paleosols, in part of alluvial or swamp type, with some lake beds including cold-water diatoms. Both at In Eker (about 1,000 m.) and Tarhenanet (1,900 m.), NAP is strongly dominant in corresponding organic alluvium (see Table 15). At the latter site, the 12 per cent arboreal pollen consists of 1 to 5 grains for each of various mediterranean genera; more interesting is the great abundance of fern spores (322 per cent), certainly indicative of a moist micro-setting (see Van Campo *et al.,* 1964). This period of geomorphological stability and cool, semiarid climate was probably contemporary with early Saourian sedimentation.

4) *Cold-pluvial climate.* The climax of the Würm Glacial was marked by the creation of firn patches in the high Atakor at 2,400 to 2,600 m., with formation of nivational niches, firn moraines, and so-called protalus ramparts, deposited by debris-laden avalanches after sliding across snowdrifts or perennial snow patches. Rognon's evidence, including also massive frost-shattering, is convincing, although some of his "periglacial" slope deposits may also be attributed to landsliding and combined colluvial-gravity processes. At lower elevations, carbonates were deposited in swamps or lakes, including a rich molluscan fauna (93 per cent Palearctic species; see also Sparks and Grove, 1961), some Aterian artifacts, and the Palearctic buffalo *Homoioceros.* Correlation with Saourian units II and IV is probable, and Conrad (1969) obtained C^{14} dates of 35,500, 33,700 and 18,800 B.P. from lacustrine deposits with *Cardium* from the lowlands northwest of the Hoggar, and of 24,200 and 17,700 B.P. from ostracod limestones and sandy tufas west of the Hoggar foothills. Unfortunately, no local pollen data are available to fill in this picture of a cold, semiarid environment. Some evidence of late glacial and early Holocene moisture may be given by C^{14} dates from upland lake carbonates: 11,580 $\pm$350 and 8380 $\pm$300 B.P. (Délibras and Dutil, 1966). Post-Pleistocene conditions were, eventually, dry, i.e., Sahelian, and comparable to those of today, despite temporary soil and swamp formation during a moister interval at the time of Neolithic settlement after 6,500 B.P. (see ch. 33).

Although the Hoggar sequence does not bring out the details of each climatic phase, it serves to emphasize the different possible patterns of Pleistocene paleoclimates. Comparable features are emerging from continuing studies of the Tibesti Massif (3,360 m.) (see Ergenzinger, 1968; Messerli and Indermühle, 1968; Hagedorn and Jäkel, 1969).

In the Libyan Desert the Hoggar pattern can be compared with the late Pleistocene record of the Kurkur Oasis (300 m.) (Butzer and Hansen, 1968, ch. 7), a wadi draining the high limestone tableland west of the Nile Valley. This succession can be partly correlated with the less satisfactory and undated sequence from the Kharga escarpment (Caton-Thompson, 1951; also Caton-Thompson and Gardner, 1932) that is frequently alluded to in the literature: (*a*) *Wadi Tufa IIIa.* Intensified denudation of uplands with alluviation of 5 m. or more of gravel, marl, and organic tufas in the valleys, and possible accumulation of derived red soil sediments in pans of the plateau. Two small handaxes *in situ,* with a C^{14} date on carbonate of "greater than 39,900 B.P." (I-2063). (*b*) *Wadi Tufa IIIb.* Accelerated spring activity and a higher water table accompanied further valley-floor aggradation of 3–5 m. of travertines and organic tufas with leaf impressions of *Ficus.* (c) Stream dissection

(6–10 m.), followed by a dry climate with little stream activity but accumulation of aeolian sands. (*d*) *Wadi Tufa IIIc*. Accelerated spring activity, with extensive valley ponding and deposition of marls or sands with a freshwater molluscan fauna. C^{14} date on carbonate of "greater than 39,900 B.P." (I-2064). (*e*) Stream dissection (6–9 m.), under drier conditions. (*f*) *Wadi Tufa IV*. Some upland denudation and contemporary spring activity, leading to alluviation of gravel, sand, and marl in the valleys, of calcareous slope wash with *Zootecus* snails on the hillsides. This last significant "pluvial" phase is probably a little younger than a C^{14} date on carbonate of 31,800 ± 1700 B.P. (I-2062). (*g*) Stream dissection (3 m.), under drier conditions. (*h*) Accumulation of soil sediments ("Red Silt") in shallow depressions by occasional, strong rains. Possibly contemporary with some spring activity and tufa accumulation ca. 10,300 B.P. Subsequently, climate was generally quite arid, with limited geomorphological activity.

This sequence of late Pleistocene environmental change lacks the evidence of intense cold common to the mountains; nonetheless, as a result of greater winter or summer rains, the climate was repeatedly wetter than today, particularly during the first half of the Würm Glacial. This recalls the two contemporary periods of accelerated fluvial activity recorded on the Red Sea coast of Egypt: the corresponding alluvial terraces are younger than Eem coral reefs but older than the main Würm regression of sea level (Butzer and Hansen, 1968, ch. 8).

THE LATE PLEISTOCENE RECORD OF THE SENEGAL DELTA AND THE CHAD BASIN

The complicated train of late Pleistocene events along the southern margins of the Sahara can best be illustrated from the Senegal Delta and the Chad Basin, both areas well within the tropical summer rainfall belt. An interesting sequence has been established along the lowermost Senegal River by Michel (1968; Michel *et al.*, 1968; Faure and Elouard, 1967; also Grove and Warren, 1968; Tricart, 1961):

a) The first datable horizon is a marine transgression (the Upper Inchirian), dating from about 31,000 B.P. to "greater than 40,000 B.P.," recording either the Paudorf Interstadial or the Tyrrhenian III. The Senegal River aggraded a +2 m. Low Terrace of gravel at this time, suggesting greater stream competence. The earlier +15 m. Middle Terrace, with ferricreted conglomerates, is, by inference, correlated with the Lower Inchirian (Tyrrhenian II or III). A rather moist interval was responsible for ferruginization (presumably during the transition from interglacial to glacial, or from interstadial to stadial). Bedrock cutting after both of the Inchirian transgressions, in response to a lower sea level, also suggests a competent stream with effective flood crests—unlike dissection of fill, which usually accompanies declining flood discharge.

b) At the height of the Würm regression, continental dunes with a general southwest orientation were deposited right across the present Senegal Delta, where they attain a local relief of 30 m. and extend to below modern sea level. At the time of their formation the Senegal River did not reach the coast, indicating a much drier climate during the later Würm pleniglacial.

c) A red paleosol, with a *(B)*-horizon some 2 m. deep, formed on these dunes during a warm, moist period of intensive biochemical weathering. At the same time the Senegal and other rivers once more cut across the dunes to the sea.

d) A marine transgression (to 2 m. above present sea level) flooded the new Senegal Delta, about 5500–4000 B.P., favoring deposition of a sandy river terrace as much as 250 km. upstream, merging in the delta with beach deposits including *Arca senilis, Dosinia isocardia,* and *Cardium edule.*

e) A protracted regression, with a positive oscillation of sea level about 3000–1650 B.P., subsequently led to the development of a series of beach ridges and associated littoral dunes. The Senegal River followed a regime similar to that of today.

As it is now understood, the late Pleistocene history of the Senegal Delta is quite comparable to that of the Nile Valley and other Saharan regions that are properly dated and interpreted. Here, again, there have been several rather significant changes of climate, with the terminal Pleistocene quite dry, the early Würm moderately wet, and one or more moist phases during the Holocene (and the late glacial?). Stratigraphy and overall interpretation of the middle Niger area and its deposits will also require a thorough revision in the light of current understanding of the Senegal Delta, as already outlined by Grove and Warren (1968).

A last area of major interest in the Saharan late Pleistocene is the great basin formerly occupied by an inland sea ancestral to modern Lake Chad (Grove and Warren, 1968; Grove and Pullan, 1963; Pias, 1962; Faure, 1966; Franz, 1967; Servant *et al.,* 1969; Servant and Servant, 1970).

The surface of the present Lake Chad stands at 280 m., with an area fluctuating between 10,000 and 25,000 sq. km., a mean depth varying between 3 and 7 m., and a maximum depth of 11 m. The lake is separated from two extensive depressions—the Bodélé and Djourab—by a low divide, breached by the dry valley of the Bahr el-Ghazal. The lowest strandline of Lake Chad, at +4 to 6 m., would already permit water to overflow into the Bodélé depression, 500 km. away. At its highest stand (320 m.), the Pleistocene lake ancestral to the Chad formed conspicuous shorelines at +40 to 50 m: and so occupied an area of about 400,000 sq. km. with direct links, via seepage waters and groundwater tables, into many of the intradunal depressions of the Ténéré Desert to the northwest. The great evaporative losses of the present lake are largely balanced by the inflow of the Logone and Shari from the south. Grove and Pullan (1963) estimate that the evaporation of the expanded lake may have been sixteen times greater, so that it must have received annually a volume of water equal to one-third the annual discharge of the Congo.

The ancient Chad Sea is, then, an excellent piece of evidence in favor of greater moisture in adjacent subhumid tropical latitudes. Unfortunately there has been no correlation of shorelines between different parts of the basin. The complexity and age of lacustrine sedimentation is illustrated by the 600 m. of Pleistocene beds known to underlie parts of the basin. One high lake stage—preceded and followed by aeolian sands—is dated in the mid-Würm by four C^{14} dates on carbonate: 41,000; 38,000; 30,400; and 21,900 B.P. A second high stand has now been dated ca. 12,400-10,600 B.P. (C^{14} dates of 12,060; 11,950; and 10,900 B.P.). However, the major shorelines and related surficial deposits date from 10,300-6850 B.P. (ten C^{14} dates), roughly contemporary with various high lake levels in East Africa (see below). Approximately contemporaneous are pollen spectra from Djourab and Borkou (Quézel and Martinez, 1958; see Table 15), with a fair proportion of arboreal pollen, almost entirely from Mediterranean species, dominated by pine and cypress. After a major but temporary regression, with extensive dune formation on the former lake plains, a moderately high Chad level is once again indicated for about 5500 to 2350 B.P. (five C^{14} dates). This last phase is, once again, common to several lake basins in East Africa (see chapter 33). Subsequently, conditions have generally remained arid, with one or two moister oscillations. Of particular interest along the southern margins of the Sahara are the stabilized or immobile dunes in what are now semiarid regions (see Grove, 1958; Grove and Warren, 1968). In northern Nigeria the Hausaland sand fields would imply 150 mm. annual precipitation instead of the modern 750 mm. As in the Senegal Delta, most of this aeolian activity appears to date from the second Würm pleniglacial.

THE NATURE OF PLUVIALS IN THE SAHARAN AREA

In overview, the regional sequences selected here allow the following generalizations:

a) *The semiarid, northern margins of the Sahara.* Accelerated fluvial activity and a greater rainfall intensity—but not necessarily higher rainfall totals—marked the early Würm. In the Lower Pleniglacial there is evidence of accelerated frost-weathering, with moderately intensive fluvial activity, declining with time; submediterranean floras substantiate a cooler climate. The Brörup and Paudorf intervals were apparently times of soil formation and geomorphologic stability. Accelerated weathering and lower rainfalls characterized the Upper Pleniglacial, the late glacial period and, apparently, early Holocene times; forest-steppe was probably widespread along the present arid margins of the dry subtropical woodlands. There is no indication of a moist interval during either the late glacial or the early Holocene. Whereas the phenomenon of post-Roman alluviation is common, if not universal, there is no substantial evidence for a moist interval in mid-Holocene times, with the possible exception of western Morocco.

b) *The arid, central axis of the Sahara.* Two periods of accelerated fluvial activity can be recognized prior to the Upper Pleniglacial that in Egypt, and probably in the Hoggar and Saoura as well, pertain to the early Würm; these periods presuppose significantly greater rainfall totals in such hyperarid settings. Immigration of mediterranean or submediterranean floras along the major drainage lines and into the moister high country can be inferred, and open woodlands were probably to be found among the mountains. The nature of the Brörup and Paudorf environments is obscure. The Upper Pleniglacial was initially marked by a minor phase of accelerated fluvial activity in Egypt, then by hyperarid conditions (ca. 27,000–17,000 B.P.); the highlands above 2,000 m. were intensely cold in winter, with cold-climate processes paramount among the summits of the Hoggar and Tibesti; the nature of the upland vegetation is unknown. There are unmistakeable indications of several moister interludes during the late glacial and Holocene; accelerated wadi or spring activity in Egypt ca. 16,000–13,000 B.P., ca. 11,000–8000 B.P., and again ca. 5000 B.P., with soil formation about 7000 B.P.; lake formation in the Hoggar ca. 11,000 and again ca. 8500 B.P., with accelerated fluvial activity and soil formation in the Saoura valley ca. 7500–5000 B.P. These "late" moist interludes have no evident parallels on the northern margins of the Sahara, although the moist phase in Egypt about 5000 B.P. can be related to winter rains. Pollen evidence from the southern Tibesti foothills suggest that open tropical woodlands of mediterranean type were still widespread in favorable environments during these "late" moist interludes.

c) *The semiarid, southern margins of the Sahara.* The record of the first half of the late Pleistocene is fragmentary, but suggests two periods of accelerated fluvial activity in the lower Senegal, leading to alluviation during the course of two marine transgressions, with effective bedrock dissection during the initial regressional stages. A high stand of Lake Chad terminated a little after 21,000 B.P., and the second part of the Upper Pleniglacial was quite arid: dunes invaded the savanna across a broad belt from the Atlantic coast to the Nile valley, suggesting a drastic decline in rainfall, of a magnitude never experienced along the Mediterranean borderlands. A brief but very moist pluvial situation is evident ca. 12,400–10,600 and 10,300–6850 B.P., with tropical soil formation on the now immobile dunes, and the creation or expansion of lakes. After a drier interval, with reactivation of some dune fields, more restricted pluvial conditions returned ca. 5500–2350 B.P. Mediterranean floras were widespread in the Chad region during both of these moist intervals. Finally, there is some evidence for slightly moister conditions during the last 2,000 years.

The sum total of the evidence suggests a number of conclusions: (1) There are both similarities and differences in the late Pleistocene history of the three major Saharan climatic provinces. (2) Pluvial climates, in the restricted sense, are not apparent along the semiarid Mediterranean borderlands. (3) Maximum moisture in the arid Sahara is recorded during the early Würm, with less significant, short-term pluvial episodes during the late glacial and Holocene; these "late" pluvial phases are paramount in the southern Sahara, where they coincide with the early and mid-Holocene; there is no record of "late" periods of accelerated fluvial activity in the northern Sahara. (4) The climax of the Upper Pleniglacial was dry everywhere, but the intensification of aridity in the southern Sahara was phenomenal. (5) The pluvial episodes, in a narrow sense, were very brief (2,000 to 5,000 years duration) in all areas. (6) There is no simple one-to-one correlation of glacials or interglacials with long-term pluvials: both the earlier, now obsolete views of a universal glacial-pluvial correspondence, as well as the more recent hypotheses of a "shifting Sahara" with alternating glacial-pluvials in the northern Sahara and interglacial-pluvials in the southern and central Sahara (see Balout, 1955; Tricart, 1961; Tricart and Cailleux, 1965b; Büdel, 1963b; Fairbridge, 1965), must now all be regarded as gross oversimplifications. (7) It is essential to distinguish pluvial phases characterized by accelerated fluvial activity (denudation and general alluviation) from those marked by geomorphologic stability but intensive biochemical weathering (soil formation).

Morphodynamic phases—with accelerated runoff, denudation, colluvial deposition and general alluviation—can best be explained by periodic, intensive rains of sufficient duration and moderate frequency. Such conditions are most effective with an incomplete mat of vegetation and most common in a semiarid climate. Slope stability with weathering and soil development requires low- or moderate-intensity rains and/or an effective grass mat to inhibit runoff and erosion. Such morphostatic conditions would be favored by frequent rains of low intensity or short duration, and limited seasonality. There is no simple correlation of morphodynamic or morphostatic settings with either the westerly rains of the Mediterranean world or the tropical rains of the summer monsoonal regimes: almost all of the tropical summer rains in the southern Sahara are associated with heavy showers or violent weather; the winter rains of the northern Sahara are often light or gentle, but may also produce very heavy rainfalls. Possibly an overlap of summer and winter rains would produce the optimal vegetation cover. At the present state of our information it remains impossible to define, climatologically, the distinction of "warm" or "cold" pluvials, or of "tropical" or "subtropical" pluvials in the Saharan world. In fact, it is imperative that the term "pluvial" be used in a descriptive and nongenetic sense, and that distinct pluvial "types" be defined simply on the basis of the different physical or biological phenomena they produce.

THE HIGH MOUNTAINS OF ETHIOPIA AND EAST AFRICA

The evidence for late Pleistocene pluvial phenomena is complemented by the record of glaciation and pollen in the high mountain country of eastern Africa.

The Ethiopian highlands are not glaciated today, but the highest peaks of Semien (Ras Dashan, 4,580 m.; Sazza, 4,500 m.; Lagata, 4,490 m.; Buahit, 4,470 m.) harbored a number of small Pleistocene glaciers. Early work by Nilsson (1940) demonstrated the existence of two former glacial stages with climatic snowlines at 3,600 to 4,100 m. and at 4,200 m. Werdecker (1955) confirmed Nilsson's results, setting the snowline for the period of major glaciation at 3,600 to 3,700 m. A glacial retreat stage, correlated with the terminal Pleistocene, corresponded to a snowline at 4,400 m. Since the modern snowline is estimated at 4,700 to 4,800 m.—a little above the highest peaks—the maximum (late?) Pleistocene snowline depression is set at 1,100 m. Nilsson (1940) also describes a late Pleistocene glaciation of Mt. Kaka (4,133 m.), with a snowline elevation estimated at 3,700 m. Büdel (1954) mentions fossil solifluction deposits down to elevations of 2,700 m., which he attributes both to soil-frost activity and to greater humidity. At least some of

Büdel's solifluction features represent colluvial screes, tropical stone lines, or other phenomena resulting from mass movement of lubricated clays. Kuls and Semmel (1965) found no evidence for Pleistocene cold-climate processes in central Ethiopia.

Pleistocene glaciation is also apparent on the giant volcanic peaks of East Africa. The present snowline of Mt. Kenya (5,158 m.) is at 5,100 m., while the snowline depression (1,200 m.) of the major Würm-age glacial stage brought about the development of numerous highland glaciers (Baker, 1967). Mt. Kilimanjaro (Kibo, 5,802 m.; Mawenzi, 5,068 m.) now apparently lies just above the climatic snowline. Despite the assumption of continued Pleistocene and recent uplift made for the East African volcanoes, Kilimanjaro provides clear evidence for four major Pleistocene glaciations and several recessional stages of late glacial and Holocene age (Downie, 1964). A phonolite resting on moraines of the second glacial has a K/Ar date of 463,000 B.P. (Evernden and Curtis, 1965), showing that mid-Pleistocene mountain glaciation can be verified in East Africa. The late Pleistocene snowline can be estimated at 4,650 m., both from the elevation of cirque floors and by the Höfer method, indicating a snowline depression of about 1,200 m. Mt. Elgon (4,315 m.), which is today situated well beneath the snowline, was also glaciated during the Pleistocene. Mt Ruwenzori (5,119 m.) has a modern snowline of 4,750 m. on the eastern Uganda side. Pleistocene glaciers extended down to 2,900 m. on the western slopes, 2,000 m. on the eastern slopes (de Heinzelin, 1963; Whittow *et al.,* 1963). A depression of about 1,000 m. can be estimated for the regional climatic snowline.

The Pleistocene glaciers of Ethiopia and East Africa suggest that late Pleistocene climatic changes in tropical areas were broadly comparable with those of the Mediterranean region. Snowline depressions of 1,000 to 1,200 m. would imply a temperature lowering of 6.5–7.8° C., assuming a lapse rate of 6.5° C. per 100 m., and all other conditions remaining equal. The well-developed recessional moraines and evidence of late Holocene readvances found on Mounts Kenya, Kilimanjaro, and Ruwenzori suggest possibilities for future correlations with European glaciers (de Heinzelin, 1963; Zienert, 1968).

In East Africa the glacial record is abundantly complemented by the comprehensive palynological and ecological study of J.A.Coetzee (1967, part I). Pollen cores were taken from peat bogs or gyttja marsh on the slopes of Mt. Kenya (Sacred Lake, 2,440 m.; Lake Rutundu, 3,140 m.); from the Cherangani Hills (at 2,900 m.), a highland (3,316 m.) on the western edge of the Kenya Rift, at the latitude of Mt. Elgon; and on the southeastern slope of Kilimanjaro (at 2,650 m.). Some thirteen well-chosen C^{14} dates show that the Kilimanjaro boring is younger than

4,700 B.P., but that the Cherangani core goes back to 28,000 B.P., and the sample from Sacred Lake to 33,500 B.P. Major shifts in the vertical zonation of vegetation are indicated for the terminal Pleistocene. A moderately warm interval, contemporary with the Paudorf Interstadial, is indicated before 27,000 B.P., and mean temperatures were only 2–4°C. lower than today. A very cold period followed, between 27,000 and 14,000 B.P., when the vegetation belts shifted 1,000 to 1,100 m. downward, with alpine grasslands or open heath dominant on the middle slopes of the mountains. Temperatures can be estimated at 5–9°C. lower than today. The alpine tree-line began to rise after 14,000 B.P., oscillating strongly, until a more uniform and rapid warm-up began after 10,500 B.P. Maximum Holocene forest development and highest temperatures were attained about 4,000 B.P. A decrease in temperature is evident since about 2400 B.P. Livingstone (1967), who studied a rather similar post-Pleistocene succession from the western Ruwenzori, prefers to believe that moisture changes were responsible for the late Pleistocene shifts of vegetation belts. Athough moisture may have played a subsidiary role, the fully published ecological arguments of Coetzee (1967), in favor of significant temperature changes, cannot be faulted. So, for example, greater aridity could not produce an ericaceous vegetation on the mid-slopes of Mt. Kenya. Similarly, greater aridity does not explain the maximum expansion of the late Pleistocene glaciers, broadly contemporaneous with the downward shift of open alpine vegetation about 27,000–14,000 B.P. In fact, the glacial and palynological data provide convergent evidence that temperatures in the high country of East Africa were *at least* 5–6° C. colder during the second Würm pleniglacial.

PLUVIAL EVIDENCE FROM THE OMO-RUDOLF BASIN

Lake Rudolf, on the Kenya-Ethiopia border, is an alkaline, nonoutlet lake (370 m.), with a modern surface area of 7,500 sq. km. The major affluent is the Omo River which drains a high catchment in western Ethiopia at 2,000–4,000 m. elevation and with an area of 30,000 sq. km. The lower Omo River passes through a tectonic basin before debouching across a large delta into the northern end of Lake Rudolf. This basin records a series of sedimentary formations, dating back to beyond 4.25 million years (Butzer and Thurber, 1969; Butzer, 1971a). To the west the basin sediments extend towards a low-level divide (at about 450 m. elevation), beyond which lie a series of extensive mudflats that form the watershed to the Pibor-Sobat, a Nile tributary. The topography, the disposition and elevation of the sediment fill, and the mollusca and fish of the Omo-Rudolf system indicate intermittent hydrographic links between the Lower Omo Basin and the Nile system. In fact, each of the sedimentary units, which all indicate an expanded and higher lake, culminate in 450–460 m. elevation, i.e., at the level of the Rudolf-Nile threshold. In other words, the Omo-Rudolf Basin has alternately had and not had an outlet. This alternation is of great paleoclimatic interest, since each of the periods of high lake level indicates a long-term, positive hydrological budget in the Omo-Rudolf Basin, in response either to greater rainfall over the Ethiopian catchment area (which provides perhaps 90 per cent of the water) and/or to reduced evaporation over the surface of Lake Rudolf.[5] On the other hand, the absence of a sedimentary record in the Lower Omo Basin will normally reflect a low lake level and either a long-term negative hydrological budget or one similar to the present.

The late Pleistocene and Holocene sediments of the Lower Omo Valley form part of the Kibish Formation; they are undeformed and consequently provide a good record of changing environments in southwestern Ethiopia (after Butzer and Thurber, 1969; Butzer *et al.*, 1970):

a) "Member I." Deposition of deltaic and lake beds, as well as stream gravels, indicating a higher-competence Omo River. These sediments, with seven stratigraphic subdivisions, have a probable thickness of 40 m. and include fossils of a primitive *Homo sapiens* (Butzer, Day and Leakey, 1969), with a tentative Th/U date of 130,000 B.P. They appear to be interdigited laterally with coarse-grade piedmont alluvia. Followed by a period of stream cutting, lake regression, and drier climate.

b) "Member II." Deposition of massive deltaic beds (some 23 m.) indicating a second expansion of Lake Rudolf some 80–100 km. northwards. Followed by a long period of cutting and regression of lake level.

c) "Member III." Deposition of twelve units of deltaic or fluvial origin, together over 45 m. thick, with molluscan shell from the terminal beds dating "greater than 37,000 B.P." (by C^{14}) or 30,000 B.P. (by Th/U) (L-1203-A). Probably contemporary with some local piedmont alluviation and, later, development of a relict soil, suggesting biochemical weathering a little more intensive than today.

d) A protracted period of stream dissection and low lake level, with salt hydration and patination of surface gravel, suggesting a drier climate than today (ca. 35,000 [?]–10,000 B.P. by interpolation of C^{14} dates).

5. A rise in lake level of 4 m. from 1962 to 1970 coincided with abrupt rises in the levels of several East African lakes, while precipitation in Ethiopia remained near average Since the Kenya tributaries of Lake Rudolf are so minor, it appears that increased cloudiness, associated with a succession of wet years in Kenya, reduced evaporation over the lake surface, complemented by increased rainfall over the lake.

e) "Member IVa." Deposition of 13.5 m. of deltaic and littoral deposits, locally interdigited with piedmont alluvia. Nine C^{14} dates on shell indicate an age of 9700–7500 B.P. Followed by a brief interval of dissection, with a lower lake level.

f) "Member IVb." Deposition of similar beds (8 m. or more, ca. 6200–4400 B.P. (five C^{14} dates on shell), with contemporary beach ridges extending from the former delta mouth to the Rudolf-Nile watershed. A single C^{14} shell date of 3250 ± 150 B.P. (L-1203-H) from a +70 m. beach ridge may indicate a final high stand or brief transgression of the lake. Followed by stream dissection and lake regression under drier conditions.

g) Narok Beds. Aggradation of contemporary floodplains and delta plains, with lake generally low but fluctuating in a range from + 35 m. to −5 or −10 m.

Member I must be assigned to a late Middle Pleistocene moist phase while Members II and III indicate two major pluvial periods prior to 30,000 B.P., that can probably be correlated with the early Würm and with the interpleniglacial. The Upper Pleniglacial was exceptionally dry and truly arid. Members IVa and IVb record two brief pluvials during parts of the early and mid-Holocene time range. Whereas these Holocene wet phases were contemporary with those of the Chad Basin, and similar to those inferred from the nilotic deposits of Nubia, Member III appears to be equivalent to the Korosko Formation of Nubia as well as the "humid-pluvial" of the Hoggar. The arid period between about 35,000 and 10,000 B.P. has conspicuous parallels in the Saharan region. None of the Omo-Rudolf pluvials coincided with the glacial maximum, and so they must be attributed to greater rainfall, possibly during the interstadials. However, the climate of the Lower Omo Basin has never been truly humid during the Pleistocene; the pluvials only differed in the degree of semiaridity experienced. The nature of the climatic changes in Ethiopia—promoting analogous fluvial records in the Nile and Omo valleys—should be better understood once pollen cores from the uplands (taken by E.M. van Zinderen Bakker) have been analyzed.

PLUVIALS OF THE KENYA RIFT VALLEY

The Kenya Rift Valley and, in particular, the contiguous basins of Lakes Naivasha, Nakuru, and Elmenteita, have provided type areas for the Gamblian, Makalian and Nakuran pluvial phases, and have long been the focus of the East African "pluvial chronology." The negative verdict of Cooke (1958) and Flint (1959) concerning the factual bases of this chronology (see Bishop, 1967, for an annotated discussion of the terminologies) has more recently been followed by renewed work (McCall,

1967; McCall *et al.*, 1967; Washbourn, 1969; Richardson, 1966). The Naivasha-Nakuru area is currently the subject of a detailed Pleistocene study by G. L. Isaac and S. L. Richardson.

One of the most impressive features is the record of a great lake, occupying the Nakuru-Elmenteita basin to a depth of about 180 m., and probably contemporary with another deep and greatly expanded lake in the Naivasha Basin. The high Naivasha level can be dated ca. 9650–5650 B.P., and diatom and pollen data suggest a local climate subject to stronger seasonal variation than today. These high lake levels in nonoutlet basins are recorded by horizontal and undeformed shorelines or lacustrine beds, so that only a moister climate can explain their existence. The temporal coincidence with the high stands of Lake Rudolf in 9700–7500 and 6200–4400 B.P. (see above), and Lake Victoria ca. 9500–6500 B.P. (Kendall, 1969), is obvious. Lake Naivasha began to shrink after 5650 B.P. and, for about a century ca. 3000 B.P., dried out completely. Since then a small, fluctuating lake has occupied the floor of the basin, at no time larger than modern Lake Naivasha. These high, early Holocene shorelines are those originally used to define the allegedly late Pleistocene Gamblian Pluvial; i.e., the Gamblian is unquestionably a post-Pleistocene phenomenon, contemporary with Member IV of the Kibish Formation in the Omo-Rudolf Basin. Recent work does not substantiate the existence of the so-called Makalian and Nakuran subpluvials.

The late Pleistocene record of the Nakuru-Naivasha area is still incoherent at the moment, due to the lack of isotopic dates (and a number of patently erroneous K/Ar determinations; see discussion in Evernden and Curtis, 1965) and the necessity to establish firm stratigraphic relationships between deposits at different localities. It appears that a number of alluvial deposits as well as some of the tilted but unfaulted lake sediments with Acheulian impediments—some of which were once ascribed to the Kanjeran Pluvial—belong in a time range contemporary with the earlier units of the Kibish Formation. Significantly, Lake Victoria lacked an outlet from somewhat before 14,500 until about 12,000 B.P., and again for a sort period about 10,000 B.P. (Kendall, 1969).

The Kenya Rift record currently serves to do little more than show that potential stratigraphic links exist between Lake Rudolf and the Kalambo Falls (discussed below) and that pluvial-nonpluvial oscillations were similar over wide areas. More important, perhaps, is that the recent studies have shown that the old pluvial scheme is quite untenable, and that the entire terminology—from the Kageran to the Nakuran—should be abandoned once and for all.

THE KALAMBO FALLS PREHISTORIC SITES

One of the most complete and varied late Pleistocene to mid-Holocene records in Africa has been studied in the course of excavations at Kalambo Falls, along the border between Zambia and Tanzania (see J.D. Clark, 1969). Here a small alluvial basin (at about 1,150 m.) is developed amid igneous and metamorphic rocks of Precambrian age, at the southeastern edge of the Lake Tanganyika Rift Valley. The Kalambo River is slightly ponded before passing through a spillway gorge cut into quartzite rock and then plunging down the fault escarpment to the shores of Lake Tanganyika (760 m.). The falls themselves have a height of 200 m. Rainfall means increase from 600 mm. around Lake Tanganyika to some 1,000–1,400 mm. above the escarpment. The modern vegetation is a deciduous to semideciduous woodland, giving way to evergreen montane forest at elevations of 1,800–2,100 m. Ericaceae and tussock moor-grasslands are found in the alpine zone above 2,700 m. Moist galeria forests, locally with swamps or swamp forests, follow the riverine zone.

The alluvial basin above the Kalambo gorge lies between quartzite ridges, and the rapid decrease in the gradient reduces stream competence. Bed loads are deposited rapidly as the river begins to meander, approaching the gorge. The balance of erosion and deposition here changed repeatedly during the late Pleistocene, leaving the following sequence of deposits, grouped as the Kalambo Falls Formation (Bond, Haldemann, Kleindienst, and van Zinderen Bakker, in Clark, 1969):

a) Mkamba Member. Bed (1): "White Sands and Dark Clay Beds" (over 3 m. thick). Fine, often structureless, quartz sands, alternating with strata of organic clays and thin lenticles of pebbles. These fluvial beds are interrupted by at least one widespread erosional break. Good pollen records, with preservation of partial tree trunks, branches, leaves, seed pods, fruits, etc. indicate repeated changes of the vegetation picture. The lowest beds (Zone U) indicate swamp and galeria forest at the riverside, with a semi-deciduous dry forest beyond. Climate warmer and drier than today. The intermediate beds (Zone V) suggest a poorly developed riverine forest, with mixed semideciduous and evergreen dry forest such as is now found at elevations some 300 to 650 m. higher. Climate cooler (by 3° C.?) and wetter (by 500 mm.?). The topmost beds belong to Zone W, together with most of Bed (2); the pollen spectra suggest vegetation patterns very similar to those of today. Bed (1) has a number of Late Acheulian living floors. Two isotopically enriched C^{14} dates on wood (Grn-4896 and 2644) gave 61,700 ± 1300 and 60,300 ± 750 B.P.;

these should be considered as minimum dates, although the pollen profile indicates warm and cold fluctuations that almost certainly record the early Würm.

Bed (2):"Ochreous Sands and Grey Clay Beds" (9 m. thick). Fine-to-coarse oxidized sands with thin lenticles of gravel and, in the upper part, lenses of organic clays. A single pollen spectrum at the top (Zone X) is similar to Zone V, again suggesting cooler and moister conditions. Sangoan occupation floors; seven acceptable, finite C^{14} dates on wood and charcoal range from 46,100 to 37,450 B.P.

Bed (3):"Rubble Bed II." Thin horizon of gravel, marking basal fill of new channels cut in (1) and (2). An early Lupemban site, related to this gravel, has a C^{14} date of 31,660 ± 600 B.P. on charcoal (Grn-4648).

Bed (4):"Pits Channel Fill" (2.5 m. thick). Complex alluvial fills of gravels, sands, and clays, with some current-bedding. Pollen Zone Y (two samples only) indicates a well-developed galeria forest and wet, grassy lowlands, with a semideciduous dry woodland on higher ground. A warmer but moister climate is inferred, probably contemporary with the Paudorf Interstadial (C^{14} dates on charcoal 30,500 ± 2000 and 27,500 ± 2300 B.P. [L-3991]). Lupemban artifacts. These fine facies of the Mkamba Member grade into ferricreted sands and, ultimately, gravels. A major disconformity separates the Mkamba and Mbwilo Members.

b) *Mbwilo Member.* Bed (1):"Rubble Bed I" (6.5 m. thick). Complex of channel-bed gravels, interbedded with sands, and grading laterally into coarse lag of colluvial origin ("stone lines"). Derived older, artifacts as well as a Magosian floor, dated 9550 ± 210 B.P. from charcoal (L-395D). Pollen Zone Z (four samples) indicates a poorly developed galeria forest, with an open vegetation rich in Ericaceae and suggesting the transition from dry woodland to montane forest (at 1,600–2,000 m. today). A cooler (4° C.?) and humid climate is inferred, apparently contemporary with the Upper Dryas in Europe.

Bed (2):(No name, [?] 3.5 m. thick). Current-bedded sands, partly interdigited with clays or gravel lenses. Magosian artifacts but no pollen.

Bed (3):(No name, [?] 2.5 m. thick). Sandy clays and bedded sands. Microlithic sites, with C^{14} charcoal dates 3920 ± 40 and 3850 ± 40 B.P. (Grn-4224, 4225), appear to be *in situ*. No pollen. The relationships of beds (1), (2), and (3) are not clear from the report (see Clark, 1969). Gravels are again contemporary upstream.

c) *Chiungu Member.* At least 6 m. of sandy clays and clays; sands and gravels are developed further upstream. Iron Age sites *in situ,* with five C^{14} charcoal dates ranging from A.D. 550 to 1580.

The publication of the Kalambo Falls geology, in three different chapters and innumerable archeological sections,[6] does not allow a confident environmental interpretation for alternating erosion and alluviation. Consequently, the excellent paleobotanical work is of primary interest. Significant changes in the vegetation pattern are indicated, some of them quite relevant to local geomorphologic processes (e.g., the stone lines of Rubble Bed I). Some of these environmental changes are clearly coeval with changes known from mid-latitude Europe. So, for example, a part of the early Würm (Brörup-Amersfoort?) and again the Younger Dryas were 3–4° C. cooler at Kalambo Falls, while the warm Paudorf is also apparent in the record as a moister interval. The pleniglacial left no record of greater moisture. Equally interesting are apparent similarities with the sedimentary record of the Omo Basin, e.g., Mkamba (2) with Kibish Member III; Mwilo (1) with Kibish Member IVa and with the early Holocene ("Gamblian") lake of the Kenya Rift. As elsewhere in the Nile Basin and East Africa, there was no pluvial spanning the whole time range of the Würm Glacial, and despite brief, moist intervals, the greater part of the late Pleistocene was no wetter than the present day.

Another sedimentary sequence, also recording much of the late Pleistocene, is exposed at Isimila, near Iringa in south-central Tanzania (Howell *et al.,* 1962). This site, which has a rich Late Acheulian industry and some fauna, has recently been restudied by C. M. Keller and C. L. Hansen, so that a discussion of the deposits and their possible interpretation would be premature.

THE LATE PLEISTOCENE OF LUNDA

The Lunda area of northeastern Angola has a complex succession of Pleistocene deposits and erosional surfaces, as well as a large number of prehistoric sites. In the region of Dundo, the interfluves (at 900–1,000 m.) are formed of late Tertiary Kalahari Sands and their derivatives, while Cretaceous sandstones and Precambrian metamorphics may be exposed in the subparallel valleys (at 700–800 m.) that drain north to the Congo River. Although annual rainfall is about 1,400 mm., with only three to four dry months per year, vegetation is rather open as a result of the dry subsoil environment. The open, dry forest, with a grassy under-

6. The chapters by Bond and Kleindienst are difficult to reconcile, and the detailed vertical sections given are not properly related to either stratigraphic scheme. Geomorphologic interpretations are scattered through almost ever chapter, but a coherent overview is lacking.

story, is dominated by the broadleaf-deciduous genera *Brachystegia* and *Isoberlinia,* while a galeria forest of broadleaf-evergreen trees fringes the rivers.

A late Pleistocene to Holocene sedimentary sequence can be outlined after J. D. Clark (1963) and DePloey (1965); the limited but significant palynological data is from van Zinderen Bakker (1963):

1) "Laterite II." Deposition of 50–150 cm. of pisolitic laterite, i.e., limonitic pellets, intermixed with blocks of older laterite and subsequently cemented by iron-rich compounds (ferricrete). Probably derived from denudation of subsoil aggregates rich in iron, accumulating as a colluvium. This laterite seals at least one Late Acheulian site, with unworn artifacts. The original formation of ferruginous soil horizons, a ferrallitic paleosol, presupposes a long period of warm, humid climate.

2A) "Redistributed Kalahari Sands II." Silty sands, primarily aeolian in origin, accumulated to thicknesses of 20–30 m. on the uplands, thinning to less than 10 m. on valley sides, where colluvial reworking by sheetwash was dominant; partly contemporary with valley-floor conglomerates (2B). The sands are known to rest on an older surface with Sangoan artifacts (ca. 46,000–37,000 B.P. at Kalambo Falls), while early to late Lupemban collections have been made within the body of the reworked sands.

2B) "Flats Terrace," lower unit (4 m. thick). Well-stratified, sandy gravel, with lenses of sands and peaty clay. A pollen sample (78 grains identified) from the clay includes 52 per cent NAP, mainly grasses, with 33 per cent of the evergreen genus *Chrysophyllum,* 7 per cent *Brachystegia,* 4 per cent sedges, and an additional 3 per cent fern spores; very open vegetation and slightly drier conditions are inferred. Early Lupemban; two C^{14} dates on peat and wood: 38,000±2500 and "greater than 34,000" B.P. (UCLA-168, 169). Since the stream gravels may grade laterally into stone lines (De Ploey, 1965), the open vegetation indicated is quite compatible with accelerated denudation on poorly vegetated slopes and aeolian activity on even drier interfluves.

3) Stream equilibrium with ferricretion of Flats Terrace, possibly indicating eluviation of upland soils, with a high water table in the lowlands. Moist.

4) "Flats Terrace," upper unit (over 2 m. thick). Limonitic gravels with lenses of peaty clay, grading laterally into colluvial deposits. Pollen (290 grains) is mainly arboreal (only 30 per cent NAP), with 25 per cent *Podocarpus* and 30 per cent *Syzygium,* both evergreen genera, the former now confined to the montane forests of upland Angola (1,600–2,200 m.), the latter a swamp-forest denizen. Humid conditions

are also indicated by 12 per cent fern spores, and a cool, wet climate is suggested, with a depression of the vertical zones of vegetation by 800–1,000 m. Late Lupemban; a radiocarbon date on wood of 14,500±560 B.P. (C-581) may be somewhat too young.

5) Limited stream cutting, accompanied or followed by ferricretion of Flats Gravel, with local accumulation of pisolitic iron ("Laterite III"). High water table, moist.

6) "Redistributed Sands III." Aeolian sands, 2–10 m. thick on slopes, derived from deflation of older upland sands. Blowouts dot the interfluves right to the banks of the Congo at Leopoldville (Kinshasa); these are now inactive under a vegetation of forest and savanna, while their floors are at the water table (De Ploey, 1965). Tshitolian sites are found in the sands. A C^{14} date of 12,970± 250 B.P. (UCLA-172) on charcoal suggests a late glacial age.

7) Current-bedded fluvial sands (2–5 m.), with organic clayey lenses, in the stream valleys. Although derived from older aeolian materials, these beds have abundant pollen indicating a warm but moist climate. Two samples from Mufo (357 grains identified) gave 38 per cent NAP, with the deciduous savanna tree *Alchornea* dominant (46 per cent). The valley floor was moist, judging by galeria species, 14 per cent sedges and an additional 15 per cent fern spores. At Calunda a sample of 50 grains gave 52 per cent NAP (22 per cent grasses, 30 per cent Compositae), with 24 per cent *Berlinia,* a deciduous tree. Derivation from older beds or long-distance stream or wind transport may account for 2 per cent *Syzygium* and 4 per cent *Podocarpus.* In general, an open, deciduous dry woodland is indicated for the higher ground. Tshitolian sites and C^{14} dates of 6830±120 and 4700±100 B.P. on peat and wood (UCLA-167, 171) suggest a mid-Holocene age for this moister interval.

8) "Redistributed Sands IV." A veneer of aeolian and colluvial sands covering weathered upland sands, or lag surfaces, locally thickening to 5 m. The activities of late Tshitolian man may have been responsible for deflation and sheetwash through decimation of the vegetation cover (De Ploey, 1965).

9) Subrecent stream gravels, sands, and clays, with some pollen suggesting swampy lowlands and a deciduous dry woodland on higher ground. Iron Age potsherds and a C^{14} date of 1880±80 B.P. on wood (UCLA-170). More recently, gullying has dissected the slopes, while organic clays accumulate in the floodplains.

The Lunda sequence is not identical with that of East Africa. A warm, humid climate is indicated for some undated part of the early Upper Pleistocene, with most of the lower pleniglacial and the early Paudorf time range quite dry. A cooler (by at least 5° C.) and moister interval (phases 3, 4, and 5) is suggested for the later Paudorf and part of the Würm maximum. Late Würm and early Holocene times were again quite dry, with warm and relatively moist phases during both the mid- and late Holocene. The contrasts with East Africa may not be as great as they seem since the record is incomplete, at least in terms of isotopic dating, and unclarities persist about just how the upland aeolian phenomena and lowland fluvial beds are interrelated. The shifts of vegetation during the "cool-moist" phases at both Kalambo Falls and Lunda can be readily accounted for the a 5° C. drop in temperature with no change in the amount of rainfall (see Clark, 1967, Map 10). However, the major aeolian phases (2A and 6) presuppose a climate far drier than that of today; this intensification of aridity is quite comparable to that of the southern Sahara.

A similar succession of late Pleistocene events seems to be recorded at the Victoria Falls (Bond and Clark, 1954) and in Botswana (Grove, 1969), although a radiometric framework and pollen data are lacking.

THE LATE PLEISTOCENE OF THE VAAL-ORANJE BASIN

The best-known sequence of Pleistocene deposits in southern Africa is that of the Vaal River valley. Deposition of a suite of "Basal Older Gravels," apparently of early Pleistocene age, was followed by up-arching of the Vaal-Oranje interfluves and intensified bedrock incision. Eventually a body of "Younger Gravels," with a Middle Pleistocene fauna and late Acheulian artifacts, was aggraded (Cooke, 1947, 1967; Partridge and Brink, 1967). The sequence of late Pleistocene events, not yet dated isotopically or faunistically, can be tentatively outlined (after Butzer, 1971b; see also Söhnge *et al.,* 1937, van Riet Lowe, 1952).

The evolution, after the "Younger Gravels," of the Vaal River proper begins with the accumulation of 12 to 15 m. of alluvial sands and silts. In the tributary valleys these are contemporary with a more complex aggradation of "Older Fill": (a) A basal soil wash with subrounded gravel, 2 m. or more thick; grades upslope into coarse detritus due to sheetwash and accelerated runoff. (b) A massive accumulation of 3 to 15 m. of silts, clays, and marls, indicating marshy, waterlogged valley bottoms; fine-grained suspended sediments extend upslope and suggest a dense vegetation mat, with inhibited surface runoff. The terminal deposit may include some reworked aeolian material. (c) Limited channel dissection. (d) Accumulation of black, organic, marshy deposits along stream channels, to a thickness of 1.5 m., with formation of chernozemic grassland soils on hillslopes. (e) Limited channel dissection. (f) Accumulation of silts, with derived aeolian sands, to a thickness of 1 or 2 m. Acheulian materials, primarily derived by colluvial action, may be found at the base of the "Older Fill," while a number of "Middle Stone Age" sites can be directly linked to the main body of alluvium. After a long period of dissection, marking a disruption of the vegetation and of geomorphological equilibrium, the more homogeneous silts of the "Younger Fill" were deposited. These attain a thickness of 5 to 12 m. and were interrupted by several phases of erosion. Slow aggradation of the "Younger Fill" continued well into the nineteenth century, and may reflect post-Pleistocene equilibrium conditions.

Crucial to an understanding of late Pleistocene environments in the Vaal-Oranje Basin are pollen studies from Florisbad and Aliwal North. The archeological site of Florisbad is located at 1275 m. elevation on the southern margins of a deflation pan, eroded in weathered shale derivatives and now occupied by salt flats, a fossil calcareous crust, halophytic plants, and some sedge communities. Slightly brackish water emerges from springs along a subsurface dike. A low hill of spring deposits was excavated to a depth of 6.8 m., and the following stratigraphy can be described (from bottom to top, after van Zinderen Bakker, 1957, 1967; Butzer, unpublished) as resting over bedrock with dolerite intrusions:

(1) Mottled sand with discontinuous organic lenticles ("Basal Peat") 60 cm.
(2) Black, organic clay ("Peat I"), 75 cm. Groningen C^{14} dates of "greater than 48,000 yrs." and of "greater than 35,000 yrs. B.P." (L-271B). (See Broecker *et al.,* 1956, for discussion of earlier Libby dates.) "Middle Stone Age."
(3) Clay and loam with vegetable matter and columnar structure, possibly suggesting a gley; 105 cm. "Middle Stone Age."
(4) Organic, fine sands ("Peat II"), 60 cm. C^{14} date of 28,450 ± 2200 yrs. B.P. (L-271C).
(5) Sands, 135 cm. "Middle Stone Age."
(6) Clayey, organic deposit ("Peat III"), 10 cm. C^{14} date of 19,350 ± 650 B.P. (L-271D).
(7) Consolidated sands, 100 cm.
(8) Organic loam ("Peat IV") and diatomaceous silt, 58 cm.
(9) Loamy surface soil, 47 cm.

Pollen was sampled at 10 cm. intervals, with counts from 50 to over 150 grains per sample. Sedge, Chenopodiaceae, *Zygophyllum,* as well as most of the varia, fluctuated according to the discharge of the thermal springs, a nonclimatic cycle. However, the Compositae and Gramineae, on the basis of modern spectra and contemporary vegetational patterns in the Orange Free State, reflect regional climatic changes. Today, the native vegetation with 450 mm. annual precipitation is dry *grassveld,* and the Compositae/Gramineae pollen ratio is 1:24. To the southwest of Florisbad, beyond the Oranje River, the semidesert Karroo shrub is found in regions with 125–250 mm. precipitation; the modern pollen ratio is approximately 1:1. Without human disturbance, geomorphologic equilibrium is typical of the grassveld, with sheetwashing and accelerated runoff in the Karroo.

Van Zinderen Bakker (1957) found fairly abundant pollen in many of the Florisbad strata. The upper halves of Bed 1 and Bed 3 ("Peat I") have 40–70 per cent Compositae, after exclusion of the "local" pollen rain, indicating a typical Karroo vegetation with about 50 per cent less precipitation than today. The sediments and local pollen, including many *Tetraploa* spores, indicate considerable and accelerating spring activity. These two zones contain the bulk of the Florisbad fauna, much of it stratified beneath the base of bed (2) (Oakley, 1954). Of 32 mammalian species, 9 (36 per cent) are extinct today (Cooke, 1963, Table 7), supporting the early Upper Pleistocene age indicated by the radiocarbon dates. Bed (3) essentially lacks pollen, while Bed (4) still has 15–50 per cent Compositae when local pollen are excluded, indicating a drier climate in mid-Upper Pleistocene times. Spring activity was at a miximum. Beds (5) and (6) show a steady increase of grassland at the expense of Karroo, until the Compositae/Gramineae ratio was 1:24. Inversely, spring activity declined although the presence of fresh water is indicated by *Riccia* pollen. This suggests that at the height of the last pleniglacial, regional climate was at least as wet as it is today (van Zinderen Bakker, 1967). The final strata (7, 8, and 9) unfortunately have few pollen. The sequence is essentially completed by another profile at Aliwal North.

Investigation of the Florisbad deposits and other spring mounds in the area showed that these sequences of spring beds are laterally conformable with the "Older Fill" of the Modder River (Butzer, unpublished). The pollen studies consequently provide the key to interpreting the different sedimentary units of the Vaal tributaries: the initial soil wash (a) accumulated under a discontinuous Karroo vegetation; while the fine-grained beds (b), (d), and (f) reflect lush grassveld. It seems reasonable to assume that the latter units can be correlated with the Upper Pleniglacial.

The thermal springs near Aliwal North emerge from a dike intruding Permian shales and sandstones at −90 m. Cores were studied by Coetzee (1967, part 2) from a 10.5 m. sequence of peaty and clayey beds that fill a former lake ultimately reduced to a spring-fed swamp. The basal deposits date from about 13,200 B.P., by extrapolation from a series of consistent C^{14} dates. The site is located near the border of the modern grassveld and Karroo vegetation provinces at 1,355 m. elevation.

The sediment and pollen sequence can be summarized (from bottom to top) on the basis of four cores (Coetzee, 1967):

Zone u. Peaty clay, sandy clay, and clay with basal gravel (125 cm.), dating ca. 13,200–12,400 B.P. A colder and more humid climate is indicated by pure grassveld, with some *Stoebe plumosa* pollen, a species now found above 2,000 m. in Lesotho.

Zone v. Gray loam (50 cm.), ca. 12,400–12,200 B.P. Warm and dry as indicated by a Karroid vegetation.

Zone w. Sandy clay or clay loam (40 cm.), ca. 12,200–11,650 B.P. Cooler and moister as indicated by a pure grassveld.

Zone x. Sandy clay, peaty clay, sandy loam, or clay (120 cm.), ca. 11,750–11,250 B.P. Very warm and dry as indicated by a Karroo vegetation.

Zone y. Sandy or peaty clay (120 cm.). Cooler and moister as indicated by grassveld vegetation.

Zone z. Black peaty clay (250 cm.). A dry grassveld was gradually replaced by a Karroid vegetation, indicating increasingly warmer and drier conditions, ca. 10,000 B.P.

The similarity of the degree of environmental changes through time at Florisbad and Aliwal North is very striking, as is the close correlation of climatic fluctuations at the latter site with the well-known higher latitude chronology of the northern hemisphere. So, for example, pollen zone (x) can be readily related to the Alleröd Interstadial, and zone (y) to the Upper Dryas. To what extent these synchronisms can be extrapolated to the whole Vaal-Oranje Basin remains to be demonstrated. Seen in conjunction with the incompletely established evidence of cold-climate phenomena (solifluction, frost-shattering, nivational niches or cirques?) above 2000 m. in the Drakensberg of Lesotho and Natal (Sparrow, 1967), pollen zone (u) offers analogies to the invasion of the Saharan highlands by higher latitude plants.

The late Pleistocene sequences reviewed here indicate that the Vaal-Oranje Basin experienced significant changes in climate, vegetation, and geomorphologic equilibrium during the course of the Pleistocene. On theoretical grounds, Cooke (1964) suggested that long-term rainfall anomalies may have varied from 50–60 per cent of the present on the one hand to 140–150 per cent on the other; his vegetation reconstruction accordingly shows Karroo vegetation over the Vaal-Oranje Basin (to 2000 m. elevation) for the "dry" anomaly, grassveld for the "wet" anomaly. The pollen profiles of Florisbad and Aliwal

North appear to support this hypothesis, which does not, however, take into account reduced evapotranspiration during periods of lower temperatures. A second point that emerges once again for southern as well as for eastern Africa is that environmental changes during the late Pleistocene were rather rapid, and that they involved repeated oscillations to the dry side, alternating with conditions broadly comparable to those of today. There is little evidence for genuine "pluvial" conditions during the late Pleistocene, or for that matter, during the Holocene. In fact, the impression obtains that rainfall in southern Africa was lower during most or all of the late Pleistocene, and that the relatively moist periods simply reflect periods of cooler climate.

THE PLUVIALS OF EASTERN AND SOUTHERN AFRICA

A number of generalizations emerge from this discussion of late Pleistocene evidence from eastern and southern Africa. These can be outlined for each environment:

Semiarid and Subhumid East Africa. Two major pluvials occurred during the first half of the Upper Pleistocene, with moister and cooler (3° C.?) conditions during the early Würm, possibly contemporary with the Brörup-Amersfoort. The Paudorf time range was a trifle cooler and moister than today, but the Upper Pleniglacial as well as late glacial times were very dry, despite expansion of the mountain glaciers, and temperatures were at least 5–6° C. lower than today. Two intense but brief pluvial phases can be identified for the early and mid-Holocene, with a drier interruption ca. 6500 B.P. In other words, the interstadials and parts of the interglacials were moist, the glacial maxima dry.

The Northern Kalahari Margins. Based on one study area, with an incomplete record, one major pluvial of uncertain age can be inferred for the early Upper Pleistocene time range. Intensively dry conditions prevailed during the Lower Pleniglacial and during the late Upper Pleniglacial, with a cool-moist climate during the late Paudorf time range, probably persisting until the last glacial maximum. Finally, a cool and comparatively moist, late glacial phase was followed by another arid spell, with a warm pluvial in mid-Holocene times. Similarities with both the southern Sahara and eastern Africa are apparent.

The Semiarid, Southern Kalahari Margins. No dated record is available for the early Würm, but part or all of the Lower Pleniglacial was significantly drier than today. The Upper Pleniglacial experienced a lower rainfall compensated for by lower temperatures. A number of rapid pulsations are apparent in the late glacial record, with cool spells appearing to be comparatively moist (e.g., the Upper Dryas), warm spells apparently dry. No Holocene moist phases are evident, analogous to the semiarid, northern margins of the Sahara.

CONCLUSIONS

The available evidence from the African continent is so complex that it can only be discussed and evaluated when outlined in some detail. Similarly, isotopic dating is essential if useless polemics on matters of correlation are to be avoided. As a result, the preceding sections of this chapter have analyzed the record of a number of well-studied areas to obtain, possibly for the first time, an objective and meaningful overview. Although few broad generalizations can be offered in terms of regional environments for any given period, the patterns of relative climatic change through time are beginning to emerge for the major climatic provinces. The record of the Mediterranean borderlands has similarities with that of the southern margins of the Kalahari. On the other hand, the records of the southern Sahara and the northern Kalahari borderland are all basically similar among themselves, while those from the central axis of the Sahara and eastern Africa are each distinct in their own right. This suggests that Pleistocene climatic changes followed predictable patterns, symmetrical about the equator, and related to the planetary wind and weather belts.

The inventory of pluvial phases is so manifold that several points need emphasis or reemphasis:

1) Unlike the glacial-interglacial or even stadial-interstadial pulsations of higher latitudes, the African pluvials were all of short duration, some less than 2,000 years, few exceeding 5,000 years.
2) Throughout the continent, the maximum of the late Würm-Wisconsin was dry, although there were a number of wetter intervals prior to 25,000 B.P., varying in their intensity and duration. In South Africa and along the Mediterranean borderlands of the Sahara, the early to mid-Würm pluvials may have been the result of cooler temperatures rather than greater rainfall; there is little evidence for Holocene pluvial episodes in these areas. Across the central axis of the Sahara and through several of the tropical savanna climates there is good evidence for one or two pluvials beginning 9500 B.P. and terminating about 5000 B.P.; these include most of the acceptable phenomena once ascribed to the supposedly late Pleistocene "Gamblian" pluvial, including the type site. Within the savanna climate zone the mid-Würm interstadial complex was decidedly moist, as were one or more previous intervals of uncertain age.

3) It appears that pluvial phases in tropical Africa occurred during certain stadial and interstadial periods of the Würm-Wisconsin glacial as well as during the Holocene. It is therefore likely that comparable pluvials marked parts of the Eem and earlier interglacials. Although the African pluvials show sufficient synchronism within certain climatic provinces to suggest a genetic association with climatic changes on a continental or worldwide scale, their apparent or well-defined correspondence with a number of different high latitude anomalies remains to be investigated from a climatological point of view.
4) The African pluvial record is far too complex to serve as a basis for stratigraphic correlation. In addition, the "classical" East African pluvial chronology—and terminology—is based on false premises and incorrect deductions. It should therefore be abandoned entirely. As a corollary, the term "pluvial" loses its value as a universal or regional stratigraphic concept.
5) It is normally difficult, if not impossible, to infer whether a record of dry and wet climates reflects a simple change of total rainfall amount or more complex interrelationships between precipitation, seasonality, and evaporation. Consequently the term "pluvial" should best be used in a local and descriptive form—as an adjective rather than as a noun—to designate evidence for a comparatively wet climate. "Nonpluvial" could be used for the converse.
6) Pluvial phenomena are of different types, e.g., accelerated stream or spring activity, high lake levels, periods of soil formation, pollen records of closed humid forest, etc. It will therefore be essential to define local categories of pluvial evidence on the basis of the associated physical or biological features.

It would seem that these general conclusions, reached on the basis of the African data, are relevant for other lower latitude regions as well.

TWENTY-ONE

Late Pleistocene Environments of North America

INTRODUCTION

The Pleistocene left an impressive legacy of glacial forms and deposits in North America, reflecting at least four major stages of glaciation. Beyond the margins of the glaciers, aeolian deposits were spread far and wide, while extensive lakes formed in the basins of the intermontane country. In many ways the North American Pleistocene record resembles that of Europe, but features are generally represented on a larger scale. At the time of maximum glaciation, ice covered 15.7 million sq. km. of North America (excluding Greenland) (Figs. 56 and 57), compared with 6.3 million sq. km. for Europe or 7.7 million for Asia (see Flint, 1958, p. 53; Donn *et al.,* 1962). In fact, the Laurentide ice sheet, which attained the dimensions of the Antarctic (13.2 million sq. km.), extended much farther toward the equator: to latitude 37°30′ N., compared with 48°50′ N. for the Scandinavian glacier in Europe—a difference of about 800 km.

The four classical glaciations of North America are outlined in Table 16, together with the faunal stratigraphy. The Wisconsin Glacial and the Sangamon Interglacial can obviously be correlated with the European Würm and Eem, respectively, on the basis of absolute dating, stratigraphic sequence, and geomorphology (see also Woldstedt, 1965, ch. 8). Intercontinental correlations of the older stages are more problematical, although there is some reason to believe that the Illinoian correlates broadly with the Saale in Europe. Various suggestions have been offered that the Kansan and Elster glacials should be equated on faunal grounds, but the available data is still insufficient for such a task. The evidence for extensive glaciation in the Sierra Nevada shortly before 2.7 million years B.P. (Curry, 1966) may find counterparts in cold-climate phenomena in the Massif Central of France, associated with Villafranchian faunal beds that have been dated in a similar time range by K/Ar (see Curtis, 1967).

This chapter will consider the major late Pleistocene environments of North America as defined by either geomorphological or biological criteria. Primary attention will be focused on conditions during the maximum of the "classical" or late Wisconsin Glacial—the Woodfordian stage (ca. 23,000-12,800 B.P.) Earlier attempts to delineate late Pleistocene environments of North America have been made by Deevey (1949), Dillon (1956), Martin (1958a), Brunnschweiler (1964) and others. Syntheses of geological and stratigraphical data have also been made by Flint (1958) and Woldstedt (1965), although the most valuable

Table 16. Glacial and faunal stratigraphy of North America (after Evernden et al., *1964; Frye* et al., *1965; Hibbard* et al., *1965; Reed* et al., *1965; Wayne and Zumberge, 1965; Curry 1966; and others).*

Time-Stratigraphic units (stages)	(substages)	*Detail*	*Faunal stages*
Recent			
Wisconsinan	Valderan	Rapid retreat by 10,750 B.P.	
	Two Creekan	ca. 12,800-11,500 B.P.	
	Woodfordian	Maximum ca. 18,000 B.P.	
	Farmdalian	Terminated ca. 23,000 B.P.	Rancholabrean
	Altonian	Began before 70,000 B.P.	
Sangamonian		Interglacial	Irvingtonian (began before 1.4 million B.P.)
Illinoian		Glacial, with 2 or 3 stadials	
Yarmouthian		Interglacial	
Kansan		Glacial, with 3 stadials	
Aftonian		Interglacial	
Nebraskan		Glacial, with 2 stadials	Blancan (began before 3.5 million B.P.)
?		?	
(Glaciation in Sierra Nevada shortly before 2.7 million B.P.)			

source of regional and thematic data is the excellent compendium of Wright and Frey (1965), complemented by nine *Guidebooks for Field Conferences* of the 7th INQUA Congress (1965). However, many classes of data, particularly on periglacial phenomena and palynology, are still inadequate for a regional presentation. For this reason a thematic discussion is employed, as in the case of chapter 19.

BASIC CHRONOLOGY OF THE WISCONSIN GLACIAL

Analogous to the situation in mid-latitude Europe, the ice front in North America oscillated widely during the course of the Wisconsin Glacial, in response to long-term climatic trends. The pattern of these changes through time, now partly fixed by radiocarbon dating, is similar, although by no means identical, to that of Europe. A significant difference appears to have been the relatively limited deglaciation that accompanied the interstadials. The chronological sequence (after Dreimanis, 1960; Dreimanis *et al.,* 1966; Goldthwait *et al.,* 1965; Black and Rubin, 1968; Kempton and Hackett, 1968) can be tentatively drawn up as follows:

a) Before *70,000* B.P. End of Sangamon Interglacial and beginning of Early Wisconsin. Inception of Laurentide ice sheet. Cold climate.
b) ca. *65,000* B.P. Temperate interval within Early Wisconsin, designated as St. Pierre Interstadial. Partial deglaciation in Canada.
c) ca. *60,000 (?)–50,000 (?)* B.P. Early maximum of the Early Wisconsin, with full glacial conditions. Ice front south of Great Lakes (Lower Winnebago Tills).
d) ca. *50,000 (?)–44,000* B.P. Temperate interval at beginning of Mid-Wisconsin, designated as Port Talbot Interstadial. Extensive deglaciation in Great Lakes region.
e) ca. *44,000–41,000* B.P. First cold maximum of Mid-Wisconsin; glaciation of Great Lakes region (Middle Winnebago Tills).
f) ca. *41,000–33,000* B.P. Temperate interval, marking first part of what is sometimes designated as the Plum Point Interstadial. Partial deglaciation in Great Lakes region (Plano Soil).
g) ca. *33,000–29,000* B.P. Second cold maximum of Mid-Wisconsin, with glaciation of Great Lakes region (Upper Winnebago Tills). The period (a) to (g), from before 70,000 to about 28,500 B.P., comprises the *Altonian.*
h) ca. *29,000–23,000* B.P. Temperate interval, almost generally designated as Farmdale Interstadial, marking second part of Plum Point Interstadial. Deglaciation in Great Lakes region.
i) ca. *23,000–12,8000* B.P. Full glacial conditions of Late or "classical" Wisconsin, comprising the Woodfordian substage and including the

Peoria Loess. Various end moraines south of Great Lakes, with maximum ice advance ca. 20,000–18,000 B.P. (Tazewell interval), followed by successive recessional halts (Cary, ca. 16,000–14,000 B.P., and Port Huron, ca. 13,000 B.P., intervals).

j) ca. *12,800–11,500* B.P. Two Creeks Interstadial with temperate conditions and almost general deglaciation in Great Lakes region.

k) ca. *11,500–10,000* B.P. Cold Valders substage, with limited glacier readvance in northern Great Lakes area. Deglaciation was underway by 10,750 B.P. The Laurentide glacier had retreated from the United States by 10,000 B.P., subdividing into two residual ice caps separated by Hudson Bay, before 7500 B.P. Final disappearance of ice over Keewatin and Quebec by ca. 4500 B.P.

The details of the Altonian substage are still controversial, and information is far from satisfactory. Nonetheless, a picture emerges of initial glaciation, followed first by a significant but poorly dated interstadial, and then by a long period of essentially cold climate with an oscillating ice front. The so-called Plum Point Interstadial, as defined by the Plano and Farmdale horizons, recalls the Paudorf Interstadial of Europe, as marked by the Hengelo and Denekamp phases. Discrepancies of the radiocarbon dates may or may not be real; in particular, the Farmdale seems about 5,000 years younger than the Denekamp. Possibly, too, the different relative importance of weathering horizons and tills along an oscillating ice front may not always be related to pollen zones well away from the glacier. A case of this kind is the Two Creeks Interstadial, which clearly coincides with the Alleröd *plus* the Bölling in Europe. The boundary between the Wisconsin and the Recent has not been defined. Although the draining of the great proglacial Lake Iroquois via the St. Lawrence River (*ca.* 11,000 B.P.) provides a similar regional stratigraphic marker to the draining of the Baltic Ice Lake (*ca.* 10,200 B.P.), there seems to be no advantage to using different dates for the international Pleistocene-Holocene boundary. A marked pollen shift can be identified in the Great Lakes *ca.* 10,000 B.P. (Ogden, 1967), suggesting that this, in fact, is a reasonable boundary to retain.

THE LAURENTIDE, ARCHIPELAGO, AND GREENLAND ICE SHEETS

The Laurentide ice sheet covered an area of about 12.5 million sq. km. at the height of the Wisconsin Glacial (Fig. 56). The maximum expansion can almost generally be attributed to the Aftonian substage (see Lemke *et al.*, 1965; Wright and Ruhe, 1965; Black and Rubin, 1968), although the ice was at least equally extensive east of the 90° meridian (western Illinois) during the later Woodfordian substage. The Wood-

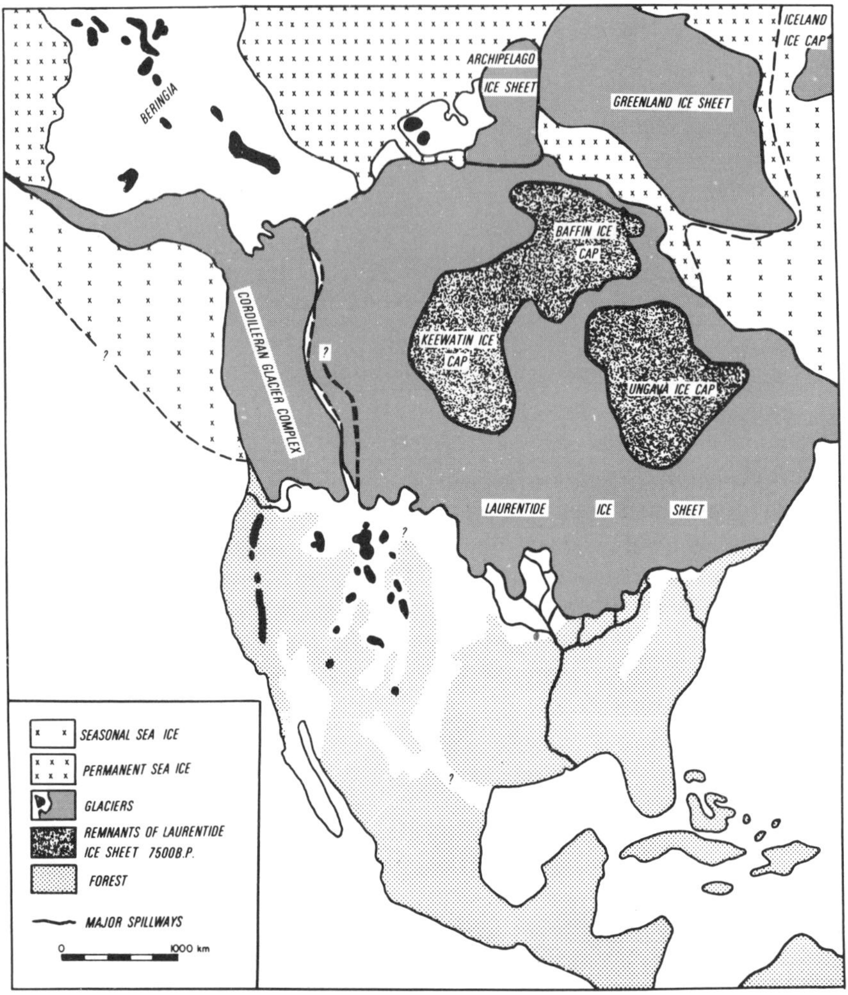

Figure 56. North America during the Wisconsin Glacial (Woodfordian substage, ca. 23,000–12,800 B.P.*). The western terminus of the Laurentide ice is uncertain, and two alternative ice margins are given. The reconstruction of forest (as opposed to parkland or grassland) is highly tentative. Shorelines approximate the – 150 m. isobath.*

fordian ice also seems to have been less extensive in the lower Mackenzie valley (see discussion comments in Bryan, 1969). The ice probably began to accumulate in three major centers—over Baffin Island, Ungava or northern Quebec, and Keewatin, areas that were still extensively glaciated 7,500 years ago (see Bryson and Wendland, 1967). As Bird (1967, p. 93) has pointed out, a drop of 3–4° C. in summer temperatures would suffice for extensive ice to develop on Baffin Island, while a fall of about 5° C. would lead to glaciation in Ungava. The growing Laurentide glacier was fed by moisture from maritime-tropical airmasses from the Gulf of Mexico and maritime-polar airmasses from the North Atlantic. In response to these southerly and southeasterly sources of moisture, the ice advanced farthest in southeastern Canada and the northeastern United States. It is here that the greatest ice gradients were eventually established, whereas the ice sheet was starved for moisture and so almost level and rather thin along its southwestern, western, and northern peripheries. The greatest isostatic readjustment took place over Hudson Bay (King, 1965), suggesting that maximum ice thickness—possibly on the order of 3,300 m. (Bird, 1967, p. 96)—occurred there.

The smaller, rather thin ice sheet over the northernmost Canadian Arctic islands was actually distinct from, although contiguous with, the Laurentide ice sheet. Much like the British ice cap, this Archipelago ice sheet was centered around several mountain massifs with radial ice drainage. In the northeast these ice caps were apparently linked with the Greenland ice sheet. Little is known about the extent of Pleistocene glaciation over Greenland, but the well-developed fiord coast and the inshore submarine topography imply more extensive glaciation around the margins of that land mass (see Fig. 56).

THE CORDILLERAN GLACIER COMPLEX

The western parts of Canada (British Columbia, the Yukon), southern Alaska, and the northwestern periphery of the continental United States were under a continuous ice sheet or an interconnected complex of valley and piedmont glaciers. This pattern of glaciers is sometimes referred to as the Cordilleran glacier complex. The more isolated clusters of mountain, valley, and piedmont glaciers, as well as the ice caps of Alaska (Brooks Range) and the western United States (Cascades, Sierra Nevada, Rocky Mountains, etc.), are frequently included under this general name. The size of the coalescent glaciated area was 2.2 million sq. km. during the Wisconsin Glacial, while the separate glaciated areas totaled about 0.9 million sq. km. (Flint, 1958, p. 53).

The Cordilleran glacier complex attained its greatest thickness in British Columbia, which lies athwart the trajectory of the moisture-bringing maritime-polar airmasses. The coastal ice streams attained thicknesses of over 1,200 m., while the coalescent piedmont glaciers in the lee of the coastal ranges were over 2,300 m. thick (Flint, 1958, p. 305). By contrast, the Rocky Mountains were in a rain shadow situation, and glacier development was correspondingly limited so that even during the Altonian maximum, the Cordilleran and Laurentide glaciers were not entirely coalescent. During much of the Woodfordian there may, in fact, have been a subcontinuous corridor along the eastern foothills of the Rocky Mountains for much of the 2000 km. stretch between the Mackenzie Delta and southern Alberta (see chapter 29).

Among the dispersed Wisconsin glaciers of the western United States (see Wright and Frey, 1965, part I, section 3), mean cirque elevation rises rapidly, in the case of the Rocky Mountains, from 1,900 m. near the Canadian border to over 3,000 m. south of latitude 45° N. (Richmond, 1965). The average snow-line depression, based on mean cirque-floor elevation, is 600 m. in the northern Rockies, 500 m. in the middle, but only 250–300 m. in the southern Rocky Mountains. In the vicinity of Mexico City, several peaks of over 5,000 m. elevation were extensively glaciated in Wisconsin times. These glaciers suggest a snowline as much as 1,000 m. lower than that of today (see White, 1962).

PROGLACIAL LAKES AND SPILLWAYS

At the height of the Wisconsin Glacial, the front of the great ice sheets ran northwest from Long Island and New York City almost to Lake Erie. From there it swung through central Ohio, Indiana and Illinois, with various tongues or lobes projecting 100 to 150 km. ahead of the general ice margin. In southwestern Wisconsin a large entrant of nonglaciated terrain, the "Driftless Hills," is recognized. Despite an almost total absence of drift, there are scattered erratics, and the erosional forms suggest an earlier, probably pre-Wisconsinan glaciation (see Black and Rubin, 1968). A similar problem exists in eastern Iowa, where the Iowan Till may be of Altonian or earlier age (see Wright and Ruhe, 1965). It is here considered as Altonian (see Fig. 57). Further west, beyond the conspicuous ice lobe in central Iowa, the glacier margin generally followed the valley of the Missouri River to the Rocky Mountains. Whereas the Altonian end moraines generally lie south of that river, the Woodfordian moraines lie north (Lemke *et al.,* 1965). The southern terminus of the Cordilleran ice sheet was intricate, with the Columbia Plateau essentially ice-free (see Richmond *et al.,* 1965).

The ice front was marked by various meltwater deposits, including outwash fans—often coalescent—and terraces, as well as proglacial lakes. Lacustrine situations developed wherever the original drainage was dammed back by the ice front, as, for example, during the Altonian stage in Montana and North Dakota (Fig. 57). Far more impressive are the much younger lake plains that developed in the central part of the continent during the glacial retreat, when extensive tracts of isostatically depressed topography were freed from the ice. Waters were impounded here, behind the higher ground and morainic ridges in the Great Lakes area. The greatest of these ice-margin features was proglacial Lake Agassiz, centered in southern Manitoba (Elson, 1967). Similar but smaller lakes existed in each of the Great Lakes basins, with glacial Lake Maumee and its successors in the Erie area the most significant (Kelley and Farrand, 1967). Equally extensive was Lake Champlain in the St. Lawrence Valley. Figure 57 shows the extent of those proglacial lakes that date from the late Pleistocene, but are no younger than the Valders readvance. Practically all of these features can be dated between the abandonment of the Cary moraines and the Valders maximum, ca. 14,000–11,000 B.P. The exit of the St. Lawrence River was opened about 11,000 years ago, and the proglacial lakes were rapidly drained. Lake Agassiz persisted until about 7,500 B.P., when the Hudson Bay exit was opened.

The glacial meltwater streams that drained the ice front during successive stages of the Wisconsin Glacial developed broad floodplains, in the main part crisscrossed by braided channels that deposited large loads of sands and gravel. The Columbia River in the northwest, the Missouri-Mississippi-Ohio system in the interior, and the Hudson River in the east performed this function. Broad sweeps of glaciofluvial "outwash" consequently accompany the modern river valleys in the form of dissected alluvial terraces. Many now unimportant rivers in Illinois, Indiana, and Ohio drained great volumes of meltwater to the Mississippi and the Ohio. These broad, Wisconsin-age spillways now harbor insignificant, "misfit" streams (see Fig. 57) and their former headwaters often drain in a reversed direction to the Great Lakes. Interpretation of the terraces of the middle and lower Mississippi River is complicated by the presence of coarse glacial age alluvia upstream, interfingering with finer interglacial alluvia and deltaic beds downstream (Woldstedt, 1960a). Also problematical are the fluvial features of the Columbia Plateau, which was apparently created by successive, catastrophic floods caused by the sudden draining of an ice-dammed proglacial lake (Richmond *et al.,* 1965).

"PLUVIAL" LAKES OF THE GREAT BASIN

The so-called pluvial lakes of the Great Basin were of different origin than the proglacial lakes along the former ice front. They are found in what are now dry, nonoutlet basins, often marked by small evaporation pans or shallow, ephemeral lakes (Fig. 57). The major body of opinion

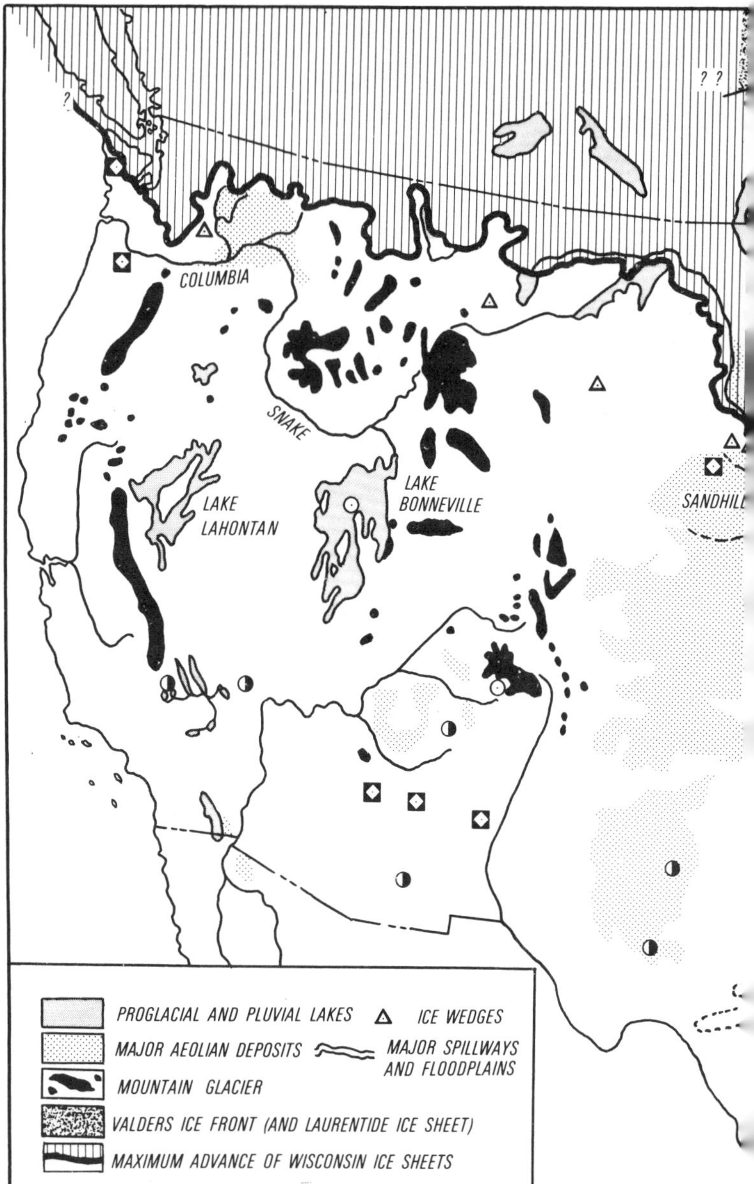

Figure 57. United States and southern Canada during the maximum of Wisconsin Glacial. The ice terminus east of the Mississippi represents Woodfordian, further west the Altonian. Those late Wisconsin lakes old

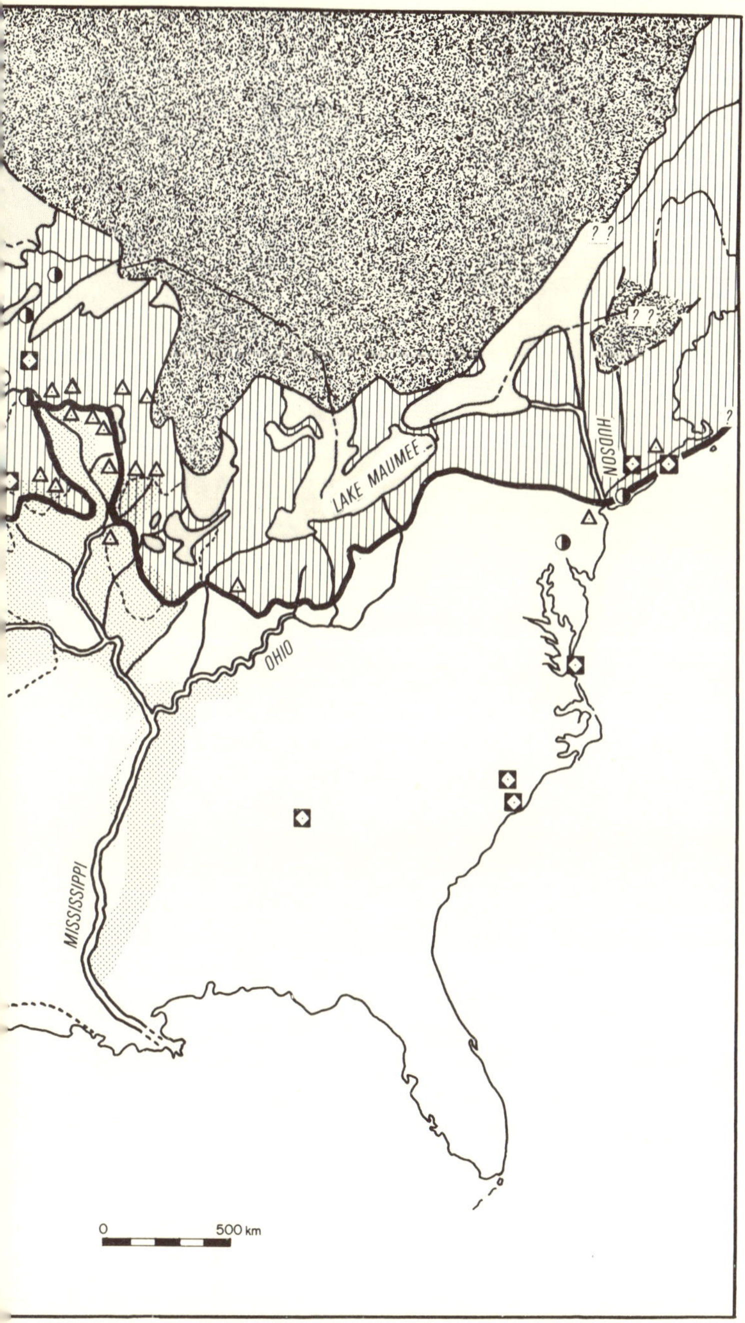

P. *are shown. Aeolian deposits refer to the Woodfordian, Two Creeks,*
ers substages, while the ice wedges date from late Altonian to
ers.

holds that the greatly expanded and deep "pluvial" lakes were made possible by a substantial increase in precipitation, a reduced evaporation accompanying glacial age temperature depression and, locally, by influx of meltwaters from adjacent mountain glaciers (see Morrison, 1965).

The largest of the "pluvial" lakes was Lake Bonneville, a predecessor of the Great Salt Lake. At its maximum it had a depth of up to 335 m. and an area of 51,700 sq. km., compared with a modern lake area that fluctuates between 2,600 and 6,500 sq. km. (Morrison, 1965). Stratigraphic study and isotopic dating (by C^{14} and Th/U) prove the existence of a long succession of "high" lakes, extending back well into the Middle Pleistocene. The last major high stand, with an overflow across to the Snake River drainage, occurred 15,000 to 12,000 B.P. A final expansion of Lake Bonneville, with shorelines at about +200 m., is dated during the Valders substage; the lake had disappeared by the early Recent. The next largest lake was Lahontan, in northwestern Nevada. At its maximum, during the Altonian substage, Lake Lahontan covered an area of 22,400 sq. km. and had a depth of up to 213 m. Its level was barely 3 m. lower early in Woodfordian times, and after temporary desiccation, Lake Lahontan attained a depth of 102 m. during the Valders. As in the case of Lake Bonneville, the Lahontan basin records a whole suite of lake shorelines and deposits, the oldest dated at about 400,000 years. Morrison and Frye (1965) have shown convincingly, by relative stratigraphy and isotopic dating, that there is a firm correlation between glaciation and lake formation, with each interglacial or interstadial accompanied by the waning or disappearance of these interior lakes.

Similar "pluvial" lakes, often interconnected, are widespread in southern California and parts of eastern Oregon (see Morrison, 1965; Flint, 1958, p. 226 ff.), and a rather comparable sequence has been established for Searles Lake, California (G. I. Smith, 1968). More restricted lacustrine deposits in western Texas are of similar age and origin (Reeves and Parry, 1965).

The hydrological budget of the "pluvial" lakes has been reconstructed by several authors, assuming a 5° C. drop in annual temperature and a corresponding 25 to 30 per cent decrease in evaporation (see Morrison, 1965, with references). This suggests an increase of precipitation on the order of 65 to 80 per cent. As a reasonable alternative suggestion, Galloway (1970) has shown that a 10–12° C. drop in temperature, accompanied by a 10 to 20 per cent decrease in precipitation, could produce similar deep lakes. This is more in keeping with the limited expansion or creation of glaciers in the southern Rocky Mountains (see Richmond, 1965). However, greater fluvial activity is verified by extensive deposits of fine-grained alluvium, with an extinct fauna (mammoth, horse, camel, bison), found throughout Arizona and New Mexico, and in adjacent areas of the Colorado Plateau. This formation is dated before 12,000 B.P., i.e., in Woodfordian times (Haynes, 1968a).

LOESS AND SAND PLAINS

A large part of the continental United States is mantled by aeolian deposits, primarily of late Pleistocene age. Figure 57 shows the major occurrences with a thickness of at least 1 m. (after Thorp *et al.,* 1952); these include (1) the Sandhills of north-central Nebraska, (2) widespread aeolian silts and more restricted sand dunes in Texas, New Mexico, and the Colorado Plateau; and (3) the broad loess mantle that extends from eastern Colorado across the central plains into Indiana, fingering down the Mississippi valley into Louisiana.

The Sandhills of Nebraska and adjacent South Dakota cover an area of 52,000 sq. km. Three generations of aeolian modeling are indicated (H. T. U. Smith, 1967). The sand itself was probably derived from extensive fluvial sands, later reworked by wind. The oldest dunes are of compound transverse type. These are most universal, reflecting a period of general aridity with little or no vegetation. Effective winds were northerly. The second generation indicates northwesterly or westerly winds, and a number of occurrences of parabolic dunes were superimposed upon the first. This would suggest more restricted deflation with a moderate vegetation cover. The third generation of wind activity is recorded by local blowouts, generally attributed to the mid-Recent. The two oldest generations of dunes are thought to be of late Pleistocene age, but recent pollen spectra and macrobotanical evidence from the northern margins of the Sandhills shows clearly that a spruce woodland *(Picea glauca)* prevailed during part of the late Woodfordian (Watts and Wright, 1966). About 12,500 B.P. this forest was abruptly replaced by grassland, although some pine *(Pinus ponderosa)* grew along the river valleys. The parabolic dunes, which Smith (1967) assigns to the late Wisconsin, therfore reflect restricted areas of deflation in a grassland environment during Two Creeks (and Valders?) time. The older, transverse fields indicate a much longer period of general and more pronounced aridity, prior to the coniferous woodlands and paleosol of the Woodfordian substage. Aeolian activity may have been sufficiently prominent at several possible times, e.g., the Sangamon Interglacial or the mid-Wisconsin interstadials.

The aeolian silts and sands of the southern High Plains can also be related to pollen sequences and lacustrine deposits. Reeves (1965) recognizes three periods of aeolian activity in western Texas and eastern New Mexico. The youngest involves minor dunes of Recent age. The next was responsible for major dunes that postdate lacustrine deposits dated between 17,400 and "greater than" 37,000 B.P. (Reeves and Parry, 1965). Since a pine parkland with some spruce covered these plains from about 24,000 to 14,000 B.P., and was succeeded by grassland and other open vegetation (Hafsten, 1961), the second dune generation must be terminal Wisconsin (see also Wendorf *et al.,* 1961). The oldest dunes may reflect a period of open vegetation ca. 33,400-24,000 B.P., or during early Wisconsin times and perhaps even before. Similar temporal precisions seem to apply to other inactive aeolian deposits of New Mexico and Arizona, and of Colorado, Oklahoma, and western Kansas. On the Colorado Plateau, now deeply weathered silts of aeolian origin rest on alluvial fills predating 12,000 B.P. (see Kottlowski *et al.,* 1965). These aeolian mantles were probably deflated from large, silt-laden rivers and need not imply an open vegetation. Widespread dune activity in Recent times suggests a drier climate.

The loess of the central plains (see also chapter 12) was accumulated at various times during the Pleistocene. Unlike the aeolian deposits further west, the great bulk of the loess—the Peoria Loess—is generally correlated with the Woodfordian (see Frye *et al.,* 1968; Reed *et al.,* 1965), although the youngest phases may correlate with renewed aeolian activity in the Sandhills (see Lugn, 1968). Its thickness almost generally exceeds 2 m. and in the major areas of accumulation averages over 5 m. The loess was derived from different sources and source regions, including the Nebraska Sandhills, but the great mass of aeolian dust was deflated from glacial meltwater streams and from river beds in general. This can be inferred from the variable mineralogy from one river system to the next, and from the maximum loess thickness found on the eastern bluffs above each stream valley, that is, downwind. Sedimentary evidence and molluscan faunas (see Taylor, 1965; also Schultz and Frye, 1968) leave little doubt that most of the loess country was moderately dry, probably with an open or parkland vegetation, at the time of aeolian accumulation. In fact the only major loess region lying within the modern zone of closed woodland is the Mississippi loess downstream of the Ohio confluence. Its localization suggests that this loess belt was not related to a regional environment, but to the abnormal abundance of suitable silts on the Mississippi floodplain (see Leighton and Willman, 1950).

All in all, the late Pleistocene aeolian deposits of the interior plains, except for the lower Mississippi valley, can best be interpreted by an open environment, with no closed woodland except in certain riverine situations. However, the aeolian deposits of the eastern Rocky Mountain foothills and the Sandhill country are not contemporaneous with the bulk of the loess. This restricts speculation about an arid climate in the central United States. Hydrological balance and density of vegetation cover in the loess belt were *not* significantly different during the Woodfordian substage than they were in historical times. However, the now semiarid western High Plains *were* drier than today during the Farmdale and Two Creeks interstadials, and possibly also during the Valders.

PERIGLACIAL PHENOMENA

The distribution of late Pleistocene periglacial phenomena in North America is incompletely understood. Ice wedges have been reported or described from nine states (Fig. 57): Rhode Island, New Jersey, central Indiana, northern Illinois, western Wisconsin, northeastern Iowa, south-central South Dakota, Montana, and Washington (H.T.U. Smith, 1962, with references; Brunnschweiler, 1964; Black, 1965; Wayne, 1967). As an indicator of past permafrost conditions, these scattered occurrences suggest that permafrost was restricted to ice proximity and to comparatively high latitudes. The southernmost extent of these features is limited by the modern 24° C. July isotherm, suggesting that a greater extension of permafrost was inhibited by summer warmth. Although the age of these ice wedges differs, with many apparently dating from the late Altonian ice advance (see Black, 1965), they do—as a group—reflect conditions characteristic of each Wisconsin glacial maximum.

Other periglacial phenomena have been recorded primarily from the Appalachian highlands, the northern Rocky Mountains, and the Cascades (see Smith, 1949, 1962; Brunnschweiler, 1964). These include blockstreams, solifluction forms, patterned ground, and allied features best developed in rough country. Involutions and patterned ground are apparently scarce beyond the former ice perimeter (see Wayne, 1967).

The available evidence, despite its fragmentary nature, leaves little doubt that there was no broad belt of permafrost ahead of the ice front, except possibly in Montana and the Dakotas. Other, minor cold-climate features in the mountains can be readily attributed to accelerated frost-weathering and soil-frost activity. The data is quite insufficient to delineate altitudinal belts with specific periglacial forms.

PALYNOLOGICAL DATA FOR FULL GLACIAL CLIMATES

The broad patterns of late Wisconsin vegetation in the unglaciated regions of the United States are only now beginning to emerge. Radiocarbon-dated pollen cores are constantly providing new pollen spectra, from the Two Creeks and Woodfordian, so that factual data are rapidly replacing the endless argumentation of the older literature. Present

Table 17. Pollen spectra and inferred vegetation of the continental United States (23,000–12,800 B.P.). (Cyperaceae excluded from NAP percentages.)

Locality	*Percentage NAP*	*Dominant Tree(s)*	*Vegetation*	*Source*
Pacific Northwest				
Humptulips, Wash.	0–25	Pine	Pine woodland	Heusser, 1965
Onion Flats, Ore.	0	Pine, fir, spruce	Coniferous forest	Heusser, 1965
Great Plains-Great Lakes				
Rosebud, Neb.	15	Spruce	Spruce woodland	Watts & Wright, 1966
McCullock Bog, Ia.	5	Fir, spruce	Fir-spruce forest	Brush, 1967
Colo Bog, Ia.	5	Fir, spruce	Fir-spruce forest	Brush, 1967
Weber Lake, Minn.	35–55	Spruce	Forest-tundra	Fries, 1962
Kotiranta Lake, Minn.	(dominant)	Spruce	Forest-tundra (?)	Cushing, 1967
North Branch, Minn.	20–30	Spruce	Spruce woodland	Fries *et al.*, 1962
Kirchner Marsh, Minn.	10–20	Spruce	Spruce parkland	Wright *et al.*, 1963
Norwood, Minn.	(dominant)		Forest-tundra (?)	Watts, 1967
Madelia, Minn.	45–85	Spruce	Forest-tundra	Jelgersma, 1962
Vandalia, Ill.	50–60	Pine, spruce	Spruce-pine parkland	Grüger, 1970
Northeast				
Chester Co., Pa.	50–75	Pine, spruce	Forest-tundra	Martin, 1958b
Kings Point, N.Y.	15–30	Pine, spruce	Spruce-pine parkland	Sirkin, 1967
Totoket, Conn.	25	Pine, spruce	Spruce-pine forest	Leopold, 1956
Rogers Lake, R.I.	15	Pine, spruce	Spruce-pine forest	Davis & Deevey, 1964
Southwest				
Searles Lake, Calif.	30–40	Pine, juniper	Juniper-pine parkland	Roosma, 1958
Tule Springs, Nev.	10–35	Pine	Pine parkland	Mehringer, 1965
Potato Lake, Ariz.	20–30	Spruce, pine	Subalpine spruce woodland	Martin and Mehringer, 1965
Laguna Salada, Ariz.	5–25	Pine, spruce	Spruce-pine forest	Hevly, 1964
Willcox Playa, Ariz.	0	Pine	Pine woodland	Martin, 1963
Dead Man Lake, N.M.	40–50	Pine, spruce	Subalpine spruce parkland	Bent & Wright, 1963
San Agustin, N.M.	0	Pine, spruce	Spruce-pine forest	Clisby *et al.*, 1957
San Juan Mtns., Colo.	40–70	Pine	Alpine	Maher, 1961
Rich Lake, Texas	0–45	Pine, spruce	Pine parkland	Hafsten, 1961
Crane Lake, Texas	10–35	Pine	Pine parkland	Hafsten, 1961
Southeast				
Cartersville, Ga.	?	Pine, spruce	Spruce-pine forest	W.A. Watts, unpub.
Singletary Lake, N.C.	20	Pine	Pine forest	Frey, 1953
Bladen Co., N.C.	5–10	Pine	Pine forest	Frey, 1953
Chesapeake Bay, Va.	15–25	Pine, spruce, fir	Spruce-pine forest	Harrison *et al.*, 1965

evidence, insofar as dated, is summarized by Table 17. Cyperaceae, dominantly represented by aquatic plants in the lake or marsh sites that are usually cored, have been excluded from the NAP percentages. Distinctive aquatics, such as *Typha,* are also omitted. Only three of the sites were located at or above the alpine tree limit during the Woodfordian: Molas Lake in the San Juan Mountains of Colorado (3,200 m.); Potato Lake, Arizona (2,340 m.); and Dead Man Lake, New Mexico (2,640 m.); the remaining spectra reflect primarily on vegetation belts of more general distribution.

Two major conclusions can be drawn from the currently available data: (a) there is no evidence of a tundra belt adjacent to the continental glaciers; and (b) there is no evidence of a mixed deciduous or deciduous forest in the southeastern United States.

Coniferous forests appear to have grown to the edge of the ice sheet in the Pacific Northwest, although parklands may possibly have been found on the drier Columbia Plateau, which has a loess mantle. Unfortunately there are as yet no records from the northern Rocky Mountains and the northern High Plains. There may have been subalpine or subarctic parklands through much of this country, as suggested by Fig. 58, although this inference must remain speculative. The Rosebud site from the Nebraska Sandhills need not indicate closed spruce woodlands for the entire region; instead, parkland may have been dictated by the great soil-moisture contrasts to be expected within that mosaic of undulating terrain. Only in Minnesota is there firm evidence for extensive tracts of parkland with dwarf-shrub communities, suggesting a forest-tundra. A host of tundra plants are indicated by the macrobotanical evidence (Watts, 1967), so that there can be no question about the floristic elements. But the poor drainage of these sites in late Wisconsin times, as well as a rather high incidence of Cyperaceae, suggests that the tundra patches within this forest-tundra mosiac were primarily a result of edaphic rather than climatic factors. Further south and east, open spruce forests bordered the ice sheet well into the Appalachian country. This can be inferred indirectly from Two Creeks profiles (see Davis, 1967) and from paleobotanical materials of full glacial age (Wayne, 1967; Grüger, 1970; Forsyth, 1965). On the east coast, mixed coniferous forests, with some parkland in Pennsylvania and New York, can also be traced along the ice front.

In more southerly latitudes, the semiarid regions of the southwestern United States were generally stocked with parkland or closed forest, mainly pine, but with much juniper in southern California and appreciable spruce at higher elevations. This shift of vegetation belts represents a vertical downward displacement of the temperate and subalpine

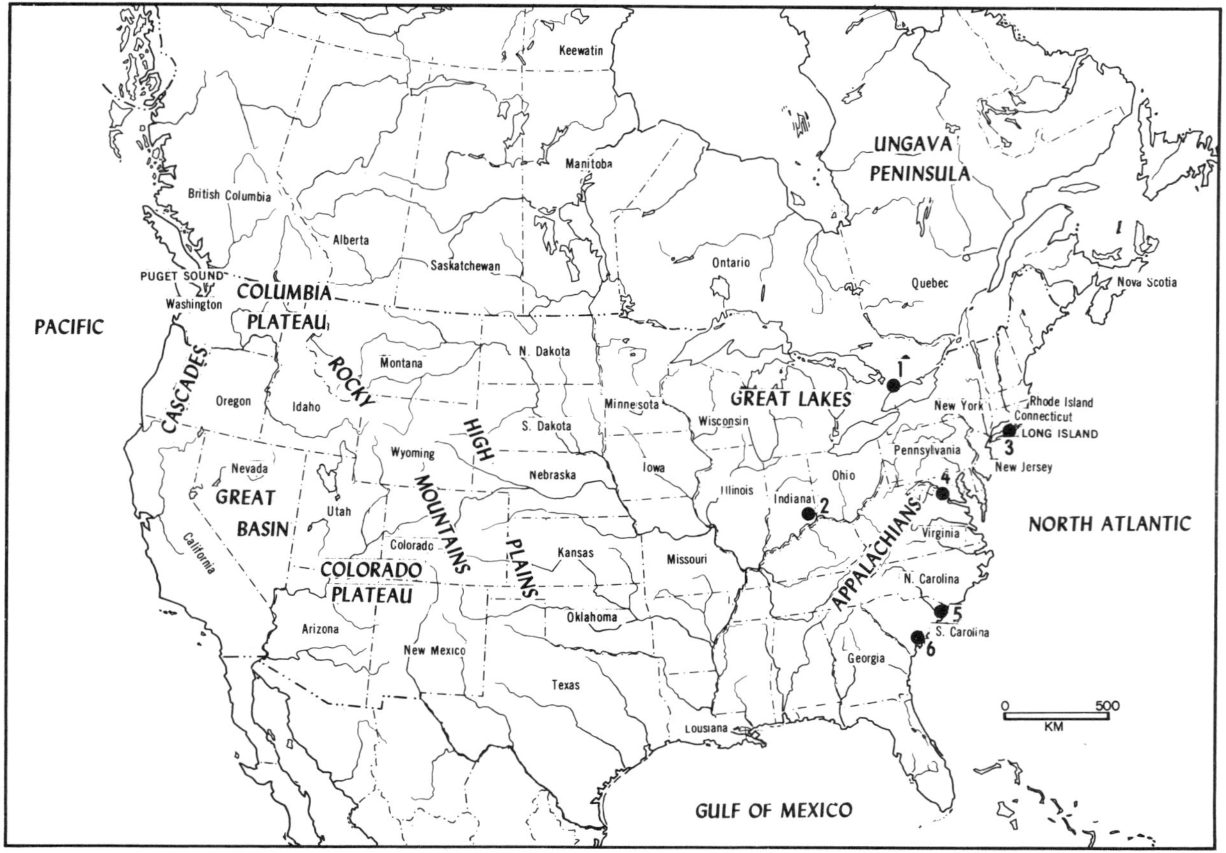

Figure 58. Locational map of part of North America. Interglacial pollen sites: (1) Toronto, (2) Richmond, Ind., (3) eastern Long Island, (4) Washington, D.C., (5) Myrtle Beach, S.C., (6) Charleston, S.C.

woodlands by 900 to 1,200 m. Further east, mixed coniferous forests are indicated for the southern states (see Whitehead, 1967). Their exact composition remains controversial, but there is certainly no evidence for warm-temperate woodlands. The small percentages of hardwood pollen found in most profiles of the eastern United States find enigmatic parallels in the full glacial spectra of tundras in Europe. They may be derived from reworking of older beds or by long-distance transport from small refugia.

The Two Creeks warm-up had a drastic effect in the Southwest, where open vegetation of a semiarid nature replaced many woodlands (Martin and Mehringer, 1965). A limited and temporary readvance of the forests in Valders time is recorded only in Texas. A similar, drastic recession of woodland *ca.* 12,500 B.P. is apparent in Nebraska. On the other hand, in Minnesota and the lower Great Lakes region, spruce forests were almost universal during the Two Creeks and the Valders. Further east, as well as in the Pacific Northwest, the terminal Wisconsin is recorded by nothing more than minor changes of forest composition.

Hardwoods, initially represented by genera such as birch, alder, or ash *(Fraxinus)*, but rapidly followed by oak and other thermophile trees, assume importance in almost all Two Creeks profiles. In the southeastern United States hardwoods rapidly replaced conifers ca. 12,500 B.P. (see Whitehead, 1967, with references), while in the southern Great Lakes region most pollen profiles of Two Creeks age have at least 25 per cent hardwood pollen. The rather sudden and ubiquitous appearance of appreciable pollen from thermophile hardwoods between about 12,500 and 12,000 B.P. suggests rapid dispersal from a number of small refugia in the southern United States. Hardwoods assumed a dominant role in most forests of the eastern United States during the initial millenia of the Recent, and there have been few systematic shifts of vegetation patterns until historical times.

Late Wisconsin vegetation of the nonglaciated parts of Alaska is discussed in chapter 29.

FAUNAL EVIDENCE

In general, faunal data has been of limited value in reconstructing past environments. There is abundant evidence that greater Pleistocene cold or aridity have affected the range and dispersal of such orders as birds, amphibians, reptiles, fish, insects, and mollusca (see Wright and Frey, 1965). A number of disjunct distributions have been explained by rapid climatic change, as, for example, cold forms isolated in now temperate environments or moisture-loving forms separated by nonpopulated semiarid zones. Speciation has also been ascribed to isolation of communities

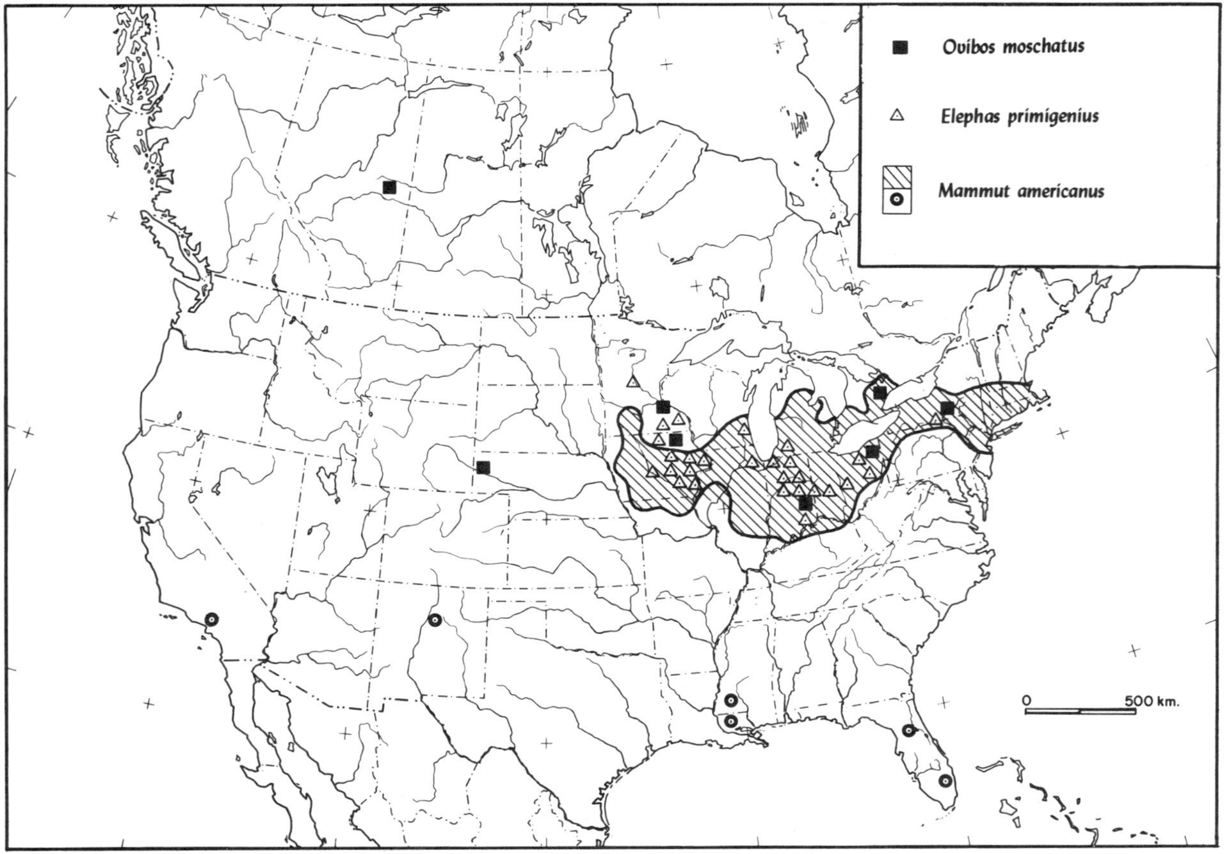

Figure 59. Distribution of musk ox, woolly mammoth and mastodon during late Wisconsin and post-Wisconsin time (compiled by William Bird). The woolly mammoth was extinct ca. 7500 B.P., mastodons by 6000 B.P.

in response to the Pleistocene glaciations or to various intensifications of aridity postulated from the geological record. However, as an independent category of evidence, only the molluscan assemblages promise to provide primary environmental information (see Taylor, 1965).

Mammalian assemblages in North America are more informative than those of the African continent, but less so than those of Eurasia. Several cold-tolerant species migrated from Asia into North America during one or more of the glacial periods, and the Rancholabrean mammalian fauna (see Table 16) has 23 such immigrants (Repenning, 1967). These include musk ox, woolly mammoth, reindeer or caribou, elk, bison, and saiga antelope, but not the woolly rhino. The late Wisconsin and early Recent distribution of a number of large mammals is given by Fig. 59, each species believed to have had a certain preference for cool or open environments. The musk ox, considered a specific tundra form, has been found only from ice proximity or in what were cold parkland or grassland environments. The woolly mammoth, believed to be a cold-steppe form that commonly ranged widely through tundra, forest-tundra and even cold woodland environments, had a distribution similar to that of the musk ox. Other mammoths, such as *Elephas (Mammuthus) columbi* and *E. imperator,* had different ecological adaptations and have been primarily recovered from lower latitudes. The mastodon *(Mammut americanus)* is generally classified as a forest browser and had a far wider range, but fossils are remarkably concentrated in the same zone as the woolly mammoth. Perhaps the availability of suitable sediments and intensity of study are also reflected in Fig. 59. Nonetheless, distribution maps for other species would probably help elucidate environmental patterns of the late Wisconsin.

ENVIRONMENTAL CONDITIONS DURING THE INTERSTADIALS

Little or no systematic information is available for the interstadials that can be recognized within the Wisconsin Glacial. Attention, so far, has been directed primarily toward the recognition and interregional correlation of weathering horizons and allied phenomena. As a result, the best-understood features of interstadial age are the paleosol horizons that separate loess or till sheets. These have been described as peat soils or as simple, humified horizons with limited oxidation or leaching, i.e., soils of *AC*-type (see Ruhe, 1968, and Forsyth, 1965). The majority of these paleosols refer to the Farmdale Interstadial; similarities with the Paudorf soil horizons of Europe are apparent.

The vegetation of the Wisconsin interstadials can be only roughly inferred from a handful of pollen profiles. In Canada the Port Talbot Interstadial was characterized by coniferous forests: pine, spruce, and larch in southwestern Ontario (Dreimanis, 1960), with alder scrub and, at a later point, spruce in Nova Scotia (Livingstone, 1968). In Ohio a peat horizons with dates ranging from "greater than 50,000" to 22,430 B.P. shows high percentages of oak, with some poplar, elm, alder, and grasses (Forsyth, 1965). In North Carolina a decline of coniferous species with a concomitant increase of thermophile hardwoods is indicated for the Farmdale and possibly also for earlier interstadials (see Whitehead, 1965, 1967; Frey, 1953). In the western United States, the Wisconsin interstadials were noted for a more open vegetation (see Hafsten, 1961; Martin and Mehringer, 1965) or lower lake levels (Morrison and Frye, 1965). From such inadequate data it is impossible to derive firm conclusions; nonetheless, the Farmdale and possibly also the Port Talbot interstadials may have approximated conditions typical of the Two Creeks interval.

FULL GLACIAL TEMPERATURES

The best estimates of Wisconsin-age temperatures are provided by the detailed work of Black (1965) on ice wedges in Wisconsin. The well-developed networks of primary and secondary ice-wedge polygons in the western part of that state indicate continuous permafrost which, by modern analogs, requires a mean annual temperature of less than −5° C. This infers a temperature depression of at least 12.5° C. In southern Wisconsin, the less well-developed polygonal networks and isolated wedges suggest discontinuous permafrost, with a mean annual temperature slightly warmer than −5° C., i.e., a depression of about 12° C. These values apply essentially to the late Altonian ("Rockian") glacial advance, but must also have been broadly applicable to the Woodfordian.

Other permafrost occurrences are inferred from scattered ice wedges and are less certain in their interpretation. However, assuming that mean annual temperatures were no higher than −2° C. in areas of discontinuous permafrost, a temperature depression of *at least* 12° C. must be assumed for Rhode Island, 13.5° C. for New Jersey, 14.5° C. for central Indiana, 12° C. for northern Illinois, 11° C. for Iowa and South Dakota, and 12° C. for Washington state. On these grounds it appears that mean annual temperatures in the northern third of the United States were at least 12° C. lower during the coldest parts of the Wisconsin Glacial.

Since the polar tree limit was at or near the ice front, it was essentially determined by the presence of the ice sheets and therefore provides no paleoclimatic information, other than that mean July temperatures within close proximity of the ice were probably at least on the order of 10° to 12° C., compared with about 24° C. today. This could be interpreted in terms of summer and winter temperature depressions of approximately equal magnitude. However, by relating present altitudinal temperature gradients to glaciation phenomena and snowline depressions, Richmond (1965) concludes that mean summer temperatures were about 8° C. lower in the southern, and 9° C. in the northern Rocky Mountains. This would appear to indicate that, as in Europe, winter temperatures were depressed more than summer temperatures, suggesting an even greater continentality than at present.

Applying these deductions to a few representative situations, locations such as Rhode Island, New Jersey, and central Indiana or Washington would have experienced full glacial temperatures as follows: mean annual temperatures in the – 2° to – 5° C. (23–28° F.) range, July means in the 10° to 15° C. (50–59° F.) range, and January means in the order of −14° C. (0° F.). Since each area was sufficiently moist to support woodland (see Table 17), a Koeppen *Dc* climate can be inferred, probably similar to climates now prevailing at latitudes 10° further north, in the boreal forest belt of Quebec and Ontario. There is a broad similarity between these climate parameters and Würm-age conditions in southern Russia, and it seems that the more southerly terminus of the ice sheets in North America eliminated the broad belt of *ET* tundra climates so conspicuous in Europe. Presumably the advance of the Laurentide and Cordilleran ice sheets was eventually brought to a halt by high summer temperatures, in a zone where July temperatures now average near 24° C. (74° F.). On the other hand, the advance of the Laurentide ice to latitude 39° N. must be attributed to the great abundance of moist and relatively warm air moving northwards from the Gulf of Mexico. These maritime-tropical airmasses would be drawn into cyclonic disturbances, accelerated and intensified by cold Canadian air, thus contributing large quantities of snow in the central and eastern United States. In addition, the topographic relief of the advancing ice front—greater than that of the Appalachian Mountains—would favor orographic intensification of precipitation.

No reliable temperature estimates can be made for the southern parts of the United States.

FULL GLACIAL MOISTURE CONDITIONS

The impression obtains that there was more abundant moisture during the colder parts of the Wisconsin Glacial. So, for example, lake levels were higher in the Great Basin, and forests were far more extensive in the southwestern United States, while fluvial activity was accelerated and stream competence greater in the Missouri-Mississippi-Ohio system. In each instance, however, there is no need to assume greater precipitation. Expanded lakes and forests in the semiarid regions of the

United States can be adequately explained by a general lowering of temperature on the order of 9° to 12° C. In fact, as Galloway (1970) points out, there is sufficient latitude, as well as some evidence, to suggest a certain decrease in actual precipitation. Elsewhere, most of the massive and coarse-grade alluvial fills reflect meltwater discharge emanating from ice sheets and major mountain ice caps. Rivers such as the Ohio or Columbia drained great volumes of glacial meltwaters during the warmer part of the year, dwindling abruptly in the winter.

There is no way to assess actual precipitation amounts or regimes for the glacial maxima or for the interstadials of the Wisconsin. It would seem, however, that the overall hydrological balance was similar to that of today—or even on the dry side—during the interstadials, while the water balance was generally more favorable during periods of maximum cold.

SUMMARY OF CLIMATIC ZONATION DURING THE WISCONSIN GLACIAL MAXIMA

In overview, available information allows a number of tentative conclusions concerning climatic distributions in North America at the maximum of the Wisconsin Glacial (see Figs. 56 and 57).

a) Glaciers covered almost all of Canada, parts of Alaska, and the northern periphery of the continental United States. This immense area of coalescent ice sheets and ice caps exceeded Antarctica in size and may have rivaled it as a climatic factor: a semipermanent, cold anticyclone in the lower atmosphere and a source of almost unlimited cold, continental airmasses, at least during winter. In the Koeppen classification these regions must have had *EF* frost climates and, except for the southern ice terminus, July mean temperatures were probably under 0° C.

b) A belt of tundra climate *(ET)* is not verified at low latitudes, except possibly for a very narrow belt ahead of the ice front where summer temperature gradients, affected by the secondary cooling of the ice mass, must have been considerable. A more typical tundra climate can be deduced for western Alaska and Beringia (see chapter 29).

c) A zone of parkland with permafrost—narrow or even nonexistent in the eastern United States—followed the ice front west of the Great Lakes. This belt may have included broad areas of the northern Great Plains, e.g., the Dakotas and Montana. Parklands or montane grasslands without permafrost extended southward along the Rocky Mountain piedmont and may have been extensive on the high ground of the cordilleran belt in Wyoming, Idaho, Washington, and eastern Oregon. The north-south variation of climate in this belt may have been similar to that found in Alberta today. Unfortunately little is known about Wisconsin environments in this part of the United States.

d) The interior loess belt, centered in northern Kansas and Missouri, Iowa and Illinois, is still difficult to interpret with confidence in default of pollen profiles. Parklands or grasslands, with some riverine woodland, seem probable. Climate was cold and presumably semiarid *(BSk?)*. Except for the absence of permafrost, the contemporary loess steppes and parkland of southern Russia, also located well south of the arctic tree-line (Frenzel, 1968a, 1968b), suggest a close analogy.

e) In the eastern United States, closed, coniferous woodlands with a subarctic climate, probably Koeppen's *Dc* type, stretched across the Ohio valley through to the mid-Atlantic states, possibly fingering southward along the Appalachian ranges. Warmer coniferous woodlands, with small refugia of mixed hardwoods in warmer locales, probably covered the southeastern United States which may have enjoyed a *Db* climate. Similar conditions can be postulated for the southwestern United States, where spruce and pine forests mantled the uplands *(Db?)* and parklands were characteristic of lower ground *(Db* or *BSk)*.

TWENTY-TWO

Some Paleoclimatic Problems of the Interglacials

INTRODUCTION

Considering that high latitude glaciers were not "normal" during the greater part of geological history, the contemporary glaciation of Greenland and Antarctica implies that the Holocene is a comparatively cool era of geological history. In fact, the Holocene is presumably nothing else but an interglacial period. For obvious reasons the Pleistocene interglacials are thought to have had a climatic character similar to that of the Holocene. However, there are differences involved. For one, the time intervals are quite different. Marine stratigraphy (ch. 3) suggests that the Eem lasted about 40,000 years, compared with 10,000 years for the Holocene so far.

Within the broad pattern of physical conditions that recall those of the Holocene, the various interglacials may conceivably have experienced phases of climate warmer or moister, as well as cooler or drier, than the average for the Holocene. Judging by the evidence, such deviations did occur and may have had considerable ecological significance. The following chapter is devoted to a brief discussion of phenomena suggesting the existence of "anomalous" interglacial climates. It is not intended to be comprehensive, since the evidence does not yet warrant such treatment.

GEOMORPHOLOGIC EVIDENCE FROM EUROPE AND AFRICA

Geomorphic processes were analogous to those of the present during the Pleistocene interglacials of both Europe and the Mediterranean area.

Both erosion and deposition were quite limited, with soil development proceeding with little interruption.

In the Saharan area little definite information is available about the interglacial intervals. The great fossil dune fields of southern Algeria and Tunisia were supposedly activated during especially dry interglacial periods (see Tricart and Cailleux, 1960–61), although no exact dating is available. In fact the occurrence of fossil dunes in much of the completely arid Sahara is rather perplexing. Possibly the pluvial phases provided large expanses of wadi and lacustrine sands that were later available for deflation under quite arid conditions. They were finally fixed by soil development under moister conditions. Consequently, dunes in true desert country do not necessarily indicate exceptionally arid conditions. Another problem of the Saharan area is the broad belt of fossil dunes extending through the Sudanese grassland belt between the White Nile and the Atlantic Ocean (see Grove and Warren, 1968). These dunes must be explained by greater aridity. But the age of the dunal complex, which has apparently been subjected to repeated deflation and deposition, is primarily glacial (see chapter 20), so that there is no sound evidence to assign any major period of aeolian activity to interglacial stages.

Although evidence of abnormally dry interglacial climates is not proven in the case of East Africa (Flint, 1959), southern Africa promises to be more rewarding in terms of "abnormal" interglacial deposits (Flint and Bond, 1968; Bond, 1957, 1963). Of particular interest here are wind-blown sands derived from the late Tertiary Kalahari beds. These sands are largely under vegetation today, but on account of their arid subsoil environment they respond rapidly to a rainfall decrease by active deflation. Such sands form a useful index of greater aridity when found in stratigraphic context in nearby areas (e.g., the Rhodesias, Transvaal). A part, but certainly not all (see chapter 20), of these "Kalahari sands" appears to be of interglacial age, but dating is still comparatively uncertain (Grove, 1969).

More informative than continental sedimentation are the marine deposits associated with high interglacial sea levels. Thermophile molluscan faunas, such as those formerly found in the North Sea or Mediterranean Sea, are useful as an index of greater warmth. On this evidence, parts of the Eem interglacial must have been warmer than at present. The controversial O^{18}/O^{16} isotope measurements of various deep-sea cores of the North Atlantic and eastern Mediterranean (see Emiliani, 1955a, 1964, 1966; Rona and Emiliani, 1969) may also suggest that surface water temperatures were a little warmer, perhaps by 1° or 2° C., at the thermal maximum of the last interglacial.

Contrary to a once widely held opinion, high interglacial sea levels do not by themselves establish the existence of warmer climates by arguing for a greater melting of the residual ice caps and mountain glaciers. Short, comparatively important oscillations about the interglacial "mean," such as are evident on the island of Mallorca (Butzer and Cuerda, 1962a), must, however, be attributed to glacio-eustasy.

THE EVIDENCE OF PALEOSOLS IN EUROPE AND AFRICA

African, European, and Mediterranean paleosols dating from the Holstein and Eem suggest somewhat different conditions of soil development. Profiles are invariably deeper, implying a considerably greater duration of soil development. Also the degree of chemical alteration is slightly greater, suggesting greater intensity of weathering than during the Holocene. The great profile depth of Holstein or Eem soils can be particularly well observed in the case of buried soils, for which further soil development after burial is more or less excluded. For example, soils developed on Riss moraines or terraces are often buried under Würm moraines or loess, providing comparable soil profiles of Eem and Holocene age.

To illustrate the differential of soil development a few cases can be cited: *(a)* In northern Italy the *in situ (B)*-horizon of buried Holstein soils which developed on morainic deposits varies between 4 and 6 m. in depth; for the Eem the depth is 120–150 cm.; for the Holocene, 50–90 cm. (Fränzle, 1965). The local climax soil of the interglacials is a braunlehm or rotlehm, of the Holocene a podsolic braunerde. *(b)* In Catalonia the *in situ (B)*-horizon of buried Holstein soils which developed in river terrace deposits varies between 1 and 5 m. in depth, with red colors (2.5 YR on the Munsell scale); in the case of Eem soils this value is 60–100 cm. with reddish-yellow colors dominant (5 YR); in the case of Holocene soils a comparatively weak soil profile of only 30–50 cm. depth and brownish color (10 YR) is developed. The Holstein climax soil approaches a rotlehm, the Eem a braunlehm, the Holocene a meridional braunerde (Butzer, 1964a). Numerous examples can also be cited from temperate Europe where deep braunlehms were particularly characteristic for the interglacials, compared with the more modest Holocene climax soil development. None of these features can be entirely explained by a greater duration of soil development.

Of particular interest are the ecological conditions associated with the development of the terra rossa or rotlehm paleosols of the Mediterranean region. Approximately comparable soils today appear to be developing only in the savanna belts with high temperatures and noticeable periodicity of rainfall (see Kubiena, 1957). The intensity of chemical weathering suggested is much greater than that possible under modern climatic conditions in the area. The dry summer must have persisted, but the transitional seasons must somehow have provided an optimal combination of warmth and moisture. Consequently warm, moist phases—stratigraphically of interglacial age (see reference in ch. 6)—were at times characteristic of the Mediterranean region. These were not accompanied by alluviation, presumably on account of a very complete and luxuriant vegetative mat with a subtropical rainforest rather than an open subtropical woodland (Durand, 1959). The striking lack of geomorphic evidence makes the term "pluvial" seem rather inappropriate. It is important to realize that present climatic conditions do not represent the totality of interglacial climates in the Mediterranean region. Rather, there have also been very warm, seasonally humid, "tropical" interglacial climates. The contemporary climate suggests a subhumid, comparatively "temperate" interglacial. Just as there were both cool-moist (classical "pluvial") and cool-dry phases during the Mediterranean glacial age record, there also were warm-moist and warm-dry phases during the interglacials.

In the Saharan area, red paleosols of interglacial age were already mentioned in chapter 20. Kubiena (1955, 1957) also describes fossil and relict braunlehms from the Canary Islands and the Hoggar Mountains, where they too presumably are of interglacial age. Unfortunately too little is known of paleosols in tropical Africa.

In the case of Europe, one might presume that a moderately warmer climate during a part of certain interglacials would have sufficed to produce the intensity of weathering suggested by the braunlehm soils. However too little is known about soil development to permit estimates of the quantitative differences in climate that were involved; the qualitative conclusions are already of sufficient importance.

Despite a lack of geomorphic evidence, the European, Mediterranean, and Saharan paleosols leave no doubt that ecological conditions were different during some parts of the Holstein and Eem interglacials. Soil development records chemical weathering and is therefore capable of recording changes not reflected in the sedimentary or erosional record. Most difficult to identify are periods of climate like that of the present, since such periods would leave a minimum record of soils and geomorphic features. Such "uneventful" interglacial phases may well have dominated the greater part of both the Holstein and Eem, but within these two intervals, conditions must occasionally have been somewhat different from those of today. This recognition of anomalous interglacial climates through paleosols raises an important problem of interglacial paleoclimates.

PALEOBOTANICAL EVIDENCE FROM MID-LATITUDE EUROPE

Floral information is particularly abundant in the interglacial record of temperate Europe, although comparatively unknown in lower latitudes. Ecological interpretation of the interglacial vegetation known from middle and higher latitude Europe may be attempted on the basis of two different classes of evidence: the Tertiary relicts particularly important among the older Pleistocene flora, and the occasional northward or eastward spread of "modern" warm-temperate species beyond their present ecological boundaries.

At the close of the Pliocene the forest composition of Europe was quite unlike that of the present. Many of the species present are now limited to the tropical rainforest; others are found in the subtropics or in the temperate woodlands of other continents. Still others are extinct. During each cold climate phase, beginning in the Villafranchian, climatic conditions in middle latitudes were evidently unfavorable, thus requiring either a southward migration of thermophile species or their local extinction. Particularly in the case of the Netherlands, the gradual diminution of exotic species as *Liquidambar, Taxodium, Liriodendron, Pseudolarix, Parrotia,* and *Zelkowa* is well understood as a result of painstaking study (see Florschütz and Van Someren, 1950; Van der Vlerk and Florschütz, 1953). The terminal Pliocene Reuver flora contains species of which 79 per cent are now extinct in the Netherlands. The Tiglian flora only has 41 per cent exotic species. At an even later date, the Holstein has only 17 per cent, the Eem 9 per cent. The persistence of many species such as cedar *(Cedrus),* certain pines *(Pinus pithyusa, P. aldarica),* and the lotus in southern Europe shows that there was a progressive southward displacement of the European vegetation belts during the successive Pleistocene interglacials. In large part this can be attributed to the gradual net cooling of higher middle latitudes during the early Pleistocene.

Of particular paleo-environmental interest are the Middle and Upper Pleistocene interglacials. The different distribution of thermophile species can be informative for a study of climatic zonation. So, for example, seven species of the Polish Holstein and eight of the Eem either do not occur or do not reproduce in Poland today, although they thrive in warmer parts of Europe. The map diagrams (Figs. 60 and 61) of the modern and interglacial (Holstein or Eem) distribution of beech *(Fagus silvatica, F. orientalis),* hornbeam *(Carpinus betulus),* and holly *(Ilex aquifolium),* and the Pontic alpine rose *(Rhododendron ponticum)* (see Frenzel, 1968a, and Walter, 1954) are particularly instructive. The beech seems to be roughly delimited by the January isotherm of −2° C. (Walter, 1954, p. 30), although nine Russian localities have provided interglacial evidence for its occurrence in areas with January means as low as − 13° C. Hornbeam has been found at some twenty Russian localities beyond the present limits of the species. The holly, whose northeasterly limits approximately coincide with a January mean of 0° C. or better, with 345 days per year with maximum temperatures over the freezing point (Walter, 1954, p. 29), has been found in fossil form at several localities in eastern Germany (January, − 2° C.) as well as at a site south of Moscow (January, − 11° C.). The yew, which does not occur northeastward of the − 4° C. January isotherm today (Walter, 1954, Fig. 5), has also been found at Moscow (January, − 11° C.). *Rhododendron ponticum,* today confined to the Iberian peninsula, Turkey, and Caucasia, occurs in association with the wild grape *(Vitus silvestris)* at three Alpine localities—Hötting near Innsbruck, Calprino near Lugano, and Re near Locarno (see Woldstedt, 1958, pp. 189–91). The Hötting site is generally considered to be of Holstein age, and implies a mean temperature 3° C. warmer than today's. The Lugano and Locarno sites are apparently of Eem age.

All in all the paleobotanical evidence of the last two interglacials suggests appreciably warmer temperatures. Frenzel (1967, 1968a) has examined the Holstein and Eem ranges of various thermophile trees and plants in Europe and the Soviet Union, and tabulated implied temperature deviations for each species. This data is summarized in Table 18. There can be little doubt that the thermal maximum of both interglacials was marked by appreciably warmer winters in eastern Europe and Siberia, and by slight yet significant increases of summer warmth throughout Europe. In addition, the distribution of moisture-loving species *(Tilia platyphyllos,* holly, hickory) in drier continental environments of eastern Europe suggests a somewhat greater rainfall during the growing season. During most of the Holstein, the forest boundary in Siberia also advanced at the expense of what is now steppe. Frenzel concludes that climate at the maximum of both interglacials was moister, with appreciably milder winters. In other words, European climate was less continental. Such a climatic change would not only help explain the European paleosols but also those of the Mediterranean.

The distribution of vegetation belts in mid-latitude Europe and the Soviet Union is comparatively well understood in its rudimentary outline, thanks to the work of Frenzel (1968a, p. 100 ff.). A reconstruction for the late Holstein Interglacial, based on data from about 75 sites, is given in Fig. 62. The widespread distribution of conifers, particularly fir and, to a lesser extent, spruce, favored a very wide distribution of mixed forests, even in western Europe. Despite a broad resemblance of the pattern of boreal forests, temperate mixed forests, subtropical wood-

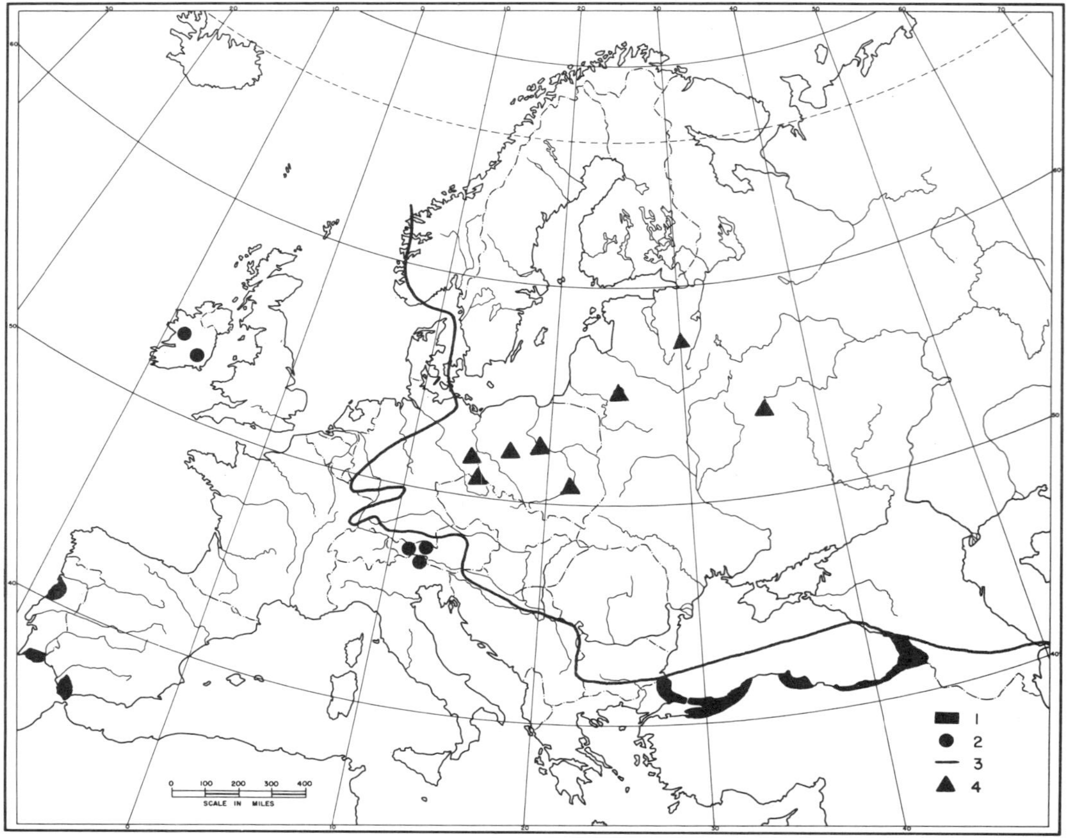

Figure 60. Interglacial shifts of European floras: (1) modern distribution of Rhododendron ponticum, (2) interglacial beds with Rhododendron ponticum, (3) modern northern and eastern limits of holly (11ex aquifolium), (4) interglacial beds with holly (after **Walter***, 1954, and Frenzel, 1968a).*

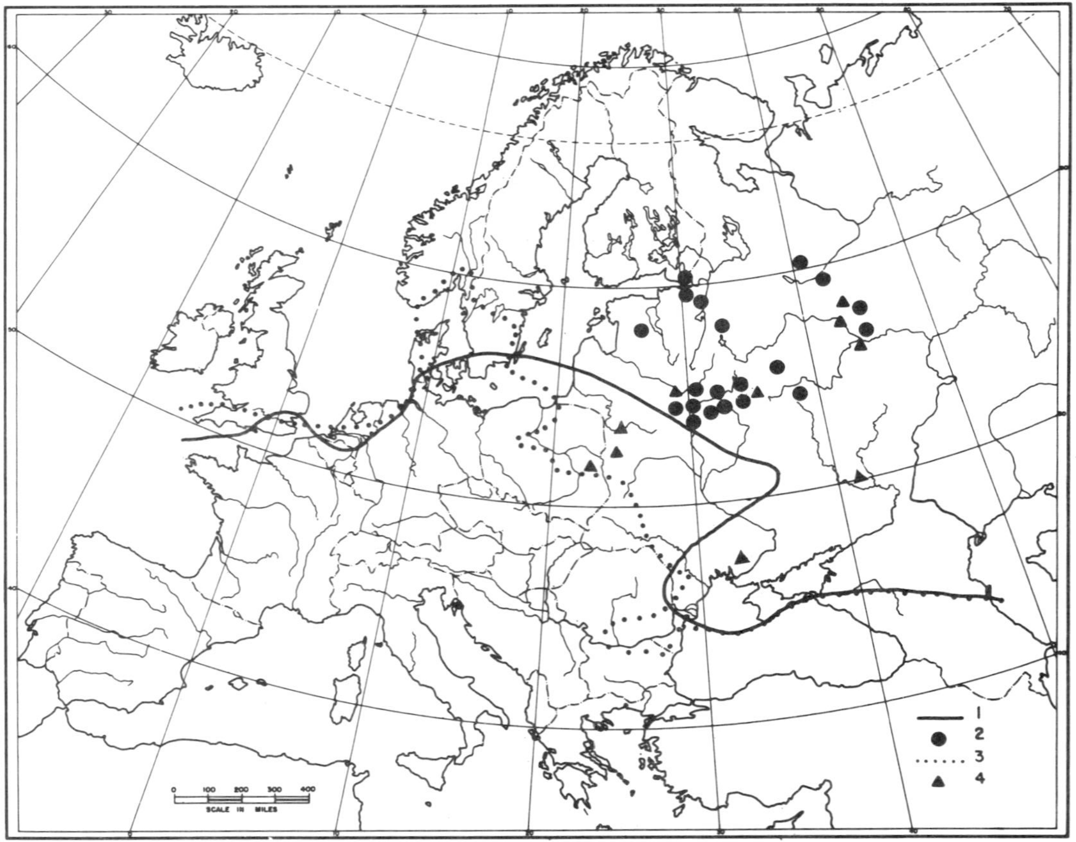

Figure 61. Interglacial shifts of European floras: (1) modern northern limit of hornbeam (Carpinus betulus), (2) interglacial beds with hornbeam, (3) modern northern and eastern limits of beech (Fagus silvatica, F. orientalis), (4) interglacial beds with beech (after Walter, 1954, and Frenzel, 1968a).

lands, and steppe to that of today, there are two readily discernible shifts: (a) subtropical woodlands and rain forests were far more extensive in Italy, the Balkans, and Caucasia, while (b) steppe and forest-steppe were partly replaced by forest in central Russia and western Siberia. Seen in detail, the dominant composition of the forests was unlike that of modern woodlands, even though the kinds of trees represented were essentially modern. Frenzel's (1968a, Plate 4; 1968b) reconstruction of a slightly earlier phase in the forest development of the Eem Interglacial is intermediate between the Holstein and recent patterns: (a) the extent of steppe and forest-steppe compares with that of today; (b) the extent and luxuriance of subtropical woodlands in southern Europe and Caucasia are more restricted than during the Holstein; and (c) the boreal forest realm of northern Europe is smaller than during either the Holstein or the Holocene.

Seen on the temporal plane, the vegetation patterns of the Holstein and Eem were subject to considerable variation, as a result of edaphic changes through time as well as the successional history of forest types. At the end of each glacial, extensive areas of mineral sediment, often disturbed by soil frost and solifluction are available for forest colonization. The earliest immigrants are those nearby, generally cold-tolerant forms that thrive with little shade on little-weathered, calcareous soils. As soil development proceeds, and more and more warmth-loving species arrive on the scene, closed forests develop. The new hardwoods

Table 18. Positive temperature anomalies for the Holstein and Eem interglacials in Eurasia (in °C.) (modified after Frenzel, 1967, 1968a).

Area	*Holstein*			*Eem*		
	January	July	Year	January	July	Year
British Isles	–	2	1	–	–	–
Denmark	1	2	2	1	1	1
Netherlands and Northern Germany	–	2	1	1	3	2
Poland	3	–	1–2	3	3	3
European Russia	8–10	2	5–6	9–10	2	4–6
Western Siberia	20–22(?)	1	11–12(?)	4	3	3

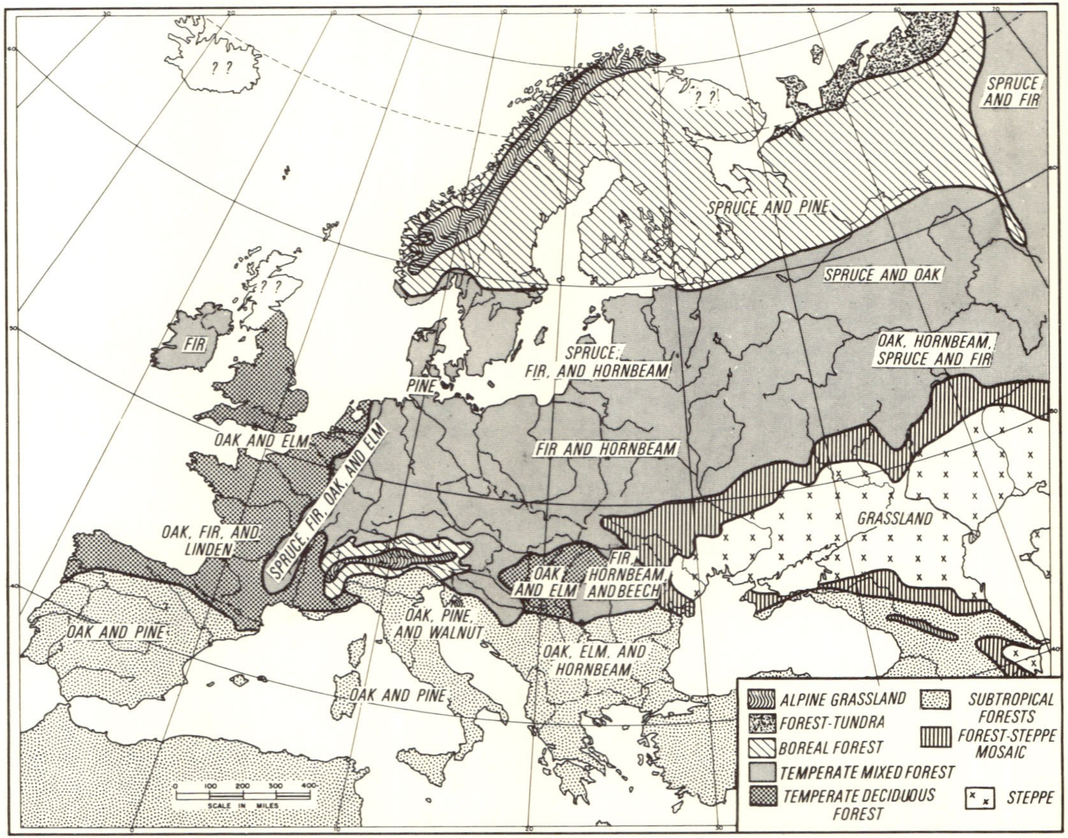

Figure 62. Environmental zonation of Europe during the thermal optimum of the Holstein Interglacial. Simplified and modified after Frenzel (1968a), with corrections in southwestern Europe.

grow up in the shade of heliotrophic species which they ultimately replace. In this way the initial grassland-woodland is replaced by climax forest. Eventually soil acidity increases, favoring forest dominants more tolerant of low base-status soils. Increasing podsolization, with the development of extensive oligotrophic environments, destroys the climax forest. Coupled with falling temperatures towards the close of the interglacial, moors with peat mosses develop and acidophile woodland-moorlands precede the onset of a new glacial. Soil development then almost comes to a halt, and increased mechanical weathering and sedimentation provide fresh soil parent material, while solifluction and cryotubation disrupt existing soil mantles. Finally, an open arctic vegetation dominated by cold-resistant forms is installed.

This picture of a cyclical development of vegetation through time is illustrated by Table 19, a model that is designed to represent conditions in the western half of Europe, and can be applied to successive interglacials from the Cromerian to the Holocene (Turner and West, 1968; Frenzel, 1967, 1968a). The reconstruction of Holstein vegetation in Fig. 62 refers to the Late Temperate phase.

Table 19. Ecological zonation for the European Interglacials (modified after Turner and West, 1968; and Iversen, 1958).

Zone names	*Characteristics of pollen assemblage*	*Soil properties*	*Vegetation type*
Early Glacial	High NAP, few or no thermophile trees	Solifluction	Open arctic-alpine
Post-Temperate	*Pinus, Betula*	Podsol, very acid	Acid woodland-moorland
Late Temperate	*Carpinus, Abies Quercus*	Brown forest soil, slightly acid	Climax forest
Early Temperate	*Quercus, Tilia Ulmus*		
Pre-Temperate	*Betula, Pinus*	Unleached, calcareous soil	Basic/neutral grassland-woodland
Late Glacial	High NAP, few or no thermopile trees	Solifluction	Open arctic-alpine

PALEONTOLOGICAL EVIDENCE FROM MID-LATITUDE EUROPE

The mammalian faunas of the interglacials do not provide much information concerning temperature conditions other than in a very general way. Possibly the specialized interglacial woodland elephant *(Elephas antiquus)* or Merck's rhino would not thrive in the more continental climates of modern Europe. But since these genera are extinct, such speculation is fruitless. One of the few animals of possible interest is the hippopotamus, a warm to warm-temperate species requiring perennial waters, and so unlikely to be found in seasonally frozen rivers. Interglacial hippos have been recorded from southern Europe, France, England, and western Germany (Frenzel, 1968a, Fig. 1). Lack of evidence from Scandinavia or eastern Europe suggests that no radical changes in winter temperatures need be assumed. The extinct water buffalo *(Buffelus murrensis)* from Germany and the monkey *(Macaca sylvana* spp.) found in different parts of temperate Europe during the Holstein may, however, be more suggestive of warmer conditions (Thenius, 1962).

Quite informative are the molluscan faunas. So, for example, occurrences of *Helix pomatia* and *Fruticicola fruticum* in Holstein sediments of Czechoslovakia suggest mediterranean influences and temperatures 3–4° C. higher than today (Ložek, 1964). A different molluscan assemblage, with abundant thermophile species, indicates a 2–4° C. temperature increase for the Eem (Ložek, 1964). This compares with the marine molluscan assemblages from the Netherlands coast, which also suggest water temperatures some 2–3° C. higher (Straaten, 1956).

THE SANGAMON INTERGLACIAL IN NORTH AMERICA

The last, or Sangamon, interglacial in North America is represented by paleosols, swampy deposits, and littoral sands or muds. The wide distribution of paleosols indicates that the Sangamon was essentially a period of geomorphologic equilibrium, as in Europe and possibly also in Africa. The only continental deposits that can be recognized with certainty are the wash and marsh accumulations found in swales of the Illinoian till sheets, and the floodplain deposits of innumerable stream valleys. Littoral phenomena include a variety of erosional and depositional forms, related to repeated marine transgressions, particularly along the coastal plains of Alaska (see Hopkins, 1967, ch. 4) and of the eastern United States (see Flint, 1966; Richards, 1962).

The paleosol of the Sangamon Interglacial can be traced between successive loess mantles or till sheets from southwestern Texas through the Great Plains to Ohio, a stretch of 3,000 km. The soil type varies in zonal progression from reddish semidesert or grassland soils in the southern plains to gray brown or reddish prairie soils in Illinois. In Texas this paleosol has comparatively deep *B* and *Ca*-horizons, more strongly developed than post-Wisconsin soils in the same areas (Frye and Leonard, 1965.) In fact, a westward shift of pine woodlands is indicated for southwestern Kansas on both paleontological and palynological grounds (see Kapp, 1965), apparently during the late Sangamon. In Indiana the profiles of buried soils are up to 2 m. deep in till, and over 3 m. in sands and gravels (Wayne and Zumberge, 1965). The *gumbotil* of Iowa and adjacent areas is an exceptional variant of Sangamon soil, with dense, compact, and poorly drained subsoil horizons, related to impeded drainage in the swales of Illinoian till plains (see Ruhe, 1965). Except where high water tables may have impeded the depth of soil leaching, paleosol profiles are deepest and exhibit slightly more intensive weathering than recent soils.

In the western United States, the Sangamon is represented by red, clayey podsolics (with profile depths of 1.8–2.4 m.) found up to an elevation of 2,400 m. in the Rocky Mountains, and to 3,150 m. in the mountains of Utah. This suggests a warmer and wetter climate since these profiles are three or four times as deep as those of the modest, post-Pleistocene soils (Richmond, 1965). The Sangamon paleosol of the Great Basin is of chestnut or brown soil type, reflecting drier conditions in the lower country (Morrison, 1965).

Swampy deposits are best recorded on top of the Illinoian till plains, and include colluvial, alluvial, and organic components. Of some interest are the so-called accretion-gleys of Illinois and neighboring states (see Frye *et al.,* 1965). These are montmorillonitic clays, up to 2 m. thick, frequently rich in organic matter, and locally perhaps including some aeolian silt. These are not groundwater soils, as the name may suggest, but colluvial wash accumulated slowly in shallow swales. Among the most informative deposits are swampy floodplain beds such as those near Richmond, Ind., and Washington, D.C. (see map, Fig. 59). At the Richmond site a progression of vegetation from coniferous to deciduous forest, and a subsequent return to coniferous woodland (at the beginning of the Wisconsin) is recorded by a long pollen profile (Kapp and Gooding, 1964). The pollen assemblages of the intermediate warm zone suggest slightly warmer conditions: oak and hickory each attain up to 40 per cent of the pollen count, while a number of genera whose present northern limits lie in southern Indiana are well represented. At the site near Washington, initial and terminal cool conditions are indicated by spruce, fir, and pine. The main interglacial is recorded by a former cypress swamp with macrofossils (including great stumps of *Taxodium*) and pollen of oak, hickory, and other warm-deciduous species, suggesting a climate slightly warmer than today (Knox, 1962; see also Grüger, 1970).

Coastal deposits of marine, deltaic, or fluvial facies have provided more than information on high interglacial sea levels. So, for example, a complete pollen profile in marine clays on Long Island records basal horizons of spruce and pine, followed by a zone of oak (up to 50 per cent of pollen), hickory and beech – similar to post-Wisconsin spectra in the area – and concluded by spruce and pine (Donner, 1964). Littoral deposits near Charleston, S.C., indicate pine, oak, and hickory in Sangamon times, suggesting conditions like those of today (Leopold, 1958). However, freshwater beds at Myrtle Bay, S.C., suggest warmer and moister conditions with pollen spectra dominated by oak, hickory, gums and cypress (Frey, 1952). Perhaps the most impressive of all the coastal profiles has been obtained from the deltaic Don Beds, near Toronto, Canada, which record a high (+ 18 m.) level of Lake Ontario (Terasmae, 1960). There are 70 species of trees and herbs, 20 of diatoms, 40 of mollusca, and 6 of mammals recorded here in a unique assemblage. A number of trees and plants with distinctive "southern" distributions (e.g., holly, *Liquidambar, Fraxinus quadrangulata)* indicate a climate at least 2 to 3° C. warmer and also somewhat wetter than today. The Don Beds are overlain by the Scarborough Beds, of Early Wisconsin age, with much improverished biota. The flora indicates a coniferous forest, with temperatures 5° C. colder than at present.

The sum total of pedological and palynological evidence in eastern North America indicates that despite a broad resemblance with the Recent, the Sangamon Interglacial appears to have favored some differences in the dominant vegetation. Until many more pollen profiles are available, these anomalies cannot be interpreted properly. However, the depth of soils and the northward shifts of some tree genera may imply warmer and slightly moister conditions during some part of the Sangamon. The similarity with Eem-age conditions in mid-latitude Europe is apparent. Unfortunately no discussion is warranted for the Yarmouth Interglacial, which is essentially known only through its paleosols.

Information from other parts of North America is too fragmentary for the Sangamon, and there is little systematic information about soil horizons. In the Puget Sound lowlands of Washington, a number of pollen records of apparent interglacial age indicate that the same kinds of trees were present as today, but both lodgepole pine and mountain hemlock were more frequent, while NAP counts were higher (Heusser, 1965). The climate may have been more continental here, with some parkland in evidence. Finally, in western Alaska, the vegetation of the Sangamon Interglacial was similar to that of modern times (Colinvaux, 1967), although a warmer climate is inferred for an earlier interglacial.

RETROSPECT ON INTERGLACIAL CLIMATES

In a general way, the paleoclimates of the Pleistocene interglacials were approximately analogous to those of the present. This is particularly substantiated by the geomorphic record, which is everywhere quite similar to that of the Holocene. But temporary anomalies in both temperature and moisture seem to have occurred during parts of the Holstein and Eem. Unequivocal evidence is lacking for the older interglacials.

Several comments can be ventured on conditions during the thermal maxima of the Eem, the Holstein, and their North American counterparts:

a) The occurrence of thermophile trees well beyond their present northern or eastern limits during the Eem and Holstein suggests appreciably warmer winters in eastern Europe and western Siberia, and slightly warmer summers throughout Europe. Similar changes are suggested for eastern North America.

b) The distribution of certain moisture-loving plants and trees in eastern Europe suggests a more humid climate, particularly for the Late Temperate of the Holstein, when forests invaded the perimeter of the Eurasian steppe.

c) The relict or buried terra rossa and rotlehm soils of the Mediterranean area imply much more intensive chemical weathering, conceivable only with a warmer rainy season (September through May). More intensive weathering is also indicated by Sangamon paleosols from the southwestern United States.

d) Thermophile molluscan assemblages in central Europe (terrestrial and aquatic) and the North Sea (marine) imply interglacial temperatures some 2° to 4° C. higher than today, while a thermophile fauna characterized the Tyrrhenian II and III stages of the Mediterranean Sea.

e) According to the less reliable O^{18}/O^{16} isotopic measurements, the surface waters of the tropical North Atlantic may have been about 1° C. warmer, those of the Mediterranean about 1–2° C. warmer.

f) No systematic data are available from Africa, where such inferred interglacial features as fossil dunes and other indices of drier climate seldom stand careful scrutiny.

The dangers involved in unqualified use of biological evidence, and the limited nature of the data, require great caution in evaluating the quantitative implications of temperature change. But there can be little doubt that temperatures were somewhat warmer than today's during the thermal maxima of the Eem and Holstein. Possibly the tropics were only a little warmer, implying that higher temperatures were pre-eminently characteristic of higher latitudes. This in turn supports a hypothesis of reduced winter cold and a reduction of latitudinal temperature gradients.

The interglacials offer a fascinating field for further research. So far, most of the literature has only considered interglacials as important stratigraphic markers within the record of repeated Pleistocene glaciation. More emphasis directed toward a fuller paleo-environmental understanding of the interglacials will surely be rewarding. Only in this way can the conventional stereotype of uncomplicated, warm-dry interglacial periods be replaced by a more realistic concept.

TWENTY-THREE

The Climatic Changes of the Pleistocene

INTRODUCTION

Discussion of the climatic changes of the Pleistocene inevitably leads to the question of ultimate causes. Theories accounting for aberrant Pleistocene climates are formulated anew every few years. They range from changes in solar radiation, solar particle emission, and sunspot activity, to terrestrial volcanic activity, ocean circulation, and mountain-building as well as to astronomical features of the earth's axis and orbit. It is basically absurd, even though frequently stimulating, to speculate on ultimate causes when the empirical paleoclimatic evidence is as scanty as it is, and when the theoretical meteorological patterns are as little understood as recent symposia on changes of climate have clearly demonstrated (see Wallén, 1963; Sawyer, 1966; Mitchell, 1968). There are then two quite distinct aspects of Pleistocene climate: the *causes* of the climatic changes, and the *patterns* of such change.

Fundamental to our whole understanding of the Pleistocene is the accumulation of well-dated, local paleoclimatic information. The foregoing chapters have attempted to outline current information of this type, a picture that is obviously far from satisfactory. Considering that our knowledge for Asia, Australia, and South America is limited or practically nil, the gathering of sound empirical paleoclimatic data remains a basic prerequisite.

The second step towards an understanding of Pleistocene changes involves synthetic analysis of paleoclimatic data: strictly contemporary data, quantitative or qualitative, must be collated and compared. Thus regional atmospheric patterns may be deduced, as they are for example in Poser's (1950) study and interpretation of late Würm wind directions and large-scale circulation features on the basis of European dunes. Theoretical reconstructions of atmospheric patterns on a continental or world-wide scale may then be possible by comparing regional paleoclimatic information with meteorological theory. Such meteorological deductions should, however, be based upon two principles:

a) Any reconstruction of atmospheric patterns must be founded upon sound empirical paleoclimatic evidence and must be fully compatible with it.

b) All meteorological deductions so allowed must have present-day counterparts in short or long-term weather patterns. In other words, it is undesirable to assume the existence, in the Pleistocene, of weather patterns unknown today.

Only when an understanding of the patterns of observed climatic change has been attained for the northern hemisphere can scientific attention be directed toward the problem of ultimate causes. So far, short-term variations of the general circulation have not been conclusively associated with extraterrestrial phenomena, as Berg (1957) has convincingly shown with particular reference to the alleged significance of sunspot cycles.

METEOROLOGICAL THEORY ON PLEISTOCENE ATMOSPHERIC CIRCULATION

Although the causes of climatic change are beyond the scope of this book, the interested reader will probably demand an elucidation of the climatic periods of Pleistocene glacials and interglacials. It should be remembered, however, that any such discussion is theoretical and incompletely founded on empirical data, and consequently to be rated as a summary of probable or possible hypotheses.

Work directed toward an understanding of the Pleistocene general atmospheric circulation has a long tradition, extending back to a first discussion by A. Penck (1914). Penck postulated that primary atmospheric cooling induced continental glaciation, with certain corollary changes. In particular, despite a general persistence of existing atmospheric circulation patterns, the continental ice sheets will have deflected traveling cyclonic disturbances equatorward. This would increase rainfall in the subtropical desert belts (pluvials) and might lead to a simultaneous equatorward shift of the lower latitude deserts at the expense of the humid tropics. Subsequent work, until 1948, did not add substantially to Penck's ideas.

The introduction of then novel meteorological tenets into Pleistocene research by H. C. Willett (1949) and by H. Flohn (1952, 1953) was of epoch-making importance. Willet, in particular, argued that the short-term patterns of the general circulation that occur over a period of several days or weeks today are essentially similar in character to those that occurred on a larger scale in historical and geological times.

According to Willett and Flohn, these short-term fluctuations can be broadly grouped into two major patterns, known as *high index* and *low index* circulation types.[1] During the former, the world wind belts tend to be well developed and aligned latitudinally, with lower and middle latitudes comparatively warm, and high latitudes comparatively cold. Middle latitudes enjoy a uniform weather succession, with lower middle latitudes remaining undisturbed by traveling lows and attendant rainy periods from higher latitudes. Willett and Flohn consider this as the prototype of an interglacial circulation. The low index circulation type, on the other hand, is characterized by massive latitudinal airmass exchange in middle latitudes, leading to frequent cold-air outbreaks into lower latitudes, and warm-air thrusts into high latitudes. Middle latitude weather is then subjected to rapid and often violent changes, while the frequent cyclonic depressions provide abundant moisture in lower middle latitudes. This was considered as the typical glacier-type circulation. Both circulation types are necessary for the mechanism of the atmospheric circulation and must have existed at all times. But the comparative frequency or intensity of the one or the other pattern may have varied in longer-term periods just as it does today in short-term periods.

Subsequent work (see Wallén, 1953, and Butzer, 1957, 1961a) has shown that the above patterns are very much oversimplified, and that changes of circulation patterns cannot be conveniently grouped into glacial and interglacial units. Instead the dominant circulations accompanying glacial advances, glacial standstills, and glacial retreats must each be considered as distinct. On a larger scale this would apply to the major last glacial subdivision of the early, main, and late Würm (see Büdel, 1950b; Viete, 1951). Similarly, allowance must be made for both moist and dry interglacial periods in lower middle latitudes.

CLIMATIC PHASES OF THE EUROPEAN UPPER PLEISTOCENE

In order to outline the broader climatic implications of the Eem and Würm in Europe and the Mediterranean area, a synthesis of the paleocli-

1. On a local scale a high index pattern is equivalent to *zonal* circulation, with little airmass transfer across the parallels of latitude, and low index is equivalent to *meridional* circulation, with considerable latitudinal transfer of airmasses.

matic data bearing on each of the climatic stages may be tentatively attempted. Turning first to the Würm glaciation, the major lines of paleoclimatic evidence can be summarized as follows (see chapters 18 and 19):

I. Europe (north of the Pyrenees, Alps, and Balkans)
  a) Periods of ice accumulation and advance and the early Würm in general were characterized by
    1) the major part of cold-climate stream alluviation (Schaefer, 1950)
    2) major large-scale solifluction and colluvial activity (Büdel, 1950b; Woldstedt, 1958; Richter, 1968)
    3) gradual lowering of temperatures
  b) Periods of glacial standstill or retreat, and the main and late Würm in general were characterized by
    1) fluvial equilibrium, eventually going over to stream downcutting (Schaefer, 1950)
    2) a general reduction of solifluction and an increase in cryoturbation or ice-wedging (Büdel, 1950; Woldstedt, 1958; Richter, 1968)
    3) major aeolian activity with loess and parabolic dune deposition (Büdel, 1950b; Poser, 1950)
    4) full glacial temperature depressions of at least 10°–12° C. on the yearly mean, of about 7°–9° C. in July
    5) an apparent temperature inversion at 1,000–2,500 m. elevation

II. Mediterranean Region (including southern Europe and northern Africa)
  a) Periods of incipient marine regression (and advance of the continental glaciers) and probably the early Würm in general were characterized by
    1) stream alluviation and slope colluviation (Butzer, 1958a, 1963a)
    2) gradual lowering of temperatures
  b) Periods of protracted low sea level (pleniglacial) and the late Würm rise of sea level were characterized by
    1) cessation of alluviation with semiarid soil development (as today), local aeolian activity, and open vegetation
    2) full glacial temperature depressions of 6°–9° C.

From the available evidence it seems that the ice advance of the early Würm was characterized by a cool maritime climate in higher latitudes with more intensive or increased rainfall in subtropical latitudes. In other words, the "pluvials" of the subtropics were roughly contemporaneous with the growth of the continental glacier. High latitudes need not have been moister than today, but it is improbable that they were drier. A local meteorological explanation can be provided to account for glacier growth in Scandinavia and effective precipitation in the Mediterranean area. It involves shifting proportions of the common large-scale weather patterns affecting the European area: many of the meridional weather types that introduce cold maritime air and winter snowfall to Scandinavia also bring cool, stormy weather to the Mediterranean Basin; other meridional types associated with cold, winter high pressures and surface inversions over higher latitude Europe also favor cool, stormy weather further south.[2]

High pressures and surface inversions were assumed to be common in middle and higher latitude Europe during the pleniglacials. These would fall into a class of meridional weather types with southerly storm tracks producing particularly heavy rainfall in the Mediterranean region. Since, however, the latter area experienced dry conditions during the pleniglacials, another meteorological explanation must be sought. A plausible solution is provided by the reduction of evaporation induced by a 5° C. lowering of ocean surface temperatures, which according to Flohn (1953) would reduce evaporation by at the very least 20 per cent. This would decelerate the evaporation-precipitation cycle. However, the glacial retreat must have been associated with frequent warm airmass transfer into higher latitude Europe,[3] while Poser's (1950) reconstruction of moderate or strong wind directions in summer suggests a predominance of zonal and mixed classes for that season at least. Simultaneously, those weather situations favoring both glacier growth and

2. Of the twenty-eight European large-scale weather patterns identified by Hess and Brezowsky (1952), Wallén (1953) has shown that six "north meridional" types and most of the nine zonal or mixed types are responsible for heavy winter snowfalls in the Scandinavian highlands, while seven "south meridional" types lead to warm air advection from southerly, southwesterly, or southeasterly quadrants, and five "warm meridional" types produce clear, cool winter weather in Fennoscandia. In summer the zonal situations favor cool, cloudy weather, the warm meridional situations steady, warm weather. In the Mediterranean region the north meridional types are associated with a 13 per cent increase in cyclonic activity during the winter half-year (October through March); the warm meridional, a 37 per cent increase; the zonal and mixed types, a 37 per cent decrease (Butzer, 1960c). Consequently, in wintertime, only the north and warm meridional types would simultaneously favor glaciation in Scandinavia and greater rainfall in the subtropics. These classes occur with frequencies of 12.2 per cent and 11.4 per cent in the winter half-year today, and presumably experienced increased winter frequency during the early Würm. Summer circulation patterns are less significant for the Mediterranean area.

3. These south meridional circulation patterns are partly favorable, partly unfavorable for cyclonic activity in the Mediterranean area. The over-all average is only 10 per cent above normal (Butzer, 1960c).

Mediterranean storminess must have decreased. Pleniglacial circulation patterns were then probably quite different from those of the early Würm. This over-all decrease of meridional types in favor of zonal circulation was probably unfavorable for cyclonic activity in the subtropics, a suggestion which may explain pleniglacial arid phases in northern Africa. A climatological outline of the glacial retreat in Europe and North America is given by Manley (1951, 1955; see also Bryson and Wendland, 1967b).

The problem of moist and dry interglacials in the Mediterranean region and northern Africa appears to be complex. A very simple solution can be offered to explain the aberrant, moist interglacial type however. Four meridional circulation patterns responsible for much of the warm air advection to higher latitude Europe today are combined with great cyclonic activity in the Old World subtropics.[4] An abnormally high frequency of these situations could explain the local evidence well.

In summary, Table 20 includes some tentative suggestions concerning climatic phases of the Upper Pleistocene.

Table 20. Upper Pleistocene climatic phases in Europe and the Mediterranean region.

Phase	*Middle Latitudes*	*Subtropical Latitudes*	*Increased Frequency of Circulation Class*
Dry Interglacial	As today	As today	As today
Moist Interglacial	Warmer	Moister, warmer	Moist South Meridional
Early Glacial	Cooler, moist	Moister, cooler	North and Warm Meridional
Full Glacial	Colder, drier	Drier, colder	Warm Meridional, Zonal (?)
Late Glacial	Cooler, drier	Drier, cooler	Zonal, South Meridional

4. This "moist south meridional" class has an index of cyclonic activity 27 per cent above average, and occurs with a frequency of 11.3 per cent during the winter half-year today (Butzer, 1960c). It is a major rain-bringer in the Mediterranean region.

CURRENT PROBLEMS IN PALEOMETEOROLOGICAL RESEARCH

Those few meteorologists with an interest in the Pleistocene are unfortunately hampered by difficulties of two kinds: (*a*) the empirical paleoclimatic data are painfully inadequate so far, and (*b*) meteorological theory concerning the modern general circulation is not considered to be satisfactory as it is. In fact, as Sutcliffe (1963) has emphasized, it has not yet been convincingly shown that the climate of the earth should be distributed as it actually is.

The meteorologist obviously cannot be expected to collect and evaluate the specialized paleoclimatic data he is to analyze. This must be provided by the Pleistocene specialist who should attempt to interpret his data with great caution, so that better catalogued materials with qualified evaluations such as "reliable," "probable," or "possible" may be accessible to the meteorologist. A great deal of more satisfactory paleoclimatic information must be available before this major barrier to paleometeorological study is removed. But there can be little doubt that the information will indeed be forthcoming.

Another obstacle of a purely meteorological kind must also be overcome. Further work such as that on the secondary effects of particularly snowy winters on radiation, temperature, pressure distributions, etc., seems highly promising, especially since the valuable study of Namias (1963). Indeed, it remains to be determined whether a relative change of circulation patterns might by itself induce higher latitude glaciation, with planetary temperature depression as an indirect result rather than a primary cause. Sutcliffe (1963) has suggested that the difference between the winter and summer general atmospheric circulation today is greater than that between a glacial and an interglacial. Consequently, Sutcliffe raises the question whether the climatic changes of the Pleistocene might still be within the normal, built-in range of variability of the general circulation. Obviously these are considerations of prime significance for a fuller understanding of the Pleistocene.

It is to be hoped that more meteorologists will devote attention to related problems. Only then can the present impasse in paleometeorological interpretation be overcome. When and if this is accomplished, an answer to the burning question of ultimate causes for the phenomenal Pleistocene era will probably not be far away.

PART SIX

Man-Land Relationships in Prehistory

TWENTY-FOUR

Early Subsistence and Settlement Patterns

THE SCOPE OF ENVIRONMENTAL ANALYSIS

A study of man-land relationships in prehistory requires a good understanding of the many aspects of the geographic environment. Such environmental analysis would include contributions by the earth and biological sciences as well as by ethnology. Study should be directed toward three major goals:

1) Understanding the *regional environment,* including the climate, vegetation, soils, and geomorphic agencies. This has been the topic of the preceding chapters and needs no further detailed discussion here. This class of information can often be readily obtained as a result of geomorphological and palynological work.
2) Understanding of the regional food resource base or *economic area.* In the case of hunter-gatherer populations this requires:
   a) analysis of fossil faunas from several archeological sites or natural sediments;
   b) estimation of the biomass existing in the region, based on paleontological data and an understanding of vegetation and animal ecology;
   c) identification of preserved vegetable foods or at least pollen of species with edible fruits, bark, roots, etc;

d) understanding of the nutritional patterns of modern "primitive" groups having comparable technology and living in comparable environments; and

e) over-all assessment of the human resource base in terms of potential population level.

This class of material is usually difficult, if not impossible to obtain. Studies of animal ecology in relation to vegetation and abiotic environmental features are still rudimentary in most areas. Comparative ethnological studies available are often inadequate. And, the common absence of vegetable remains in archeological assemblages may distort our image of Pleistocene hunter-gatherer economies.

In the case of agricultural populations, understanding of the economic area would entail:

a) assessment of how much of the human diet was based upon agricultural plants, domesticated animals, and native food resources, including game and wild vegetable foods;

b) assessment of area of local arable land as bounded by topographic features (coasts, rivers, swamps, mountains) and conditioned by vegetation cover (grassland, parkland, open or dense forest), soil depth, and terrain slope;

c) assessment of native vegetation as an obstacle to clearing and tilling;

d) assessment of soils as to friability and fertility;

e) assessment of available grazing for herd animals; and

f) assessment of game and fishing resources.

3) Understanding of the *local setting* of a site, i.e., location with respect to the terrain, hydrography, groundwater, and other local features. This would include:

a) factors other than food supply that would influence selection of a settlement site on a seasonal or perennial basis (water availability, natural shelter, exposure, dry ground, fuel resources);

b) factors impeding or facilitating human movements (coasts, rivers, swamps, steep slopes, vegetation patterns);

c) factors affecting game movements, such as availability and localization of drinking water, topographic barriers impeding or channeling movements into select areas; and

d) factors providing an added marine or aquatic food element, such as location near the seashore, lakes, rivers, or streams.

This general class of information is usually available, at least for later Pleistocene and Holocene sites.

Obviously, total environmental analysis is seldom possible. But the complexity of the geographic movement must be understood, and a complete understanding should always be attempted. It is important to know

> precisely what geographic conditions obtained at each state of human settlement; the extent to which the economic activities of any particular community were limited by the external environment; and above all how far the economic activities of the people [studied] are reflected in and can be reconstructed from changes in the geographic surroundings (J. G. D. Clark, 1957, p. 20).

MATERIAL MANIFESTATION OF HUMAN CULTURE

Turning from the environment *per se* to the activities of man *per se,* we find that the full complexity of human culture—in the widest sense of the word—is far more difficult to outline than the geographical environment. But for a discussion of man-land relationships among preliterate peoples, the material aspects of culture are paramount. There is, in fact, little choice other than to place major emphasis on an empirical economic approach such as suggested by J. G. D. Clark (1953).

A few of the more significant material aspects of culture are as follows:

1) *Economy.* The basic economic patterns in prehistoric times include (1) unspecialized food-collecting, (2) specialized hunting-gathering, (3) primitive agriculture, based in some part upon crop planting and limited keeping of domesticated animals. Nomadic herding and more advanced agricultural or mixed urban-argicultural economies had also evolved by the beginning of the historical era in the Near East. Apart from functional tool interpretation, archeological evidence on economics is largely confined to analysis of biological remains and refuse. In the case of agricultural groups such analysis includes:

   a) Composition of the faunal remains as to orders and species giving evidence as to the range of ecologic niches exploited;
   b) Indications of possible selectivity of species, suggesting deliberate choice, particular hunting methods, seasonal availability of certain species, etc.;
   c) Age and sex composition of the fauna, giving further information on hunting techniques and seasonal activities;
   d) Disposition of faunal remains in a site, providing evidence of human behavioral patterns, methods of butchery, etc.;
   e) Determination and over-all interpretation of any vegetable remains.

In the case of agricultural populations study of biological remains is directed toward the determination of these factors:

a) Proportions of wild to domestic animal species and individuals, possibly suggesting relative importance of hunting;
b) Presence of the "normal" domesticated animals (dog, goat, sheep, pig, cattle, horse) and in which proportions; absence or poor representation of some species may have cultural significance;
c) Age and sex composition of domesticated animals, giving indications of relative and seasonal aspects of animal food use;
d) Kinds and species of cultivated plants present, and in what proportions;
e) Evidence of wild vegetable foods.

2) *Technology.* Technology includes a wide variety of features: tools (in the broad sense of the word), fire, clothing, shelter, transport, hunting techniques, etc. Unfortunately, stone tools, by virtue of their preservation, are by far the most common evidence of human technology in the Paleolithic culture range. It is therefore reasonable but nonetheless unfortunate that stone tool typology has long been equated with both culture and economy. Style and function are, however, not identical, and stone typology exemplifies only a limited part of human technological skills. In the case of nonagricultural groups, the kinds and proportions of stone tools, and the site associations of tools, bones, and other vestiges of human activity are all vital to functional interpretation. Careful study of worked bone and wood is equally important. In more advanced cultures, the kinds, functions, and sources of pottery and metal objects can be equally informative. The materials employed in tool-making frequently provide information about group movements or trade connections.

3) *Settlement.* Settlement includes architecture, settlement morphology, location and permanence, population distribution and density, etc. A more detailed discussion of settlement patterns is given below, so that brief reference to the question of settlement duration will be sufficient. Müller-Wille (1954) recognizes the following settlement types based on duration of site occupation:

a) Ephemeral settlements of a few days duration;
b) Temporary settlements of several weeks duration;
c) Seasonal settlements of some months duration;
d) Semipermanent settlements of some years duration;
e) Permanent settlements lasting for several generations.

Although these settlement classes cannot be freely equated with economic traits or cultural level—either in modern primitive groups or in prehistoric cultures—consideration of these criteria in archeological evaluation can be rather useful.

4) *Land use,* i.e., human impact on the environment. Direct modification of the biological environment was possible since the earliest times through hunting, the use of fire for hunting or plot clearance, and ultimately, field cultivation and herd grazing. Such environmental modifications are another material manifestation of man's activity.

CLASSIFICATION OF PREHISTORIC CULTURES

The standard European nomenclature employed for prehistoric cultures is primarily based on stone tool typology. Cultures are divided into an old (Paleolithic), a middle (Mesolithic), and a new stone age (Neolithic). The Paleolithic is further subdivided into lower, middle, and upper units, each composed of a constellation of implement assemblages. Chronological criteria have also slipped into this classification: the break between Lower and Middle Paleolithic is frequently defined by the Riss-Eem boundary, while the Mesolithic is usually separated from the Paleolithic by the Würm-Holocene boundary. Clearly, the existing terminology can only be justified as a matter of convenience.

Braidwood (1960a; Braidwood and Howe, 1962) has proposed a more economic classification of prehistoric cultures, based primarily on levels of subsistence:

1. *Unspecialized Food-Collecting.*
   a) Naturally determined mammalian subsistence and free-wandering,[1] with tools fashioned but not yet standardized. This is thought to include the australopithecine groups and the very crude and typologically variable early pebble tools.
   b) Food-gathering, with free-wandering, hunting, and earliest standardized tool-making traditions. Subsistence patterns are significantly determined culturally. Tools of the early standardized traditions of core bifaces, flakes, and choppers appear, with broad distribution for a given tool type.
   c) Food-gathering, with elemental restricted wandering, hunting, and some variety in standardized tool forms within regions. This includes a great number of mixed industrial assemblages.
2. *Specialized Hunter-Gatherers.*
   d) Food-collecting, with selective hunting and seasonal collecting patterns for restricted wandering groups. Considerable typological

1. Presumably this should not indicate areally unlimited wandering but rather shifting, ephemeral settlements of a few days duration within limited territories.

variety and "tools to make tools," with rather marked regional restriction of any given industry, although a generalized tool-preparing tradition such as blade tools may be widespread.

e) Food-collecting, with intensified hunting and collecting with seasonally different activities by restricted wandering or center-based wandering groups. Beginning of plant manipulation and greater concentration on the taking of fish, fowl, mollusks, and fleeter mammals.

3. *Primary Food-Producers.*

f) Incipient plant cultivation and animal domestication, within the natural habitat of potential plant and animal domesticates. This elusive stage involves experimental manipulation within a subsistence milieu of the food-collecting type, among restricted wandering to semipermanent groups.

g) Food-producing with the appearance and diffusion of the primary village-farming community—in which a marked proportion of the dietary intake is of produced food. Settlement is semipermanent or permanent.

h) Food-producing, with expanded village-farming and, possibly, incipient urbanization. Permanent settlement is possible on a subsistence pattern of predominantly produced food. Plow and draft animals in wider use, craft specialization and metallurgy common.

Many anthropologists will disagree with this scheme, partly because of the apparent equation of culture and technology, partly because of its emphasis on cultural rather than technological evolution. Furthermore, both Pleistocene cultures and technology are far too elusive to fit any preconceived system. But Braidwood's classification provides a more meaningful synopsis for nonarcheologists than do the conventional tool typologies.

MODERN SETTLEMENT PATTERNS OF PRIMITIVE FOOD-GATHERERS

No modern "primitive" group provides an accurate picture of prehistoric populations. In particular, the technological traits vary strongly between dispersed groups of a similar economic level, emphasizing their acculturation to nearby higher cultures. But certain aspects of the economic and social structure may provide a fair analogy to prehistoric communities. Geographically, economic and social attributes are possibly expressed most usefully in such settlement patterns. Accordingly, characteristic settlement features observed among contemporary "primitive" peoples are outlined below as a valuable aid toward understanding prehistoric settlement and, indirectly, man-land relationships in prehistory. The discussion is freely based upon a study of settlement geography by Gabriele Schwarz (1961, part 2; for useful complementary data, see Lee and De Vore, 1968; Service, 1966).

Modern or subcontemporary ethnological groups of primitive gatherers are found in tropical rainforests (the Negritos of the Philippines, the Kubus of Sumatra, the Toala of Celebes, the Tapiros of New Guinea, the Semang of Malaya, the Congo Pygmies, and small bands of the Amazon Basin), in dry grasslands or semideserts (the Australian aborigines and the Kalahari Bushmen), and in cool temperate forests (the Alacaluf of southern Chile). Probably each of these environments was too marginal and unattractive for the Pleistocene populations to be considered, but the information provided transcends these limitations (see Chang, 1967, with discussion comments).

Such groups as fall under the category of primitive gatherers consist of groups of families or bands, socially differentiated by age and sex only. A leader, if present at all, owes his advisory functions to strength or experience, and has no vested authority, certainly none that may be inherited from father to son. In the economic sphere the group's needs are modest, and the necessary food, clothing, and shelter can all be provided by the community itself, without recourse to any special craftsmanship or organized exchange of goods. In other words, each band is a self-supporting entity except where recent trade relationships have been established with neighboring agricultural populations. Since technological skills are poorly developed, the environment has comparatively little to offer for exploitation at this low cultural level. Consequently the subsistence area is large, and 10–250 square kilometers per person seems a fair estimate of population densities. This requires considerable shifting of camp sites.

The topographic location of such food-gatherers' (ephemeral or temporary) campsites is not chosen with much deliberation. Natural glades along or close to a stream, and hence near a water supply, are preferred in forest country. In case of danger such groups retreat into the dense forest. In dry country, watering places are selected, although such water holes may be evacuated to attract scarce mammalian faunas on which the livelihood depends. Obviously then, the suitability of a location is not a strong motive of choice, since if the site is unsatisfactory the group can move on with little trouble.

Both natural and artificial shelters are employed. Hollow trees were widely used by the Australians and Tasmanians, and in part, fire was used to enlarge existing cavities. Caves were commonly occupied after being lined with leaves, grass, and bark. Rock crevices and overhangs, amplified by branches, grass, and moss, were used by the Bushmen.

Where natural shelter is available, various artificial constructions are used. Most elementary is the *wind shelter* or *lean-to.* The principle is that of two upright stems with forked tops, a crosspole, and various sticks supported against this at 45-degree angles. For use as a rain protection in tropical areas, the roof framework is then covered with large palm or banana leaves. Part of the structure may be derived from growing saplings. In Australia the lean-to was often closed in on both faces. In other areas, a crude type of *basket hut* is employed as shelter. Branches are stuck into the ground in a crude semicircle and the tops bound together. This framework is then covered, with palm leaves in rainforest areas; with grass, reeds, or bark in other areas; with hides or furs in cold regions. In the latter case, earth may be thrown up around the base.

The size of the settlements is characteristically one of small group settlements with anywhere from 4 to 24 families and 20 to 100 people. Each individual family occupies one shelter. As community effort is vital, dispersed settlement is unusual. As social organization is not rigid, individual families may leave one band and join another, so that the size of the shifting settlements is rather variable. Larger settlements in Australia attained 800 people, but in such cases the settlement was subdivided according to bands. Lack of social differentiation implies that no special structures exist. The size of the group(s) is essentially determined by availability of nourishment in an area.

The morphology of settlements is variable. In the forests, irregular, circular, or semicircular arrangements are common, with the wind shelters facing inward. This is partly to take advantage of the natural configuration of the forest glades, and partly to provide protection against wild animals.

It is questionable whether the wind shelters or basket huts described here were known in the Lower or Middle Pleistocene. But many attributes of the subsistence-settlement pattern are pertinent to an understanding of early Paleolithic cultures.

MODERN SETTLEMENT PATTERNS OF ADVANCED FOOD-GATHERERS

Contemporary groups frequently considered as "advanced" food-gatherers include the Eskimo, the northern woodland Indians of Canada, and some Paleoasiatic groups of Siberia.

Each of these arctic or subarctic groups shows distinct preferences in settlement location. The coastal Eskimo choose coastal headlands or islands in the center of good hunting areas. The Indians of the tundra or forest prefer locations along streams and lakes. Reasons for the latter choice are manifold: rivers and lakes provide an extra source of food supply in the form of fish; hunting can be done by boat; many mammals may be conveniently attacked while they are crossing rivers; and finally, rivers provide the principal transportation lines (by boat).

Construction among the Eskimo is widely characterized by different summer and winter abodes, occupied on a temporary basis for several weeks or occasionally several months. Primitive huts may in some cases be dug into the ground during winter, and sunken dwellings are frequent. Where little wood is available, stones are built up with decreasing circumference in a roundish, domed structure. Where wood is available, a rectangular plan with pyramidal shape is employed, wooden beams or animal bones supporting the roof. In summer the same people usually exchange such sunken dwellings for tents of basket or conic structure. The forest Indians frequently live in tents the year around and do not share the architectural tastes of the Eskimo.

Settlement size is usually not greater than that of the more primitive gatherers, although seasonal concentrations of several groups may occur for the purpose of common organized băttue hunts, e.g., buffalo hunting. Among such temporary or seasonal settlements it is not uncommon to find a community or council hut for ritual purposes, around which the other huts are grouped. Other analogies to the primitive food-gatherers' settlements are applicable.

Archeological evidence suggests that such subsistence-settlement patterns were common, although not necessarily characteristic, in late Pleistocene Europe.

MODERN SETTLEMENT PATTERNS OF SPECIALIZED HUNTERS AND FISHERS

Recent or contemporary populations of specialized hunters or fishers that may be analogous to terminal Paleolithic and Mesolithic groups are certain hunting and fishing communities of the tundra and subarctic forests, as well as the Yuki and Maidu Indians of north-central California.

Some of the few changes from previous patterns, for example, include more carefully chosen topographic locations, with security as a dominant motive. Summer settlements are temporary, with primitive huts or tents used for shelter. Winter dwellings are occupied on a seasonal basis so that care is taken to provide maximum protection against the cold. *Sunken dwellings* are most characteristic, although rectangular, gabled houses of crude planks were used in some areas. In California, basket or conic huts were most commonly used in the winter seasonal settlement.

To give an example of settlement size and morphology: the Yuki and Maidu of California (population density: 1 person per 2.2 sq. km.; see

Baumhoff, 1963) settled in groups of about 100 individuals. The groups maintained a major settlement of some permanence, with auxiliary sites of sporadic occupation. Huts were loosely arranged according to topographic considerations so that many settlements tended to display linear morphology. As the chieftain has little authority or permanence, his hut was generally not singled out in terms of noticeably careful construction.

Compared with a settlement density of one person per every 10–250 square kilometers, as in the case of the food-gathering populations, these specialized food-collectors lived in comparatively dense concentrations. Furthermore, the seasonal character of settlement reflects more effective food acquisition. Some of the late Pleistocene and early Holocene food-collecting cultures of Europe and the Near East appear to have approached similar efficiency at the close of the Pleistocene.

MODERN SETTLEMENT PATTERNS OF PRIMITIVE AGRICULTURISTS

Primitive agriculture as practiced by farmers employing the hoe, or the digging or planting stick, is largely synonymous with the practice of shifting agriculture in the tropics of the Old and New World. Used in this sense, primitive agricultural settlements show adaptations to their peculiar economic pattern. The gradual reduction of fertility of the surface soil requires continuous shifting of plots. Similarly, the temporary invasion of cultivated lands by tough weed grasses may also make the soil too difficult to work. When all land in an area has thus been "used" within a decade, for example, a fallow period of thirty years or more would mean that another decade or two would have to elapse before any of the local fields could be cultivated again. In the meantime it might be preferable to abandon the settlement. Plow agriculture, permitting easy working and regular mixing of a deep soil layer, can better guarantee a permanent basis for settlement—at least in extratropical areas.

With primitive agriculture the "economic area" is much more limited in size than in the case of a collecting economy. At first the settlement is immediately surrounded by land in cultivation, but as these fields are left in fallow the belt of cropped plots lies farther and farther afield, with unproductive land in between. Field huts are often set up to provide shelter, while sowing or harvesting takes place in very distant fields that may eventually lie as much as 10 or 20 kilometers from the settlement. When this inconvenience becomes too great or when concentric expansion is limited by geographic or social barriers, the settlement may be abandoned. Consequently, a rather loose association between site and "economic area" is characteristic for the settlements of primitive agriculturists. Needless to say, there are important modern exceptions where primitive agricultural settlements are permanent.

The topographic location of sites of primitive tropical agriculturists today is primarily dictated by terrain. River levees or terraces are highly favored in the rainforest because of the better alluvial soils, the possibility of river transport, and the added food resource of fish. Steeper slopes are avoided because of the danger of slumping. In the savannas, small elevations are favored because of greater air movement and safety from sheetflooding. Invariably, alluvial flats, swampy terrain, or flood-periled savanna plains are avoided. Elevated locations or mountain spurs are also more secure, and riverine sites between complex meander systems may be occupied as a defensive measure. The security factor of many primitive agricultural settlements reflects a need for safety in a semipermanent settlement as well as a tribal social structure without central authority.

The typical shelter form employed is that of one-room houses. Huts and houses with conic roofs are constructed with a circular row of posts joined by wickerwork of plant fiber, usually coated with mud, mixed mud and dung, or mud with chaff or straw ("daub-and-wattle"). The roof is supported by vertical as well as oblique beams, and closed off with grass, straw, or reeds. These house-types are characteristic in open-tropical landscapes.

A gabled house-type, designed to permit better ventilation, is more common in the rainforest. Vertical posts in rectangular array support the walls and a saddle roof. The latter is covered by palm leaves, palm leaves and ribs, or sugar cane. Wickerwork of leaves, sliced bamboo, or bark is used for the walls, or possibly crude planks. The posts are obtained from palm stems, breadfruit trees, bamboo, etc. Raised pile dwellings are common.

Household or small herd animals are kept in the house (or, in the case of pile dwellings, under the house) at night, while enclosures of some form are used for larger herd animals. Storage facilities are not required in rainforest areas where agricultural products are available continuously, but they are necessary where seasonal drought impedes cultivation for part of the year. Storage houses or huts, often on posts, are common in the seasonal tropics; large pottery storage urns or mud silos in the drier subtropics. Community buildings serving complex social or religious functions are usual, and stand out by their greater size, better craftsmanship, and ornamentation. The same applies to the house of the local chieftain.

The size of such settlements is generally that of the clan group—

hamlets or small villages of fifty to a few hundred inhabitants. Large villages with several thousand inhabitants may develop when (*a*) protection requires greater concentrations of population, (*b*) advanced political and social organization has been developed, (*c*) strong leaders require settlement amalgamation for military purposes, and (*d*) natural resources are particularly good. Population density estimates in the order of 5–10 persons per square kilometer (Braidwood and Reed, 1957) seem well founded as an average, although there is considerable variability.

The morphology of these primitive agricultural settlements is oriented along topographic lines in the rainforest: a linear orientation along streams or along pre-existing forest trails. In open landscapes the morphology is related to social or economic features, and in both cases circular forms dominate. In the latter instance, larger herd animals may be enclosed within a circle of huts. Where social rather than economic factors prevail, the living huts are located around a core area that includes the chieftain's house, community buildings (for school, social, council, religious, or craft purposes), and storage facilities. Protection is commonly obtained through construction of palisades, thorn fences, and the like.

Settlement pattterns in Egypt appear to have been of analogous type as late as the Old Kingdom (during the third millennium B.C.), but the archeological evidence from southwestern Asia suggests that many settlements of an early agricultural type already had considerable permanence. Construction in the mediterranean climate zone of the Near East was also more elaborate, long before the advent of plow cultivation. Architecture was commonly rectangular in plan, making use of adobe mud or even sun-dried bricks. But the general analogy is still useful for a better understanding of late prehistoric agricultural settlement.

TWENTY-FIVE

Human Origins in Sub-Saharan Africa

INTRODUCTION

Men, by the definition of "Man the Toolmaker," are first known from early Pleistocene deposits of the African continent. In 1957 a number of crude, man-made, stone artifacts were first found in association with teeth and jaw fragments of the fossil ape-man *Australopithecus* at Sterkfontein, near Krugersdorp in the Transvaal (Robinson and Mason, 1962). Unfortunately, isotopic dating has not been possible at Sterkfontein, and the fauna does not allow precise stratigraphic dating (see Cooke, 1963; Brain, 1958). Subsequently, between 1960 and 1963, several cultural levels, representing occupation floors and associated with hominid fossils, were excavated in Bed I of Olduvai Gorge, Tanzania (M. D. Leakey, 1967). A series of potassium-argon dates from these levels indicate an age of about 1.85-1.65 million years (Evernden and Curtis, 1965). Of comparable or possibly greater age are certain sites with fresh pebble tools[1] found in basal Pleistocene units (Moulouyian, pre-Sicilian regression?) of Morocco (Biberson, 1961a, 1961b).

The first appearance and the subsequent dispersal of toolmakers in Africa during the early Pleistocene raises the fundamental question of human origins: Who were the toolmakers and what were their antecedents? At this point we are restricted to the biological evidence of

1. Made from pebbles by removing two or three flakes from opposite faces to obtain a jagged cutting edge on one side of the pebble.

hominid[2] fossils, and matters of definition and recognition become difficult and controversial. Paleontologists generally agree that two principal sets of criteria must be satisfied if a fossil primate is to be classified as a hominid: (1) evidence of habitual, upright bipedalism as the normal form of locomotion, and (2) presence of teeth essentially human in form, with small and vertically implanted anterior teeth (incisors and canines), and a particular occlusal pattern of the posterior teeth (molars and premolars) (Pilbeam, 1968). The earliest known fossil of hominid type is *Ramapithecus,* an advanced primate found in late Miocene beds (ca. 14 million B.P.) of Kenya and northern India (Pilbeam and Simons, 1965; Howell, 1967; Pilbeam, 1968). No fossils of the pelvis or lower extremities are available, but the dental structure and morphology suggest that at least one of the two major adaptive "boundaries" that define a hominid had already been crossed. Pending further fossil discoveries from the early and mid-Pliocene, the earliest unquestioned hominid remains pertain to the extinct subfamily of the australopithecines. These are now known to range in age from 3.5 million B.P. in the Lower Omo Basin (Howell, 1968b, 1969) to about 1 million B.P. in Olduvai Gorge (Bed II). What is badly needed at the moment in order to determine whether *Ramapithecus* was a hominid is a series of fossils linking it through gradual morphological change to the earliest, undoubted hominids—the australopithecines.

APE-MEN OF THE LATE PLIOCENE AND EARLY PLEISTOCENE

The fossil ape-men of the australopithecine subfamily are divisible into two lineages, the first including gracile forms and the second, robustly built forms. In the past the two lineages were distinguished at the generic level (Robinson, 1967), but recent studies (especially Tobias, 1967; see also Howell, 1968a, 1969; Simons, 1968) have shown that specific or subgeneric distinction is the most that is warranted. Subgeneric distinction (as suggested, for example, by Howell, 1967) has certain advantages: the close phylogenetic relationship between two species belonging to the same lineage may then be implied in the full nomina of the species. The two subgenera are appropriately known as *Australopithecus* (gracile) and *Paranthropus* (robust) (see Fig. 63). To date, two species of the "robust" subgenus have been recognized. Their

2. According to a widely favored taxonomic classification, the Old World higher primates are subdivided into the hominoids and cercopithecoids (Old World monkeys). The hominoids in turn are subdivided into the hominids and pongids (anthropoid apes). The pongids include two living subfamilies, the pongines (including gorilla, chimpanzee, and orangutan) and the hylobatines (gibbons). The hominids include two subfamilies, the australopithecines (extinct) and the hominines *(Homo erectus, H. sapiens neanderthalensis,* and *H. sapiens sapiens),*

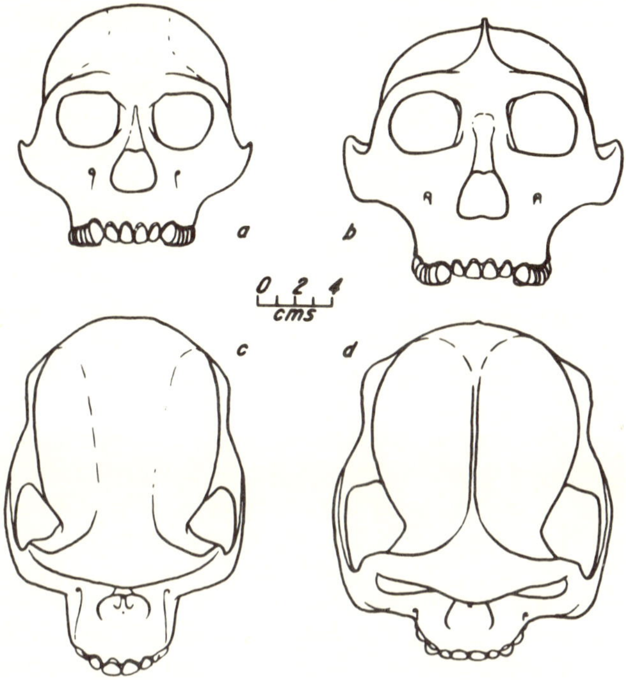

Figure 63. Facial and top views of skulls of Australopithecus *(a, c) and* Paranthropus *(b, d). Both skulls are of females. (From J. T. Robinson, 1963, by permission of the author and of the Wenner-Gren Foundation for Anthropological Research.)*

formal nomina are *Australopithecus (Paranthropus) boisei* and *Australopithecus (Paranthropus) robustus,* although abbreviated forms or descriptive terms will be used in the subsequent text as a matter of convenience. Both have skulls without foreheads, but with sagittal crests. Both have small incisors and canines in combination with very large premolars and molars. The premolars amd molars of *A. boisei* are especially large—outside the range of those of *A. robustus.* There is evidence to suggest that *A. boisei* is geologically older than *A. robustus* and may be ancestral to it.

Only one species of the second subgenus has been recognized so far. It is formally designated as *Australopithecus (Australopithecus) africanus.* Others may be named in the future when studies are completed of some fossils that have recently been found. Unlike the members of the "robust" subgenus, members of the "gracile" one had foreheads and generally lacked sagittal crests. The contrast in size between the premolars and the canines was also far less marked.

Members of both subgenera shared a number of important features. In both cases, average cranial capacity, a rough estimate of brain size, seems to have been about 500 cc. Members of both subgenera were habitual bipeds, although the available bones of the pelvic (hip) girdle and lower limbs suggest that they may not have possessed the unique striding gait of modern man (Howell, 1967). There also appear to have been differences in gait between the gracile and robust forms, although the footbones necessary to document this distinction conclusively are not available.

The postcranial skeleton (that is, the body bones) of the robust form is fairly well known, and suggests a weight of 120 to 150 pounds with a height of close to 5 feet for adult individuals. Few postcranial bones are available from the gracile form, which was about 4 feet tall; weight estimates vary considerably, from as little as 40 or 50 pounds to as much as 100 to 120 pounds for adult individuals.

In addition to the typical australopithecines, a number of evolved forms from Olduvai Gorge Bed I appear to be intermediate between *A. africanus* and *Homo erectus,* the typical mid-Pleistocene species of man. These intermediate fossils have been assigned to the species *Homo habilis* (see Tobias, 1965), though some authorities would prefer to regard them as representatives of a very advanced, gracile australopithecine. Their taxonomic designation, however, is not as important as the fact that the Bed I *Homo habilis* fossils constitute a veritable "missing link" between *A. (Australopithecus)* and later *Homo.* They are nicely intermediate both in terms of tooth size and shape and in terms of cranial capacity (estimated to be about 660 cc.). Until a definitive study of the related fossils has been made, it would be premature to decide upon the status of *Homo habilis.*

The many questions of australopithecine taxonomy can hardly be resolved with the available fossil evidence, since the various differences and resemblances may be the result of one or more of several factors: (1) different ecological adaptations; (2) different time ranges represented; and (3) variability within populations. Brief consideration of these possibilities and implications here is of more than taxonomic interest.

The differences in body size and cranial morphology, but particularly the contrasts of the dentition certainly appear to suggest fundamental distinction between the gracile and the robust australopiths. Robinson (1963, 1967) has interpreted these according to dietary adaptations, suggesting that the small anterior teeth and the massive posterior teeth of *A. robustus* would be most suited to a vegetarian. On the other hand, the dentition of *A. africanus,* with well-developed incisors and canines, may indicate an omnivorous diet, including substantial proportions of meat and other animal protein. To what degree this generalization holds is open to dispute (see Tobias, 1967), but there are no systematic differences in the degree of tooth wear and chipping or in the patterns of dental abrasion. The problem of ecological adaptation of the major groups of australopiths will be considered further below.

The question of temporal differences and of normal population variability can only be discussed after an outline of the known fossil material according to locality and stratigraphy or age (see Fig. 64):

I. *East Africa.*

a) Omo-Rudolf Basin (Howell, 1968b, 1969; Butzer, 1971a; Patterson and Howells, 1967 and unpublished; R. E. Leakey *et al.*, 1970). (1) Shungura Formation, near Omo Delta. Numerous teeth of one or more gracile australopithecines from various strata ranging from 3.5 million to 1.9 million B.P.; possible artifacts *in situ.* (2) Shungura Formation. Six partial or complete mandibles of hyperrobust type, tentatively referred to *A. boisei,* as well as a partial skull of a juvenile, from a similar time range. (3) Usno

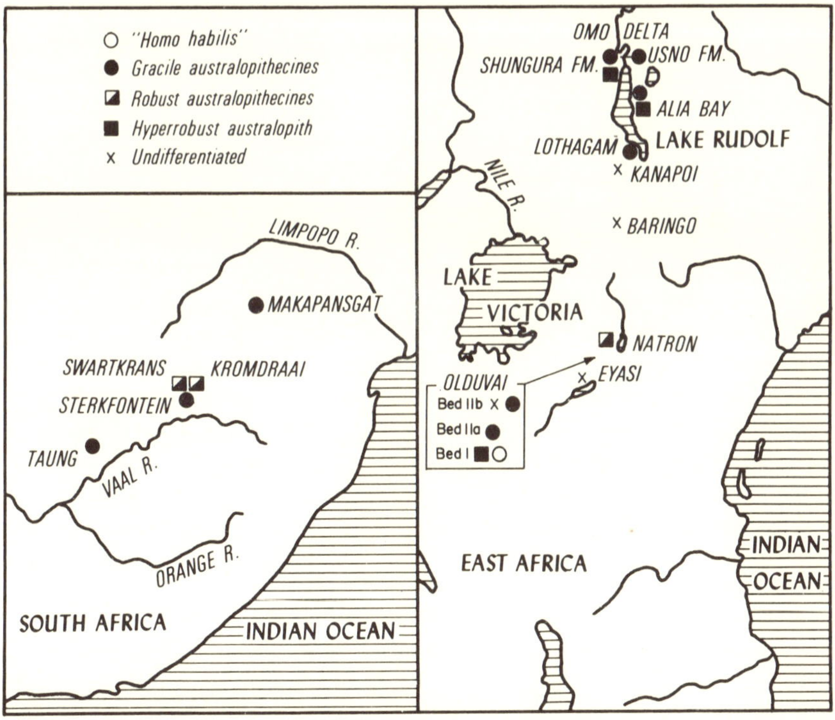

Figure 64. Australopithecine sites of East and South Africa.

Formation, north of Omo Delta. Numerous teeth of a gracile australopithecine from strata a little younger than 2.6 million B.P. (4) Kanapoi, southwest of Lake Rudolf. A rather robust arm bone older than 2.9 million B.P. (5) Lothagam, southwest coastal sector of Lake Rudolf. Partial mandible of a gracile australopithecine, age uncertain. (6) Alia Bay–Koobi Fora, east of Lake Rudolf. Two partial mandibles, a palate and a skull of hyperrobust australopithecines, as well as partial skulls of gracile type, from early Pleistocene beds currently under study; probable artifacts *in situ*.

b) Lake Baringo, central Kenya (Martyn and Tobias, 1966). An undiagnostic skull fragment of an australopithecine in the Chemeron beds, of "early" Pleistocene age.

c) Lake Natron, northern Tanzania (Issac, 1967). Complete mandible of a typical *A. robustus* from Humbu Formation, possibly dated ca. 1.5 million B.P.[3]

d) Olduvai Gorge, northern Tanzania (Tobias, 1967 and unpublished; Leakey *et al.,* 1967). (1) Bed I. Mandibles, teeth, skull or limb bones of six individuals ascribed to *Homo habilis,* dating ca. 1.85–1.65 million B.P. Various cultural levels. (2) Bed I. Skull and upper dentition of *A. boisei,* ca. 1.65 million B.P., on an occupation floor. (3) Lower Bed II. Mandible and partial skull of *A. africanus.* Cultural levels. (4) Upper Bed II. Teeth of robust type at two localities, as well as teeth of a gracile hominid. Skull of *Homo* cf. *erectus,* all much older than 500,000 B.P. Various cultural levels. (5) Bed IV. *Homo erectus* fossils from mid-Pleistocene strata, not yet described.

e) Garusi Korongo, near Lake Eyasi, northern Tanzania (Howell, 1967). Maxillary fragment of a gracile australopith or *Homo* cf. *erectus* in Laetolil beds, probably contemporary with upper Bed II of Olduvai.

II. *South Africa.*

a) Taung-Buxton, northern Cape Province (Peabody, 1954). Skull of australopith child, in "early" Pleistocene cave fill.

b) Sterkfontein ("Main" and "Extension" sites), southern Transvaal (Robinson, 1967; Robinson and Mason, 1962; Brain, 1958, 1967a, 1967b; Tobias and Hughes, 1969). A large collection of *A. afri-*

3. Earlier speculation that this date may be far too old was based on the contradiction of a normal polarity of the Humbu basalts with the then known paleomagnetic stratigraphy. Many more brief magnetic "events" are currently known (see Cox, 1969), and several magnetic reversals took place between 1.64 and 1.61 million B.P., which is within the stated range of error of the Humbu Formation date. Since no satisfactory K/Ar dates have been obtained from Bed II at Olduvai either, the tentative faunal correlation between the Humbu and upper Bed II poses no contradiction.

canus fossils in cave breccias of "early" Pleistocene age. Primitive stone artifacts in youngest deposits.

c) Swartkrans, southern Transvaal (Brain, 1958, 1967a; Robinson, 1967; Mason, 1962). Various fossils of *A. robustus,* in cave breccias of "early" Pleistocene age, associated with stone artifacts and a mandible and partial cranium of *Homo erectus* (Clarke *et al.,* 1970).

d) Kromdraai, southern Transvaal (Brain, 1958; Robinson, 1967). Various fossils of *A. robustus,* in cave breccias apparently younger than those of Swartkrans. On faunal grounds, both Swartkrans and Kromdraai are younger than the other australopithecine breccias of South Africa.

e) Makapansgat, central Transvaal (Mason, 1962; Tobias, 1967, and unpublished). Various fossils of *A. africanus,* as well as of types intermediate between *africanus* and *robustus,* in "early" Pleistocene cave breccias.

This survey of australopith sites allows several inferences: (1) the subfamily is only known from eastern and southern Africa, and the distribution of all three australopith forms may have been restricted to what are now semiarid or subhumid environments of sub-Saharan Africa. (2) Although the South African sites remain undated, typical forms of *A. africanus* are everywhere older than typical forms of *A. robustus*; in East Africa the former are verified ca. 3.5-1.6 million B.P., approximately contemporary with *A. boisei.* (3) The evolved gracile form, designated *Homo habilis,* appears to replace typical *A. africanus* a little after 2 million B.P., at least in East Africa. (4) Larger but possibly less specialized hyperrobust forms, such as *A. boisei,* appear to precede the characteristic *A. robustus* prior to 1.6 million B.P. in East Africa. Finally, (5) gracile and robust (or hyperrobust) forms, despite their individual variations through time, appear in apposition throughout the late Pliocene to early Pleistocene time range. Thus, although many of the polytypic aspects of the two australopithecine lineages may reflect temporal differences, variation through time cannot explain the two major strains.

A final question is that of "normal" variability within populations. Sexual dimorphism is rather greater among apes than among modern man, and it has been emphasized that nearly as much variation as exists in all the known australopithecine material can be found in a single subspecies of living ape, the lowland gorilla (Simons, 1968). In fact, many of the basic differences of skull form between *A. africanus* and *A. boisei* are quite analogous to male-female contrasts among lowland gorillas. Yet, despite the partial overlap between anatomical traits of *A.*

africanus and *A. robustus* or *A. boisei,* at any one time, few authorities accept less than a specific distinction between these two main lineages. In addition, the almost exclusive concentration of *A. africanus* at Sterkfontein and of *A. robustus* at Swartkrans and Kromdraai argues convincingly against a simple case of sexual dimorphism, since there is no reason why only males or only females should have died at these adjacent localities.

TOOLMAKING AND THE CASE FOR "HOMO HABILIS"

The anatomical case for the distinctiveness of *Homo habilis* is inconclusive (Howell, 1967; Simons, 1968), since the mandible and lower teeth are hardly different from those of *A. africanus.* However, the shape of the cranial vault is different and the intracranial capacity is susbtantially greater than that known or inferred for any South African *A. africanus.* Similarly, the foot and lower leg bones are remarkably man-like in overall structure and proportions.

Far more relevant is the evidence of toolmaking and culturally patterned behavior. Except possibly at Sterkfontein, the typical finds of *A. africanus* are nowhere associated with artifactual remains or butchering sites. On the other hand, *A. robustus* has been found within cultural-archeological contexts but, since this creature persisted until well after the appearance of true men on the *Homo erectus* grade, it is widely believed that these robust australopiths were hunted by man. This leaves open the question of who made the stone implements or brought in the natural stones that litter the butchering sites recorded by successive sites at Bed I at Olduvai Gorge (see M. D. Leakey, 1967). These carefully excavated occupation floors leave no question about culturally patterned behavior, ranging from stone-tool manufacture to specific butchery activity-areas, with one rather curious, irregular circular pattern of heaped-up rocks. The game bag included substantial quantities of tortoise, snakes, birds and fish, with many small mammals (rodents, porcupine, hare, shrews, small carnivores) and some larger genera (gazelles, antelopes, pigs, carnivores such as the sabertooth, and, as it now seems, *A. boisei*). Clearly, man was present, and he was already a proficient hunter as well as a scavenger of "minor" food animals.

Unless another and distinct hominid is found in Bed I—a discovery becoming increasingly less probable—the toolmaking activities and other culturally patterned behavior evident in Bed I of Olduvai must be attributed to *Homo habilis.* As further anatomical and contextual data become available, there may well be sufficient grounds to define this evolved australopithecine as a true man of the genus *Homo.* In fact, once the cultural associations are unequivocal, the title of *Homo* should be applicable regardless of the anatomical criteria.

Opinions diverge slightly as to the cultural status of the remaining, nonhominine australopiths, but Tobias (1967, p. 240) states the broad case for tool-using and limited "object utilization and fabrication" particularly well (see also Brain, 1969a), insisting that "all were more proficient than the apes in manipulating and manufacturing, and all had come to depend on their implemental activities, whether with sticks, stones, bones, horns, or stalactites, for survival," Although much remains to be clarified and established, the emergence of man from among the basic stock of the australopithecine hominids seems to be indicated. The fascinating story of human origins now being unraveled in Africa suggests an age of about 2.5–3.0 million B.P. for this decisive event, in view of artifacts tentatively identified from the Omo Valley and, more convincingly, from east Rudolf (R. E. Leakey *et al.*, 1970; G. L. Isaac, in press).

HABITAT AND ECOLOGY OF THE AUSTRALOPITHECINES

The protohuman australopithecines were presumably restricted in their geographic distribution by temperature—depending on their hair-covering and metabolism, by the availability of food and water, and possibly also by competitive primates. As we have seen, it is possible, although by no means proven, that the robust australopiths were mainly vegetarians, and that this group was as little specialized in its dietary requirements as are modern chimpanzees, gorillas, or baboons. It is also possible, if not probable, that the gracile australopiths were dependent on animal protein for a substantial part of their diet. Precisely this ability to eat both animals and vegetables would be an advantage in times of food shortage. Whatever our interpretation of the meager evidence, it is difficult not to assume that the two groups exploited different ecological niches.

The contextual data of the australopith sites, obtained from sediment and fossil studies, provides a general but still unsatisfactory ecological setting for the two divergent groups. Information of this type is currently restricted to the Omo Valley, Olduvai Gorge, and the Transvaal cave sites.

In the Lower Omo Basin, the gracile and robust australopith fossils are occasionally found at the same sites and show a broad temporal overlap (between 3.5 and 1.9 million B.P.). The geological study of the related depositional environments remains to be completed (by J. de Heinzelin and F. H. Brown), but it appears that most of the fossiliferous deposits are related to riverine or lakeshore settings. In fact, the bulk of the deltaic or lacustrine clays and tuffs of the Plio-Pleistocene Shungura and Usno formations are sterile, and bone beds are concentrated in sandy or silty strata. These are primarily related to river or distributary channels of the primeval Omo, or to the lakeside margins of sandy alluvium deposited by local intermittent streams (see Butzer, 1970b). The channel banks would have attracted many open-country animals seeking water, as well as those genera inhabiting the fringe of riverine forest. Ungulates would fall prey to terrestrial or aquatic predators such as the crocodile or possibly the extinct alligator *Euthecodon,* and their bones would be incorporated into the river bed. In fact, the mandibles of *A. boisei* show certain well-developed cracking patterns, related to bone structure, that develop when fossils are exposed at the surface for a while before burial in a sediment (N. C. Tappen, personal communication). Some animals may also have been swept into the Omo River by flood surges or bank-undermining. The former lakeside situation finds contemporary parallels where cattle today graze on aquatic emergents well out into the shallow parts of Lake Rudolf, in the same way that elephants were observed to do during the nineteenth century. Consequently, numerous kinds of animals would have sought out the lake fringe for water and fodder, sometimes falling prey to carnivores and afterward being preserved in the muddy-sandy depositional environment between lake floor and alluvial fan. Associations of the different australopith fossils with certain other primates and ungulates are currently being studied (by G. Eck). It may thereby be possible to determine whether the gracile or robust australopiths preferred the dense, fringing vegetation or the more open country beyond the floodplain and lakeshore.

Elsewhere in East Africa, the *Homo habilis* occupation floors of Bed I at Olduvai include the one classic specimen of *A. boisei,* suggesting that the robust australopiths were sometimes hunted by their more advanced biological cousins. The sites are found within or under lacustrine clays or tuffs, or in the water-worked volcanic ash. This indicates site locations in proximity to or along the shores of a perennial lake, which was probably subject to seasonal fluctuations in size and depth. The deeper parts were saline, the surface and marginal waters were probably fresh (Hay, 1964, 1967). There is little evidence for a substantially wetter climate, and the presence of both open upland vegetation and fringing woodlands is suggested by the great diversity of the mammalian and microfaunas (see L. S. B. Leakey, 1967; Howell, 1967; Cooke, 1963). Bone preservation was favored by repeated ash falls in the wake of continuing volcanic activity.

In South Africa, the australopithecine sites of the Transvaal form part of old cave fills. The caves are developed in dolomitic limestone as a result of karstic solution, while the deposits within are derived from several sources: (a) blocks of dolomite, fallen from the roof and walls of interior caverns, vertical shafts, or sinkholes; (b) insoluble residues of oxides, quartz, and chert left by the breakdown of dolomite; (c) carbonate precipitates, mainly flowstones; (d) colluvial detritus, derived from outside of the cave and by rainwash and mass movements, ranging from reddish decalcified soils to crude dolomite scree, accumulating by gravity at the base of rock outcrops or washed down the hillsides. The great bulk of the fossiliferous deposits consists of colluvial detritus admixed with collapsed blocks of dolomite. The resulting sediments were impregnated by carbonates during and after deposition and now form typical *limons rouges* (see chapter 19). Only at Makapansgat are there appreciable deposits consisting mainly of insoluble cave residues and precipitates. At or near each of the sites, karstic pools, seeps, or springs were probably available in early prehistoric times, providing sources of water for hominids and other animals and presumably favoring a more luxuriant plant cover at the local level.

Brain (1958, 1967b) has made detailed analyses of the Transvaal cave breccias. First, he was able to show conclusively that the calcareous cement was precipitated from percolating solutions, and that it had not been introduced as a fine, aeolian lime dust with the original soil wash. Furthermore, true aeolian quartz or loess-like components were found to be absent. In fact, all of Brain's textural data on the noncalcareous soil residual indicate sandy silts poorly sorted according to grain size. This conforms well with the known sedimentology of analogous *limons rouges* developed in limestones of the Mediterranean region (see, for example, Butzer and Cuerda, 1962a). However, Brain (1958) does deduce a number of climatic changes on the basis of three major criteria: (a) Angularity of quartz sand grains. Examination of such grains in dolomite soils from different South African environments indicated that quartz sand tends to be most rounded in shape with a rainfall of 400–750 mm., most angular with rainfall in excess of 900 mm., and intermediate in areas with less than 350 mm. rainfall. The apparent reason for this pattern is that the quartz residue within the dolomitic rock is angular, and a wet climate rapidly releases large quantities of fresh quartz. In drier settings, weathering of the dolomite is slower, so that the relased grains are altered and smoothed with time. In a very dry climate, the rounding process is retarded by calcite cement adhering to the quartz grains. (b) Ratio of chert to quartz sand. After determining the ratio of chert and quartz grains in the bedrock and in the local soil residues today, any "excess" of quartz grains in the sedimentary column is attributed to foreign quartz sand of aeolian origin, admixed with the soils. (c) Sediment color. Humid hillside dolomite soils in South Africa tend to be brownish, drier counterparts reddish.

From this kind of evidence, Brain (1958) suggested that during the australopithecine time range, the local rainfall at the cave sites varied from less than 500 mm. to greater than 1,000 mm. per year, compared with a central value of about 750 mm. today. In fact, a hypothetical rainfall curve was constructed, inferring dry conditions for accumulation of the Sterkfontein and Makapansgat fills (with the *A. africanus* fossils), and moist conditions for Swartkrans and Kromdraai (with *A. robustus*). This curve has been modified and reproduced in the literature (e.g., Cooke, 1963), and widely used to infer that *A. africanus* thrived in a dry grassland or thornbush, *A. robustus* in a lush savanna with riverine woodlands. Finally, different authors (Brain, 1958, 1967b; Cooke, 1963; Robinson and Mason, 1962; Robinson, 1963) have even attempted to employ this hypothetical curve to establish the relative stratigraphy of the Transvaal australopithecine sites, sometimes in relation to the discredited East African pluvial chronology.

The author is convinced that Brain's analyses have little bearing on the environments immediately contemporary with deposition at the different sites. All of the detrital breccias were deposited under identical geomorphologic conditions: a rupture of the equilibrium between soil formation and soil erosion. Under such circumstances, residual products of many ages, reflecting accumulation on hillsides and footslopes over long periods of time, are all swept into the same colluvial wash. Consequently, the statistical results would seem to have limited environmental significance and no true quantitative applications. In each case the *limons rouges* indicate no more than an incomplete or ineffective vegetation mat, since rainfall intensity today is far greater than anywhere in the Mediterranean region. Such conditions must, therefore, in the South African case, be interpreted to mean either a drier climate or a less reliable or less well-distributed rainfall. The same applies for the Taung site, where the red, sandy cave-fill was thought to be essentially aeolian and derived from the Kalahari sands (Peabody, 1954). The soil matrix from the Taung skull is, however, similar to that of the *limons rouges* from Sterkfontein or Swartkrans.

In conclustion, the australopithecine cave breccias do not appear to provide evidence of fluctuations of climate through time, nor do the sediments themselves have any special implications for dating. If stratigraphic deductions are to be made, they must be made from faunal criteria. Thus the presence of the extinct Cape horse *(Equus capensis)* as well as the common zebra at Sterkfontein (topmost strata only),

Makapansgat, and Swartkrans (C. K. Brain, personal communication; also Robinson and Mason, 1957) indicate that the very oldest South African sites are little older than Olduvai Bed I (see also formal comparisons by Cooke, 1963, 1967; Howell, 1967; L. S. B. Leakey, 1967). Similarly, paleoclimatic deductions will have to be based on biological data. Some 80 to 90 per cent of the fauna represented at all of the sites is gazelle or antelope (Brain, personal communication), suggesting fairly uniform grassland settings. Until these bovidae have been studied in detail, inferences based on other rare animal types are essentially meaningless. The cave breccias, and the question of how the bones accumulated, will be discussed in more detail later, using the type case of Sterkfontein. In overview, it appears that the australopithecines and the earliest hominines of eastern and southern Africa lived in grassland or parkland environments in close proximity to streams, springs, or lakeshores with their more dense and luxuriant vegetation. Both gracile and robust australopiths are occasionally found at identical sites and the geographical distribution of the two forms overlapped substantially. The probable differences of ecological adaptation will remain enigmatic until more substantive data becomes available. But it is possible that *A. africanus* and, even more so, *Homo habilis* ranged through a wider variety of meso-habitats as a result of greater versatility and adaptability to different ecological opportunities.

ENVIRONMENTAL CHANGES AND AUSTRALOPITHECINE EVOLUTION

The possibility that environmental factors have influenced hominid evolution cannot be ignored, and several authors have drawn attention to climatic change as a motivating force both in the selection of bipedal locomotion and in the adoption of an increasingly omnivorous diet. Central to such hypotheses (e.g., Robinson, 1963) is the assumption of late Tertiary desiccation in Africa, accentuated by the fragmentation and restriction of forested environments in the wake of rift-tectonics and vulcanism. Although of potential interest, the lack of an early Pliocene hominid record is unfortunate, while the lack of evidence for Pliocene desiccation or environmental disruptions in the Omo Valley suggests that these universal theories are premature.

More relevant are the immediate considerations of how repeated environmental changes may have affected the forces of natural selection in eastern and southern Africa during the late Pliocene and early Pleistocene. Gene flow or cross-breeding may be accelerated by environmental changes (Hiernaux, 1963). Seasonal migrations resulting from alternating wet and dry seasons, or migrations favored by long-term climatic changes, may result in gene exchange between different groups. Similarly, the greater the number of hominid bands supported by a given area during periods of favorable climate the better the chances for gene flow and acculturation. Other ecological conditions may, according to Hiernaux (1963), favor the reverse process of random genetic drift. So, for example, wide dispersal of small bands in dry country, or increased isolation as a result of desiccation, will favor inbreeding and genetic drift. Through the premature death of certain individuals of a breeding isolate, certain blood groups or genes may be lost to a small population, whereas other biological traits may be accentuated. The same result may be obtained by the dispersal of a group's offspring into daughter communities. In other words, certain environmental changes may enforce or favor group isolation and hence genetic drift, while other modifications may facilitate gene exchange and fusion of population aggregates. It is possible that alternating cross-breeding and genetic drift, motivated by alternating pluvial and nonpluvial climates at the dawn of African prehistory, may have influenced and accelerated human evolution.

Bourlière (1963, comments on p. 642) had distinguished the possible influence of environmental changes on early hominids at several levels:

a) Individual level; effects on stature, body build, and fecundity;
b) Population level: effects on size of breeding populations, on over-all population density, isolation and, possibly, on population cycles (see also Bartholomew and Birdsell, 1953);
c) Continental level: At the carnivorous hominid and early hominine level, dense populations would be found in open woodlands, savannas, and grasslands, with little or no occupation of dense rainforests or desert country. In mountainous terrain a number of distinct ecological niches are provided by moist, wooded uplands, drier open lowlands, fringing forests along streams and lakes, etc. Isolated groups could grow up and evolve within such niches.

These perspectives help illustrate the ways in which environmental changes can affect the density and general distribution of primitive populations.

As tool users, early Pleistocene hominids certainly had no mastery over their environment. Without knowledge of clothing (which one must assume) or of fire (which seems to be empirically evident; Oakley, 1961), they probably could not operate comfortably outside of warm climates because of the tropical or subtropical adaptations of most monkeys, pongids, and hominids. Within their preferred thermal environment, carnivorous hominids were again restricted to areas with available water and animal foods. The latter are scarce both in the desert and in dense forest. In association with the archeological and paleontological evidence, it seems that the decisive biological, cultural, and intellectual evolution of the hominid line took place in the tropical woodlands and savannas of Africa during the earliest phases of the Pleistocene. And, with the increasing intellectual capacity of the most progressive hominid lineage, the first hominines were less and less restricted to a relatively rigid pattern of responses to the environment.

SIGNIFICANCE OF EARLY PLEISTOCENE HOMINIDS FOR THE ENVIRONMENT

The hominids of the early Pleistocene were unable to modify or noticeably effect the environment. Hunting activities were probably confined to the gracile australopiths, and although the cultural evidence from Olduvai points to successful taking of a variety of large mammals, a considerable emphasis was still given to "minor" animal food sources: rodents, insectivores, bats, birds, chameleons, lizards, crabs, fish, and tortoise. Although we know nothing of the social organization or numbers of such hunters, their hunting techniques can be inferred to some degree. Apparently they drove large animals into swamps or open water and then despatched them by clubbing or stoning the bemired prey. In fact, at Olduvai (in lower Bed II) the disjunct lower limb bones of a primitive elephant *(Dinotherium)* were found standing upright in a once-boggy sediment, while the scattered bones of the upper body had been dismembered nearby and the meat devoured by hominid carnivores (see M. D. Leakey, 1967). Scavenging the kills of larger carnivores, a well-known practice among some primitive groups today, may also have been common. Whatever their proficiency, it is evident that the impact of early hominids upon wildlife did not compare with that of the major animal predators.

The vegetable-gathering habits of the small dispersed bands of australopithecines were, at most, seasonally and locally important for the vegetation. However, lacking control of fire, it is highly improbable that early hominids had a greater impact on the plant cover than do chimpanzees or gorillas today. Man-land relationships during the early Pleistocene were, then, entirely one-sided, with the hominid populations but a minor ecological factor in their environment. If anywhere in human history, environmental determinism would have been significant at this stage of biological and cultural evolution.

THE NATURE OF AN AUSTRALOPITHECINE SITE: STERKFONTEIN

The many problems of interpreting australopithecine bone accumulations are well illustrated by the Sterkfontein site, which is situated at 1,450 m. elevation some 9 km. northwest of Krugersdorp in the southern

Transvaal. The first australopithecine fossil was recovered there in 1936, and excavations were conducted off and on between 1947 and 1958 by R. Broom, J. T. Robinson, and R. J. Mason (Robinson and Masoh, 1962; Brain, 1958, ch. 12). An interdisciplinary restudy of the site was begun in 1966 (see Tobias and Hughes, 1969). Various geologists have visited the site, but C. K. Brain has provided the only detailed study (Brain, 1958, also 1967b). The origin of the former cavern and its deposits will be described here, taken in part from Brain's fundamental work at the site, and in part from personal observations.

Sterkfontein refers to a complex of small and large caves, sinkholes, or deep shafts resulting from karst solution and corrosion in Precambrian dolomites. Denudation has lowered the landscape and has removed part of the dolomite roof that once sheltered the australopithecine cavern, so that the site is now situated on a small hill some 30 m. above the valley floor. The nearest stream, the Blaauwbank, drains a series of quartzite and shale cuestas at the continental divide, and forms a minor tributary to the Crocodile River, ultimately emptying into the Indian Ocean via the Limpopo. The local topography is today characterized by high rolling plains, studded with steep ridges of more resistant rock. A number of dissected erosional surfaces, possibly reflecting former stream-cut platforms, follow the Blaauwbank. Of rather younger age are remnants of fluvial conglomerates at +8 m. in the valley floor (see Robinson and Mason, 1962)

The karstic patterns of the Sterkfontein area are strongly influenced by two characteristics of the dolomite rock: First, sinkholes are generally found along the natural joint-patterns of the rock, particularly where such crack-systems intersect (commonly at 75° angles). Others may be related to fracture lines that cut through the dolomite; such faults may be marked by dolomitic breccias ("shatter rock") as well as by veins of milky quartz. Second, sinkholes are tied in with one or more interconnected underground caverns, many filled with water at depth. The geometry of these underground passages reflects, again, on the joint-patterns as well as on the inclination of the rock. At Sterkfontein the dolomite and the interbedded chert bands dip 20–30° in a northerly or northwesterly direction; the shape of the underground cavern is closely related to the alignment of the bedding planes of the bedrock.

Karstic caverns are originally formed *beneath* the water table by the cumulative effects of solution as waters move through the rock fissures. The former caverns at Sterkfontein, and at the neighboring sites of Swartkrans and Kromdraai, are all found at median elevations of 1,452 to 1,454 m., suggesting a broadly contemporaneous origin in relation to essentially a single water table. Since the free-standing water is now about 42 m. lower (Robinson and Mason, 1962; but 60 m. according to Oakley, 1954), well below the bed of the Blaauwbank, the stream must have been at least 40 m. higher at the time the australopithecine caves were first dissolved out of the rock.

The earliest deposits at Sterkfontein began to accumulate after the water table had fallen, vacating a deep cavern that was apparently connected to the surface by two sinkholes. The final emergence of the cavern floor above the water table was probably a result of the Blaauwbank cutting down its valley, thus lowering the water table.

The sedimentary sequence (see Fig. 65) begins with a great mass of dolomite slabs and blocks, set in a matrix of very stony, yellowish-red, sandy silt. This Lower Breccia, well over 8 m. thick, appears to be sterile (Brain, 1968; Tobias and Hughes, 1969), and it may owe its origin to rock collapse around a sinkhole. Masses of red surface soil were washed in, where they intermingled with the jumble of rocks sliding or falling from above. One or more horizons of banded white travertine interrupt the Lower Breccia, indicating flowstone development in the deep cavern, and assuring thorough impregnation and cementation of the soil matrix with calcium carbonate. A massive travertine horizon, as much as 60 cm. thick, terminates the Lower Breccia. A longer period of nondeposition followed, with some removal of breccia by solution, erosion, or collapse of undermined sections (Fig. 65).

At a somewhat later date, soil from adjacent hillslopes began to wash freely into the cave, and there are indications that there was a fairly direct surface entrance near the highest preserved breccias, on the south side of the cave. The amount of soil carried in suggests that a long slope, with a northerly dip, still existed immediately behind this entrance, probably linking it to the hill (on which a modern restaurant is located). At the same time, free rock faces must have existed to the north of the cave, since dolomite talus was constantly being added to the soil wash. Some of this fossiliferous Upper Breccia (following the informal nomenclature of Brain, including part of the "Lower" and all of the "Middle Breccia" of J. T. Robinson) slid rapidly down inclines of 20° to 30° into the depths of the cave. Other parts were washed in gradually at angles of 10° to 15°, preserving distinct bedding that suggests the cave mouth was nearby. Some units also clearly represent collapse breccias, with masses of cave earth with fresh as well as corroded rock (dolomite, chert and cemented Lower Breccia) falling into lower subcaverns as a result of undermining. The Upper Breccia thus consists of several distinct facies, with many generations of active accumulation recorded—each followed by longer periods of nondeposition and limited ero-

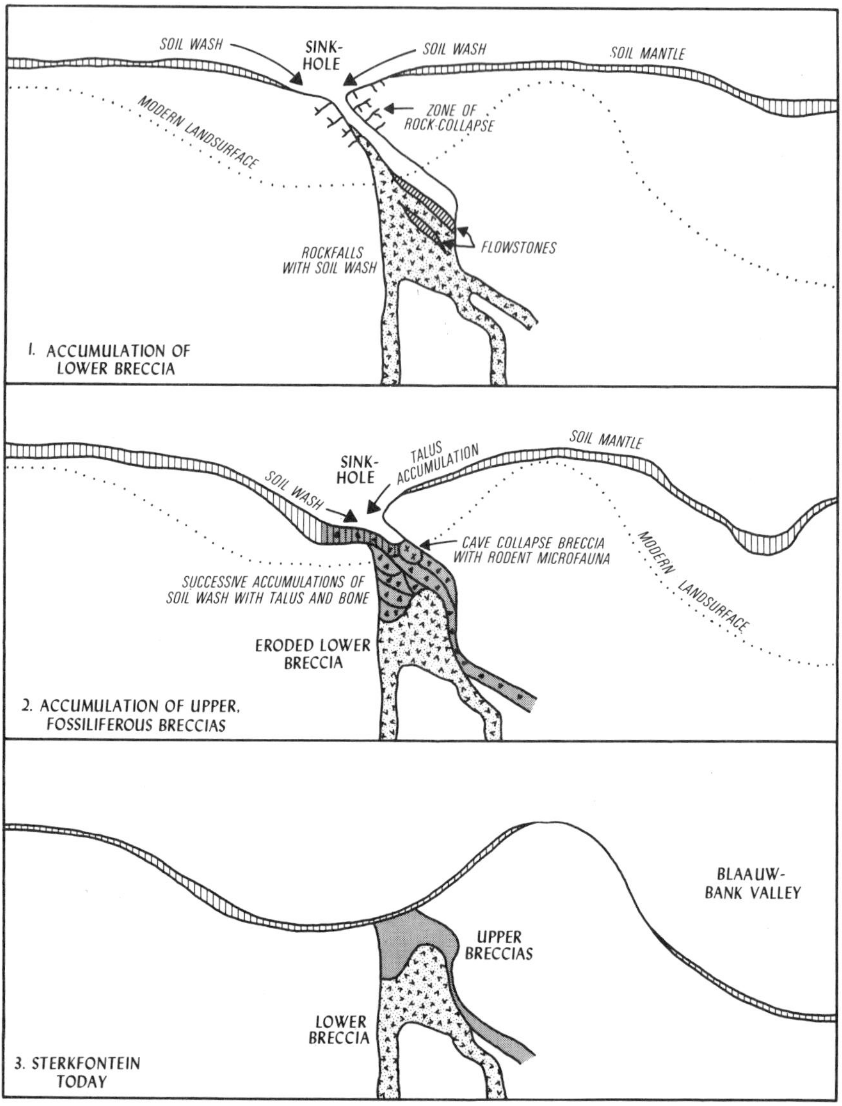

Figure 65. Possible geomorphologic evolution of the Sterkfontein Cave. (Schematic and not to scale.) Note: long periods of erosion and corrosion separate each stage.

sion. The typical matrix is a reddish-brown sandy silt, also cemented by calcium carbonate, but with infrequent laminae or bands of discontinuous flowstone. The cave earth of the collapse breccias is very rich in small rodent bones (the "rodent breccia" of Tobias and Hughes,

1969), while the more typical soilwash includes a variety of mammalian bone, including over 170 fragments pertaining to *A. africanus*. The stone artifacts, associated with three adult teeth and a juvenile palate of a gracile australopithecine (see Robinson and Mason, 1962), come from what this author feels is one of the youngest generations of undisturbed wash.

During the long period of time that has elapsed, the dolomite hill has been denuded and most of the cave superstructure destroyed. Decalcification of the uppermost breccia on one or more occasions has allowed mixing of younger artifacts, of late Pleistocene age, into what was the contemporary soil zone.[4]

The depositional suite at Sterkfontein is complex and multigenerational, as evidenced by the gross units of the Lower Breccia and by the many facies discontinuities and planes of nondeposition in the Upper Breccia. Consequently a time span of 100,000 years or more may well be recorded between the emergence of the cavern and the deposition of the artifact-bearing horizon. In terms of their bedding and disposition, the sediments vary a great deal more than Brain's (1958) study would suggest. However, in keeping with Brain's findings, these variations reflect on the depth and slope of the former cave interior, and on the changing size and nature of the external entrance. The sediment matrix remains typical of *limons rouges* throughout. The fact that such reddish soil products were repeatedly washed into the cave indicates that the vegetation mat was, at times, incomplete, and that high intensity rains were no less common than today. At such times the grass cover was certainly less dense than it was at the arrival of the European settlers in the mid-nineteenth century. Consequently, the Sterkfontein sediments suggest drier conditions, either with a lower rainfall, or with a less equitable distribution of the rains. Size and shape of the dolomite and chert detritus of the Upper Breccia indicate primary derivation from outside of the cave, pointing to scree or talus accumulations. These may well reflect on periods of mechanical weathering (and greater cold?).

Attention can now be given to the nature of the bone accumulations. The faunal inventory (R. F. Ewer in Brain, 1958, appendix A; A. N. Pocock in Tobias and Hughes, 1969; Cooke, 1963) includes the following genera or species: 2 bats; 4 shrews and moles; 1 monkey; 3 baboons; 1 hare; 16 rodents; 3 hyenas; 2 dogs; 1 leopard; 2 saber-tooths; 2 hyrax; 2 horses; 1 pig; 2 antelopes; 1 or 2 gazelles; 2 lizards; and various songbirds, in addition to *A. africanus*. The question is, just what is the

4. The "Upper Breccia" of Robinson (Robinson and Mason, 1962) is simply a partly decalcified horizon developed in the older deposits.

origin of these bones? The two basic possibilities are that (a) the bones were derived from the exterior and brought in by running water or mass movements; or (b) the bones accumulated in place due to the activity of hominids or other carnivores. Although the fossiliferous breccia has a minimum of 200 bone fragments per cubic foot (30-cm. cube) (Tobias and Hughes, 1969), the bone is not generally concentrated or disposed in such a fashion as to suggest a primary accumulation. Instead the great mass of fossils appears to have been washed in from the outside (Brain, 1971).

Only the "rodent breccia," at the western end of the site, appears to be a primary accumulation, probably representing owl droppings mixed with cave earth, later involved in a minor internal cave collapse. Owls frequently perch in dark caves during the day, emerging to feed in the evenings. The small rodents on which they usually subsist are swallowed whole, and the bones later excreted, intact and wrapped in fur (Brain, 1971). In this way, rodent bones accumulate on cave floors, sometimes amounting to 20 per cent of the sediment by weight. Owl pellets also include remains of beetles, millipedes, lizards, and chameleons.

The great bulk of the Sterkfontein fossils appear to reflect on activities at or within the former surface entrance of the cavern. Brain (1971, 1969b) has discussed various possibilities on the basis of comparative observations of different animal activities:

1) Porcupines are compulsive collectors of bones, some 60 per cent of which show evidence of gnawing. A porcupine is recorded in the Sterkfontein inventory, but at Swartkrans only 2 per cent of the bone shows evidence of gnawing, suggesting that porcupines played a minor role at best.
2) Hyenas were once widely held to accumulate bone in their lairs. However, they eat their food where they find it and rarely, if ever, drag it back to caves. Furthermore, they ingest and digest bone fairly completely, leaving little or no recognizable bone.
3) Saber-tooth cats, of which two extinct species are recorded at Sterkfontein, are adapted to tear, rip, or slice meat. They are hardly able to damage bone, whereas most of the Sterkfontein fossils are strongly fragmented.
4) Leopards drag their kills into trees in order to avoid the competition of hyenas and other scavengers. Interestingly enough, some of the few large trees of the rolling South African grasslands are found in front of cave entrances or in sinkholes. There the trees, such as white stinkwood *(Celtis kraussiana)*, are sheltered from wind and grass fires, while they have access to extra moisture. Such sites would repeatedly attract leopards. At Swartkrans the types of bones pre-

served most frequently (cranial vault, face, jaws, horn cores, and certain long bones) are precisely those left intact at leopard kill sites. In fact, the imprint of leopard canine teeth as been found on the cranial vault of a juvenile *A. robustus* at Swartkrans.

5) Some smaller animals, such as rabbit, hare, and occasionally gazelle, seek out caves to die. Rabbit and hare bones are uncommon, however.

6) Primitive hunters can be recognized by their practice of smashing long bones to get at the marrow. Yet at Swartkrans, even with its evidence of human artifacts, only about 7 per cent of the bone suggests the possibility of human activity. There is no charred bone at either site.

In overview, it would seem that Brain's conclusions for Swartkrans are also applicable in a general way to Sterkfontein, that "although [the cave entrance] may have provided intermittent refuge to primitive man and various animals, it was essentially a leopard lair, used over many thousands of years" (1969b, p. 138).

The next matter of interest at Sterkfontein is the nature of the artifacts found in the youngest breccias in the western part of the site ("Extension Site"). A total of 286 stone objects were excavated in 1957–58 (Robinson and Mason, 1962), and many more have been uncovered by the current excavations. In terms of raw material the original collection is:

57 per cent quartzite (available in the Blaauwbank valley at 500 m.)

38.4 per cent quartz (vein quartz is found at the surface near various faults)

2.5 per cent chert (locally interbedded with dolomite)

2.1 per cent diabase (also found in the Blaauwbank valley gravels).

This indicates that 97.5 per cent of the rock is "foreign," and could therefore have been carried in by man. However, the possibility must also be considered that the floodplain of the Blaauwbank may have been 30 m. or so higher at the time the site breccias accumulated. In this way "foreign" rock could have been introduced directly by fluvial transport. This pertinent question can only be answered by an intensive geomorphologic study of the area, currently being undertaken by A. B. A. Brink and T. C. Partridge.

R. J. Mason (Robinson and Mason, 1962) classified 98 of the stones as artifacts, the remainder as natural pebbles (32) or objects which may have been naturally fractured (156). Of the 98 artifacts, 92 are claimed to be in mint condition, indicating little or no subsequent transport. M.D. Leakey (1970), on the other hand, recognizes only 73 as artifacts, while in a more recent restudy of essentially the same collection, R.G.

Klein (unpublished) identified even fewer undoubted artifacts (35), and placed a greater number into the category of pieces that may have been naturally flaked. Further, traces of abrasion or weathering were observed on a larger number of the artifacts than Mason indicated. Vertical and horizontal plots of the artifacts by Robinson and Mason (1962, Fig. 6) indicate a fairly random scatter through this particular breccia unit. A far more detailed understanding of the artifacts should result from the eventual study of the large collection which has been acquired since 1966. For the moment it is possible to say that the typology is rather nondescript: the great majority are irregular hunks of rock, from which a few flakes have been struck, or oval pebbles on which rough cutting edges have been made on one or two sides. There are three rudimentary "hand-axes." In addition to the stone tools, Mason identified a single split-bone point with evidence of polish as an artifact (Robinson and Mason, 1962).

The Sterkfontein stone industry is unusual in several ways. There are far fewer stone flakes than have obviously been detached from the artifacts present. Although some of the small quartz debris may have resulted from shattering of the brittle vein quartz during toolmaking (Robinson and Mason, 1962), the stone chips typically produced in the process of toolmaking ("waste flakes") are essentially lacking. This lack strengthens the argument, provided by the random scatter of artifacts, that this was not an occupation floor. The many difficulties in assessing the Sterkfontein stone industry illustrate the overall problem of recognizing and analyzing early artifact collections that lack distinctive, fashioned tools. The same applies to the rather primitive stoneworking evident on the Olduvai Bed I occupation floors, and even more so to the apparent artifacts from the Omo Valley and east Rudolf.

A final question raised by the Sterkfontein material is, who made these artifacts? The total absence of artifacts in the main body of the Upper Breccia, with its mass of *A. africanus* fossils, suggests that *A. africanus* did not make such tools (Robinson and Mason, 1962). In fact, the gracile australopiths probably ended up as leopard prey. Unfortunately the three adult teeth and the juvenile palate of a hominid recovered together with the artifacts are insufficiently diagnostic: they may belong to *A. africanus,* but they may also belong to *Homo habilis* or even *H. erectus* (Howell, 1967). But whatever the final verdict of the fossils, man was present in the Sterkfontein area during the final period of breccia accumulation.

In overview, the australopithecine breccia at Sterkfontein records a long period of cave filling, contemporaneous with accelerated soil erosion and a drier (and colder?) climate. The sediments offer no clues as to relative dating, and no material suitable for absolute dating has been recovered. However, the presence of 2 equids (*Equus capensis* and zebra) only in the top units of the Upper Breccia indicates that Sterkfontein is, in part, older than Olduvai Bed I, i.e., greater than 1.8 million years. Bones related to leopard kills near the cave entrance were washed into the cave interior over long periods of time, over tens of thousands of years. These fossils included *A. africanus,* who apparently fitted the menu for other, more successful carnivores. There was little or no occupation of the cavern itself by either hominids or other animals, except for cave owls. Finally, during the terminal millenia of sedimentation, prehistoric man, of unknown physical type, made or used artifacts near the mouth of the cave. The period of occupation was probably brief, but some of the artifacts were gradually swept into the cave with soil wash from adjacent hillslopes.

Sterkfontein serves to illustrate the precarious existence of the early Pleistocene ape-men, and to emphasize the incomplete nature of the archeological record for this fascinating but poorly understood range of time.

TWENTY-SIX

Man-Land Relationships of Acheulian Hunter-Gatherers

TOOLMAKING TRADITIONS OF THE EARLY AND MIDDLE PLEISTOCENE

The primary record of prehistoric man during the long span of mid-Pleistocene time is provided by stone artifacts. The earliest standardized toolmaking tradition known today is represented among the middle and upper strata of Bed II at Olduvai Gorge (M.D. Leakey, 1967). The most notable implements, which are replicated in considerable numbers, are hand-axes. These are made from large flakes or pebbles, on which a significant portion of the circumference has been trimmed from two surfaces to produce a sharp edge. Trimming usually extends far back from the edges of such tools and often covers most or all of both opposed surfaces. As a consequence, hand-axes are frequently known as "bifaces" (= two-faced implements). Generally, when a hand-axe is viewed in plan, one end will be relatively pointed (see Fig. 66, [1 and 2]), while the other (called the "butt") may be rounded and sometimes completely unworked. Most hand-axes are easily recognized as man-made artifacts, and have therefore been subject to uncontrolled collecting for almost a century. For better or for worse these implements have become the index fossil of the Lower Paleolithic of Europe, and of the Earlier Stone Ages of Africa and India. It is, in fact, common to speak of "hand-axe industries," almost all of which are generally grouped within the Acheulian "culture" or—better—industrial tradition.

It is not hard to imagine how gradual change through time in the extent to which a stone knapper modified a pebble or large flake would have led to the development of hand-axes by populations whose ancestors had made simpler Oldowan "choppers" and "chopping tools" on pebbles. In fact, there are pieces from Sterkfontein which could be regarded either as particularly elaborate chopping tools or as very crude hand-axes (R.G. Klein, unpublished). In this sense, the Acheulian can be considered a logical outgrowth of the Oldowan. While Oldowan toolmakers simply modified the circumference of pebbles in their efforts to produce a tool, Acheulian craftsmen usually completely transformed pebbles, chunks, or large flakes so that it is now often impossible to determine on what kind or shape of object a finished hand-axe was made. In addition to hand-axes, distinguished generally by being more or less pointed at one end, Acheulian knappers also made another basic kind of biface, called a "cleaver" (Fig. 66, [8]). The essential element here is an axe-like straight edge at the end where a point might be expected on a hand-axe. The major use of hand-axes and cleavers appears to have been as mattocks and flensers for removing hide, cutting ligaments, parceling meat, and removing flesh from the bone or hide of large game (J. D. Clark, 1960). Hand-axes could be used specifically to pierce hide, while cleavers were more suited for general chopping purposes or for hacking through or prying apart the joints of large animals.

The chips or flakes trimmed from bifacial tools were commonly detached by striking the rock face, from an angle, with a suitable stone "hammer" or a piece of bone or wood. The latter provides a "soft" hammer particularly useful for the fine trimming of cutting edges. Some of the intact or battered, shapeless stones at Acheulian sites probably pertain to such hammer-stones, or may have been otherwise employed as missiles or as general chopping and pounding equipment. The diverse flakes, formed as a by-product of biface preparation, were frequently employed as tools, with or without a little edge-trimming or retouch. Depending on their shape, flakes could be held in the hand and used as a scraper (see "side-scraper," Fig. 66, [7]) or chisel (see "end-scraper," Fig. 66, [6] to remove meat from hide or bone. Notching a flake (Fig. 66, [4]) can also produce a notch or hook of possible use for trimming hide or wood. Whether stone tools were also used to secure vegetable food is quite uncertain.

The Acheulian of East Africa ranges in time from Olduvai middle Bed II, perhaps 500,000 or even 1 million years ago,[1] to about 60,000 B.P. at

1. The single K/Ar date from upper Bed II (0.5 million years) is unacceptable since sample provenance is uncertain (R. L. Hay, personal communication), while the 1.1-million-year date from lower Bed II may represent a contaminated sample from Bed V (Hay, comments on Evernden and Curtis, 1965).

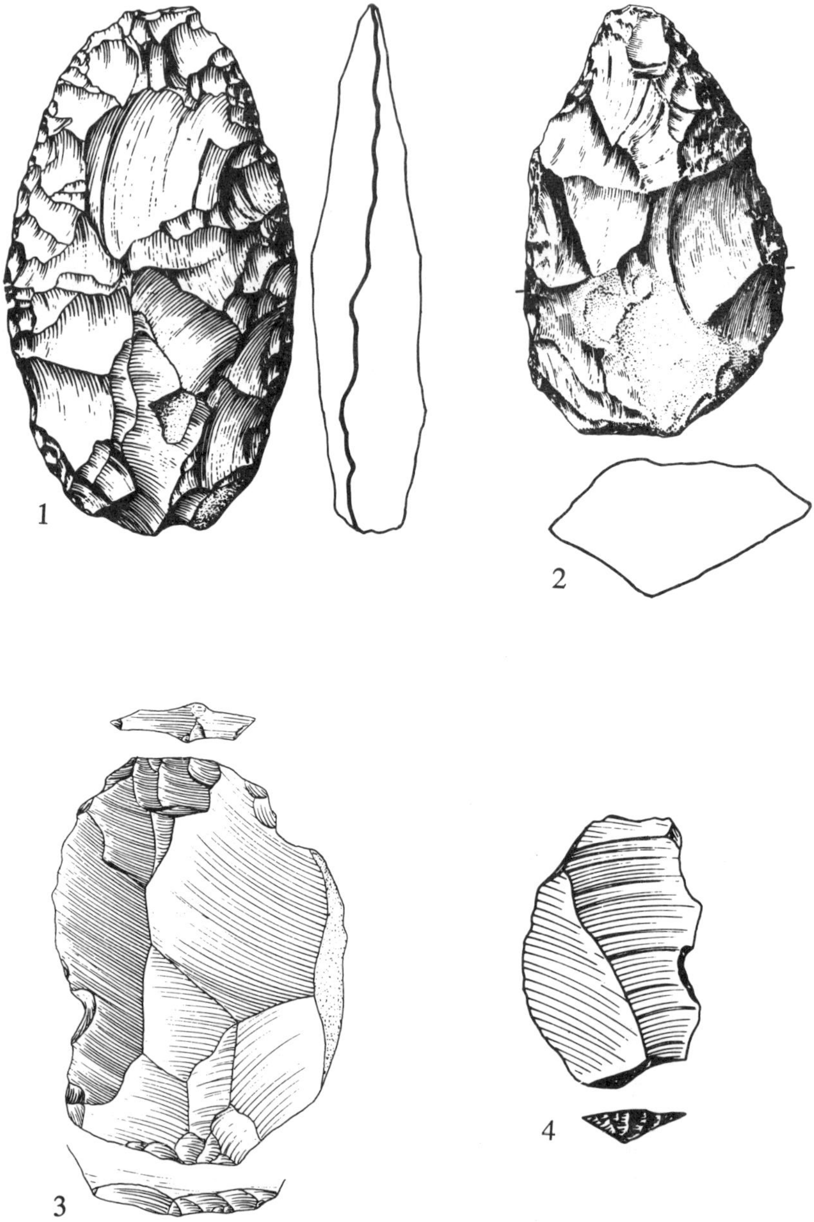

Figure 66. Some Middle Acheulian stone implements from the Somme Valley (Cagny) (from F. Bordes, 1961a, by permission of the author): (1,2) different

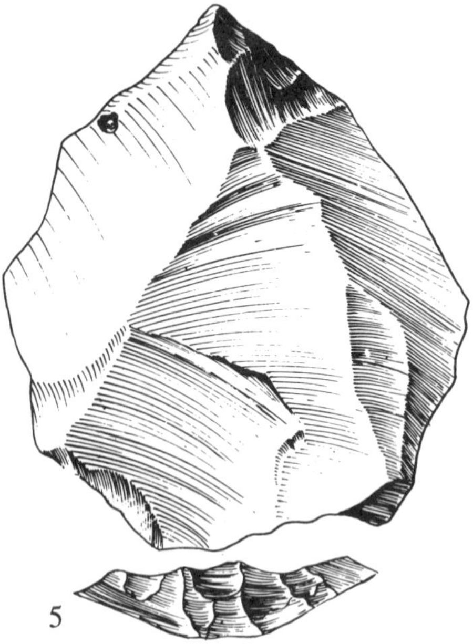

5

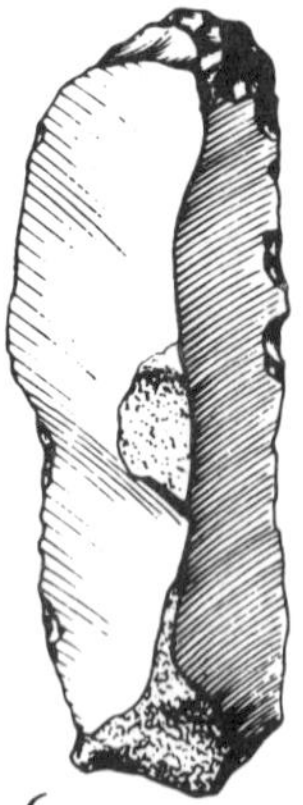

6

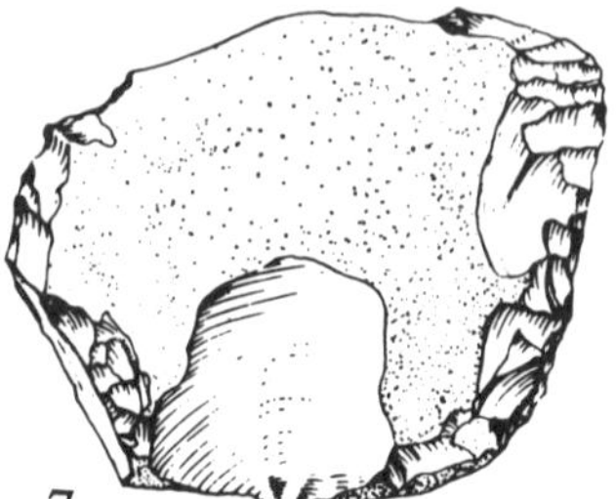

7

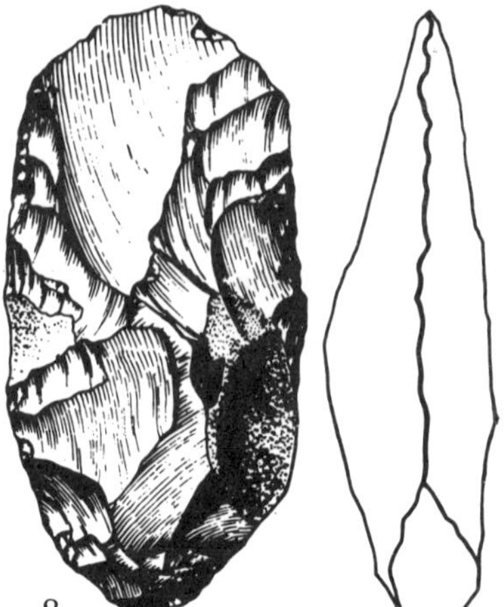

8

bifaces or "hand-axes," (3) flake, (4) notched flake, (5) core, (6) end-scraper, (7) side-scraper, (8) cleaver. (Reduced to one-half natural size.)

Kalambo Falls J. D. Clark, 1969). In Europe the Acheulian is first recorded from Elster-age deposits and lasted through the Eem Interglacial. To be sure, there are developments through time. Bifacial tools tended to become more sophisticated and were frequently characterized by fine symmetry in plan and cross-section. At some point in time the soft-hammer technique was also invented. Nonetheless, hand-axes with unprepared butts or with irregular and assymetrical shapes continued to be made in all Acheulian workshops, often side by side with handsome, blade-thin pieces. Frequently the absence of attractive bifaces was a result of the absence of suitable raw material: extremely hard rock may only fracture under very heavy blows, while the fractures of many rocks with geometrical jointing or fracture lines cannot be fully controlled by the workman.

Consequently, the presence or absence of refined bifaces is *not necessarily* a reflection of temporal development. On the other hand, one trend that does seem to have been more or less universal within the Acheulian was an increasing emphasis on flakes as time passed. Eventually, an elaborate variety of tools came to be made of flakes. This increasing reliance on flakes led to an increased sophistication in the techniques whereby flakes were produced. In the earlier Oldowan, most flakes, whether used or not, were probably simply by-products of the manufacture of other tools. Already in the late Oldowan and early Acheulian, however, flakes had come to be specially produced from pieces that we call "cores" (Fig. 66, [5]). Cores were used principally for the production of flakes, and whether or not an "exhausted" core was later used as a tool was in most cases incidental to its initial purpose. In the middle and later Acheulian, a major innovation was the preparation of a core in such a way that the size and shape of a flake struck from it were predetermined (see Fig. 66, [3] for a flake struck from a prepared core of "Levallois" type). Sophisticated prepared-core techniques were particularly prominent in Europe during the later Middle Pleistocene and thereafter, but were also present in Africa.

In general, then, there is evidence to suggest that the Acheulian changed through time. At this stage in our knowledge, however, it would be dangerous to generalize on this any further. Thus, while Howell (1966), Bordes (1968), and Collins (1969) believe that they can trace the detailed evolution of the Acheulian in Europe, Isaac (1968b, 1969) has so far been able to distinguish only two main stages of the Acheulian in East Africa.

The almost incredible time-span of 500,000 years or more for the Acheulian industries is matched by a distribution over three continents, from South Africa to England, and from Morocco to India. Apparent stylistic elements, such as biface size or shape, e.g., almond, pear, twisted, lanceolate, ovate, etc., remain enigmatic in terms of interpretation, and no systematic variation through either space or time has yet been demonstrated. Possible functional attributes, such as the frequency and diversity of flake tools, also show more variation among contemporaneous or sequential occupations in one area than on an interregional basis. As a result, it is difficult to subdivide the geographic continuum of Acheulian industries. Regional variants appear to be present in northwestern Europe, where typical cleavers are very rare, and on the Indian subcontinent, where non-Acheulian components are prominent (see below). But throughout Africa, Spain, and the Near East there is an amazing continuity of traditions.

The Acheulian is not the sole industrial tradition of the mid-Pleistocene. Oldowan craftsmanship was still practiced at Olduvai, throughout the deposition of Bed II, and Mary Leakey (1967) assumes the coexistence of two distinct industrial traditions with a limited, but nonetheless discernible, degree of contact with one another. It is just as possible, however, that assemblages both with hand-axes and without them were left by people of the same culture. Similar precursors of Oldowan type, possibly also representing archaic survivals, are recorded in the Sterkfontein and Vaal valleys (see chapter 25, and Mason, 1963) as well as in western Morocco (Biberson, 1961b). Of a rather different nature are the non-Acheulian traditions evident along the peripheries of Acheulian distribution in Europe and Asia (Fig. 67). So, for example, the Elster-age Buda industry of Vertésszöllös, near Budapest, has a very low proportion of shaped tools, and even these are rather unimpressive (see Kretzoi and Vertes, 1965; also Howell, 1966). Apart from a good number of small, well-flaked, scraper-like flakes, there are many small choppers and chopping tools, pieces on which trimming was much less extensive than on hand-axes and cleavers. Of comparable or greater age are possibly similar industries in Czechoslovakia (Stranska) and southern France (Vallonnet). Hand-axes and cleavers are also absent from the important, Holstein-age Clactonian industry and its Saale-age offshoots in western Europe (see Howell, 1966; and Collins, 1969). Here the emphasis is on flake-tools, many of them originally produced by smashing pebbles against large rocks. Chopping tools are of less and less importance.

Equally significant are the non-Acheulian industries of eastern and southern Asia (see Movius, 1943, 1948, 1955; Heekeren, 1957). These are now known from northwest India (Soan), central Burma (Anyathian), northern (and southern?) China (Choukoutienian), Malaya (Tampanian), and Java (Patjitanian). These industries are characterized principally by choppers and chopping tools. Choppers have cutting edges flaked from one side (direction) only, while the edges of chopping tools are struck from two sides (directions). One or more edges may be worked, and an occasional example even approaches a hand-axe in its general appearance. No purpose would be served by calling these artifacts hand-axes, however, since they are clearly part of another spectrum of artifactual variability than that containing the true hand-axes of contemporary Europe and Africa. In addition to choppers and chopping tools, the non-Acheulian industries of Asia were characterized by a variety of other large tools (e.g., pieces called "hand-adzes") and by flake tools. Geological evidence suggests that the general chopper/chopping-tool tradition characterized eastern and southeastern (and possibly central) Asia during the early and middle Pleistocene, persisting well into the late Pleistocene.

The origins of the Eurasiatic chopping-tool traditions is quite obscure, as are their mutual interrelationships. Equally enigmatic is the nature of interaction between contiguous populations practicing Acheulian and non-Acheulian industrial traditions, and there is the possibility that in some cases Acheulian and non-Acheulian assemblages may have been left behind by one and the same people doing different things at different sites. A crucial study area will be the Indian subcontinent, where both traditions overlap. In the Narmada Valley, where both are found together, chopper-chopping tools are found only in a stratigraphic context older than the Acheulian (A. Khatri, personal communication). In fact the Indian Acheulian seems relatively late and may well have displaced an older chopping-tool tradition.

FOSSIL REMAINS OF MID-PLEISTOCENE MEN

A number of localities have produced early and middle Pleistocene strata with human fossils, some of them associated with artifacts of either the hand-axe or chopping-tool traditions. These fossils are best referred to the genus *Homo,* and they appear to represent the general stock ancestral to modern man. However, variability is so great that, with the rather sporadic and incomplete fossil record, more questions are raised than are answered. Certainly the evidence does not warrant the mass of generic names that has been applied to each fossil, nor does it support a generic distinction from later human types (see Howell, 1960). But even the differences among some contemporary fossils is so great that subspecific or specific distinctions may be necessary. So, for example, the Elster-age hominines from Heidelberg and Vertésszöllös are more advanced than those from Ternifine or Java. However, we should expect at least as much geographical variation among early or mid-Pleistocene men as among populations of modern man. For this reason a number of authors prefer to place all fossils of Lower Pleistocene age into a single taxon, *Homo erectus* (see Howell, 1960). *Homo sapiens,* as represented at Swanscombe and Steinheim, appears to have developed out of *Homo erectus* at or near the beginning of the Middle Pleistocene. It is of course possible that the *sapiens* threshold was crossed earlier in some places than in others, perhaps earliest of all in Europe, as appears to be suggested by the remarkable *sapiens*-like fragment from Vertésszöllös.

Some of the more important localities with early hominid remains can be listed here (see Fig. 67) in stratigraphic order, with their identification—often tentative—and, in quotation marks, their superfluous generic or descriptive names:

A. Early Lower Pleistocene
   - *a)* Olduvai; upper Bed II. *Homo* cf. *erectus* ("Chellean Man") and one or more australopithecines.
   - *b)* Swartkrans. *Homo erectus* ("Telanthropus") (see Clarke *et al.,* 1969) and *Australopithecus robustus.*
   - *c)* Sangiran and Modjokerto, Java; Djetis fauna. *Homo* cf. *erectus* ("Meganthropus palaeojavanicus") and *Homo erectus* ("Pithecanthropus") (see Tobias and Koenigswald, 1964).

B. Late Lower Pleistocene
   - *a)* Heidelberg-Mauer, Germany. *Homo cf. erectus* ("Palaeanthropus"), of post-Cromerian but pre-Elster age (see Howell, 1960).
   - *b)* Vertésszöllös, Hungary. *Homo* cf. *sapiens* of Elster age (see Kretzoi and Vertes, 1965).
   - *c)* Ternifine, Algeria. *Homo erectus* ("Atlanthropus") see Howell, 1960).
   - *d)* Koro Toro, Chad. *Homo* cf. *erectus* ("Tchadanthropus"), a poorly preserved fossil of uncertain age (see Tobias, 1968).
   - *e)* Sangiran and Trinil, Java; Trinil fauna. *Homo erectus* ("Pithecanthropus") (see Koenigswald, 1949, 1968).

C. Uncertain, early to mid-Pleistocene
   - *a)* Olduvai; Bed IV. *Homo erectus* (see Leakey, Tobias, and Isaac, 1967).
   - *b)* Choukoutien I, near Peking, China. *Homo erectus* ("Sinanthropus Pekinensis") of Holstein age or from an earlier interglacial or interstadial (see Chang, 1968; Kahlke, 1968).
   - *c)* Lantian, China. *Homo erectus* ("Sinanthropus lantianensis"), of Holstein age or more probably from an earlier interglacial (see Kahlke, 1968).

D. Early Middle Pleistocene
  a) Swanscombe, England. *Homo* cf. *sapiens* of late Holstein age (see Howell, 1960).
  b) Steinheim, Germany. *Homo* cf. *sapiens* of late Holstein age (see Howell, 1960).
  c) Montmaurin, France. *Homo* cf. *sapiens* of late Holstein or Riss interstadial age (see Howell, 1960).

E. Late Middle Pleistocene
  a) Rabat and Sidi Abderrahman (Littorina Cave), Morocco. *Homo* cf. *sapiens,* of early post-Tyrrhenian I age (see Howell, 1960; Biberson, 1961a, 1961b).
  b) Lower Omo Basin, Ethiopia; Kibish Formation, Member I (130,000 B.P.?). *Homo sapiens* (Butzer, Day and Leakey, 1969; Butzer, Brown and Thurber, 1970).

Omitted from this list are the primitive *Homo sapiens* from Kanjera and Kanam (see Tobias, 1968) on the shores of Lake Victoria. Although originally supposed to be of early or mid-Pleistocene age, they may well be early Upper Pleistocene.

A survey of these fossils – of their specific attributes and localities as well as their industrial associations – shows that few biological generalizations can be made for the carriers of the different tools. The Acheulian at Swanscombe is associated with a proto-*sapiens* fossil, while *Homo erectus* appears to be related to the earlier hand-axe industries of Olduvai and Ternifine. The nonbiface industries of Europe may possibly have been associated with proto-*sapiens* types, while at least the earlier chopper/chopping-tool industries of eastern Asia were clearly the work of *Homo erectus.* On the other hand, it is also possible to infer that, during the early Pleistocene, *Homo erectus* made Acheulian artifacts in Africa and Europe as well as non-Acheulian (chopper/chopping tool) artifacts in eastern Asia. In mid-Pleistocene times, early *Homo sapiens* continued the Acheulian tradition in Europe and Africa, and there is as yet no compelling reason to suppose that the makers of the mid-Pleistocene variants of the Asian chopper/chopping-tool tradition were not also members of the modern species. At the moment, there simply are too few human remains in suitable archeological contexts to resolve the problem of the biological identity – if any – of the Acheulian craftsmen.

FURTHER EVIDENCE OF ACHEULIAN TECHNOLOGY

Many Acheulian sites have provided cultural items other than stone tools.

In addition to stone, bone and wood were widely employed for artifactual purposes. Although comparatively rare, worked bone and wood have usually been found at well-excavated sites that do favor bone preservation. In the Elster-age Acheulian occupation horizons of Torralba and Ambrona, there are substantial quantities of bone with clear evidence both of deliberate fashioning and trimming and of use, such as scratches, polish, and wear (Biberson and Aguirre, 1965). In particular, the persistent patterns of bone-splitting have been duplicated by modern experiments, using hand-axes and fresh elephant bone. The Torralba evidence shows that long bones and ribs of various larger animals were deliberately fractured longitudinally down the center. The resulting fragments were often flaked by use of a hammer, producing some tool types analogous in shape to picks, hand-axes, or cleavers. Other implements find no analogies among stone artifact assemblages and may have had functions not duplicated by any stone tools. It seems that bone-working at Torralba and Ambrona was a consequence of the local scarcity of good raw material for stone implements. However, worked bone has also been found on the Acheulian living-floors of Bed II at Olduvai Gorge (M. D. Leakey, 1967).

An increasing number of wood implements has come to light in recent years, although wood preservation is usually poor from early and mid-Pleistocene time ranges. Perhaps the most interesting collection was made at Torralba, where several pieces of pine wood with various shapes show clear evidence of cut marks, trimming, or sharpening. The pointed pieces are suggestive of offensive weapons. More convincing evidence of wooden spears is available from the Eem-age site of Lehringen, where a 2-meter spear, made of yew, was found between the ribs of an elephant. It is quite possible that the meager record of offensive weapons is a result of the widespread use of wooden implements such as spears, clubs, or throwing-sticks. Wooden sticks or bone fragments may also have been used to dig for roots or bulbs of wild vegetables, and a number of obliquely truncated and pointed sticks from the early Würm site of Kalambo Falls, Zambia, probably belong in this category.

The use of fire is already recorded by charred bone at Vertésszöllös (Kretzoi and Vertes, 1965), by several good hearths at a cave near St. Estève in southeastern France, and by charcoal fragments, carbon, and charred bone at Torralba (Howell, 1966), each of these sites dating from the Elster. The early use of controlled fire through much of middle latitude Eurasia is further suggested by abundant hearths at Choukoutien (Holstein age or earlier) (see Oakley, 1961). This practice seems to have spread slowly into warmer latitudes, and may possibly have remained unknown in Africa until late Pleistocene times (J. D. Clark,

1960). Whether or not mid-Pleistocene man knew how to use fire as a hunting device remains unknown.

Other technological information is provided by structures commonly interpreted as simple shelters. Already in Bed I at Olduvai Gorge there is a circle of loosely piled lava blocks which appears to have been artificially constructed (M. D. Leakey, 1967). A similar interpretation must be given to a linear arrangement of elephant bones and rocks at Ambrona (Howell, 1966), and stone circles from Acheulian sites in Nubia and Syria also suggest structures. More impressive are arrangements from the Holstein-age (early Tyrrhenian I), Acheulian site of Terra Amata, at Nice (de Lumley, 1967, 1969a). A number of structures, interpreted as huts, are indicated by: *(a)* imprints of a series of stakes, each averaging about 8 cm. in diameter, that were driven into sand to form the walls of a shelter; *(b)* lines of stones, paralleling the stake imprints, and occasionally stacked one on the other, apparently to brace the walls of the shelter; and *(c)* impressions of thick posts along the central axis of the structures. Floor plans range from 8 to 15 m. in length and 4 to 6 m. in width, and the oval floors are covered with organic matter, ash, artifacts, and waste flakes. A similar but less convincing structure is indicated within the nearby Lazaret cave, apparently occupied during late Riss time (de Lumley, 1969b). In a later time range, ca. 60,000 B.P., a rough arc of intentionally placed stones at Kalambo Falls, Zambia, may also have formed the base of a wind shelter (J. D. Clark, 1960). From this fragmentary evidence, it nonetheless seems abundantly clear that mid-Pleistocene men knew how to construct windbreaks and other crude shelters in open situations as well as within caves. The fact that such structures are not represented at all sites suggests that they were only assembled when weather made them necessary and, possibly, where longer periods of occupation were contemplated.

The sum total of Acheulian technology is simple and perhaps rudimentary, but the available data serves to show how biased and unrepresentative stone tools are as an index of prehistoric culture.

ACHEULIAN CULTURE VERSUS TECHNOCOMPLEX

Comprehensive studies of the European Lower Paleolithic or of the African Earlier Stone Age invariably emphasize the low degree of systematic cultural differentiation in mid-Pleistocene assemblages. Isaac, in particular, has shown that there are few apparent long-term trends of change in the Acheulian of East Africa (Isaac, 1968b, 1969). Instead, the great variability of samples that are close together in time shows that these long-term trends account for an insignificant part of the over-all diversity of Acheulian craft norms. A similar case could be made for geographical variation, namely, that variation within one ecologically defined region is at least as prominent as variability between regions or even continents.

In view of the apparent homogeneity of Acheulian artifact assemblages through time and space, one may ask to what extent the Acheulian was *a* culture or culture complex. As Bowman (1971) points out, many archeologists have explicitly or implicitly defined a "culture" as an assemblage of material objects (so, for example, Childe, 1956, p. 123). In fact, Grahame Clark (1957, p. 169) feels that criteria reflecting choice or style are more reliable for defining prehistoric cultures than criteria influenced by ecological or economic factors. If this view were accepted, then the Acheulian could be identified as *a* culture that persisted over 500,000 years and spanned two and a half continents (Bowman, 1971). In terms of our contemporary understanding of the stability and dispersal of identity-conscious sociocultural groups, this would, of course, be absurd. The problem remains to find a reasonable interpretation for the apparent homogeneity of Acheulian industries. Two approaches appear profitable in this regard.

It is quite possible that different groups with Acheulian traits did in fact have specific ecological adaptations, and that their sociocultural patterns varied far more than their tool inventory. Unfortunately, the only statistically significant sample of Acheulian manifestations is that of the stone industries. Existing tool classifications emphasize—almost exclusively—form and style, rather than function (Bowman, 1971). It may be that a functionally oriented typology of artifact assemblages will provide more relevant insights into Acheulian tool-kits, reflecting on subsistence patterns, ecological adaptations, or dietary habits. However, until a successful functional typology has in fact been devised, it must remain questionable whether tangible data of this kind will be forthcoming.

The problem of Acheulian "identity" may also be approached by analogy from modern ethnological information (Isaac, 1968b, also 1968a). It appears that among low-density populations of hunter-gatherers, individual "patrilocal" bands tend to show marked, if trivial, cultural idiosyncrasy, while larger groupings into tribes or linguistic groups are often arbitrary and seldom meaningful (see Owen, 1965). As a result, the bands, on the one hand, and the "culture-area," on the other, seem to be the only useful cultural taxa. Isaac (1968b) suggests that this pattern may have close analogies with the Acheulian artifact assemblages, as there are no recognizable entities between the taxonomic level of the individual camp-site occurrence and that of the industrial complex. In such hunter-gatherer societies, there may also be an adaptive continuity of idiosyncratic male culture and locality-specific lore, with the dispersal and recombination of women into scattered bands favoring a continual diffusion and reinforcement mechanism for basic cultural and technological traits (Owen, 1965). Again, it seems that such a system of cultural transmission should have a high inertia against radical change. In this way an adequately adapted body of technological and socioeconomic traditions might be maintained over wide areas during a period of perhaps a half million years. Such stability must, of course, also be viewed within the perspective of less evolved brains and relatively specialized ecological-economic relationships that is suggested by the paleontological and archeological evidence (Isaac, 1968b).

Perhaps the concept of the technocomplex, introduced by D. L. Clark (1968, ch. 8), offers the most viable approach to the technical and the adaptive manifestations of the Acheulians. The technocomplex embraces a huge system of loosely related culture groups, cultures, assemblages and artifact types. The sociocultural basis of such a system could be completely heterogeneous in terms of social organizations, languages, and the like, although it would inevitably contain small homogeneous units of cultural status. The technocomplex represents "the partly independent arrival of diverse developing culture systems at the same general equilibrium format based on a similar economic strategy, operating in similar environments, with a similar technology and similar past trajectories" (Clarke, 1968, p. 355). Thus heterogeneity is broadly constrained by both the organizational format and the operational ecosystem. The technocomplex does not mask the variable composition of the Acheulian tool-kit that suggests the practice of several discrete activities, each requiring a special set of tools (J. D. Clark, 1960). Similarly it does not assume a single socio-organizational pattern or undirectional cultural development. In fact, the technocomplex offers an ideal framework for considering the evidence for the increasing complexity of mid-Pleistocene hominid behavior.

INFERENCES ABOUT SUBSISTENCE AND SETTLEMENT PATTERNS

Wherever bone has been preserved, the archeological inventory of Acheulian sites suggests that large mammals provided an important, if not the major, food source. Almost all of the mammalian bone is disarticulated, broken, or battered, and sometimes reduced to small splinters, so that there can be little question that the animals were butchered and used for food. Numbers of individuals indicate that certain genera are heavily represented, as for example, elephant at Torralba and Ambrona (Howell, 1966), baboon at Olorgesailie in the Kenya Rift (Isaac, 1968a). A wide spectrum of large animals was hunted, however, including such diverse forms as hippo, horse, bovids, pigs, elephant, and rhino at Olorgesailie, as well as wolf and a large lion at Ambrona. Furthermore, the hunting bag was not limited to large mammals: rodent, bird, and reptile bones are commonly present including crocodile at Olorgesailie and a weasel at Ambrona. Whether some of these animals were scavenged from animal kills is difficult to decide, but the great bulk was certainly killed by man. The general impression obtains that all forms of protein available with given hunting techniques were utilized. Sporadically, as in the case of the one baboon concentration at Olorgesailie, a whole troop was probably ambushed and killed on a fortuitous occasion (Isaac, 1968a). In other instances, such as at Torralba and Ambrona, a long-term concentration on elephant hunting is apparent, possibly reflecting on a successful local hunting adaptation or on the exceptionally high meat yield per effort expended.

The nature of Acheulian hunting techniques is still imperfectly understood. Evidence from both Olduvai Gorge and Torralba-Ambrona suggests that animals were frequently driven into swamps or onto boggy grounds, so reducing their mobility and making individual kills possible. Fire may possibly have been used to drive animals for this purpose, but no empirical evidence is available to that effect. Bola stones, attached to hide ropes, may also have been used to bring down running animals: the stone would be hurled between the animal's legs, so that the rope would wrap itself around the limbs. Some of the many unbattered stones found in and around occupation floors may well have been used for this purpose. Nighttime attacks by large groups of hunters armed with spears, clubs, and heavy stones may also have been used to ambush sleeping animals (see Isaac, 1968a). With only flimsy weapons available, it seems unlikely that large animals like the elephant, rhino, or hippo would ever have been confronted directly. However, the possible use of vegetable poisons—such as those known to the modern Bushman—would have lent new potency to a spear-point.

Next to nothing is known about the extent to which wild vegetable foods—such as berries, fruits, nuts, stems, roots or bulbs—were exploited. The empirical evidence is simply not preserved except in exceptional cases, e.g., various fruits at Kalambo Falls. However, modern hunting groups depend for most of their subsistence on sources other than meat and, except in subarctic environments, mammal hunting provides only 20 to 40 per cent of the diet (Lee and DeVore, 1968, chap. 4). Hunting, as opposed to vegetable-gathering, would probably have become increasingly important as populations migrated out of the tropics into areas with a cold season, where plant foods are scarce for at least part of the year.

The demography of mid-Pleistocene hunter-gatherers is obviously beyond the possibilities of reconstruction. Nonetheless, the matter deserves thought, since even a speculative consideration based on modern analogies has its value. Thus, unspecialized hunter-gatherers typically aggregate in local groups or bands of 25 to 50 persons, with over-all population density seldom exceeding 4 to 100 persons per 1000 square kilometers (see Lee and DeVore, 1968). Disease, malnutrition, and possibly infanticide would provide obvious mechanisms of population control, and severe ecological reverses once in a generation or in every century would cut back the population level drastically. Lee and DeVore (1968, p. 11) suggest that the most successful long-term situation would be a population stabilized at 20 to 30 per cent of the carrying capacity. Such data is compatible with the rather fragmentary archeological record (see Howell and Clark, 1963; Howell, 1966; Isaac, 1968a). For example, Acheulian hunter-gatherers at Olorgesailie apparently lived in groups varying in size from 4 to 30 adults, the smaller groups possibly reflecting temporary splitting of relatively stable bands with 20 or 30 adults (Isaac, 1968a).

Acheulian living sites take the form of areal tool and bone concentrations that record both the toolmaking and meat-eating activities of man. Some parts of such "cultural floors" were primarily used for tool-flaking, others for the dismemberment and efficient use of animal game for food. Various foci of archeological concentration may suggest that a band reoccupied a site on several occasions, or that several groups may have temporarily occupied it together. In either case the lack of appreciable thickness to such cultural floors seems to indicate that sites were only ephemeral camps, possibly in a seasonal movement within a hunting territory.

> Some of these sites can have been no more than stopping places for consuming a single large food animal. Others, however, provide signs of deliberate and more prolonged occupation and are believed to represent butchery sites where a number of animals, on more than one occasion, were killed, cut up, and eaten, or where a seasonal crop of vegetable foods determined a stay of several days (J. D. Clark, 1960, p. 314).

Such sites appear to qualify as ephemeral settlements. Nonetheless, the evidence for simple shelters, some built within caves, suggests that at least a few settlements had greater permanence and were intended for the duration of several weeks or even months. The Lazaret cave shelter may possibly have been occupied between mid-November and mid-April, judging by bones of 5-month-old ibex (born in spring, killed in autumn) and of marmot (available only after the winter hibernation) (de Lumley, 1969b). On the other hand, pollen recovered from human coprolites included *Genista* and some other plants, indicating that one or more of the Terra Amata sites was occupied for an uncertain length of time during late spring or early summer (de Lumley, 1967, 1969a). Such inferences involve a number of questionable assumptions although, in general, sites with obvious advantages, such as suitable shelter, water, and accessible game, should have favored longer occupation.

The mobility and range of Acheulian hunter-gatherers is difficult to assess. At Torralba and Ambrona, some rocks used for tool manufacture were brought in from distances tens of kilometers away, while at Olorgesailie the absence of many types of exotic rocks suggests a more restricted range. Modern hunter-gathering groups do, however, move about a great deal, and each local group is associated with a home base or camp and a geographical range—although such groups do not function as closed social systems (Lee and DeVore, 1968, p. 11). As a result of this mobility, personal property would need to be kept at a minimum. At the same time, the nature of the food supply restricts the size of the living groups, since large concentrations of population would rapidly exhaust local food resources. Since food supplies vary from region to region with the season or year, local groups probably did not ordinarily maintain exclusive rights to resources (see Lee and De Vore, 1968).

In their culture-ecological model of "elemental man," Watson and Watson (1969, ch. 5) illustrate how social organization, subsistence, and ecology operate as interacting factors for people at the level of Acheulian hunter-gatherers.Several sociocultural phenomena appear to have been particularly significant: *(a)* The development of a complex language as a prerequisite to effective social organization and the transmission of knowledge. *(b)* The development of family structures and social institutions as a means of securing internal cohesiveness, by formalizing and extending patterns of sharing and cooperation. In fact, economic success and survival depend on the cooperative activity of every individual and family within a group. *(c)* The development of culturally determined food preferences, as groups develop tastes for only certain foods—primarily meat—without fully exploiting every possible food source. In this way, despite more efficient social patterns and improved tool techniques, Acheulian man may have been supported in smaller numbers in a given area than australopithecines were, since populations could only increase to the limits of the preferred rather than the possible food supply.

MAN-LAND RELATIONSHIPS

The distribution of artifactual sites and human fossils documents the fairly rapid dispersal of man during the early Pleistocene. Vallonnet, in southeastern France, indicates human occupancy in Europe during the youngest pre-Cromerian cold phase (de Lumley *et al.*, 1963),[2] while the earliest hominines of Java are certainly no younger (see Koenigswald, 1968). This picture of the primary dispersal of mankind seems completed by the close of the Lower Pleistocene, when man occupied much or most of Africa, the warmer and temperate zones of Europe, as well as southern, southeastern, and eastern Asia (see Fig. 67). It remains impossible to reconstruct the routes and stages of this dispersal in any meaningful detail, and a possible ecological motivation or influence must remain entirely speculative. Man could enter Asia by the rather ancient isthmus of Suez and cross to Java dry-shod during a glacial regression, and he may have first reached Europe via western Asia.[3]

On the African continent, Acheulian sites (see distribution maps in J. D. Clark, 1967) are

> most concentrated in regions which today are grass and park savanna in East and South Africa and are invariably within easy distance of water or of places where it is evident that water would be available under a slightly increased rainfall. Lakes, pans, swamps, permanent and seasonal watercourses, springs, and deep water in limestone caves were all favored camping places. There is no evidence that the earlier Acheulian populations occupied country that is today arid or often waterless, on the one hand, or evergreen forest on the other. During the later stage, however, the populations spread into the semi-arid and arid regions of (Somalia and southern Africa) as well as establishing themselves in the Sahara. At the same time they made inroads on the peripheral parts of the Congo Basin into country that is today . . . savanna . . . and penetrated along corridors into the forest itself opened by its recession from the main interfluves (Howell and Clark, 1963, p. 525-26).

In Asia, Acheulian industries have not yet been found north of the mountain barrier formed by the Caucasus, the Elburz, the Hindu Kush, and the Himalayas, and it is possible that the cold plains or steppes of

2. The sea-level stratigraphy of Vallonnet is inconclusive, but the late Villafranchian fauna, with *Elephas meridionalis, Equus stenonis, Euctenoceros senezensis, Histrix refossa* and *Crocuta perrieri,* is no younger than Cromerian (see Kurtén, 1968). The available evidence of frost-fractured rock may allow a tentative correlation with a cold phase such as the Menapian.

3. The Bosporus-Dardanelles would be reduced to a river during a glacial regression (see Pfannenstiel, 1944) and could have been forded far more easily than the Straits of Tunis or Gibraltar, both of which are very deep (thresholds at − 324 m. and about − 250 m. respectively) and subject to complex currents. The faunal isolation of the Mediterranean islands since early Pleistocene times, as well as the absence of Paleolithic archeology from all but Sicily—and there of late Pleistocene age (see Vaufrey, 1928, 1929)—rules out land bridges anywhere in the Mediterranean Sea. However, stylistic similarities between the Acheulian of Spain and of Morocco suggest that the restriction of the Acheulian to western Europe may not be fortuitous. Consequently, accidental or deliberate crossing of the Straits of Gibraltar by light craft during the Elster must be considered as a definite possibility.

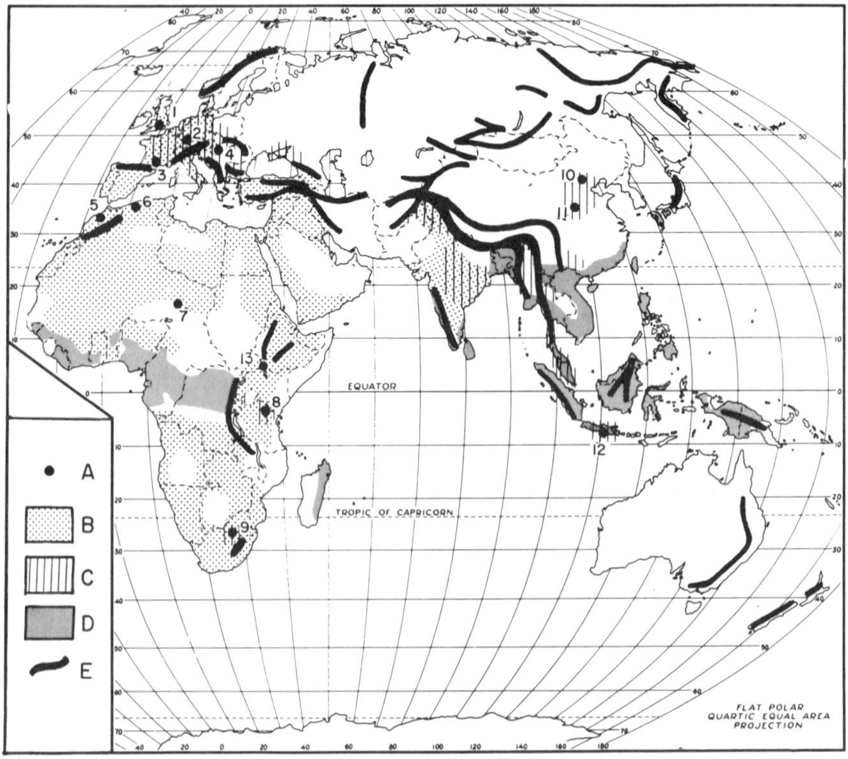

Figure 67. Distribution of early to mid-Pleistocene men: (A) fossil sites, (B) acheulian industries, (C) chopper/chopping-tool industries, (D) rainforests, (E) major mountain chains. Key to fossil sites: (1) Swanscombe, (2)Heidelberg and Steinheim, (3) Montmaurin, (4) Verteszöllös, (5) Rabat and Sidi Abderrahman, (6) Ternifine, (7) Koro Toro, (8) Olduvai, (9) Swartkrans, (10) Choukoutien, (11) Lantian, (12) Modjokerto, Sangiran, and Trinil, (13) Lower Omo Basin.

the Ukraine, the Iranian Plateau and Central Asia proved uncongenial. Similarly, typical Acheulian industries have not been found beyond the monsoon forests of East Pakistan and the India-Burma border.[4] In India and West Pakistan, Acheulian sites are found along the river valleys of the Himalaya foothills and across the open hill and plateau country of the Deccan, in what are now semiarid or subhumid environments, not unlike those of tropical Africa. On the other hand, the Acheulian sites of western Asia are found in settings analogous to those of northern Africa.

The situation is somewhat different in Europe (see distribution maps

4. Movius, in discussion to Collins (1969), does not accept the Patjitanian of Java as Acheulian, despite the presence of hand-axes.

in Collins, 1969) where the Acheulian is prominent in what were cold and open environments during the glacial stages. Occupation at Torralba and Ambrona was contemporary with an alpine grassland in Elster times, while the majority of the English, French, Italian, and German sites were situated amid the loess steppes or grasslands of the Saale Glacial, a time when open vegetation appears to have reached the Mediterranean shores. In fact, the only sites that must be linked with woodland environments are those of Holstein age, when Acheulian settlement was concentrated in northern France and southern England. Even here it might be argued that open woodlands were selected,[5] and a case could be made for an almost general regression of settlement.

There can be little question that Acheulian populations favored country with open rather than closed vegetation. This finds a ready explanation in their subsistence-base of big-game hunting, which would be optimal in grassland or savanna environments (see ch. 10). It is also hardly fortuitous that riverine, lakeshore, or spring settings, often with local development of thickets or fringing forest, were typically selected at the level of the mesohabitat. This is the sum total of the ecological adaptation of the Acheulian technocomplex. Beyond this limited degree of specialization, divergences become discernible. In particular, there was apparently no settlement on the colder plains of Asia, despite successful adaptation to equally severe environments in western Europe during the Elster and Saale Glacials. The Holstein adaptation of Acheulian populations to woodland environments in northwestern Europe may be another case in point. These examples suggest an increasing diversification of regional adaptation through time, especially along the peripheries of Acheulian settlement.

Whether or not the environmental changes evident in the geological record exerted an influence on ecological adaptations is difficult to establish, at least until a typology of functional attributes has been devised and applied to Acheulian assemblages from diverse contexts. It is reasonable to postulate that the economic structure of society would be upset by environmental change, so that readjustment might be necessary (see J. D. Clark, 1958, p. 2). This may well have been the case in western Europe, where rather sweeping shifts of vegetation accompanied the alternation of glacial and interglacial climates, in a circumscribed area bounded on three sides by open seas. Change through time was less significant but nonethless important in northern Africa, though

5. But the detailed palynological evidence (see West, 1968, with references) certainly does not support the extreme formulation of "open country with pine woodlands" (Collins, 1969, Fig. 2).

less so in the eastern and southern parts of that continent. Tropical Africa is far more diversified at the level of meso-environments than is Europe, and the absence of major barriers suggests that prehistoric groups could adjust their range more easily than their subsistence base in response to environmental change. It is even more tenuous to suggest cultural speed-ups in Africa during "interpluvials" (see Clark, 1960), particularly when mid-Pleistocene stratigraphy is not understood. It is theoretically plausible, however, that pluvial climates would facilitate population movement, gene flow, and cultural contacts in the semiarid and subhumid tropics, while dry phases would favor isolation, restrict gene flow, and favor cultural specialization (Clark, 1960).

Acheulian hunter-gatherers lacked the technology and numbers to modify their environment in any significant way. Lack of specialization in hunting techniques precludes any significant modification of the fauna. However, the use of fire as a possible device for hunting or regeneration of food plants requires attention. West and McBurney (1954) found evidence of a sudden appearance of pioneer weed plants in the Acheulian cultural layer of the Holstein-age pollen profile of Hoxne, Suffolk. This suggests forest burning with subsequent recolonization by lower, light-loving plants and ultimately tree species. Deliberate burning by man or an accidental forest fire may have been responsible. Certainly the ethnological, historical, and archeological work of Sauer (1944, 1947) and Stewart (1956) make it impossible to ignore the potential effects of the burning of the vegetative cover by man. It is improbable, however, that any appreciable, large-scale influence was exerted on the natural vegetation during the course of the early and middle Pleistocene.

All the evidence suggests, in agreement with the model of "elemental man" (Watson and Watson, 1969, p. 83), that the Acheulian hunter-gatherer lived in ecologic balance with the natural environment, affecting it in ways and degrees no different from other large animals. Despite his improved social organization and technology, he was by no means always the dominant member of his ecological community, and he was hardly so successful as to push out other large predators. The physical environment set the problems; man's solutions to these problems did not yet disrupt the ecological balance of the system, and population numbers were controlled by the basic food supply (Watson and Watson, 1969, p. 83). Nonetheless Acheulian man was a successful hunter, no less successful than other large predators. Not all groups were under constant ecological pressure, and many modern hunting populations show a remarkable lack of concern about the problem of finding food, actually working short hours while exploiting abundant food resources (Lee and DeVore, 1968, p. 6, and chs. 4, 5, 9). Almost certainly Acheulian man enjoyed better health, a more balanced diet, and more leisure than most agricultural populations do today.

AN EXAMPLE OF AN ACHEULIAN LIVING SITE: TORRALBA, SPAIN

The early Acheulian site of Torralba is located 156 km. northeast of Madrid at an elevation of 1,115 m. Originally studied by the Marqués de Cerralbo from 1907 to 1911, the site was nearly completely excavated by F. C. Howell and L. G. Freeman from 1961 to 1963 (see Howell, 1966). In the following discussion the geology and paleogeography is based on Butzer (1965), the paleontology on E. Aguirre (unpublished), the palynology on F. Florschütz and J. Menéndez-Amor (unpublished).

The geographical setting consists of a broad, steep-sided valley incised into a limestone plateau. The lower part of the valley cuts into impermeable strata of gypsum and siltstone, so that springs are frequently found at the base of the porous and permeable limestone bedrock above. A spring of this kind almost certainly existed at the site during the time of occupation. Running water is now available in a local stream at 500 m. distance, and was certainly once accessible in nearby torrents at 300 m. during most of the year. The topography itself is such that the steep slopes accompanying the valley margins would have impeded free passage of some animal species from the valley to the uplands. Furthermore, this valley represents one of the few low level passes joining Old Castile to the north of the central sierras, and New Castile to the south (Fig. 68). Torralba is therefore located along what may have been a seasonal migratory route for larger game moving northward in the late spring, southward in the late summer. Also of significance is the absence of surface water on the limestone plateau as a result of very rapid percolation. This means that man and animals have at all times been required to obtain water in the valleys.

The geologic setting of the Torralba site, and its twin at nearby Ambrona, is within a river terrace at 40 m. above the local stream, part of the headwaters of the Jalón, a major tributary of the Ebro. The stratigraphic column of this 40-meter terrace deposit is listed as the "Lower Complex" of Fig. 69. The over-all climate was both cooler and relatively moister than today's, although subject to repeated oscillations. Human occupation is first recorded here during a cold interval. Subsequently conditions were temperate but then once more reverted to a cold climate. Occupation is no longer recorded during a later warm oscillation. Geomorphic phenomena include evidence of considerable frost-weathering, solifluction (partly as a result of soil frost, partly due to lubrication of clayey beds), and valley alluviation. Most of the occupation levels are found in a series of intercalated slope deposits and stream

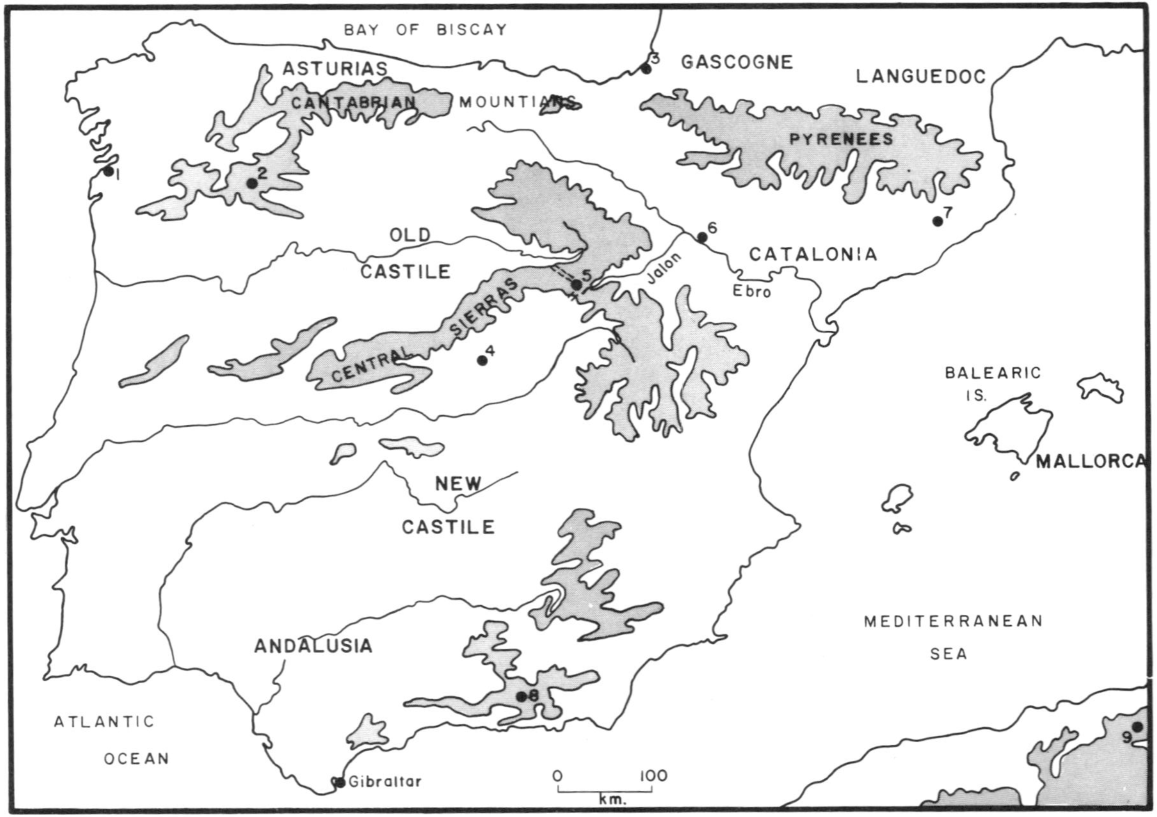

Figure 68. The Iberian Peninsula. Areas over 1,000 m. elevation shaded. (1) Vigo, (2) Laguna de las Sanguijuelas, (3) Biarritz, (4) Madrid, (5) location of Torralba and Ambrona with route of low-level pass, (6) Zaragoza, (7) Moya, (8) Sierra Nevada, (9) Jurjura Mountains, (Topography modified after Prentice-Hall World Atlas [2nd ed.], by permission of Geographischer Verlag Ed. Hölzel, Vienna.)

alluvia. One major cultural level is found within and on top of a stratum of coarse, subrounded, frost-weathered gravel (Bed IIb) which mantles the upland surfaces and hillsides of the area. Stone rings and garlands found at the Torralba site indicate significant soil-frost phenomena. All in all, the geomorphic record suggests a marginal periglacial environment with some evidence of accelerated fluvial activity. The latter includes well-rounded, coarse gravel in the stream bed deposits, considerable spring activity, widespread fine valley alluviation on the floodplain peripheries, and typical colluvail deposits.

Four major alluvial terraces were observed in the upper Jalón and Henares valleys, the second of which is equivalent to the mixed alluvial-slope deposit sequence at Torralba and Ambrona (see Gladfelter, 1971). Evidence of significant frost-weathering and some frost-heaving is limited to the second terrace. Similarly, evidence of subsequent, intensive reddish soil development is confined to the second terrace, whereas the next younger terrace was less intensively weathered. Instead, it may contain derived red sediments of the older soil at its base. This fossil red soil, stratigraphically fixed in the time interval separating the second and third terraces, is distinctive, particularly as developed *in situ* on the sediments at Torralba and Ambrona. External correlation with the equally striking rotlehm found developed on the "high terrace" of several Catalonian coastal streams is permissible. The Catalonian rotlehm is contemporary with the Tyrrhenian I or Holstein (Butzer, 1964a). A younger interglacial, presumably the Eem, is recorded by swamp marls at the base of the valley floor. These beds record a warm, dry climate as indicated by abundant grass and 10 to 15 per cent *Castanea* pollen—although the chestnut is now absent from the Spanish sierras. Above and below this marl are sandy fills, the lower with frost-weathered gravel, the upper with pollen suggesting a cool grassland environment. Thus the Torralba "Lower Complex" can be correlated with a pre-Holstein cold phase.

Palynological work corroborates the geomorphologic evidence. Corresponding to the cool, moist phases are pollen spectra with 10–60 per cent pine pollen, and a high proportion of grasses and some sedges. During the moist, temperate oscillations, pine pollen attains over 75 per cent, frequently with sedges accounting for most of the remainder of the spectrum. In conjunction with the geomorphology this suggests that the cold phases were characterized by seasonally inundated grasslands or swamps in the valley, and by a high-altitude grassland on the uplands with some open pine scrub or parkland in areas of broken ground. Much bare soil was probably exposed here with an incomplete mat of lower vegetation, including some *Artemisia*. During the more temperate

PLEISTOCENE STRATIGRAPHY AT TORRALBA & AMBRONA (SORIA), SPAIN

UNIT	SEDIMENT	THICKNESS (cm) TORRALBA	THICKNESS (cm) AMBRONA	ASSOCIATED PROCESSES	CLIMATIC INFERENCES	MAMMAL FAUNA	HUMAN OCCUPATION TORRALBA	HUMAN OCCUPATION AMBRONA	TENTATIVE CORRELATION
	IV. Reddish colluvium	120	200	Soil colluviation with local fan alluviation at valley margins	(Human interference)				Historical
				EROSION					
UPPER COMPLEX	III. Fine dark alluvium	80		Valley alluviation	Moist, temperate				Middle Holocene
	II. Coarse brown alluvium	70		Alluviation at valley margins by lateral tributaries; tufa deposits locally. Some solifluction initially.	Cool, moist		Upper Paleolithic (surface traces)		Würm Glacial
	I. Reddish colluvium	125							
				EROSION					
MIDDLE COMPLEX	II. Yellowish sands	10		Colluviation and valley alluviation, following intensive frost-weathering. Slumping of subsurface, lubricated Keuper silts producing faulting at both sites. Some solifluction.	Cold, moist				Riss Glacial Complex
	I. d. Reddish colluvium	55							
	I. c. Reddish alluvium	60							
	I. b. Reddish colluvium	30							
	I. a. Cryoclastic detritus	20							
				EROSION					
PEDO-GENESIS	*Terra fusca* soil developed on Lower Complex IV and V exclusively. B	160	150		Warm seasonally very moist				Great (Holsteinian) Interglacial (= Tyrrhenian I stage)
	Bc	10	35						
	Ca	10-20	10-60						
				EROSION					
LOWER COMPLEX	V. d. Coarse reddish alluvium	165	95	Shallow alluviation at valley margins by lateral tributaries.	Very cold		Hiatus	EARLY MIDDLE ACHEULIAN OCCUPATION	Stadial
	V. c. Fine reddish alluvium		85						
	V. b. C-gravels		60						
	V. a. Gritty marl		90						
								Hiatus	Interstadial
	IV. b. Gray marl	150	200	Valley back swamps filled with homogeneous fine silts from sluggish flood waters; pseudo-gley conditions indicated by limonitic Fe-horizons.	Moist, temperate		EARLY MIDDLE ACHEULIAN OCCUPATION		Stadial
	IV. a. Marl with channel beds		220	Valley flood-plaining dominated by very fine alluviation, but with coarse, moderate cryoclastic channel beds locally.	Moist, cool				
				EROSION					
	III. b. B-gravel	–	15	Coarse valley alluviation with reduced soil frost.	Moist, cool				
	III. a. Upper gray colluvium	80	80	Fine valley alluviation with some solifluction.	Cold, moist				
				EROSION & CONGELIFLUCTION					Interstadial
	II. d. Sandy marl	90	150	Fine valley filling.	Moist, temperate				Stadial
	II. c. Lower gray colluvium	100	?	Well-stratified gritty sands with intercalated gravels. Valley alluviation.	Cold, moist				
	II. b. A-gravel	30	60	Coarse cryoclastic gravels on slopes, partly calcrete, partly interbedded with grey silts. Some congelifluction.	Cold, moist				
	II. a. Light sand	70⁺	300	Fine valley filling of homogeneous sands, partly silty at top.	Cool, moist				
				EROSION					Interstadial ?
	I. Red colluvium	400⁺	?40	Medium, highly cryoclastic detritus at base of slopes.	Very cold		STERILE		Stadial
	O. Redeposited Keuper (several phases)	100⁺	200⁺	Congelifluction and earth flows of lubricated clays, silts and marls.	Moist, cold.				

(Tentative correlation for Lower Complex: LATE ELSTER)

Figure 69. Stratigraphic columns of Torralba and Ambrona.

phases, pine woodlands dominated the uplands while open country in the valley was reduced to a few sedge swamps. A fringe of deciduous trees accompanied the stream throughout this period. Macrobotanical remains confirm that Scot's pine *(Pinus silvestris)* was the dominant tree type, although a few pieces of hardwood have also been identified. The cold intervals represented a transitional forest-steppe or grassland environment such as found above the tree line at about 2,000 m. in the central sierras today (see Welten, 1954). This suggests a depression of the altitudinal vegetation belts by at least 900 m. Modern January averages lie at about +2° C., July averages at 19° C. The January means must have been at the very least 5° to 6° C. lower in Elster times in order to account for the significance of frost action.

The abundant mammalian fauna is limited in terms of species. The inventory of the 1961–63 excavations is dominated by elephant, with at least 30 and possibly as many as 55 individuals represented, exclusively of a grassland-adapted variety of *Elephas antiquus* (Aguirre, 1969). Next in importance is a primitive horse (*Equus caballus torralbae*), with at least 26 individuals. Deer, primarily red deer (*Cervus elaphus*) but including also fallow deer (a large form of *Dama* cf. *clactonia*) and roe deer (*Capreolus* sp.), are represented by at least 25 individuals, the aurochs (*Bos primigenius*) by at least 10, the steppe rhinoceros (*Dicerorhinus hemitoechus*) by at least 6, and unidentified carnivores by 4. The faunal list of the contemporary occupation at Ambrona further includes a big lion, the wolf (*Canis lupus*), a weasel, a mouse, a hare, a cercopithecoid monkey (*Macaca* sp.), two species of duck, grouse, various falcons or kites, a lizard, and a toad. Although some of these animals are considered as woodland or indifferent forms, the majority of the individuals represented are grassland-specific. This is in keeping with the pollen evidence.

In overview, the regional environment was that of a high-altitude grassland or forest-steppe during the cold phases, a subboreal woodland with open, swampy lowlands during the more temperate oscillations. The local setting of the site at the time of Acheulian occupation was a stony footslope located at the side of broad, grassy valley with local sedge swamps. Some stunted pine scrub may have been found in sheltered draws, with grasses and herbs also characteristic of the uplands.

Turning to the cultural inventory, a great variety of stone implements and waste chips, as well as deliberately worked wood, bone, and tusk tips are interspersed with butchered faunal remains. A few distinct occupation levels with articulated or semiarticulated elephant bones were found, almost completely undisturbed by subsequent sedimentation. In other horizons, disarticulated remains of several species occur in association, suggesting a certain amount of derivation by slope wash or solifluction. Nevertheless, the distribution of disjunct associations or true "floors" through a vertical column exceeding 3.5 m. at Torralba and 6.5 m. at Ambrona indicates the repeated use of both sites for kill or butchering purposes during a protracted period of many millennia. Most of the fauna shows distinct evidence of butchery practices, and there is an unusually high frequency of juvenile animals. There can be little doubt that, with a few possible exceptions, all of the animals were killed by man. They were eaten more or less where killed, and the bones were dismembered and in part scattered considerably. Bones with marrow were frequently cracked and split open. No traces of shelters or the like have been found, and it is possible that Torralba served repeatedly as an ephemeral camp-settlement of a few days duration.

The stone implements include numerous flake tools of several kinds (especially scrapers of different sorts) and bifacial tools, including cleavers and hand-axes. These are specifically adapted for butchering purposes. Raw material was provided by quartzite, flint, and limestone, most of which was either available locally in the stream gravel or within a few kilometers distance. Bone fragments and elephant tusk tips with traces of cutting and purposeful trimming, as well as worked wood, have been found. Some of the wood may belong to weapons, others may have been used as pointed tools. Several pieces of charcoal, charred wood, much carbon, as well as charred bone have been found, suggesting human control of fire even though there is no direct evidence of hearths. In default of preserved vegetable foods, it would seem that hunting was the mainstay of the economy, with elephants representing over four-fifths of the total meat obtained. This was probably a matter of deliberate selection, partly dictated by the facility of hunting proboscidians in the often swampy terrain of a narrow, steep-sided valley.

In retrospect, the small or moderate sized, simple hunting group or groups that occasionally preyed on migrating herds of herbivores on the swampy river floodplain at Torralba are probably quite characteristic of the Acheulian hunter-gatherers. Seasonal abundance of animals was assured each spring and autumn as the herds moved between their winter grazing grounds, in the cool, open woodlands of southern Spain, to their summer habitats in the colder grasslands of the northern plains and plateaus. Man was only one of the many elements of the biological environment.

TWENTY-SEVEN

The European Mousterians and Their Environment

NEW CENTERS OF CULTURAL ACTIVITY IN HIGHER LATITUDES

In the course of the later Pleistocene, several major cultural innovations preceded and accompanied a major expansion of hominid population into the Eurasian subarctic and into the Americas and Australia. But equally intriguing is the more subtle, geographically significant appearance of new population centers in Eurasia, the continent which had eclipsed the African heartland by the late Pleistocene. If tool-workmanship be an index of cultural progressiveness, and site tool density an index of population size, then Africa would probably qualify as the major center of population and cultural innovation of early and middle Pleistocene times. But during the last interglacial, tool-craftsmanship found a new focus in Europe, at least judging by the artful hand-axes of the final Acheulian, or the fine flaking techniques of the Mousterian. And population density, insofar as can be inferred from the evidence, achieved a new high during the European late Pleistocene. By this time Africa may have become a cultural backwater (see J. D. Clark, 1960).

The transfer of man's cultural and biological focus from the tropics to the middle latitudes was possibly largely a result of improved environmental control. Knowledge of fire had apparently first opened the middle latitudes to man during the Elster. Other advances in technology that are not recorded in the inventory of stone implements may have included hides or furs used as clothing. Whatever the case, by the onset of the

Würm, man insured for himself a suitable microclimate through clothing and shelter so as to be comfortable even in subarctic climates. The rapid subsequent cultural development of these "new" lands was possibly facilitated by a sound ecologic subsistence.

Ironically, the European center of accelerated cultural innovation during the last glacial was found in the then prevailing forest-tundra and cold loess steppes, rather than in the warmer, temperate woodlands of the Mediterranean region. Assuming that the Franco-Cantabrian cave art marked the heart of the European culture area, one could suggest that the forest-tundra of Europe had displaced the African savannas as a center of innovation. This poses new ecologic problems. The clue seems to lie in the economic basis of food-collecting societies, namely, their hunting resources. Animal meat need not necessarily have provided the bulk of Paleolithic man's diet, but in most cases it probably provided the greatest part of the calorie intake. Dahlberg and Carbonell (1961) studied the dentition of a late Paleolithic skeleton of a young woman from Cap Blanc (Dordogne), emphasizing the lack of wear. This implies a grit-free, i.e., mainly carnivorous, diet. Relevant to the argument of hunting resources is the high biomass and carrying capacity of the tundra and probably also the forest-tundra (see ch. 9). In addition, the low latitude Pleistocene tundra must have enjoyed rather favorable radiation conditions[1] while the better-drained loess environments must also have had a greater carrying capacity than the modern, high latitude tundras. The masses of animal fossils retrieved at such late Pleistocene sites as Solutré, France (many thousands of wild horses; Woldstedt, 1954, p. 279) or Předmost, Czechoslovakia (600 woolly mammoths) supports this suggestion. The low latitude Pleistocene tundras possibly provided an environment as favorable as that of the temperate grasslands today. Glacial-age Europe was consequently not a marginal resource base for sufficiently advanced hunting populations.

CULTURAL PATTERNS OF THE MOUSTERIAN

Neanderthal Man (*Homo sapiens neanderthalensis*) appears to have evolved from *Homo erectus* early in the Upper Pleistocene. In Europe, fossils of Neanderthaler type are associated with the Mousterian "cul-

1. The length of day varies from 24 hours on June 21 to 0 hours on Dec. 21 at the arctic circle in Lapland. In southern France this range is only from 16 to 8 hours. Such differences in the length of the photoperiod may have been significant for animal grazing activity. Similarly the midsummer angle of incidence of solar radiation is much higher in France than in Lapland, implying that, given identical conditions of air temperature, moisture, and cloudiness, plant photosynthesis would be more effective in glacial-age France than in modern Lapland. The food resources of herbivores would be correspondingly greater in a low latitude tundra.

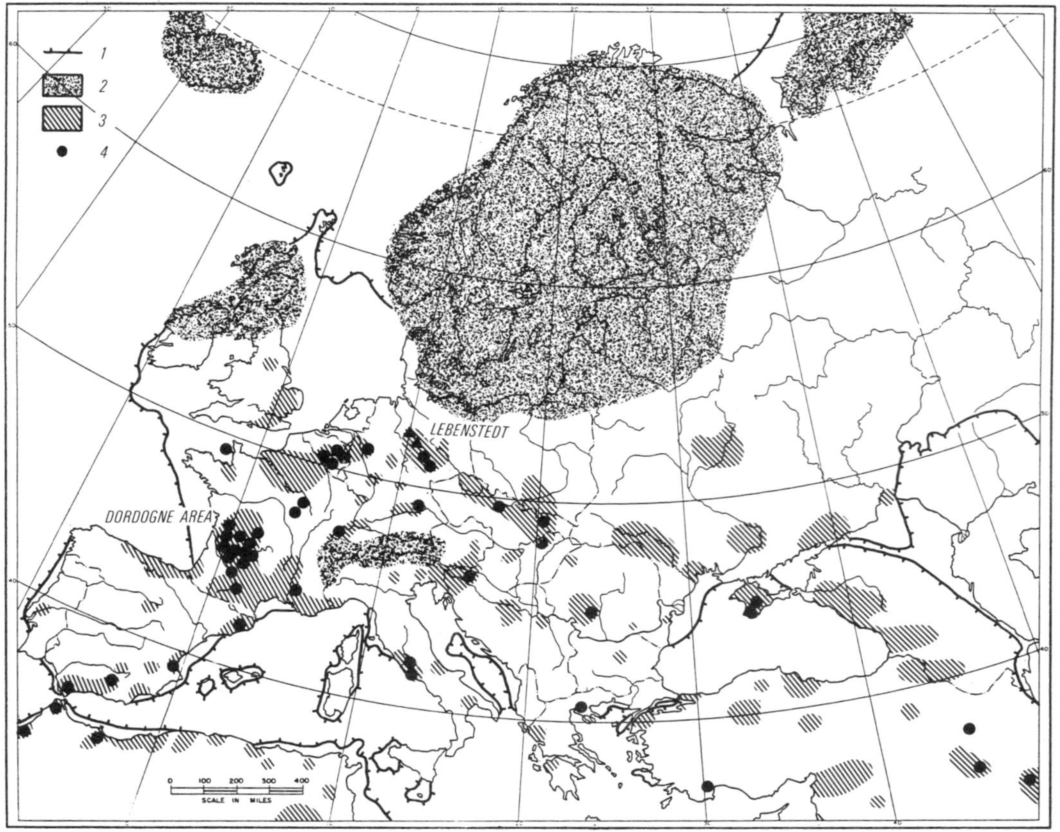

Figure 70. European settlement during the early Würm: (1) approximate position of the coastlines; (2) glaciers, borders tentatively approximated by analogy to the Pomeranian stage; (3) distribution of Mousterian and related industries during the early Würm; (4) sites with Neanderthal skeletal remains dating from the late Eem and early Würm.

ture" (or Middle Paleolithic), which is thought to have developed out of the late Acheulian. Although the classical anatomical traits of the Neanderthalers and the typical attributes of the Mousterian are first evident during the early phases of the Würm Glacial, there is reason to believe that a number of Eem-age sites already record early manifestations of Neanderthal man and his evolving "culture." Consequently, the Mousterian in its widest sense spans the late Eem and early Würm, until about 35,000 B.P.

The "classical" Neanderthalers of early Würm-age Europe were characterized anatomically by large, massive skulls, with a record endocranial capacity averaging 1,450 cc. Body-build was short but heavy, and mean stature is estimated at 157 cm. Since the early Neanderthalers of the Eem as well as the later neanderthaloid populations of Asia and Africa do not show equally extreme cranial specializations, it is believed that the "classical" Neanderthalers developed as a result of close intermarriage in the European forest-tundras (see Howell, 1958). However, the characteristics of Neanderthal-like skeletal materials (Fig. 70) such as are found outside of Europe in Uzbekistan (Teshik-Tash and Aman Kutan), Iraq (Shanidar), Palestine (Mt. Carmel), Cyrenaica (Haua Fteah), Morocco (Jebel Irhoud and Tangier), Sudan (Singa), South Africa (Makapansgat and Saldanha) (see Tobias, 1968), and Java (Ngandong) leave no doubt about the broad continuity of related human stock.

The Mousterian was a culture-complex with multi-functional tools, although both regional and ecological specialization had begun. For the first time the technological uniformity of the Old World, excepting the Far East, is interrupted. The industrial assemblages of contemporary sites in different areas show appreciable differentiation, even within the framework of broad cultural affinity (Klein, 1969a). The several Mousterian cultures of the European Neanderthalers (Fig. 71) consist of variable assemblages of flake tools, in part finely retouched, in possible combination with older tool types such as small hand-axes (see Bordes, 1953, 1961a, 1961b; Klein, 1969a). Regional specialization or differentiation is also noticeable among contemporary cultures in other continents, as for example in Africa (J. D. Clark, 1960).

The greater part of this incipient regional specialization is probably a matter of ecological adaptation to new environments: *(a)* the European tundras of the Würm, the first known case of human occupation of a subarctic or arctic environment, *(b)* the African rainforest, now first colonized by man, and *(c)* local cave environments, which had previously only been occupied sporadically, but were now in vogue.

An example of colonization of new environments can be cited from the case of the African rainforest. There the contemporaneous Sangoan

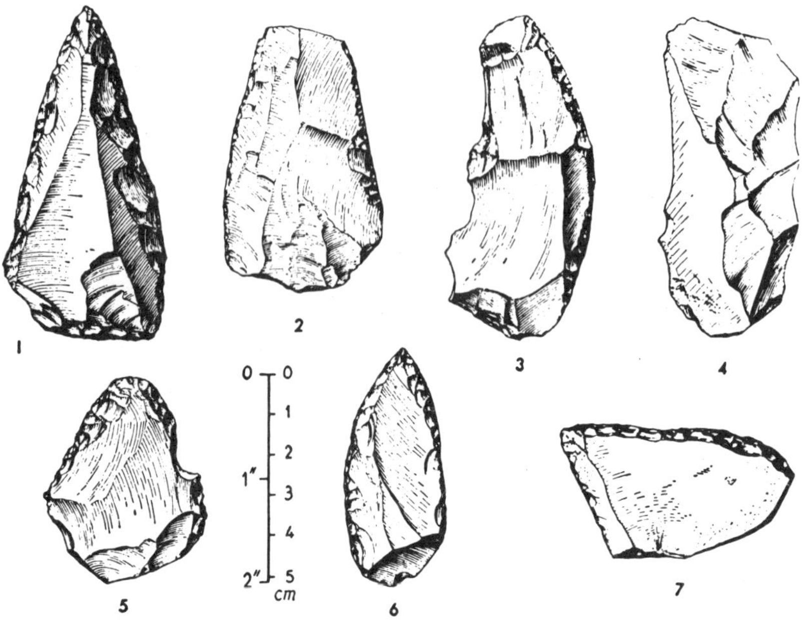

Figure 71. Some stone implements of the "typical" Mousterian from the Dordogne (from F. Bordes, 196la, by permission of the author): (1,6) points, (2,3) side-scrapers, (4) flake, (5) end-scraper on a flake, (7) transverse scraper.

and Lupemban emphasized wood-working tools and equipment, suggesting a greater dependence on, or greater use of, vegetable foods (J. D. Clark, 1960). Among the changes involved are replacement of the hand-axe by a picklike tool, and the appearance of chisels, gouges, planes, lanceolate points, etc.

ENVIRONMENTAL ADAPTATION OF THE MOUSTERIANS

Little is known about the specific adaptations of the Mousterians to an arctic environment. Their tools represent specialization in the direction of meat and possibly skin preparation. It may be justified to conclude from this that they effectively prepared skins, hides, and furs as clothing. For a convincing argument, however, bodkins or belt-fasteners should be present, but are not. Although their offensive weapons are better known, e.g., the alleged "bola" stones of various sites, the wooden spear of Lehringen, or the antler "clubs" of Lebenstedt, similar features are known from earlier sites so that it is unlikely that much innovation distinguished the Mousterian inventory in that sphere. An exception may be provided by the barbed bone points of Lebenstedt (Tode *et al.*, 1953).

The archeological evidence of open-air sites such as Lebenstedt supports the opinion that some Mousterians followed the gregarious herds northward into the open tundra in summer, during which time they at least occasionally constructed shelters at ephemeral or temporary camp sites, such as that of Molodova in southwestern Russia (Klein, 1969a). According to Bordes (1953) the fine Mousterian flakes may have been worked as tools for longer-term use by seasonal or permanent cave occupants, while more simple Levallois flakes without lateral retouch may have been frequently manufactured for brief use by groups wandering on the tundra during the summer. In parts of southwestern Asia the Middle Paleolithic of cave sites is also more carefully worked than that of open-air sites (Braidwood and Howe, 1962). Some of the Mousterians may then have adapted to a seasonal movement, following the herds of reindeer, mammoth, horse, and bison into the forest-tundra during the winter, and returning with them onto the broad expanse of herbaceous tundra in summer. Vallois (1961) suggests that the extreme morphological resemblance between certain Neanderthal groups of France and Italy implies widespread group relationships, possibly a result of seasonal migrations.

In apparent contradiction to seasonal migration is the evidence from the Mousterian and late Paleolithic cave sites of southwestern France. On the basis of reindeer antlers and dentition,[2] Bouchud (1954) showed that most of these sites must have been occupied all year round. In addition to the inference of semipermanent settlement, the regional complexity of Mousterian industries in southern France has been interpreted as a matter of several distinct cultural entities (Bordes, 1953, 1961b). However, statistical analysis suggests that these are little more than tool-kits used for specific activities, some of them on a seasonal basis (Freeman, 1966).

In overview it would seem that seasonal migration was probably not a general phenomenon. Possibly it was confined to marginal populations at the edge of the tundra, while the Mousterians of the less severe forest-tundra environments were able to settle on a basis of more permanent animal resources. The fuller archeological picture of the Eu-

2. Adult male reindeer shed their antlers in November or December, females and juveniles during the late winter or early spring. The relative frequencies of shed and broken antlers and the sex composition within a group can be used to determine the season of hunting. Fawns are usually born in the spring so that age, as determined from the dentition, is also useful in this regard (for discussion see Bouchud, 1954, and Zeuner, 1963, p. 123 ff.).

ropean Upper Paleolithic sheds more light on the complexities of settlement patterns (see chapter 28). At any rate the existence of seasonal movements may be considered as a possible—localized—adaptation to the more rigorous environments of Würm-age Europe. It is extremely important that more work be done to establish the true pattern of settlement.

Good evidence for a tent or wind shelter at an open-air site comes from Molodova (Klein, 1969a). Cave interiors were also rendered comfortable by fires, and possibly by branches and hides suspended across the cave entrance. In fact, a posthole at the entrance of Combe Grenal (Dordogne) suggests a row of shafts used to support skins or woven branches (Bordes, 1961b).

The animal booty of Neanderthal man suggests that he was a courageous as well as an efficient hunter. Woolly mammoths and rhinos were successfully hunted, and the presence of various fish and fowl at the Lebenstedt site underscores his proficiency. Neanderthal man was not the patient, slow-witted, and muscular hillbilly he is often made out to be. Although there is uncertainty as to the exact nature of the adaptations involved, the Mousterian Neanderthalers certainly did open up a new environment for human colonization, since the arctic tundra seems to have been the "last frontier." For a while at least the Neanderthalers were successful in their new environment. Their ultimate failure may have been due to biological factors, although the causes remain obscure. Between about 40,000 and 35,000 B.P. the Neanderthalers were abruptly displaced by anatomically modern men of *Homo sapiens sapiens* type, carriers of a more advanced culture. Skeletal remains dating from the later Würm have so far shown no evidence of cross-breeding between these two populations, although logically the remnants of the Neanderthalers would have been absorbed by the biologically more progressive and more numerous *Homo sapiens* peoples. Possibly the Neanderthalers were pushed out into the more marginal environments by the encroaching newcomers, and ultimately succumbed to the inhospitable environment during the maximum cold of the full glacial period (for discussion see Narr, 1963, p. 74).

AN EXAMPLE OF MOUSTERIAN HABITAT AND ECONOMY: SALZGITTER-LEBENSTEDT

The Mousterian site of Salzgitter-Lebenstedt in northern Germany was excavated in 1952 by Tode, Preul, Richter, Kleinschmidt *et al.,* (1953; also Tode, 1954), and is particularly important on account of the thorough excavation techniques employed. This open-air summer camp-site has been dated at 55,000 B.P.±1,000 (Grn. 2,083) and belongs to a slightly warmer interval in the early Würm. The location is on a small tributary stream of the Aller-Weser drainage system, some 15 km. south of the city of Braunschweig, at an elevation of 82 m. above sea level.

The geological setting is provided by a cold-climate stream terrace immediately overlain by the ground moraine of the Riss-Saale glaciation. The stream has since dissected the Riss deposits, and the Würm-age sediments, which contain the site, are embanked against the foot of the Riss terrace (Fig. 72). These Würm beds consist of alternating stream and slope deposits with the following stratigraphy, beginning at the base:

a) 50 cm. fluvial sand with evidence of patterned ground or frost cracks, overlying an older, eluviated layer of stones.
b) 150 cm. stream gravel grading laterally into solifluction mantles at base and top, generally showing evidence of cryoturbation and patterned ground or frost cracks. The cultural layer is associated with peaty beds in the stream gravels.
c) 350 cm. alternating silts and sands showing considerable evidence of cryoturbation, in part stream-bedded, in part bedded downslope by solifluction. Another peaty layer is located at the base.
d) 0–100 cm. solifluction mantle derived from the Riss terrace, eluviated at its outer edge.
e) 100–200 cm. silts with evidence of cryoturbation, disconformably underlying fluvial peat with Holocene pollen.

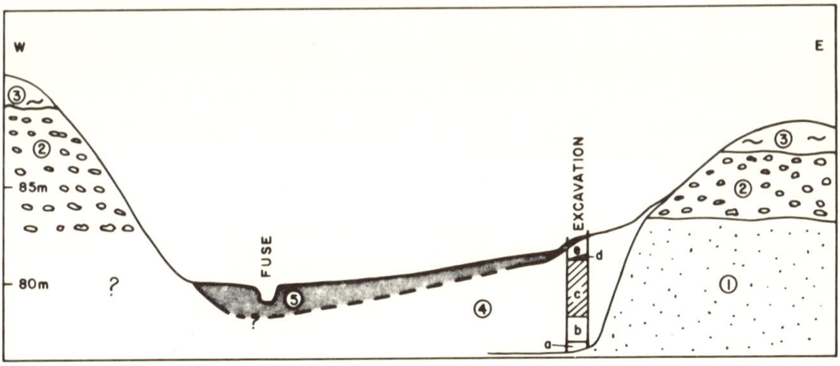

Figure 72. Geographical and geological setting of the Salzgitter-Lebenstedt site: (1) Lower Pleistocene (?) sands and gravel, (2) Riss-Saale periglacial stream terrace, (3) Riss-Saale ground moraine, (4) Würm stream terrace intercalated with solifluction beds from slope (the letters in the excavation column refer to text description); (5) Holocene alluvium and peat. Data obtained from 1:25,000 topographic map and geological sections by F. Preul (Tode et al., *1953). Vertical exaggeration approximately 30 ×.*

The interpretations given were obtained after careful study of pebble orientation, gravel morphometry, particle-size spectra and micro-stratigraphy.

Paleobotanical studies were carried out in the peaty layers of (*b*) and at the base of (*c*). The pollen spectrum of the peat in (*b*) was dominated by NAP, particularly sedges. The arboreal pollen was mainly that of pine, with some birch, willow, spruce, and a trace of alder. The macro-botanical remains included various mosses and lichens (mainly determined by the stems), alpine or arctic herbaceous plants (determined by seeds, fruits, or stones), as well as leaves of *Salix polaris* and *S. herbacea*. The interpretation is one of low willow brush in moist depressions, with sporadic stands of pine at a distance. The locality was probably very close to the northern margins of the forest-tundra, i.e., near the arctic tree-limit. Considering that the modern July temperature is 17° C., July temperatures were 5°–7° C. cooler at the time. The apparent presence of permafrost further suggests that the annual mean (+8.5° C. today) was at least 10.5° C. lower, i.e., the major temperature depression occurred during the winter season.

The highly organic bed at the base of horizon (*c*) contains only a little pollen of willow, birch and pine, being almost exclusively a spectrum of NAP. The macrobotanical remains are confined to mosses and sedges, indicating that the local environment was that of a moist, drained tundra lowland.

The faunal remains from Lebenstedt were studied in minutest detail, and contained the following species:

reindeer	approx. 80 individuals (19% juveniles), 72% of animals present
woolly mammoth	approx. 16 individuals, 14%
bison	6–7 individuals, 5.4%
horse	4–6 individuals, 4.6%
woolly rhino	approx. 2 individuals, 2%

Further single specimens included wolf, muskrat (*Desmana moschata*), crane or swan, duck, an extinct vulture, perch, pike, other unidentified fish, crabs, aquatic mollusca, and insects. The horses are of the heavy-headed, woodland type, so that three species (reindeer, mammoth, rhino), totalling 88 per cent of the mammalian fauna, are by preference tundra forms. Two characteristic tundra forms, the musk-ox and lemming, are conspicuously absent. It is thought that the wolf and vulture were killed (by stones?) while scavenging at the fringes of the camp; the muskrat was probably killed while insect-hunting in the refuse heaps.

Careful recording of the location, orientation and level of every bone fragment subsequently enabled a reconstruction of summer conditions at the site: the cultural floor was seasonally flooded by rapidly moving waters, presumably the spring meltwaters. At other times tiny rivulets, ponding as stagnant pools with peat development, flowed around the small alluvial bench occupied by the hunters in the valley bottom.

The cultural inventory itself is classified as Mousterian of Acheulian tradition, and about 10 per cent of the 2,000 stone tools are well-worked hand-axes of late Acheulian or Micoquian typology. The remainder are flake tools, mainly scrapers and points, with careful Mousterian retouching on both surfaces. The stone tools are functionally related to butchering practices. Of further interest are worked reindeer antlers (clubs?), bone points of mammoth ribs (digging sticks?). and barbed bone points (spearheads?). These items may represent a part of the weapons inventory. On the basis of the animal remains it is believed that a moderately large band, not exceeding 40 or 50 individuals, occupied the site for a few weeks during several summer seasons.

Essentially the temporary camp of these Mousterian hunters—located in a small, sheltered valley bottom following the late spring thaw—was not unlike that of Torralba. But the cultural floor is better defined and of greater concentration and depth, suggesting longer occupation. The actual site may also have been better chosen. Although evidence of fishing is known from the early Paleolithic of Africa (J. D. Clark, 1960), the birds of Lebenstedt are of interest from the perspective of hunting proficiency. From all available evidence these hunters of the early Würm must have managed comparatively well in the European tundras. Man-land relationships had become more complex, even while remaining in the mold of the Acheulian.

TWENTY-EIGHT

Upper Paleolithic Man-Land Relationships in Europe

THE APPEARANCE OF ADVANCED HUNTERS IN EUROPE

No single culture group of the Pleistocene stands out as distinctly as the Upper Paleolithic hunters of Europe. The picture of man-land relationships at the Mousterian culture level probably also does justice—at the present state of information—to Africa and tropical Asia for the closing millenia of the late Pleistocene. But during the course of the mid-Würm interstadial, rather specialized hunter-gatherers occupied the European stage, requiring a new assessment of man-land relationships. For the first time an acceleration of cultural and technological innovation is noticeable. Compared with a half million years of early Paleolithic occupation in the Old World, the European Mousterian comprised a time span of 40,000 to 60,000 years, the Upper Paleolithic about 25,000.

There is a common tendency to equate several distinct phenomena: the appearance of modern man, the advent of the Upper Paleolithic blade-tool industries, and the colonization of Australia and the Americas during a "secondary dispersal" of man. The apparent association of these late Pleistocene events is misleading and probably fortuitous. Only in Europe does the equivalence of *Homo sapiens sapiens* and the Upper Paleolithic appear to be real, because both phenomena were introduced from without. In Asia and Africa, on the other hand, the temporal and anatomical distinctions of *Homo sapiens neanderthalensis* and *H. sapiens sapiens* are rather fluid. And above all, there is no necessary association of the new racial types with the Upper Paleolithic blade industries. The biological evolution and cultural innovation first evident during the second half of the late Pleistocene is at best incompletely understood. It is therefore necessary to limit discussion to empirical manifestation of the advanced hunter-gatherer economies.

The appearance of the Upper Paleolithic hunters in western Europe can be dated to a little before 35,000 B.P. (Movius, 1960), although the replacement of the European Neanderthal populations may have taken several millenia. The exact temporal relationships of the Mousterian and Upper Paleolithic remain vague because of the inaccuracies of radiocarbon dating, but the replacement probably took place during the early or middle phases of the Paudorf Interstadial (see chapter 18). Present evidence suggests a sharp biological and technological break throughout Europe (see Narr, 1963, p. 50 ff.), although two seemingly hybrid artifact assemblages—the Chatelperronian in France and the Emiran in the Levant—may possibly indicate some technological evolution from local roots (Delporte, 1955; Bordes, 1961b; Howell, 1959a). There is some reason to believe that the Upper Paleolithic as such originated somewhere in the Near East, since certain blade industries of late Eemian age in Cyrenaica and the Levant (see Rust, 1950; McBurney, 1967) have been considered as precursors of the Upper Paleolithic. Possibly, then, the Upper Paleolithic developed among more progressive stock of general *Homo sapiens* type, at the same time that the European Neanderthalers specialized in their evolutionary *cul-de-sac*. As a result, the advent of *Homo sapiens sapiens* in Europe between 40,000 and 35,000 B.P. gives the impression of a vast, sweeping movement. According to Bates (1953), migration at the food-gatherer's level would be an infiltration process, an accumulation of small territorial readjustments and perhaps, under environmental stress, the complete displacement of small tribal groups. Narr (1963, p. 74) argues that the distribution of the Neanderthalers and the new, infiltrating groups formed a complex mosaic of semicontiguous but generally distinct territories.

UPPER PALEOLITHIC TECHNOLOGY

The Upper Paleolithic of mid-latitude Eurasia was essentially a blade-tool assemblage, characterized by an abundance and variety of long, parallel-sided flakes known as "blades." Comparable blade tools are almost or wholly absent from contemporary late Paleolithic industries in most of Africa and in southern and eastern Asia. Although existing toolmaking techniques were still maintained in some instances, the Eurasian Upper Paleolithic was novel and specialized. The blade-tool inventory, usefully illustrated and discussed by Klein (1969b, ch. 2), was in part especially devised for bone and wood-working (Fig. 73). *Burins*

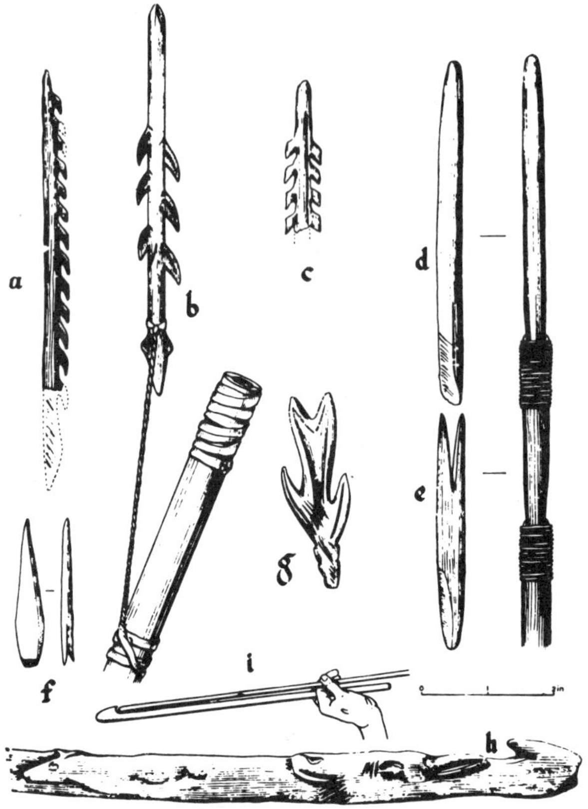

Figure 73. Bone and antler weapons from the European Upper Paleolithic (from K. P. Oakley, 1958, by permission of the Trustees of the British Museum, Natural History): (a) barbed point of antler, (b,c) harpoons of antler, (d, e) spear-point and link-shaft of antler, (f) split-base bone point, (g) bone fish-gorge (?), (h) antler spear-thrower, (i) wooden spear-thrower of modern Australian Aborigine. (Not to scale.)

are chisel-shaped blades for engraving or working in wood, bone, or antler, and are in turn often used as handles or shafts for other implements. *Borers* are blades worked to a narrow, sharp point and used for drilling holes into wood, bone, shell, skin, etc. *Notched* or strangulated blades were probably employed for preparing arrow or spear shafts by scraping or shaving wood. *End-scrapers* were prepared with a blunted end and probably used to hollow out wood or bone, scrape hides, or remove bark from wood. Round or *nosed* scrapers have a well-worked, sharp-edged semicircular end or nose, presumably used to plane and scrape wood and bone. Other tools were more generally employed for cutting or scraping purposes. These include the *backed blade,* with one purposely blunted edge, presumably to facilitate holding. Other stone implements were used for offensive purposes. *Tanged* or shouldered *points* were designed as projectile points, to be fixed to a wooden shaft. *Laurel-leaf* blades were carefully worked to thin, sharp-edged knives or arrowheads, possibly used as daggers.

Eiseley has emphasized the economy of late Upper Paleolithic stone workmanship in general. In the pebble-tool cultures a pound of flint provided about 5 cm. of cutting edge; in the hand-axe technique 20 cm.; in the Mousterian, 100 cm.; and in the Upper Paleolithic, 300 to 1,200 cm. This exponential rate of flint-working economy probably provides an index of the technological advance of the period.

Upper Paleolithic bone and antler tools include new items such as polished pins or awls, and split-based and other types of points to be fixed onto spear shafts. The later, Magdalenian inventory of western Europe (now dated ca. 19,000–13,000 B.P., with late derivatives lasting until 10,000 B.P.) further includes hooked rods employed as spear-throwers, barbed points and harpoons for fishing purposes, fishhooks, needles with eyes, bone and ivory bodkins, belt-fasteners, as well as other tools of less clearly understood use. Such implements were in many cases artistically engraved, depicting various animals of the hunt.

The use of bow and arrow is first verified from the late Paleolithic. Stone arrowheads are claimed from the late Pleistocene Aterian industry of North Africa. More convincing are the archeological remains unearthed from a former lake at Stellmoor near Hamburg, dating from the Ahrensburg culture ca. 10,500 B.P. (Rust, 1943, 1962). Some 100 wooden arrows were found. They were made from pine splints varying between 20 and 80 cm. in length. About a quarter of these arrows were designed for use without flint arrowheads. One such untipped arrow was in fact found in a wolf vertebra. Rust (1943) believes that flint arrowheads may have been used for large game only, while the untipped type of arrow was used for smaller animals. Obviously the antiquity of the untipped arrow may be much greater. Two bows of pine woods were also found at the Stellmoor site. Gaping holes in reindeer shoulder blades bear mute testimony to the effectiveness of flint arrowheads.

UPPER PALEOLITHIC ECONOMY

In the sphere of daily life, Upper Paleolithic hunting techniques were still comparable to those of earlier times. Spears, javelins, harpoons, clubs, stone missiles, bow and arrow, boomerangs or throwing-sticks all counted among the offensive weapons. Bolas, consisting of stone balls joined by thongs and thrown among the legs of running animals, may also have been known. Snares and pitfalls were almost certainly used for hunting some large game, while gregarious herbivores were stampeded into enclosures or defiles or over precipices in organized battue hunts. The bones of over 100,000 horses in the Upper Perigordian occupation horizon, at the foot of the cliffs at Solutré in the Rhone Valley, attest to the hunting efficiency of contemporary man. Modern analogies of the Scandinavian Lapps bear out the practice of organized hunts designed to drive reindeer herds over precipices or into pitfalls (see Broegger, 1926), Kühn (1929, p. 314) believes that cave drawings at Montespan, Marsoulas, Niaux, Font-de-Gaume, Bernifal, and Les Combarelles in southwestern France and at La Pileta, La Pasiega, Castillo, and Buxu in Spain, represent various forms of snares, traps, pitfalls, and enclosures. Soergel (1922) pointed out that juveniles formed a large percentage (25–35 per cent) of the animals caught at many stations, suggesting destruction of entire herds, particularly in autumn, probably in order to provide a winter meat store. But individual attack on animals in the open is also confirmed—e.g., by a drawing at La Colombière of the formidable woolly rhinoceros, showing a number of arrows penetrating the body.

The taking of fish was already practiced in the late Villafranchian of Africa (J. D. Clark, 1960). It is fairly easy to catch fish in the shallow lakes or ponded rivers during the low-water season. Weirs may also have been built to strand fish in shallow waters. Possibly the same simple method was employed by the Mousterian people at Lebenstedt. During the late Paleolithic, however, fishing was revolutionized by the invention of harpoons and fish-gorges. Aquatic foods already had considerable dietary significance in some terminal Paleolithic economies.

Little direct information is available concerning late Paleolithic clothing habits in Europe. However the frequency of bone sewing needles, bodkins, and belt-fasteners suggests that elaborate wearing apparel, presumably of tanned skins and furs, was in common use. The material culture has, in fact, frequently been compared with that of the Eskimo, undoubtedly the result of ecological convergence rather than direct ethnic or cultural association.

Some authors have argued that the reindeer was semidomesticated and that some form of reindeer nomadism had been invented during the terminal Paleolithic. There is, however, a complete lack of supporting empirical evidence to this effect. Smolla (1960, pp. 84-87), who has presented both sides of the case in detail, concludes that neither the ethnological nor archeological evidence supports this hypothesis.

UPPER PALEOLITHIC SETTLEMENT PATTERNS IN EURASIA

Blade-tool industries of main and late Würm age are widespread, occurring through most of the unglaciated parts of Europe, the Near East, western Siberia, and parts of northern Africa. Despite their over-all similarity, these industries are diversified and suggest strong regional differentiation of cultural traditions and, presumably, breeding populations. None of the cultural areas was large—and the "eastern Gravettian," ranging from Germany through southern Russia into Siberia, is nothing but a catchall for a host of distinct traditions.

Settlement patterns were variable, as can be seen from open-air habitation sites excavated at Fourneau-du-Diable (Dordogne), in the Usselo area of the Netherlands, at Poggenwisch and Borneck near Hamburg, at Moravany and Unter-Wisternitz (Dolní Vestoniče) in Czechoslovakia, at Langmannersdorf in eastern Austria, and at Pushkari, Gagarino, and Kostenki-Borshevo in Russia (Fig. 74). To these must be added a large number of well-excavated cave habitation sites ranging from Spain into Central Asia. Wherever the state of archeological investigation is good, it seems that both cave and open-air habitation sites were occupied in one area, although the relative importance of the one or the other varies regionally. Some areas suggest seasonal, winter cave occupation with temporary summer habitations in the open. Others suggest semipermanent open-air settlements. Settlement patterns were complex, and the regional diversification contrasts strongly with the simpler patterns apparent from the earlier Paleolithic.

Group economies were specialized on certain animals. So, for example, the reindeer was the most important meat source in western Europe, mammoth in eastern Central Europe, reindeer and horse in central Russia, and bison in the southern Ukraine. In Spain, where gregarious ungulates were few or absent, game resources were even more diversified. Specialization, wherever suitable animals were available, was not unique to the late Paleolithic. It had been practiced at Torralba and Ambrona (elephants) during the early Paleolithic. But it was specialization that dictated the settlement type.

If the staple game species was a seasonal migrant in a particular environment, specialized hunters would probably have resorted to seasonal changes of settlement. On the other hand, if the gregarious herds were locally present all year around, semipermanent settlement would have been possible. The Gravettian mammoth hunters of the forest steppe and loess steppe of central Europe obviously belonged to the latter category. The Magdalenian, Hamburgian, and Ahrensburgian reindeer hunters of the central European tundras belonged to the former, judging by antler studies carried out on several thousand animals found in Swiss caves at Kesslerloch and Schweizersbild (for autumn and winter occupation, see Soergel, 1922) and at open-air sites in the Hamburg area (for summer or winter occupation, see Rust, 1962). On the other

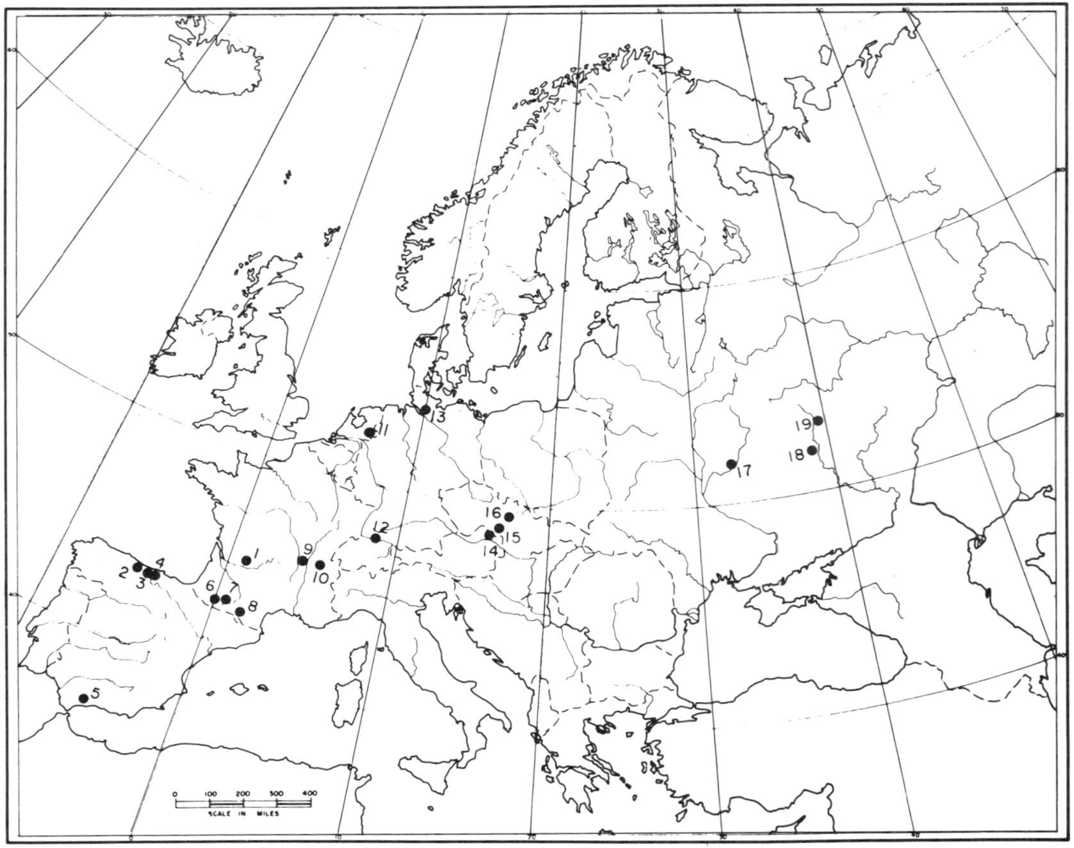

Figure 74. Some European Upper Paleolithic sites: (1) Dordogne area (La Mouthe, Font-de-Gaume, Les Combarelles, Bernifal, Fourneau-du Diable, Cap Blanc), (2) Buxu, (3) La Pasiega, (4) Castillo, (5) La Pileta, (6) Montespan, (7) Marsoulas, (8) Niaux, (9) Solutré, (10) La Columbière, (11) Usselo, (12) Kesslerloch and Schweizersbild, (13) Hamburg area (Ahrensburg, Meiendorf, Borneck, Stellmoor, Poggenwisch), (14) Langmannersdorf, (15) Unter-Wisternitz, Pollau, Moravany, (16) Předmost, (17) Pushkari, (18) Kostenki-Borshevo, (19) Gagarino.

hand, in the forest-tundra of southwestern France both the antlers and dentition of late Pleistocene reindeer remains suggest that these animals were locally available as well as hunted throughout the year (Bouchud, 1954).

Whether cave shelters or open-air tents and huts were occupied probably depended on (*a*) whether suitable caves were available locally and (*b*) whether such caves were suitably located with respect to animal trails or grazing areas. Although little is preserved to prove the point, habitation sites at cave entrances or under overhangs were probably improved by use of branches, hides, or other insulating materials.

Of greatest interest are the open-air shelters, first recognized on cave drawings from Font-de-Gaume, Les Combarelles, and La Mouthe. In recent years a conscious search for such features in archeological situations uncovered ample evidence of elaborate tents and huts. The amount of information is so great that only two examples can be outlined here, the semipermanent settlement of Pollau (Pavlov) at Unter-Wisternitz (Dolní Vestoniče) (Klíma, 1954, 1962), and the seasonal tent camps of the Hamburgians at Borneck and Poggenwisch (Rust, 1958, 1962).

At Pollau several hut plans were found along the banks of a small stream and surrounded by heaps of animal bones representing about one hundred mammoths, mostly juveniles. One hut had a 9 by 15 m. ground plan, containing five hearths in regular distribution. It is thought to have been a tentlike, roofless summer enclosure. More impressive was a second winter or permanent hut with a circular plan 6 m. in diameter. The floor was dug into the slope on one side, where it was retained by means of a carefully laid wall of limestone slabs. Vertical post holes suggest that a roof of bones or branches with animal skins spanned the dwelling pit. In the general area, mammoth tusks were rammed into the ground to form the framework of simple defensive walls, presumably filled out with brush. Fires, both within the huts and in the general camp area, were probably fed by the drippings of fat-laden animal bones. A radiocarbon date of 24,800±150 B.P. (Gro. 1325) places the small settlement in the period of maximum cold during the main Würm. On the basis of ethnographic parallels and the amount of faunal remains, Klíma (1962) suggests that each hut housed 20 to 25 persons and that the community may have numbered about 100 to 125 members. This assumes that all or most of the huts were occupied at one time. The Pollau patterns are repeated at Langmannersdorf in nearby Austria and in the more elaborate Upper Paleolithic settlements of southern Russia (Klein, 1969b). All in all, the impression of semipermanent habitation, at least of a part of the population, is convincing.

Figure 75. Reconstruction of a Hamburgian (Borneck I) summer tent (from A. Rust, 1962, by permission of K. Wachholtz Verlag, Neumünster).

Some of the more important habitation sites at Ahrensburg, near Hamburg, include summer tents at Borneck and Poggenwisch, and a winter tent-complex at Borneck. The summer tents belong to the Hamburgian cultures, dated about 15,500 B.P. (W-93, 15,200±800), the winter tent to a later culture dating from about 11,500 B.P. on pollen-stratigraphic grounds. The summer tent at Borneck (I) consisted of an interior oval tent with a ground plan measuring 2.5 by 3.5 m., and a horseshoe-shaped exterior tent measuring 5.5 m. across (Fig. 75). The furs or hides presumably used were weighted down by heavy rocks, while earth was thrown up around the base of the outer tent. The fire was built at the entrance. Most of the flint-working was done outside, implying summer occupation. The later, winter tent at Borneck (II) consisted of two circular tents, joined by a narrow, covered corridor (Fig. 76). The larger of the two circular tents was 4 m. in diameter, set on a base of laid rocks, with a deep hearth in the middle and insulated by earthworks around the base. The smaller tent probably served for storage. Built during the warmer Alleröd interval, this tent-complex suggests winter occupation. Neither of the Borneck habitation sites is completely understood in a community context, but they give an impression of seasonal shelters as constructed during the terminal Pleistocene.

In general, the more complete archeological record of the European Upper Paleolithic suggests that the population size of individual bands was not much larger than for the Mousterian (Vallois, 1961; Narr, 1963, p. 51 f.), but the density of sites and the immense faunal accumulations

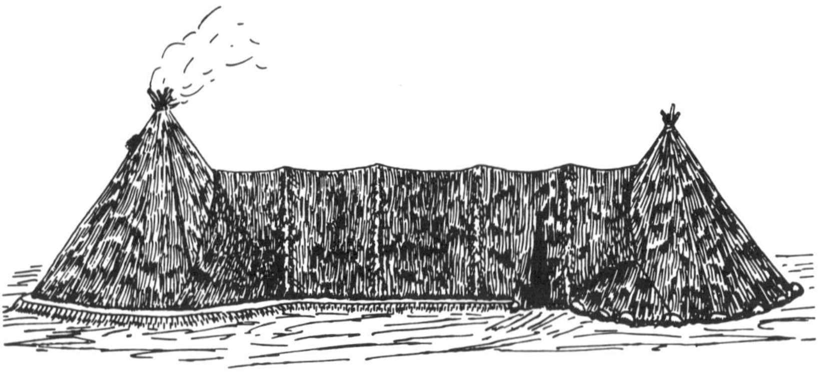

Figure 76. Reconstruction of a Magdalenian (Borneck II) winter tent complex (from A. Rust, 1962, by permission of K. Wachholtz Verlag, Neumünster).

at the individual localities suggest a much greater over-all population. Possibly, although there is no way to substantiate it, the European population had increased by as much as a factor of ten between the early and late Würm. Cultural variability through space was much more marked than for the Mousterian.

MAN-LAND RELATIONSHIPS OF UPPER PALEOLITHIC POPULATIONS

The foregoing discussion of technology and settlement suggests that man had, by the terminal Pleistocene, learned to live successfully in all environments except the high arctic barrens and the true deserts. In fact, the subsistence patterns of tundra, forest-tundra, and steppe residence had possibly developed into the highest standard of living known at any time and at any place prior to the advent of food-producing economies. Ecologic adaptations are increasingly evident in many phases of human culture, ranging all the way from sewn clothing to shelter construction and a variety of hunting techniques. Even the European mastery of bone-working may reflect on the scarcity of wood in tundra environments.

An ample food supply may have encouraged Upper Paleolithic man to develop the astounding degree of artistic perfection particularly evident in southern France and in Spain. Not only were bone and antlers skillfully engraved, but large semirelief sculptures were done in clay, and outstanding line or polychrome drawings affixed to the walls and ceilings of interior cave galleries.[1] The indication of the animal's heart, or the

1. Possibly one of the most thoughtful of the many evaluations of Paleolithic art has been given by Bandi and Maringer (1952); more recently, see Ucko and Rosenfeld (1968).

association of arrows or spears with an animal drawing, suggest sympathetic magic as a primary motive.

Yet despite man's comparative technological efficiency in hunting, fishing, clothing himself, and providing shelter, mortality rates were frightfully high. Less than 50 per cent of some 76 Upper Paleolithic skeletons studied from Eurasia are those of people who had attained the age of 21, while only 12 per cent had passed the age of 40. Practically no women had reached the age of 30 (Vallois, 1961). This probably represents no improvement on the mortality rate of the Mousterian populations.

The comparative skill and diversity of Upper Paleolithic man seriously poses, for the first time, the possibility of human impact on the environment. Any such modifications would almost certainly have been confined to modifications of the biological world.

The question of fire, used as a hunting tool or as a device to facilitate removal or regeneration of vegetation, is equally enigmatic as in earlier Paleolithic times. However, one important exception can be cited. The humic paleosol horizons of the Alleröd are commonly developed on aeolian sand or loess of late Würm age, and buried by further aeolian beds of the Younger Dryas. In Belgium, the Netherlands, and adjacent parts of northwestern Germany, fire horizons, particularly rich in pinewood charcoal, are commonly found in the horizons of the Alleröd rankers and micropodsols. (Hijszeler, 1957; Van der Hammen, 1957a, 1957b, and earlier references). There is no question that these charcoal strata were the result of forest fires during the Alleröd. In one case crowberry (*Empetrum nigra*) was strongly favored by such burning; in other cases high pollen values of the pioneer fire weed *Epilobium angustifolium* also substantiate a conflagration. The radiocarbon date of the charcoal horizons is 11,000 B.P., and such horizons frequently coincide with archeological strata belonging to the Tongrian culture (Late Magdalenian or Federmesser), a hunting population inhabiting northern Germany, the Low Countries, and adjacent parts of England.

Van der Hammen (1957a) suggests that many of these forest fires had human rather than natural causes. Narr (1961, p. 105; 1963, p. 100), with reason, raises the question that these fire strata may represent deliberate burning by man rather than accidental forest fires in dead pine woodland during the climatic deterioration heralding the Younger Dryas. Narr cites several possible incentives for deliberate burning, in part following Stewart (1956): (*a*) to permit easy sighting of game, (*b*) to favor the growth of plants and grasses required by the desirable game species, (*c*) to increase the size of game pasture at the expense of the forest, or to localize grazing areas and so facilitate hunting, or (*d*) to favor the growth of wild vegetable foods and berries for human consumption. It is possible and probable that the Tongrians did indeed practice burning on a large scale. These widespread charcoal horizons in conjunction with the isolated evidence of a forest fire from the cultural horizon at Hoxne (West and McBurney, 1954) emphasize the need for systematic restudying of Pleistocene pollen profiles for possible evidence of human disturbance of vegetation.

The second potential impact of Upper Paleolithic man on the biological world is through indiscriminate hunting of large, breeding animals. The question of late Pleistocene or early Holocene "overkill" is currently more controversial in the New than in the Old World, but the problem is equally pertinent.

By the end of the Alleröd oscillation (see discussion in chapter 31) the woolly mammoth and rhino, the steppe bison, the giant "elk," the cave bear, the European wild ass (*Equus hydruntinus*), a genus of goat (*Hemitragus bonali*), a porcupine, and two genera of rodents became extinct in the Old World (see Kurtén, 1968; Vereshchagin, 1967). Several other species, including the spotted cave hyena, the saiga antelope, Arctic lemming and musk ox, simultaneously disappeared from mid-latitude Europe, surviving further north or east in Eurasia. In the same connection it is also relevant that several species or genera had already become totally or regionally extinct during the first half of the Würm Glacial. The totally extinct forms include the straight-tusked elephant (*E. antiquus*), Merck's rhino, the steppe rhino (*D. hemitoechus*), and a woodland bison (*B. schoetensacki*), while the regional extinctions include the hippopotamus. The disappearance of certain woodland forms early during the Würm has traditionally been attributed to the onset of cold glacial climates, although one may, of course, question why these genera did not survive in the Mediterranean lowlands. The terminal Pleistocene extinctions have, on the other hand, been widely ascribed to the rapid recolonization of the mid-latitude tundras by post-glacial forests, as all of the extinct forms were linked with the steppe-tundra biotope (Kowalski, 1967).

The recent debate over man's possible role in animal extinctions in North America has served to lead some authors to reconsider the European case. Kurtén (1968, p. 272) points out that the mass deaths can hardly be ascribed to climatic causes alone, for there was no comparable mass extinction at the beginning of previous interglacials. In fact, several genera or families were eradicated for the first time since the Villafranchian, and elephants and rhinos are now absent from temperate Eurasia for the first time since the Tertiary. Yet, it must also be remembered that the wild horse, aurochs, and woodland bison managed to survive man's predations until modern times, while the wolf and brown bear were only displaced from western central Europe less than two centuries ago. Furthermore, many of the extinct species were not among the usual or even the common hunting booty of Upper Paleolithic man. Most of the larger carnivores involved were probably hunted incidentally, so that the disappearance of such species should be at best indirectly linked to a human competitor.

It is possible, as Kurtén (1968, p. 273 f.) argues, that man played an indirect as well as a direct role in accelerating the demise of species or genera that were earmarked for eventual extinction or were undergoing severe crisis. There is, for example, a general trend of size decrease through the course of the Pleistocene, with extinction of the larger mammals. Similarly, climatic and ecological stress may have reduced the size of some populations to such a low number of individuals that they became especially vulnerable. Hunting pressures or competition for similar ecological niches may have been critical, if not lethal, for such species. At the present stage of investigation there is no proof that any one species became locally or totally extinct through the agency of Pleistocene man. On the other hand, however, the massive slaughtering of game evident at some sites must have had serious repercussions on both the biomass and the equilibrium of animal populations in Europe and North America. Advanced hunters such as the Upper Paleolithic peoples of Europe do not have direct control over the natural environment, but they do exercise a considerable effect on it. They were the major predators in the ecological communities of which they were a part, and the balance of animal energy exchange can hardly be understood without regarding them as a basic factor (Watson and Watson, 1969, p. 83). All in all, it would therefore seem justified to say that the terminal Pleistocene marks the beginning of significant environmental modification by man.

TWENTY-NINE

The Paleo-Indians: Origins and Man-Land Relationships

INTRODUCTION

The earliest cultural manifestations of man in Africa suggest a long period of "hominizations" within the original geographical range of the australopithecines. After the integral developments of culture, technology, subsistence, intellect, and biology had been completed, man dispersed rapidly across Africa and temperate Europe as well as tropical and temperate Asia. This primary dispersal of mankind, during the early Pleistocene, was not immediately followed by any major extension of settlement into new environments. Then, during the late Pleistocene, man began to exploit the last, major economically feasible territories, expanding into the tropical rainforests and the boreal woodlands, and crossing over sea or by land-bridge into the New World—the Americas and Australia. This secondary dispersal presumes an improvement of the technological subsistence base that made man more adaptable to new ecological opportunities and stresses: possibly the diversification and specialization of late Paleolithic stone tools is a symptom of this new versatility and flexibility. It also presumes some form of centrifugal stimulus, favoring an extension of territoriality along the fringes of the inhabited world, the *oikoumene*. Whether denser settlement patterns, with incipient population pressures, provided such a stimulus is unknown. Conceivably, population pressure was no greater than during earlier times of ecological stress and, instead, man's response to environmental change may have been different.

Whatever the underlying motivation, the first colonization of the New World opens vistas of ecological interest, rivaled only during an earlier era by the dispersal of Acheulian hunter-gatherers, and at a later time, by the invention and diffusion of agricultural traits. In each instance, migration to the Americas and Australia involved mastery of various environmental obstacles: trekking through endless tracts of peripheral lands in the one case, navigating open seas in the other. Neither venture can have involved foresight or long-range planning, but each opened to rather small groups a veritable promised land in which to expand and multiply under optimal conditions.

A geographical study of these early colonizations involves an understanding of the Old World environments and subsistence economies from which the immigrants came, a discussion of the technical and ecological problems encountered in coping with land bridges or barriers and, finally, an examination of the colonists in their new surroundings. Although the available information is far from satisfactory, a great deal is known about many of the material aspects of New World colonization. This by itself lends relevance to a theme of great intrinsic interest, since the case of the Americas or Australia has direct bearing on the earlier dispersal of man out of tropical Africa.

LATE PLEISTOCENE ENVIRONMENTS OF EASTERN SIBERIA

The eastern half of Siberia presents a harsh, subarctic environment even today (see Suslov, 1961). Some of the world's lowest temperatures are recorded from the middle Lena Basin, and islands of sporadic permafrost are found as far south as the Amur Valley and beyond. With exception of the Arctic coastal plain, the terrain consists primarily of hill country, crisscrossed by rugged mountain chains with dominant elevations of over 2,000 m.

During the late Pleistocene, glaciation was widespread in the form of mountain and valley glaciers in the high country. The climatic snowline was situated below 500 m. in coastal proximity, rising to over 2,000 m. in the continental interior (Frenzel, 1959, p. 57 ff.); the only major ice cap was that of the Taimyr Peninsula (see Fig. 77). The overall chronology of glacial events, largely fixed by radiocarbon dating of related fluvial deposits (Kind, 1967; Kind *et al.,* 1969; Klein, 1971), is similar to that of both North America and Europe:

Beginning of *Holocene:* ca. 10,500 B.P. (Recent)

Sartan Glacial: ca. 27,000–10,500 B.P., with a significant warm oscillation ca. 11,500 B.P. (Late Wisconsin, Two Creeks, Valders)

Karginsky Interstadial Complex: ca. 50,000–27,000 B.P., interrupted by

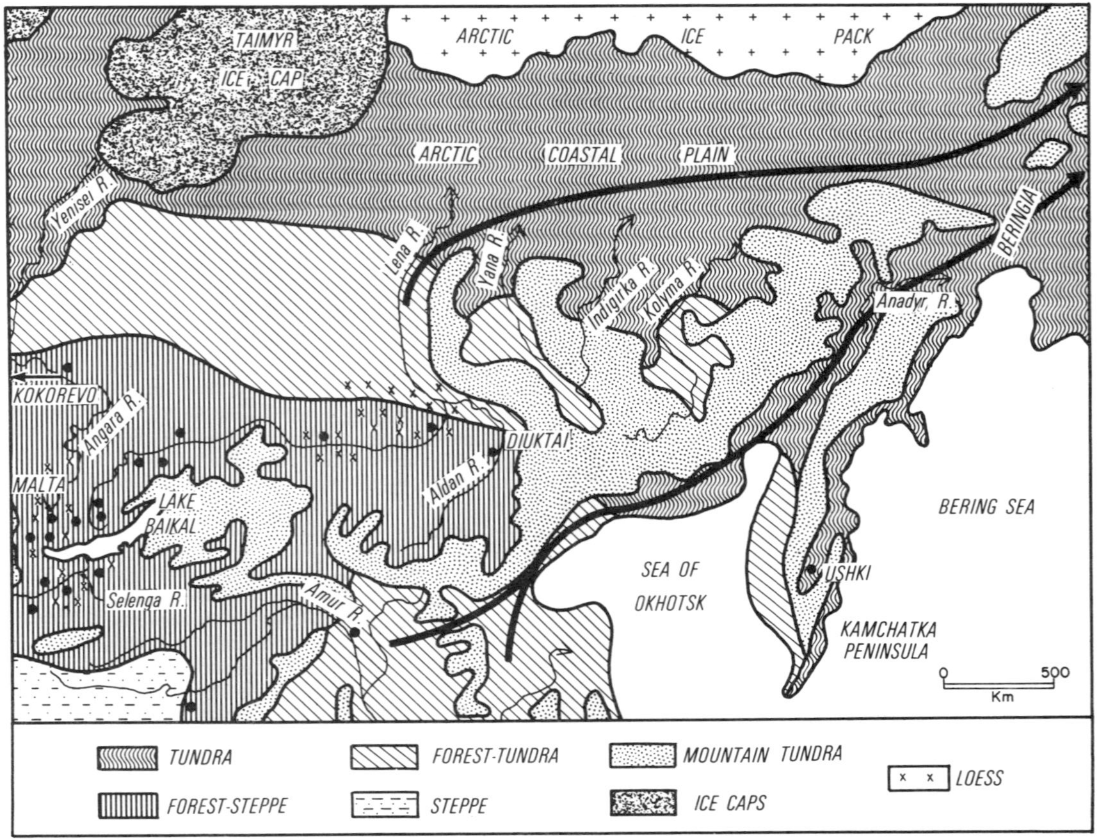

Figure 77. Late Pleistocene environments and archeological sites of eastern Siberia. Arrows indicate possible migration routes to the New World; circles mark major sites.

cold intervals ca. 45,000–41,000 and 37,000–32,000 B.P. (Plum Point Interstadial complex)

Zyriankan Glacial: before 70,000 to ca. 50,000 B.P.

Kazantsevo Interglacial: (Sangamon)

Little systematic information is available on the exact stratigraphy of the Siberian mountain glaciers.

Alluvial terraces, mainly aggraded in response to cold climate, are well-developed along most of the major river systems (see various chapters in Michael, 1964). Only some of these fluvial beds were directly associated with glacial meltwaters. Loess deposits, of full glacial age, are found in the Lake Baikal area and in the middle Lena Basin (Frenzel, 1959, p. 122 ff, map 3) (see Fig. 77). Since permafrost is so widespread today, and a great variety of periglacial processes are actively at work (see Frenzel, 1959), it can be safely assumed that the entire region had continuous or discontinuous permafrost during the Zyriankan and Sartan glacial stages. Mean annual temperatures were presumably well below −2° C. throughout the late Pleistocene, and more extensive glaciation was inhibited only by a semiarid climate.

The late Pleistocene vegetation of eastern Siberia has been reconstructed by Frenzel (1968a, plate 10; regional detail in Frenzel, 1960, pp. 73-84; see also Giterman and Golubeva, 1967; M. P. Gričuk and A. P. Vaskovskiy in Michael, 1964). The bulk of the palynological data comes from the Lake Baikal region, the Lena Basin, and the Indigirka Valley (see Frenzel, 1968a, plate 14, with sources), while little is known about the northeastern peripheries. The broad pattern of vegetation can be outlined as follows (Fig. 77):

a) *Tundra.* The lower Yenisei, the Arctic coastal plain, the exposed Beringian land bridge, the Anadyr Valley, and most of the Kamchatkan coastal plains.

b) *Mountain Tundra.* The mountain ranges of eastern Siberia and Transbaikalia, above 1,500 m. or so, with widespread but restricted mountain glaciation. Much of this rough country undoubtedly bare rock, intensively sculptured by periglacial processes.

c) *Forest-Tundra.* A broad belt of forest-tundra mosaic, parkland, or open scrub, with dwarf birch and larch in the hill country west of the middle Lena, and refugia of pine and larch in the upper Yana, Indigirka and Kolyma valleys. Similar patterns are inferred for the middle and lower Amur Basin.

d) *Forest-Steppe.* Another broad belt of parkland and forest-steppe mosaic, grading into forest-tundra nearer the poles, into open grasslands to the south. Primary forest component pine, with *Artemisia,*

Chenopodiaceae and *Ephedra* dominant among the herbaceous genera. Some extended loess tracts.

e) Cold Steppe. Open grasslands, extending across Mongolia and fingering northward into the lowlands of the Baikal and upper Amur regions. Significantly, the boreal forests of today were replaced by open vegetation that may have supported a larger variety of big game as well as a much greater biomass.

The late Pleistocene faunas of eastern Siberia are compatible with this picture of an open parkland vegetation and an intensely cold climate with little snow cover in winter (Vangengeim *et al.*, 1966). The dominant faunal elements are woolly mammoth and rhino, horse, and steppe bison *(Bison priscus)*. In the middle Lena Basin and, in general, north of latitude 58° N., these species occur together with typical arctic forms such as reindeer, arctic fox, lemming, and musk ox. Reindeer ranged farther south, to Lake Baikal, where the faunal spectrum of the upper Lena and the Angara basins included wolverine, brown bear, kulan *(Equus hemoinus)*, yak *(Poephagus)*, *Bos primigenius*, saiga, horned antelope *(Spiroceros)*, and some steppe rodents (see also Ermolova, 1963). Arctic marine mammals were presumably found in coastal areas.

From the viewpoint of the hunter-gatherer, eastern Siberia was probably more attractive during the late Pleistocene than it is today. With a considerable biomass of cold-steppe mammals, the carrying capacity was greater for late Paleolithic groups than for modern hunter-gatherers. Consequently, once the means had been devised to withstand the exceptionally severe winter cold, settlement of northeastern Siberia and Beringia was possible—perhaps even attractive—for late Paleolithic man.

SIBERIAN ANTECEDENTS TO NEW WORLD COLONIZATION

Archeological sites of Pleistocene age are found primarily in the alluvial terraces or loess mantles of the upper Lena, the Angara, the upper Yenisei, and the Selenga rivers. Cave sites are also known from the mountains northeast and west of Lake Baikal. Little or nothing is reported by way of older materials from the Arctic coastal plain or northeastern Siberia. The reader is referred to Vangengeim *et al.* (1966), Chard (1969) and Klein (1971), with references to regional studies, for the background material to the subsequent outline.

There are no sites in eastern Siberia of early or mid-Pleistocene age, and the oldest sites that have been dated or stratigraphically fixed are younger than the Karginsky Interstadial. Nonetheless, potentially older sites are reported from the middle and lower Amur Basin and from the

Altai Mountains west of Lake Baikal. The Amur sites and perhaps the comparable levels from Choukoutien (locality 15) include choppers and chopping tools, miscellaneous flakes, and in some instances, large knives and scrapers. The Altai sites and some Mongolian localities have a more Mousterian flavor, with points, side-scrapers, Levalloisian flakes, and some evidence of bifacial working. These peripheral industries may be a little older than the typical Siberian late Paleolithic, but they may also be broadly contemporary.

Best known of the Siberian Paleolithic occurrences are the site clusters of the Baikal region (Fig. 77), including Malta (14,750 B.P.) and Buret. Artifacts include points, end-scrapers, side-scrapers, knives and notches—all made on blades, as well as perforators, burins, and gravers. Semisubterranean structures, employing mammoth bone, a variety of Venus figurines, and bone needles and awls suggest a broad similarity with some Upper Paleolithic traits of Europe. This individualistic yet unspecialized late Paleolithic industry is locally followed by horizons with bifacial, laurel-leaf points. The late Paleolithic sites of the middle Yenisei and lower Angara rivers have a stronger Mousterian flavor; for the most part, they lack burins but include such diverse items as bifacial tools, Mousterian points, and microblades. The Kokorevo site complex has dates ranging from 15,500 to 12,600 B.P., while the archeological levels of Afontova Gora appear to be limited by dates of 20,900 and 11,300 B.P. Different still are the blade-tool and scraper industries of the middle Lena, which may date ca. 12,000 B.P. Of particular interest are two unique sites, the Diuktai Cave on the middle Aldan, and Ushki Lake in central Kamchatka (Chard, 1969b). The Diuktai Cave, with a single date of 13,070 B.P. includes bifacial, pressure-flaked projectile points and knives, in association with a mammoth fauna. Levels VI and V at Ushki, with a date of 10,360 B.P. (14,300 and 13,600 B.P. for underlying level VII), have bifacial foliate points and knives.

All in all, the available evidence indicates that late Paleolithic blade-tool traditions were widely dispersed through the more mesic environments of eastern Siberia by about 15,000 B.P. These have been described as "Aurignacian-like" by Müller-Beck (1967), but they represent distinctive regional industries with no necessary cultural relationships to the big-game hunters of western Europe. Traces of bifacial working are not exceptional at most sites, but elaborate bifacial points, produced by indirect percussion or pressure-flaking, first appear in the record about 13,000 B.P. This suggests that bifacial projectile-points may have appeared in eastern Siberia and North America simultaneously, and that industrial traditions on either side of Beringia may well have some common roots. However, the late Paleolithic of eastern

Siberia includes an amalgam of diverse traditions, the exact temporal and spatial relationships of which are far from understood. Also relevant to early colonization of the Americas is the absence of documented sites greater than 20,000 years old in eastern Siberia.

THE BERINGIAN LAND BRIDGE

The Bering Strait separates the land masses of Asia and America across a broad but very shallow continental shelf. At its narrowest point the strait is about 80 km. across, and a sea-level drop of only 46 m. would expose a narrow land connection between Siberia and Alaska by way of St. Lawrence Island. A lowering of 50 m. would create a second narrow connection north of the Bering Strait, and a reduction to – 100 or – 150 m. would expose most of the continental platform to form a flat and almost featureless plain (Hopkins, 1967, chs. 2 and 24). Study of the bottom sediments of the now-submerged shelf is compatible with emergence during most of the late Pleistocene (Hopkins, 1967, chs. 2 and 3).

On the basis of the regional evidence (including submerged shorelines and the nature of certain bottom sediments) and general information on glacial-eustatic fluctuations of sea level, Hopkins (1967, ch. 4 and pp. 460-65) has outlined the late Pleistocene history of the land bridge as follows:

1) The land bridge was exposed for most of the Zyriankan-Altonian substage, prior to perhaps 35,000 B.P.
2) The Woronzofian transgression, coinciding in part with the Karginsky-Farmdale substage, interrupted the land connection as late as 25,000 B.P.
3) The subsequent regression exposed the shelf until mid-Sartan-Woodfordian times, probably until shortly after 15,000 B.P. (Cary interval).
4) The terminal substages of the Wisconsin were marked by several major oscillations of sea level, during the course of the Krusensternian transgression. In a critical range of elevations, they periodically severed and connected the land bridge. In particular, the land bridge was probably open about 13,000 (Port Huron interval) and again about 11,000 B.P. (Valders).

In other words, man could have crossed the Bering land bridge dry-shod during much of the late Pleistocene. In addition, he could have crossed on the winter pack ice even at times when the land bridge was essentially severed. Finally, the prehistoric groups that first arrived in Beringia may well have had knowledge of light craft that could have negotiated the open straits: if Acheulians crossed the Straits of Gibraltar, why might not late Paleolithic groups have navigated the Bering Strait? Regardless of the answer to the last question, small groups may have crossed the land bridge at several times during the late Pleistocene. The date of their arrival in temperate North America would ultimately be determined by ice conditions in western Canada.

The regional environment of Beringia during the late Pleistocene does not appear to have differed significantly from that of northeastern Siberia (see Colinvaux, 1967). NAP is dominant and almost exclusive in pollen profiles, and a dry tundra-steppe is inferred. The abundance of grazers such as horse, bison, woolly mammoth, saiga antelope, and yak supports the interpretation of extensive grassy tracts within the mosaic of poorly drained herbaceous tundra and marsh vegetation (Hopkins, 1967, ch. 24). Forest-tundra or forest-steppe appears to have been confined to some of the interior basins of central Alaska. In general, the big-game carrying capacity of Beringia, including western Alaska, appears to have been far better during the late Pleistocene than it is at present.

Unfortunately, there is next to no archeological record of human settlement in Alaska prior to about 8500 B.P. (Hopkins and Giddings, 1965; Laughlin, 1967). Most of the sites that must have existed may now be submerged under the Bering Sea. Recently, however, two significant finds have begun to fill the gap: a convincing biface at Healy Lake in east-central Alaska (11,090 B.P.) and a bifacial fragment with large choppers at Glacier Bay, southeastern Alaska (10,180 B.P.) (R. E. Ackerman, comments in Bryan, 1969).

AN ICE-FREE CORRIDOR IN WESTERN CANADA?

The really critical factor in the first arrival of man in temperate North America seems to have been the existence of an ice-free corridor in western Canada, to permit small late Paleolithic groups in Beringia to move southward. From the Aleutian Islands to Puget Sound, the Cordilleran ice sheet calved into open Pacific waters over a stretch of 4,000 km. for most or all of the late Pleistocene. This seems to preclude the possibility of migration along the coast. Similarly, ice merged with the open ocean along the whole Atlantic coast of Canada and New England. This leaves only one realistic probability—an ice-free corridor along the eastern foothills of the Rocky Mountains.

There is general consensus that the Cordilleran and Laurentide glaciers merged, on a broad but discontinuous front, during the maximum of the Wisconsin Glacial (see the glacial map of Wilson *et al.,* 1958). The question is, when was this contact achieved and how long did it persist? In the Dakotas, Montana, and the Columbia Plateau, the late Altonian ice advance (Bull Lake stage) extended much farther than the

Woodfordian (Pinedale stage) (see Richmond *et al.*, 1965; Lemke *et al.*, 1965). Similarly, in the lower Mackenzie Basin, maximum glaciation was achieved prior to the mid-Wisconsin interstadial (R. S. McNeish, comments in Bryan, 1969). Consequently, at the present state of our knowledge, it is by no means proved that the Cordilleran and Laurentide ice sheets merged during the Woodfordian (see Fig. 56 for alternative positions of the Laurentide ice front), except possibly in a discontinuous fashion for a few millenia. It seems that the oft-stated generalization that no ice-free corridor existed from approximately 25,000 to 12,000 B.P. (or even 9,000 B.P.) is a hypothesis, not a demonstrated fact. The only immediate support for this point of view is indirect—the restriction of faunal migration down such a corridor (see Bryan, 1968, 1969).

Whatever the position of the ice termini at the maximum of the Woodfordian substage, the Canadian corridor must have been opened shortly after the Cary interval (ca. 15,000 ± 500 B.P.). Figure 78 attempts to show the position of the ice sheets at that time, as delimited by the Vashon stade in western Washington, the middle Pinedale in eastern Washington and Idaho, the Alamo moraines in North Dakota, and their northern extension into Saskatchewan and eastern Alberta. Radiocarbon dates from southern Saskatchewan suggest a temporary glacial readvance shortly after 13,000 B.P., possibly coincident with the Port Huron stade (see W. M. Hlady, comments in Bryan, 1969); the same areas were ice free as late as 20,000 B.P.

Available information on the timing of the ice-free corridor in western Canada is unsatisfactory. At the moment it would seem, however, that this corridor was probably essentially open after about 14,500 B.P. On the other hand, it is possible, although not probable, that migration along the eastern Rocky Mountain foothills was completely restricted only during parts of the Altonian and again ca. 20,000–18,000 B.P. A conclusive answer must await comprehensive study of glacial stratigraphy and dating in Alberta and the Mackenzie Valley. In the meanwhile it seems pointless to relate the first arrival of man in temperate North America to a complex time schedule dictated by the opening and severing of the Bering land bridge on the one hand, and the closing and opening of the Canadian corridor on the other. Once in Alaska and northwestern Canada, groups could thrive there on the tundra-steppe game until a time when the crucial ice-free corridor was available.

THE PALEO-INDIANS

The first advent of man in the temperate and tropical Americas is the subject of a controversy that is far from a satisfactory solution. A number of authors, including Krieger (1964) and Bryan (1969), feel that

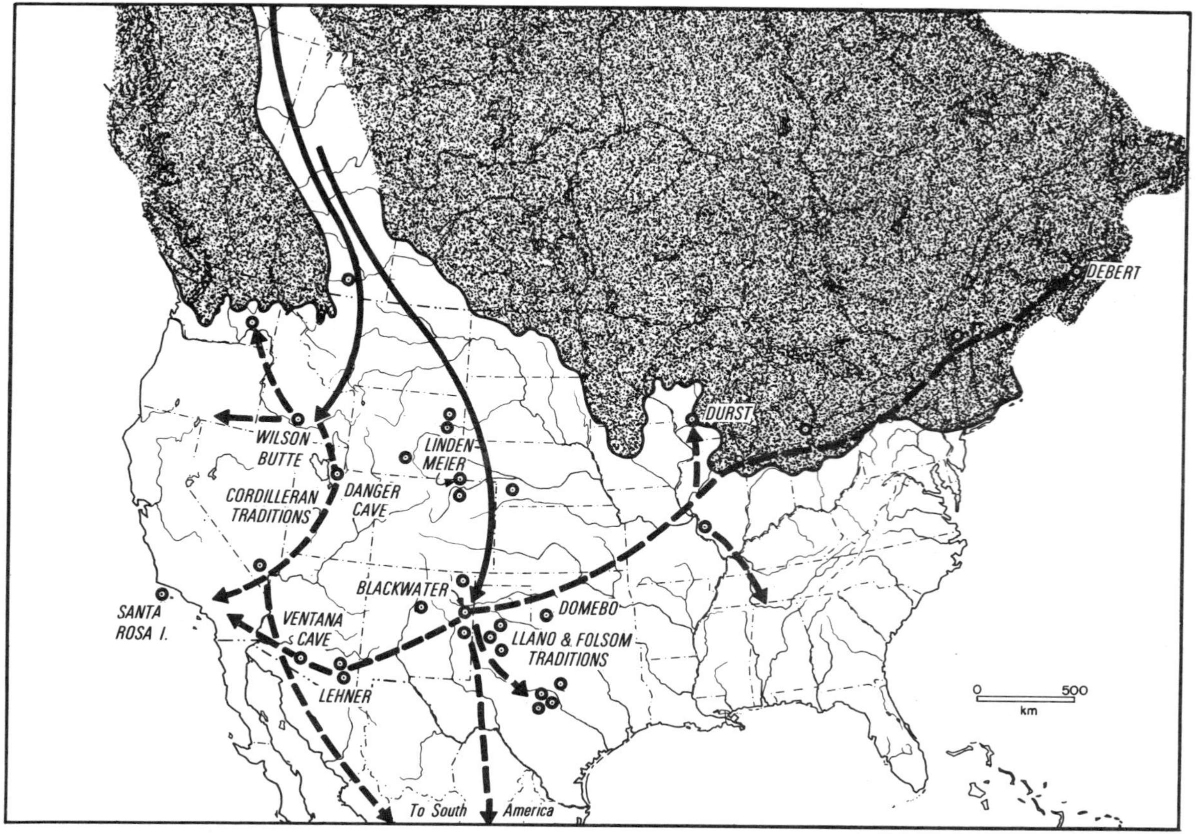

Figure 78. Late Pleistocene archeological sites (ca. 14,500–10,000 B.P.) and extent of ice sheets (ca. 15,000 B.P.) in North America. Hypothetical migration routes and principal axes of dispersal suggested by solid and broken arrows.

the first immigrants arrived south of the ice sheets during the Farmdale or even earlier, i.e., somewhat before 20,000 B.P. This point of view may eventually find support in sites near Puebla (Valsequillo) and Mexico City (Tlapacoya) that appear to confirm the presence of man about 24,000–22,000 B.P. (see Irwin-Williams, 1967; Szabo *et al.,* 1969).[1] Another group of authors, whose views have recently been formulated by Graham and Heizer (1968), believe that there is no reasonable evidence for man in America before 12,000 B.P., the oldest possible date for the fluted projectile industries. Still others prefer an intermediate position, lending greater credence to sites as old as 15,000 B.P. The net result is a confusing image of ideological writings, discredited claims, and conflicting evidence, set against a mass of sites that have seldom been adequately analyzed or published in full.

At the present state of investigation, the oldest verified site in the Americas appears to be Wilson Butte Cave in the Snake River basin of southern Idaho (Gruhn, 1961). Lower stratum C, with extinct camel, contains a small, leaf-shaped biface, a blade, and a flake with a burin blow (Crabtree, 1969). There are two radiocarbon dates on animal bone: 14,500 $\pm$500 and 15,000 $\pm$800 B.P. (M-1409, 1410) (see comments by Haynes and reply by Bryan, in Bryan, 1969). It seems inevitable that other convincing sites in a 12,000 to 14- or 15,000-year time range will find more general recognition within the foreseeable future. This implies that the Indian cultural traditions of the Americas go back only to the opening of the ice-free corridor in western Canada (ca. 14,500 B.P.?),[2] which seems reasonable in view of the date of characteristic late Paleolithic settlements in eastern Siberia. If, as also seems probable, earlier groups did penetrate into the United States and Mexico before 20,000 B.P.,[3] they may not have survived, for whatever reason. At any rate, there seems to be a curious lack of even potential archeological vestiges for the time span 20,000 to 15,000 B.P.

There is little general agreement on terminologies and cultural

1. Although these mid-Wisconsin sites are not sufficiently convincing to find general acceptance, "it appears to be only a matter of time before a convincing find is made" (Haynes, 1969, p. 712).

2. Suggestions have been made bringing early man from Siberia by water craft around the North Pacific (Bushnell, 1961), or from Europe around the North Atlantic (Greenman, 1963). During the Wisconsin Glacial such feats of determination would have involved traverses of 4,000 and 10,000 km., respectively, along unbroken stretches of calving ice sheets, shelf ice or pack ice. Apart from the extreme unlikelihood of such odysseys in savage polar waters, the Paleo-Indians were specifically adapted to the hunting of land herbivores. There is no evidence of any early coastal cultures prior to the appearance of the Aleut and Eskimo in Alaska during early post-Pleistocene times (Laughlin, 1967).

3. In addition to the sites of central Mexico, the problematical nature of the possible mid-Pleistocene site at Calico Hills, in the Mohave Desert of southern California, has been outlined objectively by Haynes (1969).

affinities for the earliest accepted industries in the Americas. More than anything else, this reflects an almost total lack of systematic typological study of lithic assemblages. The one outstanding exception to this chaos and subjectivity is the incompletely published work of Wilmsen (1968). A comprehensive source of information on sites, as of the 1950s, is given by Wormington (1957), while many of the traditional interpretations may be consulted in Jennings and Norbeck (1964). The following outline is based in part on the analyses of Mason (1962) and Müller-Beck (1966), with radiocarbon chronology modified after Haynes (1964, 1967, 1969) and Stuckenrath (1966).

The late Pleistocene to early Holocene industrial complexes of North and South America are commonly designated as Paleo-Indian, although some authors use a more restricted definition, confining the term to industries with certain categories of projectile-points. These points were probably affixed to short spears, hurled from special spear-throwers or attached to long lances. Differences were partly stylistic and partly functional, and the rather handsome fluted varieties mark a high point of stone workmanship (see Fig. 81).

The Paleo-Indian projectile-point traditions of North America are usually subdivided into three successive horizons on the basis of projectile-point style: Llano, Folsom, and Plano. The Llano tradition, including Clovis and Sandia points, spans a time-range of 11,600–10,900 B.P.[4] It appears suddenly and already well developed, so that some authors suppose that incipient Llano industries should eventually be found in Alaska or the lower Mackenzie Valley. The distribution of Clovis points from western Texas to southern California, and from Wyoming to Nova Scotia (Fig. 78), suggests a very rapid dispersal. The offshoots on the eastern seaboard persisted longer, to about 10,500 B.P. Folsom points, represented primarily in the area just east of the Rocky Mountains, presumably mark a regional development of the Llano tradition about 10,800–10,300 B.P. A shift from mammoth hunting to bison hunting is evident between the Llano and Folsom sites, but there seems to be no structural change between the artifact assemblages. The Plano tradition encompasses an increasingly diverse assembly of projectile-point types, initially including Plainview, Midland, and Agate Basin (ca. 10,200–9600 B.P.), and superseded by Scottsbluff, Eden, Allen, Angostura, Frederick, Infiernello, and others (ca. 9600–7900 B.P.). Sites are distributed between Mexico and Alberta, ranging from the Rocky Mountains to the mid-Atlantic coast. Within the Great

4. The Llano, with a date of 11,600–10,900 B.P., is contemporary with the Valders readvance (see ch. 21), not with the Two Creeks, as some authors have stated. The subsequent Folsom dates from the early part of the Valders retreat.

Plains-Rocky Mountain focal area, the technological and functional processes associated with hunting appear to have remained stable from Llano to Plano times, over a period of almost 4,000 years (Wilmsen, 1968), although this does not exclude the possibility that collective bison hunting was first employed by Plano groups, e.g., at the Plainview site.

The projectile-point sites of the semiarid western United States are normally associated with big game, mainly mammoth in the case of the Llano, and bison in the case of the Folsom and the Plano. Many, if not the majority, represent kill and butchering sites, with few artifacts but relatively many projectile-points and low proportions of waste flakes and chips (Wilmsen, 1968). This and the frequency of imported raw material suggests that artifacts or unfinished "blanks" were carried along on hunting forays, and that such kill sites with low artifact concentrations were used very briefly by small bands. Other sites, such as Lindenmeier, involve large collections of over 7,000 pieces, with much higher proportions of stone waste and a greater variety of tool types (Wilmsen, 1968). More diverse animal types, including small mammals and carnivores, were utilized, and there is reason to suspect longer-term occupancy.

The Paleo-Indians of the Llano, Folsom and Plano traditions were hunter-gatherers, moving within a more or less well-defined territory.

> Bands appear to have broken up periodically either under the stress of seasonal fluctuations in resources or to take more efficient advantage of ecological opportunities. Surface quarrying and plant collecting do not require large numbers to be carried out effectively. It may be that one segment of a band exploited one set of resources while other segments directed their attention to different parts of the environment. Band segments regathered periodically and, in fact, bands themselves may have joined with other bands (as at Lindenmeier) in order to exploit the larger environment and to maintain socio-economic integration. Hunting parties as well as raw material and plant collection parties may have voluntarily moved out from these larger units and returned to distribute the products of their activities to the group as a whole. It may be that hunts for mammoth and bison were carried out principally and perhaps only by groups such as these, at times when large band units were assembled (Wilmsen, 1968, p. 987).

Paleo-Indian colonization of the eastern woodlands resulted in exploitation of new opportunities offered by a different environment. Technologies were readapted to the hunting of small mammals and deer, so that those elements most useful in the new environment were emphasized. Prolonged emphasis in one direction gradually produced a technology among the eastern Plano groups that was distinctive from that of the parent, other-directed technology of the Great Plains (Wilmsen, 1968; also Mason, 1962; Fitting *et al.,* 1966). Structurally related changes in other sectors of the cultural system must have paralleled the technological adaptations, and the fact that eastern sites such as Shoop,

Williamson, Bull Brook, and Quad all appear to be small habitation sites seems to underscore their divergence from the Llano kill sites of the Great Plains.

Other projectile-point traditions are found in South America, within the Andean ranges and in Patagonia. They include the Toldense and its apparent offshoots (see Müller-Beck, 1966, with references; Lanning, 1967; Bryan, 1969 discussion), dating from before 10,700 to after 8,000 B.P. These archeological complexes are rapidly proving to be of great interest.

The Paleo-Indian projectile-point traditions do not represent the total spectrum of late Pleistocene man in the Americas. A number of coeval industrial variants, quite possibly representing distinct technological and cultural entities, can be recognized in the intermontane basins and Pacific coastal sectors of both the North and South American cordilleran belts. A chopper–chopping tool assemblage, with prepared flakes and no projectile-points, has recently been excavated at Tagua, central Chile (Mostny, 1968). Mastodon, horse, and deer are present, indicating big-game hunting. The radiocarbon date is 11,380 B.P. (see Lynch and Kennedy, 1970). In North America there is an incoherent complex of industries, probably not homogeneous and, in part, mixed, that has been regionally or locally described by various nonequivalent terms; e.g., Paleowestern, Desert Culture, Old Cordilleran, Ventana, San Dieguito. Some consensus exists that one or more traditions—with or without bifacial techniques, and distinct from the Llano-Folsom-Plano complex—are represented here (e.g., Jennings, 1957; Butler, 1961; Mason, 1961; Bryan, 1969).

These Cordilleran traditions, as they are descriptively and tentatively labeled on Fig. 78, may either reflect distinct Old World roots or different ecological adaptations. Until the different assemblages of the Cordilleran traditions have been properly sorted out, such adaptations cannot be specified. Some regional forms, possibly a seasonal phase only, suggest nonspecialized exploitation of the entire environment, with activities ranging from hunting to small-seed harvesting (see Mason, 1962). However, there also is evidence for big-game hunting (see Butler, 1961; Jennings, 1966, 1968). In view of the complexity of local traditions in Siberia it would not seem unreasonable to assume that more than one industrial tradition was introduced to the New World during the late Pleistocene. Possibly there were two (or more) waves of migration, one introducing the Llano tradition to the United States shortly before 11,600 B.P., the other (or others) bringing a less specialized form of late Paleolithic lithic tradition, with some elements of bifacial workmanship, at an earlier date. This as yet hypothetical wave of earlier immigrants appears to have left no certain record in alluvial or cave deposits east of the Rocky Mountains, although the possibilities of such less specialized lithic traditions as the controversial "Lively Complex" of northern Alabama must be carefully explored.

A PALEO-INDIAN SITE: LEHNER

The Lehner Mammoth Site is a small but well-studied example of a Llano kill site. It was partially excavated in 1955–56, with further excavation for palynological and geological studies in 1962–65. The following outline is based on Haury *et al.* (1959) for archeology, Lance (1959) for paleontology, Mehringer and Haynes (1965) for palynology, Haynes (1968b, 1965) for stratigraphy and geomorphology, and Haynes (1967, 1968a) for radiocarbon dating.

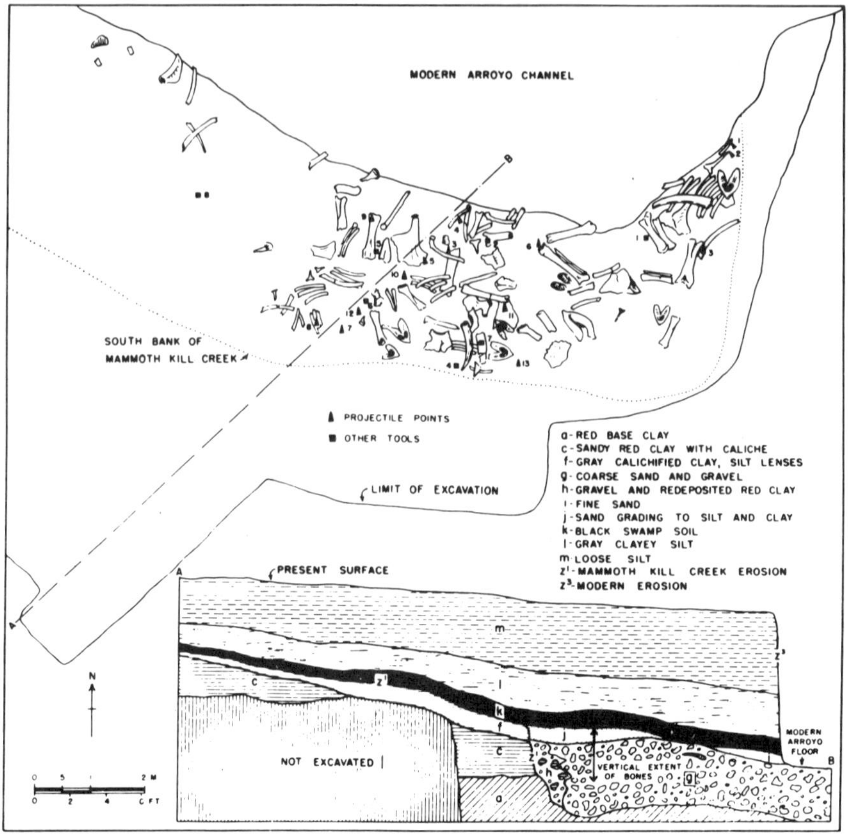

Figure 79. Plan and stratigraphic profile of the Lehner Mammoth Site. The Lehner Formation begins over erosion Z^1 with beds (g)–(j). Erosion Z^2 (not shown) separates (l) from (m). From E. W. Haury and others (1959), by permission of American Antiquity.

The Lehner site is situated at 1,280 m. elevation, in the San Pedro Valley of southeastern Arizona. Deposits with mammal bones and artifacts were first exposed along the banks of a broad, intermittent stream (2.5 m. deep) that is tributary to the San Pedro. This arroyo, with its sandy-gravelly bed, was cut during the 1920s, before which the local watercourses followed a series of grassy swales. The vegetation has deteriorated since the 1890s from a grassland with riverine trees to semidesert mesquite bush today. Local rainfall, about 290 mm., comes primarily in July and August (50 per cent) and, more evenly distributed, during the winter months.

The San Pedro Valley is of tectonic origin and is filled with several generations of Pleistocene alluvium. The youngest widespread deposit is a white, lacustrine marl, with abundant ostracod and snail shells, deposited about 22,000–12,000 B.P. Shortly thereafter the lake was lowered or drained, and the marl was dissected as new stream channels were cut toward the central valley axis. About 11,300 B.P., bedded sandy gravels began to accumulate in these arroyos, at a time when Llano hunters killed and butchered mammoth at several sites in the area. This first unit ("g" and "h" of Fig. 79) of the Lehner Formation is up to 90 cm. thick at the site, and includes several kills from repeated occupations. The pollen is dominated by grasses (60–65 per cent), with pine (16–22 per cent), juniper (3–4 per cent) and oak (3–9 per cent), while some terminal silty sands ("i" and "j") contained up to 12 per cent sedge pollen (see Fig. 80). The tree pollen was probably derived from the foothills some distance away, although macroremains of *Fraxinus* indicate the presence of riverine trees. The second unit ("k") is a black organic clay or silt, deposited after 10,800 B.P. to a thickness of 10–20 cm. in channels and draws. Although snail shells and some rodent bone are present, there is no evidence of human occupation. There are calcerous root casts, and the pollen includes more grass and less trees than (g). This bed, suggestive of a floodplain soil, marks an end to torrential stream discharge. Instead, there was a dominance of gentle washing with a dense mat of grassy vegetation along the channel floor as well as on the upland. The third unit ("1") is a light gray silt or sand, locally grading into marl, with a thickness of 60 cm. These flood silts and sands still suggest a well-vegetated channel, with seasonal wet spots, but rainfall may have been less evenly distributed. Shortly after 7900 B.P. these beds were dissected as gully-cutting began, with an incomplete vegetation mat and sporadic, torrential runoff. Subsequent channel fills (including "m") suggest drier conditions, similar to those of the nineteenth century.

At the time of Llano occupation, rainfall may have been 25–35 per

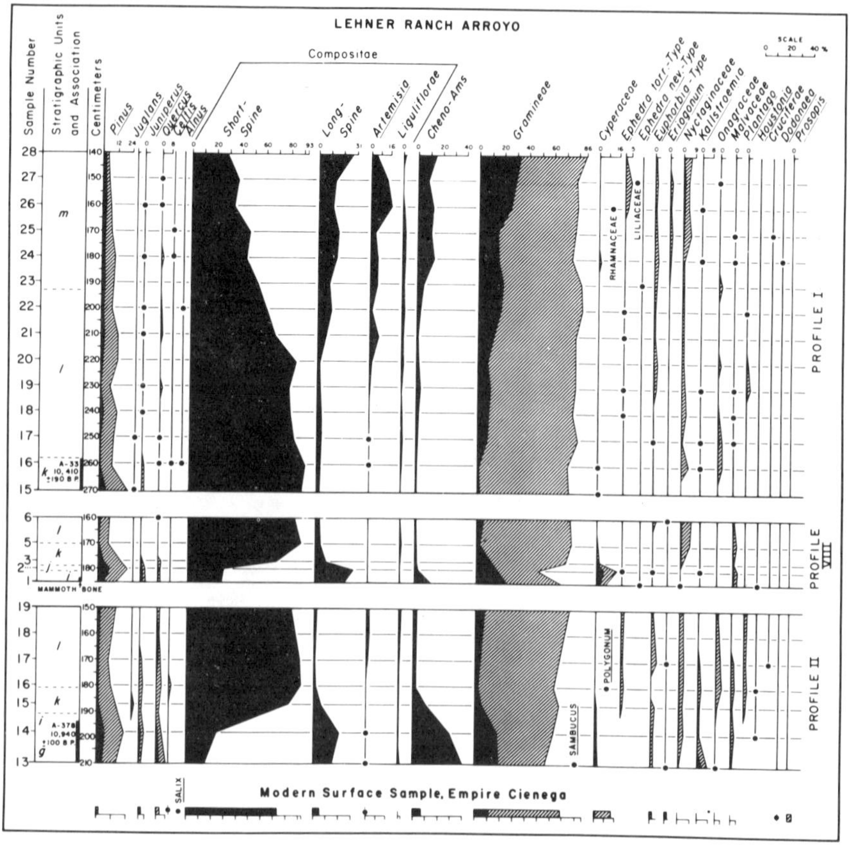

Figure 80. Pollen profiles from the Lehner Mammoth Site. From P. J. Mehringer and C. V. Haynes (1965), by permission of American Antiquity.

cent greater and temperatures 1–2° C. lower than today. The occupation levels and hearths have five accepted dates ranging from 11,290 to 10,900 B.P., so that horizons (g) to (k) were deposited at the time of the Valders readvance. The previous period of lake drainage, dissection, and an inferred drier climate corresponds to the Two Creeks substage. The molluscan data suggests a cool climate contemporary with the lacustrine phase, prior to 12,000 B.P. in Woodfordian times.

The archeological site itself has yielded bones of at least nine immature individuals of the extinct *Mammuthus columbi,* together with incidental remains of bison, tapir, and horse. Other localities in the same formation have also produced wolf and extinct camel. The faunal associations indicate several successive kills, and differences in preservation and degree of abrasion suggest unequal periods of exposure or transport.

Some 13 fluted projectile-points of Clovis type, together with 8 cutting and scraping tools (see Fig. 81), were found in intimate association with the animal bone and two hearths, through a vertical depth of 90 cm. However, parts of the site had already been eroded by the gully, while others remain to be excavated.

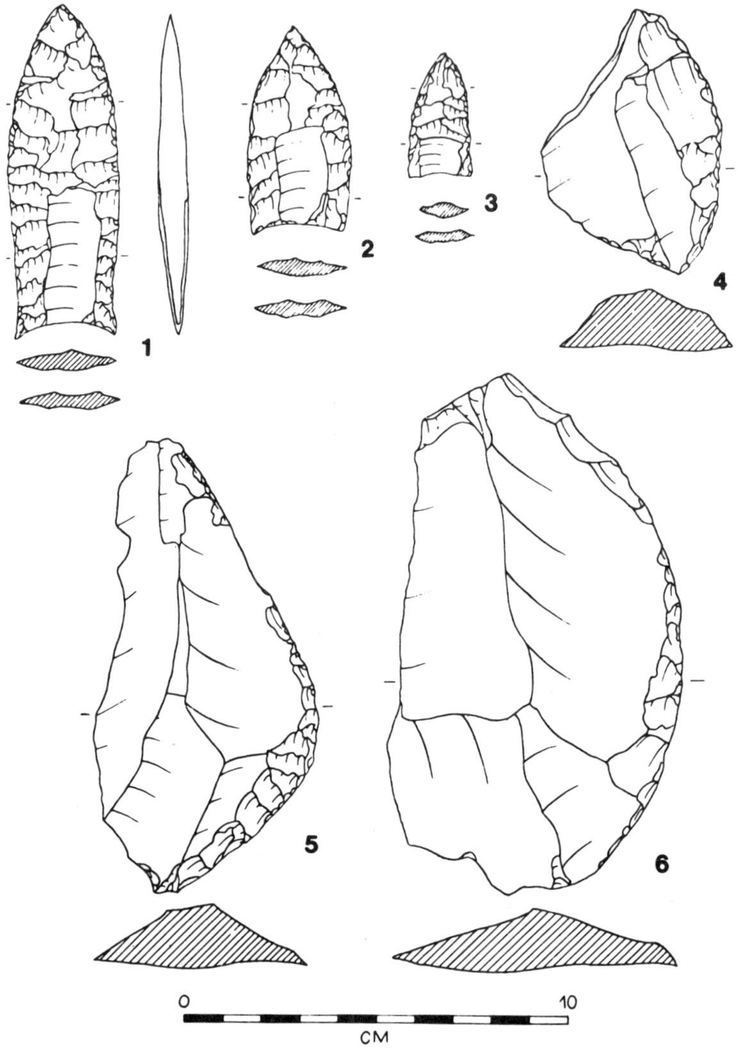

Figure 81. Stone implements from the Lehner Mammoth Site: (1)–(3) Clovis points, (4)–(6) "Mousteroid" side-scrapers. From H. J. Müller-Beck in Science *152, 27 May 1966, 1191–1210, copyright 1966 by the American Association for the Advancement of Science.*

In retrospect, the Lehner site represents a typical kill and butchering site, with ephemeral occupation at various times over a period of perhaps several centuries. The cultural patterns are identical to those recorded at innumerable Paleolithic kill sites of the Old World, and the analogies with Torralba are quite apparent.

MAMMALIAN EXTINCTIONS AND MAN

At the close of the Pleistocene, North America was the scene of wholesale animal extinctions, affecting large terrestrial herbivores, their ecologically dependent carnivores, and their scavengers. This wave of extinctions struck animals with quite different environmental preferences and adaptions, many of them versatile rather than specialized in their food requirements. Many genera or families were totally eradicated, amounting to approximately 30 per cent of the mammalian genera of North America. A good number of these have no ecological replacements.

Extinctions of organisms are normally attributed to disease or loss of adaptation, to loss of the ability to readapt to changing environments, to an inability to compete successfully with ecological competitors, or to an inability to maintain a stable equilibrium with respect to predators, animal, or man. The late Pleistocene extinctions of North America remain enigmatic, and the specific cause or variety of causes are of considerable interest. Numerous factors have been suggested, including climatic change, disease, poor synchronization of breeding habits with climate, and predation by man. Most points of view in this rather pertinent controversy have found a record in the symposium volume of Martin and Wright (1967), a comprehensive source for any interested reader. At a time when modern man has been blamed, justly, for widespread and irrevocable destruction of the natural environment and its ecological balance, it seems pertinent to take a careful look at the evidence and the arguments.

A partial list of the North American extinctions (see Martin and Wright, 1967; Hibbard *et al.,* 1965) includes mastodon, mammoth (5 species or subspecies), tapir, equids (kulan, ass, and several species of horse), pigs (3 genera or species of peccary), camelids (camel and llama), bison (6 species or subspecies), cervids (3 genera or species), a mountain goat, musk oxen (3 genera), antelopes (4 genera), shrub oxen (2 genera), yak, giant beaver, capybara (2 genera), a pocket gopher, a tropical gopher, a deer mouse, a grasshopper mouse, glyptodon (2 genera), giant armadillo (2 genera), ground sloth (4 genera), wolves and coyotes (6 species), cat (3 species or genera), a jaguar, saber-tooth (2 genera), and bear (4 species or genera).

Some of these are "primitive" forms that were already confined to the New World by late Pleistocene times and are now totally extinct: mastodon, two extinct genera of deer *(Cervalces, Sangamona),* woodland and musk oxen *(Bootherium, Symbos),* four extinct genera of pronghorn antelope *(Capromeryx, Breameryx, Stockoceros, Tetrameryx),* shrub oxen *(Euceratherium, Preptoceras),* giant beaver, the armor-plated glyptodons, the ground sloths, a wolf *(Aeonocyon),* short-faced bear *(Arctodus),* and the saber-tooths *(Smilodon, Dinobastis).* Only mammoth extinction was a world-wide, not strictly a New World, phenomenon. Other forms represented by different species or genera, are now restricted to warmer environments in the Americas (tapir, peccaries, tropical gophers, capybaras, spectacled bears) or to arctic or mountain regions (tundra musk ox, llamas). Others still are represented by closely allied species in the Old World: the equids, camel, saiga antelope, and yak. The remaining forms, it could be argued, experienced extinction of one or more members in a polymorphic population through intraspecies replacement: bison, caribou, mountain goat, pronghorn, the various rodents, armadillo, the canids and felids, and the grizzly bear. Almost all of these phyletic replacements involve ecologically equivalent species.

The ecological adaptations (see Martin and Guilday, 1967) of this rather inhomogeneous array of evolutionary and geographical forms were equally diverse. The greater part were grazers, generally with preferences for open country or parkland: most of the species of mammoths; kulan, ass, and most or all of the horse species; camel and llama; most of the bison species; mountain goat; musk ox, antelopes, and yak. Some were browsers, preferring parkland or forest: mastodon; the cervids; woodland musk oxen; spectacled bear; and possibly the ground sloths and shrub oxen. Others, such as the peccaries, armadillos, and glyptodons, were omnivorous, probably ranging through varied environments, while modern tapirs, beavers, and capybaras are semiaquatic. Some of the modern counterparts are now restricted to warmer latitudes, but extinct species may well have had different temperature tolerances or preferences (see Slaughter, 1967). All in all, with such a range of biotopes represented, the spectrum of mammalian extinctions in North America is quite different from that in Europe, where only steppe-tundra elements were affected. It is consequently impossible to relate the North American problem to large-scale climatic or environmental shifts. In fact, the preferred habitats of the different species are still widespread in North America today, and there were no major physical barriers to migration (see Mehringer, 1967).

In considering the problem of extinction, population size and density are of the utmost importance. Most mammals have optimal and minimal population sizes and densities, and animal populations tend to fluctuate, under normal conditions, between the optimal and minimal levels, thus establishing a stable equilibrium. Fluctuation below the minimal point because of unfavorable environmental conditions will lead to extinction, due primarily to the reduction of the breeding population. A greatly reduced breeding population leads to a reduction of the amount of genetic variabilility, with the possibility of increasing homozygosity and a more or less random fixation of genes without reference to their adaptive value (Simpson, 1944, p. 67). The unstable equilibrium point may be rather high for gregarious animals, and extinction of populations may ensue even after a negative external factor ceases to operate or serves only to reduce the size of a population threshold (Simpson, 1953a, p. 297). The basic problem, therefore, is not whether climate change, disease, or man the hunter caused an already decimated and doomed population to become extinct, but rather what initial factor or factors caused such herbivore populations to be reduced below the minimal population level.

There are several reasons to suggest that many, if not most, of the extinct species were at or below the point of unstable population equilibrium well before the advent of man in North America: (1) At least 20 of the extinct genera were archaic forms that—by the late Pleistocene—survived only in the New World, with its less efficient herbivores and a more limited range of predator types. No New World forms invaded Eurasia during the middle or late Pleistocene (see Repenning, 1967), suggesting that Nearctic faunas were either not very expansive or not competitive. (2) A progressive diminution of size of many mammalian species, including mammoth and bison, is evident throughout the later Pleistocene (see Reed *et al.,* 1965; Flerow, 1967; Edwards, 1967). This may be interpreted as an adaptation to keep population numbers high in situations of deteriorating resources (see Kurtén, 1965) or of increased competition. (3) The great majority of the extinct mammals were large animals, and in the case of phyletic replacement, the extinct species were generally larger. Big animals, and particularly "giant" forms, can only be supported in limited numbers by a given habitat, since they place greater demands on space, food, and cover (Guilday, 1967). At the same time, their generation lengths are great, sexual maturity is attained late, and litter sizes are small. As a result, such populations are less flexible, reach the point of unstable equilibrium more easily and recover an adequate population level more slowly and with greater difficulty than do smaller mammals. (4) Over a quarter of the extinct species appear to have disappeared by 12,000 B.P., before the rapid expansion of the Llano populations (see Table 21). This is

Table 21. Stages of late Pleistocene and early Holocene extinctions in North America (data based on Hester, 1967; Martin, 1967; Lundelius, 1967; and others).

A. Last dated record prior to 12,000 B.P.
- *Bison latifrons* (32,000 B.P.)
- *Bison alleni* (31,000 B.P.)
- *Boreostracon* (glyptodon) (23,000 B.P.)
- *Holmesina* (armadillo) (16,000 B.P.)
- *Dinobastis* (saber-tooth) (Friesenhahn Cave)
- *Felis bituminosa* (cat)
- *Felis daggetti* (cat)
- *Smilodon* (saber-tooth)
- *Canis furlongi* (coyote)
- *Canis milleri* (coyote)
- *Ursus optimus* (bear)

(*Felis bituminosa* through *Ursus optimus*: La Brea Tar Pits, [?] 13,890 B.P.)

B. Last record contemporary with Llano (ca. 11,600–10,900 B.P.)
- *Panther atrox* (jaguar)
- *Arctodus* (short-faced bear)
- *Mammuthus exilis, M. imperator* (mammoths)
- *Equus hemionus* (kulan)
- *Sangamona* (deer)
- *Bison priscus crassicornis*
- *Rangifer fricki* (caribou)
- *Bootherium, Symbos* (woodland musk ox)
- *Breameryx, Capromeryx, Stockoceros, Tetrameryx* (antelopes)
- *Euceratherium collinum, Preptoceras sinclairi* (shrub oxen)

C. Last record contemporary with Folsom (ca. 10,800–10,300 B.P.)
- (?) *Paramylodon* (ground sloth)

D. Last record contemporary with Plano (ca. 10,300–7900 B.P.)
- *Canis dirus, Aeonocyon* (wolves)
- *Mammuthus columbi, M. primigenius* (mammoth)
- *Tapirus* (tapir)
- *Equus* (horse)
- *Mylohyus, Platygonus* (peccaries)
- *Camelops* (camel)
- *Tanupolama* (llama)
- *Bison bison antiquus*
- *Oreamnos harringtoni* (mountain sheep)
- *Ovibos* (musk ox)
- *Castoroides* (giant beaver)
- *Dasypus bellus* (armadillo)
- *Megalonyx, Nothrotherium* (ground sloths)

E. Last record after 7900 B.P.
- *Mammut americanus* (mastodon) (?5950 B.P.)
- *Bison occidentalis* (7350 B.P.)

---

symptomatic of declining populations as a result of environmental stress, disease, or competition. (5) A large number of predators, last recorded from the La Brea tar pits near Los Angeles, were extinct before man could have begun to decimate the herbivore populations. This can be explained only by reduced populations of herbivores during the late Wisconsin.

It is difficult to determine the exact time of extinction, but Hester (1967) and Martin and Guilday (1967) have assembled a large amount of data to this point. Table 21 attempts to break this down into temporal units, reflecting probable extinction prior to the Llano and then, successively, during the Llano, Folsom, Plano, and early Archaic "stages." There is some dispute about a number of these species, so that the table is tentative and necessarily incomplete. Data from different parts of the United States are differentially represented at different times, and the majority of the faunas of stages (B) through (E) come from archeological sites. Despite these uncertainties, the three apparently distinct waves of extinction are probably real. The first of these waves affected animals from the Rancholabrean type site, prior to the Llano period. These extinctions may have taken place during the Two Creeks interval (ca. 12,800–11,600 B.P.)—a warm oscillation. The second wave, affecting large herbivores that are generally scarce in the paleontological record, must be dated during the Llano-Folsom transition, ca. 11,000–10,500 B.P., i.e., during the Valders maximum—a cold phase. The third wave, involving a more diversified faunal spectrum, and including well-represented forms such as woolly and Columbus mammoth, bison, horse, and peccary, must be dated early in the Holocene, ca. 8,000–7,000 B.P., when postglacial readjustments were completed.

A number of authors have marshaled arguments to the effect that "man the predator" was the initial factor responsible for mammalian extinctions at the close of the Pleistocene. Martin (1967) has been the most articulate spokesman for this view, and some of his most cogent points can be summarized: (a) The terminal Pleistocene extinctions are relatively unique in that they affected mainly one class of organisms (the mammalia), and even here were confined to the megafauna. (b) Many of the species found no phyletic or ecological replacement. So, for example, the horse was reintroduced during the sixteenth century and thrived in the same environments where other, late Pleistocene, species had met their demise. (c) There is no obvious relationship between the patterns of extinction and climatic change. Similar environments are available in North America today as were present, in different locations, during Wisconsin times. Habitats that had been congenial to mammoth, sloth, or horse during the Sangamon Interglacial are available today. (d)

The extinctions took place rapidly during a brief interval (ca. 11,000–7,500 B.P. in North America), and are comparable to waves of extinctions on other continents that are in apparent relation to early man (after ca. 40–50,000 B.P. in Africa, after 13,000 B.P. in northern Eurasia and Australia, after 10,000 B.P. in South America).

Without wishing to minimize the peculiarity and unprecedented magnitude of these extinctions, it must be pointed out that none of these arguments for prehistoric "overkill" is conclusive, and that none is acceptable without qualification. So, for example, most of the successive stages of late Cenozoic extinctions involved a large proportion of mammals, and the majority of these were large or even "giant" forms. At the same time, it should be clarified that bird extinctions were unaccountably high in the terminal Pleistocene of North America (Selander, 1965), while many nonmarine mollusca have become extinct without evolutionary replacements (Taylor, 1965). As already indicated above, a fair proportion of the extinct forms, such as bison or pronghorn, did find phyletic replacement, and only a few herbivorous species demonstrably lack ecological replacements. The history of animal evolution also shows that ecological replacement has sometimes taken millions of years (Simpson, 1953b, p. 17).

Environmental stress can create severe stress without actually eliminating a particular habitat. Deterioration would initiate interspecific competition between animals of similar, yet not identical, ecological roles, resulting in possible elimination of the less adaptive species of this fauna (W. L. Hylander, unpublished). Re-establishment of the original environmental conditions at lower or higher altitudes or latitudes would also entail the fragmentation of ranges and a complicated process of ecological shifts and readjustments for both plants and animals. Final reoccupation of the restored habitat would be restricted to the more successful members of the original fauna. Thus, environmental changes can create stresses of many kinds that affect different forms in different ways at different times. The elimination or expulsion of the tundra-steppe fauna in Europe can be related to obvious ecological pressures; the more complex pattern of extinctions in the Americas may simply reflect on more subtle stresses in a more diversified environment.

Martin's last argument demands some restrictions. As discussed already, the North American extinctions took place in three waves, beginning well before 11,000 B.P. and spanning some 5 millenia—compared with 4 millenia in Europe (see chs. 28 and 31). Considering the great number of extinctions or quasi-extinctions of the last hundred years by modern man with firearms (see Hester, 1967), this is a very long time. Indeed, the earliest American immigrants intruded into a vulnerable animal community that had evolved independent of human contact and may have had many forms that did not flee or seek concealment at the sight or scent of man (Jelinek, 1967). Tame animals, showing only curiosity at the presence of man, were in fact encountered in Antarctica and on several isolated island groups during the nineteenth century. However, it can hardly be claimed that 4 millenia of Paleo-Indian "slaughtering" were necessary to eliminate those unwary elements of the megafauna that were not already extinct by 11,500 B.P.

There are some equally pertinent arguments why man should not have played a critical role in late Pleistocene extinctions in North America:

a) There are relatively few Paleo-Indian sites over an immense area, and the majority of these have a very limited cultural inventory, suggesting small and rather dispersed populations that used individual sites briefly and sporadically. In addition, Paleo-Indian subsistence was not necessarily based, in the main part, on big-game hunting, and sites in the eastern United States show little evidence of such activities (Wilmsen, 1968).

b) Only two genera were hunted intensively: mammoth and bison. Hester (1967) has tabulated the faunal inventory from 34 archeological sites. Camel and horse are present at three sites only, pronghorn at two, tapir, peccary, musk ox, caribou, cervid, and bear at one each. On the other hand, although some 160 specimens of mastodon have been reported from Michigan alone, not a single specimen from Michigan—or any other part of the eastern United States—has yet been found in clear association with human artifacts or with evidence of human alteration (Jelinek, 1967).

c) Total faunal counts at Paleo-Indian sites are small. Some of the Llano sites represent individual kills of single animals, others record kills of several animals (up to 12 mammoths) at one time (Hester, 1967). Whereas the Llano hunters selected young mammoth, Folsom hunters took bison, without apparent preference as to age. In many instances 2 or 3 individual animals were butchered at one time, and the largest site (Lindenmeier) has 23 individuals, representing several kills over an extended period of time. The earliest evidence of mass slaughter comes from Plano sites of Holocene age: large herds were driven across unfavorable terrain, such as arroyos or stream channels, and as many as 100–200 bison were trampled to death or otherwise killed (Hester, 1967), with only a few animals being butchered and utilized. Such stampedes do not seem to have been used until after 9,000 B.P., during the final wave of extinctions. However, none of the 16 Plano sites listed by Hester has any fauna other than bison.

d) A last point that militates against the argument that man was primarily responsible for the waves of extinction is the survival of dozens of big-game species well into the nineteenth century, despite much larger and more efficient Indian populations. At a time when the total number of Great Plains buffalo *(Bison bison bison)* numbered 30 to 40 million, the Plains Indians, equipped with horse and firearms, for their own needs killed only about 320,000 animals per year during 1872–74 (see Hester, 1967). This is far less than the annual increase, and only the indiscriminate slaughter by white hunters for commercial markets drove the buffalo to the brink of extinction a decade later.

Thus there is no simple answer to the question of "why." There was no single causative factor, but rather an interacting fabric of multiple factors that confuse cause and effect. The following sequence of events is tentatively proposed as a possible schedule for the late Pleistocene extinctions:

1) Many of the archaic genera of North America were subject to extraordinary competition during the Wisconsin Glacial, partly as a result of rapid and repeated climatic changes and ecological stress, partly as a result of more adaptible and rapidly evolving competitors. Many populations dwindled and some approached an unstable equilibrium.
2) The apparent opening of the Canadian corridor after about 14,500 B.P. brought a new influx of animal populations from Alaska and Beringia into an animal community already in a state of precarious balance.[5] In addition to adverse competition, communicable disease may have decimated nonadapted American herbivore populations after diffusion of differentiated parasites, at least until new parasite-host equilibria were formed (see Edwards, 1967).
3) The drastic diminution of woodlands in the semiarid parts of North America at the onset of the Two Creeks Interstadial created sufficient ecological stress among unstable populations to accelerate phyletic replacement, primarily among dependent carnivores, with total extinction of some less adaptible forms.
4) Severe cold and violent weather accompanying the Valders readvance, perhaps the most drastic climatic reversal of the later Pleistocene, would have taxed all large mammals greatly. Heavy, early snows on the Great Plains and among the intermontane basins of the Cordilleras may have periodically eliminated winter grazing, ex-

5. The introduction of a new predator, the mongoose, into Jamaica in 1872 led to the extinction of several species of rodents, numerous snakes, lizards, turtles and terrestrial crabs as well as three genera of birds (Rensch, 1966, p. 238).

terminating large populations of herbivores. Frequent autumn blizzards at exceptionally low latitudes would have been catastrophic even to migratory herds, as modern experience shows. Intensive cold-air outbreaks during the spring and early summer would also have decimated the young of mammals with long periods of gestation and a late winter or spring birth season (Slaughter, 1967). As a consequence, a dozen or so species – most of them archaic – with minimal population levels disappeared. These included both woodland browsers and plains grazers. Equally or more significant was the reduction of mammoth, horse, and camel to near or below the minimum population level at this time.

5) World-wide extinction of the several species of mammoth may pose a special problem. Diminution of size during the course of the late Wisconsin – in both America and Eurasia – suggests severe ecological stress. It is suggested here that the New World bison was undergoing rapid evolution at about the same time, and that the repeated extinctions of its species or subspecies were only a matter of phyletic replacement by increasingly competitive forms. In fact, it is quite possible that the different subspecies of New World bison, with their diverse ecological adaptations, were responsible for the demise of the mammoth and possibly also the mastodon. This matter deserves careful study. It may cast light on the total disappearance of elephants and rhinos from the Holarctic faunal region.

6) During the final retreat of the American glaciers modern vegetation patterns were established. *(a)* The coniferous forests of the southeastern United States had shifted into the Great Lakes area during the Two Creeks, and now pushed northward into central Canada. From an open, relatively dry and summer-warm forest in the southern Great Lakes region during late Pleistocene times, these coniferous woodlands now became the dense and damp boreal forests of their present northerly situation. Short, cool summers in a high latitude setting favored oligotrophic conditions and widespread bog development (see Flerow, 1967). Browsers adapted to the eastern woodlands, including mastodon, would not all have migrated further north, even if they could not and did not compete successfully in the new animal communities of the deciduous woodlands. *(b)* In semiarid regions, the restriction of lush grasslands and forest-steppe in favor of drier grassland types would have reduced the carrying capacity for many grazers and unspecialized browsers. Although the major retreat of the forests can be dated ca. 12,000 B.P., the evidence from Lehner and western Texas (Hafsten, 1961) indicates another intensification of aridity after ca. 10,500 B.P. The severe restriction of surface

waters, also implicit from the contemporary period of gully-cutting (see Haynes, 1968a), may have eliminated semiaquatic genera such as the tapir. As a result, the bison was probably favored over the American horses.

7) Man may well have played a secondary role during this last wave of early Recent extinctions. There is no evidence or indirect argument to this effect in the eastern woodlands. On the other hand, in semiarid regions, where Plano populations were most numerous, competition for water holes and springs may have developed significant proportions. The widespread use of fire for animal drives or to destroy grazing, in addition to inadvertent burning, may also have affected some species adversely (Jelinek, 1967). Yet even here man did no more than accelerate the phyletic replacement of *Bison bison antiquus* by *B. bison bison.*

In retrospect, late Pleistocene extinctions in North America offer a wide field for pertinent research. Each of the many ecological problems and interacting factors is of great interest. But at the present level of understanding, we have no reason to blame prehistoric man.

PLEISTOCENE AND RECENT ADAPTATIONS IN NORTH AMERICA

The earliest verified inhabitants of the New World, the Paleo-Indians, carried one or more cultural and technological traditions derived ultimately from the Old World. These roots are to be sought among the late Paleolithic populations of eastern Siberia, although several millenia of transit through Beringia, Alaska, and the lower Mackenzie Valley may well have allowed a considerable degree of independent development and specialization. The major wave of immigrants came through the ice-free corridor in western Canada during the Two Creeks Interstadial, and the big-game hunters of Llano tradition are first recorded among the Rocky Mountain foothills at the beginning of the Valders, ca. 11,600 B.P. The Llano culture appeared in the United States fully developed, and remained technologically stable through successive phases of development spanning some 3,500 years. The Llano hunters were preceded by one or more small groups, of similar but less specialized tradition(s), that first appeared in the United States about 14,500 B.P. or even earlier. Equipped with a tool-kit that contained choppers, chopping tools, scrapers and, in many instances, bifacial artifacts, these groups lacked the elaborate projectile-points of the Llano tradition.

Once south of the Canadian corridor, the Paleo-Indians dispersed rapidly. The Llano tradition was first established in the semiarid western United States, probably in areas of large mammoth herds. It spread rapidly northeastward through the more open spruce forests of the southern Great Lakes and along the ice front to the east coast—all mammoth country (see Fig. 59). The demise of the mammoth on the High Plains led to a shift of hunting emphasis, and the smaller Folsom points may be functionally related to the taking of the somewhat smaller bison. Bison hunting, with the same basic tool-kit, continued until after 8,000 B.P. in the grassland setting of the Great Plains and the Rocky Mountain valleys. In the eastern woodlands, deer and smaller mammals replaced the mammoth, and hunting was by necessity deemphasized. Here the later Paleo-Indian groups—such as those of Quad, Shoop, and Williamson (Wilmsen, 1968)—took to a more intensive utilization of vegetable and aquatic foods, a trait that had been minimized during the long sojourn of the earliest Paleo-Indians in the high arctic of Beringia. A similar accentuation of food-collecting, in a diversified and seasonally alternating economy, is also apparent in the mixed environments of the Cordilleran basins and ranges.

Moving equally rapidly through Central America into the Andean country and the plains and foothills of Patagonia, the Paleo-Indians arrived at the southern tip of South America by about 11,000 B.P. Here two traditions are in evidence, initially based on game hunting, but presumably diversifying and adapting within new environments well before 8,000 B.P.

There are no specific post-Pleistocene readaptions apparent in the New World: environmental adaptation seems to have been initiated, if not largely accomplished, during the original dispersal of the Paleo-Indians (see Wilmsen, 1968). Nonetheless, many archeologists claim to recognize what might be described as a "Meso-Indian" complex, marking a shift from big-game hunting to the taking of small mammals, fish and mollusca, with intensified use of vegetable foods. This culture complex is widely labeled as "Archaic" in the United States; it follows the Paleo-Indian and precedes the local introduction of agricultural traits. A very brief survey of Meso-Indian "adaptations" is warranted here, both to keep the Paleo-Indian traditions in a proper perspective and to show up the parallels and differences with post-Pleistocene adaptations in the Old World. A basic source with a full bibliography is given by Willey (1966), and Jennings (1968), while more recent radiocarbon dates can be found in *Radiocarbon*.

In the Great Plains area, the Archaic is fixed by radiocarbon dates from shortly after 8,000 B.P. into the early Christian era. A different category of projectile-points was used, and the hunt shifted to modern bison (buffalo), deer, and antelope. Fishing and the collecting of nuts, berries, and tubers are inferred. It is well-known that the buffalo herds of historical times were no less abundant than those met by late Paleo-Indian hunters. This writer fails to see either any reason for or any evidence of any ecological readaptions in these early Archaic assemblages: the differences of projectile-points appear to be essentially stylistic,[6] and there seem to be no grounds for hunting *Bison bison bison* any differently than *B. bison antiquus.*

In the intermontane basins and coastal perimeter of the Cordilleras, the Archaic represents a diversification of traditions after 7,500 B.P., adapted to specialized ecological niches, including riverine fishing and bird-hunting in the moister northwest, plant-using and ultimately seed-grinding in the drier interior. Deer, elk, and pronghorn were the principal game. Once again, this is a direct continuation of late Paleo-Indian subsistence patterns. Grinding stones had diffused northwards from Mexico to the northwestern United States by 3500 B.P., and agriculture was introduced to Arizona after about 2300 B.P.

The eastern woodlands also witnessed a continuation and intensification of the late Paleo-Indian patterns of technology and subsistence. The earliest Archaic is here dated about 8500 B.P. (at Russell Cave). Hunting with projectile-points and spearthrowers concentrated on deer, while riverine food resources were increasingly utilized. Intensive exploitation of shell foods and wild grasses, complemented by ground-stone implements, allowed a high degree of permanence to large riverine settlements after 5000 B.P. Agriculture was first introduced about 2500 B.P., but only achieved widespread importance about 1200 to 1500 years ago.

Most significant to this problem of post-Pleistocene adaptations and innovations is the sequence of cultures identified from Tehuacán in south-central Mexico by MacNeish (1964) and his associates (Flannery, 1966; West, 1964, chs. 12-14). In the earliest, Paleo-Indian horizons here, there is documentation for hunting of small game, such as deer, pronghorn, rabbit, rat, birds, turtle and the like, in addition to occasional horse or antelope. Within that time range there is no evidence for a shift from larger to smaller game, but from larger to smaller hunting bands, in accord with the habits of the new animals being pursued (Flannery, 1966). Between 8750 and 7000 B.P. the collecting of plant foods, such as wild beans and amaranths, assumed ever greater importance, while squashes and avocados may have been first domesticated. Grinders, mortars, pestles, and pounders of polished stone became prominent elements of the tool inventory. Further domesticates after 7000 B.P. included corn, amaranths, zapotes, and various beans, even though the

6. Perhaps the apparent increased regionalization in projectile-point styles from the Llano onward reflects increased regionalization of adaptation. It might prove useful to search for spatial correlations between projectile-point styles and meso-environments.

economy was still based very largely on hunting and gathering. Animal domestication is evident after 5400 B.P., agricultural villages after 5000 B.P., and pottery after 4300 B.P. When these local sequences can be understood in a broader regional context, they should rival those of the Near East (see ch. 32) in terms of their significance for man-land relationships.

The Meso-Indian or Archaic developments of the United States are similar in many ways to those of post-Pleistocene hunter-gatherers in Europe (see ch. 31), although more informative sites are available in North America. But the Archaic differed from earlier, regional variants of the late Paleo-Indian traditions only in matters of detail, mainly stylistic elements involving projectile-points. Within the three great provinces of the Cordilleran basin-and-range country, the Great Plains, and the eastern woodlands, ecological adaptations and adjustments of the subsistence economy had already been initiated successfully by the Paleo-Indians. The specialized food-collecting patterns were but a development of trends well established. Hopefully, comprehensive analyses of Archaic assemblages will be forthcoming, so that more objective comparisons and contrasts can be drawn between different archeological horizons, as well as between environmental provinces. At that point it may become possible to search for possible repercussions of Recent climatic changes, such as have been inferred for the western United States (see Martin, 1963; Malde, 1964; Heusser, 1965; Morrison, 1965; Mehringer *et al.*, 1967; Haynes, 1968a; Irwin-Williams and Haynes, 1970).

The Americas provide a unique test case for ecological adaptations and cultural divergence among local groups little disturbed by external immigration. The initial settlement, by a fairly homogeneous population of proto-Mongoloid stock (Laughlin, 1967; Jennings, 1968, p. 46 ff.), took place very rapidly. Subsequent development was little disturbed by Eskimo and Aleut colonization of the Arctic littoral, and by possible Siberian influences with some new immigration into the arctic and boreal zones of Alaska and Canada during the later stages of the Archaic. With further problem-oriented studies, man-land relationships in the New World promise to be an even more stimulating theme for research.

THIRTY

Early Colonization of Australia

SOUTHEASTERN ASIA: POTENTIAL ANTECEDENTS

The earliest colonists of Australia presumably came from southeastern or southern Asia, most probably from the Indonesian island group. Java, Sumatra, and Borneo are attached to the Malay Peninsula by the shallow Sunda Shelf (Fig. 82). This broad area was partly emerged during several of the Pleistocene glacial-eustatic regressions, allowing man to reach Java at a very early date. However, deep waters separate the remaining islands from the Sunda Shelf on the one hand, and from the land mass of Australia and New Guinea on the other. There has been no land bridge here since early Tertiary times, judging by the archaic and highly distinctive flora and fauna of Australia in contact times. In fact, over half of the native Australian mammals were marsupials, which give birth to partly developed, almost embryonic young. Marsupials include the kangaroos and over 100 other "pouched" species of a group that died out in the Palearctic region some 25 million years ago. Most of the modern mammalian families never reached this periphery of the Asiatic continent.

A glacial-eustatic regression of 100 to 150 m. would link New Guinea and Australia by the shallow Sahul Shelf, over a land bridge some 1,100 km. wide. Another land bridge, about 90 km. in width, would link Australia with the island of Tasmania across Bass Strait. Yet, although a broad continental shelf was in fact exposed along the northwest coast of Australia about 19,000 B.P. (Van Andel *et al.,* 1967), the Timor Straits

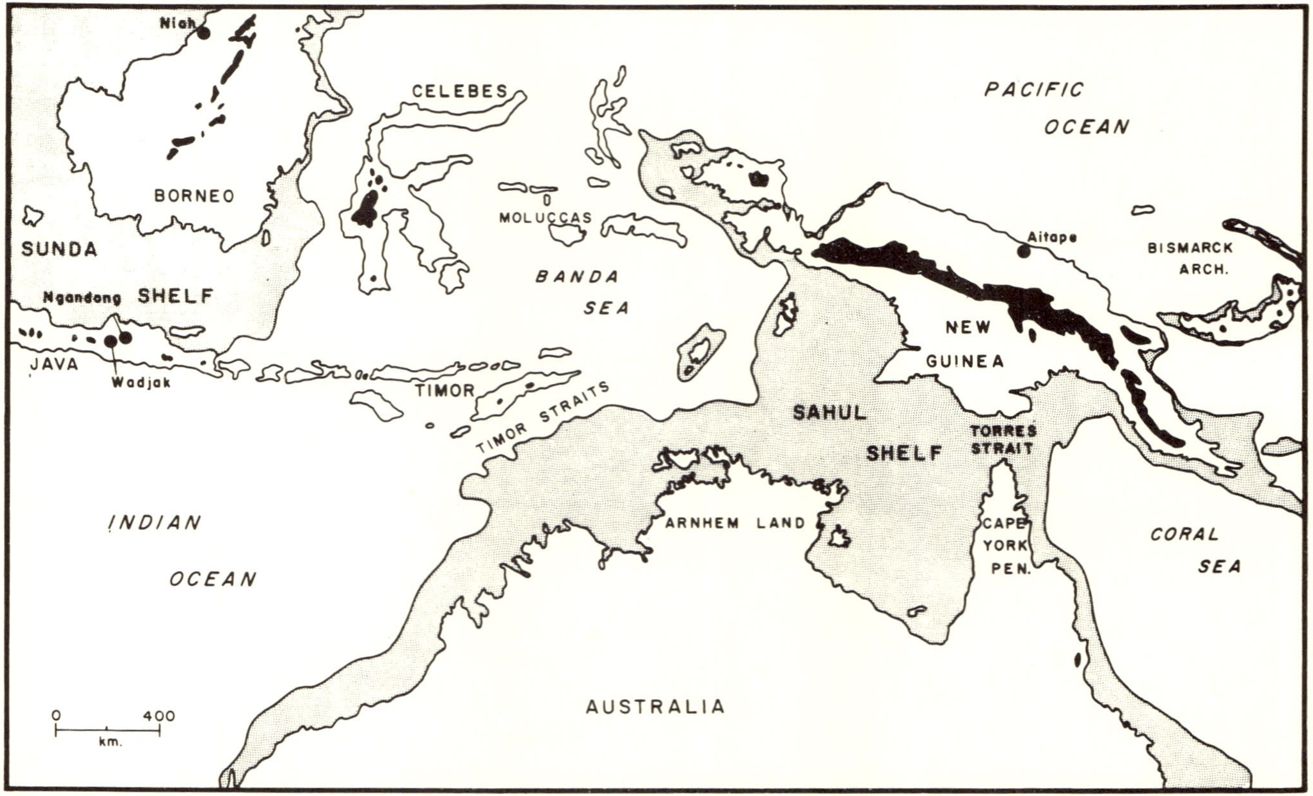

Figure 82. The East Indies, New Guinea, and northern Australia, with major late Pleistocene fossil sites and location of the Sunda and Sahul shelves. Continental shelf (0 to 150 m. depth) stippled, elevations over 1.500 m. black. (Base map modified after Goode's World Atlas [11th ed.], copyright 1960 by Rand McNally & Co., Chicago.)

(Fig. 82) remained a deep trough, some 200 km. wide. Other islands such as the Celebes and the Moluccas also remained isolated, although several were sufficiently close to be visible in clear weather. There is no evidence of significant tectonic movements in recent geological times, and man could not have crossed dry-shod from the Sunda to the Sahul land masses. This presupposes seaworthy water craft. Two major island routes may have served as steppingstones for such a migration: Java to Timor, or Borneo to New Guinea via Celebes and the Moluccas. For the last link of such trips, the minimum stretch of open water would be 180 km. in the Timor Straits, 95 km. between the Moluccas and New Guinea. Southeasterly winds and currents are dominant in these regions during the southern winter (July), with westerly winds in summer (January). Consequently, crossings would probably have been made during the southern summer. The oceanic and atmospheric circulation patterns do not favor a landfall on the western coast of Australia at any season (see Jones, 1968). A much longer movement, from China via Taiwan to the Philippines and Celebes (Mulvaney, 1969, p. 58), is a possibility but demands far greater navigational skills and ambitions.

Next to nothing is known about the late Pleistocene environment of Indonesia. However, the emerged parts of the Sahul Shelf indicate a comparatively dry climate ca. 16,000 B.P., with salt lakes and semiarid calcareous soils (Van Andel *et al.,* 1967). As late as 11,000 B.P., the waters of the Timor Straits were slightly cooler than today. The late Pleistocene Ngandong fauna of Java includes elephant, *Stegodon,* hippo, and a number of cervids and giant water buffalo; its affinities lie in the Indian subcontinent (see Koenigswald, 1938; van Heekeren, 1957; Movius, 1955). A crane *(Grus grus)* that now winters in central China may indicate a colder climate, but there is no indication of "pluvial" conditions.

The archeological record of Indonesia from the late Pleistocene time range is poorly dated, with most collections coming from the surface (see Movius, 1955; van Heekeren, 1957). Flake industries found at Ngandong and nearby Sangiran, on Java, include scrapers, core-scrapers, flake blades and some bone tools. These collections may be related to the eleven neanderthaloid crania found in late Pleistocene beds at Ngandong ("Solo Man") or otherwise to two more modern *sapiens* crania from a breccia at Wadjak ("Wadjak Man") (see Coon, 1962, p. 399 ff). A similar flake industry, without context, has been found in southern Celebes, probably indicating a late Pleistocene crossing to that island – a minimum distance of 40 km. (see van Heekeren, 1957). More firmly dated is the late Pleistocene depositional sequence from the Niah Cave in Sarawak, northern Borneo. A series of consistent radiocarbon dates between about 12,000 and 40,000 B.P. cover horizons with nondescript flakes and some pebble choppers and bone tools. A fragmentary cranium of a *Homo sapiens sapiens* was found under a charcoal horizon dated 40,000 B.P. (Brothwell, 1960).

AUSTRALIA AND TASMANIA: LATE PLEISTOCENE ARCHEOLOGICAL EVIDENCE

The Australian continent includes New Guinea, Australia in the restricted sense, and Tasmania—all linked by shallow shelves—as well as New Zealand, separated and apart beyond a distant sea. It is, therefore, not surprising that New Zealand was first settled a little after A.D. 900 by Polynesians. The record of habitation of the other three land masses is far longer.

In the Eastern Highlands of New Guinea, stone artifacts have been found in a buried soil with dates of 16,300 and 19,350 B.P. (Jones, 1968), although a suspected late Pleistocene fossil site at Aitape was dated at younger than 5000 B.P. (four dates ranging from 4400 to 5070, Gill, 1967). The earlier industries of New Guinea are in some respects analogous to those of Java and Celebes, with a dominance of flake tools; they include scrapers, choppers, and bone tools. Younger industries, dating back to 10,730 B.P. (see Jones, 1968, with references), include ground adzes that have striking parallels on the Australian shores. Land connections with Australia were probably severed at the beginning of the Holocene.

There is a substantial record of early settlement on the Australian mainland (Fig. 83) prior to 20,000 B.P. (see Jones, 1968; Bowler *et al.*, 1970; and Mulvaney, 1969, ch. 5 and p. 16, 145 f., 178 ff., with references; also Jones, personal communication):

Malangangerr and Nawamoyn, Arnhem Land (North Australia): neighboring stratified sites with ground adzes, flake and core tools; four dates for basal horizon from 24,800 to 21,450 B.P.

Koonalda Cave, Nullarbor Plain (South Australia): a quarry-workshop site with waste flakes, some utilized or retouched; four dates on lower strata vary between 21,200 and 19,300 B.P.

Lake Menindee, Darling River Plain (New South Wales): dune sands and lake shore deposits with two flakes, possible hearths, and an extinct marsupial fauna—all in possible but unproven association; relevant C^{14} dates are 26,300 and 18,800 B.P.

Lake Mungo, 160 km. from Lake Menindee: former lakeshore dune sands, interbedded with archeological materials and calcrete horizons; a partially cremated human skeleton, a variety of mammalian, bird, and fish bone, abundant imported shell, 16 hearths; and a total of 92 worked

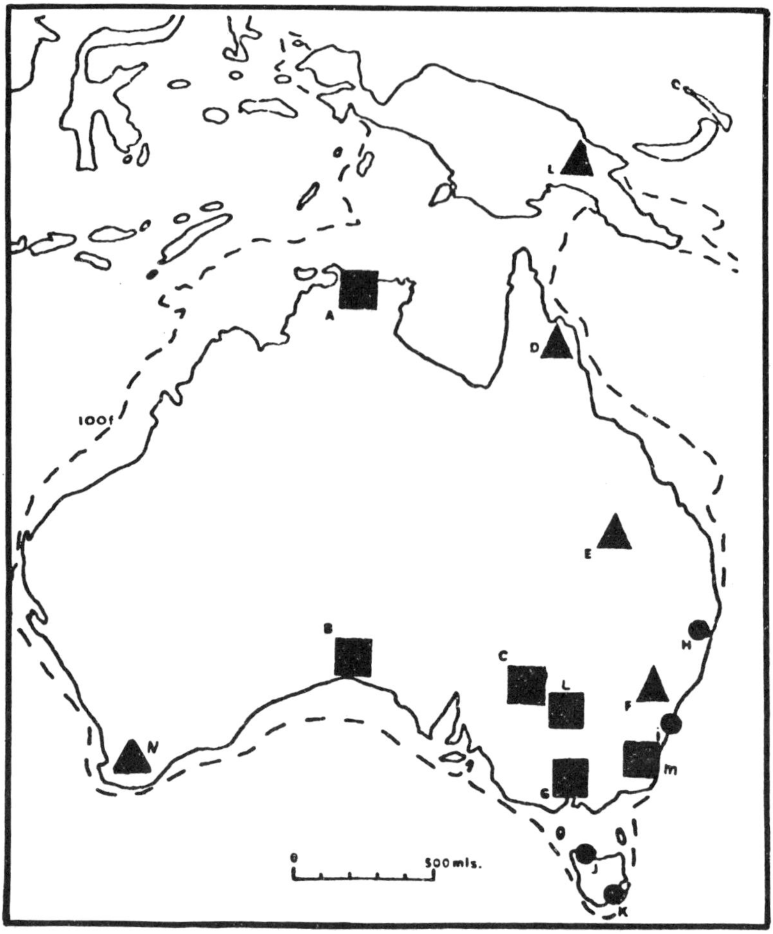

Figure 83. Early man sites in Australia, New Guinea, and Tasmania. Squares (25,000–20,000 B.P.*): (A) Malangangerr and Nawamoyn, (B) Koonalda, (C) Lake Menindee, (G) Keilor, (L) Mungo Lake, (M) Burrill; triangles (20,000–10,000* B.P.*): (D) Laura, (E) Kenniff, (F) Noola, (N) Nanruh Cave; circles (10,000–5,000* B.P.*): (H) Seelands, (I) Curracurrang, (J) Rocky Cape and Sister's Creek, (K) Carlton River, (L) Kiowa, Kafiavana. From Rhys Jones (1968), by permission of Archaeology and Physical Anthropology in Oceania.*

tools (mainly heavy core tools and scrapers) and 95 flakes are fixed with relation to a sequence of C^{14} dates between 30,250 and 32,750 B.P. (ANU-303, 331).

Keilor, near Melbourne (Victoria): alluvial silts from which a controversial cranium was obtained without excavation control; a date of 15,000 B.P. was obtained from the level of the cranium, near or at the site, although attached carbonates gave only 7360 B.P. More recently, indisputable artifacts have been recovered from this deposit well below a horizon dated 18,000 B.P.

Burrill Lake, on the New South Wales coast: shelter site with a chopper and flake-scraper industry and C^{14} date of ca. 20,000 B.P.

There are further sites dating between 20,000 and 10,000 B.P. (Fig. 83):

Laura Shelter, Cape York (Queensland): rich occupation levels with ground adzes, flake-scrapers and choppers, extrapolated to well beyond 10,000 B.P. from a date higher in the sequence.

Kenniff Cave, central Queensland: lower levels with flake-scrapers and large cores used as scrapers or chopping-tools; three inconsistent dates vary from 16,130 to 12,900 B.P. In southeastern Queensland, an archaic skull was found at Talgai; calcareous nodules from the presumed level in an alluvial deposit gave an age of 10,850 B.P.

Noola Shelter, eastern New South Wales: lower level with scrapers and corers; two inverted dates of 11,600 and 12,500 B.P.

Nanruh Cave, near the southwestern coast of Western Australia: an unpublished site of comparable antiquity, currently under study.

In addition, younger levels at Malangangerr have yielded dates of 19,600 and 18,000 B.P.; at Koonalda, 18,200 and 13,700 B.P. As yet undated is an extraordinarily primitive series of skulls, with many characteristics comparable to *Homo erectus,* that have been recovered from Kow Swamp in northern Victoria (Thorne, 1969). The site and the materials are under study.

It is still premature to seek regional patterns among the early Australian industrial traditions, until the lithic assemblages have been published. Similarly, it is not yet possible to link these industries to the Asiatic mainland. It would seem, however, that regional traditions did exist, and that these will ultimately prove critical to evaluating possible cultural roots within southeastern Asia. Superficially, the leading tool types appear to reflect on the chopper and chopping-tool traditions of that continent. Yet flaking traditions are sufficiently individualistic that Jones (1968) sees closer analogs in the late Pleistocene "Middle Stone Age" of India. Post-Pleistocene technological innovations in Australia appear to have been essentially indigenous. They include blade industries, with fine unifacially or bifacially trimmed points, as well as microlithic traditions, first appearing about 7000 B.P. Labeled as the Inventive Phase (Mulvaney, 1969, p. 107 f.), this trend to introduction and diffusion of new tools and technologies did not displace the older tool traditions, but continued alongside them.

The human fossils now known from ancient Australian contexts (see

Gill, 1968) appear to fall into two categories. Some, such as the Keilor skull, pertain to *Homo sapiens sapiens* and are representative of modern Aborigine populations. Others, such as Talgai and Kow Swamp, exhibit more primitive features (see Mulvaney, 1969, p. 158 f.; Thorne, 1969), which nonetheless seem to fall within the range of variation of modern man. These traits may eventually help relate some of the first Australians to a *Homo erectus*-derived stock in southeastern Asia.

The first archeological records in Tasmania are from a number of coastal sites: caves at Rocky Cape and Sister's Creek, with dates ranging from 8120 to 5425 B.P.; and Storm Bay, an open shell midden going back to 8700 B.P. Since the early Tasmanian cultures were oriented toward marine and littoral food sources, older sites were almost certainly submerged by the late glacial rise of sea level (Jones, 1968). The island was cut off from the mainland about 9,000 B.P., before the dispersal of the dog-like dingo into southeastern Australia. The earliest artifact collections include generalized tool types made on flakes, cores, pebbles, and bone. Subsequent technological developments indicate isolation from the Australian mainland. The backed blades, points, and adzes of the Inventive Phase found no parallels in the conservative Tasmanian assemblages, and in contact times the Tasmanians lacked the spearthrower, boomerang, shield, axe, and composite hafted tools of the Aborigines (Mulvaney, 1969, p. 133 ff.).

THE ENVIRONMENT OF THE NEW CONTINENT

Available information on late Pleistocene environments of Australia is quite unsatisfactory. In particular, there seems to be no immediate solution to the question of whether the great arid interior lowlands were moister or drier during the late Pleistocene. On the one hand, there is the evidence for expanded "pluvial" lakes in South Australia (see Wopfner and Twidale, 1967) and the broad floodplains reflecting far greater flood discharge in eastern Australia (Dury, 1967). On the other, there is the evidence for widespread aeolian activity, including stabilized longitudinal dunes (see Galloway, 1965), coastal aeolianites (Fairbridge, 1948; Woldstedt, 1962a) extending below modern sea level in northwestern and southern Australia, repeated deflation of the lacustrine sediments of South Australia, and extensive aeolian silts or clays in eastern Australia (Butler and Hutton, 1956). These apparent contradictions in part reflect inexact correlations or differences of interpretation, and in part the assumption that climatic changes were similar and simultaneous throughout the continent. There are parallels with interpretation of the "pluvial" lakes and stream alluviation of the American Southwest (chapter 18), with the coastal aeolianites of the Mediterranean region (chapter

19), and especially with the matters of exact chronology so essential in making sense out of late Pleistocene events in the Saharan world (chapter 20). Resolution of the many inherent problems must await further, detailed regional studies. It is nonetheless obvious that the interior of Australia was subject to repeated and significant environmental changes during the course of the late Pleistocene, changes to the dry as well as to the wet side.

The evidence for greater cold in the mountains of southeastern Australia (Galloway, 1965, with references) and on the island of Tasmania (Davies, 1967) is established rather more firmly. Permafrost may have been found above about 2,000 m. on the mainland and above 1,500 m. on Tasmania; and periglacial phenomena such as involutions, frost cracks, patterned ground, and solifluction mantles are widespread above 1,000- 1,500 m. and 600 m. respectively. There was some limited glaciation in New South Wales (Mt. Kosciusko, 2,195 m.) with more extensive development of glaciers on Tasmania (Mt. Ossa, 1,590 m.) as well as in New Zealand (see Suggate, 1965). Radiocarbon dating (see Jones, 1968) indicates that the maximum advance of ice occurred shortly after 26,500 B.P., with the last major retreat beginning about 14,000 B.P. The last stages of deglaciation in western Tasmania are associated with a date of 8720 B.P. An estimated temperature depression of 8° to 10° C., at the maximum of the last glacial, would have been significant for the southward dispersal of the first Aborigines, coming as they did out of a tropical environment; on the other hand, cooler temperatures would have affected the hydrological balance of interior Australia in a positive sense. In Tasmania, both a snowline elevation below 1,000 m. and the abundant evidence of frost imply a raw and harsh climate during late Pleistocene times.

MAN-LAND RELATIONSHIPS

When the first European settlement was established near Sydney in 1788, the aboriginal populations of Australia are estimated to have totaled about 300,000. These were commonly aggregated in bands of 25 to 40 persons; and the 7.6 million sq. km. area of the continent supported from 5,000 to 10,000 territorial groups of this size (Mulvaney, 1969, ch. 2; Birdsell, 1953; Elkin, 1954). There were about 500 real tribal units and some 700 linguistic groups, including approximately 220 languages. Birdsell (1967) has proposed three broad racial categories, presumably reflecting successive waves of immigration: (a) a short, dark-skinned population, including the extinct Tasmanians and remote groups ("Negritoids") from the rain forests of northeastern Queensland; (b) the thickset, light-colored hirsute Aborigines of southern and southeastern Australia, sometimes compared with the Ainu; and (c) the tall, dark, less hairy Aborigines of the northern, tropical half of the country, with some affinities to the so-called Veddoid hunters of Indonesia, Malaya, and Ceylon. These racial classes are controversial but, like the basic demographic, tribal and linguistic data, they do throw some light on the heterogeneity of what is a single technocomplex in terms of the recent archeological record.

The subsistence base was a nomadic form of hunting and gathering, and the only domesticated animal was the dog or dingo, presumably introduced from Asia. Wherever available, marine and aquatic resources were exploited, and only 30 to 40 per cent of the subsistence of contemporary Australian groups is based on hunting (Lee and DeVore, 1968, ch. 4). A more congenial environment with more diverse food sources was responsible for greatest population densities along the northern and eastern coasts as well as in the Murray River basin of the southeastern interior. These coastal and riverine environments were rich in game and fish, and provided abundant vegetable foods. Settlement of the semi-desert interior was always thin, and the widely scattered bands depended largely on collecting sparse vegetable food, rodents, reptiles, and insects.

Economy and technology prior to 1788 were adapted to the vegetation and dependent fauna within each habitat, and the populations lived in ecological adjustment with their many environments (Mulvaney, 1969, pp. 48 f., 57). It appears that population levels were adjusted to the minimum carrying capacity of lean years, not only by natural factors but by an intricate social structure. This system served to regulate man's organic interrelationships with the land by developing systems of taboo and enforcing dietary restrictions on social groups and age grades. Unfortunately, anthropological work has been concentrated on marriage and kinship patterns, with little attention devoted to man's attitude toward and conscious utilization of the environment. The opportunities for such study are rapidly disappearing as even the most remote tribes are being exposed to the outside world.

There is no reason to assume that man-land relationships in Australia were substantially different during the late Pleistocene than they were in the eighteenth century. Nonetheless, the impact of man on the vegetation has been appreciable and it may have originally upset the ecological balance. The Aborigines burned wide tracts of country—seemingly indiscriminately—to promote the growth of the next season's edible shoots, as well as to hunt game or clear tracks through tall grass or forests (Mulvaney, 1969, p. 48 f.; Jones, 1969). The term "peripatetic pyromaniacs" (see Merrilees, 1968) seems unfair, but the effect of constant burning on the natural vegetation has been fundamental, changing forest or thick scrub into open savannas (Jones, 1968). It is also conceivable that these modifications were instrumental in large-scale marsupial extinctions during the late Pleistocene.

Animal extinctions have recently assumed considerable interest in Australia. Two large quadrupeds (*Diprotodon* and *Nototherium*), a number of giant kangaroos, a marsupial "lion" (*Thyacoleo*), and various flightless birds are involved, including two entire families and at least 20 genera (Merrilees, 1968). As in the case of North America, there is little or no evidence that these animals were hunted. Of the sites predating 5000 B.P. discussed above, only one (Lake Menindee) includes bones of extinct genera, although there is no lack of archeological association with modern faunal elements (see Merrilees, 1968). There is reason to believe that the extinct species were already scarce or absent by 25,000 B.P., and there are only two younger dates on extinct marsupials: 13,700 B.P. from Victoria, 11,100 B.P. from South Australia (see list of radiocarbon dates in Mulvaney, 1969, p. 178 ff.). In fact, Jones (1968) now believes that the major wave of late Pleistocene extinctions took place between 30,000 and 20,000 B.P., not after 13,000 B.P., as postulated by Martin (1967). Although modification of the vegetation cover by fire may have aided in the process of extinction, the environmental stress inevitably created by late Pleistocene oscillations of climate seems an equally plausible explanation at the moment. The dingo, first recorded between strata dated at 8600 and 7450 B.P. from a South Australian site (see Mulvaney, 1969, p. 65 and 179), can hardly be blamed for these extinctions, although it may well have played a role in the more recent demise of certain smaller marsupials.

The native Tasmanians, before their brutal eradication early in the nineteenth century, did not eat fish. However, the archeological record shows that subsistence was originally oriented toward the sea: marine mollusca provided 50 per cent of the meat, while seal and fish accounted for 95 per cent of the remaining protein (Jones, 1968). The interior of the island was probably inhospitable and uninhabited. Eventually, land marsupials and sea birds assumed greater importance, and part of the population moved inland. Whether this was due to a drastic decline of the chief food animal, the elephant seal (*Mirounga leonina*), or to a deliberate diversification of the economy is uncertain. Trails were kept open in now-impenetrable rain forest by burning, and constant firing was also employed to maintain tracts of open sedgeland in the rainforest environment (Jones, 1968). Thus, as the elephant seal was driven to the brink of extinction by excessive hunting, the Tasmanian economy shifted from a coastal hunting-fishing-gathering to a generalized hunting-gathering subsistence.

THIRTY-ONE

Environmental Change and Cultural Adaptation in Early Postglacial Europe

INTRODUCTION

The onset of the Holocene in Europe brought climatic and ecological conditions analogous to those of the present. During the subsequent 10,000 years there evidently were numerous short-term fluctuations as well as longer-term trends to a cooler or warmer, moister or drier climate. But none of the changes involved approached the importance or possible ecologic significance of the environmental changes during and at the close of the Pleistocene proper.

In the cultural realm, the Holocene and the last millenia of the Pleistocene mark a period of rapid cultural evolution culminating in the literate civilizations of the Near East, Europe, and Asia. Some archeologists have found reason to see environmental adaptation or responses as the root of several cultural changes. In particular, the termination of the Pleistocene is thought to have had a profound impact on certain populations in both Europe and the Old World subtropics. In the case of Europe, continued specialization led to persistence of food-collecting subsistence in a somewhat different environmental setting. In the Near East, however, more spectacular results followed upon the invention of agriculture and animal domestication during the last millenia of the late Pleistocene.

ENVIRONMENTAL CHANGES OF THE TERMINAL PLEISTOCENE

Throughout most of the late glacial, accompanying the gradual retreat of the Scandinavian, British, and Alpine glaciers, climatic conditions remained cold and rigorous in mid-latitude Europe. This is shown by widespread aeolian activity as well as by tundra vegetation in northwestern, central, and northeastern Europe.

First heralded by a brief warm oscillation, known as the Bölling, the warm *Alleröd* interval or interstadial began some 12,200 years ago. The Alleröd resulted in forest recolonization of most of the Würmian tundras (Lemée, 1954; Lang, 1963; Firbas, 1949–52; Gams, 1950; Iversen, 1954; Godwin and Willis, 1959). According to numerous pollen spectra, pine woodlands are known to have dominated the southern half of France, southern Germany, and northern Poland, with spruce and pine dominant farther to the southeast (Fig. 84). A birch woodland with some pine occupied northern France and northern Germany, a birch parkland most of England and Denmark. Tundra was confined to small areas marginal to the ice sheets in southern Sweden and Finland.

Most of Fennoscandia was still glaciated during the Alleröd, the ice margin being located a little inland of the Norwegian coast and extending through south-central Sweden into southern Finland. The British glacier had been reduced to a highland glaciation of northern Scotland. In the Alps, glaciers were confined to the inner mountain valleys. Deglaciation had proceeded so far that the world sea level was about 40 m. lower than today's (see Farrand, 1964), still, however, leaving the Straits of Dover and the southern North Sea dry. The basin of the Baltic Sea was partly occupied by a large freshwater lake.

Various estimates of average July temperatures have been made. On the basis of remains of various thermophile plants, Iversen (1954) suggests a July temperature of 13°–14° C. for Denmark (15°–16° C. today); on the basis of the altitudinal limit of pine in the Black Forest, Firbas (1949–52, vol. I, p. 287) suggests a July temperature depression of 2.5° C. for southern Germany. The upper tree-line of the Alps was 500 m. lower than today's, according to Lüdi (1955), possibly suggesting an analogous temperature depression. This would imply that the July temperatures were 5°–6° C. warmer during the Alleröd than during the main Würm. Climate was rather continental and possibly on the dry side (Iversen, 1954; Schweitzer, 1958).

The climatic deterioration of the Younger Dryas (ca. 9400–8300 B.C.: see Hammen *et al.*, 1967; Nilsson, 1964a, 1964b) was as spectacular as the forest advance of the Alleröd. Glaciers readvanced in Europe and North America, and world sea level dropped by as much as 10 m. (see Farrand, 1964). NAP pollen dominates the pollen spectra of north-

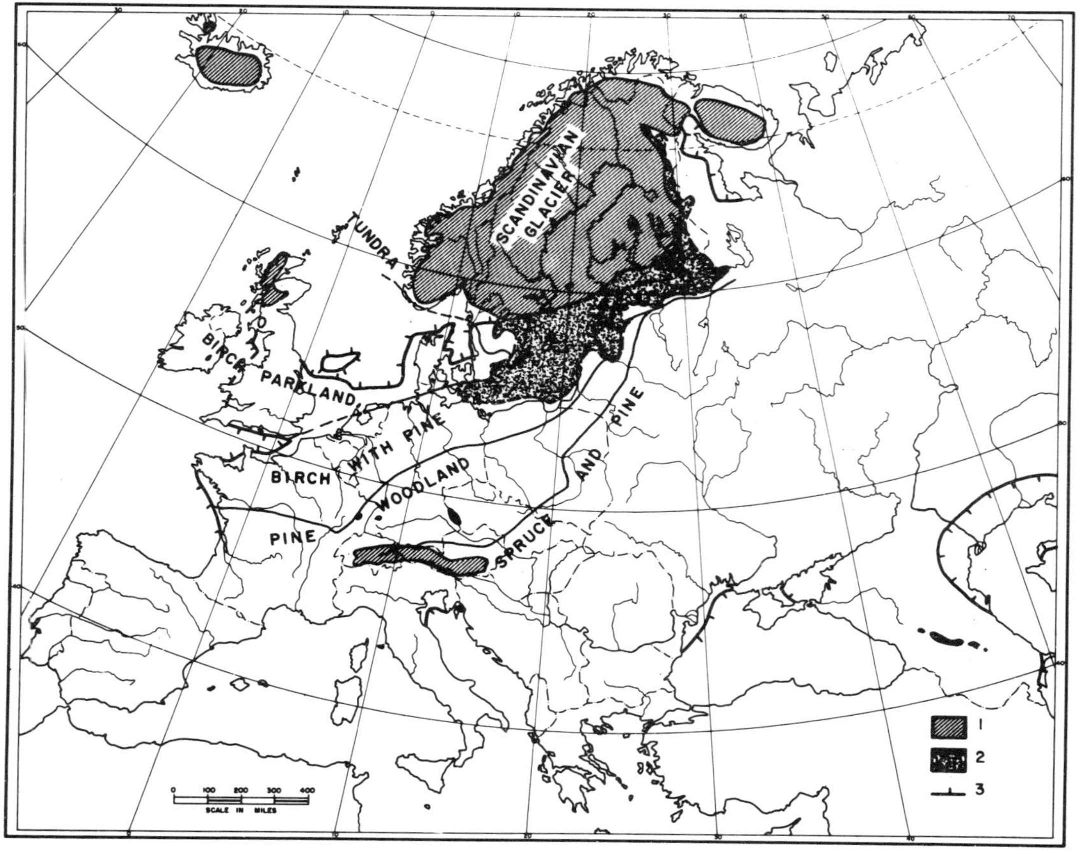

Figure 84. Environment of northwestern Europe during the Alleröd: (1) glaciers, (2) Baltic ice lake, (3) coastlines. Vegetation lines after Firbas (1949); Gams (1950); Iversen (1954); and Godwin (1956).

western Europe, and forest-tundra probably dominated in the former Würm tundras of Europe (Fig. 85). North of the Alps, birch and pine woodland were first encountered in the warmer lowlands of central France and southern Central Europe, with the arctic tree-limit located somewhere in Denmark and Britain. The altitudinal tree-limit in the Alps was 800 m. lower than today's (Lüdi, 1955). From various lines of botanical evidence Firbas (1949–52, vol. 1, p. 288) and Iversen (1954) suggest sudden and drastic July temperature depressions of 5.5°–7° C. for Germany and 5°–6° for Denmark. Similar values probably apply to other parts of mid-latitude Europe as well.

During the Preboreal (ca. 8300–7500 B.C.), the first phase of the Holocene, environmental conditions were once more reversed. Rapid forest recolonization of the ground lost during the Younger Dryas reduced the forest-tundra to small areas in Britain and southern Scandinavia. The remnants of the British glacier in the Scottish Highlands disappeared, while the Alpine glaciers were reduced to approximately

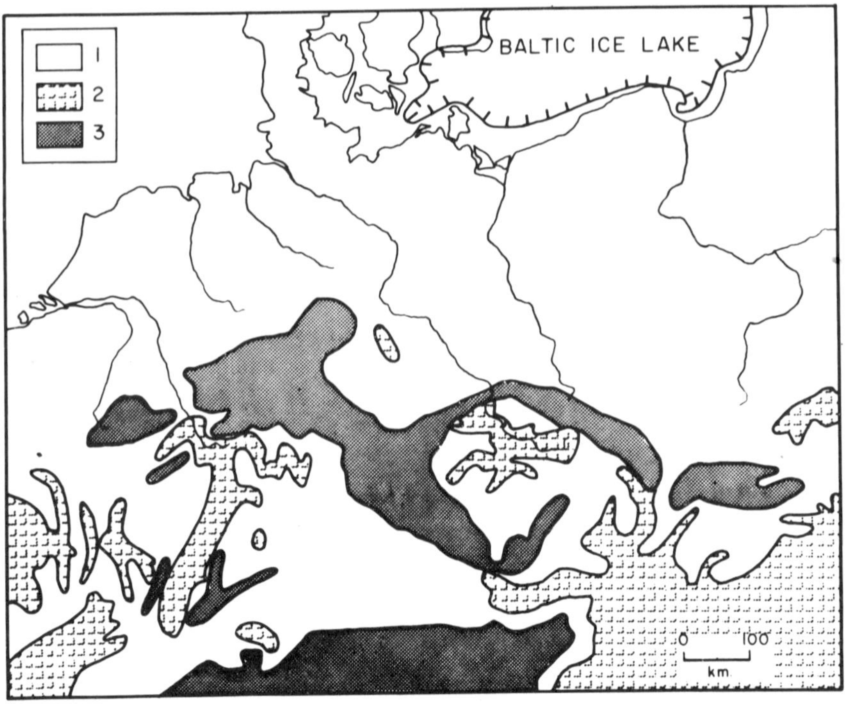

Figure 85. Vegetation zones in central Europe during the Younger Dryas (modified after Firbas, 1949): (1) forest-tundra, (2) birch and pine woodland, (3) alpine meadows.

their present dimensions. The Scandinavian glacier began its final retreat and dissipation. This retreat was comparatively rapid, averaging a little over 300 m. per year in Sweden and Finland. Each yearly halt is indicated by lines of small "annual" moraines. At the close of the Preboreal, much of Norway and northern Sweden were still covered by residual ice masses, and world sea level was still 10 or 15 m. below that of the present (see Farrand, 1964; Jelgersma, 1966). Yet the presence of various thermophile water plants in Denmark (Iversen, 1954) and England (see J. G. D. Clark, 1954), and a tree-line elevation coinciding with that of today in the Alps (Lüdi, 1955) all suggest that July temperatures were quite similar to those of the present. Similarly, O^{18} isotopic concentration in the Greenland ice sheet indicates that air temperatures were almost as warm in 8300 B.C. as they are now (Dansgaard and Tauber, 1969).

The environmental changes marking the transition of the Older Dryas to the Alleröd and the Younger Dryas-Preboreal are similar in magnitude and kind. In each case a few centuries apparently sufficed to enable the forests to migrate several hundred kilometers northward. The warming trends were probably gradual, so that the local appearance of certain species provides rather arbitrary stratigraphic dates within the transitional period. Equally arbitrary is the original Pleistocene-Holocene boundary, defined by the draining of the Baltic ice lake and establishment of intercommunication with the North Sea.

Climatically then, there was no sudden changeover from glacial to postglacial conditions at about 8300 B.C., but rather an irregular, violent, oscillation of warmer and colder conditions, producing an over-all warm-up, the gradual ablation of the continental glaciers, and a climate ultimately warmer than that of the present. In fact, the mountain glaciers of Alaska and the Alps readvanced briefly about 8000 B.C. (see Frenzel, 1966), while a major halt or regrowth of the North American glacier is recorded about 6500 B.C. (Cochrane stade). Altogether some 15,000 years elapsed between the maximum extent of the ice during the Brandenburg stage and the final dissipation of the ice when world sea level once more reached its present *niveau,* by about 3000 B.C. (see Bryson and Wendland, 1967; Lind, 1969). The composite picture of environmental change can therefore not be arbitrarily broken down into oversimplified units delimited by the date 8300 B.C. A survey of the floral and faunal changes can provide further information toward a more realistic understanding of this problem.

EARLY POSTGLACIAL VEGETATION CHANGE IN EUROPE

The standard pollen zones of Europe were established by L. von Post in 1916, and with some modifications are still accepted today (Table 22

Table 22. Late glacial and Holocene pollen zones of northwestern Europe.

Zone	*Date* (B.C.)	*Name*	*Dominant vegetation*	*Inferred climate*
VIII	After 300	Subatlantic	Beech	Maritime
VII	3300–300	Subboreal	Oak-beech	More continental
VI	6200–3300	Atlantic	Oak-elm	Warmer and maritime
V	(7500)–6200	Boreal	Hazel-pine-oak	Warmer and continental
IV	8300–(7500)	Preboreal	Birch-pine	Warm-continental
III	9400–8300	Younger Dryas	Forest-tundra	Arctic
II	10,200–9400	Alleröd	Birch-pine	Temperate-continental
Ic	10,400–10,200	Older Dryas	Tundra	Arctic
Ib	10,800–10,400	Bölling	Birch parkland	Subarctic
Ia	before 10,800	Oldest Dryas	Tundra	Arctic

and Fig. 86). Apart from the addition of new detail to the late glacial sequence, considerable reinterpretation has been necessary of the ecological implications of the vegetational changes. It was originally thought that the Preboreal, with its birch-pine vegetation, provided an analogy to present-day vegetation in Lapland. As a result, the Preboreal was thought to be a period of subarctic climate in northwestern Europe. Similarly, the hazel maximum of the pollen profiles was thought to be evidence of increasing temperatures. In 1954 Iversen indicated that botanical evidence in Denmark implied a warm climate during the Preboreal, an opinion seconded by Godwin (1956) for the case of Britain. In 1960 Iversen was able to show convincingly that most of the climatic interpretation of the Holocene pollen zones requires considerable modification.

According to Iversen (1960), much of the vegetation change recorded by early postglacial pollen profiles represents no more than a succession

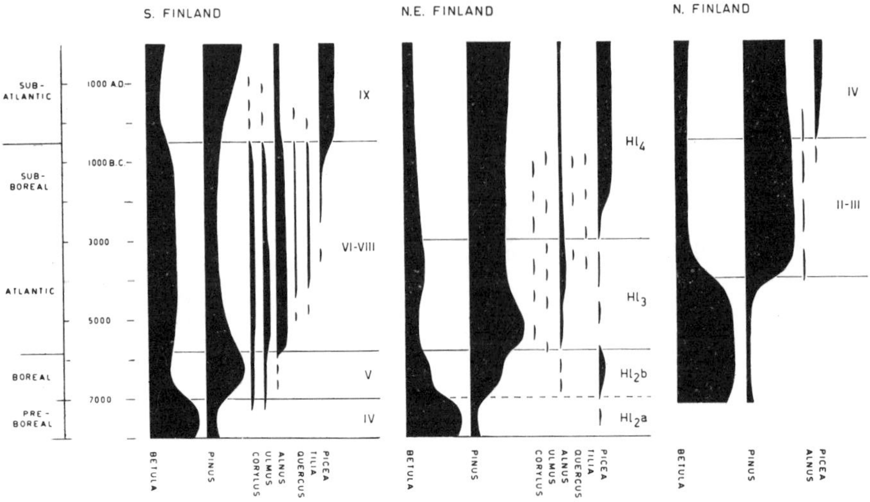

Figure 86. Composite postglacial pollen diagrams for southern, northeastern, and northern Finland (from J. J. Donner, 1963, by permission of the author).

from "pioneer" to "climax" species. Pioneer species have a high rate of reproduction, effective seed production, and dispersal, attaining maturity at an early age. Consequently, such species, which include willow, aspen, and birch, are particularly adapted to rapid colonization of new ground left by retreat of the ice, by the filling-in of lakes, the abandonment of cultivated land, or the destruction of primary woodland through forest fires. The weakness of pioneer species is their high light requirement and short life span. Consequently, they succumb easily in competition with so-called climax species. The latter reproduce slowly, but are notably tenacious, and they tolerate as well as provide ample shade. Species such as elm, lime, oak, and beech gradually follow the pioneer species and ultimately displace them.

Iversen and Godwin showed that the Preboreal and Boreal intervals record exactly such a forest succession. So, for example, in the case of Denmark the Preboreal began with a juniper *(Juniperus)* maximum, soon followed up by a brief aspen *(Populus)* maximum, then by an extended birch *(Betula)* maximum, during which time pine *(Pinus)* gradually increased. Juniper and aspen were locally present, and hence marked the first maxima. Juniper, a small tree, was easily overshadowed and soon displaced by aspen. Birch and pine then rapidly invaded the area and displaced the shade-intolerant and short-lived aspen. Other climax species were located at greater distances and travel slowly, so that their immigration required several millennia. In the meantime the birch forest of northwestern Europe did not imply a cool, dry climate as formerly postulated, but simply represented a stage in colonization. Vegetational changes are not as rapid as the primary climatic changes, and colonization of many different soil environments on till, outwash, loess, or marshy ground had to be completed first. Persistence of tundra patches as a result of edaphic factors is substantiated by the importance of *Artemisia, Hippophae,* and *Ephedra.*

The Boreal is now usually defined by the first appearance of hazel (*Corylus*), a shade-tolerant shrub which first had to migrate from southern France before it could take its place in the northern forests. Hazel thrives well under birch and pine, but prevents their regeneration, particularly that of birch. As the first tolerant scrub, *Corylus* remained unchallenged until the arrival of elm and lime, which then crowded it out. As a result, neither the extended maximum of *Corylus* nor the disappearance of *Betula* have climatic significance (Iversen, 1960). Similarly, the Preboreal-Boreal border is not climatic, and cannot be synchronous, as it simply marks the local arrival of hazel. Consequently, the very broad approximation of such a date as 7500 B.C. (see Godwin and Willis, 1959), indicated in parentheses in Table 22, should be regarded with caution.

The same laws governing the stages of the Preboreal are also involved in the sequence of maxima found in the Boreal and the early Atlantic. Sequence of immigration—depending on rate of travel and distance from the nearest refuge location—and species competition were entirely responsible for the vegetational change. Already during the Preboreal the presence of a thermophile water plant such as *Cladium* indicated a warm climate in Denmark. During the Boreal, mistletoe *(Viscum)* and ivy *(Hedera)* indicate warmer summers and warmer winters than is the case today (Frenzel, 1966).

From the perspective of vegetation, the environmental changes of the early Holocene proceeded gradually. Forest did not abruptly replace tundra at the close of the Pleistocene. Rather, forest-tundras and parklands, succeeded by open and irregular woodlands—dotted by numerous undrained tundra lowlands, and reflecting a multitude of edaphic factors—superseded the herbaceous tundra only gradually.

FAUNAL CHANGE AT THE CLOSE OF THE PLEISTOCENE

The ecological impact of the final climatic oscillations of the terminal Pleistocene on animal life was indirect, through the intermediary of available food. The previous discussion has shown that the picture of climatic and vegetational change was complex and that there was no sudden disappearance of the tundra environment at the close of the

Pleistocene. The woodlands of the forest-tundras had probably been preferred by reindeer and bison in winter. Consequently, the former range of the gregarious tundra animals would not appear to have been suddenly eliminated at 8300 B.C. or 10,200 B.C. Instead, conditions must have spasmodically deteriorated during a 3,500 year interval, from the beginning of the Bölling to the close of the Preboreal. Only by the eighth millenium had the tundra of temperate Europe been fully eliminated.

Faunal evidence documenting the local disappearance of individual species in western Europe is fairly good. Both the late Magdalenian cave art and corresponding paleontological evidence in cave strata show that reindeer, bison, horse, and woolly mammoth were still present in southwestern France in considerable numbers as late as 11,000 B.C., before the Bölling. The absence of musk ox from these later drawings may be significant, but such representations were always rather scarce. The woolly rhino may still have been present, judging by a single late drawing from La Mouthe (see Breuil, 1950). The contemporary Hamburgian sites of Meiendorf, dated ca. 11,500 B.C., yielded a rich faunal inventory with 127 reindeer, 3–4 hares, 3 ground squirrels, and single specimens of lemming, horse, badger, fox, polecat, and wolverine (Rust, 1937, 1962). There is no ready explanation why bison and woolly mammoth should be completely absent in the Hamburg area other than by reason of deliberate specialization on reindeer by prehistoric man.

Woolly mammoth and rhino, as well as musk ox, giant "elk," bison, and the cave predators, are conspicuously absent from west European faunal assemblages of the Younger Dryas, while reindeer remained the staple food. It seems that the local extinction of most of the tundra fauna took place between 11,000 and 9,000 B.C., i.e., during the Alleröd at the latest. Possibly the warmer climate and the reduction of suitable grazing, and in some instances, possibly man were responsible (see ch. 28). Only the more versatile reindeer survived. Of the 668 mammalian individuals represented in the Younger Dryas cultural stratum of Stellmoor (Rust, 1943), 650 are reindeer. This suggests that reindeer were still very plentiful in the northern and eastern parts of Europe during the terminal Pleistocene. Yet among the faunal remains of early Postglacial cultures such as Star Carr, ca. 7600 B.C., and the Maglemosian of the Baltic and North Sea areas (during the Boreal), reindeer and horse are both absent. The reindeer must have withdrawn from temperate Europe during the first half of the Preboreal, although there is reason to believe that woodland reindeer persisted in Scotland and eastern Europe into the Subatlantic. Wild horse and forest bison *(Bison bonassus)* lived in the European woodlands until recently, but not the steppe horses of the tarpan and Przewalski type, nor the extinct Pleistocene steppe bison *(Bison priscus)*.

Summing up, the Pleistocene tundra fauna of western Europe became locally extinct in stages. Certain larger species, including the mammoth, woolly rhino, giant "elk," and musk ox appear to have disappeared or become extremely scarce during the Bölling or Alleröd. This was at least in part a result of environmental change. Throughout the terminal Pleistocene, reindeer formed the overwhelming bulk of man's food supply, this dietary pattern persisting into the Younger Dryas. The reindeer appears to have withdrawn to northeastern parts of Europe during the Preboreal, while a new animal spectrum was established in the birch and pine woodlands. Only the very end of some four millenia of ecological transition from tundra to forest (ca. 11,000–7500 B.C.) seems to have driven the remaining tundra fauna from temperate Europe.

CULTURAL ADAPTATION TO NEW ENVIRONMENTS AT THE CLOSE OF THE PLEISTOCENE

The original terminology employed for the European stone age cultures did not include a Mesolithic, but rather employed terms such as "Final Paleolithic" to describe terminal collecting groups in the early Holocene. The recognition of a Mesolithic phase was primarily due to V. G. Childe (1925). Childe believed that the environmental changes accompanying the onset of the Holocene in Europe had a sudden and serious impact on man. Childe realized the great herds of herbivores had been replaced by more solitary game such as deer, wild cattle, boar, and the like,

> the pursuit of which required more arduous tactics and a new equipment. That spelt the end of the cultures that had brought prosperity to Upper Paleolithic hunters. Adaptations to the novel and really sterner conditions are represented by so-called mesolithic cultures (Childe, 1958, p. 25).

One of the alleged consequences of this apparent deterioration of hunting resources was the breakdown of the organized community-hunting activities thought to be characteristic of many Upper Paleolithic groups. This change of social order is believed to have been more or less responsible for the disappearance of the Franco-Cantabrian cave art. The Mesolithic was then considered a consequence of environmental changes, a fact clearly emphasized by the arbitrary delimitation of 8000 B.C. for the beginning of this last collecting stage.

The cultural innovations of Europe during the early Holocene have since been almost generally described as a matter of readjustment to new environments. The specific nature of these changes has been recently restated by Binford (1968) and Waterbolk (1968): *(a)* A significantly smaller number of sites, a reduction of population, and eventually a major shift in the centers of "population growth." *(b)* Greater cultural diversity, both regionally and locally, suggesting more specific responses to local environmental conditions. *(c)* A marked increase in the exploitation of aquatic foods and wild fowl, with a concomitant trend to the hunting of smaller game. *(d)* A major change in the form of stone tools. The question may be raised whether each of these generalizations is warranted, at least without qualification, and whether indeed this environmental interpretation of the terminal Paleolithic cultures of Europe has not been overemphasized.

To begin with, it is doubtful whether the number of sites in western and central Europe dating from the time span 8000 to 5000 B.C. is substantially less than for any 3,000-year time span of the later Würm. Settlement continued in a large number of cave sites with the same degree of discontinuity evident in the Upper Paleolithic sequences, while good explanations can be offered for the scarcity of open-air sites. For one, streams were actively downcutting during the Würm-Holocene transition, and the Rhine incised its bed rapidly at the end of the Younger Dryas. Thus riverine sites will be largely lost at the bottom of post-Pleistocene alluvial fills. Similarly, at the coast, the rapid and continuing rise of sea level until 3000 B.C. will have submerged the great bulk of sites associated with the exploitation of littoral and marine resources. Finally, the absence of loess sedimentation or reworked loess colluvium on upland surfaces explains the lack of the depositional medium enjoyed by many Paleolithic sites; as a result, open-air sites were destroyed except for scattered stone tools. Altogether, a diminution of population has not been proven, although new centers of population growth did appear along the seashores of northern Europe, recently vacated by the ice.

Whether or not the cultural diversity of the Mesolithic in western Europe was greater than that of the Upper Paleolithic is debatable. A comparison of the Italian coastal cave sites, the Azilian of northern Spain, and the Tongrian of the Netherlands about 9000 B.C. shows no fewer contrasts than those evident between the Mediterranean "Romanellian," the Asturian of Cantabria, and the North Sea Maglemosian of about 7000 B.C. In fact, regional differentiation increased steadily throughout the Mousterian and Upper Paleolithic, and the Mesolithic falls well within the trends set earlier during the late Pleistocene. Now, as before, adaptations to local resources are, of course, apparent, but the degree of diversity was not comparably greater than that of the new environmental contrasts.

The increased exploitation of aquatic food resources by Mesolithic populations merely continued a trend evident during the later Pleistocene. The Upper Paleolithic strata of caves in Cantabrian Spain have fairly abundant shell, often transported kilometers inland. The Azilian, which corresponds to the cold Younger Dryas relapse, has considerably more shell, while the early post-Pleistocene Asturian sites are veritable shell middens (G. A. Clark, personal communication). However, mammalian faunas retained their significance for the Asturians, and the game bag (red and roe deer, as well as chamois, mountain goat, and boar) is not much different than it was for the Magdelenians or Aurignacians. In fact, the midden sites of the Asturians in Cantabrian Spain, and of similarly adapted populations in northwestern Iberia and central Portugal (all dating ca. 8000–4000 B.C.), form a direct cultural continuum with the regional Upper Paleolithic. Contrary to what Binford (1968) says, wild fowl are not verified in significant frequencies from Mesolithic sites, and fishing at riverine sites was no more common than during the Magdalenian. In this same context, it should also be emphasized that the tool inventory does not change drastically at any point, despite continuing innovations. Technologically, the Mesolithic forms a part of the dynamic Upper Paleolithic traditions.

It would nonetheless be incorrect to minimize the change in biomass between the European tundras and the closed woodlands. There can be little doubt that the disappearance or emigration of the great mammoth and bison herds in eastern Europe during the Bölling or Alleröd, and of the reindeer herds from western Europe during the Preboreal, removed the cornerstone of the European meat supply. Such changes would have introduced a major food crisis for specialized hunters, since the hunting of isolated deer, wild cattle, or boar was difficult if large meat reserves were to be established. But the individual hunt did not require greater skills than a reindeer battue. Since the winters were now mild, it was possible to hunt locally abundant forest game throughout the year, and a winter larder was unnecessary. New hunting techniques did not have to be learned, for deer and wild cattle had been successfully hunted throughout late Pleistocene times. Yet, regardless of technological skill, the disappearance of reindeer, for example, meant a catastrophic decline in game resources. Possibilities of following the retreating herds were limited, due both to the reduction of suitable reindeer habitats and the territoriality of other hunting peoples. However, reindeer was the staple of only a small segment of the late Paleolithic populations of Europe. Elsewhere, other dominant resources became scarce at various times between 11,000 and 8,000 B.C.; even the reindeer appears to have disappeared from its winter haunts in southwestern France some two millenia before its decline in northern Germany, Denmark, and Britain.

Consequently, although the European hunters north of the Pyrenees, Alps, Carpathians, and Caucasus underwent a severe resource crisis, with a significant decline of game biomass in the encroaching boreal and deciduous woodlands, this crisis varied in time—from place to place, and according to the staple game resources—between 11,000 and 7500 B.C. Considering the size and diversity of the European continent, the differences of regional specialization and adaptation and, above all, the length of the time interval involved, one can hardly speak of an incisive end-of-the-Pleistocene resource crisis in Europe. In the same light, one cannot help but feel a certain overemphasis on environmental readaptation in the thinking of many archeologists. The following discussion of a proto-Maglemosian site in Preboreal England may illustrate the point.

AN EXAMPLE OF MESOLITHIC HABITAT AND ECONOMY: STAR CARR, YORKSHIRE

The early Mesolithic site of Star Carr, located 8 km. south-southeast of Scarborough, near the English North Sea coast, was excavated by J. G. D. Clark in 1949–51. Through painstaking evaluation of the paleo-ecological evidence, particularly by H. Godwin and D. Walker, the subsistence patterns of the group are quite well understood. The locality is at 23 m. above sea level on the margins of a poorly drained peat lowland once occupied by a lake in early postglacial times.

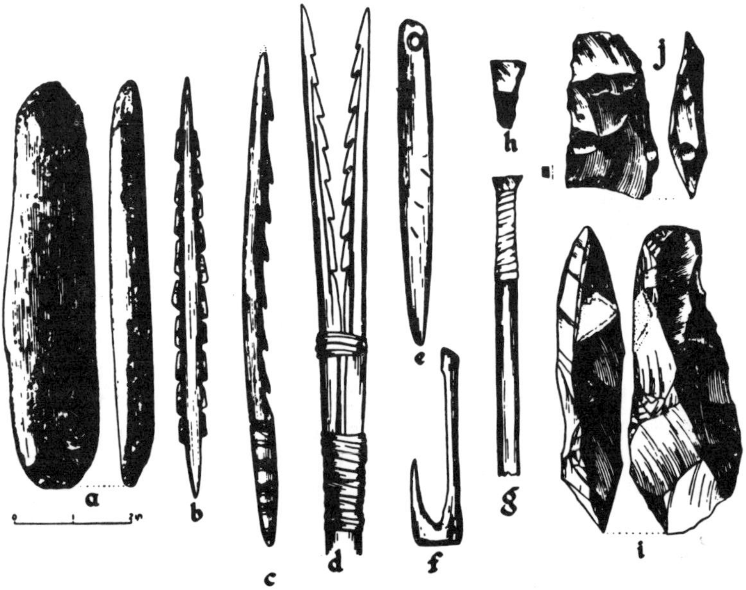

Figure 87. Mesolithic fishing and hunting equipment (from K. P. Oakley, 1958, by permission of the Trustees of the British Museum, Natural History): (a) a hammer for clam and mussel bashing, (b) bone fish-spear with microlith barbs, (c) barbed point in deer antler, (d) leister prongs of modern Eskimo fisherman, (e) net-making needle (?), (f) bone fish-hook, (g,h) microlith (transverse arrowhead) hafted in wood with sinew binding, (i,j) axes.

The cultural strata are found over solifluction deposits of Younger Dryas age, and under a meter or so of peat. Pollen analysis and macro-botanical evidence showed that the horizon is within the upper part of pollen zone IV, i.e., the late Preboreal. Two C^{14} dates of 7535±350 B.C. (C-353) and 7600 B.C. ±210 (Q-14) confirm this stratigraphic position.

The dominant tree species of the land areas was birch (*Betula pubescens*), and a closed birch forest with some pine (*Pinus silvestris*)—probably covered the hillsides and the drier parts of the valley bottom. Willow and a little alder fringed the reed swamps, indicating the presence of the lake edge. Although a number of aquatic birds record the presence of water nearby, the mammalian fauna is entirely one of woodland forms: at least 80 individuals of red deer (*Cervus elaphus*), 33 of roe deer (*Capreolus*), 11 of elk (*Alces*), 9 of aurochs (*Bos primigenius*), 5 of boar (*Sus scrofa*), together with a varying number of beaver, marten, red fox, wolf, badger, hare, and hedgehog. Fish or invertebrates are absent. Both the aquatic plants and mammalian fauna suggest a warm-temperate forest environment, not unlike that found in Britain today.

The areal extent of the settlement area is about 200 square meters, the vertical depth of cultural deposits about 15–40 cm. There are no remains of huts or or any kind of structure, although these are known from contemporary cultures in Denmark, northern Germany, and the Netherlands. The occupation area suggests that the group did not exceed a total of 16 to 25 individuals of whom 5 were adult men able to hunt large game. The state of the deer antlers indicates that the site was occupied during winter and spring, and abandoned during the summer.

The artifactual materials include some 2,500 finished or utilized flints, dominated by scrapers, burins, and microliths—minute, carefully retouched bladelets employed as arrowheads or projectile barbs. There is a large number of axes, adzes, and gouges, best interpreted as wood-working tools. Considerable use of bone and antler was made, particularly for barbed antler points, mattock heads, batons (for working flint?), bodkins, fastening pins, and in one case, as a harpoon head. Apart from various bone ornaments a 26-cm. long wooden paddle was found, possibly suggesting the presence of boats. On the basis of the meat represented by the faunal collection, 20 people might have been supported for 6.2 years. In view of seasonal occupation from October through April, and the use of some vegetable foods, even in winter, the site may have been repeatedly occupied for at least 12–15 years, possibly with some interruptions.

The general subsistence pattern of the Star Carr people is thought to have been one of seasonal winter settlement, with almost total emphasis on game hunting during that season. During the summer, the group possibly spent its time fishing in nearby lakes or rivers or food-collecting at the coast. Considerably more vegetable food could be obtained then, so that camps were possibly occupied on a temporary basis only. The general proficiency or cultural level of this population is not obviously advanced beyond that of the Upper Paleolithic, although the settlement patterns are adapted to a different environment. The later Maglemosian culture has a few better-worked tools, and indicates an appreciable calorie source from fish and invertebrates. In particular, fish spears, leisters, lines with barbless fish hooks, and nets provided with floats and stone sinkers all underscore the significance of fishing. This picture is complemented by further wooden paddles and a dugout canoe. But man's modification of or mastery over the environment was no more conspicuous than during the late Pleistocene.

THIRTY-TWO

Agricultural Origins in the Near East as a Geographical Problem

INTRODUCTION

The previous chapters attempted to appraise man-land relationships during the slow process of cultural innovation characterizing the Paleolithic and the Mesolithic. These hunter-gatherer populations had all been very sparsely settled and technologically simple, with a limited or even negligible impact on the natural environment. However, the same transition of Pleistocene and Holocene that left Europe at the cultural level of advanced food-collecting, witnessed the dramatic beginnings of agriculture in the Near East.

The culture groups of the Near Eastern late Pleistocene were specialized hunter-gatherers (Hole and Flannery, 1967; Flannery, 1969). But, at least as far as their tool industry is concerned, these Upper Paleolithic people were comparatively uninteresting and not remarkably progressive or specialized. Then about 11,000 years ago two cultures appear in the Levant and northeastern Iraq; the Natufian and the Karim Shahirian. Both assemblages were characterized by so-called agricultural implements such as sickle blades, grinding stones, and polished stone axes known as celts, presumed to have been used as hoes in many cases. None of these tools as such necessarily indicates agricultural activity, but the combination suggest partial subsistence on either wild grains or cultivated cereals. And at Zawi Chemi Shanidar, one site of the Karim

Shahirian assemblage, there is fairly good proof of the presence of domesticated sheep (Perkins, 1964) about 8900 B.C. (C^{14} dates 8910 and 8640 B.C.; Solecki and Rubin, 1958). By 7000 B.C. subsistence farming had become a common economic trait in parts of the Near East.

A hearth of agricultural origins in this particular area and at this particular time is of environmental and geographical interest. First, the localization of early domestication is to some extent circumscribed by environmental factors. Suitable biological resources must be present if local domestication is to be possible. A second problem concerns possible environmental influences on the cultural processes implied by agricultural origins. And third, the invention of agriculture is of great physical import, marking a drastic change in man-land relationships. The following chapter attempts to outline some of these environmental problems in relation to the hearth of domestication in the Near East. Beyond doubt the environmental problems related to first domestication in other culture areas, for example in the New World, are quite distinct. But their consideration lies beyond the scope of a selective survey of man-land interactions at different cultural and technological levels.

THE NEAR EAST AS A HEARTH OF DOMESTICATION

There have been several hearth areas in which domestication of specific associations of plants and animals was apparently first carried out. Basically such areas are habitats where a number of wild plants and animals are suitable for domestication, and presumably where such species could first be domesticated in the habitat of their wild ancestors (Braidwood, 1958). There were at least two independent hearths of domestication (in the Old and New World) and probably three; specifically *(a)* Mesoamerica and the Andean highlands, *(b)* the Near East, particularly the hill country of southwestern Asia and the Aegean area, and *(c)* southeastern Asia, probably along the margins of the Bay of Bengal and in Burma (see Sauer, 1952; Gorman, 1971; Solheim, 1969). The first two hearths at least had no obvious cultural intercommunications; in the case of areas *(b)* and *(c)*, one in a subhumid winter rainfall belt, the other in the humid tropics, techniques and cultural backgrounds are so different that any *initial* contact would be rather difficult to establish. Finally, there *may* have been minor hearths of domestication in which single species were first domesticated before an agricultural economy had been introduced from without. North China (Ho, 1969, and Watson, 1969), Ethiopia, and West Africa (Alexander and Coursey, 1969) provide possibilities of this kind.

The Near Eastern hearth region provided the biological materials, intellectual achievements, and cultural associations that underlie the civilizations of western Asia, northern Africa, and Europe. The basic biological inventory includes seed plants (cereals) and herd (as opposed to household) animals. More specifically, the food-producing cultures of these areas have from the very beginning depended primarily on the cultivation of wheat and barley for subsistence (Helbaek, 1959; Harlan and Zohary, 1966).

THE NATURAL HABITAT OF THE CEREALS

According to Helbaek (1959) the locus of domestication of a wild plant would presumably be within its area of original distribution in the wild state. Consequently, a prehistoric group dependent upon wild wheat as its main food should have developed its subsistence pattern within the original area of natural distribution of that species. The same should apply to a culture primarily dependent upon barley.

The wild ancestor of domesticated barley *(Hordeum spontaneum)* is now distributed across the Near East and in several parts of southern

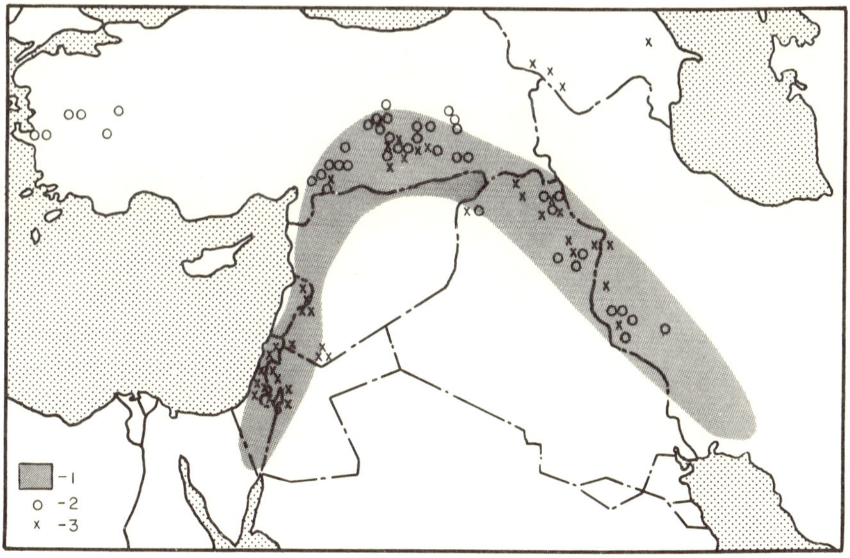

Figure 88. Natural distribution of the wild wheats and wild barley (modified after Harlan and Zohary, 1966). Native habitats of (1) Hordeum spontaneum, (2) Triticum aegilopoides, and (3) Triticum dicoccoides.

Europe and northern Africa (Fig. 88).[1] On the other hand, the two wild wheats, from which all domestic wheats have been derived directly or by complex hybridization, are more restricted in range. The large-grained *Triticum dicoccoides,* direct ancestor of emmer wheat *(T. dicoccum),* has its natural distribution in the Zagros Mountains of Iraq and Iran, the Taurus of southeastern Turkey, and much of the Levant. The wild small-grained *T. aegilopoides,* straight-line ancestor of einkorn *T. monococcum),* occurs through much of Turkey and the Zagros (see Harlan and Zohary, 1966). If one can assume that 12,000 years ago the distribution of wild wheat and barley was as it is today,[2] it would seem that the cradle of the "western" plant husbandry cultures lies in the winter rainfall zone of the Near East (Helbaek, 1959).

Helbaek (1959) considers that cereal domestication proceeded in several stages. The first essential change from reaping of wild cereals to planting may have included concentration of the desired plant by sowing, improvement of growth by tilling, exclusion or removal of unwanted plants from the tilled plot, and protection of the crop against animals and birds.

Another major step was to select particular types of grain and thus begin the process of selection, specialization, and ultimately adaptation to peculiar environments outside of the limited natural range of distribution. In the case of wheat it meant moving down the domesticant to, first, the plains or, later on, into the artificial ecology of the irrigated floodplains, and then into more northerly zones or higher altitudes (Helbaek, 1960b; Flannery, 1965).

The last major step was the hybridization of the wheats into more advanced, specialized types such as club wheat, bread wheat, spelt, and naked wheat, and the apparent evolution of barley into another, six-rowed type (Helbaek, 1966).

Besides conscious "primary" domestication, Helbaek distinguishes a "secondary" domestication, namely the segregation, for intentional cultivation, of a weed growing in cultivated soil which already was unintentionally subjected to selecion through being reaped along with the intended crop. Either wheat or barley was probably so introduced, and rye and oats are typical examples. Both of the latter were introduced as weeds in wheat fields, rye from west-central Asia and oats from the Near East or eastern Europe. Both "appear" very late in the arche-

1. Harlan and Zohary (1966) feel that the wild barley of northern Iran and Afghanistan on the one hand, and of Cyrenaica and the Aegean area on the other, have spread as a result of the disturbances of agricultural settlement.

2. Different climatic conditions during the terminal Würm may have modified the natural distribution, while man may since have eradicated the wild species in some areas. Barley, in particular, is cold-sensitive (Harlan and Zohary, 1966).

ological record and were probably never primarily planted anywhere but in cooler latitudes, where they proved to be particularly hardy plants. They play no role whatever in the Near East.

Regarding other plants, the various millets have an obscure history (see von Wissmann, 1957). These are summer rainfall plants, so that it is unlikely that they were first cultivated in the Near East with its mediterranean-type climate. Of further note is the wild flax plant, *Linum bienne,* used for fiber and oil, which has the same habitat and cultural context as wheat and barley. Together with starchy vegetables of Near Eastern origin, the wine grape, olive, date, fig, apple, pear, cherry, etc. also seem to originate somewhere in the Near East. In overview, winter-rainfall *cereal cultivation, orchard husbandry, and viticulture* are characteristic of early plant domestication in the Near Eastern hearth (Helbaek, 1959, 1960a).

THE NATURAL HABITAT OF THE HERD ANIMALS OF THE NEAR EAST

Present knowledge on the locus of the first domestication of herd animals is far less satisfactory than on that of the "western" cereals. The former range of the wild ancestors is usually extensive; the wild progenitor(s) is frequently a matter of strong controversy, often due to rather muddled taxonomic situations; and the archeologic-osteologic material is far less complete. The most up-to-date survey of the problem has been made by Zeuner (1963), Reed (1969), and Higgs and Jarman (1969).

The dog *(Canis familiaris)* is generally considered to be descended from the wolf, although later interbreeding with jackals may have taken place in the semiarid subtropics. The natural habitat of the wolf includes the greater part of the forest zone of Eurasia and North America. As the domestication of the dog took place rather early among European Mesolithic groups during the Preboreal or Boreal, the dog has no necessary association with agriculturists. There is, however, evidence of domesticated dogs in the Near East by 7000 B.C. (Reed, 1969).

The goat *(Capra hircus)* is most generally thought to be descended from the bezoar goat *(Capra aegagrus),* ranging from Palestine to the Caucasus, from Greece to the Indus. Fossil bezoar goats are also known from the late Pleistocene of the Levant. The actual habitat of the wild goat is somewhat more limited as a result of the ecological niche to which the goat is adapted, i.e., rough ground with rocky slopes which enable this agile climber to escape possible predators.

The sheep *(Ovis aries)* is probably mainly descended from the urial *(Ovis orientalis),* although other species of wild sheep may have contributed to certain breeds of domesticated sheep. The urial occurs in northern Iran, Afghanistan, northwestern India, and adjacent parts of Central

Asia. Another possible wild ancestor, the eatern moufflon *(Ovis musimon* ssp.*)* inhabited Anatolia, Caucasia and western Iran. Yet another, the argali *(Ovis ammon),* is found in Central Asia. Sheep are adapted to open, rolling country, avoiding open plains or dense forest.

Cattle *(Bos taurus)* are in all probability descended from the large, long-horned, wild *Bos primigenius* or aurochs once distributed throughout the forested regions of Europe, southwestern Asia, and northern Africa. A shorthorned species called *B. longifrons* or *B. brachyceros* has been postulated, but these animals were probably females of *B. primigenius.* Wild cattle favored woodland or forest as a habitat.

Originally there were several subspecies of wild pig *(Sus scrofa)* native to the woodlands of Eurasia and North Africa. The European domesticated pigs are essentially descendants of the wild boar *(Sus scrofa scrofa),* and the Chinese ones of the banded pig *(Sus vittatus)* native to southeastern Asia.

Domestication of horse, reindeer, and camel came relatively late and played no role in the original transition to food production, so that these genera are of peripheral interest in this discussion.

The natural habitats of the western Asiatic herd animals overlap in a broad way with the native distribution of the wild wheats and barley in the Near Eastern highlands. The range of the wild ancestors of the herd animals is very much greater than that of the wild wheats and barley, however. Although the boar and the aurochs, and possibly barley, were native to the alluvial floodplains of Mesopotamia and Egypt, sheep, goat, and wild wheats were absent. The Syrian, Iranian, and Central Asian deserts fall outside of this natural habitat zone.

THE NATURAL HABITAT ZONE

If there was sound reason to believe that cereal domestication preceded animal domestication, the Near Eastern hearth of agricultural origins could be more or less localized into a zone of preference—the Near Eastern highlands, and possibly a more peripheral zone, the alluvial floodplains of the Nile and the Tigris-Euphrates. So far, archeological evidence of animal domestication predates the earliest proven domesticated grains by as much as two millennia. It is only as a matter of convenience that the zone of overlap of the wild cereals and wild herd animals is emphasized here, even though the available archeological evidence suggests that the evidence may not be fortuitous.

A brief examination of the physical geography of the modern natural habitat of the wild wheats can be rather informative. The areas involved are characterized by irregular and diversified terrain and a minimum annual precipitation of 300–500 mm.; they coincide with the subtropical mediterranean-type woodlands of the Fertile Crescent and the temperate forests of Anatolia (Fig. 89). Significant is the exclusion of this particular habitat from the steppe or semidesert areas. Equally interesting is the location of known agricultural communities predating about 6000 B.C. These were all found within or at the peripheries of the woodland belt.

The alluvial valleys enjoy somewhat different environmental conditions. Apart from the peculiar terrain features of floodplains, neither the lower Nile Valley nor the Tigris-Euphrates lowlands have sufficient rainfall for nonirrigated agriculture. But crops could be planted as the annual floods receded (October in Egypt, June in Mesopotamia), and the moisture retained in the soil would normally be sufficient to bring one crop to maturity. The ecologic patterns of these alluvial floodplains were generally quite distinct from those of the highlands, even though a winter growing season would be common both to Egypt and the wooded hill country.

The geographical traits and subsistence economy of the earliest known Near Eastern farming communities speak for agricultural origins in the winter rainfall belt. This region corresponds closely to that ideal physical environment envisaged for first agriculture by C. O. Sauer (1952). From a different premise, Sauer argued that agriculture began in wooded lands rather than in grasslands with deep and continuous sod. This argument is based on the difficulty of cultivating heavy sod with primeval agricultural tools. Rather, a varied, open woodland could be more easily cleared by deadening the trees, so providing open spaces with looser topsoil for easy sowing. Dense forests were also inimical to primitive hoe agriculture. Sauer emphasizes that diversity of terrain is optimal in providing numerous ecologic niches—"a land of hills and valleys, of streams and springs, with alluvial reaches and rock shelters in cliffs" (1952, pp. 5–6). For it is here that the greatest diversity of plants and animals and suitable genetic reservoirs are to be found.

POSSIBLE ENVIRONMENTAL CHANGES IN THE NEAR EASTERN AREA AT THE CLOSE OF THE PLEISTOCENE

Climatic conditions in western Asia during the main Würm were discussed briefly in chapter 19. Any specific changes that may have occurred at the close of the Pleistocene appear to have been confined to the highlands, with little evidence for ecologically significant change in the more mesic low country.

The fauna of the terminal Paleolithic Zarzian culture of Iraq, dated 10,450 B.C., is not considered indicative of a different climate than today's, according to Braidwood, Howe *et al.* (1960, pp. 167-70). The fauna at Palegawra (965 m. elevation) includes gazelle, wild goat and

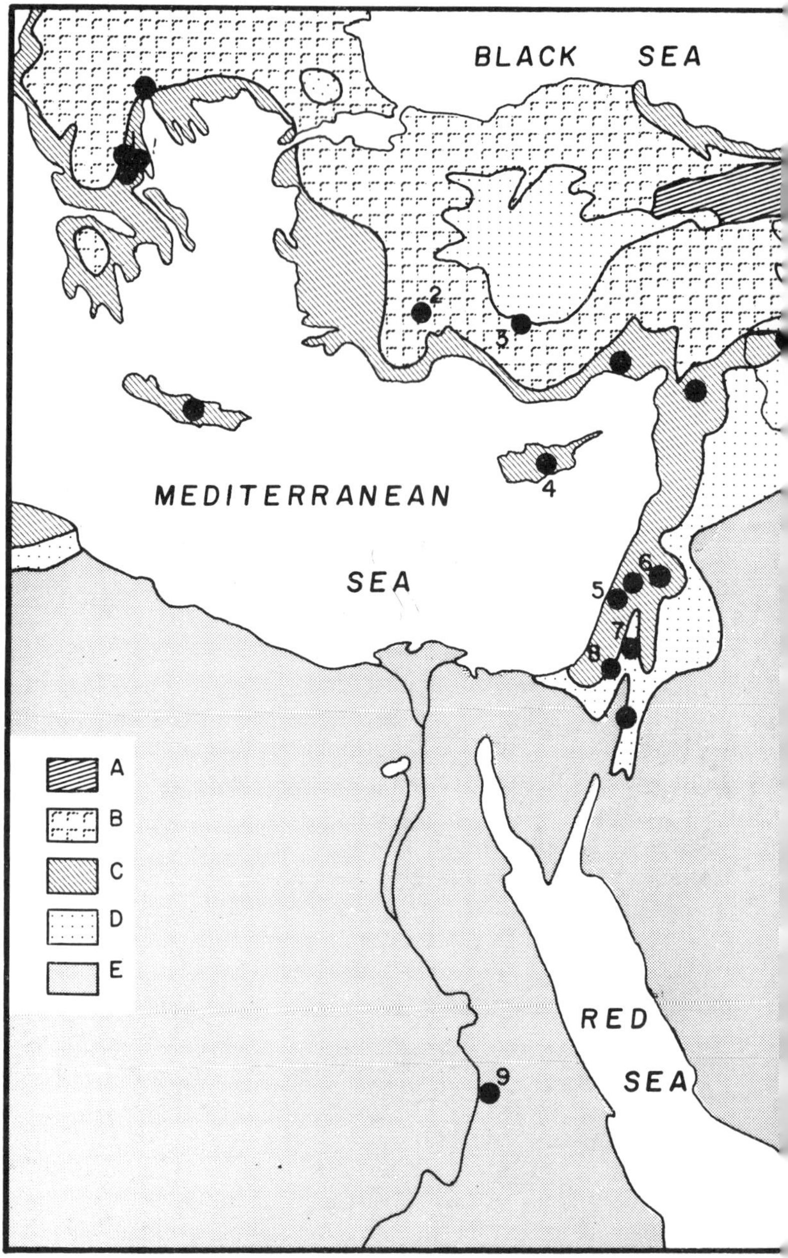

Figure 89. Natural postglacial vegetation of the Near East and location agricultural and proto-agricultural sites (ca. 9000–5750 B.C.): (A) conife forests, (B) deciduous and mixed forests, (C) subtropical woodlands, (D) grassland, (E) desert-grassland, semidesert, and desert. The galeria woo of the major rivers are not shown. Sites: (1) Sesklo, (2) Haçilar, (3) Çata

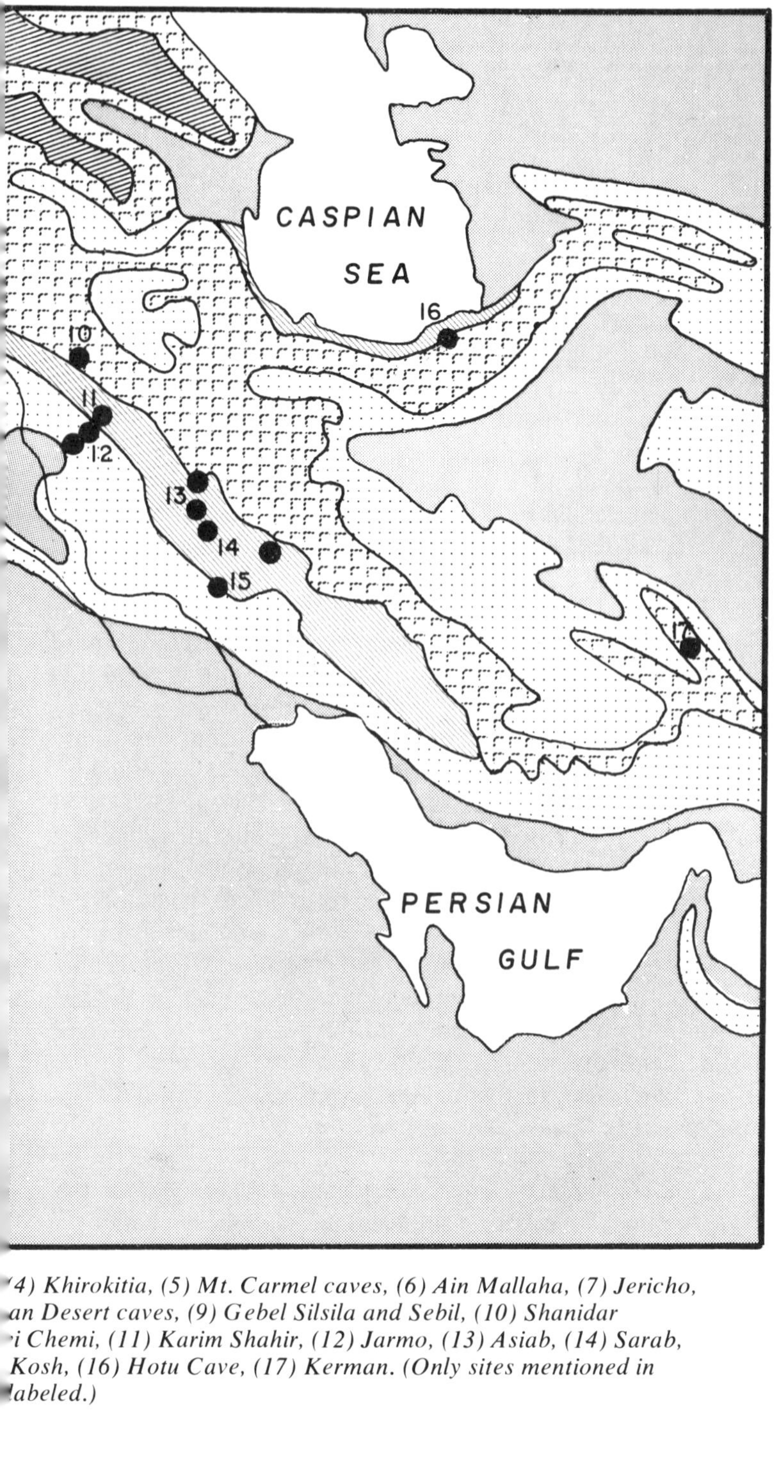

'4) Khirokitia, (5) Mt. Carmel caves, (6) Ain Mallaha, (7) Jericho, ·an Desert caves, (9) Gebel Silsila and Sebil, (10) Shanidar ·i Chemi, (11) Karim Shahir, (12) Jarmo, (13) Asiab, (14) Sarab, Kosh, (16) Hotu Cave, (17) Kerman. (Only sites mentioned in ·labeled.)

sheep, wild cattle, red and roe deer, boar, onager (?), red fox, wolf, lynx (?) and a hedgehog (Braidwood, Howe *et al.,* 1960, pp. 58-59) while the fauna of the corresponding Shanidar level B-2 (730 m. elevation) is dominated by wild goat (60 per cent) and red deer (20 per cent) together with bear, wild sheep, and boar (Perkins, 1964). The late Upper Paleolithic fauna of Ksar Akil, Lebanon (Hooijer, 1961), is equally indicative of more or less contemporary conditions. From this, one would conclude that local ecological conditions during the late Würm were similar to those of the present. Nonetheless, pollen studies from Lake Zeribar, near Marivan (1,300 m.) in the Zagros, indicate a change from almost 100 per cent NAP to about 15 per cent oak ca. 10,000 B.C. Oak then increases in an irregular fashion, while Chenopodiaceae decrease from 70 to 50 per cent, with *Artemisia* fluctuating around 10 per cent (Zeist, 1967). An open oak parkland is postulated (see Wright *et al.,* 1967). Northern microfaunal elements are gradually replaced by southern ones, documenting a rise in temperatures (Megard, 1967). Consequently, the high country was rather different in appearance prior to the Alleröd, and environmental prerequisites only approached modern conditions early in Holocene times. Similar patterns are apparent in northern Greece where closed mediterranean woodlands were reestablished as late as 6000 B.C. (see Bottema, 1967).

The cold relapse of the Younger Dryas possibly did not pass quite unnoticed in this part of the world. *Éboulis secs* horizons found in contemporary horizons of Ksar Akil, Lebanon (see Ewing, 1951), and the Haua Fteah cave, Cyrenaica (McBurney, 1967), offer possible suggestions but no proof for a cooler and moister climate at the close of the Pleistocene. Similarly there is evidence for recessional stages of the Würm glaciers of the Caucasus, eastern Anatolia, and northwestern Iran, some of which have been compared with the final Würm oscillations of the Alpine glaciers (see references and discussion in Butzer, 1972). It is quite probable, although beyond the possibility of accurate dating at the moment, that a small glacial readvance occurred in the highlands at this time. Seen in this perspective it would, therefore, be unjustified to say that ecologic conditions were truly "modern" prior to 8000 B.C.

Although the rather modest temperature changes suggested for the Younger Dryas cannot have been significant for human habitation, a posssible depression of 1°–3° C. would have had an effect on the distribution of the wild cereals. Wild wheats are now found to over 2,000 m. elevation in southeastern Turkey and Iran, while wild barley is rarely found above 1,500 m. (Harlan and Zohary, 1966; Helbaek, 1959). Colder late glacial climates may therefore have excluded these species from parts of the Near Eastern highlands during the late Würm.

Locally, in Palestine and Sinai, conditions may have been somewhat moister during a part or all of the Natufian period (ca. 9500–7500 B.C.). The gazelle, a characteristic open-country biotype, is comparatively infrequent at this level in the Mt. Carmel caves of Palestine, and a half dozen species of this genus disappeared at the time (Bate, 1940). Complementing the faunal record is archeological evidence of fishing in the dry wadis of the arid south Judean highlands. This suggests permanent pools of water available throughout the year. The presence of hunting populations in the Negeb and Sinai deserts, as indicated by plentiful distribution of Natufian flints, also seems relevant. Corroboration is provided by contemporary spring deposits and alluvia in the Jordan valley (Picard, 1963; Vita-Finzi, 1969; Nir and Ben-Arieh, 1965). And in Egypt there is good evidence of local wadi alluviation during the terminal late Pleistocene (Butzer and Hansen, 1968). These seem to be the available indications of greater moisture during the last millennium or so of the Pleistocene. The evidence appears to be limited to the lowland areas peripheral to the subtropical deserts. Such a "moist spell" probably did not have ecological significance in the mesic woodlands or cool high country.

ARCHEOLOGICAL EVIDENCE OF EARLY AGRICULTURE AND LIVESTOCK-RAISING

In reviewing the archeological record it is often difficult to determine whether a particular community practiced food production or whether agriculture and livestock-herding were entirely unknown. Smolla (1960) has devoted considerable attention to this problem of archeological evidence for early agriculture and animal domestication.

The stone artifacts commonly associated with agricultural operations are not unequivocal.

Sickle blades, consisting of rectangular flint blades, were designed to be mounted into a wooden or bone haft. Such bone hafts have been found on numerous occasions. However the sickles need not have been used to reap cereal crops, but may just as well have been employed on certain wild grasses or on reeds used for matting and hut construction. The sheen or lustre frequently developed on such blades may be a silicon deposit derived from straw or grasses (Smolla, 1960, p. 109 ff., with references). Since wild cereals "shatter" upon touch, it is questionable whether sickle-reaping would be possible at all. In fact, the ethnological record shows that the simplest primitive reaping of wild cereals is performed by plucking the ears or by beating the plants and catching the grain or seeds in a basket (Smolla, 1960, p. 110). Sickle-harvesting in the unripe state would not produce sickle-sheen, while the seeds may not be reproductive.

Mortars, consisting of hollowed stone vessels, and querns or pestles used as handstones, are pre-eminently effective as grinding stones for crushing grain or seeds to make flour. However *some* Natufian mortars were used to grind pigment (Garrod, 1958), while mortars and pestles are sometimes used for meat-grinding today (L. Binford and R. J. Braidwood, personal communication), and could also be employed for grinding acorns, wild grains, or bone grease.

Stone celts, resembling polished axes or hoes, may have been used as axes or hoes. There are, however, no good ethnological parallels for stone hoes (Smolla, 1960, p. 53).

All in all, the so-called agricultural tools are difficult to interpret, although when found in association and in large numbers they strongly suggest the intensive use and preparation of vegetable foods and probably of domesticated crops. Unfortunately there is no archeological record of more meaningful items such as digging sticks.

Botanical evidence of plant domestication can be recognized, but many of the morphological changes resulting from domestication take place very slowly. Theoretically, a single mutation will produce a "nonshattering" grain, so that selection of "nonshattering" mutants could rapidly produce a new domesticated stock with new morphological characteristics (J. D. Sauer, personal communication).

Osteological evidence for earliest domestication would be difficult or impossible to demonstrate by bone anatomy alone. An interesting example of circumventing this problem has been made by Perkins (1964) at Zawi Chemi Shanidar, the site of the earliest evidence of animal domestication to date. Here the faunal compositions of the Middle and Upper Paleolithic strata were quite uniform, with wild goat outnumbering wild sheep by 3:1, and constituting about 60 per cent of the fauna. About 25 per cent of the animals were juveniles under a year of age. Suddenly, in the Zawi Chemi horizon, sheep bones jumped to 75 per cent, of which 60 per cent were immature. Goat dropped down to 10 per cent, still with 25 per cent juveniles. It is concluded that the sheep must have been domesticated at this stage, and that the larger part of each year's young were killed for food and skins before the end of the year. The hunting of wild goats had consequently become relatively unimportant.

As a result of these difficulties in accurate assessment of the archeological record, the absence of evidently domesticated cereals or animals from many sites need not prove that agriculture was unknown. Equally so, the presence of so-called agricultural implements does not necessarily prove knowledge of crop planting.

THE NEAR EASTERN ARCHEOLOGICAL RECORD PERTAINING TO EARLY AGRICULTURE

The Near Eastern tool inventory of various Upper Paleolithic cultures, culminating with the Kebaran assemblage in Palestine, the Nebekian in Syria, and the Zarzian in northeastern Iraq (Howell, 1959a; Hole and Flannery, 1967), is broadly comparable to the European counterparts, although showing early microlithic traits. Settlement was largely concentrated in caves, although some Zarzian open-air sites have been tentatively identified (Briadwood, Howe *et al.,* 1960, pp. 155–57). The only contemporary cultural group that falls out of this framework is the Sebilian complex of the Egyptian Nile Valley. The Sebilian groups are of particular interest since they were semisedentary, occupying campsites on the banks of the Nile, where they intensively used the aquatic and riverine food resources of their localized environment. Modest kitchen middens in the Kom Ombo area of Upper Egypt testify to considerable use of freshwater mollusks, fish, and more rarely, turtle and crocodile; in addition, a wide range of woodland and steppe mammals were hunted. Grinding stones are already present, often in great numbers. Geologically, the Sebilian complex has been dated ca. 15,000–10,500 B.C.(Butzer and Hansen, 1968), i.e., no later in time than the Kebaran or Zarzian.[3]

Rather abruptly, archeological indications of agriculture appear in the Levant and Iraq ca. 9500 B.C.,suggesting a very early diffusion of agriculture in the Near Eastern highlands. Sickle blades and pounding and milling stones appear more or less simultaneously in both the Natufian assemblage of Palestine, Lebanon, and Syria (Garrod, 1958), and the Karim Shahirian of Iraqi Kurdistan (Braidwood, Howe *et al.,* 1960). The contemporary Asiab assemblage of northwestern Iran (Braidwood *et al.,* 1961) does not yet appear to have sickles, grinding stones, or celts. An analogous culture with microliths, sickle blades, and grinding stones has also been discovered at Kerman, in southeastern Iran (Huckriede, 1962). No evidence of cereals is available from the Natufian but cereal pollen is found at Zawi Chemi Shanidar (Leroi-Gourhan, 1969), so that plant domestication was at least well underway. Domesticated sheep are

3. A full ecological interpretation of the Sebilian complex and other contemporary industries will undoubtedly contribute to understanding agricultural origins in the Near East. This will only be possible after complete publication of the archeological results by P. E. L. Smith and M. A. Baumhoff. For further discussion, see chapter 33.

present in the Karim Shahirian. These two cultures, which possibly extend through most of the ninth and eighth millenia, precede a bona fide agricultural economy, certainly established in parts of western Asia by 7000 B.C. Both assemblages are essentially found within the natural habitats of wild wheat, barley, sheep, and goat. This may be the elusive stage of "incipient agriculture and animal domestication"—which Braidwood (1960a) describes as experimental manipulation of potential domesticates within a dominant food-collecting economy, at first still within the ecological niche to which the wild ancestor of the domesticate was adapted. However, successful adaptation to the lowland steppes began very early, as is shown by the Bus Mordeh assemblage of Ali Kosh in the Khuzistan foothills (Hole and Flannery, 1967). Emmer and barley were cultivated and goat and sheep kept.

By 7000 B.C. agriculture had become the primary subsistence of village farmers found in the Levant, the Zagros area, and southwestern Anatolia. These people grew einkorn, emmer, and barley, and kept domesticated goats and sheep. The domesticated pig also appears in archeological context somewhere in the seventh millennium in the pottery levels of Jarmo, northeastern Iraq. By the beginning of the sixth millennium, village-farming communites are verified in Thessaly (see Renfrew, 1969), Crete (see Higgs and Jarman, 1969), Cyprus (Dikaios, 1953), Anatolia (Mellaart, 1965; Renfrew, 1969, Reed, 1969), a good range of sites in the Levant, northern Iraq, and adjacent parts of Iran as well as in the Belt Cave on the Caspian shores of Iran (see Ralph, 1955).

The regional appearance of the various achievements of cultural innovation and evolution in the Near East are summarized in Table 23. The major expansion of food-producing populations at the Neolithic level into the cooler environments of temperate Europe and into the different environment of the Tigris-Euphrates and Nile floodplains, appears to postdate 5500 B.C. These later aspects will be considered in chapters 33 and 34.

THE ECOLOGY OF THE NATUFIAN IN PALESTINE

Remains of the Natufian cultural assemblage, dating from approximately 9500–7500 B.C., are widely distributed in the southern Levant (Fig. 90).[4]

4. Sites have been found in the Jabrud cave of Syria (Rust, 1950), at Beirut, in three caves of the Mt. Carmel area of Palestine (Garrod and Bate, 1937; Garrod, 1958), at the base of Jericho (Kenyon, 1959), at Ain Mallaha near Lake Huleh (Perrot, 1962, 1966, 1968), as well as in a number of caves in the wadis of the Judean hills both northwest and southeast of Jerusalem (Neuville, 1951). Surface finds have been made east of the Jordan river, in the Negeb and Sinai deserts, and at el-Omari and Helwan, near Cairo, although their affinities are less certain.

Table 23. Archeological evidence of early cereal cultivation and animal herding in the Near East (based on Hole and Flannery, 1967; Mellaart, 1965; Reed, 1959, 1961, 1969; Renfrew, 1969; and others).

Sites and stratigraphy	*Approximate dates* (B.C.)	Barley	Einkorn	Emmer	Bread Wheat	Sheep	Goat	Cattle	Pig	Dog
Aegean Area										
Argissa (Thessaly), Aceramic	6500	x	x	x		x	x	x	x	?
Nea Nikomedeia (Macedonia)	6200	x		x		x	x	x	x	?
Knossos (Crete), stratum X	6100	x		x	x					
Khirokitia (Cyprus), Aceramic	6000					x	x			
Sesklo (Thessaly), Aceramic	6000-5000	x		x						
Ghediki (Thessaly), Aceramic	6000-5000	x	x	x						
Anatolia										
Haçilar, Aceramic	7000			x						?
Haçilar, Ceramic	5800-5000	x	x	x	x					?
Çayönü	7000					x			x	x
Çatal Hüyük, VI-II	5850-5600		x	x	x	x		?	x	
Levant										
Tell Ramad (Syria)	7000	x	x	x	x					
Jericho, Prepottery Neol. A.	7000-6500	x		x						
Jericho, Prepottery Neol. B.	6500-5500	x	x	x						
Beidha (Jordan), Prepottery	7000			x			x			
Amouq (Antioch), A.	5750	x		x			x			
Mesopotamia-Khuzistan										
Ali Kosh, Bus Mordeh	7500-6750		x	x		x	x			
Ali Kosh, Ali Kosh	6750-6000	x		x		x	x			
Ali Kosh, M. Jaffar	6000-5600	x		x		x	x			
Tepe Sabz, Sabz	5500-5000	x			x	x	x	x		x
Tell es-Sawwan (Samarra)	5800-5600	x	x	x	x					
Hassuna	5800	x		x			x			
Kurdistan-Luristan										
Zawi Chemi, Karim Shahir	8900					x				
Jarmo	6750-6500		x	x		x	x		x	x
Tepe Sarab	(?) 6500					x	x			
Tepe Guran	6200-5500	x								
Matarrah	5800	x		x					x	

One of the best published and culturally important sites of the Natufian is found in the Mugharet el-Wad cave of Mt. Carmel, at an elevation of 45 m. on the southern face of a small wadi.[5] The base of the cave is 12.5 m. above the wadi floor, and extends for some 85 m. with an average height of 10 m. The Natufian strata underlie 0.3–1.2 m. of a consolidated brown earth and limestone rubble with early Bronze Age and later remains. About 0.2–3.0 m. thick, these beds consist of unconsolidated, stony red earth with limestone talus in the sections located in front of the cave entrance. The underlying deposits of the interior cave contain Upper and Middle Paleolithic industries. Interpretive geomorphological work has not yet been carried out, so that the implications of the beds are obscure.

In the further absence of known botanical remains, the rich faunal collection of the Mugharet el-Wad is ecologically important. It includes rodents and insectivores with two species of hedgehog, mole rat *(Spalax)*, a vole, squirrel, hare, a gerbil, and hyrax (*Procavia*). Spotted hyena, red fox, wolf (not dog; Clutton-Brock, 1963), badger, marten, musteline, the Syrian bear (?), wild cat, and leopard number among the carnivores, while the bulk of the animals represented are various ungulates: fallow deer *(Dama mesopotamica),* gazelle, wild goat, wild cattle, onager, and boar. Ecologically these species are partly woodland, partly open country, and partly even desert or cliff forms (the gerbil and hyrax). They corroborate the local situation of wooded upland to the northeast, and perennial streams or ponds with fringing forests and widespread open country on the Pleistocene dunes of the coastal plain to the south. They also show that diversified hunting played an important role in the Natufian economy. There is no trace of domesticated forms (Perrot, 1968).

The cave floor included a mass of flint implements, waste materials, broken and occasionally charred animal bones, burials, and some crude stonework, possibly associated with the interments. Although architecture is lacking at this site, house foundations have been uncovered at Ain Mallaha (Perrot, 1966). Some thirty-nine burials have been found, the dentition of which shows excessive wear and a very high frequency of abscesses of the premolars. Patricia Smith (personal communication; also Dahlberg, 1960), who has studied the dental pathology of the Ain Mallaha and el-Wad burials in detail, finds the severe attrition and tooth loss in older individuals indicative of a gritty diet, probably with a dominance of cereals or other coarse vegetable foods. Interestingly, the

5. The present-day climate has a January mean temperature of 13° C., a July mean of 27° C., and an annual precipitation of 625 mm. falling almost exclusively during the three winter months. The natural vegetation of the area is mediterranean woodland.

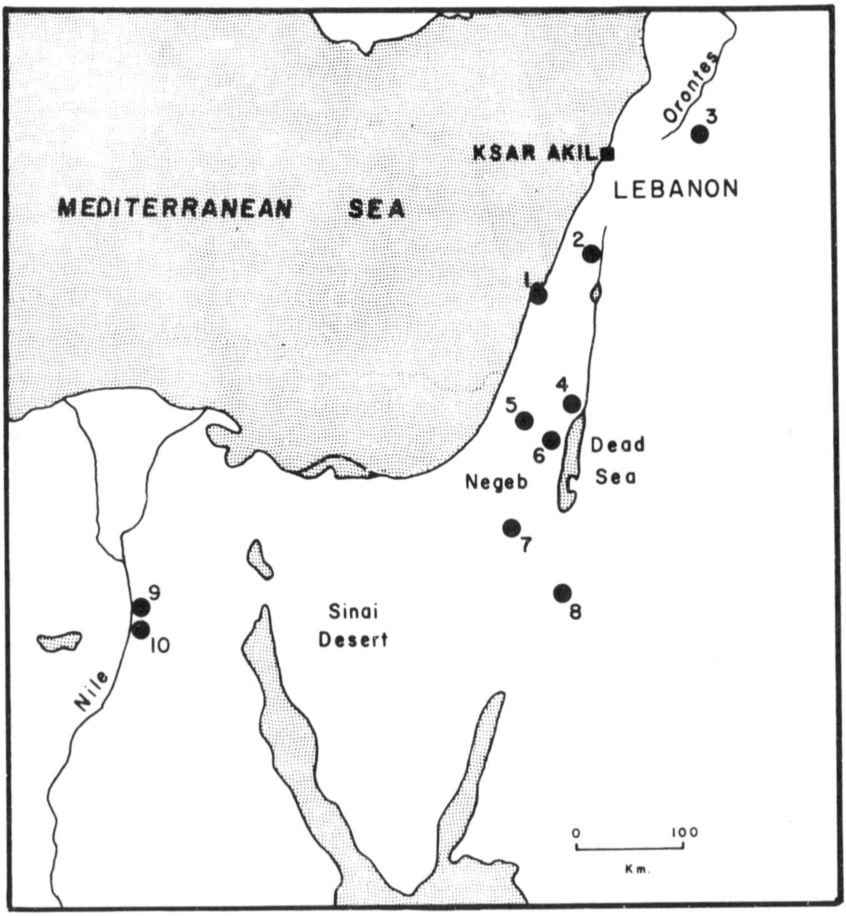

Figure 90. Distribution of the Natufian industries (ca. 9500-7500 B.C.): (1) Mt. Carmel (Wadi Fallah, el-Wad, Kebara), (2) Ain Mallaha, (3) Jabrud, (4) Jericho, (5) Shuqba, (6) Tor Abu Sif, Erq el-Ahmar, and el-Khiam, (7) Nahal Rimon, (8) Beidha, (9) el-Omari, (10) Helwan.

preceding Kebaran people appear to have utilized meat rather than fibrous and starchy plant foods.

The technological inventory contains, in part, a number of implements common to the Upper Paleolithic: backed blades, burins, massive scrapers, and rough picks, together with naturalistic carving in bone and stone. Great numbers of microliths, and at certain other sites, bone spear points, harpoons, fishhooks, pins, needles, and awls recall certain Mesolithic innovations. Novel for the Natufian in general, however, are lustrous sickle blades (sawtoothed varieties appearing in the Upper Natufian), some blades with bone hafts, as well as celts, mortars, and pestles. Flint arrowheads figure among the few innovations of the Upper Natufian. However the total assemblage shows a distinct shift of relative proportions between a dominance of Paleolithic artifacts in the Lower Natufian, and one of more characteristically Neolithic artifacts in the Upper Natufian.

The Natufian culture represents one of the most interesting transitional assemblages of the Near East. Partly dated by C^{14} dates of 9220, 8840, 7900 and 7850 B.C. (see Perrot, 1968; Zeuner, 1963, p. 31) from the Middle Natufian underlying Jericho, the populations in question were at least semisedentary, judging by cave occupancy or house habitation in the open. Intensive exploitation of the different ecological niches of the natural environment is a well-established characteristic, recalling both the earlier Middle Sebilian and the later European Mesolithic. Simultaneously, cereal agriculture was very probably known, judging by the abundant presence of all so-called agricultural implements, and the dietary value of gritty foods as suggested by the dentition. Unfortunately no plant foods have been found so far. But the beginnings of plant and animal domestication must be conceived of at a stage and in a setting such as that of the Natufian or the broadly contemporary Karim Shahirian of Iraq.

ECOLOGY OF A VILLAGE-FARMING COMMUNITY: JARMO, NORTHEASTERN IRAQ

The townsite of prehistoric Jarmo is located on a bluff at some 770 m. above sea level in the rolling hill country of the Kurdish foothills of northeastern Iraq.[6] The village appears to have been occupied more or less continuously for about a quarter of a millennium shortly after 7000 B.C., judging by a wide scatter of radiocarbon dates (Braidwood, Howe *et al.,* 1960; see also the ecological synthesis of Braidwood and Reed, 1957).

The irregular terrain is a consequence of dissection of late Pleistocene silts by steep-sided stream valleys and gullies. An intermittent stream, probably perennial before the destruction of the natural vegetation, has partially destroyed the western end of the site by undercutting. During the period of settlement (Wright, 1952), the site was located about 36 m. above this stream bed, which probably formed the major water supply of the village.

6. By extrapolation from other climatic stations in the area, the January mean temperature is about 6.5° C., the July mean 29° C., the annual precipitation about 630 mm., falling predominantly in winter. The natural vegetation is that of a mediterranean woodland.

Botanical remains at Jarmo include both domesticated and wild emmer and einkorn wheat, domesticated two-row barley, as well as acorns, pistachio nuts, lentils, the field pea, and blue vetchling.

Faunal materials include the remains of domesticated goat and pig (the latter in the upper strata of the site only; Reed, 1969), as well as a fair number of wild animals representing the hunting booty of the community. Species listed are red fox, wolf, gazelle, wild cattle, red and roe deer, wild sheep, boar, and onager (?). Great masses of terrestrial snails *(Helix salomonica)* are present together with some freshwater crabs and fish. The faunal selection suggests a woodland environment with some areas of open plain or rough country.

The village covered a total area of about 12,500 square meters, and the cultural materials attain about 7 m. in depth. A good third of this area was never occupied by houses, and a total of 25 houses is estimated as the maximum size of Jarmo. This includes a guess on how much of the site has been destroyed by gullying. Each house presumably represented a family unit. Assuming a family-household size of 5 to 7 people, 25 houses would indicate a population of 125 to 175 people. The lower figure is probably closer to the truth. This is, incidentally, the average size of villages in the area today.

The architecture itself, although well-defined, was not pretentious. Sun-dried mud was employed, being laid in successive 10 to 15 cm. tiers, often set on foundations of unmortared stone. The resulting mud-walled house had several rectangular rooms and was not unlike the local houses of today. The village had no regular plan, and consisted of simple houses, animal shelters, and storage buildings, without evidence of community buildings or social structure.

The technological inventory of Jarmo contains various flint implements, among which great quantities of sickle blades and microliths made of a glassy volcanic rock, obsidian, are of most interest. The obsidian was quarried some 500 km. to the north in the Lake Van area, indicating commercial contacts. Together with the celts are various grinding stones and bowls. Pottery appears only in the upper third of the settlement strata. Other items include bone needles, awls and the like, as well as evidence of reed matting. The technology is then a complex of domestic, hunting, and agricultural equipment. The dentitions of seven skeletons show signs of only moderate wear (Dahlberg, 1960), implying a less coarse diet than was common for the Natufian population. This probably points toward better preparation of vegetable foods, and possibly also to a fair proportion of meat in the dietary economy.

All in all, the farmers of the village of Jarmo appear to have established a well-balanced economy which, even at the stage of primitive subsistence agriculture, ensured adequate local food resources for permanent habitation over two centuries. The absence of the plow, or for that matter plow animals, means that some form of hoe agriculture was practiced. Although cereals dominated in the sown fields, a number of vegetable crops may also have been grown. Domesticated animals, apparently present in good numbers after the local introduction of the pig, supplied a dependable and possibly appreciable meat fraction to the diet. Hunting was still an important economic trait, while gathering of wild plant and animal foods is substantiated by finds of acorns, pistachio nuts, and snails. Jarmo is indeed the prototype of agricultural villages which already dotted the moister hill country of the Near East by the close of the seventh millenium. The origins of the cultural landscape and the expression of man-land relationships at the food-producing level will be considered in the subsequent chapters.

THE DESICCATION THEORY OF AGRICULTURAL ORIGINS

Although the cultural and intellectual processes basic to the economic transition from food-collecting to food-producing are of no direct concern to the natural scientist, the abundant environmentalist theories on that topic certainly are. These theories are based on the belief that late glacial or early Holocene desiccation affected wide areas of the subtropics that had enjoyed pluvial conditions earlier during the Pleistocene. As a result, the former hunting populations of the deserts of northern Africa, Arabia, Iran, India, and central Asia were allegedly expelled or forced to concentrate along sources of permanent water at springs or along permanent streams.

The oldest of these hypotheses can be associated with R. Pumpelly (1908, pp. 65–66), who excavated at the Neolithic site of Anau, southern Turkmenistan:

> With the gradual shrinking in dimensions of habitable areas and the disappearance of herds of wild animals, man, concentrating on the oases and forced to conquer new means of support, began to utilize the native plants; and from among these he learned to use seeds of different grasses growing on the dry land and in marshes at the mouths of larger streams on the desert. With the increase of population and its necessities, he learned to plant the seeds, thus making, by conscious or unconscious selection, the first step in the evolution of the whole series of cereals.

In the same sense Peake and Fleure (1927, p. 14) write:

> . . . men naturally turned their attention back to the old habit of collecting food as their hunting became less successful. In certain regions however, men were led towards a new idea; it occurred to them to produce food by the cultivation of edible plants.

Or as Childe (1929, p. 42) describes the same process in more detail:

> Enforced concentration in oases or by the banks of ever more precarious springs and streams would require an intensified search for means of nourishment. Animals and man would be herded together round pools and wadis that were growing increasingly isolated by desert tracts and such enforced juxtaposition might almost of itself promote that sort of symbiosis between man and beast signified in the word domestication.

For Childe, the resulting "emancipation from dependence on the whims of the environment" (1929, p. 46) was *the* impetus for the economic revolution ("Neolithic revolution") heralded by the invention of food-production. Toynbee (1935, vol. I, pp. 304–5) adopted the same economic revolution and the same impetus as the "physical challenge" at the root of ancient Egyptian and Mesopotamian civilizations, as well as for the origin of nomadic pastoralism (1935, vol. III, pp. 10–12). Similar ideas persist in more recent revisions of both Childe and Toynbee.

There is no doubt today that the simple patterns envisaged by the theories of Pumpelly, Peake and Fleure, Childe and Toynbee are archeologically untenable, since the food-producing revolution does not seem to have taken place in the deserts. In the hill country of western Asia, where the decisive steps of local agricultural invention were probably undertaken, the desiccation theory loses all meaning. These are well-watered regions where pluvial-interpluvial oscillations would not seriously reduce wild game resources. The native vegetation of the Near Eastern highlands is a subtropical or warm-temperate woodland under modern climatic conditions. Streams from the higher country provide abundant, perennial waters, or at least did so before the catastrophic impact of deforestation and soil erosion in historical times. Even if rainfall changes had occurred, they would only have carried limited ecological implications in an area of varied topography and with numerous local ecological niches. Instead, temperature changes may have had greater importance, particularly in late glacial times when the cold highlands once more became habitable. Such changes would therefore have enlarged the area of suitable lands at about the time of agricultural origins.

In conclusion, the previous review of paleoclimatic information does not suggest any incisive changes in the late glacial and early Holocene record of western Asia, and the climatic changes that did take place certainly did not follow a simple pattern of progressive desiccation. It seems unlikely that the cultural innovation of the Near Eastern hearth of domestication was associated with any dramatic ecological changes at the close of the Pleistocene. Instead, a bountiful natural environment with a fortuitous assembly of suitable domesticates presumably favored the geographic location of the Near Eastern hearth.

POSTSCRIPT

It has become apparent that the concept of a Near Eastern hearth of domestication requires modification. During the late 1940s and early 1950s, the literature dealing with agricultural origins in the Old World was highly speculative. Little factual material was available, and wide-ranging hypotheses were formulated on the basis of limited evidence. Today, some fifteen years later, the wealth of available archeological and biological data favors a more empirical approach. There is no lack of speculative writing, but serious attempts to interpret the evidence have become unduely restricted. New finds are often categorized within the increasingly rigid framework of a single "nuclear area" and a single cultural-ecological association. In fact, some workers have adopted a new form of environmentalism that obscures the fact that the origin of agriculture is, in the first place, a cultural phenomenon.

The archeological record remains very incomplete, despite the increasing number of excellent sites that span a long range of time and appear to reflect on different stages in the development of food-producing economies. Anatolia has already brought many surprises, and wider exploration is bound to reveal further evidence that the warm-temperate environments of modern Turkey were far from being a cultural hinterland of the Taurus-Zagros area. The amazing Neolithic sites of Macedonia and Thessaly now show that southeastern Europe was one of the core areas of early agriculture, and studies further afield in Iran and Transcaucasia are bound to extend the concept of a "nuclear area."

Patterns of sedentary or semisedentary settlement—believed by some to be vital for the earliest agricultural innovations—were common in late Pleistocene and early post-Pleistocene times. They must already be inferred for some Acheulian populations, and both the open-air and cave sites of Upper Paleolithic groups in the tundra-steppes of Europe frequently indicate seasonal, if not semipermanent, occupation of suitable localities. A similar record of semisedentary settlement is suggested for the Sebilians and other groups that were settled along the Nile River as early as 15,000 B.C. Long-term residence at one or several closely adjacent sites may reflect the availability and reliability of food sources more clearly than technology. Such prerequisites were present at many times and at many places during the course of the Pleistocene; they were not unique to the Near Eastern "nuclear area" at the close of the Pleistocene.

Given suitable food resources, intensified food-collecting provided the basis for large settlements of considerable permanence. Recent excavations have adequately demonstrated this proposition for the

Tehuacán Valley in south-central Mexico and for the Tennessee and Wabash valleys in the central United States. At Tehuacán, intensified food-collecting permitted a significant increase of population over several millenia, at a time when cultigens provided less than 10 per cent of the food intake. In the case of the eastern woodlands of North America, efficient exploitation of riverine environments permitted large and stable villages for some 3,000 years prior to the local adoption of agriculture. The Sebilians and other folk of Egypt and Nubia are another case in point, with some large cemeteries speaking for populations of at least moderate size. Here intensive utilization of a river-oasis remained practible or preferable for ten millenia, until agriculture first penetrated the Nile Valley—presumably with fresh populations—at a surprisingly late date. Much more recently, certain Indian populations of California and the Pacific Northwest provide examples of the same process.

These cases should serve as reminders that farming and livestock raising were not the only means of supporting relatively large populations in suitable meso-environments. Situations of this kind were the exception rather than the rule, but they are almost certainly relevant to agricultural origins and diffusions.

The first steps to agriculture marked no sharp break in subsistence patterns or population level among semisedentary groups with an intensive food-collecting economy. When agricultural traits spread and were adapted by preference or necessity, there probably were few discontinuities between the subsistence forms of adjacent agricultural and nonagricultural populations. The strong distinction made between farmers and hunters today reflects western cultural attitudes that have intensified over millenia, reinforced by an increasing technological gap. Initially, however, the convergence of unlike economies would not have been considerable within any one meso-environment. On the other hand, regional specializations must have been conspicuous. Each environment provided an individual set of potential resources that were managed and exploited distinctively, both before and after the introduction of agriculture. Consequently, the early stages of agricultural innovation must have been marked by strong regional contrasts, reflecting both different resources and different traditions. Barring violent conquest or displacement, traditional methods, attitudes, and preferences may have persisted over centuries or millenia, long after the introduction of agricultural traits.

Viewed within this perspective, wheat-and-barley farming can reflect but a part of the spectrum of advanced subsistence patterns in the Near Eastern area before 5000 B.C. One set of questions that can be raised concerns the variable role played by one or more domesticated animals in different areas. Were there food-collectors who herded animals? Were there herders who cultivated some grains during part of the year? Did or did not herding precede farming initially? Or locally? Did herding and farming originally have different roots among regionally specialized food-collectors, or did they spring from a single regional tradition? Did herding and farming traits diffuse at similar or different rates? In their entirety or selectively? Another group of questions could be formulated about the relative role of legumes and certain other vegetables. Such plants were cultivated in Mexico for almost 3,000 years before the first domestication of maize; and grains may also have been preceded by other cultigens such as peas and lentils in the Near East. Our cultural bias has favored an overemphasis of grain-farming in the current Near Eastern literature, possibly obscuring the significance of other domesticates.

In retrospect, it appears that archeological research must be directed at a wider range of problems. Present understanding of the "terminal food-collecting stage" in North America shows how much more we need to know about the intensive food-collecting economies of a broader area in northern Africa, southeastern Europe, and western Asia. We have learned little new about the "level of incipient agriculture" in the Near East during the past decade, despite ongoing excavations. Perhaps there has been too much attention to house structures, burials, and the identification of habitation residues—with only rudimentary analysis of total archeological associations. Only conscious effort will serve to demonstrate dietary and subsistence patterns and to allow inferences about different regional traditions. On presently available evidence, the "nuclear area" of the Near East must be extended into southeastern Europe, to account for evidence there of advanced, domesticated cattle a millenium earlier than anywhere in western Asia. Future excavations will probably reveal that the "nuclear area" included parts of Transcaucasia and Iran, beyond the Zagros ranges to Turkmenistan.

The basic environmental requisites for the complex of agricultural traditions of the Near Eastern hearth area are fairly simple: a winter growing season with sufficient moisture for dry farming. This submediterranean environment now extends through Anatolia into the Aegean world, and into certain uplands of the central and western Mediterranean Basin. The essential restricting factor would seem to have been the availability of suitable domesticates. In the case of potential herd animals, the situation is still rather fluid, with few limitations. In fact, it now seems probable that cattle were first domesicated in the Aegean area and possibly also in some part of Africa. The progenitors of our sheep and goats remain a puzzle, and the last word has not yet been said on the locus of domestication of the first farmyard pigs. If, indeed, wheat was the original cultigen in the Near East, the locus of first domestication was more restricted. But the abundance of wild wheats in natural habitats may be exaggerated, for the wild wheats now thrive in deforested areas first made available by human interference in historical times. In undisturbed woodlands, wild wheat would hardly be so abundant as to permit a subsistence based primarily on the harvesting of such primitive stands.

In concluding, we make a plea for the primary relevance of cultural traditions in agricultural origins. The Near Eastern-Aegean "nuclear area" must have contributed a number of local subsistence patterns that were ultimately adapted and fused to a hybrid, food-producing economy, with exchange and competition between neighboring groups.[7] Just as the subtropical forests of Transcaucasia may have provided orchard trees and the temperate woodlands of Macedonia the domesticated cow, other meso-environments may have contributed not only to the array of cultigens but also to other facets, such as manipulation, preparation, patterns of complementary cultivation, and above all, dietary preferences. Hopefully, renewed archeological search will extend beyond the established village farmers to those groups of more diversified food-collectors who made the first steps to agriculture possible in the Near East, and who each put their stamp upon the subsistence economy that subsequently diffused through mid-latitude Eurasia.

7. Recent dating of bone collagen from unquestionably domesticated sheep and cattle in southeastern Europe provides substantial grounds for reconsidering the role of the Aegean region in the origins of animal domestication (Reiner Protsch, *Radiocarbon Dates for some of the earliest domesticated animals in Europe*, unpublished M.A. Thesis, University of California, Los Angeles, 1970). Meticulous processing of domesticated sheep and cattle from Argissa-Magula (Aceramic level) gave unadjusted dates of 6180 ± 100 B.C. (UCLA-1657A) and 6040 ± 50 B.C. (UCLA-1657D) respectively. Similarly, domesticated cattle from the Neolithic site of Obre I, near Sarajevo in Jugoslavia, yielded C^{14} ages of 5290, 4760 and 4200 ±60 B.C. (UCLA-1605 H, G, I). Further dating of early domesticated animal fossils from both the Near East and Greece should help to resolve many of the remaining problems of where and when the herd animals were first domesticated.

THIRTY-THREE

Agricultural Dispersals into Europe and Northern Africa

INTRODUCTION

Village farming communities had been established through much of the mediterranean woodlands of western Asia and Greece shortly after 7000 B.C., and a millenium later a rapid dispersal of agricultural techniques was well under way. Food-production had been introduced into the Danube Basin as far upstream as Vienna by 5000 B.C., at which time the indirect archeological evidence points to agricultural or herding populations in the western Sahara, in the central Mediterranean region, and probably also in parts of temperate Asia. Simultaneously the Hassunan farmers of Mesopotamia (ca. 5900–5200 B.C.) pushed out beyond the grassland fringe of the Near Eastern woodlands into the river floodplains. By 4000 B.C. food-production was a widespread phenomenon throughout Northern Africa and temperate Europe, with offshoots of the western agricultural conplex almost certainly established as far east as the Indian subcontinent and northern China.

This dispersal of agricultural traits was not simply a matter of cultural diffusion. Each new environment required a number of new ecological adaptations: *(a)* The winter-cold, summer-moist lands of temperate Europe were unsuitable for mediterranean farming techniques and, initially, for the available cultigens; cereal mutants had to be developed that were tolerant of the new climate and new parasites; farm animals had to be selected according to new criteria, and the whole schedule of farming revised. *(b)* The arid Saharan world was clearly not suited to cereal-farming, except in an incidental way. How, then, did food-producing traits disperse so early and so rapidly in this inhospitable region? The answer seems to lie in a nomadic way of life, emphasizing herding, and probably based largely on older food-collecting habits. *(c)* The Hassunan villages of Mesopotamia, and their contemporaries and successors, are often found beyond the possible limits of dry-farming. The close correlation of such sites with riverine locations speaks for irrigation farming, probably developed between 5500 and 4000 B.C., when the major centers of population shifted from the wooded foothills onto the lowland floodplains. Here again the whole way of life had to be readapted to a summer growing season in the wake of spring floods, to heavy, and often saline, alluvial soils, and to concentrated settlement in an environment liable to pollution and devastating epidemics.

Diffusion of even a single trait is a complicated process that is difficult, if not impossible, to understand in any prehistoric context. The nature of agricultural dispersals is by far the most complex of any such problem. Whether we are dealing with migrations of races or of ethnic groups, of subsistence patterns or of ideas, is wholly unclear at this time. Archeological research is still preoccupied with tracing, recognizing, and dating the movement of agricultural traits across the face of the Old World, and any comparison of the cultural inventory between sites of cultures separated in time or space remains casual and subjective. Consequently the real character of agricultural dispersals continues to elude us.

ECONOMIC MOTIVES FOR AGRICULTURAL DISPERSAL

It is widely implied or accepted that agricultural dispersals were a result of economic stress in one form or another: environmental change, depletion of agricultural resources by overuse, or over-population.

One of the earliest theories links agricultural dispersals to that ancient scapegoat, "postglacial desiccation" in the Old World subtropics. In particular, Childe (1925, 1929) thought that impending "post-pluvial" desiccation in the Near East continued after the first general and successful steps to plant and animal domestication. The food-producing peoples expanded rapidly in numbers but were faced with a deteriorating environment. Desiccation eventually caused or (as others have put it more cautiously) "played a part in" the rapid expansion of Neolithic peoples and cultures into the moister lands of Europe. So, for example, Coon (1939, pp. 60–65) suggested that the dispersal of the Mediterranean race from the Near East (apparently in part associated with early agriculture) was a consequence of desiccation incident upon the close of the Pleistocene. Even at a much more recent date, Childe (1958, p. 54) still suggests that postglacial desiccation of the Sahara promoted ethnic and cultural movements from North Africa into Spain in the fifth millenium. In practice these arguments have no foundation in fact. As is discussed further below, the dispersal of herding groups through the Sahara coincided with one or more intervals of moister climate, so that movements were facilitated by a climatic amelioration, rather than triggered by environmental deterioration. Elsewhere, as in western Asia, agricultural traits diffused from humid into semiarid or even arid environments, which is inconsonant with a hypothesis of desiccation.

A second theory about agricultural dispersals can be formulated by implication from primitive, shifting agriculture (see Watters, 1960, with references). Slash-and-burn farming, with digging sticks or hoes, is today largely confined to tropical woodlands, and it would, of course, be misleading to explain early prehistoric farming in the Near East and Europe by analogy with Bantu subsistence in Africa or Quechua agriculture in South America. But there may be some relevant points. The earliest farmers necessarily practiced a very extensive land use that, at least initially, must have led to a depletion in the fertility of the more workable soils. Before the concepts of fallow and of soil replenishment were understood, it is possible that the best lands adjacent to a settlement were soon exhausted within the limits of existing technology and organization. Although shifting agriculture cannot be verified or even inferred with any confidence for the Near East and Greece, settlements in temperate Europe were far less stable. Under such circumstances, the colonizing of fresh lands should have been an appealing solution for dynamic groups living near the margins of the more densely settled agricultural lands.

Perhaps the most popular concept of early agricultural dispersals is associated with expanding agricultural populations, incipient overpopulation at a new level of carrying capacity, and with draining off of the surplus population into daughter colonies in unsettled lands. Presumably drawing its inspiration from European expansion into the thinly populated New World, with the technologically superior invaders rapidly displacing, absorbing, or dominating the indigenous peoples, this view was repeatedly emphasized by Gordon Childe (1924, 1929 and later syntheses; see also Sauer, 1947; Watson and Watson, 1969, pp. 96 f., 100 f.). The invention and adoption of new tools and a new economic subsistence is assumed to promote a great increase in population, made possible by the increased and more reliable food supply. Food production per unit area would be much greater, and even during a bad crop year a certain amount of food would be available. There would not be complete dependence on the seemingly erratic movements and biological cycles of wild game. Life and death would no longer be so precariously balanced; birth rates would increase and infant mortality decline. However, when a settlement reached its new carrying capacity at agricultural subsistence, the rate of increase had to level off, either by emigration or by higher mortality rates. As long as fresh lands remained that could be cleared and planted by fire and wooden or stone tools, the agriculturalists would probably send out daughter colonies that supplanted or absorbed the sparsely settled food-gathering populations.

There is no substantive reason to question this attractive hypothesis of chronic overpopulation, but there are grounds for a more cautious interpretation of the archeological evidence. For one, the prehistoric record of the eastern woodlands of the United States shows local, intensive settlement of great stability almost entirely independent of agriculture, with subsistence based on intensive "harvest" exploitation of select natural food resources (Struever, 1968, and personal communication). This warns against *a priori* assumptions that all indigenous Mesolithic peoples were thinly settled. At the same time, our picture of dense peasant populations in western Asia or Europe today may have biased our estimates of population densities among early agriculturalists. During the first few millenia, natural food resources commonly remained significant, and it is unlikely that the earliest farmers subsisted primarily on cultivated crops and herd animals. As a means of subsistence, early agriculture was neither intensive nor efficient, and populations probably increased far more slowly than has been suggested. The size and density of full agricultural settlements in the Near East (after 6000 B.C.) and in temperate Europe (after 4500 B.C.) are sufficiently great—by all standards—to prove that, ultimately, agricultural populations were remarkably numerous. However, the evidence also indicates that most villages had decidedly mixed economies before 6000 B.C., and that in the areas of recent dispersal in Europe and, to a far greater degree, in the Mediterranean Basin and the Sahara, mixed subsistence forms were the rule to a much later date. A last point of caution can be drawn from a more searching look at New World colonization after A.D. 1500. Recent European settlement of the woodlands west of the Appalachian watershed (after 1790), of the plains west of the Mississippi (after 1840), of the Campos of Brazil and the Pampas of Argentina (since the eighteenth century) were not the result of overpopulation in areas of older European colonization. Expansion was, to a degree, influenced by a continuing influx of European colonists into the established lands, but most of the new settlement was undertaken by small, frontier-oriented groups for a great variety of personal reasons and ambitions.

With these reservations in mind, let us now turn to early agricultural dispersals in the Old World, emphasizing the patterns of settlement and the ecological problems.

SAUER'S THEORY OF AGRICULTURAL DIFFUSION FROM SOUTHEAST ASIA

C. O. Sauer (1952) has suggested that the primary hearth of first domestication was found in Southeast Asia, while several minor or "derivative centers of additional domestications" are postulated for India, the Near East, Ethiopia, and West Africa. Following E. Hahn, Sauer believes that vegetative planting of tropical tuber plants may have been the easiest and earliest step to domestication, and that this abstract concept subsequently spread through the Old World. Characteristic of this Southeast Asian hearth in Burma and adjacent areas were household animals such as dog, pig, fowl, duck, and goose; nonseed, vegetative root plants such as banana, aroids, yams, sago, pandans, bamboo, sugar cane, and breadfruits. Postulated for the derived Indian-Himalayan center are plants such as the millets, pulses, gourd, jute, and other fiber plants, as well as some herd animals: goat, sheep, zebu, buffalo and yak. The only herd animals allotted to the Near East are cattle, together with seed plants such as the wheats, grape, olive, fig, and flax. For Ethiopia, these additional domesticants are thought to include teff, sorghum, cotton, and sesame; for West Africa, the guinea hen, yam, and bush pig.

Although Sauer's ideas are only presented as a suggestive sketch, the sequence of archeological events presently available from the Near East (see chapter 32) and India (see Sankalia, 1962) suggest that agricultural origins in the former area were an essentially independent innovation. However, recent work indicates strong possibilities that various leguminous plants were domesticated in northern Thailand by 7000 B.C. (Gorman, 1971; Solheim, 1969), thus lending support to Sauer's concept of advanced fishing and planting populations in southeastern Asia.

An elaboration of Sauer's dispersal concepts is due to H. von Wissman (1957), who outlined several successive nuclei of cultural diffusion in their geographical characteristics: *(a)* the tropical forests along the rivers and coasts of the Bay of Bengal: fishers and planters; *(b)* the forest-steppe and savanna of India: seed-planters with millets and oil plants; *(c)* the subtropical highlands of Afghanistan: sheep and goat farmers; *(d)* the small oases of the highlands and deserts of western Iran and Armenia: wheat and barley farmers. From here the alleged wave of dissemination entered Mesopotamia, which is not considered a center of agricultural origins but rather of technological invention.

Several elements stressed by Wissman are: *(a)* Each nucleus sent out waves of dissemination that may have caught up with each other or may have lost some cultural elements upon entering a different climatic region. Such waves were taken over, transformed, or rejected depending on physical or human factors. *(b)* Major movement of cultures is postulated in the wooded steppes where the soil is rich and supposedly easy to work. *(c)* The movements are compared with postglacial climatic fluctuations: (1) the Holocene thermal maximum (ca. 6200–1000 B.C.) may have permitted the spread of food production over the cold Central Asian mountain zone; (2) a moist spell in the third millennium may have established agricultural contacts, across the Central Asian deserts, possibly leading to the origins of horse nomadism. *(d)* The postglacial rise in world sea level was responsible for "burying" the archeological remains of the presumed late Pleistocene fishers and shell gatherers of southeastern Asia through marine submergence or intensive alluviation in lower stream courses. Reduced floodplain alluviation after 3000 B.C., when modern sea level was attained, may have been related to the beginnings of settlement and rapid technological advance in the lower valleys of the Tigris-Euphrates and the Nile.

Although Wissman's views are interesting and deserving of attention, they go far beyond the available archeological evidence and can therefore only be rated as a hypothesis.

EUROPEAN CLIMATE DURING THE ATLANTIC PHASE

The original dispersal of agricultural traits in Europe coincides with the warm, moist Atlantic phase (ca. 6200–3300 B.C.). The Scandinavian glacier had completely disappeared, and many mountain glaciers of the Alps were smaller than they are today while others disappeared. The botanical evidence suggests a considerably warmer summer climate than today's (see Firbas, 1949–52; Frenzel, 1960b, 1966; Iversen, 1960). So, for example, the altitudinal tree limit was 200–300 m. higher than today's in the Scandinavian highlands and in the Sudeten ranges, 300 m. higher in the northern and southern Alps. Various water plants and trees requiring considerable summer warmth occurred at higher elevations or at higher latitudes than is the case today. Tree pollen occurs in certain strata of bogs in the north European tundra, while plant fruits of now sterile perennials have been found on the Arctic islands. In fact a third of the 125 species of Spitsbergen do not reproduce under present climatic conditions. Massive oaks grew beyond the present limit of oak in northeastern Russia, while the hazel was found considerably north of its present distribution in Scandinavia (Fig. 91), and even the submediterranean wild grape *(Vitus silvestris)* thrived in southern Sweden. Particularly illuminating is a comparison of growing season temperatures at the northern limits of hazel *(Corylus avellana)* distribution in Scandinavia, as shown in the tabulation:

Mean Temperature °C.	Apr.	May	June	July	Aug.	Sept.	Oct.
Former limit	0.3	5.5	11.7	13.7	11.8	7.8	1.7
Present limit	2.5	8.2	14.0	15.8	14.1	10.1	4.5
Difference	2.2	2.7	2.3	2.1	2.3	2.3	2.7

From this it may be concluded that in comparison with today's temperatures, summer temperatures in mid-latitude Europe were at least 2° C. warmer during the Atlantic (ca. 6200–3300 B.C., but with a minor readvance of the Alpine glaciers ca. 5500 B.C.) and, following a major cold oscillation ca. 3300–2500 B.C. (see Frenzel, 1966, with references), again during the second part of the Subboreal (ca. 2500–1000 B.C.). During this last phase the Alpine tree-line reached its highest elevation, about 300–400 m. above the modern tree limit (Lüdi, 1955). A greater melting of the world's residual glaciers is also indicated for parts of the Subboreal, with sea level up by as much as 3 m. about 2000 B.C., and again by almost 2 m. about 1000 B.C. (Lind, 1969). Conditions were analogous to those of an interglacial maximum, but the time interval was shorter.

The first part of the Atlantic was comparatively moist, with colonization of the drier lowland basins in central Europe by moisture-demanding trees such as alder, spruce, and fir, and with the appearance of raised bogs in former lakes and marshes (Firbas, 1949–52, vol. I, p. 290 f.). Comparable patterns of vegetation change occurred in Scandinavia, where the early Atlantic (ca. 6200–4600 B.C.) was relatively moist, the later Atlantic (ca. 4600–3300 B.C.) drier (Nilsson, 1964b). Furthermore, a temporary extension of the forest into both the present tundras and the steppes is indicated by Frenzel's (1960b) palynological reconstruction of the Atlantic vegetation of Russia. All in all, the increased evaporation reduced stream discharge by at least 30 per cent in northern Europe (Aario, 1969).

ECOLOGICAL ASPECTS OF EARLY AGRICULTURAL SETTLEMENT IN EUROPE

Agricultural settlement of temperate Europe coincided with the Atlantic, at a time when the forest composition was mainly that of a mixed oak woodland, with oak, elm, lime, ivy, and alder dominant in the west, and pine more prominent in the east. Waterbolk (1968) suggests that the Boreal-Atlantic transition brought about an environmental deterioration

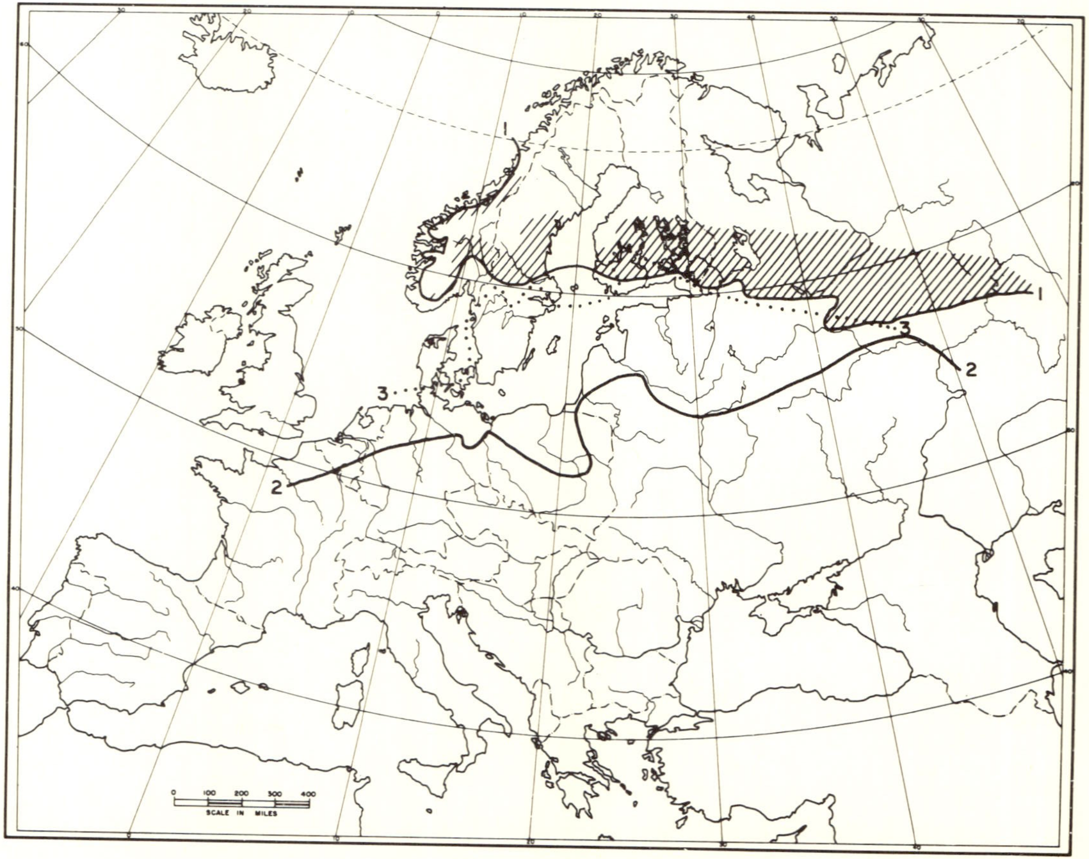

Figure 91. Temporary northward migration of European plant species during the Atlantic phase (after Fries, 1963, and Walter, 1954): (1) present northern limits of hazel (Corylus avellana), with extended Atlantic range, (2) present northern limits of the water chestnut Trapa natans, (3) Atlantic limits of Trapa natans.

unfavorable to Mesolithic populations. In western Europe the fairly open pine-hazel woodlands were replaced by more shady, mixed oak forests, reducing the grazing resources and biomass of larger game animals such as wild cattle, elk, and red deer. At the same time the formation of acid bogs in former lakes and marshes was disastrous to fish and waterfowl. The North Sea was enlarged by the encroaching marine transgression that brought salt water into the Baltic. These changes appear to have favored a shift of settlement to riverine and particularly to coastal areas, where waterfowl, fish, and sea mammals were available and allowed a higher degree of sedentary occupation (Waterbolk, 1968). Whether the continental interiors had been largely abandoned prior to the advent of Neolithic colonists is quite another matter, favored only by negative evidence of a circumstantial nature.

The early agricultural peoples in southeastern Europe settled in fairly compact villages, with a large number of single-room structures presumably housing "restricted" families; construction involved baked mud, with a light framing of poles and wickerwork, gabled or pitched roofs, and household furniture including chairs, benches, and couches (see Piggott, 1965, with references). Larger villages on this pattern were occupied over long periods, as indicated by stratigraphic sequences spanning several millenia, much like the Near Eastern tells. This permanence of settlement appears reasonable in view of the effort invested in settlement construction. Herding of cattle, pig, goat, and sheep was a prominent activity, while barley, several wheats, peas, lentils, and vetch were cultivated. The basic patterns of subsistence and settlement here in the mediterranean and submediterranean environments of southeastern Europe were comparable to those of western Asia and, at a later date, of other parts of the Mediterranean Basin. Agriculture was still largely in the initial "hearth zone," although it would be presumptuous to draw simple ethnic interrelationships between, for example, the Aegean world and interior Anatolia.

Farther north, in the loess plains and in the valleys of the Danube Basin, there are considerable differences. Settlement forms change, and differing house sizes and functions indicate other social patterns, while the relative proportions of domesticated animals change, as does the importance of game and fish. The contrasts of these new agricultural groups become crystallized ca. 4500 B.C. in the so-called early Danubian culture (also referred to as Linear Pottery or Bandkeramik settlements) (see Buttler, 1938; Narr, 1956; Piggott, 1965; Waterbolk, 1968, with references).

The Danubians were village farmers with a subsistence economy based on shifting agriculture. Three species of wheat, as well as barley, lentils, flax, beans, and peas, were cultivated and presumably formed the staple diet, judging by the quantity of milling and pounding stones. Stone adzes were probably used for felling trees. The cow was the most common domesticated animal, with the pig in second place. Sheep, goat, and dog were of minor importance. The refuse pits show evidence of hunting activity, with red and roe deer, boar, aurochs, and woodland bison as favored game, although wild animals account for less than 10 per cent of the total bone counts. Lacking agricultural implements, it is no more possible to reconstruct the patterns of cultivation than those of herding. However, a Danish experiment indicated that stone axes are adequate to clear lightly stocked oak woodlands with the aid of fire (Iversen, 1956). Seed can subsequently be sown into the warm ash and raked over with a forked stick. This method was found to yield good crops of naked barley, einkorn and emmer wheat on leached, podsolic soils; presumably more fertile soils could provide more sustained yields, particularly if the grain were planted in holes within the topsoil, rather than broadcast (see Steensburg, 1957).

The Danubians occupied long, rectangular, gabled houses of wood and wickerware, measuring 5 to 6 m. wide, and 15 to 40 m. long. Vertical posts supported the walls and roof. In addition to these timber-built longhouses, which numbered 10 to 20 per village, settlement areas were dotted with storage buildings, ditches, and pits, all loosely arranged in subgeometrical patterns, and enclosed by a fence. The Danubian longhouses probably served in part as animal stalls and storage areas, but even so must have housed "extended" families of four or five units with 20 people or so. Consequently these villages suggest population groups of 200 to 400 or more. After 10 years or so, the village sites were abandoned, with repeated occupation after fallow intervals of perhaps 50 years, although the degree of permanence of villages seems to have varied from one region to another.

The sites of the Danubian culture are very strictly limited to loess areas in the Low Countries, Germany, Poland, Austria, Czechoslovakia and Hungary (Fig. 92). No sites occur north of the margins of the Würm till. For the most part the warm, dry lowland plains or river terraces were selected, and within these, the loess areas (Gradmann, 1906, 1936). The natural vegetation of the central European loess lowlands has long been the subject of controversy. Gradmann (1933) argued that grasslands, parklands, or open woodlands were still widespread in late prehistoric times, and that such lands were optimal in terms of better soils, easier cultivation, good pasture, and more bountiful game. Others, including Nietsch (1939) and Schott (1939), have argued that more or less closed forests dominated even the drier basins, requiring clearance

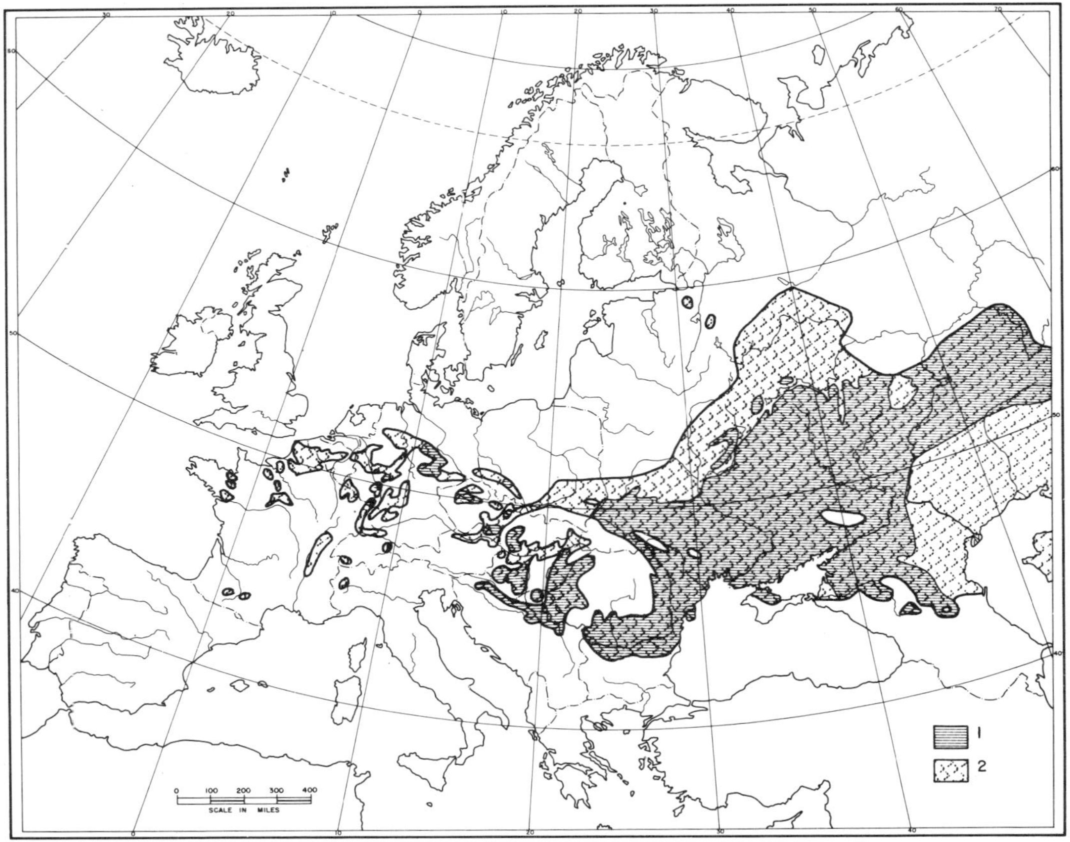

Figure 92. Extent of loess and chernozemic soils in Europe: (1) Chernozems, degraded chernozems, and kastanozems, (2) Loess; where not shaded, with braunerde, podsolic, or semidesert soils. Loess simplified after Frenzel (1959), soils after Ganssen and Hädrich (1965).

by felling or burning. Godwin (1944) was able to verify this second point of view in the case of England.

Palynological evidence (Firbas, 1949–52, vol. I, p. 356 ff.) does not support widespread grassland or parkland during the Atlantic, even though the mixed oak woodlands on comparatively dry loess soils may have been lightly stocked. On account of the gradual decrease of *Artemisia* in the pollen record, Firbas believes that all but exposed bedrock, talus slopes, and stony gravel or sand surfaces were colonized by tree vegetation early in Holocene times. But such remaining natural gaps in the forest cover would obviously not have attracted settlers. Firbas concludes that the moister loess lowlands (wherever annual precipitation exceeds 500 mm. today) were occupied by closed mixed oak forests during the Atlantic, although the drier basins probably had a parkland or open woodland vegetation. Areas qualifying as comparatively dry are the interior basin of Bohemia-Moravia, the Elbe-Saale plain, the Upper Rhine basin, and the Hungarian plain. In fact, the Hungarian lowlands have remained an oak parkland or steppe throughout historical times.

Soil studies appear to substantiate Firbas' conclusions. Loess sediments are highly permeable and evaporate more soil moisture than any other sediment, so that loess soils are comparatively dry in the edaphic sense and do not favor tree growth. The climatically drier loess lowlands commonly have soils of the "degraded" chernozem type. Such chernozems originally developed under grassy vegetation with dry, warm summers—presumably during the continental climate of the Preboreal and Boreal. Subsequent woodland invasion during the moist, maritime Atlantic led to carbonate solution, increased acidity, and chemical weathering, with oxidation and some leaching (Scheffer and Schachtshabel, 1960, p. 275 f.; Wilhelmy, 1950). These soils prove the former existence of grasslands in certain dry basins, at least until the beginning of the Atlantic. Consequently, a fair amount of parkland or open woodland was available to the Danubian colonists in the south and east (see extent of chernozemic soils shown in Fig. 92).

In short, the earliest agricultural colonists entered the central European area during a period of optimal warmth and a comparatively dry climate during the second part of the Atlantic, when moisture and runoff were further diminished because of increased evaporation. The settlements of the Danubian farmers are sharply restricted to loess sediments (Fig. 93), which obviously provided greater soil fertility. At the same time, these often were areas with parkland vegetation and calcareous, chernozemic soils, or otherwise they had base-saturated forest soils under closely stocked, mixed oak woodland. It is no mere coincidence that primeval settlement, loess, calcareous or basic soils, dry

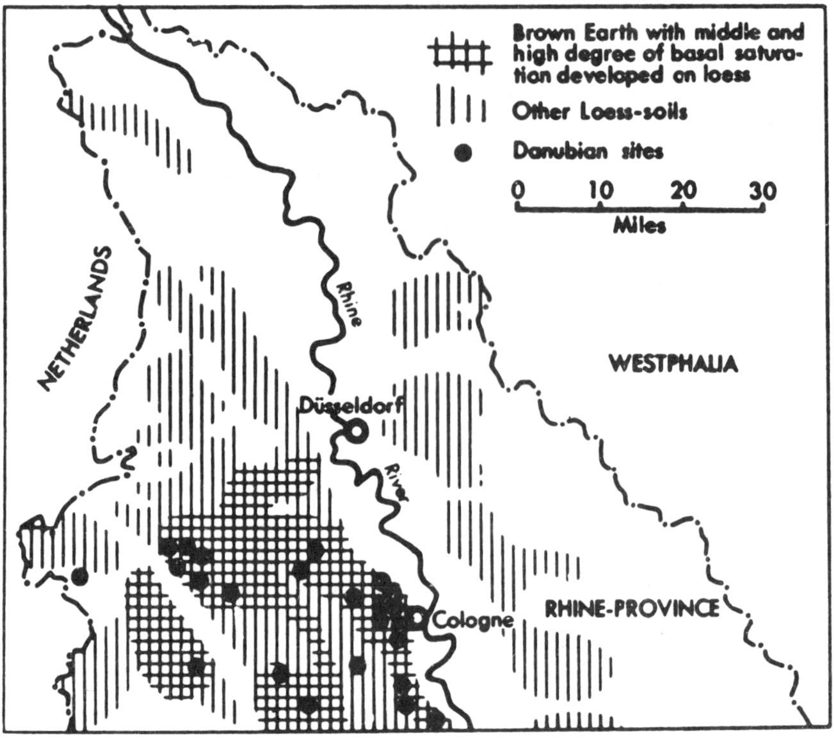

Figure 93. Danubian sites and soil types in the northern Rhineland (from K. J. Narr, 1956, copyright 1956 by the University of Chicago Press, with permission).

lowland basins, and comparatively open, oak parklands or woodlands should provide a common denominator for the earliest agricultural lands of mid latitude Europe. Only in later times, when less demanding crops such as rye, oats, or spelt had been developed, was colonization extended to the hilly tracts with more acidic and partly leached forest soils. Swampy terrain and heavy waterlogged soils were only occupied at a somewhat later date. To the north, settlement stopped abruptly at the margins of the clayey moraines and infertile outwash sands of the Würm glaciers. This suggests that natural soil fertility and easy workability were prime factors in determining Danubian colonization. Perhaps even the light woodland cover of the loess plains was deliberately used as an index of suitable farm land.

The new agricultural lands of mid-latitude Europe (Fig. 94) were not radically different from the subtropical or temperate woodlands of Asia Minor and Greece, particularly during the warm Atlantic and Subboreal

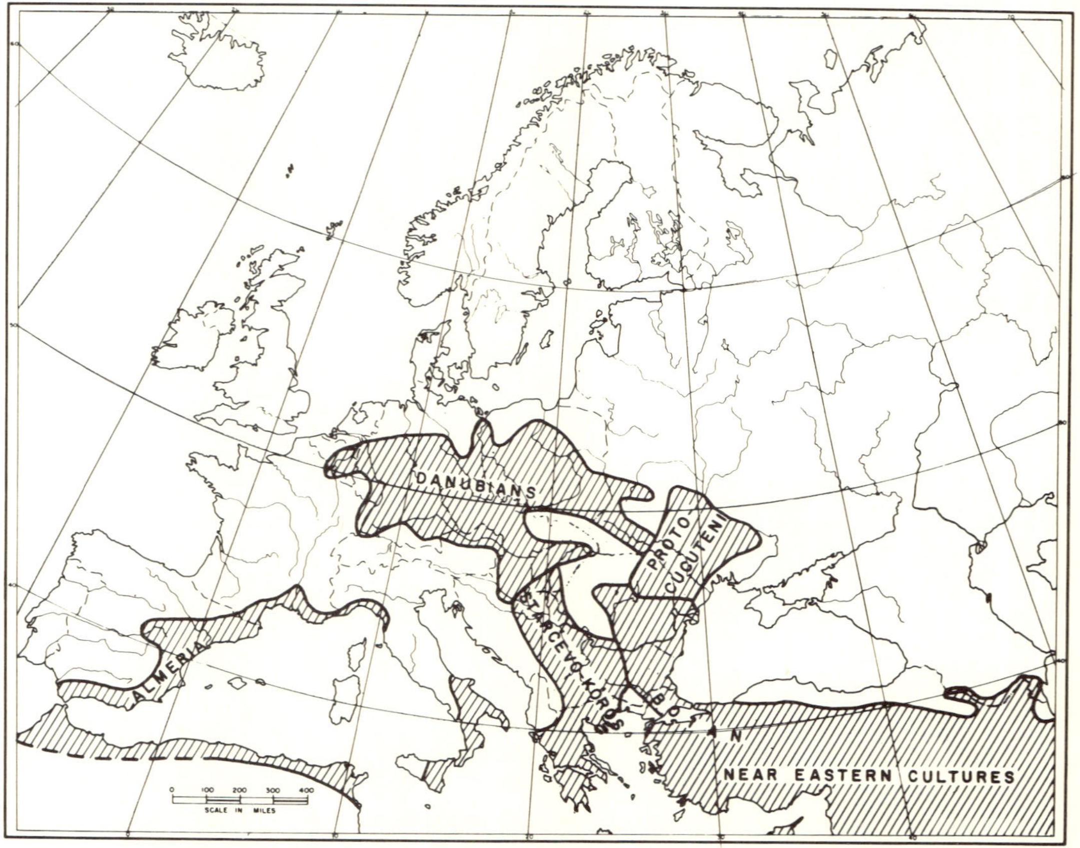

Figure 94. Early village farming cultures in Europe and adjacent areas ca. 4200 B.C. *Modified after Milojčić (1958) and Waterbolk (1968).*

phases. Despite an increase in winter cold and summer moisture, the landscape of the new environments was different in degree rather than in kind. It was probably not accidental that the pioneer farmers of Europe should select the environment most like that of their cultural antecedents: not the humid lands of the west, nor the cool, poorly drained till plains of the north, nor yet the snowy plains or open steppes of the east. Rather, the more intermediate environment of the Balkan peninsula and central Europe provided the most compatible solution in terms both of climate and edaphic factors. Just as open woodlands had probably witnessed the birth of agriculture in the Near East, they also provided the setting to the first agricultural venture into higher latitudes.

However, the change in crop ecology was important. Winters were cool rather than mild, whereas summers were decidedly moist. Severe and protracted frosts may have eliminated some winter crops from the array of domesticated plants, although "winter" wheat and barley are still commonly planted in autumn in much of central Europe today. Somewhere, however, the idea of spring sowing of mediterranean crops must have been experimented with and found to be expedient. Some of the evolution of new mutants and rapid hybridization of wheat species in temperate Europe may have resulted from deliberate changes in plant ecology at the hands of man—just like those accompanying the deliberate cultivation of oats and rye on marginal soils and in cooler climates a few millennia later. In other words, the Danubians with their different settlement forms and modified subsistence patterns represent a population with a new outlook and a social order adapted to a new environment.

Later agricultural dispersals initially show comparable differences that can be associated with the peculiarities of new meso-environments, as, for example, the Proto-Cucuteni and Tripolye in the loess steppes of Moldavia and the Ukraine, and the different Western Traditions along the wooded, maritime coastlands of western Europe (see Waterbolk, 1968, with references). In fact, the different rates of dispersal and the apparent termini to expansion may have been influenced primarily by environmental preferences and technological adaptations.

Returning to the initial question about the nature of agricultural dispersal in temperate Europe, we find that at any locality, wheat, barley, cattle, pigs, sheep, and goats seem to appear all at once. As David Asch (unpublished) has phrased it, "these domesticants seem to have been part of a tightly dependent system which 'diffused' largely as a unit." This would speak for population movements. However, intensive efforts still need to be devoted to detailed studies of animal bone and to the recovery of carbonized plant residues from Mesolithic and hybrid Neolithic sites before this picture is validated. The exceptional site of Lepenski vir, in an almost inaccessible Danube tributary valley near the Iron Gate, provides a fascinating example of apparent acculturation (see Bökönyi, 1970). The two lowest strata have abundant remains of red deer, aurochs, boar, fish, and domesticated dogs—presumed to have been eaten. The bones of the topmost stratum—still lacking in pottery—include 15 per cent domesticated cattle, with some sheep/goat, dog, and rare pig. The cattle indicate local domestications from wild stock, as well as hybridization between domestic cows and aurochs. It appears that all three occupation levels were contemporary with farming villages at no great distance.

SAHARAN CLIMATE DURING THE EARLY AND MID-HOLOCENE

Agricultural colonization of the Mediterranean Basin, in particular of the coasts of southern Europe and northwestern Africa, did not encounter appreciable environmental differences anywhere in the summer-dry subtropical woodland belt. The settlement of truly arid lands, such as the Sahara, did, however, require considerable ecological adaptation. Fortunately for the early agricultural colonists, the Saharan area enjoyed an abnormally moist climate during a time interval roughly synchronous with the Atlantic phase in Europe. The evidence in favor of several moist intervals during the early and mid-Holocene may be subdivided into three categories: faunal evidence, chiefly on the basis of rock-drawings; botanical evidence, both macrobotanical and palynological; and geological evidence, generally of a rather specific and detailed type.

A moister climate during late prehistoric times was first inferred from the widespread distribution of human artifacts and rock-drawings in desert areas, often many miles from existing waterholes. The wild animals shown in the rock art included gazelles, antelopes, and ostrich, as well as species associated with more luxuriant savanna vegetation: elephant, both the single-horned and the two-horned rhino, hippo, and giraffe. Certain paleoclimatic inferences can be attempted on the basis of these animal representations and their distribution and frequency (see Mauny, 1956; Butzer, 1958b), but indirect data of this kind is not conclusive. Consequently, it is fortunate that much geological and palynological data has been collected in recent years, often fixed by radiocarbon dating. As a result, the archeological evidence can now be seen from a new perspective.

The evidence for post-Pleistocene moist interludes is unsatisfactory and unconvincing along the Mediterranean borderlands, in Morocco, Tunisia, or the Cyrenaica. It is best developed in the Saharan highlands and along the major wadis systems or depressions that drain the higher country. So, for example, the Guirian alluvial terrace of the Saoura Valley and contemporary basin fills of the upland plains include evidence for accelerated alluvial activity and repeated, widespread lake or swamp formation, with dates of 7000, 4380, 4210, 3660, 3540, 2980, 2220, 1420, and 990 B.C. (see Chavaillon, 1964; Conrad, 1969). Pollen was examined from an exposure of gray sands, interdigited with organic horizons and capped by a travertine layer (Beucher, 1963). The limited pollen includes pine (*halepensis*?) and acacia (*raddiana*?), but few grasses. The travertine, however, is dominated by pollen of xerophytic species (*Ephedra,* Chenopodiaceae) with some dubious traces of birch, alder, hornbeam, and pine.

In the Hoggar Mountains (Rognon, 1967), silty-swampy fill accumulated in many of the valleys during Neolithic times, with lake carbonates dated 9630 and 6430 B.C. (Délibrias and Dutil, 1966). Pediment-cutting proceeded in the uplands, and organic swamp beds formed in some intradunal hollows of the lowland "sand seas." Dating here can be derived from the Erg Chech, due west of the Hoggar, where lacustrine limestones, organic sands, and tufas have been dated in three groups comparable to those of the Saoura region: 8150, 4520, 4470, 4410, and 1060 B.C. (Conrad, 1969). The rock shelter at Meniet in the Hoggar contains strata from which 87 pollen grains were identified by Quézel and Martinez (1958). Of these pollen 56 per cent belong to arboreal species. In order of numerical importance they include cypress, Aleppo pine, evergreen oak, wild olive, hackberry (*Celtis australis*), the thorn bush *Ziziphus,* juniper, and tamarisk. Macroremains of *Ziziphus,* lotus, and hackberry help substantiate the pollen record, which further includes cereal, grass, sedge, and *Artemisia.* The uppermost stratum at Meniet has a C^{14} date of 3450±300 B.C. (Délibrias et al., 1959) and contains bones of an extinct buffalo. Similar results were obtained from a sample of hyrax dung from Taessa in the Hoggar, at some 2,200 m. elevation. Pollen (3,000 grains) included similar genera as at Meniet, together with pistachio and walnut. A C^{14} date of 2730±300 B.C. was obtained for the dung. Limited numbers of pollen grains were also studied from sediments with cattle bones found below rock-drawings in the Tassili Mountains (Quézel and Martinez, 1961). These grains included Aleppo pine, evergreen oak, and cypress. A more impressive but incompletely published sequence is available from the nearby Acacus hill country, near Ghat (1,370 m). Here the rock shelter of Uan Muhuggiag (Mori, 1965, pp. 218-41) indicates two periods of occupation by food-producing people with domesticated cattle and sheep. In the earlier horizon (ca. 5600–3800 B.C.), *Typha,* a swamp plant, accounts for over

50 per cent of four samples in the horizon, with the desert shrubs *Aristida* and *Artemisia herba-alba* accounting for about 10 per cent each. Macrobotanical remains include abundant *Typha* together with acacia and some desert trees or shrubs. In the upper horizon (ca. 3500–2500 B.C.) there is little or no evidence of swamp plants, but *Artemisia herba-alba* attains over 50 per cent, *Aristida* 25 per cent, and *Artemisia campestris* up to 10 per cent.

As many other parts of the Sahara, the Tibesti area has a wealth of naturalistic rock-drawings pertaining to prehistoric hunting groups. Giraffe, elephant and, to a lesser extent, rhinoceros are fairly common among these drawings, both in the highlands and in the Borkou foothills. Although dated geological evidence is not yet available, elephant fossils have been reported, and palynological data are available from "Neolithic" cave sediments at Mosséi (Quézel and Martinez, 1958). The pollen count here was 60 per cent NAP and included 35 per cent *Acacia flava*. However, over 80 per cent of the macroremains pertained to pine and juniper.

Near the Djado Oasis, southwest of Tibesti, several lacustrine deposits with subfossil mollusca have been studied by Llabador (1962). All appear to have been laid down during the moister interval in "Neolithic" times. The Neolithic lacustrine beds of Adrar Bous, northeast of the Air Massiv, have yielded valuable paleo-ecological information concerning the local mid-Holocene environment. The fauna includes equids, cattle, antelope, wart hog, hippopotamus, tortoise, and crocodile. Two genera of fish, ostrich, and three species of gastropods complete the preliminary faunal inventory (Llabador 1962). Macrobotanical remains include the common reed and live oak. Two distinct palynological horizons are defined by Quézel and Martinez (1962). The lower consists of diatomite, attaining 3 m. in thickness, with 66 per cent of 251 identified pollen grains from NAP species. The upper horizon is a blackish swamp sediment of 10–30 cm. thickness, with 82 per cent of 192 grains NAP, almost exclusively chenopods. The arboreal species present in the diatomite include 13 per cent juniper, 12 per cent cypress, 5 per cent an extinct pine, 4 per cent myrtle, and 2 per cent pistachio. The limited arboreal species of the upper, swampy beds, dated 3180 ± 300 B.C. (Délibrias and Hugot, 1962), contain 12 per cent myrtle, 3 per cent cypress and 2 per cent juniper. A severe degradation of the open mediterranean-type vegetation is evident between the two horizons that mark the mid-Holocene moist interval.

Faure (1966) has studied the lacustrine deposits of the intradunal depressions of the Ténéré, phenomena having a wider distribution along the southern margins of the Sahara. Freshwater lakes, possibly interconnected with Lake Chad, were well developed ca. 7250–5000 B.C. (8 C^{14} dates); and a last lake episode is dated ca. 3550–1150 B.C. (4 dates). The Ténéré evidence largely parallels the sequence of post-Pleistocene lake levels of Chad (see chapter 20), with one high stand fixed by dates of 8210, 8150 (2), 7660, 7520, 7310, 7050, 6800, 5050 and 5040 B.C., the second by dates of 1550, 1430, 1210, 550 and 510 B.C. (Servant *et al.*, 1969). In the Senegal Delta the earlier wet phase appears to have been contemporary with the development of the deep red paleosol under moist, warm conditions, while the second wet phase coincided approximately with silt alluviation of the Senegal.

The Nile Valley provides further details and confirmation of several moist intervals (Butzer and Hansen, 1968). A period of accelerated wadi activity that began 9200 B.C. terminated by 6000 B.C. Shell proliferations suggest rather more vegetation in the wadis. A little later, ca. 5000 B.C., a red paleosol suggests a mat of vegetation and more frequent gentle rains. Finally, after a second dry interlude, accelerated wadi activity and extensive sheet washing—in the wake of sporadic but heavy and protracted rains—are indicated ca. 4000–3000 B.C. Historical and archeological documents suggest that the desert wadi vegetation of northern and eastern Egypt was more abundant as late as 2350 B.C., when the prevailing aridity was established (Butzer, 1959b). At the same time, spring activity in the Kharga Oasis was greater, allowing agricultural subsistence by Neolithic settlers, while the static groundwater table was higher in much of the Libyan Desert, presumably facilitating cattle-herding in now desolate areas (Murray, 1951; also Knetsch *et al.*, 1963, on the depletion of "fossil" water resources).

All in all, the Saharan evidence indicates two or three moister interludes during the early and middle Holocene. Dates are slightly at variance from place to place and since few of the local sequences are firmly dated, long-range correlations are difficult. Possibly three moisture peaks are indicated—ca. 8250 (9700?)–6400, ca. 5100–2200, and ca. 1600–500 B.C.—separated by drier interruptions, and followed, during the last 3 millenia, by conditions quite comparable to those of today. None of these changes in precipitation or effective moisture can be quantitatively estimated, and there is no evidence concerning possible changes of temperature. The impression obtains that the increase in moisture at the height of these moist intervals was ecologically significant, although not sufficient to qualify the "arid" nature of the climate. It appears that open woodland or parkland was present at edaphically favored localities in the high country, while fringing savanna-scrub and local swampy ground accompanied the major wadi lines and depressions. It is generally agreed that the paucity of modern vegetation in the Saharan highlands and wadis is due to human activity such as overgrazing and use of woody plants for fuel. However, the Sahara is exceptionally arid by standards of other world deserts, and this climatic aridity has not been accentuated by man. If left undisturbed over many generations, the plant life of the Sahara would indeed regenerate somewhat, but hardly in such a way as to permit wholesale faunal migrations or to provide resources for diverse agricultural and herding populations. The many categories of evidence for moist interludes during the early and mid-Holocene do provide evidence for real climatic variations.

MESOLITHIC AND NEOLITHIC GROUPS OF THE EASTERN ATLAS AND THE NUBIAN NILE

Northern Africa provides two interesting cases of Mesolithic groups that were resistant to the introduction of agricultural traits, and who subsequently maintained much of their cultural identity. Yet, in part, both of these groups are contemporary with the earliest food-producers in the Sahara. The first of these, the Capsians, occupied the mediterranean woodlands and steppes of the eastern Atlas, the second the well-watered Nile floodplain of Nubia. Both groups were apparently numerous and enjoyed a subsistence-economy based on intensive utilization of natural food resources. In their own, rather different ways they represent specialized food-collectors comparable to some of the eastern woodland cultures in the United States who "resisted" the innovations of food production.

The Capsians are recorded by literally thousands of sites in the hill country and on the Saharan slope of the Atlas Mountains in Tunisia and Algeria (Fig. 95) (see Vaufrey, 1933; Balout, 1952; Tixier, 1968, and personal communication; Camps *et al.,* 1968). The outstanding characteristics of these sites are great, open-air kitchen middens (*rammadiyat*) consisting of ash, terrestrial shells (*Helix* sp., *Rumina decollata*), flaking debris and artifacts, and limited amounts of animal bone. These "mounds" of ash and debris are found on hillsides and may be several meters thick, commonly extending over hundreds of square meters. In at least one instance a mound was later covered by 2 to 3 m. of soil wash before being dissected. Lacking palynological or systematic paleontological study, site distributions and the masses of terrestrial snails must tell their own story. The Capsian mounds are never found any closer than 70 km. to the coast, and the well-studied lithic assemblages show no relationships whatsoever with the coastal cultures (Ibero-Maurusian). Instead, sites are concentrated in the interior hill country, at elevations of 400 to 1200 m., in what were mediterranean parklands as late as Roman times. Noteworthy is that wood of juniper and Aleppo pine has been identified from the ash mounds. However, the southern range of the Capsian extended down into the Saharan lowlands, to below 200 m., and in one instance, to the southern side of the string of salt lakes, or chotts, that extend through central Tunisia and Algeria.

The early "Typical" Capsian has no acceptable C^{14} dates, although the estimated age ranges from about 8500 to 7000 B.C. The industry consists of small, relatively specialized blade tools (burins, backet bladelets and blades, end-scrapers), and sites are confined to the Algerian-Tunisian border country and the adjacent Saharan plains. The "Upper" Capsian has different lithic components, dominated by small notches and denticulates, truncated bladelets, and geometric microliths; five regional variants or facies are identified. This phase is dated by nineteen acceptable C^{14} dates ranging from 6530 to 4390 B.C., and probably covers a time span of 6850–4100 B.C. The Upper Capsians extended their range through most of the wooded Atlas of Algeria, almost to the Moroccan border. Skeletal finds point to robust members of the Mechta or Afalou type of *Homo sapiens sapiens* (Briggs, 1955).

The Capsians appear to have been intensive collectors, obtaining a major part of their food supply—at least seasonally—from terrestrial snails, supplemented by lizards, tortoise, small desert rodents, and occasional large mammals (*Equus, Gazella, Bos primigenius,* and the extinct buffalo *Homoioceros*). It is presumed that they also gathered nuts, acorns, fruits, and other vegetable foods in the mixed oak, pine and juniper forests of the high country during the drier part of the year. In fact, some cave sites are known there.

Obviously the Capsians did very well, only accepting a limited array of agricultural traits 1,000 years later than the adjacent coastal folk ("Mediterranean Neolithic") and 1,000 to 2,000 years after peoples in the Algerian Sahara. In fact, the earliest "Neolithic of Capsian Tradition" is found in the northwestern Sahara, as far south as latitude 27° N (Fig. 95). This Saharan variant now has seven C^{14} dates ranging between 5350 and 2980 B.C., with an inferred time span of 5550–2750 B.C.—corresponding almost precisely with one of the mid-Holocene moist intervals. The Neolithic successor in old Capsian territory has four dates spanning the period 4000–1250 B.C.

The Neolithic of Capsian Tradition retained the typical Upper Capsian tool inventory, but added stone celts or mullers (used to buff leather and to hammer ?), pottery, a variety of arrowheads, and a great range of utilized ostrich shell. The birds belonged to both the surviving *Struthio camelus* and the regionally extinct, smaller, blue-necked subspecies (*S.c. molybdophanes*), but it appears that only eggs, rather than the birds themselves, were taken. Drilling a small hole in an ostrich egg makes a perfect bottle, many of which were artistically engraved. Ostrich egg was further broken up and then retouched and polished or ground to form saucers, discs, or bottle-stoppers, tool-like edges, pendants, and beads. Sea shells, imported from the coast, were also used as ornaments. Clearly the Neolithic Capsians were very adept at conserving and transporting water, which may explain their expansion into and long persistence in the Saharan region. They have been linked with rock-drawings of domesticated cattle and sheep, and some cow bone has been recovered. Whether agriculture was practiced is still unknown, but intensive food-collecting must have remained the basis of their subsistence. Altogether, the Neolithic Capsians represent a Mesolithic group that, at a fairly late point, adapted a number of Neolithic traits without losing their basic cultural identity. These Capsians provide a fascinating counterexample to the intrusive Danubian Neolithic.

Turning to the Nubian Nile Valley, we find equally specialized late Paleolithic and Mesolithic populations of considerable regional diversity, intensively occupying the Nile floodplain in relatively large groups and fairly dense concentrations. These riverine people exploited the fish, invertebrates, mammals and waterfowl of the river, as well as the large "savanna" mammals that periodically grazed on the vegetation of the alluvial flats and watered at the banks of the Nile and its summer overflow channels. In some instances, as on the Kom Ombo Plain north of Aswan, there even are sizable shell middens (dated ca. 15,000–10,500 B.C.), and grinding stones were common with some groups, although their use is problematical (see Butzer and Hansen, 1968, ch. 4). A post-Pleistocene record, preceding the Dynastic period, is unfortunately not preserved north of Aswan, where the Nile floodplain appears to have been low at that time. But in southernmost Egyptian and northern Sudanese Nubia there is a rich archeological record of Mesolithic and early Neolithic groups (see Fig. 95).

There are several Mesolithic industrial complexes in the Nubian Nile Valley, the most widespread of which is the Qadan or Wadi complex (Wendorf *et al.*, 1968, vol. 1, p. 98; vol. 2, pp. 564 ff., 954 ff., 1050 f.; Irwin *et al.*, 1968), which has C^{14} dates of 12,550, 9460, 9250, and 4480 B.C. The oldest sites are found within or eroded from nilotic silts at + 30 m. relative to the modern floodplain, the youngest in relation to silts at less than + 10 m. In addition to these extensive riverine campsites, there are abundant burials, in shallow oval pits, including some cemeteries with a high proportion of victims of a violent death. The people clearly belong to the Mechta variety, with a long and robust skull, a short, broad face, and well-developed supraorbital ridges with low, rectangular orbits. The teeth show a high degree of attrition but a low incidence of caries, the characteristic pattern of a nonagricultural popu-

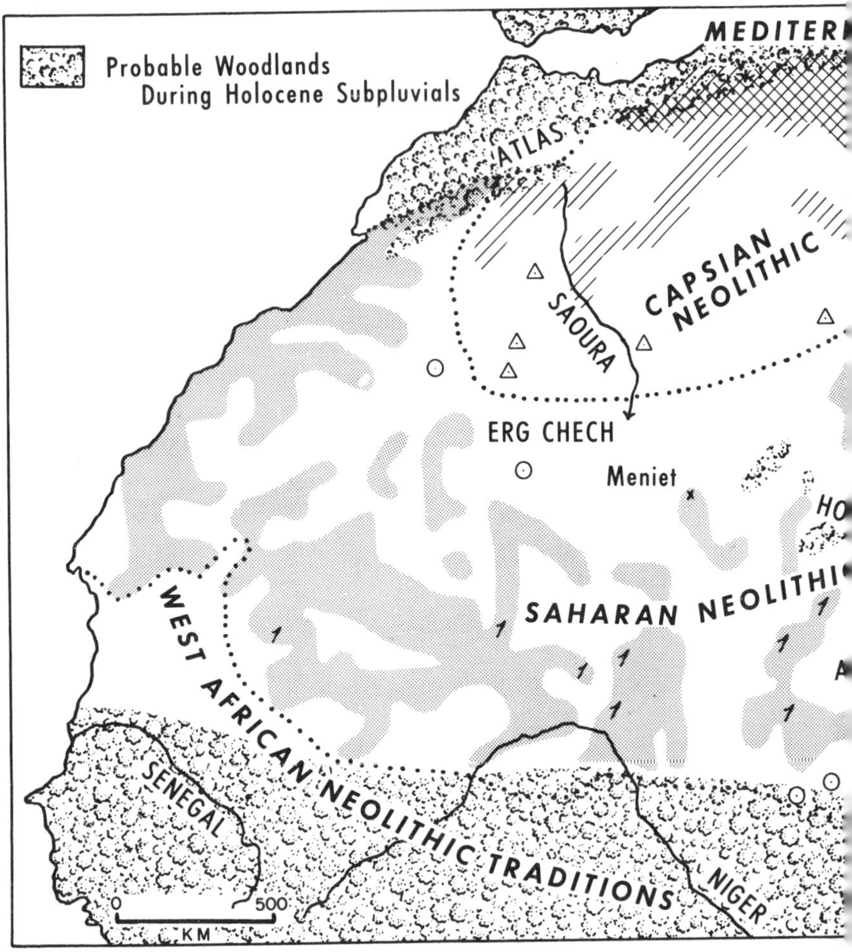

Figure 95. Mesolithic and Neolithic settlement in the Sahara (archeological data after Clark, 1967a; Wendorf et al., *1968; and others).*

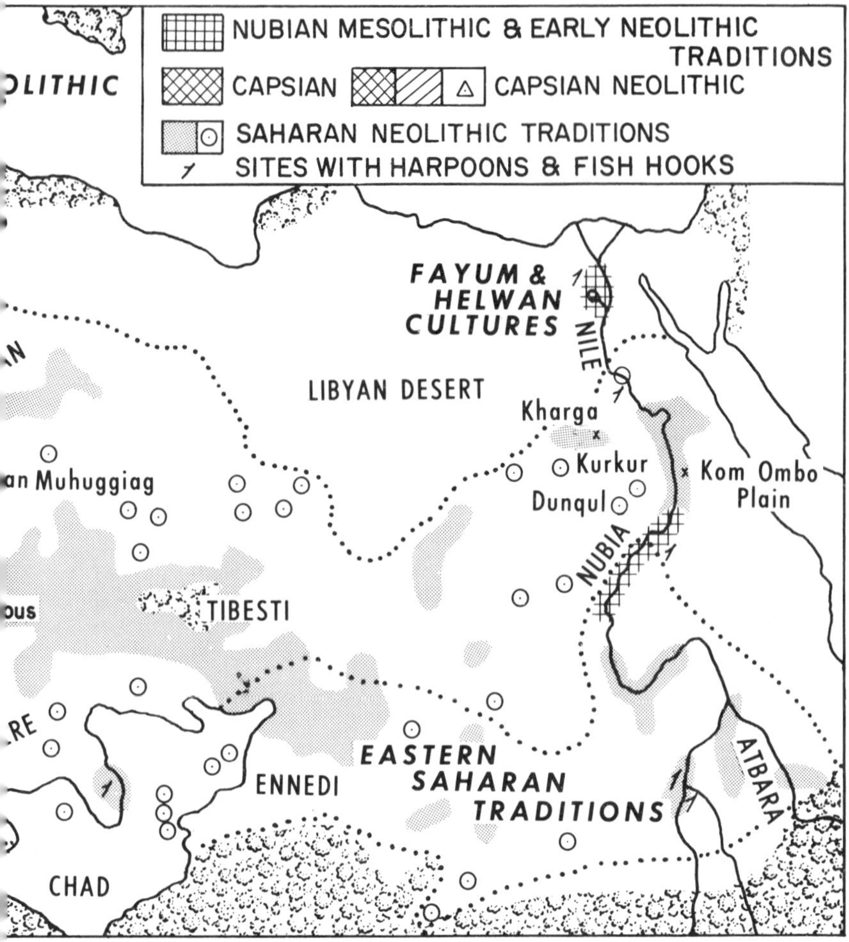
NUBIAN MESOLITHIC & EARLY NEOLITHIC TRADITIONS
CAPSIAN
CAPSIAN NEOLITHIC
SAHARAN NEOLITHIC TRADITIONS
SITES WITH HARPOONS & FISH HOOKS
OLITHIC
FAYUM & HELWAN CULTURES
NILE
LIBYAN DESERT
Kharga
Kurkur
Kom Ombo Plain
Dunqul
an Muhuggiag
TIBESTI
ous
NUBIA
RE
ENNEDI
EASTERN SAHARAN TRADITIONS
ATBARA
CHAD

lation. The lithic assemblage is dominated by scrapers, geometric microliths, burins, and backed and truncated bladelets, with very generalized affinities to the Magosian industry of eastern Africa. In addition to fish, *Bos primigenius,* hartebeest (*Alcelaphus buselaphus*), and gazelle were hunted in large numbers, and to a lesser extent, wild ass, hippo and jackal—all animals locally available in early historical times (Butzer, 1959b). Far more localized is the Arkinian industry, confined to one extended group of small sites with a date of 7440 B.C. (Wendorf *et al.,* 1968, vol. 1, p. 98; vol. 2, pp. 651 ff., 1051 f.). These were originally located on the banks of the Nile, within the Arminna Member (Gebel Silsila Formation; Butzer and Hansen, 1968, p. 276 ff.) at 11 m. or so above the modern floodplain. The industry is dominated by end-scrapers, backed bladelets, geometric microliths, and grinding stones, with affinities to contemporary coastal cultures in Algeria. The game bag includes hartebeest and wild cattle, with some hippo, gazelle, and jackal. Only a little less restricted is the Shamarkian Complex (Wendorf *et al.,* 1968, vol. I; p. 98, vol. 2, pp. 695 ff., p. 1052 f.). The compact camps and workshop sites are found on the surface of older nilotic beds, at short distances from the Nile, or in overbank silt and sand strata of the Nile at 5 or 6 m. above the modern floodplain (Kibdi Member, Gebel Silsila Formation; Butzer and Hansen, 1968, p. 277 f.). There are two dates of 5750 and 3650 B.C. The Shamarkian tool-kit was characterized by backed bladelets, with some end-scrapers, burins, geometric microliths, chopping tools, and worked ostrich shell (pendants, beads, engraved fragments), showing broad affinities with the Capsian. Fish bone is not preserved, and there is only rare mammalian bone (*Bos,* hartebeest, gazelle, hippo).

These industries would suggest a mosaic of local variants, with broad affinities both to northern and eastern Africa, intensively exploiting the riverine environment of the Nile floodplain or banks—fish and invertebrates on the one hand, large savanna mammals on the other. Neolithic archeological traits are not represented, even at a time (3600 B.C., corrected to about 4720 B.C. in terms of calendar years; see chapter 3) when food-production had been introduced to the Mediterranean coast some 1,500 years earlier, and to the west-central Sahara at least 2,000 years before.

The appearance of the Neolithic, unfortunately recognized only by pottery and arrowheads, is equally complex. The Shamarkian type sites are succeeded by a Neolithic variant of strong localization, and dated by a single C^{14} determination of 3270 B.C. (corrected to 4250 B.C. by the Stuiver-Suess formula). The major site has an area of 14,000 square meters, with both sites—averaging four or five times larger than the

Shamarkian sites—related to + 5 m. nilotic silts of the Kibdi Member. The lithic industry, which may contain some Shamarkian influences, is dominated by notches, denticulated tools and perforators, with a number of different arrowheads, and pottery. Comparisons with the Neolithic of Capsian Tradition appear warranted. The only paleontological evidence are some scraps of cattle and hartebeest bone. Far more extensive is the second, Neolithic facies, known as the Abkan (Wendorf *et al.,* 1968, vol. 2, pp. 611 ff. and p. 1053), which appears to be related to both the earlier Qadan and the succeeding A-Group culture. Numerous riverine campsites are indicated, with a larger population than during the Qadan occupation. Qadan-like artifacts, plus groovers and plain clay-tempered pottery, rare grinding stones, and possible stone fish traps, are noteworthy, but fish and mammalian bone are almost never preserved. The final Neolithic variant is found near the former floodplain margins or even out in the desert. It is clearly intrusive and related to the Khartum (Shaheinab) Neolithic of the central Nile area; it appears to be broadly contemporary with the Shamarkian and Abkan Neolithic (Wendorf *et al.,* 1968, vol. 1, p. 98; vol. 2, pp. 768 ff. and p. 1053 f.). Apart from scrapers, borers, backed bladelets, groovers, and ostrich-egg shell beads, there are the characteristic Shaheinab pottery types (surface-combed, impressed, and punctate). Identified animal bone is limited to gazelle and hare. Altogether these archeological complexes are not highly informative, but they do show a marked increase in population with a strong reserve of Mesolithic subsistence traits, despite the possibilities of some cattle-herding and limited grain-growing (by inference from the proto-historic A-Group). Thus the Nubian Nile Valley provides a close parallel—but a distinct ecological variant—to the culture-resistant but far more homogeneous Capsian populations of Algeria and Tunisia. As antecedents to later Nile Valley civilizations, they are also well worth bearing in mind.

In sum, the Capsians and the late prehistoric "Nubians" provide striking examples of populations, with intensive and successful food-collecting economies, who were resistant to Neolithic intrusions and who only adopted Neolithic traits very late and even then, rather selectively.

CATTLE NOMADS OF THE SAHARAN NEOLITHIC

The rock paintings and engravings of the interior Sahara bear testimony to two amazing archeological complexes: an Epi-Paleolithic hunting culture, and an early Neolithic food-producing culture with cattle, sheep, and local cultivation of cereals (see Rhotert, 1952; Lhote, 1959, 1965; Forde-Johnston, 1959; Mori, 1965; Clark, 1967a, 1967b; Hugot, 1968;

Camps *et al.,* 1968; Hobler and Hester, 1969, all with references). Unfortunately, it has not been possible to link conclusively the different groups responsible for the rock-drawings with the various stone industries and pottery types vaguely labeled as "Saharan Neolithic" (or "Neolithic of Sudanese Tradition"). In fact, the standard archeological inventory offers no proof for the distinctiveness of the hunters and herders at any one site, nor does it support the concept of a general ethnic identity throughout the interior Sahara at any one time. Nonetheless, the bulk of the "Neolithic" inventory—ground and polished stone axes and adzes; occasional sickle blades or grinding stones; large, bifacial tools; stone weights, platters, and dishes; different types or traditions of arrowheads and pottery; bone harpoons—can be assigned to late prehistoric populations with an economy based at least in part on livestock herding. Such associations have been established at critical sites in the Hoggar and Air, in the Tassili and Fezzan, and possibly in the Sudanese Nile Valley (see Fig. 95). However, in dealing with an area as vast as the Sahara and a time range of at least 3 and possibly as much as 6 millenia, it would be simplistic to assume cultural or economic uniformity, let alone ethnic continuity through space or time.

Despite the difficulties of resolving the broad patterns of the Saharan Neolithic with the available information, the Saharan data do provide the first verified example of nomadic pastoralism. The origins of the cattle nomads of the central and eastern Sahara, and of the cattle-and-sheep nomads of the western Sahara remain obscure. The earliest manifestation of food-producing in Egypt, the Sudan, Cyrenaica, and Tunisia are all younger than 5000 B.C., so that the Uan Muhuggiag date of 5590 B.C. and its apparent association with domesticated animals have come as a surprise, as have other dates on pottery Neolithic in the western Sahara in excess of 6000 B.C. (Fezzan: 6120; Hoggar: 6100 and 6720 B.C.) (Camps *et al.,* 1968). Consequently, whatever their origin, the early pastoralists of the western and central Sahara are probably dated with some accuracy by numerous radiocarbon dates between 6000 and 2320 B.C. (in terms of corrected calendar ages, ca. 7100–2900 B.C.), with comparable Neolithic cultures persisting on the savanna margins of the western Sudan until about 1000 B.C.

The basic economic traits of the Saharan pastoralists have been discussed by Rhotert (1952), Lhote (1959), and Clark (1967b). Subsistence was primarily based on cattle herds derived from local domestication of *Bos primigenius* (= *africanus*)—possibly somewhere in the Nile Valley. Lack of emphasis of the animals' udders in the pictorial art suggests that meat rather than milk was the major form of exploitation. Domesticated sheep overlap with cattle-raising in the Tassili region. Much game was hunted, probably reflecting local fusion with autochthonous food-collecting groups. Possibly, although not necessarily, these cattle nomads were the users of the occasional grinding stones reported from different parts of the Sahara. This may indicate that cereals were known locally, a point suggested also by the pollen from Meniet. Sorghum and bullrush millet grow wild on wadi banks in the southern Sahara today, and must have extended farther north during the subpluvials. The strong concentration of archeological sites in wadi valleys and at existing or former groundwater localities also suggests that settlement was largely confined to areas with available water—for both human and animal use. Caves were also occupied in some areas of the Tassili, Hoggar, and the Fezzan.

No direct proof of nomadism is available for the Saharan cattle-herders, but cultural associations have been shown with nomadic Kushitic or Eastern Hamitic groups of the Red Sea coasts of Egypt, the Sudan, and Ethiopia (Rhotert, 1952). The typical composite drawings of large cattle herds strongly suggest organized pastoralism. Also, the rapid dispersal of this culture through the Sahara may reflect a nomadic subsistence. With the erratic nature of the rainfall (even during the Holocene subpluvial) and the sporadic distribution of water and pasture, it is unlikely that permanent or semipermanent habitation *could* have been practiced in any one area, a conclusion not in contradiction with the windbreaks of reeds, woven mats, or skins suggested by some of the Tassili and Uweinat rock paintings. It seems necessary to assume that adequate water and fodder could only be guaranteed by periodic movements, possibly into the better watered highlands or to permanent water holes during the dry season, and to ephemeral pastures among the foothills and on nearby alluvial plains during the episodic rains.

One may suspect that this very obvious case of adaptation of food production to an adverse environment had its origins in an agricultural community that gradually expanded or was displaced into marginal arid country where livestock-raising was more economical than cereal agriculture. Yet this source region remains enigmatic, with the first introduction of agricultural traits into the Nile Valley and along the Mediterranean littoral some 1,000 to 2,500 years later. Interestingly enough, the lithic affinities of the Saharan Neolithic lie in the late Pleistocene traditions of the central or southern Sahara (J. D. Clark, unpublished). At any rate, planting played a very small role in an economy based primarily on meat animals. This contrasts with the contemporary village farmers of the Near East and Europe, among whom subsistence was primarily based on cereals.

In retrospect, the diffusion of food-producing traits into the arid zone, and in particular into the Sahara, was a case of cultural or technological adaptation to a new environment. Yet this dispersal was only made possible by the temporary improvement of the environment and resource base of the Saharan highlands during the Holocene subpluvial. Not only did desiccation play no role whatever in agricultural dispersals after 5000 B.C., but instead the prevailing moister climate must have facilitated and perhaps motivated man's expansion over the world's greatest desert. In fact the spread of food-producing populations through the arid zone of the Old World followed close upon the migration of the Ethiopian faunas through the Sahara.

The dispersal of agriculture into sub-Saharan Africa at a later date is beyond the scope of this study, as is diffusion into central and eastern Asia.

POTENTIAL IMPACTS OF FOOD PRODUCTION ON MAN-LAND RELATIONSHIPS

The impact of the new food-producing economies on the environment marks a rather significant change in man-land relationships. Two million years of Pleistocene time had witnessed a very gradual development of technology and economic patterns permitting existence of the human species under most environmental conditions. Man had also begun to modify the biological world, even if only on a local scale. Now, with the spread of ecologically potent farming communities across the Old World, transformation of the natural environment began to leave great scars in the landscape—the areal importance of which almost everywhere increased with time, and the continuity of which was assured by the persistence of human populations at ever higher technological levels. The major aspects of geographical interest include *(a)* an explosion of population, made possible by an improved subsistence economy, *(b)* physical transformation of the environment, particularly through decimation of the native flora and fauna and their partial replacement by nonindigenous domesticated species, and *(c)* the creation of a cultural landscape.

Population is essentially controlled by available food. Rapid demographic expansion has ensued upon several major technological improvements of the food supply: (1) after the first invention of tool manufacture, (2) after the invention of agriculture, (3) with the intensification of agricultural production accompanying urbanization, and in more recent times, (4) with the industrial revolution. C. O. Sauer (1947) described the history of man as a succession of higher and higher levels, each one brought about by discovery of more food, either through occupation of new territory or through increase in food-producing skill. When the maximum possible population is reached, population must level off, either by gradual convergence of birth and death rates, or by draining off the surplus into daughter colonies.

The introduction of a subsistence economy based on farming and herding would provide a greater and more stable food supply. A much smaller economic area could provide sufficient food for much larger communities. Domestic animals could be used for meat at most times of the year, while the highly productive cereal crops could be stored for the whole year following the harvest. There was no longer any need to move when the local supply of wild plant foods or of game was exhausted. Starvation no longer ensued when biological cycles reduced the local game population. Above all, the food supply was far more reliable, both in the course of the seasons, as well as during the passing of the years, so maintaining a much higher population level. Of course, exceptionally cold winters, drought years, crop and animal plagues, etc. would still exert a noticeable influence on the population curve. But man was becoming conspicuously less dependent on the vagaries of the environment.

Braidwood and Reed (1957) have discussed subsistence levels and modern ethnographic parallels, and suggested typical population densities of approximately 1 person per 100 sq. km. at the unspecialized Pleistocene food-gathering level; 5 per 100 sq. km. at the specialized late Pleistocene-early Holocene food-gathering level; 1,000 per 100 sq. km. at the early agricultural level; and 2,000 per 100 sq. km. at the early urban level. Obviously these are only meant to be orders of approximation, but the values help illustrate the degree of change involved.

The physical transformation of the natural environment was primarily the result of man's agricultural activities. Changes were originally confined to the biological sphere. The natural woodland or grassland vegetation was partly replaced by fields of wheat, barley, and vegetables. Such crops, originally native to a restricted area of western Asia, were to spread through most of the world, into lands where their very existence was often possible only through the caring hand of man. Species that in natural competition shared minute ecological niches with countless other plants now dominated acre upon acre of monocultures. Unconsciously, agricultural or grazing activities favored certain local herbaceous plants by creating open spaces in woodlands, so increasing the importance of fire-tolerant plants in the course of slash-and-burn clearance. Similarly, new ecological niches were provided for a rash of new weed plants, whose original habitats and specific niches had been as insignificant as those of the cereals or vegetables.

The same can be said for the animal world. The wild fauna, with some exceptions, was decimated through a reduction of the natural habitat by cultivation and a disturbance of breeding haunts, as well as the improved hunting techniques of ever larger populations. Instead, the new farming populations tended select domesticated animals, thus enabling dispersal of certain species on a continental scale and causing drastic changes in the composition of the fauna. Certainly these qualitative and quantitative changes of flora and fauna required millennia, and the face of the earth was at first altered only locally. The cumulative effect over several millennia has, however, been significant and sometimes catastrophic.

The cultural landscape reflects intensive settlement with effective transformation of the biological environment through agricultural land use. With the introduction of village farming into an area, cultivated fields and biologically altered grazing areas began to dot the landscape. Architectural skills had improved, and shelter requirements were met by construction of houses, stables, and storage huts. Individual farmsteads coagulated to form villages dispersed over the countryside. With incipient urbanization, these man-made structures increased in size, number, and importance as towns and cities, market places, roads, bridges, fences, and the like were added. Irrigation and drainage schemes were implemented in marginal environments. Forests were removed for land clearance and timber, and grasslands plowed up. These innovations were often followed by such unpleasant corollaries as soil deterioration and soil erosion.

Although the record of man's early transformation of the physical into a cultural landscape is poorly preserved in the Old World subtropics, the case of mid-latitude Europe is better understood. The significance of forest clearance and crop cultivation by village-farming communities was first recognized in Denmark from the pollen records of the Subboreal by Iversen (1949; also 1956). The earliest appearance of cereal pollen was accompanied by a rapid increase in the appearance of weed colonists such as *Artemisia, Rumex, Plantago,* and Chenopodiaceae, with a corresponding decrease in mixed oak forest (Fig. 46). Such discontinuities were followed by temporary birch pollen maxima—common after forest fires—with subsequent increase of alder, hazel, and finally, oak. Evidence of burning is occasionally visible in the peat stratigraphy. Iversen explained these features through forest clearance by burning and felling, with subsequent livestock-grazing or crop planting in the "opened" woodland. The fields were soon abandoned in the course of shifting cultivation, and so allowed to revert back to forest. Interestingly, open woodlands such as oak-birch forests on sandy soils showed little or no pollen discontinuity other than the presence of cereal and *Plantago* pollen. These show the existence of cultivation on plots available without recourse to intensive clearing.

This original picture of common, but not exclusive, slash-and-burn agriculture seems to be substantiated by the over-all archeological and palynological evidence in mid-latitude Europe (J. G. D. Clark, 1945; Firbas, 1949–52, vol. I, p. 363 ff; Iversen, 1956). Fire was apparently not necessary in the more open landscapes. Tree-felling with stone axes was quite feasible, as recent experiments by Steensburg (1957) showed. Bark-peeling or girdling of trees was probably also an effective clearance method, particularly after brush and lighter growth had been removed through burning. An interesting form of semiagriculture preceded true agricultural colonization in Denmark (Iversen, 1960, Troels-Smith, 1960). This Erteboelle culture may represent a contact culture, as resistant to intrusive colonists and cultural change as the North African Capsians, with an economy based largely upon stalled or tethered cattle. The animals were almost entirely fed with the foliage of elm, mistletoe, ivy, and ash. As a result there was a sharp reduction in elm pollen, formerly interpreted as a climatic change at the transition of the Atlantic to the Subboreal. A little wheat and barley was apparently grown, but there was no forest clearance worth speaking of. This example illustrates that the methods and significance of forest clearance by early agriculturists can hardly be generalized.

Prior to the first introduction of the ox-drawn plow from Mesopotamia into temperate Europe during the second half of the third millennium, soil preparation was done by hoe or digging stick. With such tools it is unlikely that most of the woodland soils yielded well for more than a year or two, requiring twenty or more years of fallow thereafter. Fertility must have been more enduring on the chernozemic soils, since recent plow agriculture without fertilization on the Ukrainian chernozems only required one fallow year in three. The exact nature of rapid soil depletion or yield reduction is complex, reflecting actual mineral depletion, rate of weed colonization, erosion resulting from soil structure changes, or humus destruction. The common symptom of sharply reduced yields probably results from a number of interacting factors.

Soil erosion was probably unimportant since cultivation was limited to the more productive lowland soils. Clearance and cultivation of hillsides was a late innovation in mid-latitude Europe. Deforestation or moderate grazing would not leave bare soil exposed for very long. Even in the Mediterranean region, in such an ancient land as Lebanon, the commercial importance of lumber in historical times suggests that widespread deforestation was rather uncommon until the end of the Roman period (Mikesell, 1969). In fact Heichelheim (1956) and Darby (1956) emphasize that general deforestation and land deterioration even in the Mediterranean region fall largely within the two millennia of our own era. It would therefore seem that early agricultural land use did not yet provoke its more unpleasant side-effects such as accelerated runoff, seasonally accentuated stream discharge, soil erosion, gullying, and gradual loss of soil moisture attendant upon the destruction or removal of humus. At any rate, both archeological and geological evidence to this effect is absent.

The preceding discussion of man-land relationships assumed that human populations automatically expand to the limit of resources available within a given technological framework. Such an assumption is, of course, questionable. Although less significant at the food-collecting level, efficiency of exploitation among technologically equivalent groups assumes considerable importance at the food-producing level. Was there a fundamental stability in the relationships of man to the exploitable resources of his habitat? Or did local overexploitation of resources already lead to temporary or semipermanent environmental crises? At the early agricultural level it would seem that a basic stability persisted, and with so much new land to occupy, it is possible that local overexploitation was still uncommon. In addition, a primitive technology and the prevalent natural hazards must often have placed both the relative security and the productivity of early agriculture in jeopardy. And, as a way of life, farming was "more continuously demanding than hunting and gathering, with the participation of all but the very young and very old being required in subsistence pursuits" (Adams, 1968, p. 44).

THIRTY-FOUR

Agricultural Settlement and Urban Origins in the Near Eastern Floodplains

THE NEAR EASTERN FLOODPLAINS DURING THE EARLY AGRICULTURAL SETTLEMENT

A new environmental situation for agricultural colonization was present in the alluvial valleys of the Tigris-Euphrates, the Nile, and at a later date, the Indus. In each of these areas rainfall is insufficient for crop planting, but a regime of periodic annual floods provides ample moisture. In each case a hostile desert intervenes between the early village-farming communities of the wooded hill country. The seasons of available moisture vary as well: instead of mediterranean-type winter rainfall, the flood season of the Tigris-Euphrates comes in late spring and early summer, those of the Nile and the Indus in late summer and early autumn. Consequently crop planting must take place at different seasons, during the recession of the floods.

The original agricultural settlement of the Near Eastern floodplains was no easy step for prehistoric man. Although the agricultural villages of the seventh and sixth millennia had sometimes tapped stream waters in the lower courses of highland streams or at oases, total readaptation was required for settlement on the floodplains. Planting had to be done at a different season, and the farmers had to learn to cope with seasonally declining water resources. Hoe or digging-stick cultivation was probably more difficult on muddy, heavy alluvial soils than it was with light, friable woodland soils. Possibly, the first colonists learned to broadcast seeds on wet alluvium, without prior working of the soil.

Not all of the staple crops that had been developed under a winter rainfall regime did well with a floodwater regime on the alluvial plains. In fact, einkorn wheat has not yet been found in ancient floodplain sites, while a new, six-rowed mutation of barley proved to be a successful adaptation to the same ecological change (Helbaek, 1960b). The new barley did very well on the rich alluvial soils and was not affected by the gradual salinization of the Mesopotamian lowlands following the introduction of irrigation. Emmer wheat and flax were salt intolerant, declined in importance, and became insignificant in lowland Iraq by the beginning of the historical era (Helbaek, 1960b). Similarly, many of the standard Near Eastern orchard crops did not thrive with the artificial ecology of the floodplains. Instead the date palm was cultivated, providing a (novel?) source of fruit and wood. In general, considerable economic adaptation was achieved.

Another problem concerns the environment encountered by the first farmers on the floodplains of the Tigris-Euphrates (after 6000 B.C.) and the Nile (after 4500 B.C.—calendar years). Were the alluvial plains covered with jungle-like thickets and extensive swamps that required major clearance or drainage? Or could the floodplains be settled without technological improvement?

A combination of natural levees and seasonally inundated alluvial flats mark the lower courses of the Tigris and Euphrates and their many branches. Annual high water occurs in May and June when the snow meltwaters of Armenia combine with the spring rainfall maximum in Turkey. Depending on the synchronization of precipitation and temperature conditions in the high country, there may be appreciable floods. But the floods are generally irregular and highly variable from year to year, and consequently not dependable. Unlike the Nile, the Mesopotamian waters contain salt in solution which is concentrated in the soil through evaporation on irrigated land. In Egypt the hazard of salinization is largely confined to the coastal marshlands of the Delta.

With perennial water available along the river banks and in backswamp areas, the Mesopotamian floodplains would support a woodland vegetation without human interference. In the natural state tamarisk and poplar were probably characteristic trees of the levee embankments, in association with oleander, acacias, and thorny *Zizyphus* species. The seasonally inundated flats would be parched by late summer, permitting a vegetation of brush and grasses only. Beyond the floodplain, on the desert uplands or on the ancient river terraces, plant growth was limited to desert or semidesert shrub.

Like the Tigris-Euphrates lowlands, the Nile north of the Egyptian-Sudanese border moves across an alluvial floodplain. The summer monsoon begins in Ethiopia during May, and the lower Nile reaches its flood peak in September. When the floodwaters recede, the levees are immediately left dry, although the lower-lying basins may remain inundated well into November. Backswamps were always localized in their occurrence, except in the northern delta where swamps merge into brackish coastal lagoons, cut off from the open sea by sand bars. These characteristics of the Nile Valley have little in common with the jungle-thickets and endless papyrus swamps commonly postulated in the archeological literature (Passarge, 1940; Butzer, 1959b). The extent of perennial swamps and lakes in the Nile Valley was limited in prehistoric times, and as today, the greatest part of the plain consisted of seasonally flooded basins. Groves of acacia, tamarisk, sycamore, and Egyptian willow crowned the levees, which were always available for permanent settlement. Papyrus swamps and quiet expanses of standing water with lotus, sedge, and reeds, teeming with hippopotamus, crocodile, and a host of aquatic birds, were available. But these constituted only a small segment of prehistoric Egypt.

From the very beginning, man could take up his abode on the levees or on the desert margins of the Near Eastern floodplains. When the floodwaters receded he could throw his crop seeds on the wet mud of the alluvial flats or graze cattle and other herd animals on the lush grasses and brush vegetation. By the time the waters rose again, the harvest had long been gathered, and livestock could be pastured on the levees or on the outer peripheries of the alluvium. Physical conditions in the Nile Delta were equally good (Butzer, 1959b), and the myth of a delta rapidly expanding seaward in historical times is as unfounded in the case of the Nile as in the case of the Tigris-Euphrates (Lees and Falcon, 1952).

Granting that drainage was not an acute problem for the prehistoric colonists of the Near Eastern floodplains, irrigation certainly was. The technological aspects of simple basin irrigation appear to have been mastered in the fourth millennium, and similar methods have remained in use until the construction of dams and barrages within the last century. The technique is as follows: The alluvial flats are naturally divided into a maze of irregular basins of variable size, resulting from abandoned levee ridges, small distributary branches, and abandoned stream channels. Depending on their elevation, these natural basins drain off successively as the flood begins to recede, unless the waters are temporarily retained through human efforts. Dry season reservoirs can be artificially created by breaching the levees, allowing uncontrolled local flooding, and then subsequently damming the waters. These can be released later on and allowed to flow into the fields as required. Inscriptions dating from the earliest historical times confirm the significance of irrigation by the third millennium.

Settlement in the Near Eastern floodplains had been essentially completed during the fourth millennium B.C. In the case of Egypt, the archeological evidence is confined to cemeteries, villages, and towns located on the desert edge. However settlement sites are far too few when compared with the many cemeteries. This shows that by 3500 B.C. most of the Egyptian population already lived on sites located within the floodplain. In Mesopotamia, sites are known both from the desert edge and the floodplains.

EARLY URBANIZATION IN THE NEAR EASTERN FLOODPLAINS

Agricultural colonization of the alluvial valleys of the Tigris-Euphrates and the Nile was followed by a rapidly increasing population and the development of an intensive cultural landscape. Sporadic evidence suggests the presence of occasional overgrown villages with as many as 5000 inhabitants, e.g., Hierakonpolis on the western edge of the Nile Valley, during the later fourth millennium (Butzer, 1960b, with references). The first copper tools are verified from Mesopotamia during the Ubaid culture (ca. 4300–3500 B.C.), from Egypt during the Badarian (ca. 4000 B.C.). Wooden, ox-drawn plows had been introduced in both areas during the late fourth millennium, and are referred to by some of the earliest written documents. Urban communities—with differentiation of labor and function—appeared in both of these areas and in the Indus Valley by 3000 B.C. All in all, the centuries preceding the dawn of history witnessed the establishment of metallurgy, irrigation, plow agriculture, and the first towns. This was the technological prelude to the literate floodplain civilizations of the Near East.

The innovations of the fourth millennium are probably epitomized in urban origins, so much so that Childe (1929, 1958) interjected some unfortunate socio-historical ideas into the archeological interpretation, introducing the concept of an "urban revolution." Childe correlates urbanization with the establishment of metallurgical industries, on the grounds that the manufacture of copper or bronze tools would require a specialized craftsman. In this way the self-sufficiency of small farming communities would break down, while new economic pursuits would concentrate on mining, manufacturing, and distributing finished products. The new class of part-time or full-time artisans would require support by other members of the community, so that the local agricultural surplus had to be made available to the families of the non-food-producing metalworkers. Childe envisages this evolution as the crucial step from a classless "barbarian" social order, based on kinship, to a class society, based on division of labor or function.

Objectively reviewed, the "urban revolution" was not a revolution but a stage of economic transition. It was not necessarily associated with metalworking, and urbanization itself was but one symptom of a complex cultural change. The beautifully decorated and well-fired pottery of the Halafian culture of northern Iraq (ca. 5200–4300 B.C.) leaves no doubt of craft specialization. One can then envisage part-time specialization and incipient division of labor at a much earlier date. The subsequent 1,500 years witnessed a number of technological innovations in the floodplains of the Near East, which possibly accompanied a considerable cultural and intellectual ferment. Some of the economic aspects of this change may have been as follows:

a) Production of food surplus, made possible by (1) more intensive, irrigation agriculture, (2) more fertile, perennially productive alluvial soils, and (3) improved sowing and harvesting techniques, culminating with the innovation of the animal-drawn plow.

b) Development of commercial contacts to obtain raw materials (flint, stone, metal in lower Mesopotamia, metal in the Nile Valley) from more distant areas, so requiring an exchange of commodities. Possibly food products and pottery craftsmanship formed the major export items.

c) Development of functional centers in some villages, designed for food storage and exchange, and possibly for manufacture or redistribution of finished products.

If these assumptions are correct, craftsmen, tradesmen, and clerks would be a necessary corollary. The craftsmen would include potters, metal and woodworkers, builders, and artists. The tradesmen, on the other hand, would perform the vital function of transferring raw materials or finished products from mines or quarries, town or village, on foot or on donkey, by land, river, or sea. The clerks would presumably handle the bureaucratic aspects of granary administration, regulation of raw material import, marketing, and outside commercial relationships. Given such prerequisites some villages—by accident, location, or ambition—outstripped their neighbors and assumed the role of functional centers. These increasingly important foci of bureaucratic organization were necessary in an economy with large-scale, planned surplus farming, commerce, and industry. Urbanization would therefore be the culmination of the technological and economic changes outlined. The terms "town," "city," or "urban" would be applicable to such a center of organized and complex activity whenever an appreciable segment of the population was employed in crafts or industries. The role of the bureaucracy would increase gradually, ultimately including community organization of drainage and irrigation, of planting and harvesting, of town, harbor, and road-building, of property demarcation and judicial settlements. From this evolved a complex social elite of scribes, politically or militarily endowed secular officials, and artistic and ritually oriented priests. At the same time, the residents of larger settlements were more easily subject to political control, taxation, military service and forced labor (Adams, 1968). Thus the functional role of the new cities was threefold: *(a)* The appropriation and redistribution of agricultural surpluses for large agricultural communities living within a small but intensively utilized radius. *(b)* A permanent institutional base that authoritatively administered the interrelationships between specialized segments of the population, both local and external. *(c)* A concentration of wealth, with the self-glorification of a new elite through ostentatious public building programs and the development of military power (Adams, 1968). In this more restricted sense, Childe's original emphasis of the significance of urban civilization is justified.

THE SIGNIFICANCE OF URBANIZATION FOR MAN-LAND RELATIONSHIPS

Urban origins on the Near Eastern floodplains had a greater effect on man's relations with his fellow beings than on his relations with his environment (Adams, 1966). The earlier steps of agricultural invention were accompanied by an increasing breadth or intensity of environmental exploitation, associated with an improved technological inventory and a new means of subsistence. Urbanization, although based on changes in exploitation of the environment, was accompanied by the development of new institutions and a complex social structure, freeing man even further from dependence on his immediate environment (Adams, 1965, 1966; Hole, 1966).

One of the leading results of urbanization was the development of more balanced economies, based upon food surplus and food exchange, and probably fostering diversification as well as specialization of subsistence. In this way, organizational activity could offset aperiodic calamities such as killing frosts, low or destructive floods, or locust plagues. Hence the vagaries of the environment were of ever less influence on the human population level.

A second major effect of urbanization was the development of trade as an important enterprise. Some forms of commerce must have existed even during the Paleolithic, but organized economic exploitation of raw materials in distant lands now became significant. Surface ore deposits were mined extensively, and raw materials were frequently transported a thousand kilometers or more to distant centers of manufacturing. Complex marketing arrangements arose between town and village, farmer and herder, leading to a high degree of economic interdependence, and in some areas, political "unification."

A third ecological impact stemmed from the accelerated technological and cultural progress resulting from a division of labor and function. Manufactured products became widely accessible, while better tools permitted greater agricultural production. Community efforts such as irrigation and drainage, distribution of water, and engineering projects of various kinds all became more elaborate. In some areas, such as the Nile Valley, complete political and economic centralization ensued. But the maze of city-states in lowland Mesopotamia provides an excellent counter-example to the concept that irrigation civilizations require centralized control (Oppenheim, 1964).

As a fourth factor, Adams (1968) rightly emphasizes that cities have been sources of severe population loss, not population growth. Urban mortality rates, particularly of infants and children, were much higher than rural ones, because of the primitive or nonexistent techniques of sanitation and the increased density and aggregate size of settlement. In fact, the maintenance of urban populations, let alone their growth, has long been dependent on continued immigration from rural areas.

Finally, urbanization has had severe negative repercussions in terms of environmental ecology (Adams, 1968). Intensified urban pressure on agricultural resources favored soil exhaustion and, in the alluvial lowlands of Mesopotamia, salinization as a result of overirrigation. The substitution of monocrop cultivation for diversified agriculture compromised the economic self-sufficiency of even the rural population, while excessive taxation repeatedly led to the abandonment of villages and farmlands. Equally precarious was the swallowing up of smaller, locally maintained systems of irrigation into state-run enterprises. The agricultural masses were unable to administer and maintain these gargantuan affairs during periods of political decline or depopulation, so that warfare, invasions, and epidemics periodically wreaked chaos on the rural landscape, bringing starvation to broad segments of the population.

Each of these immediate consequences of urbanization contributed to man's capacity to alter the face of the earth. The cultural landscape of the floodplain kingdoms had almost attained the status of a man-made environment by the beginning of the historical era. Wide adjacent tracts were exploited for mineral resources such as gold, silver, copper, tin, and amber, or forest products such as lumber, firewood, and charcoal. With the advent of political megalomaniacs, wars of conquest, slave-hunting, and revolutions led to recurrent devastations of both the cultural and physical landscape. This, then, set the patterns that have remained characteristic to the present day: man's interactions with the environment have been largely obscured by the social and cultural products of man.

RETROSPECT ON THE SIGNIFICANCE OF THE ENVIRONMENT FOR PREHISTORIC MAN

The preceding chapters have attempted to study man-land relationships during the course of prehistory. By way of review we can turn first to the potential spheres of environmental influence on man:

a) Human distribution. (1) Warmth and clothing: Early hominids may have been limited to the tropics through physiological adaptations of the species to such environments. Low temperatures, which could only be counteracted by some sort of clothing and the use of fire may have excluded early hominids from higher latitudes. Late Pleistocene colonization of subarctic environments may have followed technological improvements guaranteeing better heat control. It seems possible to observe a step-by-step settlement of higher latitudes in the course of the Pleistocene. (2) Water and food resources: Prehistoric man was never found very far from sources of permanent surface water and animal or vegetable food. This partially excluded him from desert environments. (3) Environmental change: Major changes of the environment, such as the expansion of deserts or ice sheets, certainly promoted changes in the distribution of population. Less significant ecological changes in the range forest-woodland-grassland-semidesert may have had similar, although less well-defined effects.

b) Population level. The size of Pleistocene communities was probably limited by the food supply available within their hunting territory. Climatic fluctuations or other biological factors controlling local animal population or plant productivity should have strong repercussions on the human population level. Although proof is obviously next to impossible, it remains likely that environmental control (in terms of average carrying capacity) and environmental vicissitudes (in terms of nonseasonal fluctuations of the food supply) were rather significant in determining the regional population level. Similarly, human population levels must have varied strongly from one environment to another.

c) Biological and cultural evolution. Much has been said but little convincingly proven about environmental influences on biological and cultural evolution. Environmental changes may possibly have accelerated biological evolution during the early Pleistocene. But it has not been possible to demonstrate *general* validity for the analogous concept of environmental change and cultural innovation. None of the climatic changes attending the close of the last glacial can be genetically related with cultural innovation, either in the case of the Mesolithic or in the case of Near Eastern agricultural origins. Certainly a more careful distinction should be made between cultural innovation and economic adaptation, at least for the better understood phases of prehistory.

d) Economic traits. The significance of the environmental setting of a community for its technological inventory, way of life, and subsistence economy must have been enormous in prehistoric times. It would be fascinating to know just how dietary economy, clothing, habitation, raw materials, etc. were adapted to particular environments at different cultural levels. Yet, at least during the Pleistocene, the archeological record does not provide a sufficiently coherent picture of man's economic life. Only with greatly improved cultural interpretation of archeological sites will it be possible to say more about man-land interactions at the Paleolithic level. Although modern ethnology has contributed significantly in this direction, much more needs to be done. The preceding discussions of prehistoric subsistence patterns and of adaptations to new environments show how incomplete and tentative much of our present information is. But the tentative picture that emerges is nevertheless interesting and often provocative. It is here that much remains to be done and gained.

RETROSPECT ON THE IMPACT OF MAN ON THE ENVIRONMENT

At the food-collecting level, man was present in small numbers only, with limited technological skills. Combining the results of the paleo-ecological survey of the last chapters with Heizer's (1955) review of relevant ethnological data, the following possible patterns of environmental modification can be outlined:

a) Vegetation change through human use of fire. Fire may have been used as an aid in hunting or to promote vegetative growth, both of species palatable for man and of species favored by game. In areas of vigorous forest growth, such burnings may have favored the dominance of certain species, or at least temporarily disturbed the climax forest patterns. In areas of marginal tree growth, frequent burning may have replaced woodland by grassland vegetation, or at least favored grassy vegetation out of all proportion to the available climatic and soil resources. Such regional or local vegetation changes may have affected the humus content and type of soils, and they would certainly alter the water balance of the soil. But they could hardly have had any effect on regional climates – even today man-made modifications of rural climate are quite negligible on air temperatures and precipitation except at the microclimatic level (see Thornthwaite, 1956).

b) Vegetation change through accidental spread of plant species. Unintentional and accidental spread of uncultivated plants in the wake of man's wanderings may have led to minor and localized changes in vegetation composition. Certain heliophytic local or foreign species, particularly certain weed colonists, would find new ecological niches as a result of vegetation clearance near a settlement or in any area affected by deliberate burning. Similarly, plant seeds or nuts carried as food could become scattered and grow spontaneously near settlement sites.

c) Faunal changes through hunting activity. Selective hunting or conscious sparing of certain animals for dietary or cultural reasons could lead to changes of faunal composition and equilibrium. Hunting pressure may also have led to local movements or even to migration of animal groups, and man will certainly have altered the animal biomass significantly on occasion. Whether or not local or total extinction of species was indeed the result of prehistoric hunting practices remains uncertain.

At the food-producing level, man's impact on the environment became more significant and tangible. His activities were no longer confined to modification of the biological world but also began to affect the abiotic surroundings. These spheres of influence can be enumerated as follows:

a) Forest clearance. With the establishment of agricultural economies, primary forest clearance, plot-burning, and livestock-pasturing assumed a functional role in settlement. Conscious large-scale destruction of the native vegetation was now no longer incidental or local in occurrence, but characteristic. Deliberate eradication of nonuseful plant species by farmers and herders probably began to assume importance as cultural biases toward the environment were formed.

b) Deliberate introduction of domesticates. Little needs to be added here to the discussion of how cultivated plants and domesticated animals were introduced to new lands or favored in their natural habitats. Selective favoring of economically useful plants may also have been locally extended to species not traditionally included among the domesticates: oak, ash, elm, etc. used for animal feeding.

c) Soil depletion and soil erosion through land use. Agricultural use of the soil inevitably led to changes in soil humus, structure, and over-all chemistry—changes partly matched by addition of fertilizer. Soil erosion was presumably less important in late prehistoric times than today, although it remained a potential danger wherever deforestation, cultivation, or overgrazing were practiced on hillsides and uplands.

The cultural transformation of the environment that followed in the wake of agricultural dispersal was concomitant with the beginnings of the cultural landscape. The new patterns of land use were gradually intensified as a result of increasing population density and technological advancement. And with the establishment of urban civilizations on the Near Eastern floodplains, we cross the threshold of the historical era.

The relations of man to the land in prehistory are as yet poorly understood. But, with the growth of interdisciplinary interest for the many kinds of problems involved, a fuller understanding is bound to emerge—an understanding of man, his evolution, his cultural traits, and his environment.

GENERAL OVERVIEW

A number of observable phenomena emerge from the course of the preceding chapters:

a) Despite the many uncertainties shrouding the issue of human origins, hominization in both a biological and a cultural sense can be observed during the early Pleistocene. It is unimportant whether we define hominization by taxonomic criteria applied to the range of fossils from *Australopithecus* to *Homo erectus,* or by the establishment of a peculiarly human way of life, such as that of the Oldowans.

b) During the course of the Pleistocene there was an over-all increase in human numbers, although man does not automatically expand to his carrying capacity—average or maximum—at any particular level of technological and organizational skills.

c) The dispersal of the human species is evident, even if still poorly understood. From an original home in the tropical "savanna" environments of eastern and southern Africa, man expanded into northern Africa, temperate Eurasia, and tropical Asia early in Pleistocene times. At a much later point, the tropical rainforests of Africa and the subarctic environments of Eurasia were settled, only little earlier than the first colonization of Australia and the Americas near the end of the Pleistocene.

d) An increase in technological complexity is apparent, although the primary criterion—the number of stone-tool "types"—is unsatisfactory. However, with the beginnings of food-production the record becomes far more representative, permitting a balanced picture of human activity.

e) To some extent parallel with the increase in technological complexity there appears a greater diversity, both in the technological and—implicitly—in the cultural realm. Multiplying at a geometric rate are the number of industrial assemblages, variants and complexes, as well as the culture groups. Ultimately this diversity ranges from unspecialized food-gatherers to urbanized food-producers.

f) It is difficult to define an equilibrium concept for man and the environment, or to formulate a threshold value of where "disturbance" begins. Nonetheless, there is an increased intensity of environmental modification apparent during the time span of prehistory. This ranges from simple and incidental exploitation by food-gatherers at the one end of the spectrum to intensive utilization, modification, and destruction by urban man at the other.

g) Lastly, once a "definition" of man has been accepted, all of the preceding phenomena appear to have developed at an exponential rate—exceedingly slowly over the first million years or so, then accelerating rapidly during the last few millenia prior to the historical era.

The underlying processes to the observable phenomena are at best partially and tentatively understood. But at least the problems have been formulated and a number of relevant processes are liable to inference or speculation:

i) Initial hominization seems to have been accompanied by the elimination of other hominids closely related to man, possibly those most similarly adapted and therefore most competitive. This applies to the robust australopithecines and probably plays a part in the evolution of *Homo* from the broader background phylum of the gracile australopithecines.

ii) The development of the basic human behavioral patterns took place over millions of years, with the passing on of acquired "culture" and knowledge from generation to generation. Included here are such diverse phenomena as technological skills, language, group structuring—including such features as cooperation, sharing, and division of labor,—and last, but not least, the organizational patterns of functional activities that range the spectrum from hunting or vegetable-gathering to food-preparation.

iii) In the midst of the progression in human technology, organization, and numbers, there clearly appears to be an increasing complexity to man-land relationships. Most of the details remain elusive, but here, certainly, is one of the most fascinating goals for the student of prehistory.

iv) Concomitant with the increased intensity and complexity of environmental modification, man progressed from an ecologically insignificant member of the ecosystem to a prominent and ultimately the dominant one.

v) Finally, in the wake of his increasing environmental "control," man has acquired ever more potent skills for over-exploitation. After a million or two years of slow change and balanced subsistence within the constraints of the environment, a few short millenia carried man to successive apexes of "civilization." Suddenly the conscious system of restraints has been removed and, with a lethal ability to destroy his environment as well as his own kind, man has unwittingly damaged his environment almost to the point of no return. The human species evolved to the psychology of a food-collecting way of life, with small groups living in basic harmony with their environment. Only a new consciousness of our place in the world can hope to provide us with a perspective that promises a future.

Bibliography

Aario, L., and H. Janus

1958. *Biologische Geographie.* Braunschweig: G. Westermann.

Aario, R.

1969. "The Northern Discharge Channel of the Ancient Päijänne and the Paleohydrology of the Atlantic Period," *Bull. Geol. Soc. Finland,* 41: 3-20.

Adam, K. D.

1951. "Der Waldelefant von Lehringen," *Quartär,* 5: 79-92.

1953. "Die Bedeutung der altpleistozänen Säugetierfaunen Südwestdeutschlands für die Gliederung des Eiszeitalters," *Geol. Bavarica,* 19: 357-63.

1954. "Die zeitliche Stellung der Urmenschen-Fundschicht von Steinheim an der Murr innerhalb des Pleistozäns," *Eiszeitalter und Gegenw.,* 4-5: 18-21.

1961. "Die Bedeutung der pleistozänen Säugetier-Faunen Mitteleuropas für die Geschichte des Eiszeitalters," *Stuttgarter Beitr. zur Naturkunde,* 78: 1-34.

1964. "Die Grossgliederung des Pleistozäns in Mitteleuropa," *Stuttgarter Beiträge zur Naturkunde,* 132: 1-12.

1966. "Zur Grossgliederung des mitteleuropäischen Pleistozäns," *Zeit. deut. geol. Ges.,* 115 (1963): 751-57.

Adams, R. M.

1965. *Land Behind Baghdad: A History of Settlement on the Diyala Plains.* Chicago: University of Chicago Press.

1966. *The Evolution of Urban Society.* Chicago: Aldine.

1968. "The Natural History of Urbanism," *Smithsonian Annual,* 2: 40-59.

Ager, D. V.

1963. *Principles of Paleoecology.* New York: McGraw-Hill.

AGUIRRE, E. DE
1969. "Evolutionary History of the Elephant," *Science,* 164: 1366-76.
AHLMANN, H. W.
1924. "Le niveau de glaciation comme fonction de l'accumulation d'humidité sous forme solide," *Geografiska Ann.,* 6: 223-72.
AITKEN, M. J.
1961. *Physics and Archaeology.* London and New York: Interscience.
ALEXANDER, J., and D. G. COURSEY
1969. "The Origins of Yam Cultivation." In UCKO and DIMBLEBEY, 1969, pp. 405-26.
ALIMEN, H.
1957. "Secteur Français." In "Livret Guide de l'Excursion N_1 (Pyrénées)," *5th Int. Congr. INQUA* (Madrif-Barcelona), pp. 78-89.
1967. "The Quaternary of France." In K. Rankama (ed.), *The Quaternary,* New York: Interscience, vol. 2, pp. 89-238.
ANDEL, T. H. VAN, *et al.*
1967. "Late Quaternary History, Climate, and Oceanography of the Timor Sea, Northwestern Australia," *Amer. Jour. Sci.,* 265: 737-58.
ANDERSEN, S. T.
1961. "Vegetation and its Environment in Denmark in the Early Weichselian Glacial," *Danmarks Geol. Undersoegelse,* series II, no. 75.
ANDERSEN, S. T., H. DE VRIES, and W. H. ZAGWIJN
1960. "Climatic Change and Radiocarbon Dating in the Weichselian Glacial of Denmark and the Netherlands," *Geol. en Mijnbouw,* 39: 38-42.
ANDREW, G.
1948. "Geology of the Sudan." In J. D. TOTHILL, 1948, pp. 84-128.
ANTEVS, E.
1925. "Retreat of the Last Ice Sheet in Eastern Canada," *Geol. Survey Canada,* Mem. 146.
1952. "Valley Filling and Cutting," *Jour. Geol.,* 60: 375-85.
ARAMBOURG, C.
1952a. "La paléontologie des vertébrés en Afrique du Nord française," *19th Int. Geol. Congr. (Algiers), Monographie région.*
1952b. "The Red Beds of the Mediterranean Basin," *Proc. Pan-Afr. Congr. Prehist.* (Nairobi, 1947): 39-45.
1955. "Le gisement de Ternifine et l'Atlanthropus," *Bull. Soc. Préhist. Franç.,* 52: 90-95.
1962. "Les faunes mammalogiques du Pleistocène circumméditerranéen," *Quaternaria,* 6: 97-109.
ARAMBOURG, C., J. ARÈNES, and G. DEPAPE
1952. "Sur deux floures fossiles quaternaires d'Afrique du Nord," *C. R. Acad. Sci.,* 234: 128-30.
ARAMBOURG, C., and M. ARNOULD
1949. "Note sur les fouilles paléontologiques exécutées en 1947-8 et 1949 dans le gisement villafranchien de la Garet Ichkeul," *Bull. Soc. Sci. Nat. de Tunisie,* 2: 149-57.

ARAMBOURG, C., and L. BALOUT
1952. "Du nouveau à l'Ain Hanech," *Bull. Soc. Hist. Nat. Afrique Nord,* 43: 152-59.
ARKELL, A. J.
1949a. "The Old Stone Age in the Anglo-Egyptian Sudan," *Occ. Papers Sudan Antiq. Serv.,* No. 1.
1949b. *Early Khartoum.* London: Oxford University Press.
1953. *Shaheinab.* London: Oxford University Press.
ARMSTRONG, R. L., W. HAMILTON, and G. H. DENTON
1968. "Glaciation in Taylor Valley, Antarctica, Older than 2.7 Million Years," *Science,* 159: 187-9.
ARNEMAN, H. F., and H. E. WRIGHT
1959. "Petrography of Some Minnesota Tills," *Jour. Sedimentary Petrol.,* 29: 540-54.
ARRHENIUS, G.
1952. "Sediment Cores from the East Pacific." In H. Pettersson (ed.), *Reports of the Swedish Deep Sea Expedition 1947-1948,* Goeteborg: Elanders, Vol. 5, fasc. 1.
ASTRE, G.
1937. "Faune des steppes froide à spermophile et climats du Pleistocène supérieur aux Pyrénées," *Bull. Soc. Géol. Franç.,* 7: 59-68.
AUMEN, H., *et al.*
1965. "Pétrographie des limons de Provence," *Bull. Assoc. Franc. Etude Quat.,* 2: 35-49.
AXELROD, D. I.
1960. "The Evolution of Flowering Plants." In S. Tax (ed.), *Evolution after Darwin,* Chicago: University of Chicago Press, Vol. 1, pp. 227-305.
BAAS, J.
1932. "Eine frühdiluviale Flora im Mainzer Becken," *Zeitschr. Botanik,* 25: 289-371.
BAGNOLD, R. A.
1954. *The Physics of Blown Sand and Desert Dunes* (2d ed.). London: Methuen.
BAKER, B. H.
1967. *Geology of the Mount Kenya Area.* Nairobi: Geol. Survey of Kenya, Report 79.
BAKER, F. S.
1944. "Mountain Climates of the Western United States," *Ecol. Mon.,* 14: 229-43.
BAKKER, J. P.
1957. "Quelques aspects du problème des sédiments corrélatifs en climat tropical humide," *Zeitschr. Geomorph.,* 1: 3-43.
BALOUT, L.
1955. *Préhistoire de l'Afrique du Nord.* Paris: Arts et Métiers Graphiques.
BANDI, H. G.
1944. "Nochmals die Frage: Überwinterte das Rentier?" *Ann. Soc. suisse*

Préhist., 35: 113-18.

BANDI, H. G., and J. MARINGER
1952. *Kunst der Eiszeit.* Basel: Holbein.

BANDY, O. L., E. A. BUTLER, and R. C. WRIGHT
1969. "Alaskan Upper Miocene Marine Glacial Deposits," *Science,* 166: 607-09.

BANNISTER, B.
1963. "Dendrochronology." In BROTHWELL and HIGGS, 1963, pp. 162-76.

BARAT, C.
1957. "Pluviologie et aquiolimétrie dans la zone intertropicale," Dakar: *Mém. Inst. Franç. d'Afrique Noire,* 49.

BARTHOLOMEW, G. A., and J. B. BIRDSELL
1953. "Ecology and the Protohominids," *Amer. Anthropologist,* 55: 481-88.

BARTZ, JOACHIM
1959. "Zur Gliederung des Pleistozäns im Oberrheingebiet," *Zeit. deutsche geol. Ges.,* 111: 653-61.

BATE, D. M. A.
1940. "The Fossil Antelopes of Palestine in Natufian (Mesolithic) Times," *Geol. Mag.,* 77: 418-33.
1955. "Faunas of Quaternary Deposits in Cyrenaica." In MCBURNEY and HEY, 1955, pp. 274-91.

BATES, M.
1953. "Human Ecology." In A. L. Kroeber, (ed.), *Anthropology Today, an Encyclopedic Inventory,* Chicago: University of Chicago Press, pp. 700-13.

BAUD, C. A.
1960. "Dating Prehistoric Bones by Radiological and Optical Methods," *Viking Fund Publ. Anthropol.,* 28: 246-64.

BAUER, A.
1955. "Über die in der heutigen Vergletscherung der Erde als Eis gebundene Wassermasse," *Eiszeitalter und Gegenw.,* 6: 60-70.

BAULIG, H.
1935. *The Changing Sea-Level.* London: G. Philips.

BAUMHOFF, M. A.
1963. "Ecological Determinants of Aboriginal California Populations," *Univ. Calif. Publ. Amer. Arch. Ethnol.,* 49: 155-236.

BENT, A. M., and H. E. WRIGHT
1963. "Pollen Analyses of Surface Materials and Lake Sediments from the Chuska Mountains, New Mexico," *Bull. Geol. Soc. Amer.,* 74: 491-500.

BERG, H.
1949. "Beziehungen zwischen Temperatur- und Niederschlagsanomalien (Bemerkungen zur Eiszeittheorie)," *Rev. Geofisica pura e applicata,* 14: 1-16.
1957. *Solar-terrestrische Beziehungen in Meteorologie und Biologie.* Leipzig: Geest and Portig.

BERNARD, E. A.
1962. *Théorie astronomique des pluviaux et interpluviaux du Quaternaire africain. Mém. Acad. Roy. Sci. d'Outre-Mer* (Brussels), N. S., 12, 1-232.
BERRY, L., and A. J. WHITEMAN
1968. "The Nile in the Sudan," *Geog. Jour.,* 134: 1-37.
BEUCHER, F.
1963. "Flores quaternaires au Sahara nord-occidental (Saoura)," *C. R. Acad. Sci.,* 256: 2205-08.
BEUCHER, F., and G. CONRAD
1963. "L'âge du dernier Pluvial saharien. Essai sur la flore d'un épisode lacustre," *C. R. Acad. Sci.,* 256: 4465-68.
BEUG, H. J., and F. FIRBAS
1961. "Ein neues Pollendiagramm vom Monte Baldo," *Flora,* 150: 179-84.
BIBERSON, P.
1961a. "Le cadre paléogéographique de la préhistoire du Maroc atlantique," Rabat: *Publ. Service des Antiquités du Maroc,* Mém. 16.
1961b. "Le Paléolithique inférieur du Maroc atlantique," *ibid.,* Mém. 17.
BIBERSON, P., and E. AGUIRRE
1965. "Experiences de taille d'outils préhistoriques dans des os d'éléphant," *Quaternaria,* 7: 165-84.
BINFORD, L. R.
1968. "Post-Pleistocene Adaptations." In S. R. Binford and L. R. Binford (eds.), *New Perspectives in Archeology,* Chicago: Aldine, pp. 313-41.
BIRD, J. B.
1967. *The Physiography of Arctic Canada.* Baltimore: Johns Hopkins Press.
BIRDSELL, J. B.
1953. "Some Environmental and Cultural Factors Influencing the Structuring of the Australian Aboriginal Populations," *Amer. Naturalist,* 87: 171-207.
1967. "Preliminary Data on the Trihybrid Origin of the Australian Aborigines," *Archeology and Physical Anthropology in Oceania,* 2: 100-55.
BIRMAN, J. H.
1968. "Glacial Reconnaissance in Turkey," *Bull. Geol. Soc. Amer.,* 79: 1009-26.
BIROT, P.
1954. "Problèmes de morphologie karstique," *Ann. de Géog.,* 63: 161-92.
1965. *Géographie physique générale de la zone intertropicale.* Paris: Centre de Documentation Universitaire.
1968. *The Cycle of Erosion in Different Climates* (trans. C. I. Jackson and K. M. Clayton). Berkeley: University of California Press.
BISHOP, W. W.
1963. "The Later Tertiary and Pleistocene in Eastern Equatorial Africa, with Implications for Primate and Human Distributions," *Viking Fund Publ. Anthropol.,* 36: 246-75.

1967. "Annotated Lexicon of Quaternary Stratigraphical Nomenclature in East Africa." In Bishop and Clark (eds.), 1967, pp. 375-395.

Bishop, W. W., and J. D. Clark, eds.

1967. *Background to Evolution in Africa.* Chicago: University of Chicago Press.

Bishop, W. W., and M. Posnansky

1960. "Pleistocene Environments and Early Man in Uganda," *Uganda Jour.* 24: 44-61.

Black, R. F.

1950. "Permafrost." In P. D. Trask (ed.), *Applied Sedimentation,* New York: J. Wiley, pp. 247-75.

1954. "Permafrost—a Review," *Bull. Geol. Soc. Amer.,* 65: 839-56.

1959. "Geology of Raddatz Rockshelter, Sk 5, Wisconsin," *Wisconsin Archeologist* (Milwaukee), 40: 69-82.

1965. "Ice-Wedge Casts of Wisconsin," *Trans. Wisconsin Acad. Sci., Arts and Letters,* 54: 187-222.

Black, R. F., and W. S. Laughlin

1964. "Anangula: A Geologic Interpretation of the Oldest Archeologic Site in the Aleutians," *Science,* 143: 1321-22.

Black, R. F., and M. Rubin

1968. "Radiocarbon Dates of Wisconsin," *Trans. Wisconsin Acad. Sci., Arts and Letters,* 56: 99-115.

Blanc, A. C.

1935. "Sulla fauna quaternaria dell'Agro Pontino," *Atti Soc. Tosc. Sci. Nat.* (Pisa), *Proc. Verb.,* 44: 108-10.

1936. "La stratigraphie de la plaine côtière de la Basse-Versilia (Italie) et la transgression flandrienne en Méditérranée," *Rev. Geogr. phy. Geol. dyn.,* 9: 129-62.

1942. "Variazoni climatiche ed oscillazioni della linea de riva nel Mediterraneo centrale durante l'era glaciale," *Geol. der Meere u. Binnengewässer,* 5: 137-219.

1957. "On the Pleistocene Sequence of Rome. Paleocologic and Archeologic Correlations," *Quaternaria,* 4: 108-109.

Blanc, A. C., F. Lona, F. Settepassi, G. Cova, and P. Franchesi

1955. "Richerche sul Quaternario Laziale," *Quaternaria,* 2: 151-200.

Blanc, A. C., H. de Vries, and M. Follieri

1957. "A First C14 Date for the Würm Chronology on the Italian Coast," *Quaternaria,* 6: 83-93.

Blanc, G. A.

1921. "Grotta Romanelli, Stratigrafia dei depositi e natura e origine di essi," *Arch. Antropol. Ethnolog. di Firenze,* 50: 65-103.

Blenk, M.

1960. "Ein Beitrag zur morphometrischen Schotteranalyse," *Zeit. f. Geomorph.* (N. S.), 4: 202-42.

Boisse de Black, Y.

1951. *Les glaciations de l'Auvergne.* Aurillac: Imprimerie Moderne.

BÖKÖYNI, S.

1970. "Animal Remains from Lepenski Vir," *Science,* 167: 1702-04.

BONATTI, E.

1966. "North Mediterranean Climate During the Last Würm Glaciation," *Nature,* 209: 984-85.

BOND, G.

1946. "The Pleistocene Succession near Bulawayo," *Occ. Paper Nat. Mus. Southern Rhodesia,* 12: 104-15.

1957. "The Geology of the Khami Stone Age Sites," *Occ. Paper Nat. Mus. Southern Rhodesia* (Vol. 3), 21a: 44-55.

1963. "The Pleistocene in Southern Africa with Implications for Primate and Human Distribution," *Viking Fund Publ. Anthropol.,* 36: 308-34.

BOND, G., and J. D. CLARK

1954. "The Quaternary Sequence in the Middle Zambesi Valley," *South Africa Arch. Bull.,* 9: 115-30.

BONIFAY, E.

1956. "Les sédiments détritiques grossiers dans les remplissages des grottes: méthode d'étude morphologique et statistique," *Anthropologie,* 60:447-61.

1957. "Age et signification des sols rouges méditérranéens en Provence," *C. R. Acad. Sci.,* 247: 3075-77.

1962. *Les terrains quaternaires dans le Sud-Est de la France. Publ. Inst. Prehist. Univ. Bordeaux,* 2: 1-194.

BONIFAY, E., and P. MARS

1959. "Le Tyrrhénien dans le cadre de la chronologie quaternaire méditérranéenne," *Bull. Soc. Géol. Franç.* (ser. 7), 1: 62-78.

BORDES, F.

1953. "Essai de classification des industries 'Moustériennes,' " *Bull. Soc. Préhist. Franç.,* 50: 457-66.

1961a. *Typologie du paléolithique ancien et moyen. Publ. Inst. Préhist., Univ. de Bordeaux, Mém.* 1, 2 Vol.

1961b. "Mousterian Cultures in France," Science, 134: 803-10.

1968. *The Old Stone Age.* New York: World University Library.

BORDES, F., H. LAVILLE, and M. M. PAQUEREAU

1966. "Observations sur le Pleistocène supérieur du gisement de Combe-grenal," *Actes Soc. Linn. de Bordeaux.* 103 (B), 10: 1-19.

BOTTEMA, S.

1967. "A Late Quaternary Pollen Diagram from Ioannina, Northwestern Greece," *Proc. Prehist. Soc.,* 33: 26-29.

BOUCHUD, J.

1954. "Lé renne et lé problème des migrations," *Anthropologie,* 58: 79-85.

BOULE, M., E. CARTAILHAC, R. VERNEAU, and L. DE VILLENEUVE

1906-19. *Les Grottes de Grimaldi (Baoussé-Roussé).* Monaco and Berlin: Friedlaender, 2 vol.

BOULE, M., and L. DE VILLENEUVE

1927. "La Grotte de l'Observatoire à Monaco," *Arch. Inst. Paléont.*

Humaine (Paris), mém. 1.

BOURLIÈRE, F.

1963. "Observations on the Ecology of some Large African Mammals," *Viking Fund Publ. Anthropol.*, 36: 43-54.

BOWLER, J. M., R. JONES, H. ALLEN, and A. G. THORNE

1970. "Pleistocene Human Remains from Australia: A Living Site and Cremation from Lake Mungo, Western N.S.W.," *World Archeology*, 2: 39-60.

BOWMAN, D. C.

1971. "Studies of the Acheulian Industrial Tradition: A Critical Appraisal," in preparation.

BRAIDWOOD, R. J.

1957a. "Means towards an Understanding of Human Behavior before the Present." In *The Identification of Non-artifactural Archeological Materials*, Washington, D.C.: Nat. Acad. Sci.—Nat. Res. Council Publ. 565, pp. 14-16.

1957b. "The Old World: Post-Paleolithic." In *ibid.*, pp. 26-27.

1958. "Near Eastern Prehistory," *Science*, 127: 1419-30.

1960a. "Levels in Prehistory: A Model for the Consideration of the Evidence." In S. Tax (ed.), *Evolution after Darwin*, Chicago: University of Chicago Press, Vol. 2, pp. 143-51.

1960b. "Preliminary Investigations Concerning the Origins of Food-Production in Iranian Kurdistan." *Advancement of Science*, pp. 214-18.

1961. *Prehistoric Men* (5th ed.). Chicago: Chicago Nat. Hist. Mus.

BRAIDWOOD, R. J., and B. HOWE

1962. "Southwestern Asia Beyond the Lands of the Mediterranean Littoral," *Viking Fund Publ. Anthropol.*, 32: 132-46.

BRAIDWOOD, R. J., B. HOWE, *et al.*

1960. "Prehistoric Investigations in Iraqi Kurdistan," *Studies in Ancient Oriental Civilizations* (Chicago), Vol. 31.

BRAIDWOOD, R. J., B. HOWE, and C. A. REED

1961. "The Iranian Prehistoric Project," *Science*, 133: 2008-10.

BRAIDWOOD, R. J., and C. A. REED

1957. "The Achievement and Early Consequences of Food-Production: A Consideration of the Archeological and Natural-Historical Evidence," *Cold Springs Harbour Symposia on Quant. Biology*, 22: 19-31.

BRAIN, C. K.

1967a. "The Transvaal Museum's Fossil Project at Swartkrans," *South African Jour. Sci.*, 63: 378-84.

1967b. "Procedures and Some Results in the Study of Quaternary Cave Fillings." In BISHOP and CLARK (eds.), 1967, pp. 285-301.

1969a. "The Contribution of Namib Desert Hottentots to an Understanding of Australopithecine Bone Accumulations." *Sci. Papers of the Namib Desert Res. Station*, 4: 13-22.

1969b. "Who Killed the Swartkrans Ape-Man?" *Bull. South African Mus. Assoc.*, 9: 127-39.

1971. *The South African Australopithecine Bone Accumulations, Transvaal Museum Mem.*, in press.

BREUIL, H.

1950. *Four Hundred Centuries of Cave Art.* Montignac: Centre d'Études et de Doc. Préhist.

BREUIL, H., and L. KOSLOWSKI

1931-32. "Étude de stratigraphie paléolithique dans le nord de France. La Vallée de la Somme," *Anthropologie,* 41: 449-88; 41: 27-47, 299-314.

BRIGGS, L. C.

1955. *The Stone Age Races of Northwest Africa. Bull. Amer. School Preh. Res. (Peabody Museum),* No. 18.

BRINKMANN, R.

1956. "Tertiär und alt-Quartär in den nordwestlichen Keltiberischen Ketten," *Stille Festschrift, Deutsche Geol. Ges., Geol. Verein, und Paläontol Ges.,* pp. 77-84.

BRINKMANN, R. K., O. MÜNNICH, and J. C. VOGEL

1960. "Anwendung der C-14 Methode auf Bodenbildung und Grundwasserkreislauf," *Geol. Rundschau,* 49: 244-53.

BROECKER, W. S.

1965. "Isotope Geochemistry and the Pleistocene Climatic Records." In WRIGHT and FREY (eds.), 1965, pp. 737-54.

BROECKER, W. S., *et al.*

1968. "Milankovitch Hypothesis Supported by Precise Dating of Coral Reefs and Deep-Sea Sediments," *Science,* 159: 297-300.

BROECKER, W. S., and J. VAN DONK

1970. "Insolation Changes, Ice Volumes and the O^{18} Records in Deep-Sea Cores," *Reviews of Geophysics,* 8: 169-198.

BROECKER, W. S., and T. L. KU

1969. "Caribbean Cores P6304-8 and P6304-9: New Analysis of Absolute Chronology," *Science,* 166: 404-06.

BROECKER, W. S., J. L. KULP, and C. S. TUCEK

1956. "Lamont Natural Radiocarbon Measurements III," *Science,* 124: 154-65.

BROECKER, W. S., and E. A. OLSON

1960. "Radiocarbon from Nuclear Tests," *Science,* 132: 712-21.

BROECKER, W. S., and D. L. THURBER

1965. "Uranium-Series Ages of Pacific Atoll Coral," *Science,* 149: 55-60.

BROEGGER, A. W.

1926. *Kulturgeschichte des norwegischen Altertums.* Leipzig: Harrassowitz. (Translated by V. H. Günther.)

BROOKS, C. E. P.

1949. *Climate Through the Ages* (2d ed.). London: E. Benn.

BROTHWELL, D. R.
1960. "Upper Pleistocene Human Skull from Niah Caves," *Sarawak Museum Jour.*, 9: 323-49.
BROTHWELL, D. R., and E. S. HIGGS (eds.)
1963. *Science in Archeology* (rev. ed., 1970). London: Thames and Hudson.
BROWN, C. A.
1960. *Palynological Techniques.* Baton Rouge: private print.
BROWN, R. J. E.
1960. "The Distribution of Permafrost and Its Relation to Air Temperature in Canada and the U.S.S.R.," *Artic,* 13: 163-77.
BRUNNACKER, K.
1958. "Zur Parallelisierung des Jungpleistozäns in den Periglazialgebieten Bayerns und seiner östlichen Nachbarlander," *Geol. Jahrb.,* 76: 129-50.
1960. "Zur Kenntnis des Spät—und Postglazials in Bayern," *Geologica Bavarica,* 43: 74-150.
1962. "Das Schieferkohlenlager vom Pfefferbichl bei Füssen," *Jahresber. Mitt. Oberrhein. Geol. Ver.,* N.S. 44: 43-60.
1963. "Die Sedimente der Höhlenruine von Hunas," *Eiszeitalter und Gegenw.,* 14: 117-20.
1965. "Schäzungen über die Dauer des Quartärs, insbesondere auf der Grundlage seiner Paläoböden," *Geol. Rund.,* 54: 415-28.
1967a. "Grundzüge einer Löss- und Bodenstratigraphie am Niederrhein," *Eiszeitalter und Gegenw.,* 18: 142-51.
1967b. "Die Sedimente der Crvena Stijena," *Glasnik* (Sarajevo), N.S. 21-22: 31-65.
BRUNNACKER, K., and V. LOŽEK
1969. "Loess-Vorkommen in Südostspanien," *Zeit. Geomorph.,* 13: 297-316.
BRUNNSCHWEILER, D.
1964. "Der pleistozäne Periglazialbereich in Nordamerika," *Zeit. Geomorph.,* 8: 223-31.
BRUSH, G. S.
1967. "Pollen Analyses of Late-Glacial and Postglacial Sediments in Iowa." IN CUSHING and WRIGHT (eds.), 1967, pp. 99-115.
BRYAN, A. L.
1968. "Some Problems and Hypotheses Relative to the Early Entry of Man into America," *Anthropologica,* 10: 157-77.
1969. "Early Man in America and the Late Pleistocene Chronology of Western Canada and Alaska," *Current Anthropol.,* 10: 339-67.
BRYAN, K.
1941. "Pre-Columbian Agriculture in the Southwest as Conditioned by Periods of Alluviation," *Ann. Assoc. Amer. Geog.,* 31: 219-42.
1950. "The Geology and Archeology." In E. W. Haury (ed.), *Ventana Cave, Arizona,* Albuquerque: University of New Mexico Press, pp. 75-126.

BRYAN, K., and L. L. RAY

1940. "Geologic Antiquity of the Lindenmeier Site in Colorado," *Smithsonian Misc. Coll.*, Vol. 99, No. 2.

BRYSON, R. A., and J. A. DUTTON

1961. "Some Aspects of the Variance Spectra of Tree Rings and Varves," *Ann. New York Acad. Sci.*, 95: 580-604.

BRYSON, R. A., and W. M. WENDLAND

1967a. "Radiocarbon Isochrones of the Retreat of the Laurentide Ice Sheet," *Dept. of Meteorol. Tech. Rep. 35* (University of Wisconsin, Madison), 1-28.

1967b. "Tentative Climatic Patterns for Some Late Glacial and Postglacial Episodes in Central North America." In W. J. Mayer-Oakes (ed.), *Life, Land and Water*, Winnipeg University of Manitoba Press, pp. 271-98.

BÜDEL, J.

1944. "Die morphologischen Wirkungen des Eiszeitklimas im gletscherfreien Gebiet," *Geol. Rundschau*, 34: 482-519.

1950a. "Das System der klimatischen Morphologie," *Abh. deut. Geographentags* (München, 1948), München: Amt für Deutsche Landeskunde, pp. 65-100.

1950b. "Die Klimaphasen der Würmeiszeit," *Naturwiss.* 37: 438-49.

1951a. "Die Klimazonen des Eiszeitalters," *Eiszeitalter und Gegenw.*, 1: 16-26.

1951b. "Klima-morphologische Beobachtungen in Süditalien," *Erdkunde*, 5: 73-76.

1952. "Bericht über klimamorphologische und Eiszeitforschungen in Niederafrika," *ibid.*, 6: 104-32.

1953. "Die "Periglazial"–morphologische Wirkungen des Eiszeitklimas auf der ganzen Erde," *ibid.*, 7: 249-66.

1954. "Klima-morphologische Arbeiten in Äthiopien im Frühjahr 1953," *ibid.*, 8: 139-56.

1958. "Die Flächenbildung in den feuchten Tropen und die Rolle fossiler solcher Flächen in anderen Klimazonen," *Abh. deut. Geographentags* (Würzburg, 1957), Wiesbaden: F. Steiner, pp. 89-121.

1960. "Die Frostschutt-Zone Südost-Spitzbergens," *Colloquium Geographicum* (Bonn), Vol. 6.

1961. "Die Abtragungsvoränge auf Spitzbergen im Umkreis der Barentsinsel auf Grund der Stauferland-Expedition 1959/60," *Abh. deut. Geographentags* (Cologne, 1961), Wiesbaden: F. Steiner, pp. 337-75.

1963. "Die pliozänen und quartären Pluvialzeiten der Sahara," *Eiszeitalter und Gegenw.*, 14: 161-87.

1969. "Das System der klima-genetischen Geomorphologie," *Erdkunde*, 23: 165-83.

BUNTING, B. T.

1965. *The Geography of Soil*. Chicago: Aldine.

BUSHNELL, G.

1961. "An Old World View of New World Prehistory," *Amer. Antiquity,* 27: 63-70.

BUTLER, B. E., and J. T. HUTTON

1956. "Parna in the Riverina Plain of Southeastern Australia and the Soils Thereon," *Australian Jour. Agric. Res.,* 7: 536-53.

BUTLER, B. R.

1961. *The Old Cordilleran Culture in the Pacific Northwest. Occ. Papers,* Idaho State Museum, No. 5.

BUTTLER, W.

1938. "Der donauländische und der westiche Kulturkreis der jüngeren Steinzeit." In *Handbuch der Urgeschichte Deutschlands,* Berlin: W. de Gruyter, Vol. 2.

BUTZER, K. W.

1957. "Mediterranean Pluvials and the General Circulation of the Pleistocene," *Geografiska Ann.,* 37: 48-53.

1958a. *Quaternary Stratigraphy and Climate in the Near East, Bonner Geogr. Abhl.,* 24: 1-157.

1958b. "Das ökologische Problem der neolitischen Felsbilder der östlichen Sahara," *Abhl. Akad. Wiss. Lit.* (Mainz), *Math.-Naturw. Kl.,* 1: 20-49.

1958c. "Russian Climate and the Hydrological Budget of the Caspian Sea," *Rév. canadienne de Géog.,* 12: 129-39.

1959a. "Contributions to the Pleistocene Geology of the Nile Valley," *Erdkunde,* 13: 46-67.

1959b. "Die Naturlandschaft Ägyptens während der Vorgeschichte und dem dynastichen Aeitalter," *Abhl. Akad. Wiss. Lit.* (Mainz), *Math.-Naturw. Kl.,* 2: 1-80.

1960a. "On the Pleistocene Shorelines of Arabs' Gulf, Egypt," *Jour. Geol.,* 68: 622-37.

1960b. "Archeology and Geology in Ancient Egypt," *Science,* 132: 1617-24.

1960c. "Dynamic Climatology of Large-Scale European Circulation Patterns in the Mediterranean Area," *Meteor. Rundschau,* 13: 97-105.

1961a. "Climatic Change in Arid Regions since the Pliocene," *Arid Zone Research* (UNESCO), 17: 31-56.

1961b. "Archäologische Fundstellen Ober- und Mittelägyptens in ihrer geologischen Landschaft," *Mitt. Deut. Archäol. Inst., Abt. Kairo,* 17: 54-68.

1961c. "Remarks on Soil Erosion in Spain (Abstract)," *Ann. Assoc. Amer. Geog.,* 52: 405.

1962. "Coastal Geomorphology of Majorca," *ibid.,* pp. 191-212.

1963a. "The Last 'Pluvial' Phase of the Eurafrican Subtropics," *Arid Zone Research* (UNESCO), 20: 211-21.

1963b. "Climatic-Geomorphologic Interpretation of Pleistocene Sediments in the Eurafrican Subtropics," *Viking Fund Publ. Anthropol.*, 36: 1-27.

1964a. "Pleistocene Geomorphology and Stratigraphy of the Costa Brava Region, Catalonia," *Abhl. Akad. Wiss. Lit.* (Mainz), *Math.-Naturw. Kl.*, 1: 1-51.

1964b. "Pleistocene Cold-Climate Phenomena of the Island of Mallorca," *Zeitschr. Geomorph.*, 9: 7-31.

1965. "Acheulian Occupation Sites at Torralba and Ambrona, Spain: Their Geology," *Science*, 150: 1718-22.

1966. "Geologie und Paläogeographie archäologischer Fundstellen bei Sayala (Unternubien)," *Denkschr. Österr. Akad. Wiss.* (Vienna) phil-hist. Kl., 92: 89-98.

1967. "Geomorphology and stratigraphy of the Paleolithic site of Budiño," *Eiszeitalter und Gegenw.*, 18: 82-103.

1969. "Geomorphological Observations in the Lower Omo Basin, Southwestern Ethiopia," *Colloquium Geographicum, Troll-Festschrift.* 12: 177-92.

1970. "Contemporary Depositional Environments of the Omo Delta," *Nature*, 226: 425-30.

1971a. "The Lower Omo Basin: Geology, Fauna and Hominids of Plio-Pleistocene Formations," *Naturwissenschaften*, 58: 7-16.

1971b. Fine Alluvial Fills in the Orange and Vaal Basins of South Africa," *Proc. Assoc. Amer. Geog.*, 3: in press.

1972. "Environmental Changes in Southwestern Asia and Egypt during Terminal Pleistocene and Early Holocene Times," *Fundamenta* (Cologne), Series B, vol. 3, in press.

BUTZER, K. W., F. H. BROWN, and D. L. THURBER

1970. "Horizontal sediments of the Lower Omo Basin: the Kibish Formation," *Quaternaria*, 11 (1969): 15-30.

BUTZER, K. W., and J. CUERDA

1962a. "Coastal Stratigraphy of Southern Mallorca and Its Implications for the Pleistocene Chronology of the Mediterranean Sea," *Jour. Geol.*, 70: 398-416.

1962b. "Nuevos yacimientos marinos cuaternarios de las Baleares," *Notas y Comm. Inst. geol. y minero de España*, 67: 25-70.

BUTZER, K. W., M. H. DAY, and R. E. LEAKEY

1969. "Early *Homo Sapiens* Remains from the Omo River Region of Southwest Ethiopia," *Nature*, 222: 1132-38.

BUTZER, K. W., and O. FRÄNZLE

1959. "Observations on Pre-Würm Glaciations of the Iberian Peninsula," *Zeitschr. Geomorph.*, 3: 85-97.

BUTZER, K. W., and L. G. FREEMAN

1968. "Pollen Analysis at the Cueva del Toll, Catalonia: A Critical Re-appraisal," *Geol. Mijnbouw*, 47: 116-20.

BUTZER, K. W., and C. L. HANSEN

1968. *Desert and River in Nubia: Geomorphology and Prehistoric Envi-*

ronments at the Aswan Reservoir. Madison: University of Wisconsin Press. (With contributions by B. G. Gladfelter, E. G. Leigh, and M. Van Campo.)

Butzer, K. W., and D. L. Thurber

1969. "Some Late Cenozoic Sedimentary Formations of the Lower Omo Basin," *Nature,* 222: 1138-43.

Cahen, L.

1954. *Géologie du Congo Belge.* Liège: H. Vaillant-Carmanne.

Cailleux, A.

1942. "Les Actions éoliennes periglaciaires en Europe," *Mém. Soc. Géol. France* (N.S.), Vol. 21, No. 46.

Cailleux, A., and J. Tricart

1963. *Initiation à l'etude des sables et des galets.* Paris: Centre de Documentation Universitaire. Vol. I (text; preceded by Volumes II and III of tabular data, Paris, 1959.)

Cain, S. A.

1944. *Foundations of Plant Geography.* New York: Harper.

Camps, G., G. Délibrias, and J. Thommeret

1968. "Chronologie absolue et succession des civilisations préhistoriques dans le nord de l'Afrique," *Libyca,* 16: 9-28.

Carpenter, E. F.

1955. "Astronomical Aspects of Geochronology." In T. L. Smiley (ed.), *Geochronology, Univ. Ariz. Phys. Sci. Bull.,* 2: 29-74.

Castany, G., and F. Ottmann

1957. "Le Quaternaire marin de la Méditérranée occidentale," *Rev. Géog. phys. Géol. dyn.* (2d ser.), 1: 46-55.

Caton-Thompson, G.

1952. *The Kharga Oasis in Prehistory.* London: Athlone. Geology by E. W. Gardner.

Caton-Thompson, G., and E. W. Gardner

1929. "Recent Work on the Problem of Lake Moeris," *Geog. Jour.,* 73: 20-60.

1932. "The Prehistoric Geography of the Kharga Oasis," *ibid.,* 80: 369-409.

1934. *The Desert Fayum,* London: Roy. Anthropol. Soc., 2 vol.

1939. "Climate, Irrigation and Early Man in the Hadhramaut," *Geogr. Jour.,* 93 18-38.

Chaney, R. W.

1940. "Tertiary Forests and Continental History," *Bull. Geol. Soc. Amer.,* 51: 469-88.

Chang, K. C.

1962. "China," *Viking Fund Publ. Anthropol.,* 32: 177-92.

1967. "Major Aspects of the Interrelationship of Archaeology and Ethnology," *Current Anthropology,* 8: 227-43.

1968. *The Archaeology of Ancient China.* New Haven: Yale University Press.

CHARD, C. S.

1969a. *Man in Prehistory*. New York: McGraw-Hill.

1969b. "Archeology in the Soviet Union," *Science*, 163: 774-79.

CHARLESWORTH, J. K.

1957. *The Quaternary Era*. London: E. Arnold, 2 vol.

CHAVAILLON, J.

1964. *Les formations quaternaires du Sahara nord-occidental*. Paris: Centre Nat. Rech. Sci.

CHERDYNTSEV, V. V., I. V. KAZACHEVSKIJ, G. I. KISLITSINA, E. A. KUZMINA, and N. V. KIND

1966. "Non-equilibrium Uranium in Carbonate Deposits and Their Age Determination" (in Russian), *Geokhimiya*, 2: 139-145.

CHILDE, V. G.

1925. *The Dawn of European Civilization* (7th ed., 1957). London: Routledge and Kegan Paul.

1929. *The Most Ancient East* (4th ed., 1954). London: Routledge and Kegan Paul.

1956. *Piecing together the Past*. London: Routledge and Kegan Paul.

1958. *The Prehistory of European Society*. Harmondsworth: Pelican.

CHOUBERT, G.

1948a. "Sur l'âge des limons rouges superficiels du Maroc," *C. R. Acad. Sci.*, 227: 558-60.

1948b. "Sur la nature des limons rouges superficiels du Maroc," *ibid.*, 639-41.

1957. "L'étage moghrébien dans le Maroc occidentale," *Actos V. Cong. Int. INQUA* (Madrid-Barcelona, 1957), in press.

1962. "Réflexion sur les parallelismes probables des formations quaternaires atlantiques du Maroc avec celles de la Méditérranée," *Quaternaria*, 6: 137-75.

CLARK, A. H.

1954. "Historical Geography." In P. E. James and C. F. Jones (eds.), *American Geography, Inventory and Prospect*, Syracuse: Syracuse University Press (for the Assoc. Amer. Geog.), pp. 70-105.

CLARK, J. D.

1954. *The Prehistoric Cultures of the Horn of Africa*. Cambridge: Cambridge University Press.

1960. "Human Ecology During the Pleistocene and Later Times in Africa South of the Sahara," *Current Anthropology*, 1: 307-24.

1963. "Prehistoric Cultures of Northeast Angola and Their Significance for Tropical Africa" (with an appendix on pollen samples by E. M. van Zinderen Bakker), *Museu do Dundo, Publ. Culturais*, No. 62, 2 vol.

1966. "Acheulian Occupation Sites in the Middle East and Africa: A Study in Cultural Variability," *Amer. Anthropol.*, 68.2 (part 2): 202-29.

1967a. *Atlas of African Prehisotry*. Chicago: University of Chicago Press.

1967b. "The Problem of Neolithic Culture in Subsaharan Africa." In BISHOP and CLARK (eds.), 1967, pp. 601-627.

CLARK, J. D. (ed.)

1969. *Kalambo Falls Prehistoric Site,* Vol. I. Cambridge: Cambridge University Press. Geology by G. Bond (pp. 197-214), E. G. Haldemann (pp. 20-45), M. R. Kleindienst (pp. 46-56). Pleistocene vegetation by E. M. van Zinderen Bakker (pp. 57-84), with identification of macrobotanical remains by F. White, L. Chalk, and T. C. Whitmore (pp. 216-24).

CLARK, J. G. D.

1945. "Farmers and Forests in Neolithic Europe," *Antiquity,* 19: 57-71.

1952. *Prehistoric Europe: The Economic Basis.* London: Methuen.

1953. "The Economic Approach to Prehistory," *Proc. Brit. Acad.,* 39: 215-38.

1954. *Excavations at Star Carr: An Early Mesolithic Site at Seamer, near Scarborough, Yorkshire* (with palynological contributions by D. Walker and H. Godwin). Cambridge: Cambridge University Press.

1957. *Archaeology and Society* (2d ed.). London: Methuen.

CLARK, D. L.

1968. *Analytical Archaeology.* London: Methuen.

CLARKE, R. J., F. C. HOWELL, and C. K. BRAIN

1970. "More Evidence of an Advanced Hominid at Swartkrans," *Nature,* 225: 1219-21.

CLISBY, K. B., F. FOREMAN, and P. B. SEARS

1957. "Pleistocene Climatic Changes in New Mexico," *Geobot. Inst. Rübel* (Zürich), 34: 21-26.

CLUTTON-BROCK, J.

1963. "The Origins of the Dog." In BROTHWELL and HIGGS, 1963, pp. 269-74.

COETZEE, J. A.

1967. *Pollen Analytical Studies in East and Southern Africa. Palaeoecology of Africa* (Cape Town: Balkema), 3: 1-146.

COLE, SONIA

1963. *The Prehistory of East Africa.* London: Macmillan.

COLINVAUX, P. A.

1967. "Quaternary Vegetational History of Arctic Alaska." In HOPKINS (ed.), 1967, pp. 207-31.

COLLINS, D.

1969. "Culture Traditions and Environment of Early Man," *Current Anthropology,* 10: 267-316.

CONRAD, G.

1969. *L'évolution continentale post-hercynienne du Sahara algérien.* Paris: Centre Nat. Rech. Sci.

COOK, R. M.

1963. "Archaeomagnetism." In BROTHWELL and HIGGS (eds.), 1963, pp. 72-84.

COOK, S. F.

1951. "The Fossilization of Human Bone: Calcium Phosphate, and Carbonate," *Univ. Calif. Publ. Amer. Arch. Ethnol,* 40: 263-80.

1960. "Dating Prehistoric Bone by Chemical Analyses," *Viking Fund Publ. Anthropol.,* 28: 223-45.

Cook, S. F., S. T. Brooks, and H. E. Ezra-Cohn

1962. "Historical Studies on Fossil Bone," *Jour. Paleontol.,* 36: 483-94.

Cook, S. F., and R. F. Heizer

1952. "The Fossilization of Human Bone: Organic Components and Water," *Univ. Calif. Arch. Survey Dept.,* No. 17.

Cooke, H. B. S.

1947. "The Development of the Vaal River and Its Deposits." *Trans. Geol. Soc. South Africa* (1946), 243-260.

1958. "Observations Relating to Quaternary Environments in East and South Africa," *Geol. Soc. South Africa,* Annexure to Vol. 40.

1963. "Pleistocene Mammalian Faunas of Africa, with Particular Reference to Southern Africa," *Viking Fund Publ. Anthropol.,* 36: 65-116.

1964. "The Pleistocene Environment in Southern Africa." In D. H. S. Davis (ed.), *Ecological Studies in Southern Africa,* The Hague: Junk, pp. 1-23.

1967. "The Pleistocene Sequence in Southern Africa and Problems of Correlation." In Bishop and Clark (eds.), 1967 pp. 175-184.

Coon, C. S.

1939. *The Races of Europe.* New York: Macmillan.

1962. *The Origin of Races.* New York: A. A. Knopf.

Coque, R.

1962. *La Tunisie présaharienne: étude géomorphologique.* Paris: A. Colin.

Corbel, J.

1961. "Morphologie périglaciaire dans l'arctique," *Ann. de Geog.,* 70: 1-24.

Cornwall, I. W.

1956. *Bones for the Archaeologist.* London: Phoenix House.

1958. *Soils for the Archaeologist.* London: Phoenix House.

Cotton, C. A.

1942. *Climatic Accidents in Landscape Making.* London: Whitcombe and Tombs.

1945. "The Significance of Terraces due to Climate Oscillations," *Geol. Mag.* 82: 10-16.

1949. *Geomorphology.* New York: J. Wiley.

1961. "The Theory of Savanna Planation," *Geography,* 46: 89-101.

Cox, A.

1969. "Geomagnetic Reversals," *Science,* 163: 237-45.

Cox, A., R. R. Doell, and G. B. Dalrymple

1965. "Quaternary Paleomagnetic Stratigraphy." In Wright and Frey (eds.), 1965, pp. 817-30.

Crabtree, D. E.

1969. "A Technological Description of Artifacts in Assemblage 1, Wilson Butte Cave, Idaho," *Current Anthropology,* 10: 366-67.

CRUSAFONT PAIRÓ, M.
1960. "Le quaternaire espagnol et sa faune de mammifères—essai et synthèse," *Mammalia Pleistocaenica* (Brno), 1: 55-64.

CUERDA, J.
1957. "Fauna marina del Tirreniense de la Bahia de Palma," *Bol. Sci. Hist. Nat. de Baleares,* 3: 1-76.

CUERDA, J., and A. MUNTANER
1960. "Nota sobre diversos nivales tirrenienses localizados en las cercanias del Cap Orenol (Mallorca)," *Bol. Soc. Hist. Nat. Baleares,* 6: 37-47.

CUERDA, J., and J. SACARES
1965. "Nuevos yacimientos cuaternarios en la costa de Lluchmayor (Mallorca)," *Bol. Soc. Hist. Nat. Baleares,* 10: 89-130.
1966. "Nueva contribución al estudio del Pleistoceno marino del Termino de Lluchmayor (Mallorca)," *Bol. Soc. Hist. Nat. Baleares,* 12: 63-99.

CURRAY, J. R.
1965. "Late Quaternary History, Continental Shelves of the United States." In WRIGHT and FREY (eds.), 1965, pp. 723-735.

CURRY, R. R.
1966. "Glaciation about 3,000,000 Years Ago in the Sierra Nevada," *Science,* 154: 770-71.

CURTIS, G. H.
1967. "Notes on Some Miocene to Pleistocene Potassium-Argon Results." In BISHOP and CLARK (eds.), 1967, pp. 365-69.

CUSHING, E. J.
1967. "Late Wisconsin Pollen Stratigraphy and the Glacial Sequence in Minnesota." In CUSHING and WRIGHT (eds.), 1967, pp. 59-88.

CUSHING, E. J., and H. E. WRIGHT
1967. *Quaternary Paleoecology.* New Haven: Yale University Press.

DAGLEY, P., *et al.*
1967. "Geomagnetic polarity zones for Icelandic lavas," *Nature,* 216: 25-29.

DAHLBERG, A. A.
1960. "The Dentition of the First Agriculturists (Jarmo, Iraq)," *Amer. Jour. Phys. Anthropol.,* 18: 243-56.

DAHLBERG, A. A., and V. M. CARBONELL
1961. "The Dentition of the Magdalenian Female from Cap Blanc, France," *Man,* 49-50.

DALRYMPLE, G. B., and M. A. LANPHERE
1969. *Potassium-Argon Dating.* San Francisco: Freeman.

DAMON, P. E.141-54.
1968. "Radiocarbon and Climate," *Meteorological Monographs,* vol. 8, 30: 151-54.

DAMON, P. E., A. LONG, and D. C. GREY
1966. "Fluctuations of Atmospheric C^{14} During the Last Six Millenia," *Jour. Geophys. Res.,* 71: 1055-1063.

DANSGAARD, W., S. J. JOHNSEN, J. MOELLER, and C. C. LANGWAY
"One Thousand Centuries of Climatic Record from Camp Century on the Greenland Ice Sheet," *Science,* 166: 377-81.

DANSGAARD, W., and H. TAUBER
1969. "Glacier Oxygen-18 Content and Pleistocene Ocean Temperatures," *Science,* 166: 499-502.

DARBY, H. C.
1956. "The Clearing of the Woodland in Europe." In THOMAS, 1956, pp. 183-216.

DARLINGTON, P. J.
1957. *Zoogeography, the Geographical Distribution of Animals.* New York: J. Wiley.

DAVIES, J. L.
1967. "Tasmanian Landforms and Quaternary Climates," In JENNINGS and MABBUTT (eds.), 1967, pp. 1-25.
1969. *Landforms of Cold Climates.* Cambridge: M.I.T. Press.

DAVIS, M. B.
1965. "Phytogeography and Palynology of Northeastern United States." In WRIGHT and FREY (eds.), 1964, pp. 377-402.
1967. "Late-Glacial Climate in Northern United States. In CUSHING and WRIGHT (eds.), 1967, pp. 11-44.

DAVIS, M. B., and E. S. DEEVEY
1964. "Pollen Accumulation Rates: Estimates from Late-Glacial Sediment of Rogers Lake," *Science,* 145: 1293-95.

DAVIS, W. M.
1930. "Origin of Limestone Caverns," *Bull. Geol. Soc. Amer.*, 4: 475-628.

DAWSON, E. W.
1963. "Bird Remains in Archaeology." In BROTHWELL and HIGGS, 1963, pp. 279-93.

DEEVEY, E. S.
1949. "Biogeography of the Pleistocene," *Bull. Geol. Soc. Amer.,* 60: 1314-1416.

DEEVEY, E. S., M. S. GROSS, G. E. HUTCHINSON, and H. L. KRAYBILL
1954. "The Natural C^{14} Contents of Materials from Hard-Water Lakes," *Proc. Nat. Acad. Sci.* (Washington), 40: 285-88.

DEFFONTAINES, P.
1930. "Essai de géographie préhistorique de la Tchéchoslovaquie," *Anthropologie,* 40: 275-82.
1933. "Essai de géographie préhistorique du Limoussin et son pourtour sédimentaire," *Ann. de Géog.,* 42: 461-76.

DE GEER, G.
1912. "A Geochronolology of the Last 12,000 Years," *Proc. 11th Int. Geol. Congr.* (Stockholm, 1912), 1: 241-58.
1934. "Equatorial Paleolithic Varves in East Africa," *Geografiska Ann.,* 16: 75-96.

1940. "Geochronologia Suecica principles," *K. Svenska Vetensk. Handl.* (ser. 3), Vol. 18, No. 6.

DEGERBOEL, M., and H. KROG
1959. "The Reindeer in Denmark: Zoological and Geological Investigations of the Discoveries in Danish Pleistocene Deposits," *Biol. Skrifter Danske Videnskabernes Selskab* (Copenhagen), Vol. 10, No. 4.

DE JONG, J. D.
1967. "The Quaternary of the Netherlands." In K. Rankama (ed.), *The Quaternary*. New York: Interscience, vol. 2, pp. 301-426.

DÉLIBRIAS, G., and P. DUTIL
1966. "Formations calcaires lacustres du Quaternaire supérieur dans le massif central saharien et datations absolues. *C. R. Acad. Sci.*, 262: 55-58.

DÉLIBRIAS, G., and H. J. HUGOT
1962. "Datation par la méthode C^{14} de néolithique de l'Adrar-Bous." In H. J. Hugot (ed.), *Missions Berliet: Ténéré-Tchad.* Paris. Arts et Métiers Graphiques, pp. 71-72.

DELIBRIAS, G., H. GUGOT, and P. QUÉZEL
1959. "Trois datations de sédiments sahariens récents par le radiocarbone," *Libyca*, 5: 267-70.

DELPORTE, H.
1955. "L'industrie de Chatelperron et son extension géographique," *Congres Préhist. de France* (Strasbourg-Metz, 1953), pp. 233-49.

DEVEVAN, W. N.
1966. *The Aboriginal Cultural Geography of the Llanos de Mojos of Bolivia.* Berkeley: University of California Press.

DE PLOEY, J.
1965. "Position géomorphologique, génèse et chronologie de certains dépôts superficiels au Congo occidental," *Quaternaria*, 7: 131-54.

DEVORE, I., and S. L. WASHBURN
1962. "Baboon Ecology and Human Evolution," *Viking Fund Publ. Anthropol.*, 36: 335-67.

DEVRIES, H., G. W. BARENDSEN, and H. T. WATERBOLK
1958. "Groningen Radiocarbon Dates II," *Science*, 127: 129-37.

DE VRIES, H., and A. DREIMANIS
1960. "Finite Radiocarbon Dates of the Port Talbot Interstadial Deposits in Southern Ontario," *Science*, 131: 1738.

DE VRIES, H., F. FLORSCHÜTZ, and J. MENÉNDEZ-ARMOR
1960. "Un diagramme pollinique simplifié d'une couche de "gytjja" située à Poueyferré près de Lourdes," *K. Nederlandse Akad. Wetenschappen* (ser. B), 63: 498-500.

D'HOORE, J. L.
1960. "The Soils Map of Africa South of the Sahara (1:5 million)," *Proc. 7th Int. Congr. Soil Science*, Madison, vol. 2, 11-19.

DIKAIOS, P.
1953. *Khirokitia.* London: Oxford University Press.

DILLON, L. S.
1956. "Wisconsin Climate and Life Zones in North America," *Science,* 123: 167-76.
DONN, W. L., W. R. FARRAND, and M. E. EWING
1962. "Pleistocene Ice Volumes and Sea-Level Lowering," *Jour. Geol.,* 70: 206-14.
DONNER, J. J.
1963. "The Zoning of the Post-Glacial Pollen Diagrams in Finland and the Main Changes in Forest Composition," *Acta Botanica Fennica,* No. 65.
1964. "Pleistocene Geology of Eastern Long Island, New York," *Amer. Jour. Sci.,* 262: 355-76.
DONNER, J. J., and B. KURTÉN
1958. "The Floral and Faunal Succession of Cueva del Toll, Spain," *Eiszeitalter und Gegenw.,* 9: 72-82.
DORF, E.
1955. "Plants and the Geologic Time Scale," *Geol. Soc. Amer., Spec. Paper* 62: 575-92.
DOORNKAMP, J. C.
1968. "The Role of Inselbergs in the Geomorphology of Southern Uganda," *Trans. Inst. Brit. Geogr.,* 44: 151-162.
DOWNIE, C.
1964. "Glaciations of Mt. Kilimanjaro, Northeast Tanganyika," *Bull. Geol. Soc. Amer.,* 75: 1-16.
DREIMANIS, A.
1960. "Pre-classical Wisconsin in the Eastern Portion of the Great Lakes Region," *21st. Int. Geol. Congr.,* Norden, 1960, pt. 4, pp. 108-19.
DREIMANIS, A., J. TERASMAE, and G. D. MCKENZIE
1966. "The Port Talbot Interstate of the Wisconsin Glaciation," *Canad. Jour. Earth Sci.,* 3: 305-25.
DUBIEF, J.
1952. "Le vent et le déplacement du sable au Sahara," *Trav. Inst. Rech. Sahar.* (Algiers), 8: 1-44.
DUBOIS, C., and P. ZANGHERI
1957. "Palynologie de quelques sédiments tourbeux de le basse plaine du Po," *Bull. Serv. Carte Géol. (Alsace-Lorraine),* 10: 145-50.
DÜCKER, A.
1937. "Über Strukturböden im Riesengebirge. Ein Beitrag zum Bodenfrost-und Lössproblem," *Zeit. deut. Geol. Ges.,* 89: 113-29.
1951. "Über die Entstehung von Frostspalten," *Schriften Naturw. Ver. Schleswig-Holstein,* 25: 58-64.
DUIGAN, S. L., with B. W. SPARKS
1963. "Pollen Analyses of the Cromer Forest Bed Series in East Anglia," *Phil. Trans. Roy. Soc. London,* (series B), 246: 149-202.
DUNHAM, K. C., *et al.*
1968. "The Geochronological Significance of Argon-40/Argon-30 Age

Determination on White Whin from the Northern Penine Orefield," *Proc. Roy. Soc.,* Series A, 307: 251-266.

DURAND, J. H.

1959. "Les sols rouges et les croûtes en Algérie," *Serv. des Études Scient.* (Alger-Birmandreis).

DURHAM, J. W.

1950. "Cenozoic Marine Climates of the Pacific Coast," *Bull. Geol. Soc. Amer.,* 61: 1243-64.

DURY, G. H.

1959. *The Face of the Earth.* Harmondsworth: Penguin.

1967. "Climatic Change as a Geographical Backdrop," *Australian Geographer,* 10: 231-42.

DYCK, W.

1967. "Recent Developments in Radiocarbon Dating: Their Implications for Geochronology and Archaeology," *Current Anthropology,* 8: 349-51.

DYLIK, J.

1952. "The Concept of the Periglacial Cycle in Middle Poland," *Bull, Soc. Sci. Letters of Lodz, Cl. III., Sci. Math. Nat.,* 3: 1-29.

1956. "Coup d'oeil sur la Pologne périglaciaire," *Biuletyn Peryglacjalny,* 4: 195-238.

EDWARDS, W. E.

1967. "The Late Pleistocene Extinction and Diminution in Size of Many Mammalian Species." In MARTIN and WRIGHT (eds.), 1967, pp.

EISELEY, L. C.

1955. "The Paleo Indians: Their Survival and Diffusion." In "New Interpretations of Aboriginal American Culture History," 75 *Anniv. Vol. Anthropol. Soc. Washington,* pp. 1-11.

ELISSÉEFF, V.

1960. "Das Paläolithikum Nordostasiens." In A. Varagnac (ed.), *Der Mensch der Urzeit,* Düsseldorf and Cologne: E. Diederichs, pp. 116-40.

ELKIN, A. P.

1954. *The Australian Aborigines: How to Understand Them.* Sydney: Angus and Robertson.

ELSON, J. A.

1967. "Geology of Glacial Lake Agassiz." In W. J. Mayer-Oakes (ed.), *Life, Land and Water,* Winnipeg: University of Manitoba Press, pp. 37-96.

ELTON, C.

1946. The Ecology of Animals (2d ed.). London: Methuen.

EMBLETON, C., and C. A. M. KING

1968. *Glacial and Periglacial Geomorphology.* London: Arnold.

EMILIANI, C.

1955. "Pleistocene Temperature Variations in the Mediterranean," *Quaternaria,* 2: 87-98.

1961. "Cenozoic Climatic Changes as Indicated by the Stratigraphy and Chronology of Deep-Sea Cores of Globigerina-Ooze Facies," *Ann. New York Acad. Sci.,* 95: 421-36.

1964. "Paleotemperature Analysis of the Caribbean Cores A254-Br-C and

Cp-28," *Bull. Geol. Soc. Amer.*, 75: 129-44.

1966. "Paleotemperature Analysis of Caribbean Cores P6304-8 and P6304-9 and a Generalized Temperature Curve for the Past 425,000 Years," *Jour. Geol.*, 74: 109-26.

EMILIANI, C., and J. GEISS

1957. "On Glaciations and Their Causes," *Geol. Rundschau*, 47: 576-601.

EMILIANI, C., T. MAYEDA, and R. SELLI

1961. "Paleotemperature Analysis of the Plio-Pleistocene Section at Le Castella, Calabria, Southern Italy," *Bull. Geol. Soc. Amer.*, 72: 679-88.

EMILIANI, C., and T. MAYEDA

1964. "Oxygen Isotopic Analysis of Some Molluscan Shells from Fossil Littoral Deposits of Pleistocene Age," *Amer. Jour. Sci.*, 262: 107-13.

ERDTMANN, G.

1954. *An Introduction to Pollen Analysis*. Waltham: Chronica Botanica.

ERGENZINGER, P.

1968. "Beobachtungen im Gebiet des Trou au Natron, Tibestigebirge," *Die Erde*, 99: 176-83.

ERICSON, D. B., M. EWING, G. WOLLIN, and B. C. HEEZEN

1961. "Atlantic Deep-Sea Sediment Cores," *Bull. Geol. Soc. Amer.*, 72: 193-286.

ERICSON, D. B., and G. WOLLIN

1968. "Pleistocene Climates and Chronology in Deep-Sea Sediments," *Science*, 162: 1227-34.

ERMOLOVA, N. M.

1963. "O faune mlekopitaiuschchikh epokhi palelita i neolita Pribaikal'ia," *Materialy po Etnografii (Moscow)*, vyp. 3: 27-64.

EVANS, I. S.

1970. "Salt Crystallization and Rock Weathering: A Review," *Rev. Géomorph. dyn.*, 19: 153-77.

EVERNDEN, J. F., and G. H. CURTIS

1965. "The Potassium-Argon Dating of Late Cenozoic Rocks in East Africa and Italy," *Current Anthropology*, 6: 343-85.

EVERNDEN, J. F., D. E. SAVAGE, G. H. CURTIS, and G. T. JAMES

1964. "Potassium-Argon Dates and the Cenozoic Mammalian Chronology of North America," *Amer. Jour. Sci.*, 262: 145-98.

EWING, J. F.

1951. "Comments on the Report of H. E. Wright on His Study of Lebanese Marine Terraces," *Jour. Near Eastern Studies*, 10: 119-22.

EYRE, S. R.

1963. *Vegetation and Soils: A World Picture*, London: E. Arnold.

FAEGRI, K., and J. IVERSEN

1964. *Textbook of Modern Pollen Analysis* (2d ed.). Copenhagen: E. Munksgaard.

FAIRBRIDGE, R. W.

1948. "The Geology and Geomorphology of Point Peron, Western Australia," *Jour. Roy. Soc. Western Australia*, 33: 1-43.

1961. "Eustatic Changes in Sea Level," *Physics and Chemistry of the Earth,* 4: 99-185.

1965. "Eiszeitklima in Nordafrika," *Geol. Rund.,* 54: 399-414.

FARRAND, W. R.

1961. "Frozen Mammoths and Modern Geology," *Science,* 133: 729-35.

1964. "The Deglacial Hemicycle," *Geol. Rund.,* 54: 385-98.

1969. "Geology, Climate and Chronology of Yabrud Rockshelter I," *Alfred-Rust Festschrift (Fundamenta,* Series A, vol. 2), pp. 121-32.

FAURE, H.

1966. "Evolution des grands lacs sahariens à l'Holocène," *Quaternaria,* 8: 167-75.

FAURE, H., and P. ELOUARD

1967. "Schéma des variations du niveau de l'océan Atlantique sur la côte de l'Ouest de l'Afrique depuis 40,000 ans," *C. R. Acad. Sci.,* 165: 784-87.

FELIX, C. J.

1961. "Palynology." In H. N. Andrews (ed.), *Studies in Paleobotany,* New York: J. Wiley, pp. 436-62.

FINK, J.

1959. "Geologische Problemstellung." In F. Felgenhauer, J. Fink, and H. de Vries, "Studien zur absoluten and relativen Chronologie der fossilen Böden Österreichs. I.," *Archaeologica Austriaca,* 25: 35-73.

1960. "Leitlinien einer österreichischen Quartärstratigraphie," *Mitt. Geol. Ges. Wien,* 53: 249-66.

1961. "Die Südabdachung der Alpen," *Mitt. Österr. Bodenkundl. Ges.,* 1961: 123-83.

1962. "Die Gliederung des Jungpleistozäns in Österreich," *ibid,* 54: 1-25.

1965. "The Pleistocene in Eastern Austria," *Bull. Geol. Soc. Amer., Spec. Paper,* 84: 179-99.

FIBRAS, F.

1937. "Der Pollenanalytische Nachweis des Getreidebaus," *Zeitschr. Botanik,* 31: 447-78.

1949-52. *Spät-und nacheiszeitliche Waldgeschichte Mitteleuropas nordlich der Alpen.* Jena: G. Fischer, 2 vol.

1950. "The Late-Glacial Vegetation of Central Europe," *New Phytologist,* 49: 163-73.

FIRBAS, F., and B. FRENZEL

1960. "Floren-und Vegetationsgeschichte seit dem Ende des Tertiärs," *Fortschr. der Botanik,* 22: 87-111.

FIBRAS, F., and P. ZANGHERI

1954. "Über neue Funde pflanzenführender Ablagerungen in der südlichen Po-Ebene bei Forli," *Nachr. Akad. Wiss. Göttingen,* pp. 11-18.

FITTING, J. E., J. DE VISSCHER, and E. J. WAHLA

1966. *The Paleo-Indian Occupation of the Holcombe Beach. Anthropol. Papers,* Mus. Anthropol. Univ. Michigan, 27: 1-147.

FLANNERY, K. V.

1966. "The Postglacial 'Readaptation' as Viewed from Mesoamerica,"

Amer. Antiquity, 31: 800-05.

1969. "Origins and Ecological Effects of Early Domestication in Iran and the Near East." In UCKO and DIMBLEBEY (eds.), 1969, pp. 73-100.

FLEISCHLER, R. L., P. B. PRICE, R. M. WALKER, and L. S. B. LEAKEY

1965. "Fission-Track Dating of Bed I, Olduvai Gorge," *Science,* 148: 72-74.

FLEROW, C. C.

1967. "On the Origin of the Mammalian Fauna of Canada." In HOPKINS (ed.), 1967, pp. 271-80.

FLINT, R. F.

1957. *Glacial and Pleistocene Geology* (1st ed., 1947). New York: J. Wiley.

1959. "Pleistocene Climates in Eastern and Southern Africa," *Bull. Geol. Soc. Amer.,* 70: 343-74.

1965. "The Pliocene-Pleistocene Boundary," *Spec. Paper, Geol. Soc. Amer.,* 84: 497-533.

1966. "Comparison of Interglacial Marine Stratigraphy in Virginia, Alaska and Mediterranean Areas," *Amer. Jour. Sci.,* 264: 673-84.

FLINT, R. F., and G. BOND

1968. "Pleistocene Sand Ridges and Pans in Western Rhodesia," *Bull. Geol. Soc. Amer.,* 79: 299-314.

FLOHN, H.

1952. "Allgemeine atmosphärische Zirkulation und Paläoklimatologie," *Geol. Rundschau,* 40: 153-78.

1953. "Studien über die atmosphärische Zirkulation in der letzten Eiszeit," *Erdkunde,* 7: 266-75.

FLÖRSHUTZ, F., and A. M. H. VAN SOMEREN

1950. "The Paleo-Botanical Boundary Pliocene-Pleistocene in the Netherlands," *Proc. 18th Int. Geol. Congr.* (London, 1948), sec. 9: 40-46.

FÖLSTER, H.

1964. "Morphogenese der südsudanesischen Pediplane." *Zeit. f. Geomorph.,* 8: 393-423.

FORDE, C. D.

1934. *Habitat, Economy and Society.* New York: E. P. Dutton.

FORDE-JOHNSTON, J.

1959. *Neolithic Cultures of North Africa.* Liverpool: The University Press.

FORSYTH, J. L.

1965. "Age of the Buried Soil in the Sidney, Ohio Area," *Amer. Jour. Sci.,* 263: 571-97.

FRAGA TORREJON, E. DE

1958. "Catalogo bibliografico de la fauna cuaternaria asturiana," *Monograf. Geol., Inst. Geol. Aplicada,* (Oviedo), No. 8.

FRANK, A. H. E.

1969. "Pollen Stratigraphy of the Lake of Vico (Central Italy)," *Palaeogeography, Palaeoclimatology and Palaeoecology,* 6: 67-85.

FRANZ, H.

1960. *Feldbodenkunde als Grundlage der Standortsbeurteilung und Bodenwirtschaft.* Vienna and Munich: G. Fromme.

1967. "On the Stratigraphy and Evolution of Climate in the Chad Basin during the Quaternary." In BISHOP and CLARK (eds.), 1967, pp. 273-283.

FRÄNZLE, O.

1959. "Glaziale und Periglaziale Formbildung im östlichen Kastilischen Scheidegebirge," *Bonner Geogr. Abhl.*, Vol. 26.

1965. *Die pleistozäne Klima-und Landschaftsentwicklung der nördlichen Pro-Ebene. Abhandl. Akad. Wiss. Liter.* (Mainz), *Math.-Naturw. Kl.*, 1965, no. 8: 1-141.

FRECHEN, J., and G. V. D. BOOM

1959. "Die sedimentpetrographische Horizontierung der Pleistozänen Terrassenschotter im Mittelrheingebiet," *Fortschritte zur Geologie des Rheinland und Westfalen,* 4: 89-125.

FRECHEN, J., and H. J. LIPPOLT

1965. "Kalium-Argon-Daten zum Alter des Laacher Vulkanismus, der Rheinterrassen und der Eiszeiten," *Eiszeitalter und Gegenw,* 16: 5-30.

FREEMAN, L. G.

1966. "The Nature of Mousterian Facies in Cantabrian Spain," *Amer. Anthropol.,* 68, no. 2, pt. 2: 230-37.

FRENZEL, B.

1959-60. "Die Vegetations-und Landschaftszonen Nord-Eurasiens während der letzten Eiszeit und während der postglazialen Wärmezeit," *Abhl. Akad. Wiss. Lit.* (Mainz), *Math.-Naturw. Kl.,* Nr. 13; 1960, Nr. 6.

1964. "Zur Pollenanalyse von Lössen," *Eiszeitalter und Gegenw.,* 15: 5-39.

1965. "Über die offene Vegetation der letzten Eiszeit am Ostrande der Alpen," *Verhandl. Zool.-Botan. Ges. Wien,* 103-104: 110-43.

1966. "Climatic Change in the Atlantic/Sub-Boreal Transition on the Northern Hemisphere; Botanical Evidence," in *World Climate from 8000 to 0 B. C.* London: Royal Meteorological Society, pp. 99-123.

1967. *Die Klimaschwankungen des Eiszeitalters.* Braunschweig: Vieweg.

1968a. *Grundzüge der Pleistozänen Vegetationsgeschichte Nord-Eurasiens.* Wiesbaden: Steiner.

1968 b. "The Pleistocene Vegetation of Northern Eurasia," *Science,* 161: 637-49.

FRENZEL, B., and C. Troll

1952. "Die vegetationszonen des nödlichen Eurasiens während der letzten Eiszeit," *Eiszeitalter und Gegenw.,* 2: 154-67.

FREY, D. G.

1952. "Pollen Analysis of the Horry Clay and a seaside Peat Deposit near Myrtle Beach, South Carolina," *Amer. Jour. Sci.,* 250: 212-25.

1953. "Regional Aspects of the Late-Glacial and Postglacial Pollen Succession of Southeastern North Carolina," *Ecol. Monographs,* 23: 289-313.

FRIEDMAN, I., R. L. SMITH, and D. CLARK

1963. "Obsidian Dating." In BROTHWELL and HIGGS (eds.), 1963, pp. 47-58.

FRIES, M.

1962. "Pollen Profiles of Late Pleistocene and Recent Sediments from Weber Lake, Minnesota," *Ecology,* 43: 295-308.

1963. "Vad myren berättar," *Sartryck Sveriges Naturs Arsbok,* pp. 91-107.

FRIES, M., H. E. WRIGHT, and M. RUBIN

1961. "A Late-Wisconsin Buried Peat near North Branch, Minnesota," *Amer. Jour. Sci.,* 259: 670-93.

FRITTS, H. C.

1963. "Recent Advances in Dendrochronology in America with Reference to the Significance of Climatic Change," *Arid Zone Research* (UNESCO), 20: 255-63.

1965. "Dendrochronology." In WRIGHT and FREY (eds.), 1965, pp. 871-79.

1966. "Growth-Rings of Trees: Their Correlation with Climate," *Science,* 154: 973-79.

FRITTS, H. C., D. G. SMITH, and M. A. STOKES

1965. "The Biological Model for Paleoclimatic Interpretation of Mesa Verde Tree-Ring Series," *Amer. Antiquity,* 31: 101-21.

FRYE, J. C.

1962. "Comparison between Pleistocene Deep-Sea Temperatures and Glacial and Interglacial Episodes," *Bull. Geol. Soc. Amer.,* 73: 263-66.

FRYE, J. C., and A. B. LEONARD

1965. "Quaternary of the Southern Great Plains." In WRIGHT and FREY (eds.), 1965, pp. 203-16.

FRYE, J. C., H. B. WILLMAN, and R. F. BLACK

1965. "Outline of Glacial Geology of Illinois and Wisconsin." In WRIGHT and FREY (eds.), 1965, pp. 43-62.

FRYE, J. C., H. B. WILLMAN, and H. D. GLASS

1968. "Correlation of Midwestern Loesses with the Glacial Succession." In SCHULTZ and FREY (eds.), 1968, pp. 3-22.

GAILLARD, C.

1934. "Contribution à l'étude de la faune préhistorique de l'Egypte," *Arch. Mus. Hist. Nat. Lyon,* 14: 1-125.

GALLOWAY, R. W.

1965. "Late Quaternary Climates in Australia," *Jour. Geol.,* 73: 603-18.

1970. "Full-Glacial Climate in Western U.S.A.," *Ann. Assoc. Amer. Geog.,* 60: 245-56.

GAMS, H.

1950. "Die Allerödschwankung im Spätglazial," *Zeitschr. Gletscherkunde* (N.S.), 1: 162-71.

GANSSEN, R., and F. HÄDRICH

1965. *Atlas zur Bodenkunde.* Mannheim: Bibliographisches Institut.

GARDNER, E. W.
1932. "Some Lacustrine Mollusca from the Faiyum Depression: A Study in Variation," *Mém. Inst. d'Égypte,* 18: 1-123.
1935. "The Pleistocene Fauna and Flora of Kharga Oasis, Egypt," *Quart. Jour. Geol. Soc.,* 91: 479-518.
GARROD, D. A. E.
1958. "The Natufian Culture: The Life and Economy of a Mesolithic People in the Near East," *Proc. Brit. Acad.,* 43: 211-27.
GARROD, D. A. E., and D. M. A. BATE
1937. *The Stone Age of Mt. Carmel, Vol. I., Excavations at the Wady el-Mughara.* Oxford: Clarendon Press.
GARROD, D. A. E., *et al.*
1928. "Excavation of a Mousterian Rock-Shelter at Devil's Tower (Gibraltar)," *Jour. Roy. Anthropol. Soc.,* 58: 33-113.
GELLERT, J. F.
1962. "Das Lössproblem in China." *Petermann's Mitt.,* 106: 81-94.
GEORGE, W.
1962. *Animal Geography.* London: Heinemann.
GERMAIN, L.
1923. "Les climats des temps quaternaires d'après les mollusques terrestres et fluviatiles," *Anthropologie,* 33: 301-22.
GIDDINGS, J. L.
1954. "Tree Ring Dating in the American Arctic," *Tree Ring Bull.,* 20: 23-25.
GIGOUT, M.
1960. "Nouvelles recherches sur le Quaternaire marocain et comparisons avec l'Europe," *Trav. Labor. Géol. Fac. Sci. Lyon* (N.S.), No. 6.
1962. "Sur le Tyrrhénien de la Méditérranée occidentale," *Quaternaria,* 6: 209-28.
GILL, E. D.
1967. "Significance of Aitape Radiocarbon Dates for Eustasy and Tectonics," *Austr. Jour. Sci.,* 30: 142-47.
1968. "Palaeoecology of Fossil Human Skeletons," *Palaeogeography, Palaeoclimatology and Palaeoecology,* 4: 211-17.
GITTERMAN, R. E., and L. V. GOLUBEVA
1967. "Vegetation of Eastern Siberia during the Anthropogene Period." In HOPKINS (ed.), 1967, pp. 232-44.
GLADFELTER, B. G.
1971. *Meseta and Campiña Landforms: A Geomorphology of the Alto Henares Basin. Research Papers,* University of Chicago Geography Dept., 130: 1-204.
GLINKA, K. D.
1927. *The Great Soil Groups of the World and Their Development.* Translated by C. F. Marbut. Ann Arbor: Edwards Bros. (mimeographed).
GODWIN, H.
1944. "Neolithic Forest Clearance," *Nature,* 153: 511-14.

1956. *The History of the British Flora*. Cambridge: Cambridge University Press.

1962. "Half-life of Radiocarbon," *Nature*, 195: 984.

GODWIN, H., D. WALKER, and E. H. WILLIS

1957. "Radiocarbon Dating and Post-glacial Vegetational History," *Proc. Roy. Soc. London* (series B.), 147: 352-66.

GODWIN, H., and E. H. WILLIS

1959. "Radiocarbon Dating of the Late Glacial Period in Britain," *Proc. Royal Soc. London* (seriesB), 150: 199-215.

GOLDTHWAIT, R. P., A. DREIMANIS, J. L. FORSYTH, P. F. KARROW, and G. W. WHITE

1965. "Pleistocene Deposits of the Erie Lobe." In WRIGHT and FREY (eds.), 1965, pp. 85-98.

GORMAN, C. F.

1969. "Hoabinhian: A Pebble-Tool Complex with Early Plant Associations in Southeast Asia," *Science*, 163: 671-73.

1971. "Nimrods, Piscators, Pluckers and Planters," *World Arch.*, 3: in press.

GRADMANN, R.

1906. "Beziehungen zwischen Pflanzengeographie und Siedlungsgeschichte," *Geog. Zeit.*, 12: 305-25.

1933. "Die Steppenheide-theorie," *ibid.*, 39: 265-78.

1936. "Vorgeschichtliche Landwirtschaft und Besiedlung," *ibid.*, 42: 378-86.

GRAHAM, J. A., and R. F. HEIZER

1968. "Man's Antiquity in North America: Views and Facts," *Quaternaria*, 9(1967): 225-37.

GRAHMANN, R.

1932. "Der Lösz in Europa," *Mitt. Ges. Erdkunde Leipzig*, 51: 5-24.

1937. "Form und Entwässerung des nord-europhaischen Inlandeises," *ibid.*, 54: 48-70.

1955. "The Lower Palaeolithic Site of Markkleeberg and Other Comparable Localities near Leipzig," *Trans. Amer. Phil. Soc.*, 45: 507-687.

GRAUL, H.

1955. "Bemerkungen zu einer geologischen Übersichtskarte des Iller-Riss-Gebietes," *Zeit. Deut. Geol. Ges.*, 105: 517-24.

GRAUL, H. (ed.)

1968. *Beiträge zu den Exkursionen analässlich der Deuqua-Tagung* 1968 *in Biberach on der Riss. Heidelberger Geograph. Arb.*, 20: 1-124.

GRAUL, H., and K. BRUNNACKER

1962. "Eine Revision der Pleistozänen Stratigraphie des schwäbischen Alpenvorlandes," *Petermanns Mitt*, 106: 253-271.

GREENMAN, E. F.

1963. "The Upper Paleolithic and the New World," *Current Anthropology*, 4: 41-66.

GROSS, H.
1958. "Die bisherigen Ergebnisse von C-14 Messungen und paläolithischen Untersuchungen für die Gliederung und Chronologie des Jungpleistozäns in Mitteleuropa und den Nachbahrgebieten," *Eiszeitalter und Gegenw.*, 9: 155-87.

GROVE, A. T.
1958. "The Ancient Erg of Hausaland and Similar Formations on the South Side of the Sahara," *Geog. Jour.*, 124: 528-33.
1969. "Landforms and Climatic Change in the Kalahari and Ngamiland," *Geogr. Jour.*, 135: 191-212.

GROVE, A. T., and R. A. PULLAN
1963. "Some Aspects of the Pleistocene Paleogeography of the Chad Basin," *Viking Fund Publ. Anthropol.*, 36: 230-45.

GROVE, A. T., and A. WARREN
1968. "Quaternary Landforms and Climate on the South Side of the Sahara," *Geogr. Jour.*, 134: 194-208.

GRÜGER, E.
1970. "The Development of the Vegetation of Southern Illinois Since Late Illinoian Time," *Rev. Géogr. Phys. Géol. Dyn.*, 12: 143-48.

GRUHN, R.
1961. *The Archeolcgy of Wilson Butte Cave, South-central Idaho. Occ. Papers,* Idaho State College Museum, No. 6.

GRUND, A.
1903. "Die Karsthydrographie, Studien aus Westbosnien," *Pencks geog. Abhandl.*, 7: 103-200.

GUENTHER, E. W.
1961. *Sedimentpetrographische Untersuchungen von Lossen* (Part I). Cologne: Böhlau.

GUILCHER, A.
1958. *Coastal and Submarine Morphology* (trans. B. W. Sparks and R. H. W. Kneese). London: Methuen.

GUILDAY, J. E.
1967. "Differential Extinction During Late Pleistocene and Recent Times." In MARTIN and WRIGHT (eds.), 1967, pp. 121-40.

GUTENBERG, B.
1941. "Changes in Sea Level, Postglacial Uplift, and Mobility of the Earth's Interior," *Bull. Geol. Soc. Amer.*, 52: 721-72.

HAEKEL, J. (ed.)
1961. *Theorie und Praxis der Zusammenarbeit zwischen den anthropologischen Disziplinen.* (Bericht über das 2. österreichische Symposion auf Burg Wartenstein, 1959.) Horn: F. Berger.

HAFSTEN, U.
1961. "Pleistocene Development of Vegetation and Climate in the Southern High Plains as Evidenced by Pollen Analysis." In WENDORF, *et al.*, 1961, pp. 59-91.

HAGEDORN, H. J., and D. JÄKEL

1969. "Bemerkungen zur quartären Entwicklung des Reliefs im Tibesti-Gebirge," *Bull. Assoc. sénégalaise Étude Quatern. Ouest afric.*, no. 23-24: 25-43.

HALL, E. T.

1963. "Dating Pottery by Thermoluminescence." In BROTHWELL and HIGGS (eds.), 1963, pp. 90-92.

HAMILTON, W.

1968. "Cenozoic Climatic Change and its Cause." In *Causes of Climatic Change, Meteorological Monographs,* 8, no. 30: 128-133.

HAMMEN, T. VAN DER

1952. "Dating and Correlation of Periglacial Deposits in Middle and Western Europe," *Geol. en Mijinbouw,* 14: 328-36.

1957a. "The Stratigraphy of the Late-Glacial," *ibid.,* 19: 250-54.

1957b. "The Age of the Usselo Culture," *ibid.,* pp. 396-97.

HAMMEN, T. VAN DER, G. C. MAARLEVELD, J. C. VOGEL, and W. H. ZAGWIJN

1967. "Stratigraphy, Climatic Succession and Radiocarbon Dating of the Last Glacial in the Netherlands," *Geol. Mijnbouw,* 46: 79-95.

HAMMEN, T. VAN DER, T. A. WIJMSTRA, and W. H. VAN DER MOLEN

1965. "Palynological Study of a Very Thick Peat Section in Greece, and the Würm-Glacial Vegetation in the Mediterranean Region," *Geol. Mijnbouw,* 44: 37-39.

HANSEN, S.

1940. "Varvity in Danish and Scanian Late-Glacial Deposits, with Special Reference to the System of Ice-Lakes at Egernsund," *Danmarks Geol. Undersoegelse,* Vol. 2, No. 63.

HARE, F. K.

1954. "The Boreal Conifer Zone," *Geographical Studies* (Birbeck College, London), 1: 1-19.

HARING, A., A. E. DE VRIES, and H. DE VRIES

1958. "Radiocarbon Dating up to 70,000 Years by Isotopic Enrichment," *Science,* 128: 472-73.

HARLAN, J. R., and D. ZOHARY

1966. "Distribution of Wild Wheats and Barley," *Science,* 153: 1074-80.

HARRISON, W., R. J. MALLEY, G. A. RUSNAK, and J. TERASMAE

1965. "Late Pleistocene Uplift, Chesapeake Bay Entrance," *Jour. Geol.,* 73: 201-29.

HARTSHORNE, R.

1959. *Perspective on the Nature of Geography.* (Assoc. of Amer. Geographers, Monograph Series, No. 1.) Chicago: Rand McNally.

HASTENRATH, S.

1960. "Klimatische Voraussetzungen und grossräumige Verteilung der Frost-strukturböden," *Zeitschr. Geomorph.,* 4: 69-73.

1967. "The Barchans of the Arequipa Region, Southern Peru," *Zeit. f. Geomorph.,* 11: 300-331.

HAURY, E. W., E. B. SAYLES, and W. W. WASLEY
1959. "The Lehner Mammoth Site," *Amer. Antiquity,* 25: 2-30.
HAY, R. L.
1965. "Stratigraphy of Beds I through IV, Olduvai Gorge, Tanganyika," *Current Anthropology,* 6: 387-89, and discussion comments, pp. 381-83.
1967. "Revised Stratigraphy of Olduvai Gorge." In BISHOP and CLARK (eds.), 1967, pp. 221-28.
HAYNES, C. V.
1964. "Fluted Projectile Points: Their Age and Dispersal," *Science,* 145: 1408-13.
1965. "Stratigraphy of the Lehner Site, Arizona." In L. A. Heindl and E. K. Reed (eds.), *Guidebook for Inqua Field Conference. H, Southwestern Arid Lands,* Lincoln: Nebraska Acad. Sci., pp. 55-57. (With molluscan data by J. C. Bequaert.)
1967. "Carbon-14 Dates and Early Man in the New World." In MARTIN and WRIGHT (eds.), 1967, pp. 267-86.
1968a. "Geochronology of Late Quaternary Alluvium." In R. B. Morrison and H. E. Wright (eds.), *Means of Correlation of Quaternary Successions,* Salt Lake City: University of Utah Press, pp. 591-631.
1968b. "Preliminary Report on the Late Quaternary Geology of the San Pedro Valley, Arizona," *Southern Arizona Guidebook III,* Arizona Geol. Soc. pp. 79-96.
1969. "The Earliest Americans," *Science,* 166: 709-15.
HAYNES, C. V., and G. A. AGOGINO
1966. "Prehistoric Springs and Geochronology of the Clovis Site," *Amer. Antiquity,* 31: 812-21.
HAYNES, C. V., and D. C. GREY
1965. "The Sister's Hill Site and Its Bearing on the Wyoming Post-glacial Alluvial Chronology," *Plains Anthropologist,* 10: 196-207.
HEEKEREN, H. R. VAN
1957. *The Stone Age of Indonesia. Verhandl. Koninkl. Inst. Taal-, Land-, Volkenkunde,* 21.
HEICHELHEIM, F. M.
1956. "Effects of Classical Antiquity on the Land." In THOMAS, 1956, pp. 165-82.
HEINZELIN, J. DE
1957. *Les fouilles de Ishango.* Bruxelles: Institut des Parcs Nat. du Congo Belge.
1963. "Paleoecological Conditions of the Lake Albert-Lake Edward Rift," *Viking Fund Publ. Anthropol.,* 36: 276-84; discussion, pp. 602-3.
1964. "Le sous-sol du temple d'Aksha," *Kush,* 12: 102-10.
1968. "Geological History of the Nile Valley in Nubia." In F. WENDORF, *et al.,* pp. 19-55.
HEIZER, R. F.
1955. "Primitive Man as an Ecologic Factor," *Kroeber Anthropol. Papers,* 13: 1-31.

1960. "Physical Analysis of Habitation Residues," *Viking Fund Publ. Anthropol.,* 28: 93-157.

1963. "Domestic Fuel in Primitive Society," *Jour. Roy. Anthropol. Inst.,* 93: 186-194.

HEIZER, R. F., (ed.)

1958. *A Guide to Archaeological Field Methods* (3d ed.). Palo Alto: National Press.

HELBAEK, H.

1959. "Domestication of Food Plants in the Old World," *Science,* 130: 365-73.

1960a. "The Paleoethnobotany of the Near East and Europe." In BRAIDWOOD, HOWE, *et al.,* 1960, pp. 98-118.

1960b. "Ecological Effects of Irrigation in Ancient Mesopotamia," *Iraq,* 22: 186-96.

1966. "Commentary (1966) on the Phylogenesis of *Triticum* and *Hordeum," Econ. Bot.,* 20: 350-60.

HESCHELER, K., and E. KÜHN

1949. "Die Tierwelt der prähistorischen Siedlungen der Schweiz." In O. Tschumi, *Urgeschichte der Schweiz,* Frauenfeld, Vol. I, pp. 121-368.

HESS, P., and H. BREZOWSKY

1952. "Katalog der Grosswetterlagen Europas," *Ber. Deut. Wetterdienstes US-Zone,* No. 33.

HESSE, R., W. C. ALLEE, and K. P. SCHMIDT

1951. *Ecological Animal Geography* (2d ed.). New York: Wiley.

HESTER, J. J.

1967. "The Agency of Man in Animal Extinctions." In MARTIN and WRIGHT (eds.), 1967, pp. 169-92.

HEUSSER, C. J.

1965. "A Pleistocene Phytogeographical Sketch of the Pacific Northwest and Alaska." In WRIGHT and FREY (eds.), 1965, pp. 469-84.

HEVLY, R. H.

1964. "Paleoecology of Laguna Salada," *Fieldiana* (Anthropol.), 55: 171-87.

HEY, R. W.

1962. "The Quaternary and Palaeolithic of Northern Libya," *Quaternaria,* 6: 435-49.

1963. "Pleistocene Screes in Cyrenaica (Libya)," *Eiszeitalter und Gegenw.,* 14: 77-84.

1968. "The Quaternary Geology of the Jabal al-Akhdar Coast." In F. T. Barr (ed.), *Geology and Archaeology of Northern Cyrenaica,* Tripoli: Petrol. Explor. Soc. of Libya, pp. 159-65.

HIBBARD, C. W., D. E. RAY, D. E. SAVAGE, D. W. TAYLOR, and J. E. GUILDAY

1965. "Quaternary Mammals of North America." In WRIGHT and FREY (eds.), 1965, pp. 509-26.

HIERNAUX, J.

1963. "Some Ecological Factors Affecting Human Populations in

Sub-Saharan Africa," *Viking Fund Publ. Anthropol.,* 36: 534-46.

HIGGS, E. S.

1961. "Some Pleistocene Faunas of the Mediterranean Coastal Areas," *Proc. Prehist. Soc.,* 27: 144-54.

1967. "Environment and Chronology." In MCBURNEY (ed.), 1967, pp. 16-74.

HIGGS, E. S., and M. R. JARMAN

1969. "The Origins of Agriculture: A Reconsideration," *Antiquity,* 43: 31-41.

HIJSZELER, C. W. W. J.

1957. "Late-glacial Human Cultures in the Netherlands," *Geol. en Mijnbouw,* 19: 288-302.

HILLY, J.

1962. *Étude géologique du Massif de l'Edough et du Cap de Fer. Bull. Service Carte Géol. d'Algérie,* N.S. 19.

HO, P. T.

1969. "The Loess and the Origin of Chinese Agriculture," *Amer. Hist. Rev.,* 75: 1-36.

HOBLER, P. M., and J. J. HESTER

1969. "Prehistory and Environment in the Libyan Desert," *South Afr. Arch. Bull.,* 33: 120-30.

HOLE, F.

1966. "Investigating the Origins of Mesopotamian Civilization," *Science,* 153: 605-11.

HOLE, F., and K. V. FLANNERY

1962. "Excavations at Ali Kosh, Iran, 1961," *Iranica Antiqua,* 2: 97-148.

1967. "The Prehistory of Southwestern Iran: A Preliminary Report," *Proc. Prehist. Soc.,* 33: 147-206.

HOOIJER, D. A.

1961. "The Fossil Vertebrates of Ksar Akil, A Paleolithic Rock Shelter in the Lebanon," *Zoologische Verhandelingen* (Leiden), No. 49.

HOPKINS, D. M. (ed.)

1967. *The Bering Land Bridge.* Stanford: Stanford University Press.

HOPKINS, D. M., and J. L. GIDDINGS

1953. *Geologic Background of Iyatayet Archeological Site, Cape Denbeigh, Alaska. Smithsonian Misc. Collec.* 121, No. 11.

1965. "The Quaternary Geology and Archaeology of Alaska." In WRIGHT and FREY (eds.), 1965, pp. 355-76.

HOPKINS, D. M., T. N. V. KARLSTROM, *et al.*

1955. "Permafrost and Groundwater in Alaska," *U. S. Geol. Survey Prof. Paper* 264 f: 113-46.

HOPKINS, D. M., and F. S. SIGAFOS

1950. "Frost Action and Vegetation Patterns on Seward Penninsula, Alaska," *U. S. Geol. Survey Bull.* 974-C: 51-101.

HOWE, B.

1967. *The Palaeolithic of Tangier, Morocco: Excavations at Cape Ash-*

akar, 1939-1947. Bull. Amer. School of Prehist. Res., 22: 1-200. With a chapter on the geology by C. E. Stearns.

Howell, F. C.

1958. "Upper Pleistocene Men of the Southwest Asian Mousterian." In *Neanderthal Centenary,* Utrecht: Kemink, pp. 185-98.

1959a. "Upper Pleistocene Stratigraphy and Early Man in the Levant," *Proc. Amer. Philos. Soc.,* 103: 1-65.

1959b. "The Villafranchian and Human Origins," *Science,* 130: 831-44.

1960. "European and Northwest African Middle Pleistocene Hominids," *Current Anthropology,* 1: 195-232.

1961. "More on Middle Pleistocene Hominids: A Reply," *ibid.,* 2: 118-20.

1966. "Observations on the Earlier Phases of the European Lower Paleolithic," *Spec. Publ., Amer. Anthropologist* pt. 2, v. 68, no. 2: 88-201.

1967. "Recent Advances in Human Evolutionary Studies," *Quart. Rev. Biol.,* 42: 471-513.

1968a. Review of P. V. Tobias, *The Cranium and maxillary dentition of Australopithecus boisei. Amer. Anthropol.,* 70: 1029-30.

1968b. "Omo Research Expedition, 1967," *Nature,* 219: 567-572.

1969. "Some Remains of Hominidae from Pliocene/Pleistocene Formations in the Lower Omo Basin," *Nature,* 223: 1234-39.

Howell, F. C., and J. D. Clark

1963. "Acheulian Hunter-Gatherers of Sub-Saharan Africa," *Viking Fund Publ. Anthropol.,* 36: 458-533.

Howell, F. C., G. H. Cole, and M. R. Kleindienst

1962. "Isimila, an Acheulian Occupation Site in the Iringa Highlands, Southern Highlands Province, Tanganyika," *Actes IV Congr. Pan-Africain de Préhistoire et de l'Etude du Quaternaire, Musée Royal de l'Afrique Centrale Ann.* (Tervuren), 40: 42-80. (Geological section after E. G. Haldemann, pp. 45-60.)

Huckriede, R.

1962. "Jung-Quartär und End-Mesolithikum in der Provinz Kerman (Iran)," *Eiszeitalter und Gegenw.* 12: 25-42.

Hugot, H. J.

1968. "The Origins of Agriculture: Sahara," *Current Anthropol.,* 9: 483-88.

Irwin, H. T., J. B. Wheat, and L. F. Irwin

1968. *Investigations of Paleolithic and Epipaleolithic Sites in the Sudan. Univ. Utah Anthropol. Papers,* 90: 1-123.

Irwin-Williams, C.

1967. "Association of Early Man with Horse, Camel and Mastodon at Hueyatlaco, Valsequillo." In Martin and Wright (eds.), 1967, pp. 337-47.

Irwin-Williams, C., and C. V. Haynes

1970. "Climatic Change and Early Population Dynamics in the Southwestern United States," *Quat. Research,* 1: 59-71.

ISSAC, G. L.
1966. "The Geological History of the Olorgesailie Area," *Actes V. Congr. Pan-Africain de Préhistoire et de l'Étude du Quaternaire,* Teneriffa (1964), 125-33.
1967. "The Stratigraphy of the Peninj Group." In BISHOP and CLARK (eds.), 1967, pp. 229-58.
1968a. "Traces of Pleistocene Hunters: An East African Example." In LEE and DEVORE (eds.), 1968, pp. 253-261.
1968b. "Divisions Within the Acheulian of Eastern Africa: Some Interpretative Suggestions," Annual meetings of American Anthropological Association, Seattle, Nov. 21, 1968 (mimeographed).
1969. "Studies of Early Culture in East Africa," *World Archaeology,* 1: 1-28.

IVERSEN, J.
1949. "The Influence of Prehistoric Man on Vegetation," *Danmarks Geol. Undersoegelse,* Series IV, Vol. 3, No. 6.
1954. "The Late-Glacial Flora of Denmark and Its Relation to Climate and Soil," *ibid.,* Series II, 80: 87-119.
1956. "Forest Clearance in the Stone Age," *Scient. American,* 194(3): 36-41.
1958. "The Bearing of Glacial and Interglacial Epochs on the Formation and Extinction of Plant Taxa," *Uppsala Univ. Arsskr.,* 6: 210-15.
1960. "Problems of the Early Post-Glacial Forest Development in Denmark," *ibid.,* Series IV, Vol. 4, No. 3.

JACKSON, M. L.
1958. *Soil Chemical Analysis.* Englewood Cliffs, N.J.: Prentice-Hall.
1964. "Soil Clay Mineralogical Analysis." In C. I. Rich and G. W. Kunze (eds.), *Soil Clay Mineralogy,* Chapel Hill: University of North Carolina Press, pp. 245-94.

JÁNOSSY, D.
1961. "Die Entwicklung der Kleinsäugerfauna Europas im Pleistozän (Insectivora, Rodentia, Lagomorpha)," *Zeitschr. Saugetierkunde,* 26: 1-11.

JÄTZOLD, R.
1960. "Aride und humide Jahreszeiten in Nordamerika," *Stuttgarter Geog. Studien,* 71.

JELGERSMA, S.
1962. "A Late Glacial Pollen Diagram from Madelia, South-central Minnesota," *Amer. Jour. Sci.,* 260: 522-29.
1966. "Sea-Level Changes During the Last 10,000 Years," in *World Climate from 8000 to 0 B. C.* London: Roy. Meteor. Soc., pp. 54-71.

JELINEK, A. J.
1967. "Man's Role in the Extinction of Pleistocene Faunas." In MARTIN and WRIGHT (eds.), 1967, pp. 193-200.

JENNINGS, J. D.
1957. *Danger Cave. Memoirs,* Society for American Archaeology, No. 14.

1966. "Early Man in the Desert West," *Quaternaria,* 8: 81-90.
1968. *Prehistory of North America.* New York: McGraw-Hill.

JENNINGS, J. D., and E. NORBECK (eds.)
1964. *Prehistoric Man in the New World.* Chicago: University of Chicago Press.

JENNINGS, J. N., and J. A. MABBUTT (eds.)
1967. *Landform Studies from Australia and New Guinea.* Cambridge: Cambridge University Press.

JOHNSSON, G.
1960. "Cryoturbations at Zaragoza, Northern Spain," *Zeitschr. Geomorph.,* 4: 74-80.

JONES, R.
1968. "The Geographical Background to the Arrival of Man in Australia and Tasmania," *Archaeology and Physical Anthropology in Oceania,* 3: 186-215.
1969. "Fire-stick Farming," *Austr. Nat. Hist.,* September, 224-28.

JUDSON, S.
1949. "The Pleistocene Stratigraphy of Boston, Massachusetts." In E. S. Barghoorn, *et al., "The Boylston Street Fishweir II,* Peabody Found. Archaeol. Papers, Vol. 4, 1: 7-48.
1950. "Depressions of the Northern Portion of the Southern High Plains of Eastern New Mexico," *Bull. Geol. Soc. Amer.,* 61: 253-74.
1953a. "Geology of the San Jon Site, Eastern New Mexico," *Smithsonian Misc. Coll.,* Vol. 121.
1953b. "Geology of the Hodges Site, Quay County, New Mexico," *Bull. Bureau Amer. Ethnol.,* 154: 285-302.

KAHLKE, H. D.
1961. "Revision der Säugetierfaunen der klassischen deutschen Pleistozän-Fundstellen von Süssenborn, Mosbach und Taubach," *Geologie,* 10: 493-525.
1968. "Zur relativen Chronologie ostasiatischer mittelpleistozänen Faunen und Hominoidea-Funde." In G. Kurth (ed.), *Evolution und Hominisation,* Stuttgart: Fischer, pp. 91-118.

KAHLKE, H. D. (ed.)
1965. *Das Pleistozän von Voigtstedt. Palaontol. Abhandl.,* Abt. A, 2: 221-692.

KAISER, K. H.
1960. "Klimazeugen des periglazialen Dauerfrostbodens in Mittel- und Westeuropa," *Eiszeitalter und Gegenw.,* 11: 121-41.

KAISER, W.
1961. "Bericht über eine archäologisch-geologische Felduntersuchung in Ober-und Mittelägypten," *Mitt. deut. archäol. Inst., Abt. Kairo,* 17: 1-53.

KAPP, R. O.
1965. "Illinoian and Sangamon Vegetation in Southwestern Kansas and Adjacent Oklahoma," *Contributions,* Univ. Michigan Museum Paleontology, 19: 165-257.

KAPP, R. O., and A. M. GOODING
1964. "Pleistocene Vegetational Studies in the Whitewater Basin, Southeastern Indiana," *Jour. Geol.,* 72: 307-26.
KATSUI, Y., and Y. KONDO
1967. "Dating Method Using Hydration Layer of Obsidian," *Quaternary Research* (Tokyo), 6: 168-71.
KEITH, M. L., and G. M. ANDERSON
1963, "Radiocarbon Dating: Fictitious Results with Mollusk Shells," *Science,* 141: 634-36.
KELLEY, R. W., and W. R. FARRAND
1967. *The Glacial Lakes Around Michigan.* Bulletin, Michigan Geological Survey, 4: 1-23.
KELLOGG, C. E.
1941. "Climate and Soil," *Yearbook of Agriculture, 1941* (U. S. Dept. of Agriculture): 265-91.
KEMPF, E. K.
1966. "Das Holstein-Interglazial von Tonisberg im Rahmen des niederrheinischen Pleistozäns," *Eiszeitalter und Gegenw,* 17: 5-60.
KEMPTON, J. P., and J. E. HACKETT
1968. "Stratigraphy of the Woodfordian and Altonian Drifts of Central Northern Illinois." In R. E. Bergstrom (ed.), *The Quaternary of Illinois,* University of Illinois, College of Agriculture, Spec. Publ. 14, pp. 27-34.
KENDALL, R. L.
1969. "An Ecological History of the Lake Victoria Basin," *Ecol. Monographs,* 39: 121-76.
KENYON, K. M.
1959. "Earliest Jericho," *Antiquity,* 33: 5-9.
KEREKES, J.
1951. "Zur periglazialen Sedimentbildung in mitteleuropaischen Höhlen," *Quartär,* 5: 41-49.
KILMER, V. J., and L. T. ALEXANDER
1949. "Methods of Making Mechanical Analysis of Soils," *Soil Science,* 68: 15-24.
KIND, N. V.
1967. "Radiocarbon Chronology in Siberia." In HOPKINS (ed.), 1967, pp. 172-92.
KIND, N. V., *et al.*
1969. "New Materials on the Absolute Chronology on the Upper Pleistocene Glaciations of Siberia (according to C^{14} data)" (in Russian), *Doklady Akad. Nauk SSSR,* 184 (6): 1387-90.
KING, L. C.
1961. "The Palaeoclimatology of Gondwanaland During the Palaeozoic and Mesozoic Eras." In NAIRN, 1961, pp. 307-31.
KLAER, W.
1956. "Verwitterungsformen im Granit auf Korsika," *Peterm. Mitt. Ergzh.,* 261.

1962. "Untersuchungen zur klimagenetischen Geomorphologie in den Hochgebirgen Vorderasiens," *Heidelberger Geog. Arb.*, No. 11.

KLEBELSBERG, R. VON

1948-49. *Handbuch der Gletscherkunde und Glazialgeologie.* Vienna: Springer, 2 vol.

KLEIN, A.

1953. "Die Niederschläge in Europa in Maximum der letzten Eiszeit," *Peterm. Geog. Mitt.*, 97: 98-104.

KLEIN, R. G.

1969a. "The Mousterian of European Russia," *Proc. Prehist. Soc.*, 35: 77-111.

1969b. *Man and Culture in the Late Pleistocene: A Case Study.* San Francisco: Chandler.

1971. "The Prehistory of Siberia," *Quat. Research,* 1: in press.

KLÍMA, B.

1954. "Paleolithic Huts at Dolní Vestonice, Czechoslavakia," *Antiquity,* 109: 4-14.

1962. "The First Ground-Plan of an Upper Paleolithic Loess Settlement in Middle Europe and Its Meanings," *Viking Fund Publ. Anthropol.,* 32: 193-210.

KLÍMA, B., J. KUKLA, V. LOŽEK, and H. DE VRIES

1962. "Stratigraphie des Pleistozäns und Alter des paläolithischen Rastplatzes in der Ziegelei von Dolni Vestonice," *Anthropozoikum* (Praha), 11: 93-145.

KLIMASZEWSKII, M., W. SZAFER, B. SZAFRAN, and M URBANSKI

1950. "The Dryas Flora of Kroschienko on the River Dunajic," *Bull. Serv. Geol. de Pologne,* Vol. 24, No. 2.

KLINGE, H.

1958. "Eine Stellungnahme zur Altersfrage von Terra-Rossa-Vorkommen," *Zeitschr. Pflanzenernährung, Düngung, Bodenkunde.* 81: 56- 63.

1960. "Beiträge zur Kenntnis tropischer Boden (II)," *ibid,* 89: 211-16.

KLUTE, F.

1928. "Die Bedeutung der Depression der Schneegrenze für eiszeitliche Probleme," *Zeitschr. Gletscherkunde,* 16: 70-93.

KNETSCH, G.

1960. "Über aride Verwitterung unter besonderer Berücksichtigung natürlicher und künstlicher Wände in Ägypten," *Zeitschr. Geomorph.,* Suppl. Vol. 1 (1961): 190-205.

KNETSCH, G., *et al.*

1963. "Untersuchungen an pluvialen Wassern der Ost-Sahara," *Geol. Rundschau,* 52: 587-610.

KNOX, A. S.

1962. "Pollen from the Pleistocene Terrace Deposits of Washington, D.C.," *Pollen et Spores,* 4: 357-58.

KOENIGSWALD, G. H. R. VON

1940. "Neue *Pithecanthropus*-Funde 1936-1938. Ein Beitrag zur Kenntnis der Praehominiden," *Nederlandsch-Indie Mijnbouw Dienst. Wetenschap. Mededeel.,* No. 28.

1968. "Das absolute Alter des *Pithecanthropus erectus.*" In G. Kurth (ed.), *Evolution and Hominisation,* Stuttgart: Fischer, pp. 195-203.

KOEPPEN, W.

1920. "Die Lufttemperatur an der Schneegrenze," *Peterm. Mitt.,* 66: 78-80.

KOPP, K. O.

1963. "Schneegrenze und Klima der Würmeiszeit an der baskischen Küste," *Eiszeitalter und Gegenw.,* 14: 188-207.

KOTTLOWSKI, F. E., M. E. COOLEY, and R. V. RUHE

1965. "Quaternary Geology of the Southwest." In WRIGHT and FREY (eds.), 1965, pp. 287-98.

KOWALSKI, K.

1967. "The Pleistocene Extinction of Mammals in Europe." In MARTIN and WRIGHT (eds.), 1967, pp. 349-64.

KRETZOI, M.

1961. "Stratigraphie und Chronologie, Stand der ungarischen Quartärforschung," *Prace Instytut Geologiczny* (Warzawa), 34: 313-32.

KRETZOI, M., and L. VERTES

1965. "Upper Biharian (Intermindel) Pebble-Industry Occupation Site in Western Hungary," *Current Anthropology,* 6: 74-87.

KREIGER, A. D.

1964. "Early Man in the New World." In J. D. JENNINGS and E. NORBECK (eds.), 1964, pp. 23-81.

KRUMBEIN, W. C.

1941. "Measurement and Geologic Significance of Shape and Roundness of Sedimentary Particles," *Jour. Sedimentary Petrol.,* 11: 64-72.

KUBIENA, W. L.

1938. *Micropedology.* Ames: Collegiate Press.

1953. *The Soils of Europe.* London: T. Murby.

1954a. "Genesis and Micromorphology of Laterite Formation in Rio Muni," *Proc. 5th Int. Soil Science Congr.* (Léopoldville), 4:77-84.

1954b. "Über Reliktboden in Spanien," *Aichinger Festschrift,* 1 (Mitt. Inst. f. angewandte Vegetationskunde, Vienna): 213-24.

1955. Über die Braunlehmrelikte des Atakor (Zentral-Sahara)," *Erdkunde,* 9: 115-32.

1957. "Neue Beiträge zur Kenntnis des planetarischen und hypsometrischen Formenwandels der Böden Africas," *Stuttgarter Geog. Studien* (Lautensach Festschrift), 69: 50-64.

1963. "Paleosols as Indicators of Paleoclimates," *Arid Zone Research* (UNESCO), 20: 207-9.

KUBITZKI, K., and K. O. MUNNICH

1960. "Neue C-14 Datierungen zur nacheiszeitlichen Waldgeschichte Nord-westdeutschlands," *Ber. deut. botan. Ges.,* 73: 137-45.

KUHN, H.

1929. *Kunst und Kultur de Vorzeit Europas: das Paläolithikum*. Berlin and Leipzig: de Gruyter.

KUENEN, P. H., and W. G. PERDOK

1962. "Frosting and Defrosting of Quartz Grains (Experimental Abrasion, 5)," *Jour. Geol.,* 70: 648-58.

KULS, W., and A. SEMMEL

1965. "Zur Frage pluvialzeitlicher Solifluktionsvorgänge im Hochland von Godjam (Äthiopien)," *Erdkunde,* 19: 292-97.

KUNERT, R., and M. ALTERMANN

1965. "Das Pleistozän zwischen Saale und Wipper," *Geologie,* 14: 520-53.

KURTÉN, B.

1957. "The Bears and Hyenas of the Interglacials," *Quaternaria,* 4: 69-81.

1959a. "On the Longevity of Mammalian Species in the Tertiary," *Commentationes Biol. Soc. Sci. Fennica,* Vol 21, No. 4.

1959b. "On the Bears of the Holsteinian Interglacial." *Stockholm Contrib. in Geol.,* 2: 73-102.

1960a. "Chronology and Faunal Evolution of the Earlier European Glaciations," *Commentationes Biol. Soc. Sci. Fennica,* Vol. 21, No. 5.

1960b. "Faunal Turnover Dates for the Pleistocene and Late Pliocene," *ibid.,* Vol. 22, No. 5.

1965. "The Carnivora of the Palestine Caves," *Acta Zool. Fennica* (Helsinki), 107: 1-74.

1968. *Pleistocene Mammals of Europe*. Chicago: Aldine.

KUTZBACH, J. E., R. A. BRYSON, and W. C. SHEN

1968. "An Evaluation of the Thermal Rossby Number in the Pleistocene," *Meteorological Monographs,* vol. 8, no. 30: 134-38.

LAATSCH, W.

(1957). *Dynamik der mitteleuropaischen Mineralboden* (4th ed.). Dresden and Leipzig: T. Steinkopff.

LACHENBRUCH, A. H.

1962. "Mechanics of Thermal Contraction Cracks and Icewedge Polygons in Permafrost," *Geol. Soc. Amer., Spec. Paper* 70: 1-69.

LAIS, R.

1932. "Die postglazialen Sedimente einer Höhle am Isteiner Klotz in Baden," *Fortschr. der Geol. und Paläont.* (Berlin), Vol 11, No. 36.

1941. "Über Höhlensedimente," *Quartar,* 3: 56-108.

LAMING, A. (ed.)

1952. *La decouverte du passé. Progrès récents et techniques nouvelles en préhistoire et en archéologie*. Paris: A. and J. Picard.

LANCE, J. F.

1959. "Faunal remains from the Lehner Mammoth Site," *Amer. Antiquity,* 25: 35-42.

LANG, G.

1963. "Probleme der spätzeitlichen Vegetationsentwicklung in Südwestdeutschland und im französischen Zentralmassiv," *Pollen et Spores,* 5: 129-42.

LANNING, E. P.

1967. *Peru before the Incas.* Englewood Cliffs, N.J.: Prentice-Hall.

LAUER, W.

1952. "Humide und aride Jahreszeiten in Afrika und Südamerika and ihre Beziehung zu den Vegetationsgürteln," *Bonner Geogr. Abhl.,* 9: 15-98.

LAUGHLIN, W. S.

1967. "Human Migration and Permanent Occupation in the Bering Sea Area." In HOPKINS (ed.), 1967, pp. 409-50.

LAUGHLIN, W. S., and W. G. REEDER

1962. "Rationale for the Collaborative Investigation of Aleut-Konyag Prehistory and Ecology," *Arctic Anthropology,* 1: 104-8.

LAVILLE, H.

1964. "Recherches sédimentologiques sur la paléoclimatologie du würmien récent en Périgord," *L'Anthropologie,* 68: 1-48, 219-52.

LAVILLE, H., and D. DE SONNEVILLE-BORDES

1967. "Sédimentologie des niveaux moustériens et aurignaciens de Caminade-Est," *Bull Soc. Préhist. Franc.,* 64: 35-52.

LEAKEY, L. S. B., P. V. TOBIAS, and G. L. ISAAC

1967. "Tanzania." In K. P. Oakley and B. G. Campbell (eds.), *Catalogue of Fossil Hominids. I. Africa,* London: British Museum (Natural History), pp. 105-18.

LEAKEY, M. D.

1967. "Preliminary Survey of the Cultural Material from Beds I and II, Olduvai Gorge, Tanzania." In BISHOP and CLARK (eds.), 1967, pp. 417-46.

1970. "Stone Artifacts from Swartkrans," *Nature,* 225: 1221-25.

LEAKEY, R. E., *et al.*

1970. "New Hominid Remains and Early Artifacts from Northern Kenya," *Nature,* 226: 223-30.

LEE, R. B., and I. DE VORE (eds.)

1968. *Man the Hunter.* Chicago: Aldine.

LEES, G. M., and N. L. FALCON

1952. "The Geographical History of the Mesopotamian Plain," *Geogr. Jour.,* 118: 24-39.

LEFFINGWELL, E. DE K.

1919. "The Canning River Region Northern Alaska," *U. S. Geol. Survey Prof. Paper,* No. 109.

LE GROS CLARK, W. E.

1955. *The Fossil Evidence for Human Evolution.* Chicago: University of Chicago Press.

LEIGHTON, M. M., and H. B. WILLMAN

1950. "Loess Formations of the Mississippi Valley." *Jour. Geol,* 58: 599-623.

LEMÉE, G.

1954. "Observations nouvelles sur la végétation au dernier interglaciaire et au tardiglaciare en France d'après l'analyse pollinique," *Rapports et Comm. 8. Congr. Int. Botanique* (Paris, 1954), sec. 6: 263-64.

LEMKE, R. W., W. M. LAIRD, M. J. TIPTON, and R. M. LINDVALL

1965. "Quaternary geology of Northern Great Plains." In WRIGHT and FREY (eds.), 1965, pp. 15-28.

LEOPOLD, E. B.

1956. "Two Late Glacial Deposits in South Connecticut," *Proc. Nat. Acad. Sci.,* 42: 863-67.

1959. "Pollen, Spores and Marine Microfossils of the Charleston Phosphate Area, South Carolina," *Bull. U.S. Geol. Sur.,* 1079: 49-53.

LEOPOLD, L. B., M. G. WOLMAN, and J. P. MILLER

1964. *Fluvial Processes in Geomorphology.* San Francisco: W. H. Freeman.

LEROI-GOURHAN, A.

1961. "Flores quaternaires françaises," *Bull. Soc. Botan. Franç.,* 108: 244-54.

1969. "Pollen Grains of Gramineae and Cerealia from Shanidar and Zawi Chemi." In UCKO and DIMBLEBEY (eds.), 1969, pp. 143-148.

LEVERENZ, J. M.

1960. "Mammalian Fauna Depicted by Cave Drawings of Southwest Europe with Ecological Implications," Unpubl. research paper, Dept. of Geography, Univ. of Wisconsin, Madison.

LHOTE, H.

1959. *The Search for the Tassili Frescoes.* New York: Dutton.

1965. "L'évolution de la faune dans les gravures et des peintures rupestres du Sahara et ses relations avec l'évolution climatique," in *Miscelánea en Homenaje al Abate Henri Breuil,* Barcelona: Dip. Prov. de Barcelona, pp. 83-118.

LIBBY, W. F.

1955. *Radiocarbon Dating* (2nd ed.). Chicago: University of Chicago Press.

LIND, A. O.

1969. *Coastal Landforms of Cat Island, Bahamas Research Papers,* Univ. of Chicago Geography Department, No. 122: 1-156.

LIVINGSTONE, D. A.

1967. "Postglacial Vegetation of the Ruwenzori Mountains in Equatorial Africa," *Ecol. Monographs,* 37: 25-52.

1968. "Some Interstadial and Post-glacial Pollen Diagrams from Eastern Canada," *Ecol. Monographs,* 38: 87-125.

LLABADOR, F.

1962. "Résultats malacologiques de la mission scientifique du Ténéré. In *Missions Berliet: Ténéré-Tchad.* Paris: Arts et Métiers Graphiques, pp. 243-70.

LLOPIS-LLADO, N.

1957. In F. Hernandez-Pacheco, *et al.,* "Livret guide de l'excusion N2:Le

Quaternaire de la région cantabrique," *5th Int. Congr. INQUA* (Oviedo).
LORTET, V., and C. GAILLARD
1903. "La faune mommifiée de l'ancienne Egypte," *Arch. Mus. Hist. Nat. Lyon,* Mém. 8.
LOTZE, F.
1962. "Pleistozäne Vergletscherungen im Ostteil des Kantabrischen Gebirges, Spanien," *Abhandl. Akad. Wiss. Liter.* (Mainz), math.-naturw. kl. 1962: 149-69.
LOUIS, H.
1934. "Glazialmorphologische Studien in den Gebirgen der Britischen Inseln," *Berliner Geog. Arb.,* No. 6.
LOZEK, V.
1964. *Quartärmollusken der Tschechoslowakei.* Prag: *Rozpravy Ustavu Geologikého,* 31: 1-374.
1967. "Beiträge der Molluskenforschung zur prähistorischen Archäologie Mitteleuropas," *Archäol. Zeit.* (Berlin), 1: 88-138.
LUDI, W.
1955. "Die Vegetationsentwicklung seit dem Rückzug der Gletscher in den mittleren Alpen und ihrem nördlichen Vorland," *Ber. Geobotan. Forschungsinst.* Rübel (Zürich), 1954: 36-38.
LUGN, A. L.
1962. *The Origin and Sources of Loess. Univ. of Nebraska Studies,* N.S., 26: 1-105.
1968. "The Origin of Loesses and Their Relation to the Great Plains in North America." In SCHULTZ and FREY (eds.), 1968, pp. 139-84.
LUMLEY, H. DE
1963. "Les niveaux quaternaires marins des Alpes-Maritimes," *Bull. Soc. Géol. Franç.,* 7 series, 5: 562-79.
1967. "Découverte d'habitats de l'Acheuléen ancien, dans des dépôts mindéliens, sur le site de Terra Amata (Nice)," *C. R. Acad. Sci.,* 264: 801-04.
1969a. "A Paleolithic Camp at Nice," *Scientific Amer.,* 220: 42-50.
LUMLEY, H. DE (ed.)
1969b. *Une Cabane acheulienne dans la Grotte du Lazaret. Mém. Soc. Préhist. Franç.,* vol. 7: 1-235.
LUMLEY, H. DE, S. GAGNIERE, L. BARRAL, and R. PASCAL
1963. "La Grotte du Vallonet, Roquebrune—Cap Martin," *Bull. Musée d'Anthropol. Préh. de Monaco,* 10: 5-20.
LUNDELIUS, E. L.
1967. "Late Pleistocene and Holocene Faunal History of Central Texas." In MARTIN and WRIGHT (eds.), 1967, pp. 287-319.
LÜTTIG, G.
1956. "Eine neue, einfache geröllmorphometrische Methode," *Eiszeitalter und Gegenw.,* 7: 13-20.
1959. "Eiszeit-Stadium-Phase-Staffel. Eine nomenklatorische Betrachtung," *Geol. Jahrb.,* 76: 235-68.

1962a. "The Shape of Pebbles in the Continental, Fluviatile and Marine Facies," *Int. Assoc. Sci. Hydrol.*, Publ. 59: 253-58.

1962b. "Italienisches und griechiches Pliopleistozän," *Zeit. deut. Geol. Ges.*, 114: 7-31.

1965. "The Bilshausen Type Section, West Germany," *Spec. Paper, Geol. Soc. Amer.*, 84: 159-78.

LYNCH, T. F., and K. A. R. KENNEDY

1970. "Early Human Cultural and Skeletal Remains from Guitarrero Cave, Northern Peru," *Science*, 169: 1307-09.

MAARLEVELD, G. C.

1951. "De Asymmetrie van de kleine Dalen op net noordelijk Halfrand," *Tijdschr. kon Nederl. Aardrijkskundig Genootschap*, 68: 297-312.

1960. "Über die pleistozänen Ablagerungen im Südlichen Afrika," *Erdkunde*, 14: 35-46.

1964. "Periglacial phenomena in the Netherlands during different parts of Würmtime," *Biul. Peryglazjalny*, 14: 251-56.

MACKAY, J. R., W. H. MATTHEWS, and R. S. MACNEISH

1961. "Geology of the Engigstciak Archaeological Site, Yukon Territory," *Arctic*, 14: 25-52.

MACNEISH, R. S.

1964. "Ancient Mesoamerican Civilization," *Science*, 143: 531-37.

MADLER, K.

1939. "Die pliozäne Flora von Frankfurt am Main," *Abhl. Senckenberg. Naturf. Ges.*, 446: 1-36.

MAHER, L. J.

1961. *Pollen Analysis and Postglacial Vegetation History in the Animas Valley, San Juan Mountains, Colorado.* (Univ. Minnesota Ph.D. dissertation.)

MALDE, H. E.

1964. "Environment and Man in Arid America," *Science*, 145: 123-29.

MALDE, H. E., and A. P. SCHICK

1964. "Thorne Cave, Northeastern Utah: Geology," *Amer. Antiquity*, 30: 60-73.

MANLEY, G.

1951. "The Range and Variation of the British Climate," *Geog. Jour.*, 117: 43-68.

1955. "A Climatological Survey of the Retreat of the Laurentide Ice-Sheet," *Amer. Jour. Sci.*, 253: 256-73.

MARBUT, C. F.

1936. "Soils of the United States." In *Atlas of American Agriculture* (U. S. Dept. of Agriculture, Washington, D. C.). Pt. 3.

MARÉCHAL, R., and G. C. MAARLEVELD

1956. "L'extension des phénomènes périglaciaires en Belgique et aux Pays

Bas," *Med. Geol. Sticht* (Haarlem) (N. S.), 8: 77-86.

MARTIN, F.

1968. "Pleistocene Mollusks from Sudanese Nubia." In WENDORF (ed.), 1968, pp. 56-79.

MARTIN, P. S.

1958a. Pleistocene Ecology and Biogeography of North America." In C. S. Hubbs (ed.), *Zoogeography,* Amer. Assoc. Adv. Sci. Publ. 51, pp. 375-420.

1958b. "Taiga-tundra and the Full Glacial Period in Chester County, Penn.," *Amer. Jour. Sci.,* 256: 470-502.

1963a. *The Last 10,000 Years.* Tucson: Univ. of Arizona Press.

1963b. "Geochronology of Pluvial Lake Cochise, Southern Arizona, II," *Ecology,* 44: 436-44.

1967. "Prehistoric Overkill." In MARTIN and WRIGHT (eds.), 1967, pp. 75-120.

MARTIN, P. S., and J. E. GUILDAY

1967. "A Bestiary for Pleistocene Biologists." In MARTIN and WRIGHT (eds.), 1967, pp. 1-62.

MARTIN, P. S., and P. J. MEHRINGER

1965. "Pleistocene Pollen Analysis and Biogeography of the Southwest." In WRIGHT and FREY (eds.), 1965, pp. 433-52.

MARTIN, P. S., B. E. SABELS, and D. SHUTLER

1961. "Rampart Cave Coprolite and Ecology of the Shasta Ground Sloth," *Amer. Jour. Sci.,* 259: 102-27.

MARTIN, P. S., and H. E. WRIGHT (eds.)

1967. *Pleistocene Extinctions: The Search for a Cause.* New Haven: Yale University Press.

MARTYN, J., and P. V. TOBIAS

1966. "Pleistocene Deposits and New Fossil Localities in Kenya," *Nature,* 215: 476-80.

MARX, H., and C. A. REED

1957. "Observations on the Burrowing Rodent *Spalax* in Iraq," *Jour. Mammology,* 39: 386-89.

MASON, H. L.

1936. "The Principles of Geographic Distribution as Applied to Floral Analysis," *Madrono,* 3: 181-90.

MASON, R. J.

1961. "The Earliest Tool-Makers in South Africa," *South Afr. Jour. Sci.,* 57: 13-16.

1962. *Prehistory of the Transvaal: A Record of Human Activity.* Johannesburg: Witwatersrand University Press.

MASON, RONALD J.

1962. "The Paleo-Indian Tradition in Eastern North America," *Current Anthropology,* 3: 227-78.

MATTHES, F. E., and A. D. BELMONT

1950. "The Glacial Anticyclone Theory Examined in the Light of Recent

Meterological Data from Greenland," *Trans. Geophys. Union,* 31: 174-82.

MAUNY, R.

1956. "Préhistorie et zoologie: la grande 'faune éthiopienne' du Nord-Quest africaine du paléolithique à nos jours," *Bull. Inst. Franç. d'Afrique Noire* (series A), 18: 246-79.

MCBURNEY, C. B. M.

1967. *The Haua Fteah (Cyrenaica) and the Stone Age of the Southeast Mediterranean.* Cambridge: Cambridge University Press.

MCBURNEY, C. B. M., and R. W. HEY

1955. *Prehistory and Pleistocene Geology in Cyrenaican Libya.* Cambridge: Cambridge University Press.

MCCALL, G. J. H.

1967. *Geology of the Nakuru-Thompsons Falls-Lake Hannington Area.* Nairobi: Geol. Survey of Kenya, Report 78.

MCCALL, G. J. H., B. H. BAKER, and J. WALSH

1967. "Late Tertiary and Quaternary Sediments of the Kenya Rift Valley." In BISHOP and CLARK (eds.), 1967, pp. 191-228.

MEGARD, R. O.

1967. "Late Quaternary *Cladocera* of Lake Zeribar, Western Iran," *Ecology,* 48: 179-89.

MEHRINGER, P. J.

1965. "Late Pleistocene Vegetation in the Mohave Desert of Southern Nevada," *Jour. Ariz. Acad. Sci.,* 3: 172-88.

1967. "The Environment of Extinction of the Late Pleistocene Megafauna in the Arid Southwestern United States." In MARTIN and WRIGHT (eds.), 1967, pp. 247-66.

MEHRINGER, P. J., and C. V. HAYNES

1965. "The Pollen Evidence for the Environment of Early Man and Extinct Mammals at the Lehner Mammoth Site, Southeastern Arizona," *Amer. Antiquity,* 31: 17-23.

MEHRINGER, P. J., P. S. MARTIN, and C. V. HAYNES

1967. "Murray Springs: A Mid-Postglacial Pollen Record from Southern Arizona," *Amer. Jour. Sci.,* 265: 786-97.

MEIGS, P.

1952. "Distribution of Arid Homoclimates," U.N. Maps No. 392 and 393. Paris: UNESCO.

MELLAART, J.

1965. *Earliest Civilizations of the Near East.* London: Thames and Hudson.

MENÉNDEZ-AMOR, J., and F. FLORSCHÜTZ

1961. "Contribucion al conocimiento de la historia de la vegetation en España durante el Quaternario," *Estudios geológicos,* 17: 83-99.

1962. "Un aspect de la végétation en Espagne méridionale durant la dernière glaciation et l'Holocène," *Geol. Mijnbouw,* 41: 131-34.

MENSCHING, H.

1955a. "Das Quartär in den Gebirgen Marokkos," *Peterm. Mitt. Ergzh.,* Vol. 256.

1955b. "Karst und Terra rossa auf Mallorca," *Erdkunde,* 9: 188-96.

MERCER, J. H.

1969. "Glaciation in Southern Argentina More Than Two Million Years Ago," *Science,* 164: 823-25.

MERRILEES, D.

1968. "Man the Destroyer: Late Quaternary Changes in the Australian Marsupial Fauna," *Jour. Roy. Soc. Western Australia,* 51: 1-24.

MESOLELLA, M. J., *et al.*

1969. "The Astronomical Theory of Climatic Change: Barbados Data," *Jour. Geol.,* 77: 250-74.

MESSERLI, B.

1965. *Beiträge zur Geomorphologie der Sierra Nevada (Andalusien).* Zürich: Juris.

1966. "Das Problem der eiszeitlichen Vergletscherung am Libanon und Hermon," *Zeit. f. Geomorph,* 10: 37-68.

1967. "Die eiszeitliche und die gegenwärtige Vergletscherung im Mittelmeerraum," *Geographica Helvetica,* 1967: 105-228.

MESSERLI, B., and D. INDERMÜHLE

1968. "Erste Ergebrisse einer Tibesti-Expedition," *Verhandl. Schw. Naturfor. Ges.* (1968): 139-42.

MICHAEL, H. N. (ed.)

1964. *The Archaeology and Geomorphology of Northern Asia.* Toronto: University of Toronto Press.

MICHEL, P.

1968. "Genèse et évolution de la vallée du Sénégal, de Bakel à l'embouchure (Afrique occidentale)," *Zeit. Geomorph.,* 12: 318-49.

MICHEL, P., P. ELOUARD, and H. FAURE

1968. "Nouvelles recherches sur le Quaternaire récent de la région de Saint-Louis (Sénégal)," *Bull. Inst. Fond. d'Afrique Noire,* 30, Ser. A: 1-38.

MIKESELL, M. W.

1967. "Geographic Perspectives in Anthropology," *Annals Assoc. Amer. Geog.* 57: 617-34.

1969. "The Deforestation of Mount Lebanon," *Geog. Rev.,* 56: 1-28.

MILANKOVITCH, M.

1941. "Kanon der Erbdestrahlung und seine Anwendung auf das Eiszeit-problem," *Acad. Roy. Serv. Éd. Spéc.,* Vol. 133.

MILLER, D. H.

1955. "Snow Cover and Climate in the Sierra Nevada," *Univ. Calif. Publ. Geog.,* Vol. 11.

MILLER, J. A.

1967. "Problems of Dating East African Tertiary and Quaternary Volcanics by the Potassium-Argon Method." In Bishop and Clark (eds.), 1967, pp. 259-72.

Miller, R. R.

1965. "Quaternary Freshwater Fishes of North America." In Wright and Frey (eds.), 1965, pp. 569-82.

Milliman, J. D., and K. O. Emery

1968. "Sea Levels During the Past 35,000 Years," *Science,* 162: 1121-23.

Milojčic, V.

1958. In H. Bengtson, and V. Milojčic, *Grosser historischer Weltatlas,* Part I, (3d ed.), Munich: Bayerischer Schulbuch-Verlag.

Mitchell, G. F., and H. M. Parkes

1949. "The Giant Deer in Ireland," *Proc. Roy. Irish Acad.,* 52B: 291-391.

Mitchell, J. M.

1968. "Concluding Remarks." In *Causes of Climatic Change, Meteorological Monographs,* 8, No. 30: 155-59.

Mohr, E. C. J., and F. A. Van Baren

1954. *Tropical Soils.* The Hague: Van Hoeve.

Monod, T.

1963. "The Late Tertiary and Pleistocene in the Sahara and Adjacent Southerly Regions," *Viking Fund Publ. Anthropol.,* 36: 117-229.

Moreau, R. E.

1933. "Pleistocene Climatic Changes and the Distribution of Life in East Africa," *Jour. Ecology,* 211: 415-35.

1963. "The Distribution of Tropical African Birds in Relation to Past Climatic Changes," *Viking Fund Publ. Anthropol.,* 36: 28-42.

Mori, F.

1965. *Tadrart Acacus: arte rupestre e culture del Sahara preistorico.* Torino: Einaudi.

Morisawa, M.

1968. *Streams: Their Dynamics and Morphology.* New York: McGraw-Hill.

Morison, C. G. T., A. C. Hoyle, and J. F. Hope-Simpson

1948. "Tropical Soil-Vegetation Catenas and Mosaics: A Study in the Southwestern Part of the Anglo-Egyptian Sudan," *Jour. Ecology,* 34: 1-84.

Morrison, R. B.

1965. "Quaternary Geology of the Great Basin." In Wright and Frey (eds.), 1965, pp. 265-86.

Morrison, R. B., and J. C. Frye

1965. "Correlation of the Middle and Late Quaternary Successions of the Lake Lahontan, Lake Bonneville, Rocky Mountain, Southern Great Plains and Eastern Midwest Areas," *Nevada Bur. of Mines* (Univ. of Nevada, Reno), Report 9: 1-45.

Mortensen, H.

1952. "Heutiger Firnrückgang und Eiszeitklima," *Erdkunde,* 6: 145-60.

1957. "Temperaturgradient und Eiszeitklima am Beispiel der pleistozänen Schneegrenzdepression in den Rand- und Subtropen," *Zeitschr. Geomorph.* 1: 44-56.

MOSS, J. H.

1951. "Glaciation in the Wind River Mtns. and Its relation to Early Man in the Eden Valley, Wyoming," University of Pennsylvania, Museum Monographs, pp. 9-92.

MOSTNY, G.

1968. "Association of Human Industries with Pleistocene Fauna in Central Chile," *Current Anthropology,* 9: 214-15.

MOVIUS, H. L.

1943. "The Stone Age of Burma," *Trans. Amer. Phil. Soc.,* 32: 341-93.

1948. "The Lower Paleolithic Cultures of Southern and Eastern Asia," *Trans. Amer. Phil. Soc.,* 38: 329-420.

1949. "Villafranchian Stratigraphy in Southern and Southwestern Europe," *Jour. Geol.,* 57: 380-412.

1955. "Palaeolithic Archeology in Southern and Eastern Asia, Exclusive of India," *Jour. World History,* 2: 257-82, 520-53.

1960. "Radiocarbon Dates and Upper Paleolithic Archeology in Central and Western Europe," *Current Anthropology,* 1: 355-91.

MOVIUS, H. L., and S. JUDSON

1956. "The Rock-Shelter of La Colombière. Archaeological and Geological Investigation of an Upper Perigordian Site, near Poncin, Ain," *Bull. Amer. School Prehist. Research,* No. 19.

MÜLLER, H.

1965. "Eine pollenanalytische Neubearbeitung des Interglazial-Profils von Bilshausen," *Geol. Jahrb.,* 83: 327-52.

MÜLLER-BECK, H. J.

1957. "Paläolithische Kulturen und Pleistozäne Stratigraphie in Suddeutschland," *Eiszeitalter und Gegenw.,* 8: 116-40.

1964. "Zur stratigraphischen Stellung des *Homo heidelbergensis," Jahrbuch, Rom.- German, Zentralmuseum* (Mainz), 11: 15-33.

1966. "Paleo-Hunters in America: Origins and Diffusion," *Science,* 152: 1191-1210.

1967. "On Migrations of Hunters Across the Bering Land Bridge in the Upper Pleistocene." In HOPKINS (ed.), 1967, pp. 373-408.

MÜLLER-WILLE, W.

1954. "Arten der Menschlischen Siedlung" *Abhl. Akad. Raumforsch. Landesplan.* (Bremen), 28 (Mortensen Festshrift): 141-63.

MULVANEY, D. J.

1969. *The Prehistory of Australia.* London: Thames and Hudson.

MÜNNICH, K. O.

1957. "Erfahrungen mit der C-14 Datierung verschiedener Arten von Sedimenten," *Veröff. Geobotan. Inst. Rübel* (Zürich), 34: 109-17.

MÜNNICH, K. O., and J. C. VOGEL

1959. "C-14 – Altersbestimmung von Suesswasser-Kalkablagerungen," *Naturwiss.,* 46: 168-69.

MURRAY, G. W.

1951. "The Egyptian Climate: An Historical Outline," *Geogr. Jour.*, 117: 422-34.

NAIRN, A. E. M. (ed.)

1961. *Descriptive Palaeoclimatology*. New York: Interscience.

1964. *Problems in Palaeoclimatology*. New York: Interscience.

NAIRN, A. E. M., and N. THORLEY

1961. "The Application of Geophysics to Palaeoclimatology." In NAIRN, 1961, pp. 156-82.

NAMIAS, J.

1963. "Surface-Atmosphere Interactions as Fundamental Causes of Drought and Other Climatic Fluctuations," *Arid Zone Research* (UNESCO), 20: 345-60.

NANGERONI, G.

1952. "I fenomeni di morfologia periglaciale in Italia," *Proc. 17th Int. Geogr. Congr., Comm. on Periglacial Morph.* (Washington, D.C.): 7-14.

NARR, K. J.

1956. "Early Food-Producing Populations." In THOMAS, 1956, pp. 134-51.

1961. *Urgeschichte der Kultur*. Stuttgart: A. Kroner.

1963. *Kultur, Umwelt und Leiblichkeit des Eiszeitmenschens*. Stuttgart: G. Fischer.

NEUVILLE, R.

1951. "Le Paléolithique et le Mésolithique de Désert de Judée," *Arch. Inst. Paléontol. Humaine* (Paris), No. 24.

NEWBIGIN, M. I.

1936. *Plant and Animal Geography*. New York: Dutton.

NIETSCH, H.

1939. *Wald und Siedlung im vorgeschichtlichen Mitteleuropa*. Leipzig: Mannus-Bücherei, Vol. 64.

NIKIFOROFF, C. C.

1941. "Morphological Classification of Soil Structure," *Soil Science*, 52: 193-212.

NILSSON, E.

1931. "Quaternary Glaciations and Pluvial Lakes in British East Africa," *Geografiska Ann.*, 13: 249-349.

1940. "Ancient Changes of Climate in British East Africa and Abyssinia," *ibid*, 22: 1-79.

NILSSON, T.

1964a. *Standard pollendiagramme und C^{14}-Datierungen aus dem Ageröds-Mosse im mittleren Schonen. Lunds Univ. Arsskrift,* 59, No. 7.

1964b. *Entwicklungsgeschichtliche Studien im Ageröds-Mosse. Ibid.*, 59, No. 8.

NIR, D., and D. BEN-ARIEH

1965. "Relics of an Intermediate Terrace Between the Ghor and the Zor in the Central Jordan Valley," *Israel Jour. Earth Sci.*, 14: 1-8.

NISKANEN, E.
1943. "On the Deformation of the Earth's Crust Under the Weight of a Glacial Ice Load and Related Phenomena," *Ann. Acad. Sci. Fennicae,* (series A III), No. 7.

NORDENSKIJIÖLD, O., and L. MECKING
1928. "The Geography of the Polar Regions," *Amer. Geogr. Soc.,* Spec. Pub. 8.

NUSSBAUM, F.
1928. "Die diluviale Vergletscherung der östlichen Pyrenäen," *Geogr. Zeit.,* 34: 385-401.

NUSSBAUM, F., and F. GYGAX
1952. "Glazialmorphologische Untersuchungen im Kantabrischen Gebirge (Nord-Spanien)," *Jahresber. Geogr. Ges.* (Bern, 1951-52): 54-79.

NYE, P. H., and D. J. GREENLAND
1960. *The Soil under Shifting Cultivation,* Farnham Royal: Commonwealth Agr. Bur.

OAKLEY, K. P.
1954. "Study Tour of Early Hominid Sites in Southern Africa, 1953," *South African Archaeol. Bull.,* 9: 75-87.
1958. *Man the Tool-Maker* (4th ed.). London, British Mus. Nat. Hist.
1961. "On Man's Use of Fire, With Comments on Toolmaking and Hunting," *Viking Fund Publ. Anthropol.,* 31: 176-93.
1963. "Analytical Methods of Dating Bones." In BROTHWELL and HIGGS, 1963, pp. 24-34.

OAKLEY, K. P., *et al.*
1950. "The Pliocene-Pleistocene Boundary," *Report 18th Int. Geol. Congr.* (London), Pt. 9.
OGDEN, J. D.
1967. "Radiocarbon and Pollen Evidence for a Sudden Change in Climate in the Great Lakes Region Approximately 10,000 Years Ago." In CUSHING and WRIGHT (eds.), 1967, pp. 117-27.

OLAUSSON, E.
1961a. "Remarks on Some Cenozoic Core Sequences from the Central Pacific," *Medd. Oceanografiska Inst. Goeteborg,* No. 29.
1961b. "Sediment Cores from the Mediterranean Sea and the Red Sea IV. Studies of Deep-Sea Cores." In H. Pettersson (ed.), *Reports of the Swedish Deep-Sea Expedition 1947-1948,* Goeteborg: Elanders, pp. 337-91.
1965. "Evidence of Climatic Changes in North Atlantic Deep-Sea Cores," *Progress in Oceanography,* 3: 221-52.

OLDFIELD, F.
1961. "The Full and Late-glacial Period in South-West France," *Proc. Linn. Soc. London,* 172: 49-53.
1964. "Late Quaternary Deposits at Le Moura, Biarritz," *New Phytologist,* 63: 374-409.

1968. "The Quaternary Vegetational History of the French Pays Basque, I," *New Phytologist,* 67: 677-731.

OLLIER, C. D.

1969. *Weathering.* New York: Elsevier.

OLLIER, C. D., and A. J. THOMASSON

1957. "Asymmetrical Valleys of the Chiltern Hills," *Geogr. Jour.* 83: 71-80.

OPDYKE, N. D., B. GLASS, J. D. HAYS, and J. FOSTER

1966. "Paleomagnetic Study of Antarctic Deep-Sea Cores," *Science,* 154: 349-57.

OPDYKE, N. D., and S. K. RUNCORN

1959. "Palaeomagnetism and Ancient Wind Directions," *Endeavour,* 18: 26-34.

OPPENHEIM, A. L.

1964. *Ancient Mesopotamia: Portrait of a Dead Civilization.* Chicago: University of Chicago Press.

OSBORN, H. F.

1934-42. *The Proboscidea.* New York: Amer. Mus. Nat. Hist., 2 vol.

OVERBECK, F.

1958, *Die Moore, Geologie und Lagerstätten Niedersachsens III* (2d ed.). Bremen: Niedersachsisches Amt. f. Landesplanung, Series Ac, Abh. 4.

OWEN, R. C.

1965. "The Patrilocal Band: A Linguistically and Culturally Hybrid Social Unit," *Amer. Anthropol.,* 67: 675-90.

PAAS, W.

1962. "Rezente und fossile Boden auf niederrheinischen Terrassen und deren Deckschichten," *Eiszeitalter und Gegenw.,* 12: 165-230.

PANNEKOEK, A. J. (ed.)

1956. *Geological History of the Netherlands.* The Hague: Govt. Printing and Publ. Office.

PARTRIDGE, T. C., and A. B. A. BRINK

1967. "Gravels and Terraces of the Lower Vaal River Basin," *South African Geog. Jour.,* 49: 21-38.

PASCHINGER, H.

1961. "Quartäre Formenwelt im Fussgebiet der Sierra Nevada Spaniens," *Erdkunde,* 15: 201-9.

PASSARGE, S.

1940. "Die Urlandschaft Ägyptens und die Lokalisierung der Wiege der Altägyptischen Kultur," *Nova Acta Leopoldina,* 9: 77-152.

PATERSON, S. S.

1956. *The Forest Area of the World and Its Potential Productivity.* Göteborg: Royal University of Göteborg, 2 vol.

PATTERSON, B., and W. W. HOWELLS

1967. "Hominid Humeral Fragment from Early Pleistocene of Northwestern Kenya," *Science,* 156: 64-66.

PEABODY, F. E.

1954. "Travertines and Cave Deposits of the Kaap Escarpment of South Africa and the Type Locality of *Australopithecus africanus*," *Bull. Geol. Soc. Amer.,* 65: 671-706.

PEAKE, H. J., and H. FLEURE

1927-36. *The Corridors of Time.* Oxford: The University Press, 6 vols. (*Peasants and Potters,* Vol. III, 1927).

PECSI, M.

1964. "Haupttypen der periglazialen Bodenfrosterscheinungen in Ungarn," *Report VI. Int. Congr. on Quaternary* (Warsaw, 1961), vol. 4: 121-32.

PEEL, R. F.

1966. "The Landscape in Aridity," *Trans. Inst. Brit. Geog.,* 38: 1-24.

PELTIER, L. C.

1949. "Pleistocene Terraces of the Susquehanna River, Pennsylvania," *Penn. Geol. Survey,* (4th Series), Bull. G. 23.

1950. "The Geographic Cycle in Periglacial Regions as It Is Related to Climatic Geomorphology," *Ann. Assoc. Amer. Geog.,* 40: 214-36.

PENCK, A.

1914. "The Shifting of the Climatic Belts," *Scottish Geog. Mag.,* 30: 281-93.

PERKINS, D.

1964. "The Prehistoric Fauna from Shanidar, Iraq." *Science,* Vol. 14.

PERROT, J.

1962. "Palestine-Syria-Cilcia," *Viking Fund Publ. Anthropol.,* 32: 147-64.

1966. "Le gisement Natoufien de Mallaha (Eynan), Israel," *L'Anthropol.,* 70: 437-84.

1968. "Préhistoire Palestinienne," *Supplément au Dictionnaire de la Bible,* 8: 286-446.

PETTIJOHN, F. J.

1957. *Sedimentary Rocks* (2d ed). New York: Harper.

PÉWÉ, T. L.

1951. "An Observation of Wind-Blown Silt," *Jour. Geol.,* 59: 399-410.

1954. "The Geological Approach to Dating Archaeological Sites," *Amer. Antiquity,* 20: 51-61.

1966. "Paleoclimatic Significance of Fossil Ice Wedges," *Biul. Peryglacjalny,* 15: 65-73.

PFANNENSTIEL, M.

1944. "Die diluviale Entwicklungsstadien und die Urgeschichte von Dardanellen, Marmarameer, und Bosporous," *Geol. Rundschau,* 34: 342-434.

1954. "Die Entstehung der ägyptischen Oasendepressionen. Das Quartär der Levante II," *Abhl. Akad. Wiss. Lit.* (Mainz), *Math.-Natw. Kl.,* 1953, No. 7.

PFIZENMAYER, E. W.

1939. *Siberian Man and Mammoth.* London: Blackie.

PHILIPS, C. W.
1951. "The Fenland Research Committee, Its Past Achievements and Future Prospects." In W. F. Grimes (ed.), *Aspects of Archaeology in Britain and Beyond: Essays Presented to O. G. S. Crawford,* London, pp. 258-73.
PIAS, J.
1962. *Les sols des Moyen et Bas Logone, du Bas Chari, des régions Riveraines du Lac Tchad et du Bahr el-Ghazal.* Paris: *Mém. Off. Rech. Sci. Techn. Outre-Mer,* 2: 1-438.
PICARD, K.
1960. "Zur Untergliederung der Saalevereisung im Westen Schleswig-Holsteins," *Zeit. Deut. Geol. Ges.,* 112: 316-325.
PICARD, L.
1963. "The Quaternary in the Northern Jordan Valley," *Proc. Israel Acad. Sci. Hum.,* 1, No. 4: 1-34.
PILBEAM, D. R.
1968. "The Earliest Hominids," *Nature,* 219: 1335-38.
PILBEAM, D. R., and E. L. SIMONS
1965. "Some Problems of Hominid Classification," *Amer. Scientist,* 53: 237-59.
PISSART, A.
1963. "Les traces des 'pingos' du Pays de Galles et du Plateau des Hautes Fagnes," *Zeit. f. Geomorph.,* 7: 147-65.
PITTIONI, R.
1961. "Über die Zusammenarbeit der "anthropologischen Disziplinen" vom Standpunkt der Urgeschichte," In HAEKEL, 1961, pp. 10-36.
POLUTOFF, N.
1955. "Das Mammut von Taimyr. Neue Erkenntnisse zur Ökologie des sibirischen Mammuts," *Eiszeitalter und Gegenw.,* 6: 153-58.
PONS, A., and P. QUÉZEL
1957. "Première étude palynologique de quelques paléosols sahariens," *Trav. Inst. Rech. Sahar.* (Algiers), 15: 15-40.
1958. "Premières remarques sue l'étude palynologique d'un guano fossile du Hoggar," *C. R. Acad. Sci.,* 246: 2290-92.
POP, E.
1957. "Les recherches pollenanalytiques en Roumanie et leurs résultats" (Russian with French summary), *Botaniueskii Jurnal, Akad. Nauk SSSR,* 42: 363-76.
POSER, H.
1948. "Boden und Klimaverhältnisse in Mittel-und Westeuropa während der Würmeisziet," *Erdkunde,* 2: 53-68.
1950. "Zur Rekonstruktion der Spätglazialen Luftdruckverhältnisse in Mittel- und Westeuropa auf Grund der vorzeitlichen Dünen," *ibid,* 4: 81-88.

1957. "Klimamorphologische Probleme auf Kreta," *Zeitschr. Geomorph.*, 1:113-42.

POULSEN, T. L., and W. B. WHITE

1969. "The Cave Environment," *Science*, 165: 971-81.

PUMPELLY, R.

1908. "Explorations in Turkestan: Expedition of 1904: Prehistoric Civilizations of Anau," *Publ. Carnegie Inst.* (Washington), No. 73, 2 vol.

QUÉZEL, P., and C. MARTINEZ

1958. "Etude palynologique de deux diatomites du Borkou," *Bull. Soc. Hist. Nat. Afrique Nord*, 49: 230-44.

1961. "Le dernier interpluvial au Sahara Central. Essai de chronologie palynologique et paléoclimatique," *Libyca*, 6/7: 211-27.

QUÉZEL, P., and J. Y. THÉBAULT

1959. "Palynologie et datation du volcanisme récent de l'Ahaggar," *Bull. Scien. Econ. Bur. Rech. Min. Algérie*, No. 6: 59-64.

QUITZOW, W., and J. I. S. ZONNEVELD

1956. "Vorlaufiges Ergebnis der Terrassenuntersuchungen in Maas und Niederrheingebiet," *Geol. en Mijnbouw*, 18: 428-48.

RALPH, E. K.

1955. "University of Pennsylvania Radiocarbon dates I," *Science*, 121: 150-52.

1959. "University of Pennsylvania Radiocarbon dates IV," *Amer. Jour. Sci., Radiocarbon Supplement* 1: 45-53.

RATTRAY, J. M.

1960. *The Grass Cover of Africa*. Rome: FAO.

RAYNAL, R.

1956. "Les phénomènes périglaciares au Maroc et leur place dans l'évolution morphologique," *Biuletyn Peryglacjalny*, 4: 143-62.

1960a. "Les éboulis ordonnés au Maroc," *Ibid.*, 8: 21-30.

1960b. "Quelques apercus sur l'existence et l'importance des phénomènes periglaciaires préwurmiens au Maroc," *ibid.*, 9: 109-22.

REED, C. A.

1959. "Animal Domestication in the Prehistoric Near East," *Science*, 130: 1629-39.

1960. "A Review of the Archeological Evidence on Animal Domestication in the Prehistoric Near East." In BRAIDWOOD, HOWE, *et al.*, 1960, pp. 119-46.

1961. "Osteological Evidences for Prehistoric Domestication in Southwestern Asia," *Zeitschr. Tierzüchtung und Züchtungsbiologie*, 76: 31-38.

1969. "The Pattern of Animal Domestication in the Prehistoric Near East." In UCKO and DIMBLEBEY (eds.), 1969, pp. 361-80.

REED, E. C., V. H. DREESZEN, C. K. BAYNE, and C. B. SCHULTZ

1965. "The Pleistocene in Nebraska and Northern Kansas." In WRIGHT and FREY (eds.), 1965, pp. 187-262.

REEVES, C. C.

1965. "Chronology of West Texas Pluvial Lake Dunes," *Jour. Geol.*, 73: 504-08.

REEVES, C. C., and W. T. PARRY

1965. "Geology of West Texas Pluvial Lake Carbonates," *Amer. Jour. Sci.,* 263: 606-15.

REICH, H.

1953. "Die Vegetationsentwicklung der Interglaziale von Grossweil-Ohlstadt und Pfefferbichl im bayerischen Alpenvorland," *Flora,* 140: 386-443.

REID, E. M.

1921. "A Comparative Review of Pliocene Floras, Based on the Study of Fossil Seeds," *Quart. Jour. Geol. Soc.,* 76: 145-59.

REIFF, W.

1955. "Über den pleistozanen Sauerwasserkalk von Stuttgart-Münster-Bad Cannstadt," *Jahresber. u. Mitt. oberrhein, geol. Ver.* (N.S.), 37: 1-16.

REIFENBERG, A.

1947. *The Soils of Palestine.* London: T. Murby.

REMY, H.

1958. "Zur Flora und Fauna der Villafranca-Schichten von Villarroya, Prov. Logroño/Spanien," *Eiszeitalter und Gegenw.,* 9: 83-103.

RENFREW, J. M.

1969. "The Archaeological Evidence for the Domestication of Plants." In UCKO and DIMBLEBEY (eds.), 1969, pp. 149-72.

RENSCH, B.

1966. *Evolution above the Species Level.* New York: Interscience.

REPENNING, C. A.

1967. "Palearctic-Nearctic Mammalian Dispersal in the Late Cenozoic." In HOPKINS (ed.), 1967, pp. 288-311.

REVELLE, R., *et al.*

1969. *The Ocean.* San Francisco: Freeman.

RHOTERT, H.

1952. *Libysche Felsbilder.* Darmstadt: L. C. Wittich.

RICHARDS, H. G.

1962. "Studies on the Marine Pleistocene," *Trans. Amer. Phil. Soc.,* Vol. 52, Pt. 3.

RICHARDS, R. W.

1952. *The Tropical Rainforest: An Ecological Study.* Cambridge: Cambridge University Press.

RICHARDSON, J. L.

1966. "Changes in Level of Lake Naivasha, Kenya, During Postglacial Times," *Nature,* 209: 290-91.

RICHMOND, G. M.

1965. "Glaciation of the Rocky Mountains." In WRIGHT and FREY (eds.), 1965, pp. 217-30.

1970. "Comparison of the Quaternary Stratigraphy of the Alps and Rocky Mountains," *Quat. Research,* 1: 3-28.

RICHMOND, G. M. (ed.)

1968. *Quaternary Glaciation of the Alps.* Boulder: University of Colorado Press.

RICHMOND, G. M., R. FRYXELL, G. E. NEFF, and P. L. WEIS

1965. "The Cordilleran Ice Sheet of the Northern Rocky Mountains and Related Quaternary History of the Columbia Plateau." In WRIGHT and FREY (eds.), 1965, pp. 231-242.

RICHTER, K.

1956. "Klimatische Verschiedenartigkert glazialer Vorstossphasen in Nord-deutschland," *Actes IV. Congr. Int. INQUA* (Rome, 1953).

1958. "Fluorteste quartärer Knochen in ihrer Bedeutung für die absolute Chronologie des Pleistozäns," *Eiszietalter und Gegenw.*, 9: 18-27.

1968. "Klimatische Zyklen im norddeutschen Vereisungsgebiet," *Eiszeitalter und Gegenw.*, 19: 262-67.

RIET LOWE, C. VAN

1952. "The Vaal River Chronology," *South African Archaeol. Bull.*, 7: 135-49.

ROBINSON, J. T.

1963. "Adaptive Radiation in the Australopithecines and the Origin of Man," *Viking Fund Publ. Anthropol.*, 36: 385-416.

ROBINSON, J. T.

1967. "Variation and Taxonomy of the Early Hominids." In T. Dobzhansky, M. K. Hecht and W. C. Steere (eds.), *Evolutionary Biology*, New York: Appleton, vol. 1, pp. 69-100.

ROBINSON, J. T., and R. J. MASON

1962. "Australopithecines and Artefacts at Sterkfontein," *South African Arch. Bull.*, 17, no. 66: 87-125.

RODE, A. A.

1961. *The Soil Forming Process and Soil Evolution.* Jerusalem: Israel Program for Scientific Translations.

ROGNON, P.

1967. *Le Massif de l'Atakor et ses bordures (Sahara Central).* Paris: Centre Nat. Rech. Sci.

RONA, E., and C. EMILIANI

1969. "Absolute Dating of Caribbean Cores P6304-8 and P6304-9," *Science*, 163: 66-68.

RONAI, A., M. KRETZOI, and M. PECSI

1961. "Stand der ungarischen Quartarforschung," *Prace Institut Geologiczny* (Warsaw), 34: 313-34.

ROOSMA, A.

1958. "A Climatic Record from Searles Lake, California," *Science*, 128: 716.

ROSHOLT, J. N., and P. S. ANTAL

1963. "Evaluation of the Pa^{231}/U-Th^{230}/U Method for Dating Pleistocene Carbonate Rocks," *U.S. Geol. Surv. Prof. Paper*, 450-E: 108-11.

ROSHOLT, J. N., C. EMILIANI, H. GEISS, F. F. KOCZY, and P. J. WANGERSKY

1961. "Absolute Dating of Deep Sea Cores by the Pa^{231}/Th^{230} Methods," *Jour. Geol.*, 69: 162-85.

ROSHOLT, J. N., *et al.*

1962. "Pa^{231}/Th^{230} Dating and O^{18}/O^{16} Temperature Analysis of Core A254-BR-C," *Jour. Geophys. Res.*, 67:2907-11.

RUBIN, M., R. C. LIKINS, and E. G. BERRY

1963. "On the Validity of Radiocarbon Dates from Snail Shells," *Jour. Geol.*, 71: 84-89.

RUETIMEYER, L.

1862. "Die Fauna der Pfahlbauten der Schweiz," *Neue Denkschr. Schweiz. Ges. Naturw.* (Zürich), Vol. 19.

RUGGIERI, G.

1965. "A Contribution to the Marine Lower Quaternary of Italy," *Spec. Paper, Geol. Soc. Amer.*, 84: 141-52.

RUHE, R. V.

1968. "Identification of Paleosols in Loess Deposits in the United States." In SCHULTZ and FREY (eds.), 1968, pp. 49-66.

RUSKE, R.

1965. "Zur Gliederung der Holstein- und Saalezeit im östlichen Harzvorland," *Eiszeitalter und Gegenw*, 16: 88-96.

RUST, R.

1937. *Das alsteinzeitliche Rentierjagerlager Meiendorf* (with sections on geology by K. Gripp, paleontology by W. Krause, palynology by R. Schütrumpf). Neumünster: K Wachholtz.

1943. *Die alt-und mittelsteinzeitlichen Funde von Stellmoor* (with sections on paleontology by K. Gripp and W. Kollau, palynology by R. Schütrumpf, geology by K. Gripp). Neumünster: K. Wachholtz.

1950. *Die Höhlenfunde von Jabrud (Syrien)*. Neumünster: K. Wachholtz.

1958. *Die jungpaläolitischen Zeltanlagen von Ahrensburg* (with section on palynology by R. Schütrumpf). Neumünster: K. Wachholtz.

1962. *Vor 20,000 Jahren* (2d ed.). Neumünster: K. Wachholtz.

RUXTON, B. P.

1967. "Slopewash under Mature Primary Rainforest in Northern Papua." In J. N. JENNINGS and J. A. MABBUTT, (eds.), 1967.

RUXTON, B. P., and L. BERRY

1960. "Weathering Profiles and Geomorphic Position on Granite in Two Tropical Regions," *Rév. Géomorphologie Dynam.*, 12: 16-31.

RYDER, M. L.

1963. "Remains of Fishes and Other Aquatic Animals." In BROTHWELL and HIGGS, 1963, pp. 294-312.

SABBAGH, M. E.

1962. "A Preliminary Regional Dynamic Climatology of the Antarctic Continent," *Erdkunde*, 16: 94-111.

SABELS, B. E.

1960. "Trace Element Studies on Cave Sediments," *Nev. State Mus. Anthropol. Papers*, 3: 17-23.

SANDFORD, K. S.

1934. Paleolithic Man and the Nile Valley in Upper and Middle Egypt," *Univ. Chicago Oriental Inst. Publ.,* Vol. 18.

SANKALIA, H. D.

1962. "India," *Viking Fund Publ. Anthropol.,* 32: 60-83.

SAUER, C. O.

1927. "Recent Developments in Cultural Geography." In E. C. Hayes, (ed.), *Recent Developments in the Social Sciences,* Philadelphia, pp. 154-212.

1944. "A Geographic Sketch of Early Man in America," *Geog. Rev.,* 34: 529-573.

1947. "Early Relations of Man to Plants," *ibid.,* 37: 1-25.

1952. *Agricultural Origins and Dispersals.* New York: Amer. Geog. Soc.

SAURAMO, M.

1929. "The Quaternary Geology of Finland," *Bull. Comm. Géol. de Finlande,* No. 86.

SAWYER, J. S.

1966. "Possible Variations of the General Circulation of the Atmosphere," in *World Climate from 8000 to 0 B.C.,* London, Roy. Meteor. Soc., pp. 218-29.

SCHÄDEL, L., AND J. WERNER

1963. "Neue Gesichtspunkte zur Stratigraphie des mittleren und älteren Pleistozäns im Rheingletschergebiet," *Eiszeitalter und Gegenw.,* 14: 5-26.

SCHAEFER, I.

1950. "Die diluviale Erosion und Akkumulation," *Forsch. z. deut. Landeskunde,* 49.

SCHALLER, G. B., and J. T. EMLEN

1963. "Observations of the Ecology and Social Behavior of the Mountain Gorilla," *Viking Fund Publ. Anthropol.,* 36: 368-84.

SCHEFFER, F., and P. SCHACHTSCHABEL

1960. *Bodenkunde* (5th ed.). Stuttgart: F. Enke.

SCHENK, E.

1955. "Die Mechanik der periglazialen Strukturboden," *Abhl. Hessischen Landesamtes f. Bodenforschung,* Vol. 13.

SCHERY, R. W.

1952. *Plants for Man.* Englewood Cliffs, N. J.: Prentice-Hall.

SCHMID, E.

1958. "Höhlenforschung und Sedimentanalyse," *Schriften Inst. Ur-und Frühgeschichte der Schweiz* (Basel), No. 13.

1963. "Cave Sediments and Prehistory." IN BROTHWELL and HIGGS, 1963, 123-38.

SCHMIDT, W. R.

1948. "Die Steppenschluchten Sudrusslands," *Erdkunde,* 2: 213-29

SCHÖNHALS, E.

1951. Über fossile Böden im nichtvereisten Gebiet," *Eiszeitalter und Gegenw.,* 1: 109-30.

1953. "Gesetzmässigkeiten im Feinaufbau von Talrandlössen mit Bemerkungen über die Entstehung des Losses," *Ibid.,* 19-36.

SCHOTT, C.

1931. "Die Blockmeere in den deutschen Mittelgebirgen," *Forsch. deut. Landes- und Volkskunde,* 29: 1-78.

1939. "Die vorgeschichtliche Kulturlandschaft Mitteleuropas," *Zeitschr. Erdkunde,* 8: 641-50.

SCHULMAN, E.

1956. *Dendroclimatic Changes in Semiarid America.* Tucson: Univ. of Arizona Press.

SCHULTZ, C. B., and J. C. FRYE (eds.)

1968. *Loess and Related Eolian Deposits of the World.* Lincoln: University of Nebraska Press.

SCHÜTRUMPF, R.

1936. Paläobotanisch-pollenanalytische Untersuchungen der pälaolithischen Rentierjägerfundstätte von Meiendorf bei Hamburg," *Veröff. des archäolog. Reichsinst.,* 1: 1-54.

1938. "Stratigraphisch-pollenanalytische Mooruntersuchungen im Dienste der Vorgeschichtsforschung," *Prahist. Zeit.,* 28/29: 158-83.

1955. "Das Spätglazial," *Eiszeitalter und Gegnw.,* 6: 41-51.

SCHWARZ, G.

1961. *Allgemeine Siedlungsgeographie* (2d ed.). Berlin: de Gruyter.

SCHWARZBACH, M.

1963. *Climates of the Past.* New York: Van Nostrand.

SCHWEINFURTH, U.

1957. *Die horizontale und vertikale Verbreitung der Vegetation im Himalaya. Bonner Geogr. Abhandl.,* 20: 1-375.

SCHWEITZER, H. J.

1958. "Entstehung und Flora des Trasses im nördlichen Laachersee-Gebeit," *Eiszeitalter und Gegenw.,* 9: 28-48.

SEKYRA, J.

1960. "Pusobeni mrazu na pudu: Kryopedologie se Zvlast nim zretelem k CSR," *Geotechnica* (Prague), 27: 1-164.

SELANDER, R. K.

1965. "Avian Speciation in the Quaternary." In WRIGHT and FREY (eds.), 1965, pp. 527-42.

SELLI, R.

1962. "Le Quaternaire marin du versant Adriatique-Ionien de la péninsule italienne," *Quaternaria,* 6: 391-413.

SEMMEL, A.

1963. "Intramontane Ebenen im Hochland von Godjam (Äthiopien)," *Erdkunde,* 17: 173-89.

SERČELJ, A.

1963. "Die Entwicklung der Würm und der Holozänwaldvegetation in Slowenien" (Yugoslav with German summary), *Dissertationes Acad. Scient. Art. Slovenica,* Class IV: 361-418.

1966. "Pollen analytische Untersuchungen der Pleistozänen und Holozäu nen Ablagerungen von Ljubljansko Barje," (Yugoslav with German summary), *ibid.*, 9: 429-72.

Serra Rafols, J., J. F. De Villalta, and J. M. Thomas

1957. "Livret Guide des Excursions B_2-B_3 (Alentours de Barcelone et Moia)," *5th Int. Congr. INQUA* (Madrid-Barcelona).

Servant, M., and S. Servant

1970. "Les Formations lacustres et les diatomées du quaternaire récent du fond de la cuvette tchadienne, "*Rev. Geogr. Phys. Geol. Dyn.*, 12: 63-76.

Servant, M., S. Servant, and G. Delibrias

1969. "Chronologie du Quaternaire récent des basses régions du Tchad," *C. R. Acad. Sci.*, 269: 1603-06.

Service, E. R.

1966. *The Hunters*. Englewood Cliffs, N. J.: Prentice-Hall.

Shackleton, N. J.

1967. "Oxygen Isotope Analyses and Pleistocene Temperatures Re-assessed," *Nature*, 215: 15-17.

Shackleton, N. J., and C. Turner

1967. "Correlation Between Marine and Terrestial Pleistocene Successions," *Nature*, 216: 1079-82.

Sharpe, C. F. S.

1960. *Landslides and Related Phenomena*. Paterson, N. J.: Pageant Books.

Shaw, D. M., and W. L. Donn

1968. "Milankovitch Radiation Variations: A Quantitative Evaluation," *Science*, 162: 1270-72.

Shepard, F. P., and R. Young

1961. "Distinguishing Between Beach and Dune Sands," *Jour. Sedimentary Petrol.*, 31: 196-214.

Shostakovitch, V. B.

1936. "Geschichtete Bodenablagerungen der Seen als Klima-Annalen," *Meteor. Zeitschr.*, 53: 176-82.

Simons, E. L.

1968. "Assessment of a Fossil Hominid," *Science*, 160: 672-75.

Simonson, R. W.

1954. "Identification and Interpretation of Buried Soils," *Amer. Jour. Sci.*, 252: 705-22.

Simpson, G. G.

1944. *Tempo and Mode in Evolution*. New York: Columbia University Press.

1953a. *The Major Features of Evolution*. New York: Columbia University Press.

1953b. *Evolution and Geography*. Eugene: Oregon State System of Higher Education.

Sirén, G.

1961. "Skogsgränstallen som indicator for klimafluktuationera i norra Fennoskandien under historisk tid," *Communicationes Inst. Forestalis Fenniae,* Vol. 54, No. 2.

SIRKIN, L. A.

1967. "Late Pleistocene Pollen Stratigraphy of Western Long Island and Eastern Staten Island, New York." In CUSHING and WRIGHT (eds.), 1967, pp. 249-74.

SLAUGHTER, B. H.

1967. "Animal Ranges as a Clue to Late Pleistocene Extinction." In MARTIN and WRIGHT (eds.), 1967, pp. 155-67.

SMITH, G. D., and SOIL SURVEY STAFF

1961. *Soil Classification, a Comprehensive System (7th Approximation).* Washington, D.C.: U.S. Dept. of Agriculture.

SMITH, G. I.

1968. "Late Quaternary Geologic and Climatic History of Searles Lake, Southeastern California." In R. B. Morrison and H. W. Wright (eds.). *Means of Correlation of Quaternary Successions,* Salt Lake City: University of Utah Press, pp. 293-310.

SMITH, H. S.

1964. "Egypt and C^{14} Dating," *Antiquity,* 38: 32-37.

SMITH, H. T. U.

1949. "Physical Effects of Pleistocene Climatic Changes in Non-Glaciated Areas – Eolian Phenomena, Frost Action and Stream Terracing," *Bull. Geol. Soc. Amer.,* 60: 1485-1516.

1954. "Coast Dunes," *Coastal Geog. Congr., Office of Naval Research,* (Washington, Feb. 1954), pp. 51-56.

1962. "Periglacial Frost Features and Related Phenomena in the United States," *Biul. Peryglacjalny,* 11: 325-42.

1965. "Dune Morphology and Chronology in Central and Western Nebraska," *Jour. Geol.,* 73: 557-78.

SMOLLA, G.

1960. Neolithische Kulturerscheinungen: Studien zur Frage ihrer Herausbildungen," *Antiquitas* (Bonn: R. Habelt) (Series 2), 3: 1-180.

SOERGEL, W.

1921. *Die Ursachen der diluvialen Aufschotterung und Erosion.* Berlin: Borntraeger.

1922. *Die Jagd der Vorzeit.* Jena: G. Fischer.

1937. *Die Vereisungskurve.* Berlin: Borntraeger.

1940. "Der Klimacharakter des Mammuts," *Paläontol Zeitschr.,* 22: 29-55.

1941. "Rentiere des deutschen Alt-und Mitteldiluviums," *ibid.,* 22: 387-420.

1941. "Die Verbreitung des diluvialen Moschusochsen in Mitteleuropa," *Beiträge zur Geol. Thüringens,* 7: 75-95.

1943. "Der Klimacharakter der als nordisch geltenden Säugetiere des Eiszeitalters," *Sitz-Ber. Heidelberger Akad. Wiss., math.-naturw. Kl.* (1941), No. 4.

SÖHNGE, P. G., D. J. L. VISSER, and C. VAN RIET LOWE

1937. *The Geology and Archaeology of the Vaal River Basin. Geological Survey Memoir* (Pretoria), 35, Pts. 1 and 2: 1-134.

SOKOLOFF, V. P., and G. F. CARTER

1952, "Time and Trace Metals in Archeological Sites," *Science,* 116: 1-5.

SOLÉ SABARIŚ, *et al.*

1957. "Livret Guide de l'Excursion N_1 (Pyrénées)," *5th Int. Congr. INQUA,* (Madrid-Barcelona).

SOLECKI, R. S., and M. RUBIN

1958. "Dating of Zawi Chemi, an Early Village Site at Shanidar, Northern Iraq," *Science,* 127: 1446.

SOLECKI, R., and A. LEROI-GOURHAN

1961. "Paleoclimatology and Archaeology in the Near East," *Ann. New York Acad. Sci.,* 95: 729-39.

SOLHEIM, W. G.

1969. "Reworking Southeast Asian Prehistory," *Paideuma,* 15: 125-139.

SONNENFELD, J.

1962. "Prehistoric Technology: Functional Interpretations and Geographical Implications," *Professional Geographer,* Vol. 14, No. 2: 4-8.

SPARKS, B. W.

1960. *Geomorphology.* London: Longmans.

1963. "Non-Marine Mollusca and Archaeology." In BROTHWELL and HIGGS, 1960, pp. 313-24.

SPARKS, B. W., and A. T. GROVE

1961. "Some Quaternary Fossil Non-Marine Mollusca from the Central Sahara," *Jour. Linnaean Soc. London, Zoology,* 44: 355-64.

1964. "Fossil Non-Marine Mollusca from Mongonu, Northeast Nigeria," *Overseas Geol. and Mineral Resources* (London), 9: 190-95.

SPARROW, G. W. A.

1967. "Pleistocene Periglacial Topography in South Africa," *Jour. Glaciology,* 6: 551-59.

STEARNS, C. E., and D. L. THURBER

1965. "Th^{230}/U^{234} Dates of Late Pleistocene Marine Fossils from the Mediterranean and Moroccan Littorals." *Quarternaria,* 7: 29-42.

1967. "Th^{230}/U^{234} Dates of Late Pleistocene Marine Fossils from the Mediterranean and Moroccan Littorals," *Progress in Oceanography,* 4: 293-305.

STEEGER, A.

1944. "Diluviale Bodenfrosterscheinungen am Niederrhein." *Geol. Rundschau,* 34: 520-38.

STEENSBURG, A.

1957. "Some Recent Danish Experiments in Neolithic Agriculture," *Agr. Hist. Rev.,* 5: 66-73.

STEWART, O. C.

1956. "Fire as the First Great Force Employed by Man." IN THOMAS, 1956, pp. 115-31.

STIPP, J. J., J. M. A. CHAPPELL, and I. MCDOUGALL
1967. "K/Ar Age Estimate of the Pliocene-Pleistocene Boundary in New Zealand," *Amer. Jour. Sci.,* 265: 462-74.

STODDART, D. R.
1969. "Climatic Geomorphology: Review and Re-assessment," *Progress in Geography,* 1: 159-222.

STRAATEN, L. VAN
1956. "Composition of Shell Beds Formed in Tidal Flat Environments in the Netherlands and in the Bay of Arcachon," *Geol. Mijnb.,* 18: 209-26.

STRAHLER, A. N.
1969. *Physical Geography* (3d ed.). New York: J. Wiley.

STRUEVER, S.
1968. "Woodland Subsistence-Settlement Systems in the Lower Illinois Valley." In S. R. and L. R. Binford (eds.), *New Perspectives in Archeology,* Chicago: Aldine, pp. 285-312.

STUCKENRATH, R.
1966. "The Debert Archaeological Project, Nova Scotia: Radiocarbon Dating," *Quaternaria,* 8: 75-80.

STUIVER, M., and H. E. SUESS
1966. "On the Relationship Between Radiocarbon Dates and True Sample Ages," *Radiocarbon,* 8: 534-40.

SUESS, H. E.
1955. "Radiocarbon Concentration in Modern Wood," *Science,* 122: 415-17.

SUGGATE, R. P.
1965. "Late Pleistocene Geology of the Northern Part of South Island, New Zealand," *Bull. New Zealand Geol. Soc.,* 77.

SUGGATE, R. P., R. G. WEST, and B. W. SPARKS
1959. "On the Extent of the Last Glaciation in Eastern England," *Proc. Roy. Soc.* (series B), 150: 263-83.

SUSLOV, S. P.
1961. *Physical Geography of Asiatic Russia.* San Francisco: W. H. Freeman.

SUTCLIFFE, R. C.
1963. "Theories of Recent Changes of Climate," *Arid Zone Research* (UNESCO), 20: 277-80.

SZABO, B. J., H. E. MALDE, and C. IRWIN-WILLIAMS
1969. "Dilemma Posed by Uranium-Series Dates on Archaeologically Significant Bones from Valsequillo, Puebla, Mexico," *Earth and Planetary Sci. Let.,* 6: 237-44.

SZAFER, W.
1953. "Pleistocene Stratigraphy of Poland from the Floristical Point of View" (In Polish), *Ann. Soc. Geol. Pologne,* 22: 1-99.
1954. "Pliocene Flora from the Vicinity of Czorsztyn (West Carpathians) and Its Relationship to the Pleistocene," *Prace Inst. Geol.* (Warsaw), No. 11.

TABER, S. S.
1943. "Perennially Frozen Ground in Alaska: Its Origin and History," *Bull. Geol. Soc. Amer.*, 54: 1433-1584.

TATOR, B. A.
1952. "Pediment Characteristics and Terminology," *Ann. Assoc. Amer. Geog.*, 42: 293-317.

TAYLOR, D. W.
1965. "The Study of Pleistocene Non-Marine Mollusks in North America." In WRIGHT and FREY (eds.), 1965, pp. 597-612.

TEDROW, J. C. F., and H. HARRIES
1960. "Tundra Soil in Relation to Vegetation, Permafrost and Glaciation," *Oikos*, 11: 237-49.

TERRA, H. DE., and H. L. MOVIUS
1943. "Research on Early Man in Burma," *Trans. Amer. Phil. Soc.*, 32: 265-464.

TERRA, H. DE, and T. T. PATTERSON
1939. "Studies on the Ice Age in India and Associated Human Cultures," *Publ. Carnegie Inst.* (Washington), No. 493.

THENIUS, E.
1961a. "Über die Bedeutung der Paläokologie für die Anthropologie und Urgeschichte." In HAEKEL, 1961, pp. 80-103.
1961b. "Paläozoologie und Prähistorie," *Mitt. Urgeschichtl. und Anthropol. Ges.* (Vienna), 12, 3/4: 39-61.
1962. "Die Grossäugetiere des Pleistozäns von Mitteleuropa," *Zeitschr. Saugetierkunde*, 27: 65-83.

THIEL, E. C.
1962. "The Amount of Ice on Planet Earth," *Antarctic Research, Amer. Geophys. Union:* 172-175.

THOM, B. G.
1970. "Carolina Bays in Horry and Marion Counties, South Carolina," *Bull. Geol. Soc. Amer.*, 81: 783-814.

THOMAS, W. L. (ed.)
1956. *Man's Role in Changing the Face of the Earth.* Chicago: University of Chicago Press.

THOMÉ, K. N.
1958. "Die Begegnung des nordischen Inlandeises mit dem Rhein," *Geol. Jahrb.* (Hannover), 76: 261-308.

THORNBURY, W. D.
1969. *Principles of Geomorphology* (2d ed.). New York: J. Wiley.

THORNE, A. G.
1969. "Preliminary Comments on the Kow Swamp Skeleton," *Austr. Inst. Abor. Studies Newslet.*, 2(10): 6-7.

THORNTHWAITE, C. W.
1948. "An Approach Toward a Rational Classification of Climates," *Geogr. Rev.*, 38: 55-94.

1956. "Modification of Rural Microclimate." In THOMAS, 1956, pp. 567-83.

THORP, J., *et al.*

1952. *Map of Pleistocene Eolian Deposits of the United States, Alaska and Parts of Canada.* Washington: Geol. Soc. Amer.

THUN, R., R. HERRMANN, and E. KNICKMAN

1955. *Die Untersuchung von Böden* (3d ed.). Radebeul and Berlin: Neumann.

THURBER, D. L.

1962. "Anomalous U^{234}/U^{238} in Nature," *Jour. Geophys. Res.,* 67: 4518-20.

TISCHLER, W.

1955. *Synökologie der Landtiere.* Stuttgart: G. Fischer.

TIXIER, J.

1968. "Notes sur le Capsien typique." In *La Préhistoire: problèmes et tendances (Hommage à R. Vaufrey),* Paris: C.N.R.S., pp. 439-51.

TOBIAS, P. V.

1965. "*Australopithecus, Homo habilis,* Tool-Using and Tool-Making," *South Afr. Arch. Bull.,* 20, No. 80: 167-192.

1967. *The Cranium and Maxillary Dentition of Australopithecus (Zinjanthropus) boisei.* Cambridge: Cambridge University Press.

1968. "Middle and Early Upper Pleistocene Members of the Genus *Homo* in Africa." In G. Kurth (ed.), *Evolution und Hominisation.* (2d ed.), Stuttgart: Fischer, pp. 176-94.

TOBIAS, P. V., AND A. R. HUGHES

1969. "The New Witwatersrand University Excavation at Sterkfontein," *South African Arch. Bull.,* 24, no. 95: 158-69.

TOBIAS, P. V., AND G. H. R. VON KOENIGSWALD

1964. "A comparison between the Olduvai hominines and those of Java," *Nature,* 204; 515-18.

TODE, A.

1954. *Mammutjäger vor 100,000 Jahren.* Braunschweig: E. Appelhans.

TODE, A., F. PREUL, K. RICHTER, A. KLEINSCHMIDT, *et al.*

1953. "Die Untersuchung der paläolithischen Freilandstation von Salzgitter-Lebenstedt," *Eiszeitalter and Gegenw.,* 3: 144-220.

TONGIORGI, E.

1938. "Vegetation und Klima der letzten Eiszeit and des Postglazials in Mittelitalien," *Proc., 3rd, Int. Congr. INQUA* (Vienna, 1936), pp. 280-83.

TOTHILL, J. D.

1946. "The Origin of the Sudan Gezira Clay Plain," *Sudan Notes and Records,* 27: 153-83.

1948. "Origin of Soils of the Sudan." In J. D. Tothill (ed.), *Agriculture in the Sudan,* London: Oxford University Press, pp. 129-43.

TOYNBEE, A. J.

1935. *A Study of History* (2d ed.). London: Oxford University Press, Vols. 1 and 3.

TREWARTHA, G. T., A. H. ROBINSON, AND E. H. HAMMOND

1967. *Elements of Geography*. New York: McGraw-Hill.

TRICART, J.

1956a. "France," *Biuletyn Peryglacjalny*, 4: 117-38.

1956b. *Cartes des phénomènes périglaciares quaternaires en France*. Paris: Imprimerie Nationale.

1958. "Division morphoclimatique du Brésil and atlantique central," *Rev. Geomorph. Dyn*, 9: 1-22.

1961. "Notice explicative de la carte géomorphologique du delta du Sénégal," *Mém. Bur. Rech. Géol. Min.*, No. 8.

1966. "Quelques aspects des phénomènes périglaciaires quaternaires dans la péninsule ibérique," *Biul. Peryglacjalny*, 15: 313-27.

1969. *Geomorphology of Cold Environments*. London: Macmillan.

TRICART, J., and A. CAILLEUX

1956. "Action du froid Quaternaire en Italie Péninsulaire," *Actes IV. Congr. Int. Quaternaire* (Rome, 1953).

1962. *Le modelé glaciaire et nival*. Paris: Sedes.

1965a. *Introduction à la géomorphologie climatique*. Paris: Sedes.

1965b. *Le modelé des régions chaudes: forêts et savanes*. Paris: Sedes.

1967. *Le modelé des régions périglaciaires*. Paris: Sedes.

TRICART, J., and J. SCHAEFFER

1950. "L'indice d'émousse des galets. Moyen d'étude des systèmes d'érosion," *Rev. Géomorph. Dynam.*, 4: 151-79.

TRIMMEL, H.

1968. *Höhlenkunde*. Braunschweig: Vieweg.

TROELS-SMITH, J.

1960. "Ivy, Mistletoe and Elm. Climate Indicators—Fodder Plants," *Danmarks Geol. Undersoegelse*, series IV, Vol. 4, No. 4.

TROLL, C.

1944. "Strukturbüden, Solifluktion, und Frostklimate der Erde," *Geol. Rundschau*, 34: 545-694. (Also in H. E. Wright (trans.), "Structure Soils, Solifluction, and Frost Climates of the Earth," *U.S. Army Snow, Ice and Permafrost Research Establishment* [Wilmette, Ill.], Translation 43, October, 1958.)

1947. "Die Formen der Solifluktion und die periglaziale Bodenabtragung," *Erdkunde*, 1: 162-75.

1948. "Der subnivale oder periglaziale Zyklus der Denudation," *ibid.*, 2: 1-21.

1950. "Die geographische Landschaft und ihre Erforschung," *Studium generale*, 3: 163-81.

1956. "Die Klimatypen an der Schneegrenze," *Actes IV. Congr. Int. INQUA* (Rome, 1953), pp. 820-30.

1959. *Die tropischen Gebirge.*, *Bonner Geogr. Abhandl.*, 25: 1-93.

TROMBE, F.

1952. *Traité de Spéléologie*, Paris: Payot.

TURNER, C., and R. G. WEST
1968. "The Subdivision and Zonation of Interglacial Periods," *Eiszeitalter und Gegenw.,* 19: 93-101.
UCKO, P. J., and G. W. DIMBLEBEY (eds.)
1969. *The Domestication and Exploitation of Plants and Animals.* Chicago: Aldine.
UCKO, P. J., and A. ROSENFELD
1968. *Palaeolithic Cave Art.* New York: World University Library.
VALENTIN, H.
1952. *Die Küsten der Erde. Petermanns Mitt.: Erganzungs-Heft* 246.
1957. "Die Grenze der letzten Vereisung im Nordseeraum," *Abhl. deut. Geographentags* (Hamburg, 1955), Wiesbaden: F. Steiner, pp. 359-72.
VALENTINE, J. W.
1961. "Paleoecologic Molluscan Geography of the Californian Pleistocene," *Univ. California Publ. Geol. Sci.,* 34: 309-442.
VALLOIS, H. V.
1961. "The Social Life of Early Man: The Evidence of Skeletons," *Viking Fund Publ. Anthropol.,* 31: 214-35.
VAN CAMPO, M., G. AYMONIN, J. COHEN, P. DUTIL, P. GUINET, and P. ROGNON
1964-65-67. "Contribution à l'étude du peuplement végétale quaternaire des montagnes sahariennes." *Pollen et Spores,* 6: 169-94; 7: 361-71; 9: 107-20.
VAN CAMPO, M., and J. BOUCHUD
1962. "Flore accompagnent le squélette d'enfant moustérien découvert au Roc de Marsal, commune du Bugue (Dordogne) et première étude de la faune du gisement," *C. R. Acad. Sci.,* 254: 897-99.
VAN CAMPO, M., and R. COQUE
1960. "Palynologie et géomorphologie dans le sud tunésien," *Pollen et Spores,* 2: 275-84.
VANGENGEIM, E. A., A. P. OKLADNIKOV, and E. I. RAESKII
1966. "Priroda sibiri v antropogene i pervonachal' nce zaselenie severnoi azii chelovekom." In A. P. Okladnikov (ed.), *Istoriia Sibiri,* Novosibirsk: Akad. Nauk, pp. 35-49.
VAUFREY, R.
1928. "Le Paléolithique italien," *Arch. Inst. Paleont. Humaine* (Paris), Mem. 3.
1929. "Les éléphants nains des 'les méditerranéennes," *Arch. Inst. Paleontol. Humaine* (Paris), Mem. 6.
1933. "Notes sur le Caspien," *L'Anthropologie,* 43: 457-83.
VEEH, H. H.
1966. "Th^{230}/U^{238} and U^{234}/U^{238} Ages of Pleistocene High Sea Level Stand," *Jour. Geophys. Res.,* 71: 3379-86.
VEEH, H. H., and J. W. VALENTINE
1967. "Radiometric Ages of Pleistocene Fossils from Cayucos, Califor-

nia," *Bull. Geol. Soc. Amer.*, 78: 547-50.

VENZO, S.

1955. "Le attuali conoscenze sul Pleistocene Lombardo con particolare riguardo al Bergamasco," *Atti Soc. Ital. Sci. Nat.*, 94: 155-200.

VERESHCHAGIN, N. K.

1967. "Primitive Hunters and Pleistocene Extinction in the Soviet Union." In MARTIN and WRIGHT (eds.), 1967, pp. 365-98.

VÉRTES, L.

1959. "Untersuchungen an Höhlensedimenten, Methoden und Ergebnisse," *Régészeti Füzetek* (Budapest) (series 2), No. 7.

VIETE, G.

1951. "Zum Klima der Vorzeit," *Zeitschr. Meteor.*, 5: 102-10.

VILLALTA, J. F. DE

1952. "Contribución al conocimiento de la fauna de mamíferos fósiles del Plioceno de Villarroya (Logroño)," *Bol. Inst. Geol. Min. España*, 44.

VIRET, J.

1954. "Le loess à bancs durcis de Saint-Vallier (Drôme) et sa faune de mammifères villafranchiens," *Nouv. Arch. Mus. Hist. Nat. Lyon*, No. 4.

VIRGILI, C., and I. ZAMARRENO

1957. In "Livret Guide de l'Excursion B, (Environs de Barcelona et Montserrat)," *5th Int. Congr. INQUA* (Madrid-Barcelona), pp. 7-16.

VITA-FINZI, C.

1969. *The Mediterranean Valleys: Geological Changes in Historical Times*. Cambridge: Cambridge University Press.

WADELL, H.

1932. "Volume, Shape and Roundness of Rock Particles," *Jour. Geol.*, 40: 443-51.

WAGNER, P. L., and M. W. MIKSELL (eds.)

1962. *Readings in Cultural Geography*. Chicago: University of Chicago Press.

WALLÉN, C. C.

1953. "The Variability of Summer-Temperature in Sweden and Its Connection with Changes in the General Circulation," *Tellus*, 5: 157-78.

1963. "Aims and Methods in Studies of Climatic Fluctuations," *Arid Zone Research* (UNESCO), 20: 469-75.

WALKER, D., D. M. CHURCHILL, and N. T. MOAR

1966. "Commentary on Botanical Data from New Guinea, Australia and New Zealand." *In World Climate from 8000 to 0 B.C.*, London: Roy. Meteor. Soc., pp. 149-56.

WALTER, H.

1954. *Grundlagen der Pflanzenverbreitung II. Arealkunde*. Stuttgart: E. Ulmer.

1960. *Grundlagen der Pflanzenverbreitung I. Standortslehre* (2d ed.). Stuttgart: E. Ulmer.

WASHBOURN, C. K.

1967. "Lake Levels and Quaternary Climates in the Eastern Rift Valley of

Kenya," *Nature,* 216: 672-73.

WASHBOURN-KAMAU, C. K.
1970. "Late Quaternary Chronology of the Nakuru-Elementeita Basin, Kenya," *Nature,* 266: 253-54.

WASHBURN, A. L.
1956. "Classification of Patterned Ground and Review of Suggested Origins," *Bull. Geol. Soc. Amer.,* 67: 823-66.

WASYLIKOWA, K.
1967. "Late Quaternary Plant Macrofossils from Lake Zeribar, Western Iran," *Rev. Palaeobot. Palynol.,* 2: 313-18.

WATERBOLK, H. T.
1968. "Food Production in Prehistoric Europe," *Science,* 162: 1093-1102.

WATSON, R. A., and P. J. WATSON
1969. *Man and Nature.* New York: Harcourt, Brace.

WATSON, W.
1969. "Early Animal Domestication and Cereal Cultivation in China." In UCKO and DIMBLEBEY (eds.), 1969, pp. 393-402.

WATTERS, R. F.
1960. "The Nature of Shifting Cultivation, a Review of Recent Research," *Pacific Viewpoint,* 1: 59-99.

WATTS, W. A.
1967. "Late Glacial Plant Macrofossils from Minnesota." In CUSHING and WRIGHT (eds.), 1967, pp. 89-98.

WATTS, W. A., and R. C. BRIGHT
1968. "Pollen, Seed and Mollusk Analysis of a Sediment Core from Pickerel Lake, South Dakota," *Bull. Geol. Soc. Amer.,* 79: 855-76.

WATTS, W. A., and T. C. WINTER
1966. "Plant Macrofossils from Kirchner Marsh. Minnesota—A Paleoecological Study," *Bull. Geol. Soc. Amer.,* 77: 1339-60.

WATTS, W. A., *and* H. E. WRIGHT
1966. "Late-Wisconsin Pollen and Seed Analysis from the Nebraska Sandhills," *Ecology,* 47: 202-10.

WAYNE, W. J.
1967. "Periglacial Features and Climatic Gradient in Illinois, Indiana, and Western Ohio." In CUSHING and WRIGHT (eds.), 1967, pp. 393-414.

WAYNE, W. J., and J. H. ZUMBERGE
1965. "Pleistocene Geology of Indiana and Michigan." In WRIGHT and FREY (eds.), 1965, pp. 63-84.

WEBER, H.
1958. *Die Oberflächenformen des festen Landes.* Leipzig: B. G. Teubner.

WEERTMEN, J.
1961. "Equilibrium Profiles of Ice Caps," *Jour. Glaciol.,* Vol. 3: 953-64.

WEISCHET, W.
1954. "Die gegenwartige Kenntnis vom Klima in Mitteleuropa beim Maximum der letzten Vereisung," *Mitt. Geog. Ges. München,* 39: 95-116.

WELLMAN, H. W., and A. T. WILSON

1965. "Salt Weathering, A Neglected Geological Erosive Agent in Coastal and Arid Environments," *Nature,* 205: 1097-98.

WELTEN, M.

1954. "Pollenniederschlagstypen aus höhern Lagen Spaniens und ihre subrezenten Veränderungen," *Veroff. Geobotan. Inst. Rübel* (Zurich), 31: 199-216.

WENDORF, F.

1966. "Early Man in the New World: Problems of Migration," *Amer. Naturalist,* 100: 253-70.

WENDORF, F. (ed.)

1968. *The Prehistory of Nubia.* Dallas: Southern Methodist University Press, 2 vol.

WENDORF, F., *et al.*

1961. *Paleoecology of the Llano Estacado,* Vol. 1, Santa Fe: Mus. New Mexico Press.

1968. *The Prehistory of Nubia.* Dallas: Southern Methodist University Press, 2 vol. Chapters or sections by A. Gautier, J. L. Shiner, R. Schild, M. Chmielewska, H. Wieckowska, J. E. Anderson, and Wendorf are cited by page references.

WENTWORTH, C. K.

1922. "A Scale of Grade and Class Terms for Clastic Sediments," *Jour. Geol.,* 30: 277-92.

WERDECKER, J.

1955. "Beobachtungen in den Hochländern Äthiopiens auf einer Forschungsreise 1953/54," *Erdkunde,* 9: 305-17.

WEST, R. C. (ed.)

1964. *Natural Environment and Early Cultures* (Vol. 1, Handbook of Middle American Indians). Austin: University of Texas Press.

WEST, R. G.

1956. "The Quaternary Deposits at Hoxne, Suffolk," *Phil. Trans. Roy. Soc. London (Series B),* 239: 265-356.

1961. "Interglacial and Interstadial Vegetation in England," *Proc. Linn. Soc. London,* 175: 81-90.

1968. *Pleistocene Geology and Biology.* London: Longmans.

WEST, R. G., and J. J. DONNER

1956. "The Glaciations of East Anglia and the East Midlands, a Differentiation Based on Stone-Orientation Measurements of the Tills," *Quart. Jour. Geol. Soc.* 112: 69-91.

WEST, R. G., and C. B. M. MCBURNEY

1943. "The Quaternary Deposits at Hoxne, Suffolk, and Their Archaeology," *Proc. Prehis. Soc.,* 20: 131-54.

WHITE, S. E.

1962. "Late Pleistocene Glacial Sequence for the West Side of Iztaccihuatl, Mexico," *Bull. Geol. Soc. Amer.,* 73: 934-58.

WHITEHEAD, D. R.

1965. "Palynology and Pleistocene Geography of Unglaciated Eastern North America." In WRIGHT and FREY (eds.), 1965, pp. 417-32.

WHITTOW, J. B., *et al.*

1963. "Observations on the Glaciers of the Ruwenzori," *Jour. Glaciol.,* 4: 581-616.

WICHE, K.

1961. "Beiträge zur Formenentwicklung der Sierren am unteren Segura (Sudostspanien)," *Mitt. Österr. Geogr. Ges.,* 103: 125-57.

WIEGAND, G.

1965. *Fossile Pingos in Mitteleuropa. Würzburger Geogr. Arbeiten,* 16: 1-152.

WILCOX, R. E.

1965. "Volcanic-Ash Chronology." In WRIGHT and FREY (eds.), 1965, pp. 807-16.

WILHEMY, H.

1950. "Das Alter der Schwarzerde und die Steppen Mittel-und Osteuropas," *Erdkunde,* 4: 5-34.

1958. *Klimamorphologie der Massengesteine.* Braunschweig: G. Westermann.

WILLETT, H. C.

1949. "Long-Period Fluctuations of the General Circulation of the Atmosphere," *Jour. Meteor.,* 6: 34-50.

WILLETT, H. C., and F. SANDERS

1959. *Descriptive Meteorology.* New York: Academic Press.

WILLEY, G. R.

1966. *An Introduction to American Archaeology,* vol. I. Englewood Cliffs, N.J.: Prentice-Hall.

WILLIAMS, P. J.

1961. "Climatic Factors Controlling the Distribution of Certain Frozen Ground Phenomena," *Geografiska Ann.,* 43: 339-47.

WILLIS, E. H., H. TAUBER, and K. O. MÜNNICH

1960. "Variations in the Atmospheric Radiocarbon Concentration over the Past 1300 Years," *Amer. Jour. Sci., Radiocarbon Supplement,* 2: 1-4.

WILMSEN, E. N.

1968. "Lithic Analysis in Paleoanthropology," *Science,* 161: 982-87.

WILSON, J. T., *et al.*

1958. *Glacial Map of Canada.* Ottawa: Geol. Assn. Can.

WISEMAN, J. D. H.

1966. "Evidence for Recent Climatic Changes in Cores from the Ocean Bed." In *World Climate from 8000 to 0 B. C.,* London: Roy. Met. Soc., pp. 84-98.

WISSMANN, H. VON

1938. "Über Lössbildung und Würmeiszeit in China," *Geog. Zeit.,* 44: 201-20.

1957. "Ursprung und Ausbreitungswege von Pflanzen- und Tierzucht und ihre Abhängigkeit von der Klimageschichte," *Erdkunde,* 11: 81-94, 175-93.

WITTFOGEL, K. A.

1956. "The Hydraulic Civilizations." In THOMAS, 1956, pp. 152-64.

WOERKOM, A. J. J. VAN

1953. "The Astronomical Theory of Climatic Changes," In H. Shapley (ed.), *Climatic Change,* Cambridge: Harvard University Press, pp. 147-57.

WOLDSTEDT, P.

1929. *Das Eiszeitalter, Grundlinien einer Geologie des Diluviums.* Stuttgart: F. Enke.

1952. "Probleme der Terrassenbildung," *Eiszeitalter und Gegenw.,* 2: 36-44.

1954. *Die allgemeinen Erscheinungen des Eiszeitalters* (Vol. 1 of *Das Eiszeitalter*). Stuttgart: F. Enke.

1958. *Europa, Vorderasien und Nordafrika im Eiszeitalter* (Vol. 2 of *Das Eiszeitalter*). Stuttgart: F. Enke.

1960a. "Mississippi und Rhein: ein geologischer Vergleich," *Eiszeitalter und Gegenw.,* 11: 31-38.

1960b. "Alte Quartäre Strandlinien in Nordamerika und Europa," *ibid.,* 11: 12-19.

1962a. "Interglaziale marine Stände in Australien," *ibid.,* 12: 60-65.

1962b. "Uber die Gliederung des Quartärs und Pleistozäns," *ibid.,* 13: 115-24.

1965. *Afrika, Asien, Australien und Amerika im Eiszeitalter* (Vol. 3 of *Das Eiszeitalter*). Stuttgart: F. Enke.

1967. "The Quaternary of Germany." In K. Rankama (ed.), *The Quaternary,* New York: Interscience, vol. 2, pp. 239-300.

WOPFNER, H., and C. R. TWIDALE

1967. "Geomorphological History of the Lake Eyre Basin." In JENNINGS and MABBUTT (eds.), 1967, pp. 1919-43.

WORMINGTON, H. M.

1957. *Ancient Man in North America.* (4th ed.). Denver: Museum of Natural History.

WORTMANN, H.

1956. "Ein erstes sicheres Vorkommen von periglazialen Steinnetzböden im norddeutschen Flachland," *Eiszeitalter und Gegenw.,* 7: 119-26.

WRIGHT, H. E.

1951. "Geologic Setting of Ksar Akil, a Palaeolithic Site in Lebanon," *Jour. Near Eastern Studies,* 10: 115-19.

1952. "The Geological Setting of Four Prehistoric Sites in Northeastern Iraq." *Bull. Amer. Schools Oriental Research,* 128: 11-24.

1957. "Geology." In *The Identification of Non-Artifactual Archaeological Materials,* Washington, D. C.: Nat. Acad. Sci.—Nat. Res. Council, Publ. 565, pp. 50-51.

1961. "Late Pleistocene Climate of Europe: A Review," *Bull. Geol. Soc. Amer.*, 72: 933-84.

1962a. "Late Pleistocene Geology of Coastal Lebanon," *Quaternaria*, 6: 525-40.

1962b. "Pleistocene Glaciation in Kurdistan," *Eiszeitalter und Gegenw.*, 12: 131-64.

WRIGHT, H. E., J. H. ANDREWS, and W. VAN ZEIST

1967. "Modern Pollen Rain in Western Iran and Its Application to Plant Geography and Quarternary Vegetational History," *Jour. Ecol.*, 55: 415-43.

WRIGHT, H. E., and D. G. FREY (eds.)

1965. *The Quaternary of the United States*. Princeton: Princeton University Press.

WRIGHT, H. E., and R. V. RUHE

1965. "Glaciation of Minnesota and Iowa." In WRIGHT and FREY (eds.), 1965, pp. 29-42.

WRIGHT, H. E., Y. C. WINTER, and H. L. PATTEN

1963. "Two Pollen Diagrams from Southeastern Minnesota," *Bull. Geol. Soc. Amer.*, 74: 1371-96.

WUNDT, W.

1933. "Änderungen des Erdalbedo während der Eiszeit," *Meteor. Zeitschr.*, 50: 241-48.

1944. "Die Mitwirkung der Erbahnelemente bei der Entstehung der Eiszeiten," *Geol. Rundschau*, 34: 713-47.

ZAGWIJN, W. H.

1960. "Aspects of the Pliocene and Early Pleistocene Vegetation in the Netherlands," *Proefschr. Leiden* (Maastricht).

1961. "Vegetation, Climate and Radiocarbon Datings in the Late Pleistocene of the Netherlands (I)," *Meded. Geol. Stichting* (Haarlem), N.S. 14: 15-45.

1969. "Vegetation, Climate and Radiocarbon Datings in the Late Pleistocene of the Netherlands (II)," *Meded. Geol. Stichting* (Haarlem), N.S. 22: in press.

ZĚBERA, K., V. LOŽEK, V. KNEBLOVA, O. FEJFAR, and M. MAZALEK

1955. "Zpráva o II. etapé geologického výzkumu kvartéru v. Předmosti u Přerova na Moravé," *Anthropozoikum* (Praha), 4: 291-362.

ZEIST, W. VAN

1967. "Late Quaternary Vegetation History of Western Iran," *Rev. Palaeobot. Palynol.*, 2: 301-11.

ZEUNER, F. E.

1932. "Die Schotteranalyse," *Geol. Rundschau*, 24: 66-104.

1934. "Die Beziehung zwischen Schädelform und Lebensweise bei den rezenten und fossilen Nashörnern," *Ber. Naturforsch Ges. Frieburg*, 34: 21-80.

1952. "Pleistocene Shore-Lines," *Geol. Rundschau*, 40: 39-50.

1958. *Dating the Past. An Introduction to Geochronology* (1st ed., 1946).

London: Methuen.

1959. *The Pleistocene Period. Its Climate, Chronology and Faunal Successions* (2d ed.). London: Hutchinson.

1961. "Criteria for the Determination of Mean Sea-Level for Pleistocene Shoreline Features," *Quarternaria*, 5: 143-47.

1963. *A History of Domesticated Animals*. London: Hutchinson.

ZIENERT, A.

1968. "Gleiche Würm-Rückzugsstadien in den Gebirgen Mitteleuropas und Ostafrikas?" *Eiszeitalter und Gegenw.*, 19: 85-92.

ZINDEREN BAKKER, E. M. VAN

1957. "A Pollen Analytical Investigation of the Florisbad Deposits." In J. D. Clark (ed.), *Proceedings*, 3rd Panafrican Congress of Prehistory, Livingstone, 1955, pp. 56-67.

1962a. "Botanical Evidence for Quaternary Climates in Africa," *Ann. Cape Provincial Museums*, 2: 16-31.

1962b. "A Pollen Diagram from Equatorial Africa (Cherangani, Kenya)," *Nature*, 194: 201-3.

1963. "Analysis of Pollen Samples from Northeast Angola," In CLARK, 1963, pp. 213-217.

1967. "Upper Pleistocene and Holocene Stratigraphy and Ecology on the Basis of Vegetation Changes in Sub-Saharan Africa." In BISHOP and CLARK (eds.), 1967, pp. 125-47.

ZOHARY, M.

1963. *On the Geobotanical Structure of Iran. Israel Jour. Botany*, 11, supplement.

ZOLYOMI, B.

1953. "Die Entwicklungsgeschichte der Vegetation Ungarns seit dem letzten Interglacial," *Acta Biol. Acad. Sci. Hung.* (Budapest), 4: 367-430.

ZONNEVELD, J. I. S.

1968. "Quaternary Climatic Changes in the Carribbean and N. South America," *Eiszeitalter und Gegenw.*, 19: 203-08.

ZOTZ, L. F.

1951. *Altsteinzeitkunde Mitteleuropas*. Stuttgart: F. Enke.

1955. *Das Paläolithikum in den Weinberghöhlen bei Mauern* (with sections on geology by G. Freund and palynology by E. Hofman). Bonn: Quartär Bibliothek, L. Röhrscheid, Vol. 2.

1956. "Das Campignien in Süddeutschland," *Forschungen und Fortschritte* (Berlin), 30: 331-35.

Addendum

BRAIN, C. K.

1958. "The Transvaal Ape Man Bearing Cave Deposits," *Transvaal Mus. Mem.*, 11: 32-48.

OVERBECK, F.

1950. *Die Moore, Geologie und Lagerstatten Niedersachsens III* (2d ed.). Bremen: Niedersachsisches Amt. f. Landesplanung, Series Ac, Abh. 4.

Index